# Electron Microscopy and Analysis, 1985

# Electron Microscopy and Analysis, 1985

Proceedings of The Institute of Physics Electron Microscopy and Analysis Group conference held at the University of Newcastle upon Tyne, 2–5 September 1985 (EMAG 85)

Edited by G J Tatlock

Institute of Physics Conference Series Number 78
Adam Hilger Ltd, Bristol and Boston

CODEN IPHAC 78 1–592 (1985)

*British Library Cataloguing in Publication Data*

Electron microscopy and analysis, 1983: proceedings
of The Institute of Physics Electron Microscopy
and Analysis Group conference held at the University
of Newcastle upon Tyne, 2–5 September 1985 (EMAG 85).
—(Conference series/Institute of Physics; no. 78)
1. Electron microscopy
I. Institute of Physics. *Electron Microscopy and Analysis Group* II. Tatlock, G.J. III. Series
502′.8′25 QH212.E4

ISBN 0-85498-169-1
ISSN 0305-2346

The conference was organised by the Electron Microscopy and Analysis Group of The Institute of Physics.

Programme Organiser
Dr J R Fryer (University of Glasgow)

Exhibition and Local Arrangements Committee
Dr A Hendry (University of Newcastle) Chairman, Mr D Beamer (Philips), Mr H Martin (Jeol (UK) Ltd), Mr N McCormick (Link Systems), Mr A Williamson (University of Newcastle)

Electron Microscopy and Analysis Group Committee 1984–85
Dr L M Brown (Cavendish Laboratory) Chairman, Dr P J Goodhew (University of Surrey) Secretary, Mr P D Augustus (Plessey Research), Dr E D Boyes (University of Oxford), Dr D J Dingley (University of Bristol), Dr C A English (AERE Harwell), Dr P E J Flewitt (CEGB), Dr J R Fryer (University of Glasgow), Dr A Hendry (University of Newcastle), Dr F J Humphreys (Imperial College), Dr D White (BP Research Centre)

Honorary Editor
Dr G J Tatlock (University of Liverpool)

Published on behalf of The Institute of Physics by Adam Hilger Ltd
Techno House, Redcliffe Way, Bristol BS1 6NX, England
PO Box 230, Accord, MA 02018, USA

Printed in Great Britain by J W Arrowsmith Ltd, Bristol

# Preface

The new medical and dental school of the University of Newcastle upon Tyne provided the venue for EMAG '85. There were over 220 participants, 148 contributed papers and a large and well equipped trade exhibition.

For the majority of the electron optical techniques discussed, a quantitative understanding of the optics and specimen scattering relationships has been gained over the last two decades. This was reflected in the contributed papers, as the topics mainly concerned the interpretation related to a particular specimen rather than broad generalisations. It was also noticeable that the techniques were being used to their fullest extent and on specimens that a few years ago would have been considered beyond the capability of electron microscopy or analysis.

There were twelve invited speakers—four from France, two from the United States and six from Britain. After a welcoming address from the Vice-Chancellor of Newcastle University, Professor Humphreys commenced the technical proceedings with a provocative paper in which he demonstrated the electron beam drilling of 1 nm holes in alumina and other ceramics. The mechanism of this electron beam interaction was the subject of much discussion. In many ways this paper was typical of the rest of the meeting in that the papers were seeking an understanding of the specimen that was hitherto unknown because the electron optical techniques were not previously applied with such rigour or understanding. Other notable examples later in the meeting were the imaging and image simulation of pentagonal quasicrystalline structures in contributions by Portier and Knowles, and the movements of atomic columns in small gold crystals shown by Smith and Bovin.

A paper by Bourret opened the session on boundaries and together with the other papers in this session showed that high resolution electron microscopy has become a standard tool for the study of interfaces and non-periodic regions in crystals. This use of HREM was also seen in the work of Audier who discussed graphite formation on small metal particles and also in the session devoted to ceramics and surface layers. It was of interest in this latter session that the plenary paper by Lewis and many subsequent ones were using HREM as well as microanalysis to characterise technologically important materials.

Another method of structural determination is convergent beam electron diffraction. An enthusiastic session dealt with many aspects of this technique and showed that it is sensitive to strain and minor symmetry changes even in thick specimens. This type of sample is unsuitable for HREM and thus the two techniques are complementary.

Structural determination is also complementary to elemental analysis and sessions on EELS—with a plenary paper by Sevely—microanalysis of metallic systems opened by Titchmarsh, and the techniques and developments in microanalysis introduced by Goodhew, gave a well balanced assessment of the techniques and applications of analytical microscopy. The use of analysis and structural determination was highlighted in the session on surface reactions introduced by Flower. It is symptomatic of the importance of electron microscopy that complex chemical reactions at interfaces can now be examined.

The sessions described previously were concerned with the application of well developed techniques to stable specimens. One aspect of EMAG is that it is also concerned with new techniques and phenomena. The session on STEM showed that information can be obtained in reflection and microdiffraction modes that is unobtainable by TEM and that specimen excitation by the narrow, intense beam can provide spectroscopic information. Scanning Auger microscopy was the subject of several papers, notably by Prutton and his co-workers; and a workshop on high voltage microscopy chaired by Boyes examined the results and capabilities of the latest commercial instruments.

One session dealt with organic crystals and polymers. Apart from certain compounds, e.g. phthalocyanines, these materials have been considered only suitable for low resolution imaging—often by replication. Dorset and other authors in this session showed that high resolution information can be obtained by electron diffraction and lattice imaging. In a cooled specimen of paraffin, lattice resolution of 0.25 nm was obtained. The symmetry of these crystalline materials is usually low but Wittman showed how, with controlled epitaxial growth, the orientation and growth habits of the crystals can be tailored to an optimum geometry. High resolution imaging, convergent beam and microdiffraction methods were all described in successful applications to organic crystals, thus opening up an important new field for electron microscopy and analysis.

In the meeting nearly half of the papers were presented as posters which in many cases comprised the total contributions on certain subjects. My only retrospective regret of this meeting is that I wish it had been extended by a day devoted entirely to the posters.

On the social side Newcastle proved popular. The City of Newcastle and the Lord Mayor gave a lavish welcome to the delegates with a meal in the City chambers on the opening evening. The dinner was followed by a display of Morris dancing and folk-singing but despite these ancient festivities it continued to rain for much of the next four days. The following evening there was a buffet provided by the exhibitors and the conference dinner was on the final evening.

The biennial EMAG meetings have grown steadily over the years and are regarded as representing the state of the art in electron microscopy. I extend my thanks to the invited speakers, contributors and all participants for maintaining this standard.

**John R Fryer**
Programme Organiser

# Contents



†Invited paper

†Invited paper

†Invited paper

†Invited paper

†Invited paper

**Chapter 10: Metals and semiconductors**

†Invited paper

**Chapter 13: Ceramics**



†Invited paper

†Invited paper

*Inst. Phys. Conf. Ser. No 78: Chapter 1*
*Paper presented at EMAG '85, Newcastle upon Tyne, 2–5 September 1985*



# Nanometre-scale electron beam lithography

C J Humphreys, I G Salisbury, S D Berger,*
R S Timsit** and M E Mochel***

Department of Metallurgy and Materials Science, University of Liverpool, P.O.Box 147, Liverpool L69 3BX, UK.

* Cavendish Laboratory, University of Cambridge, Cambridge CB3 OHE, UK.

** Alcan Laboratories, Kingston, Ontario, Canada.

*** Materials Research Laboratory, University of Illinois at Urbana-Champaign, Illinois 61801, USA.

Abstract. Holes and lines only 1nm across may be directly cut in a variety of inorganic materials using an intense focussed beam of electrons. The mechanism involves electron stimulated desorption, and EELS spectra recorded during the cutting process reveal structural changes occurring. Material may be removed 'at a distance' without being in the path of the electron beam. The aspect ratio of the holes is so great that the Uncertainty Principle is apparently violated.

## 1. Introduction

If an intense electron beam, current density typically $10^7$ A $m^{-2}$, is focussed onto certain inorganic materials it drills holes which can be less than 1nm in diameter (see Fig.1). If the beam is moved it cuts lines, which can be less than 1nm wide.(2) We have called this process SCRIBE (Sub-nanometre Cutting and Ruling by an Intense Beam of Electrons). This writing is much narrower than other forms of electron beam lithography, which use organic resists, and it is also a one-step process, requiring no chemical development. Materials which have been drilled by the SCRIBE process include β and β"alumina (Mochel et al. 1983a), NaCl (Isaacson and Muray 1981), amorphous alumina (Mochel et al., 1984, Salisbury et al., 1984), $CaF_2$ and MgO (Salisbury et al. 1984) and $AlF_3$ (Muray et al. 1984). Patterns cut in most of these materials (not NaCl) are stable to the atmosphere. Not only may holes be drilled and lines cut, it is also possible to machine the external shapes of the above materials on a nanometre scale.

Applications of the SCRIBE technique may be far reaching, since we can use this to fabricate structures at dimensions previously inaccessible. For example, nanometre-scale 3-D electronic devices may in principle be fabricated which are extremely fast because of their small size. By drilling an array of holes of chosen size molecular sieves may be made, tailored to specific molecular dimensions. Memories may be fabricated with a density one million times greater than at present. Custom-built biotechnological materials could be synthesized by patterning a substrate in a desired form on a molecular scale, etc.

## 2. Basic features of the SCRIBE process

The resolution of conventional electron beam lithography is limited by the spreading of the primary electron beam in the resist, by backscattering from the substrate, and by the distance over which low energy secondary electrons are created and travel (organic resists are exposed mainly by secondaries). Resolution in the SCRIBE process appears to suffer from none of these factors. Fig.3 shows that a hole drilled through 1000Å of Na β-alumina maintains its straight sides (Mochel et al. 1983b) with no evidence of broadening. The hole diameter is 20Å ± 2Å: the ±2Å corresponds to surface roughness on an atomic scale. In the SCRIBE process the electron beam undoubtedly broadens and creates secondaries, however, because there is a minimum threshold beam current density for drilling we believe that only the central position of the primary beam drills - hence the very straight holes.

Do the holes drill from the top or bottom of the specimen? To answer this question we have partially drilled a hole then tilted the specimen (Mochel et al. 1983b). Fig.4 shows that for a thin specimen, the hole starts at <u>both</u> surfaces, the craters move inwards and join up to form the hole. For thick specimens the drilling is mainly from the electron entrance surface.

Is the drilling at a uniform rate? The transmitted electron intensity, I, was measured as a function of time, t, during drilling. For some materials, e.g. Na β-alumina, it is found that ln I is proportional to t, i.e. the material drills at a uniform rate (Mochel et al. 1983a) of typically 250Å $s^{-1}$. For other materials, e.g. amorphous $Al_2O_3$, the material drills rapidly at first, followed by a plateau region which may last for many seconds before perforation finally occurs (Fig.5). Some other materials drill extremely rapidly, in milliseconds. The reasons for these different behaviours are not yet clear, but it appears that somewhat different hole drilling mechanisms may be operating in different materials.

## 3. Mechanisms of the SCRIBE process

Specimens in good thermal contact with a conductor may be drilled, hence the specimen is not melting or vaporising. The threshold beam current density for drilling increases as the incident electron energy increases (Salisbury et al. 1984), hence direct electron displacement damage and sputtering are not responsible for the drilling. Since the cross section for ionisation damage increases with increasing electron energy, this suggests that ionisation damage is the primary mechanism responsible for electron beam cutting. The fact that thin specimens drill from both surfaces indicates that surface desorption is important.

We suggest that the material drills atom plane by atom plane from both surfaces, with the anion (e.g. oxygen in $Al_2O_3$) being desorbed following ionisation by the incident beam and the cation migrating to the sides of the hole. A number of different mechanisms exist for electron stimulated desorption (ESD). We will apply one of these, the Knotek-Feibelman (1979) theory, to the particular case of $Al_2O_3$. An energy level diagram of $Al_2O_3$ is given in Fig.6 : the highest occupied level of the $Al^{3+}$ ion is the

Al(2p) level having a binding energy of 73eV. If an electron ionises this level the dominant decay modc is an inter-atomic Auger process in which one O(2p) electron decays into the Al (2p) hole. This releases about 70eV of energy which is taken up by the emission of one or two Auger electrons from the O(2p) state. The $O^{2-}$ ion thus loses up to 3 electrons and is pushed out of the surface by the repulsive Coulomb force. Neutral and positive oxygen ions are, of course, being produced by the above ionisation mechanism throughout the irradiated volume of the crystal, and it is possible that very small oxygen gas bubbles and voids are formed. Surface desorption will cease, and hole drilling will stop, unless the cations (i.e. $Al^{3+}$ in $Al_2O_3$) migrate to the sides of the hole. The driving force for this migration is believed to arise from both Coulomb repulsion between neighbouring $Al^{3+}$ ions on the surface of the hole and from the induced electric field in the material. Consideration of the flux of secondary electrons leaving and arriving at a point distance r from the centre of the beam shows that material in the centre of the beam is positively charged with respect to material at the edge of the beam. The induced electric field in the specimen therefore acts radially outwards and sweeps the $Al^{3+}$ ions to the sides of the hole, thereby 'uncovering' further oxygen ions which desorb, and so on, layer by layer. If the incident beam current density is below the threshold for drilling, the cations have time to neutralise their charge by collecting slow secondary electrons and hence they do not migrate to the sides of the hole. In some circumstances a plug of metal is left behind in the hole.

## 4. Electron energy loss spectra and energy filtered images

Fig.7 shows typical EELS spectra from amorphous $Al_2O_3$. Before drilling the broad 25eV $Al_2O_3$ plasmon peak dominates the low energy region of the spectrum. During drilling a sharp peak at 9eV, and sometimes also at 15eV, arises, superimposed on the broad $Al_2O_3$ bulk plasmon peak. After drilling, these sharp peaks normally, but not always, disappear and there is an odd shaped broad peak centred at about 20eV which extends down to 0eV, i.e. the band gap has filled in. We note that metallic Al volume and surface plasmon energies are at 15eV and 11ev respectively, and it may be that the observed 9eV peak is an Al surface plasmon, energy shifted due to the proximity of the interface. We have noticed shifts in these peak energies when different forms of alumina ($\alpha$, $\beta$ and amorphous) are drilled indicating the importance of chemical and interface effects. Fig.8 shows energy filtered images. The 15eV image shows an $Al/Al_2O_3$ interfacial plasmon lying within the hole.

## 5. Drilling at a distance

Fig.9 shows various holes all drilled using a stationary 5Å diameter electron beam. The small holcs were drilled with an exposure time of a fraction of a second, the large holes with an exposure time of many seconds. After a hole is drilled with a 5Å diameter beam it continues to enlarge until it is ~200Å in diameter in amorphous alumina, even though the probe remains at the centre of the hole (the absence of drift can be seen from the fact that the holes remain reasonably circular). This 'drilling at a distance' appears to be due to the long range nature of the inelastic scattering potential for ionisation with small electron energy losses (i.e. ionisation can occur with large impact parameters). However, this interpretation causes difficulties in understanding the threshold current density necessary to commence drilling and the very straight and narrow holes drilled with short beam exposures.

## 6. Beating the Uncertainty Principle?

We can drill holes only 10Å across in at least 2000Å thick material (see Fig.3 for an example of a hole in 1000Å thick material). If we take the electron localisation $\Delta x \sim 10$Å, and the momentum uncertainty $\Delta p_x$ as fixed by the aspect ratio of the hole, we find that the hole 'measures' the electron position and momentum with an accuracy better than the Uncertainty Principle limit. In fact we are not, of course, breaking the Uncertainty Principle since not all incident electrons participate in hole drilling, and it cannot be predicted in advance whether a particular electron will be localised within the hole. However it is of interest to note that holes can be drilled with an aspect ratio such that those electrons transmitted have $\Delta x\ \Delta p_x \lesssim h$.

## References

Isaacson M and Muray A 1981 J. Vac. Sci. Technol. 19 1117

Knotek M L and Feibelman P J 1979 Surf. Sci. 90 78

Mochel M E, Humphreys C J, Eades J A, Mochel J M and Petford A K 1983a Appl. Phys. Lett. 42 392

Mochel M E, Humphreys C J, Mochel J M and Eades J A 1983b Proc. 41st Meeting of EMSA (San Francisco Press) p.100

Mochel M E, Eades J A, Metzger M, Meyer J I and Mochel J M 1984 Appl. Phys. Lett. 44 502

Muray A, Isaacson M and Adesida I 1984 Appl. Phys. Lett. 45 589

Salisbury I G, Timsit R S, Berger S D and Humphreys C J 1984 Appl. Phys. Lett. 45 1289

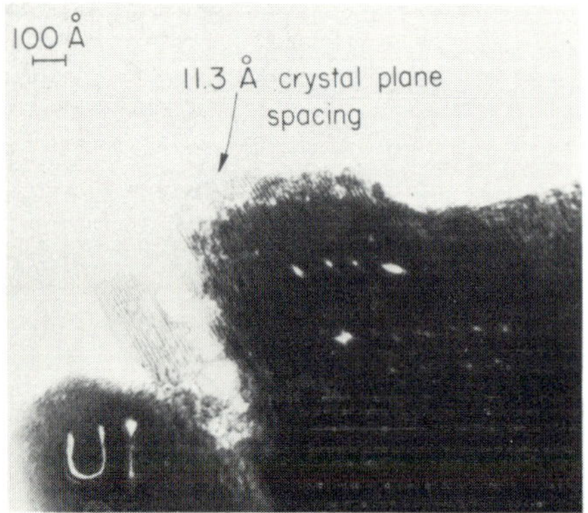


Fig. 1. Holes drilled and lines cut in sodium β-alumina. The 11.3Å spacing lattice fringes demonstrate that some of the holes have diameters less than 10Å.

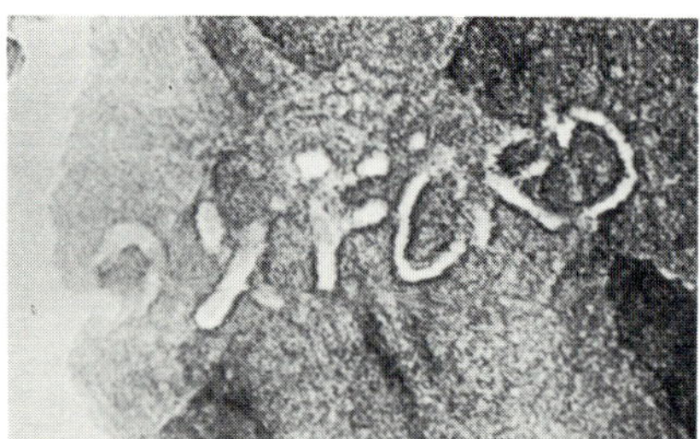

Fig. 2. The word OXFORD written in sodium β-alumina by moving the incident electron beam with a joystick. The narrowest lines are less than 10Å wide.

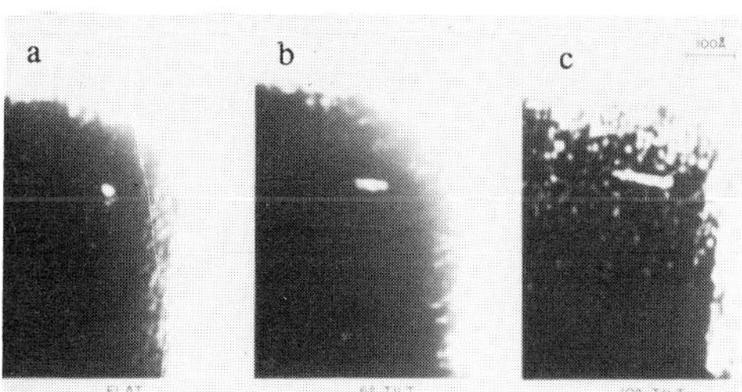


Fig. 3. Hole about 20Å across drilled through 1000Å thick sodium β-alumina: (a) as drilled; (b) sample tilted 5°; (c) sample tilted 10°. Hole is straight sided and of even width along its length.

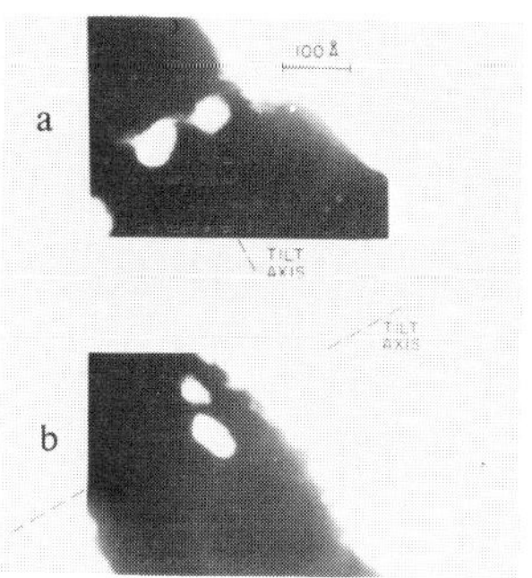


Fig. 4. Partial hole drilled by short exposure of sodium β-alumina to the beam. (a) sample tilted 5° to show top and bottom craters; (b) sample returned to flat position and then tilted by 5° about a perpendicular axis: top and bottom craters are again visible.

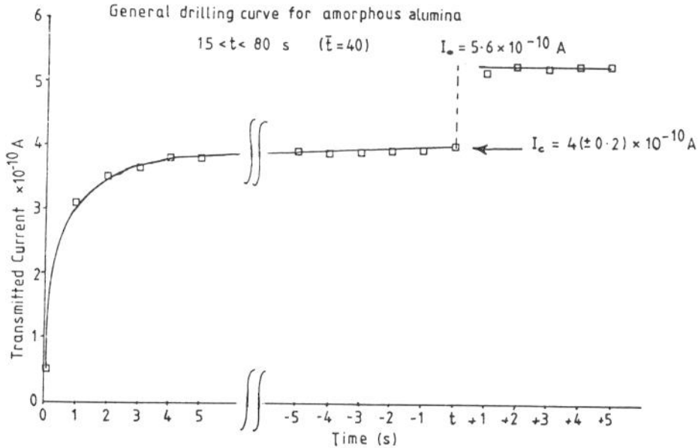


Fig. 5. Typical plot of the transmitted electron current versus time during hole drilling in amorphous alumina. The current rises rapidly at first, followed by a plateau region lasting between about 15 and 80 seconds, followed by very rapid total penetration.

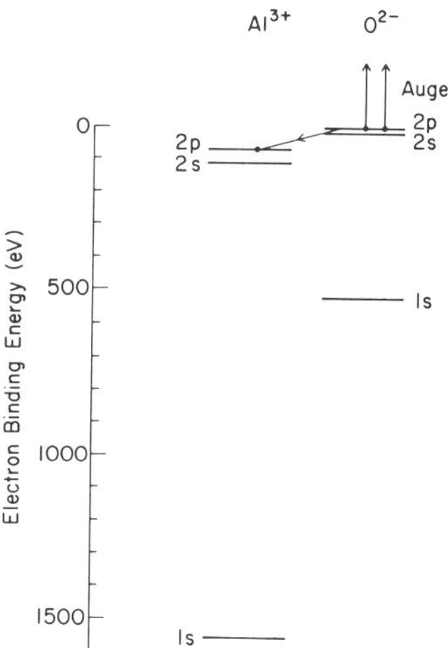


Fig. 6. Electron energy levels in $Al_2O_3$. Following ionisation of an electron from the Al(2p) level, the $O^{2-}$ ion may lose 3 electrons and hence become $O^+$.

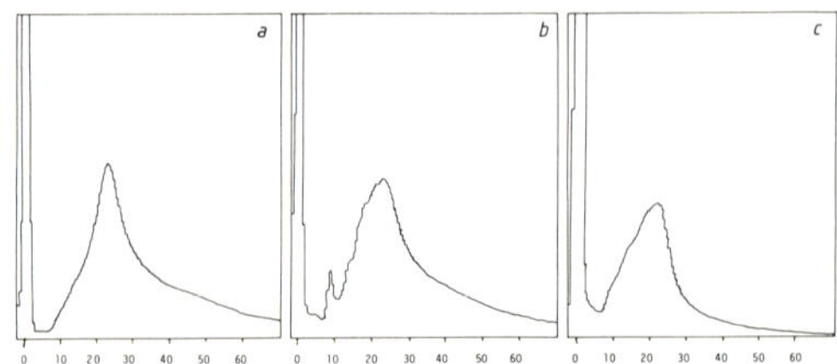


Fig. 7. Electron energy loss spectra from amorphous $Al_2O_3$: (a) before, (b) during and (c) after drilling. The band gap region in (a) has filled up in (c). The sharp 9eV peak in (b) may be an energy-shifted interface plasmon.

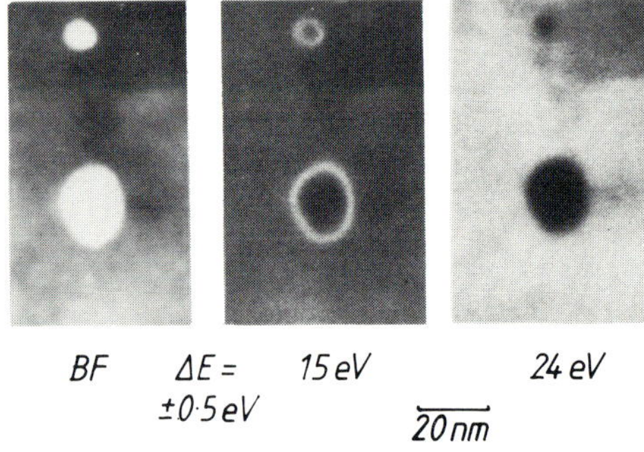


Fig. 8. Energy loss images in amorphous $Al_2O_3$ after drilling ∿200Å and ∿60Å diameter holes. The 15eV loss image shows an interfacial excitation lying within the hole. The 24eV loss image corresponds to the bulk plasmon energy of $Al_2O_3$ and hence only shows up in the material.

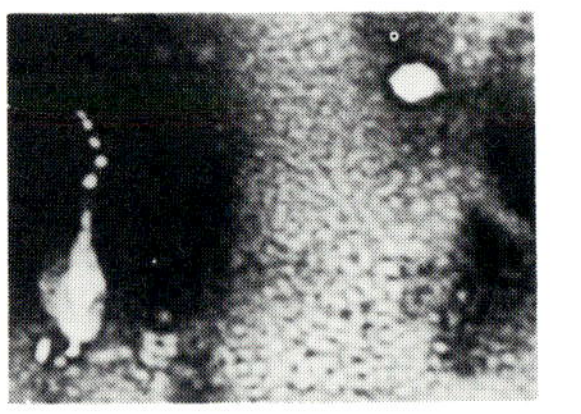

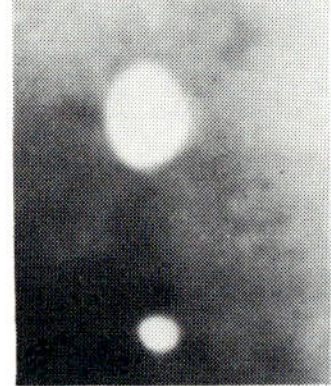

20 nm

Fig. 9. Holes of various sizes drilled in amorphous $Al_2O_3$. The small holes had short and the large holes had long exposure times to a stationary 5Å diameter probe.

# Methods for the evaluation of the electron beam damage

D Vesely and D Finch

Department of Materials Technology,Brunel University,Uxbridge,Middlesex.

## 1. Introduction.

The electron beam damage of specimens is particularly important when organic materials are being examined in the electron microscope. The loss of crystallinity, loss of elements or loss of mass may be clearly visible while many changes in the chemical composition of the specimen will not be apparent. The existing methods used for the beam damage evaluation will only partially help us to understand the complexity of the beam damage mechanism. It is the purpose of this paper to evaluate these techniques and to show that the G value (number of events for 100 eV absorbed energy), which is related to a particular change in the specimen, is the best description of the beam damage and can be easily calculated from the decay curves.

## 2. Loss of crystallinity.

The loss of crystallinity is the most clearly visible effect as it results in loss of contrast on studied features. For this reason this type of damage has been most widely studied, but it is not very easy to obtain meaningful results. The reason for this is, that the initial crystallinity for polymers is unknown and the diffracted intensity can increase in the early stages of irradiation due to some reorientation. The specimen will also lose mass and it is thus difficult to differentiate between the contribution for the Bragg and diffused scattering to the selected diffraction angle.

The best method for the crystallinity loss measurement is the use of a Faraday cage with an annular aperture at the viewing screen level, which can detect several first order diffraction spots. A typical measured curve for PE is shown in figure 1. It is obvious that the curve does not correspond to a single exponential and that the background intensity must be taken into account. The background will gradually decrease due to mass loss or due to contamination. This decay curve can be described as:

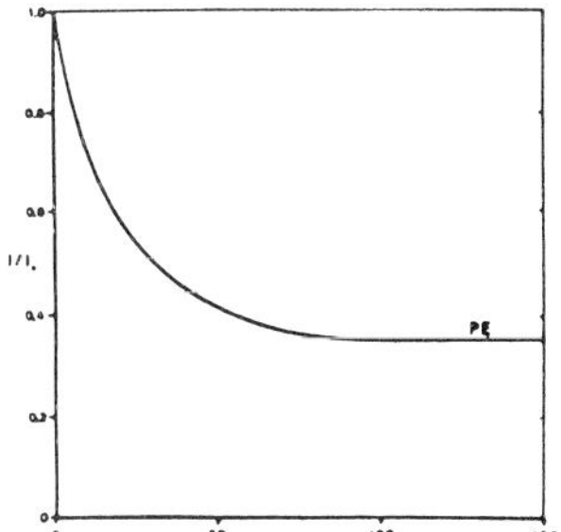

$$I/I_0=(1-A)e^{-k_1 D}+Ae^{-k_2 D} \qquad [1]$$

Fig. 1. The crystallinity loss with exposure for PE.

where $I_0$ is the initial signal from the undamaged specimen, D is the exposure and A, $k_1$, $k_2$, are material constants. The first exponential function is describing the true loss of crystallinity.

## 3. Loss of mass.

The second most visible effect is loss of mass as nearly all polymers lose some mass during irradiation. It can be measured by energy loss analysis, or more easily by monitoring the electron beam intensity of the transmitted beam, in other words, the contrast changes of the image of an amorphous polymer. The second method is more sensitive and therefore more accurate, mainly because a large area of the specimen can be utilised. The maximum contrast can be obtained by the smallest projective lens aperture and the image is reduced in size by the objective lens to fit into the Faraday cage at the screen level. The relative mass loss can be calculated for the measured intensities:

$$m/m_o = \log I/I_o (\log I_1/I_o)^{-1} \qquad [2]$$

where $m_o$ is the initial mass loss, $I_o$ the signal for zero mass and $I_1$ the initial signal for mass $m_o$. A correction for the carbon coating thickness should be made, but for thicknesses less than 50Å this is negligible. Some results are shown in figure 2. It is apparent that different polymers will lose mass at different rates and to a different residue level, depending on the formation of the volatile gases. The gases are diffusing away from the specimen and the mass loss is thus dependent on the specimen thickness and also on the irradiation rate. This effect however was small for all of the polymers investigated. The equation [1] again will describe the mass loss very well.

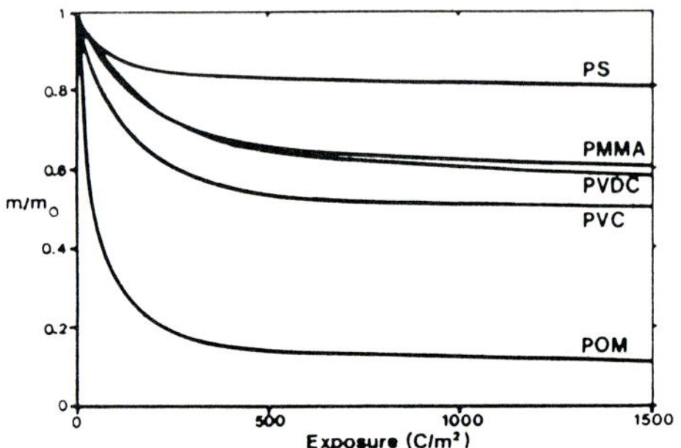


Fig. 2. Mass loss measured for different polymers.

## 4. Loss of elements.

The mass loss measurement will not provide information on elements which form the volatile compounds. An X-ray elemental analysis is very useful in this respect. An EDX signal from a selected energy window can be taken directly into a pen recorder or a computer. The time scale must be converted to exposure, from a known exposure rate. For light elements a windowless detector is required. Some measured curves are shown in figure 3. The decay curves can be best described by equation [1], but it is difficult in this case to understand why the measured curves do not correspond to a single exponential curve. The existence of two different structures in a homopolymer must be excluded, as there is no evidence to support

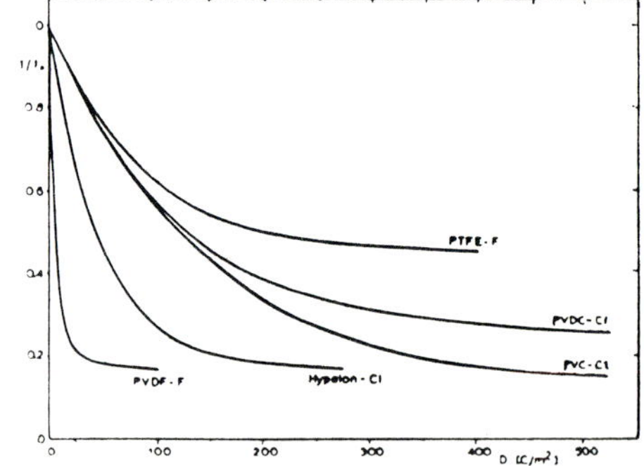


Fig. 3. Decay curves showing the loss of Cl and F from some common polymers.

this, so the only explanation could be that a certain portion of the element in question is incorporated into the new structure, formed by the beam damage. This process however would be described by three exponentials: $e^{-k_1 D}$ - the decay of the original structure, $B(1-e^{-k_3 D})$ - the formation of the new structure and $e^{-k_2 D}$ - the decay of the new structure:

$$I/I_0 = e^{-k_1 D} + B(1-e^{-k_3 D})e^{-k_2 D} \qquad [3]$$

The new structure is however formed by the decay of the original structure i.e. $k_1 = k_3$ and for a stable structure $k_1 \gg k_2$ and $k_1 + k_2 \simeq k_1$. Equation [3] will thus be identical with [1]. It can be seen that the elemental loss is more informative than the loss of crystallinity or mass, and leads to more speculations on the chemistry of the radiation process. A direct study of the chemical changes at various degrees of degradation is however most valuable.

## 5. Infrared spectroscopy.

One of the most suitable techniques for the study of chemical changes throughout various stages of irradiation is IR spectroscopy. In order to increase the signal to noise ratio, the irradiated area of the specimen must be as large as possible, however with modern techniques such as FTIR, an area obtainable in an electron microscope can be used. An ideal thickness of the specimen is about 20 microns, but with this thickness the current density at the lower levels of the specimen is slightly reduced, even at 200 keV. This makes the dosimetry more difficult, however the results will be affected more severely by poor IR spectra, than by averaging of the exposure, and specimens about 20 microns thick were therefore used. The changes in the IR spectrum with irradiation can be seen in figure 4. A quantitative analysis, which is needed for degradation studies, presents some difficulties. For example, the background, which is changing with irradiation, must be reliably established and any overlapping peaks resolved. It is thus essential to obtain spectra of highest quality for the same specimen, irradiated progressively to higher exposures. The interpretation of the spectra requires some knowledge and experience with a particular compound and additional information, e.g. from mass loss or chemical staining of double bonds, can be valuable. The chemical changes can be divided into four groups:

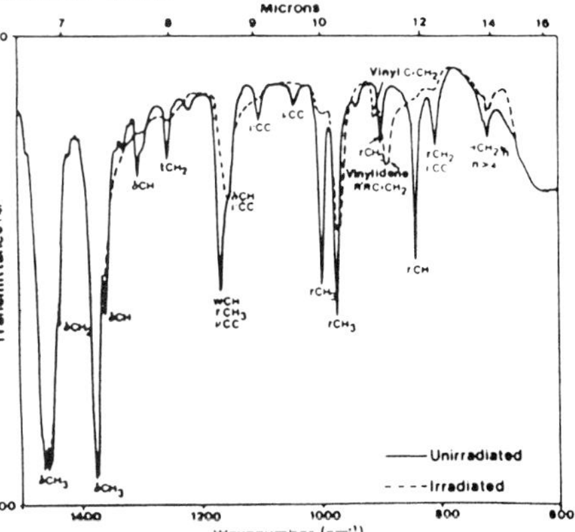


Fig. 4. An IR spectrum of PP irradiated to $100 C/m^2$.

a) changes associated with the chain conformations,(such as helix structure, cis-trans isomerism, isotactic-syndiotactic sequences). These are usually very rapid (see figure 5) and, in the case of crystalline polymers, can be related to loss of crystallinity.

b) changes associated with loss of elements. These should give identical data to X-ray measurements (see figure 6).

c) changes associated with the formation of a new chemical structure.

These are represented by peaks which appear with irradiation and are of great importance but cannot easily give quantitative information.

d) formation of intermediary structures such as vinyl groups, shown in figure 5 for PP. These have a peak at a characteristic exposure and are very valuable for polymer analysis.

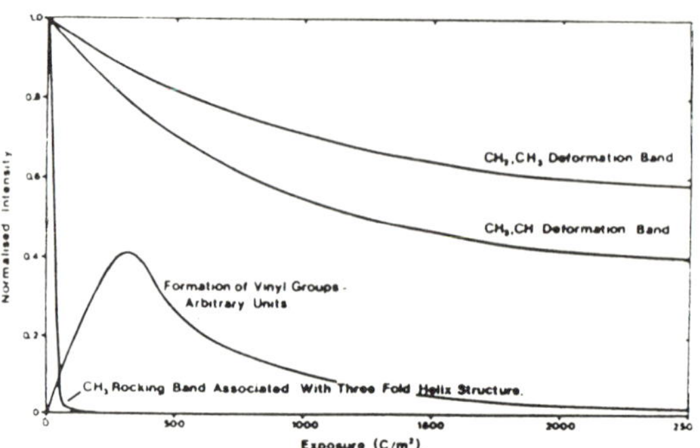


Fig. 5. Measured decay curves for PP from IR analysis.

All the decaying peaks in groups a) and b) can be related to exposure by using the equation [1], group c) cannot be described quantitatively and group d) can be described by an equation of the form:

$$I/I_0 = B(1-e^{-k_1 D})e^{-k_2 D}$$

where B, $k_1$, and $k_2$ are constants.

## 6. Conclusion.

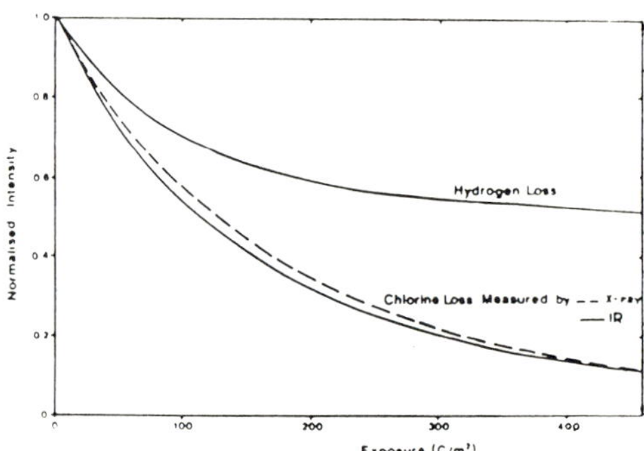


Fig. 6. Decay curves for PVC showing both Cl loss and loss of hydrogen with irradiation.

It can be concluded, that all of the processes which show a decay can be described well using an equation with two exponentials. They represent the initial fast decay of the original structure and a slow decay of another structure, which is often formed by the electron beam irradiation. Therefore, three independent constants are needed for the characterisation of a particular beam damage process in a given polymer. This is in principle, similar to the G value, but the G value is only used for the very beginning of the degredation of the original material and its dependence on the exposure is thus not considered. The G value can be calculated from a decay curve, which is represented by constants A, $k_1$, $k_2$ as follows:

$$G = \frac{e \rho Av\, a}{Mw\, Ek}\left((1-A)k_1 + Ak_2\right) \qquad [4]$$

where e is the electron charge, $\rho$ is the density of the polymer, Av - Avagadro number, a is the number of possible events in a monomer unit of molecular weight Mw and Ek is the energy absorbed from an electron with energy k in 100eV. It follows that the electron beam damage process in any particular material can only be fully understood by using the results of several techniques, but chemical analysis, e.g. by IR, must be included to obtain a detailed knowledge of the materials degredation.

*Inst. Phys. Conf. Ser. No 78: Chapter 1*
*Paper presented at EMAG '85, Newcastle upon Tyne, 2–5 September 1985*


# Use of a STEM in radiation damage measurements

F J Rocca

Cavendish Laboratory, Madingley Road, Cambridge, CB3 OHE

## 1. Introduction

The use of electron microscopy in the study of the structure of organic and biological specimens is restricted because of the damage caused by electron irradiation of the sample. In order to obtain a better understanding of the damage effects it is necessary to know how some specimen dependent properties e.g. crystallinity, vary as functions of the applied dose. Thus, the fading of the electron diffraction pattern with irradiation gives an indication of the doses at which high resolution imaging is no longer useful, owing to the destruction of the crystalline structure. Many other methods are available to assess the type and extent of the damage (Reimer 1984) but the diffraction technique is the most popular, mainly because of its relative simplicity, in situ nature and direct relevance to imaging. The study of the fading of the diffraction pattern is usually carried out on conventional fixed beam TEMs. In this work use is made of a new operating mode in a dedicated STEM (Vacuum Generators HB501) for acquiring such information and its advantages relative to the CTEM are discussed. Further details may be found in Rocca (1985).

## 2. Experimental

Two methods are available in a dedicated STEM for producing diffraction patterns: (i) selected area diffraction (SAD), in which a nearly parallel incident beam is rocked about a fixed point on the specimen (this method is very similar to conventional SAD) and (ii) microdiffraction, where a very small fixed probe (about 1 nm in diameter) is focussed onto the specimen and the diffracted beams leaving the specimen are scanned over a detector (a scintillator-photomultiplier combination situated at the exit slit of the electron energy loss spectrometer) by using post-specimen scan (Grigson) coils. In both methods the pattern is viewed on a CRT scanned synchronously with respect to the beam and whose intensity is modulated by the detector signal. A photographic record can be made but neither of these techniques is superior to those available on CTEMs, especially the method of MacLeod and Chapman (1977) where digital scanning (by using the tilt coils) and recording of spot electron diffraction patterns on a TEM are possible.

In this work, a new technique based on a combination of the two STEM operating modes is described. In particular, by interfacing a computer to the detector it is possible to acquire digitally recorded diffraction patterns or spot intensities which are immediately available for inspection and analysis. The use of a computer operated recording system enables, in principle, measurements at very different dose rates. This is not easy to carry out with photographic recording media where loading, exposing and removing the plates imposes severe constraints, not to mention the care needed to avoid saturation, etc., of the emulsion.

Fig. 1 shows a schematic of the STEM in the required mode. A nearly parallel fixed incident beam is produced on the specimen (simply by entering SAD mode and Spot mode on the VG instrument) with the beam size (typically 8 μm diameter) governed solely by the size of the selected area aperture. The diffraction pattern is then scanned over the detector using the Grigson coils. Unfortunately, it is not possible to perform this with the normal operating system of the microscope and an independent scan system, designed by Dr Denis McMullan, which allows the diffraction pattern to be scanned either linearly or circularly over the detector is used. It is, of course, possible to leave the diffraction pattern stationary and just detect a single Bragg spot or part of a diffraction ring. Moving the pattern circularly can produce the intensity variation of the first order spots, say, in a single crystal diffraction pattern. Typical results are illustrated in fig. 2.

The scintillator-photomultiplier detector is interfaced to an on-line computer (Link 860 series II) and the data stored on floppy discs. Note that it is possible to display a line scan through a diffraction pattern on the CRT but recording it photographically imposes the usual time constraints. A slight problem exists when entering spot mode viz. no display on the screen, but this can be overcome by inserting the microdiffraction screen between the Grigson coils and the detector. This screen can be viewed directly or, more conveniently, displayed on a TV screen by a camera. The beam current is controlled by the extraction voltage on the field emission source and can be varied over a fairly large range ($10^{-9}$ to $10^{-12}$ Å). By varying the condenser lens settings and the size of the selected area aperture a very large range of current density values may be achieved. The spot on the sample has a current density which is uniform to about 5%. The detector aperture (which corresponds to the condenser aperture in a TEM) controls the angular resolution. Typical acceptance semi-angles are of the order of 0.13 mrad, comparable to the characteristic inelastic angle for organic crystals at 100 keV. Smaller apertures can be used but great care must then be taken with the alignment of the microscope. Dwell times on the scintillator need only be of the order of milliseconds and the detection system is always operated in its linear response region (known from other studies). The scan/computer system can store up to 1028 channels of data. Each individual scan takes $\frac{1}{2}$ second, irrespective of how many channels per scan are being used; typically 128 channels per scan. A preset delay can be introduced between each scan; typically 2 to 16 seconds are used.

Beam currents are measured by allowing the beam to enter the electron energy loss spectrometer, and detuning this so that the beam hits the sides of the drift tube, which is insulated and acts as a Faraday cage. The beam currents can then be read off on an electrometer. Beam sizes are measured directly in the imaging mode by observing the illuminated area. The use of a beam blanking system and systematic moving of the specimen removes any pre-exposure.

## 3. Results

Measurements were carried out on two materials, p-terphenyl, an aromatic material, and ℓ-valine, an amino acid. The former was available as a self-supporting single crystal thin film and the latter as a thin polycrystalline film on a carbon support. $D_{1/e}$ values (dose for the intensity to decay to 37% i.e. 1/e, of its initial value) for first order reflections were 0.049 $Ccm^{-2}$ for p-terphenyl and $1.8 \times 10^{-3}$ $Ccm^{-2}$ for ℓ-valine at 100 keV. These are in agreement with previously published values (Parkinson, Goringe, Jones, Rees, Thomas and Williams 1976; Reimer 1984). Fig. 3 shows a decay curve for p-terphenyl first order spots, where no obvious buckling or bending of the samples occurred during recording.

## 4. Discussion

The method outlined above enables fast and efficient diffraction damage measurements to be made with immediate display of the results. This helps in determining whether buckling or bending of the sample has occurred. The only disadvantage is that it is not possible to illuminate a larger area than that from which the measurement is made (a direct consequence of having no post-specimen lenses) and any specimen and/or beam drift (which is also common to other methods) could introduce fresh sample and provide a misleading estimate of the life-time. Measurements of drift rates show, however, that they are negligible. Despite this, care was taken throughout the experiments to ensure that no such effects occurred. The reasons why the two available STEM diffraction modes are not suitable mainly have to do with either too large a current density (microdiffraction) or with recording constraints (SAD). The existing energy loss facilities are also available for use and a detailed study of how the low energy loss region i.e. the intermolecular structure, varies with applied dose and, in particular, its relation to the loss of crystallinity should be possible (Misra and Egerton 1984). This could provide a much better understanding of the damage processes occurring in beam sensitive materials.

The apparent double decay modes of p-terphenyl could be linked to visual observations which show that the spots preferentially become diffuse in an angular, rather than a radial, direction during the early stages of irradiation, but further work is necessary. However, the increase in the decay rate at greater exposures is slightly strange and could possibly be due to other effects, especially at the low normalized intensity values at which it occurs.

I would like to thank the British Petroleum Company for a research studentship, Dr D McMullan for his help with the scanning system and Prof U Valdrè and Dr A Howie F.R.S. for many helpful discussions.

## References

MacLeod A M and Chapman J N 1977 J.Phys. E: Scientific Instruments 10 37
Misra M and Egerton R F 1984 Ultramicroscopy 15 337
Parkinson G M, Goringe M J, Jones W, Rees W, Thomas J M and Williams J O 1976 Developments in Electron Microscopy and Analysis (London:Academic) pp 315-318
Reimer L 1984 Ultramicroscopy 11 291
Rocca F J 1985 Ph.D Thesis University of Cambridge

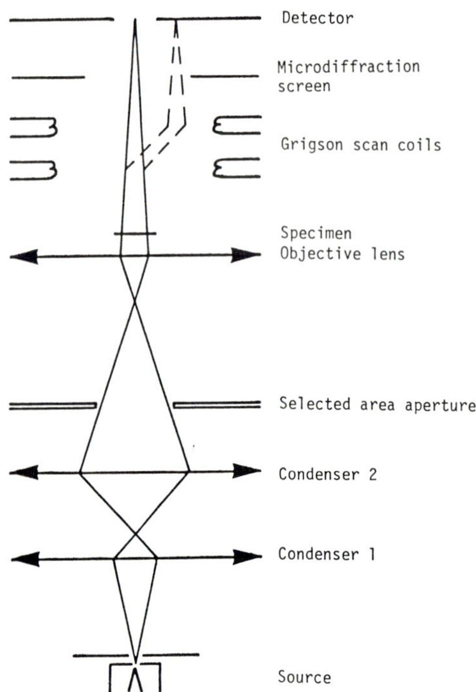


Fig.1. STEM in SAD mode with a fixed large diameter incident beam.

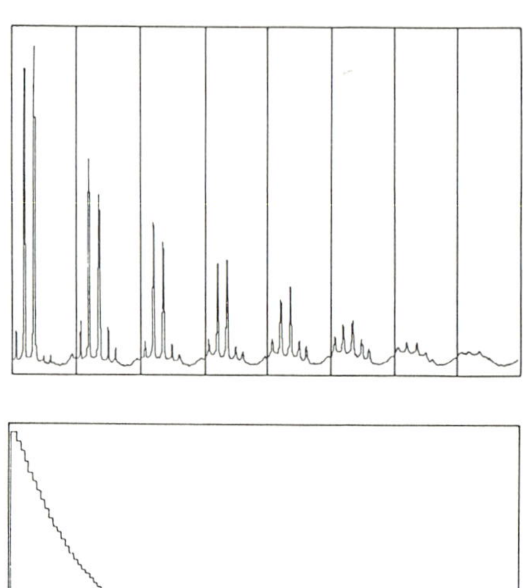

Fig.2. Top scan is for a discretely acquired pattern showing the fading of the first order reflections. Lower scan is a for a continuous measurement on a single Bragg spot.

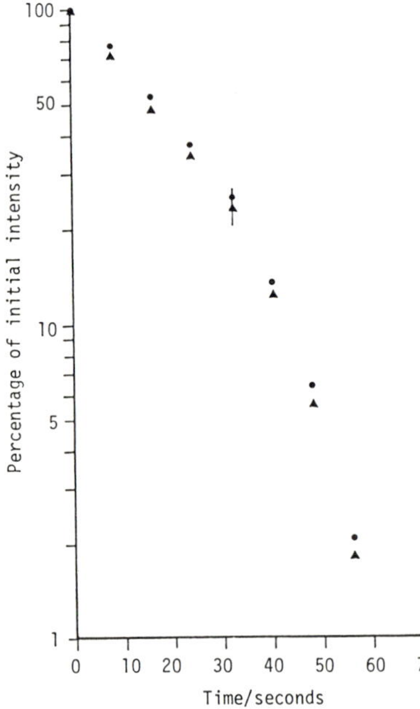


Fig.3. Plot showing decay of two separate first order spots for a p-terphenyl sample. Current density was $2.2\times10^{-3}Acm^{-2}$

Paper presented at EMAG '85, Newcastle upon Tyne, 2–5 September 1985

# The effects of electron irradiation on chalcogenide thin films

C P McHardy and A G Fitzgerald

Carnegie Laboratory of Physics, The University, Dundee DD1 4HN.

## Introduction

The use of amorphous chalcogenides coated with silver as high resolution photoresists or electron resists has been reviewed by Mizushima and Yoshikawa (1982). These applications are based on the differing solubilities of irradiated and unirradiated chalcogenide/silver films in alkaline solution and lead to the possibility of producing high resolution patterns defined by the extent of the irradiating beam.

This paper presents the results of a study of the effects of electron irradiation on films of arsenic and germanium chalcogenides coated with silver. The films have been investigated at different electron irradiation intensities and for different preparation conditions.

## Experimental Procedure

Thin films of amorphous $As_2X_3$ and GeX (X = S, Se, Te) under or overcoated with silver were prepared by vacuum evaporation. Thin films for study in the transmission electron microscope (TEM) were deposited on rocksalt and eventually mounted on aluminium grids. These grids are inert to the chalcogenides studied (Fitzgerald and McHardy 1984). Films for scanning electron microscope (SEM) analysis were deposited on aluminium or glass substrates. Film composition was determined by energy dispersive x-ray microanalysis in the TEM and SEM and surface composition was determined by Auger electron spectroscopy (AES) in an ultra-high vacuum SEM. Film thicknesses were determined by multiple beam interferometry.

## Transmission Electron Microscope Studies

The majority of amorphous chalcogenide/silver films are sensitive to the electron beam at normal electron intensities. Electron diffraction studies of the effect of contact of these materials with silver had to be carried out at low beam intensities. In almost every case the silver chalcogenide is formed without exposure to intense u.v. or electron irradiation (Fitzgerald and McHardy 1985).

The effects of electron irradiation by a focussed electron beam in the TEM upon silver coated $As_2S_3$, $As_2Se_3$, GeS and GeSe films have been described in detail (Fitzgerald and McHardy 1984 and 1985). Silver diffuses away from the irradiated area. This effect can be used to write patterns in the film. Recently films of $As_2Te_3$ and GeTe in contact with films of silver have been investigated. $As_2Te_3$ films coated with silver behave in the same way as the other chalcogenides studied (Fig. 1) whereas GeTe

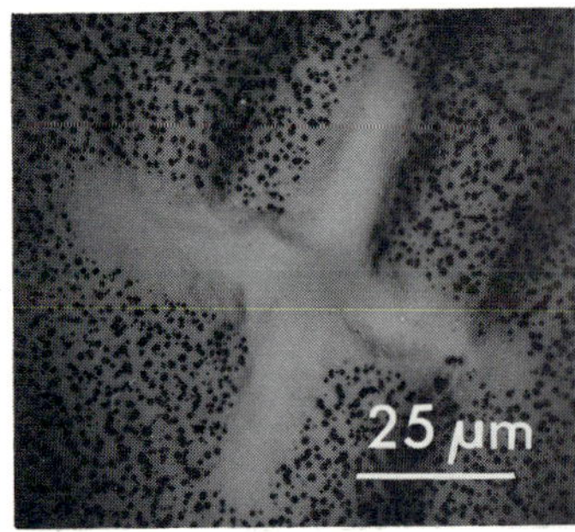


Fig. 1 Pattern produced in $As_2Te_3$/Ag film by a focussed electron beam.

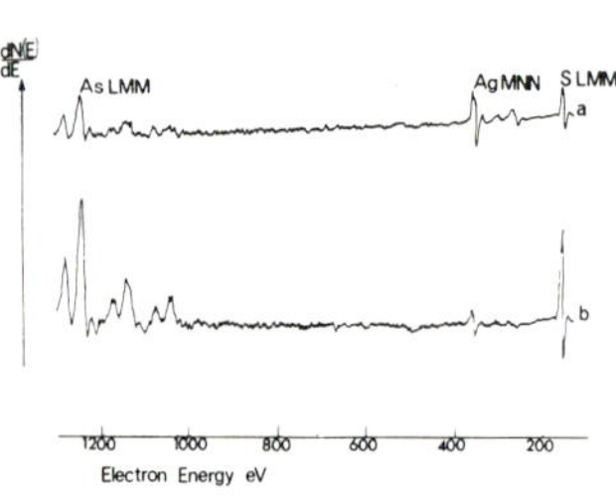


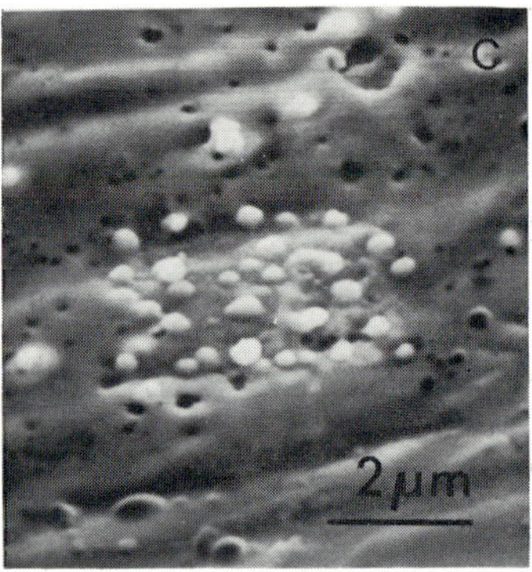


Fig. 2 Auger spectra from an area of $As_2S_3$/Ag (800 Å) (a) before high intensity electron irradiation (b) after high intensity electron irradiation. Fig. 2(c) is a scanning electron micrograph of the area irradiated at high intensity.

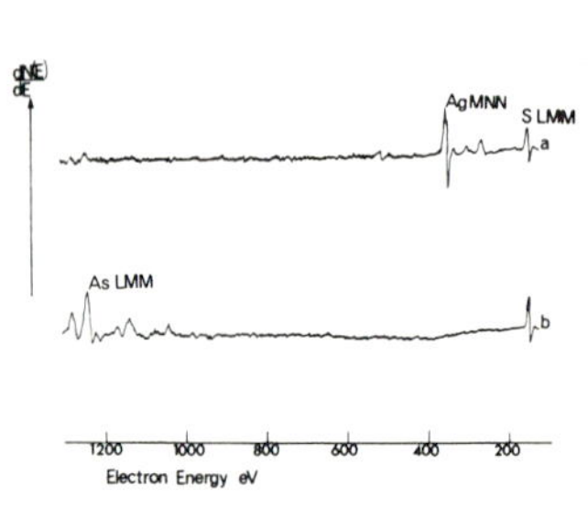


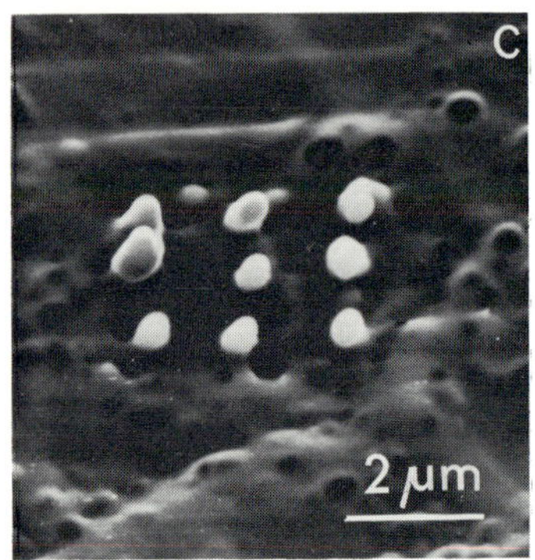


Fig. 3 Auger spectra from an area of $As_2S_3$/Ag (2400 Å) (a) before high intensity electron irradiation (b) after high intensity electron irradiation. Fig. 3(c) is a scanning electron micrograph of the area irradiated at high intensity.

films coated with silver are unaffected by a focussed electron beam. Electron diffraction studies at low beam intensity show that polycrystalline $Ag_2Te$ forms in GeTe/Ag films at room temperature whilst for $As_2Te_3$/Ag films a polycrystalline film forms with an as yet unidentified structure.

## Scanning Electron Microscope Studies

In a previous AES study of silver films overdeposited with chalcogenide, with glass as the supporting substrate (Fitzgerald and McHardy 1985), it was noted that silver diffused extensively through the chalcogenide and in some cases penetrated to the free surface. Intense electron irradiation, above a current density of the order of 0.5 nA/$\mu m^2$, resulted in the diffusion of silver away from the irradiated area. Below this current density the films were stable and consistent Auger traces were obtained on sustained electron exposure. A similar behaviour was observed for these films in the TEM.

When silver is overdeposited on the chalcogenide, which is the normal preparation sequence in producing electron resists in chalcogenides, the films were found to be unstable, even to the lowest electron irradiation intensities used to obtain Auger electron spectra. The films were however, stable to typical electron irradiation intensities used to record electron images and obtain EDX spectra in the SEM.

In order to investigate whether these effects were thickness dependent or related to the nature of the supporting substrate electron irradiation studies have been made on varying thicknesses of $As_2S_3$ overdeposited with silver at a concentration of 10wt.%. The substrates used were aluminium and glass.

The behaviour of films deposited on aluminium substrates depended on thickness. Auger traces from the thinnest films studied (800 Å) showed little variation in surface composition on electron irradiation. However, some beam damage and reduction in surface silver concentration was observed at high beam intensities (Fig. 2). For 2400 Å films exposure to an intense electron beam caused silver to diffuse away from the surface of the irradiated area and contrast changes were observed (Fig. 3). For 3400 Å films a similar behaviour was observed. However, at the highest beam intensities, instead of silver diffusing away from the irradiated area, the electron beam appeared to raise the film temperature sufficiently to cause local melting. This resulted in most of the lower melting point chalcogenide being lost from the irradiated area, and the silver concentration being enhanced (Fig. 4). These effects have also been confirmed by EDX analysis. Local melting was also observed at the irradiated area in the thickest films studied (4500 Å) and silver was not observed to diffuse away at any electron beam intensity.

The behaviour of thinner films on glass was difficult to study because of charging due to the penetration of the electron beam through to the substrate. For thicker films the effects of electron irradiation were similar to those for aluminium substrates. An additional interesting effect was the dendritic growth of silver away from the edge of the irradiated area (Figs. 5, 6).

## Conclusions

Contact reactions occur between silver and all of the amorphous chalcogenides of arsenic and germanium. Apart from GeTe/Ag these films are sensitive to the electron beam. In electron beam sensitivity studies diffusion of silver away from the irradiated area in $As_2S_3$/Ag probably arises from the temperature gradient between this area and the surrounding film. The local melting observed in thicker films results from the confinement of the electron beam within the deeper low thermal conductivity $As_2S_3$ layers. These observations suggest that careful control of current density is required in electron resist applications.

## References

Fitzgerald A G and McHardy C P 1984 Inst. Phys. Conf. Ser. 68 457.
Fitzgerald A G and McHardy C P 1985 Surface Science (in press).
Mizushima Y and Yoshikawa A 1982 Amorphous Semiconductor Technologies and Devices (Amsterdam: North Holland) pp 277-295.

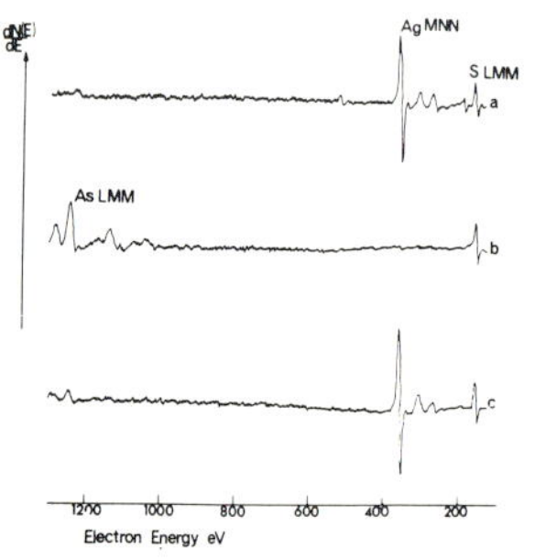


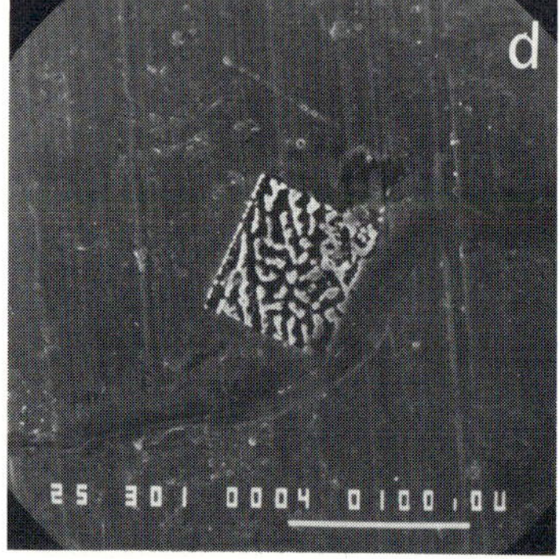


Fig. 4 Auger spectra from areas of an $As_2S_3$/Ag (3400 Å) film after irradiation at current densities of approximately (a) $2 \times 10^{-4}$ nA/μm$^2$ (b) 0.02 nA/μm$^2$ (c) 200 nA/μm$^2$. Fig. 4d is a scanning electron micrograph of the area irradiated at 0.02 nA/μm$^2$.

Fig. 5 Scanning electron micrograph from an area of $As_2S_3$/Ag deposited on glass and irradiated at high intensity.

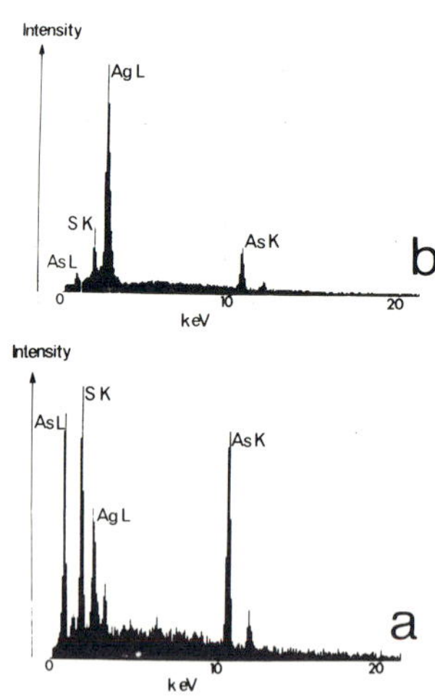


Fig. 6 EDX spectra from (a) unirradiated area and (b) area with dendritic growth shown in Fig. 5.

# The electron irradiation of silicon of varying oxygen content

W Bergholz*, I G Salisbury, J L Hutchinson and C J Humphreys

Department of Metallurgy and Science of Materials, University of Oxford, Parks Road, Oxford OX1 3PH

*Now at 1V Physikalisches Institut, Universitat Gottingen, Bunsenstrasse 11-15, 3400 Gottingen, Germany.

Elongated defects on {1 1 3} are found in Czochrälski (CZ) silicon after electron irradiation at temperatures above about 400°C, or following protracted (>100h) annealing in the 400°C to 700°C range. The structures may be over a micron long in <1 1 0>, the direction in which they are extended, and in irradiated material may be up to 30nm wide (e.g. Salisbury and Loretto, 1979). These workers favoured a self interstitial model, but high resolution studies have thus far been ambiguous; Tan et al (1981) and Pasemann et al (1983) favour a stacking fault interpretation, while Dessaux-Thibault et al (1983) have proposed an oxide model. There is a similarity between those defects and these found in annealed CZ silicon, which are generally taken to be coesite precipitates (e.g. Bergholz et al, 1985).

In the present work the defects found after 800kV irradiations at 400+30°C in two specimens were compared. The first was typical of CZ material, with an oxygen concentration of $8 \times 10^{17}$ atoms $cm^{-3}$, while the second was float zoned (FZ) silicon with less than $10^{16}$ atoms $cm^{-3}$. Each sample had a total dopant concentration of less than $10^{16}$ atoms $cm^{-3}$. This is preliminary to a more extensive investigation in which the effect of oxygen, carbon and dopant concentration upon the nature and density of the defects produced will be studied.

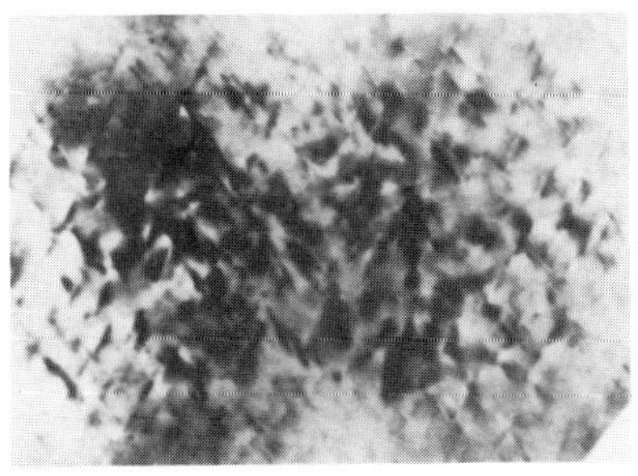


fig.1 CZ silicon after 800KV electron irradiation at 400°C

A clear difference between the behaviour of the two materials was noted. In CZ silicon the first rod-like defects were observed after about 20 seconds and the line density of these quickly built up to exceed $10^9$ $cm^{-2}$ after 3-4 minutes (fig.1). The first defects did not appear in FZ material until after 7 minutes, and, following a further 6 minutes of irradiation, the rather lower density of damage showed no sign of increasing (fig.2).

fig.2 FZ silicon after 800KV electron irradiation at 400°C

Diffraction contrast images suggest that the structure of these defects is not simple, but on the basis of inside/outside movements on reversing g in weak beam it would appear : 1) the structures are very narrow, not generally greater than 5nm and 2) the displacement, R, is not far from the x<0 0 1> reported for similar precipitates found following the annealing of CZ silicon (A and B in fig.3). There is a margin of uncertainty in the displacement direction, however, the present diffraction contrast analysis has confirmed that R is perpendicular to the long axis - the defects show total invisibility for g parallel to the axis - but does not preclude directions close to <0 0 1> for which one index is considerably larger than the other two. Relatively high index reflections are required to distinguish between an <0 0 1> displacement of unknown magnitude and, for example, the a/25<1 1 6> displacement reported by Ferreira Lima and Howie (1976) for {1 1 3} defects in irradiated germanium and, even in weak beam, these give rather poor images of such narrow structures. In this context a more detailed account of the analyses which have established an <0 0 1> displacement for the rod-like defects in annealed CZ silicon would be of interest.

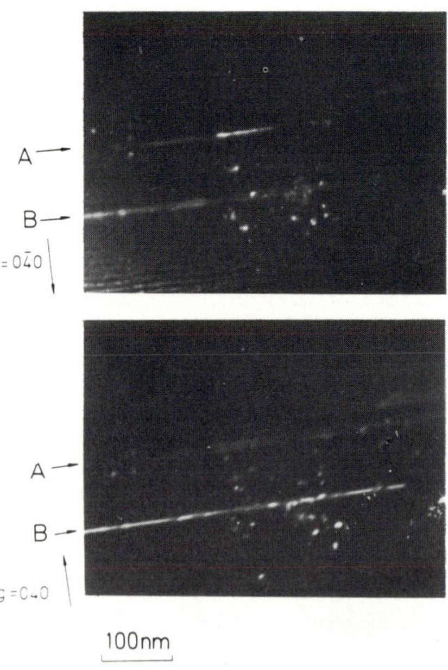


fig.3 Inside/outside contrast for rod-like defects in CZ silicon

Elongated interstitial loops with Burgers vectors of a/2<1 1 0>, perpendicular to their <1 1 0> long axes, have been found in both silicon specimens. Since the habit plane of these is usually close to {1 1 3}, Salisbury and Loretto (1979) suggested that they resulted from the unfaulting of the rod-like defects, but if the latter are taken to be oxide precipitates, an alternative explanation is required.

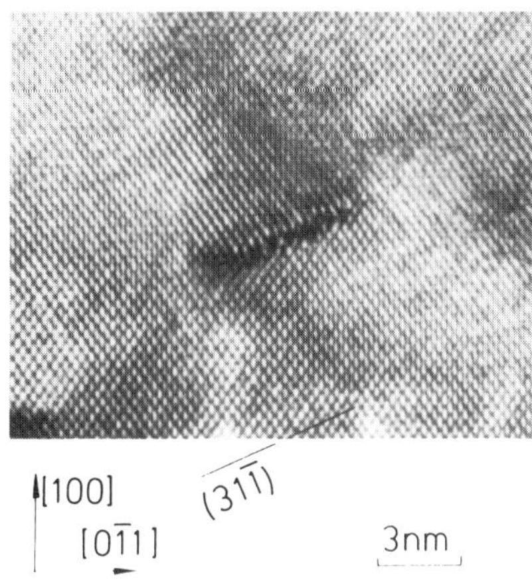


fig.4 $(3\ 1\ \bar{1})$ defect in CZ silicon

To obtain more detail it is necessary to use high resolution microscopy. Figure 4 show a typical {1 1 3} defect in irradiated CZ silicon and these structures closely resemble the rod-like defects in annealed material (e.g.Bergholz et al, 1985). One feature of note is the low level of lattice strain around the defect, a feature which makes unambiguous characterization by diffraction contrast difficult.

Not all the defects are single faults however, and figure 5 shows two examples of a complex involving precipitation on both {1 1 1} and {1 1 3} planes, with the stair rod running along the <1 1 0> long axis. While the contrast from the {1 1 3} portion remains difficult to interpret, the {1 1 1} structure is an interstitial fault with a displacement close to a/6<1 1 4>. These non-edge {1 1 1} faults, which show intrinsic fault fringes in diffraction contrast, have been reported in implanted silicon (Salisbury 1981) and several mechanisms have been proposed for their generation. Salisbury (1981) considered that their nucleation might be favoured over that of edge defects,

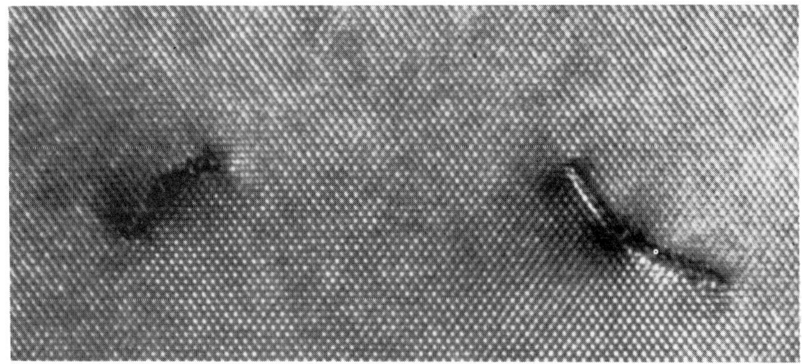

fig.5 Complex defects in CZ silicon

despite their higher boundary dislocation energy, because the interstitials can precipitate between widely spaced {1 1 1}'S({1 1 1}partial planes in the diamond structure are alternately a/4<1 1 1> and a/12<1 1 1> apart, and an edge fault must precipitate between the latter). Nucleating alongside a defect with a displacement of (approximately) <0 0 1> type, which is much closer to <1 1 4> than to <1 1 1>, would also favour the non-edge configuration.

While a detailed discussion of the present results will be given elsewhere, it should be emphasized that both the nature of the defects and the mechanism for their generation remains uncertain. The similarity between the defects in annealed and implanted CZ silicon suggests that both may be coesite precipitates and this view is supported by the difference in nucleation rate and density observed when irradiating FZ material. On the other hand, the amount of oxygen in FZ silicon is not sufficient to account for even the small number of elongated defects found after irradiation, and the density of defects found in CZ material is sometimes slightly too high to be compatible with the initial oxygen content. Some explanation for the interstitial a/2<1 1 0> loops found on {1 1 3} after irradiation (Salisbury and Loretto, 1979) is also required, and the possible role of surface oxide (e.g. Pasemann et al, 1983) requires further investigation.

Hundreds of hours of annealing are required to produce visible defects in CZ silicon (e.g. Bergholz et al, 1985), whereas similar structures are produced within minutes during irradiation. The thermal annealing results have been interpreted as implying a considerable increase in the rate of oxygen transport (Bergholz et al), possibly as a result of a rapidly diffusing oxygen/self interstitial complex, and this is clearly enhanced still further during irradiation.

Bergholz W, Hutchinson J L and Pirouz P 1985 J.Appl.Phys. in press.
Desseaux-Thibault J, Bourret A and Pennison J M 1983 Inst.Phys.Conf. Ser. (Bristol: Adam Hilger) 67 71
Ferreira Lima C A and Howie A 1976 Phil. Mag. A 6 1057
Pasemann M, Hoehl D, Aseev A L and Pchelyakov O P 1983 phys.stat.sol. (a) 80 135

Salisbury I G 1981 Acta Metall. 30 627
Salisbury I G and Loretto M H 1979 Phil.Mag.A 39 317
Tan T Y, Foll H and Krakow W 1981 Defects in Semiconductors (North Holland) pp179

# The significance of radiation damage to the determination of the composition of vanadium carbide precipitates in steels

S P Duckworth*, D G Howitt+, A J Craven° and T N Baker*

* Department of Metallurgy, University of Strathclyde, Glasgow.
+ Department of Mechanical Engineering, University of California, Davis
o Department of Natural Philosophy, University of Glasgow.

## 1. Introduction

When making a quantitative assessment of the precipitation hardening contribution to the strength of a steel, it is generally assumed that the precipitates will have chemical compositions lying in a small well defined range. Since industrial processes may lead to the formation of precipitates with compositions outside the expected range, a technique for analysis of such particles is desirable, and electron energy loss spectroscopy (EELS) has been investigated for this purpose. A study of vanadium carbide precipitates on extraction replicas, using an HB5 STEM, reported apparent carbon contents significantly lower than expected (Duckworth et al, 1983). Evidence for preferential loss of a light element during STEM microanalysis has also been observed for NiO (Crozier et al, 1984) and TiC (Thomas, 1984). The current study is part of an investigation of radiation damage effects sustained by vanadium carbide precipitates subjected to a focussed beam of electrons. In order to monitor radiation damage by observing changes in the diffraction patterns of precipitates during electron bombardment, a series of irradiation experiments were performed on vanadium carbide precipitates on extraction replicas.

## 2. Experimental details

The experiments were carried out in a Philips EM 400T with $LaB_6$ filament. Although this has a higher brightness than the conventional tungsten filament, and the EM 400T operates with a higher beam current than the HB5 STEM, the latter's higher brightness allows it to produce a far higher current density. Thus any radiation damage effects observed in the EM 400T are likely to be less drastic than those experienced in the HB5. However, the temperature rise in the specimen may be considerably greater in the 400T.

The precipitates used in these experiments were removed from a microalloyed vanadium steel and are expected to have a composition of $VC_{0.75}$ to $VC_{0.97}$ (Storms et al, 1962). Initially, sputtered silicon replicas, as used for EELS analysis of the precipitates, were examined. For comparison, carbon and aluminium replicas were subsequently investigated. All replicas were prepared from the same steel specimen and given similar washing treatments.

Since the precipitates were generally less than 100 nm in size and widely separated, a selected area diffraction (SAD) pattern obtained from the normal precipitate distribution was relatively faint. Denser clusters of precipitates were used to give a stronger diffraction pattern. During the experiments, energy dispersive X-ray spectroscopy was used to confirm that no polishing residue contaminated these denser clusters.

A micrograph and diffraction pattern were obtained from each region of interest prior to focussing the electron beam to a spot and subjecting the precipitates to a high electron dose. At various stages in the irradiation, the illumination was defocussed and a diffraction pattern and image recorded.

To maximise the electron dose at the specimen, the condenser aperture was removed during irradiation. This resulted in a relatively large focussed beam diameter of ~1 µm, and this matched the area defined at the specimen by the 30 µm SAD aperture used for the diffraction patterns.

An estimate of the beam current density at the specimen was obtained from a calibration chart of total screen current vs. exposure time (supplied by Pye Unicam). All these experiments were performed with a current density of approximately 10A $cm^{-2}$. In the HB5 STEM, current densities in a defocussed 15 nm probe are of the order of 600A $cm^{-2}$.

Experiments were carried out at 100, 80, 60 and 40 keV to detect any energy threshold for the observed damage. If such a threshold were found, it might also apply to experiments in the HB5 and permit EELS analysis to be performed with less damage.

## 3. Results and discussion

The major feature of this study is the rapid onset of damage on the silicon replica. This is not seen on the carbon replica, where evidence of damage is only apparent after a much longer irradiation period. Aluminium replicas show an intermediate behaviour. These effects have been observed both in separate experiments on the three replica types and in a direct comparison where a multispecimen holder was used to enable tests to be made on the different replicas under essentially constant microscope conditions.

A typical sequence of micrographs of a particular region during irradiation at 100 keV is shown in Fig. 1. Although some precipitates appear unchanged in shape, others agglomerate into larger particles. Additionally the silicon replica is thinned and in some regions perforated.

The corresponding diffraction pattern sequence is shown in Fig. 2. Crystallisation of the initially amorphous silicon support film under the focussed electron beam is indicated by the development of three distinct continuous narrow rings, including the strong innermost one. For control purposes, an amorphous silicon film prepared on KCl under similar conditions was given the same irradiation, and the resulting ring pattern compared with those from the replicas. The rings corresponding to the continuous rings on the replica were observed but no others. Etching and grain growth also occurred on this silicon film. The pre-dose diffraction patterns from the replicas gave ring ratios which closely matched the expected values for the FCC vanadium carbide structure. Upon irradiation these rings became broadened, with spots occurring at other distances from the centre.

Similar damage effects have been observed on silicon replicas at 80, 60 and 40 keV. For the same current density, the rate of damage appears to be greater at lower incident electron energy, but the agglomeration of the precipitates gives a smaller structure.

Carbon and aluminium replicas, under comparable electron dose conditions, do not show such marked effects. No significant change has been observed in micrographs of precipitates on a carbon replica, even after irradiation for substantially longer periods. There is no evidence of thinning of the replica or destruction of precipitates. However, the diffraction patterns show some broadening of the rings (Fig. 3), although to a far smaller extent than for the silicon replicas. The aluminium replica showed a greater tendency to become thinned and perforated after irradiation. Although the precipitates again showed no tendency to agglomerate or be destroyed, the diffraction patterns showed broadening of the rings, at a rate intermediate between silicon and carbon replicas.

It is not clear why the silicon replica differs so markedly from the others. A number of things may be responsible; increased temperature rise due to poorer thermal contact with the support grid; a reaction with the silicon and the oxygen known to be present on its surface from EELS analyses; or some damage already induced by the sputtering process.

Interpretation of the ring broadening in the diffraction patterns is not straightforward. In the early stages of damage, the broadening can be interpreted as the formation of orthorhombic $V_2C$. This would be consistent with the high V/C ratios obtained from EELS analyses, if the same process operated in the HB5 STEM. As the dose to the silicon replica is increased, particularly at lower keV, the range of spot positions increases so that the pattern is no longer entirely consistent with $V_2C$. The interpretation is unclear as the ring spacings for a range of possible phases, e.g. carbides and silicides are quite similar and no single set matches the patterns.

The lack of an energy threshold to the damage effect, and the apparent increase in damage rate at lower energy, suggest that there is an ionization process rather than a displacement process under electron irradiation in these experiments.

## 4. Conclusions

At this stage it is uncertain precisely what reaction is occurring on the silicon replica, although clearly there has been extensive radiation damage. In view of the major differences between the electron beam conditions in these experiments and those occurring during HB5 STEM EELS analysis, it is not possible to predict from these results what will happen in the latter case. The higher electron dose rate and the much lower total current in the HB5 STEM focussed probe may cause different damage mechanisms to occur. However, it seems likely that vanadium carbide precipitates extracted onto a silicon replica prepared by the present technique will lead to serious radiation damage effects which must be allowed for in any quantitative EELS analysis.

## 5. References

Crozier P A, Chapman J N, Craven A J and Titchmarsh, J M 1984 Analytical Electron Microscopy - 1984 ed D B Williams and D C Joy (San Francisco: San Francisco Press Inc.) pp79-82

Duckworth S P, Craven A J and Baker T N 1984 Electron Microscopy and Analysis 1983 (London: Inst. Phys) pp339-342

Storms E K and McNeal R J 1962 J. Phys. Chem. 66 1401
Thomas L E 1984 Analytical Electron Microscopy - 1984 ed D B Williams and D C Joy (San Francisco: San Francisco Press Inc.) pp358-362

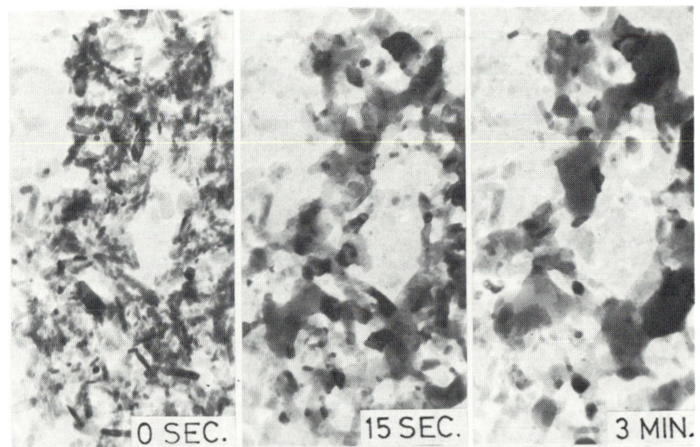


Fig. 1

Damage to VC precipitates on silicon replica at 100 kV.

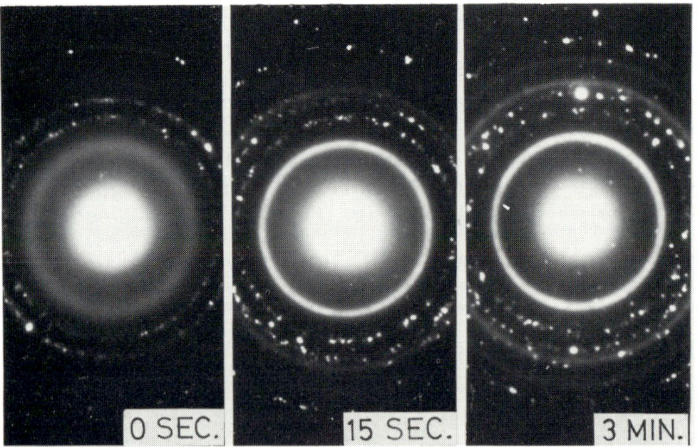


Fig. 2

SAD patterns corresponding to Fig. 1.

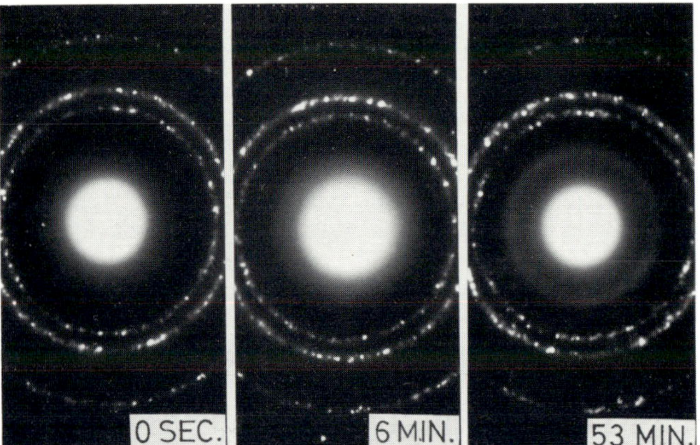


Fig. 3

SAD patterns showing lesser damage to precipitates on carbon replica at 100 kV.

*Inst. Phys. Conf. Ser. No 78: Chapter 1*
*Paper presented at EMAG '85, Newcastle upon Tyne, 2–5 September 1985* 

# STEM microanalysis of radiation induced products in ceramics

Yoshio MATSUI*
Cavendish Laboratory, University of Cambridge, Madingley Road, Cambridge CB3 OHE
*On leave from National Institute for Research in Inorganic Materials, 1-1 Namiki, Sakura, Niihari, Ibaraki 305, Japan

## 1. Introduction

Electron beam irradiation in the TEM often causes structural change in inorganic compounds (Hobbs, 1979). Recently, we reported two examples; a) formation of small particles by irradiation of hBN (Matsui, 1984) and b) decomposition of $KFe_{17}O_{25}$ with beta'''-alumina type structure by an electron beam (Matsui et al., 1985). In this paper, we report the results of microanalyses of these radiation-induced products by using dedicated FEG-STEM. Results of high-resolution TEM are also presented.

## 2. Small Particles Formed by Electron Irradiation of hBN

The crystal of hBN was irradiated by the electron beam in the TEM to form small particles around the crystal. These were shown to be particles of Cu by STEM-EELS analysis ($L_{2,3}$ edge at 930 eV), contrary to the previous assumption of cubic boron nitride (cBN) by Matsui (1984).

## 3. Electron Beam Decomposition of $KFe_{17}O_{25}$

This compound has β'''-alumina type of structure shown in Fig. 1. It consists of alternate stacking of spinel block (SB) and conduction planes (CP). Each SB has six close packed oxygen layers. Two neighbouring SBs have mutually opposite oxygen packings; (..ABCA..) and (..ACBA..). CPs are loosely packed and, therefore, imaged as lines of white dots in the HREM images taken under the Scherzer condition, as shown in Fig.5.

It is well-known that β'''-alumina itself ($NaAl_{15}Mg_2O_{25}$) is quite stable even under intense electron beam, while β''- and β''''-aluminas with rhombohedral structures easily suffer from damage during HREM observations (Matsui & Horiuchi, 1981; Matsui, 1981; De Jonghe, 1977; Bovin, 1979; Hull et al, 1981). Quite different from stable $NaAl_{15}Mg_2O_{25}$, $KFe_{17}O_{25}$ with identical structure is beam sensitive; some of CPs are lost, as can be seen in the central part of Fig. 2. This indicates that potassium atoms gradually go out of the structure, leaving defect blocks created by the direct contact of two SBs.

If we remove the second condenser aperture to irradiate strongly, significant change of structure is induced, as can be seen by the change of diffraction pattern in Fig.3, indicating the formations of a twinned cubic structure. To check

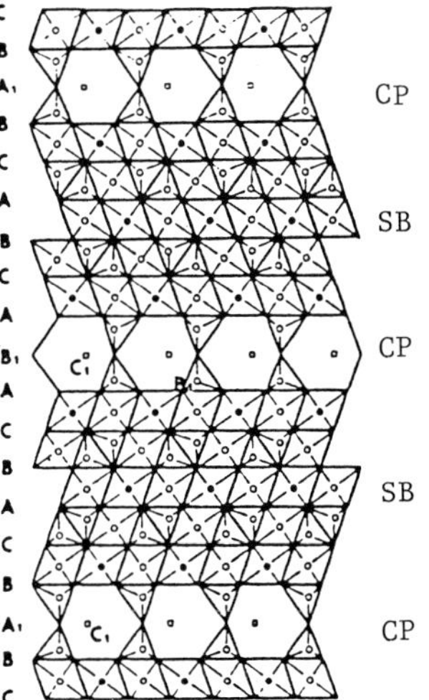


Fig. 1 Crystal structure of $KFe_{17}O_{25}$ with β''' type structure.

the change of chemical composition, EDX analyses were performed by HB5 STEM. Results are as shown in Fig. 4 where the K line of potassium detected in (a) is lost in (b) obtained after an intense electron irradiation. These diffraction and EDX data suggest that the final product of damage is magnetite ($Fe_3O_4$) with the spinel structure. The corresponding KREM image (500kV) is shown in Fig. 5.

Considerable atomic rearrangements are considered to take place to form such a twinned magnetite structure, as suggested before (Matsui et al., 1985). Briefly, when one of CP is lost, two SPs on both sides are combined directly, preserving mutual mirror relations. When the third SP is combined, rearrangements of both oxygen layers and Fe cations take place in the central part to form a spinel slab of 18 oxygen layers. This kind of process leads to the formation of twinned spinel lamella as shown in Fig. 5. Structural defects are sometimes introduced during this process as shown in Fig. 6.

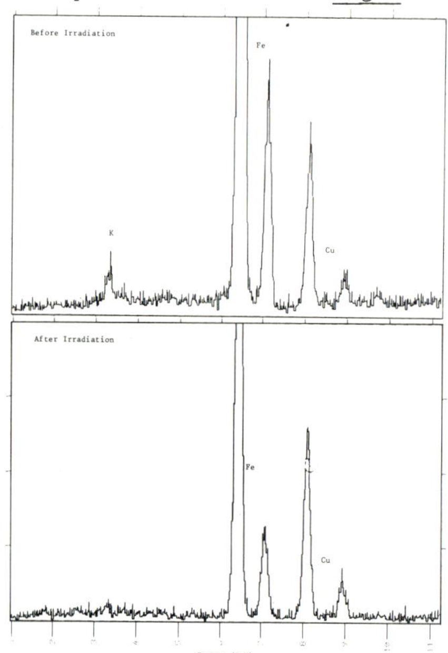


Fig. 4. EDX spectra of $KFe_{17}O_{25}$ before and after electron irradiation. (HB5 Cambridge)

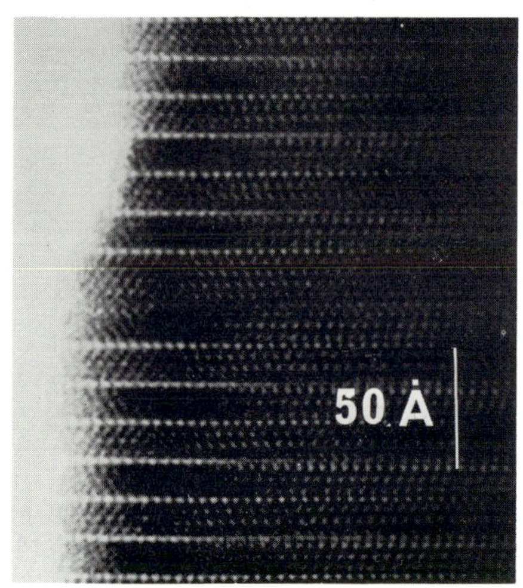


Fig. 2. HREM image of $KFe_{17}O_{25}$ with beam induced damage in the central part. (200kV Oxford)

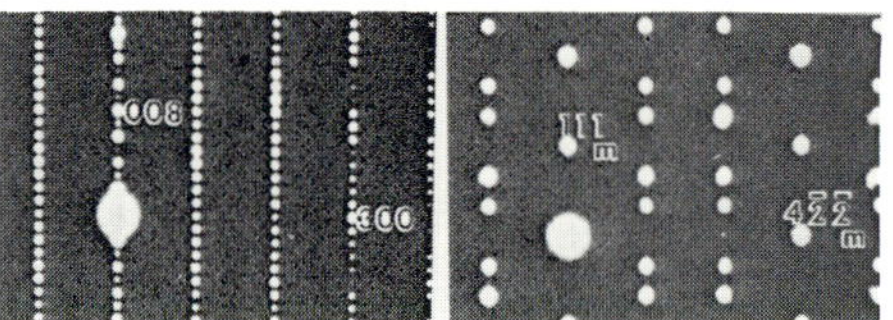


Fig. 3. Electron diffractions of $KFe_{17}O_{25}$ before (left) and after (right) electron irradiation. (100 kV Ibaraki)

Fig. 5. HREM image of radiation induced magnetite. (500 kV Cambridge)

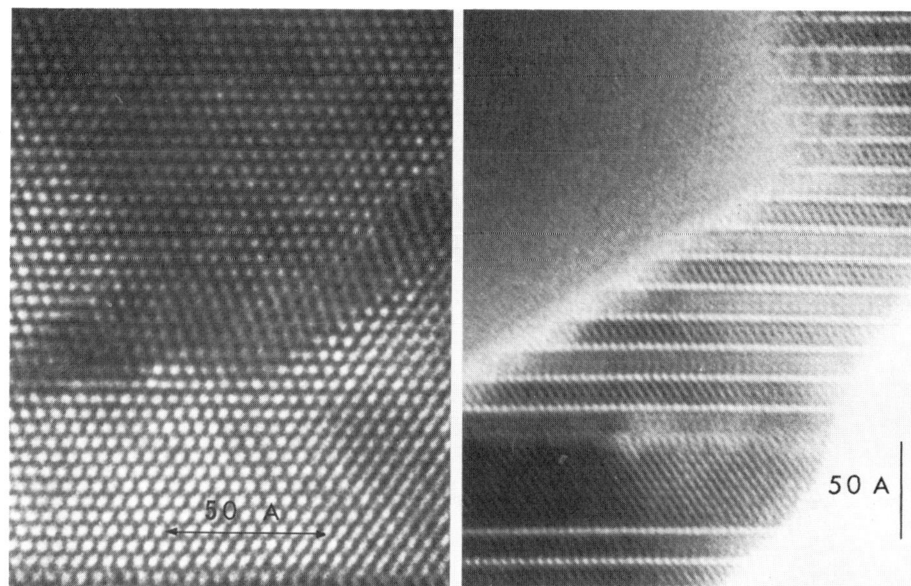


Fig. 6. Structural faults in electron induced magnetite. (200 kV Oxford)

Fig. 7 Non spinel product formed in $KFe_{17}O_{25}$ by electron beam. (200 kV Oxford)

Very occasionally, non spinel product is observed by intense electron irradiation, as shown in Fig. 7 where the radiation induced product on the top-left cannot be explained by magnetite. EDX shows that almost all potassium is lost in such a region. Formation of $Fe_2O_3$ with the corundum structure seems most probable. A detailed study by high resolution TEM and STEM is in progress.

Acknowledgements

The author thanks 1) Dr. A. Howie and all members of Metal Physics Group of Cavendish Laboratory for help in STEM experiments, 2) Drs. K. Knowles & R. Camps (HREM Cambridge) for help in 500 kV HREM experiments and 3) Drs. J.L. Hutchison & C. J. Humphreys (Oxford) for their help in experiments on JEM 200CX.

References

Bovin J 1979 Acta Cryst. A35 572 - 580
De Jonghe L C 1977 Mater. Res. Bull. 12 667 - 674
Hobbs L W 1979 Introduction to Analytical Electron Microscopy (Ed. Hren et al.) Plenum p437 - p480
Hull R et al. 1981 Proc. EMAG (Cambridge) p23 - p26
Matsui Y 1981 J. Appl. Cryst. 14 38 - 42
Matsui Y 1984 J. Cryst. Growth 66 243 - 247
Matsui Y , Bando Y , Kitami Y & Roth R S 1985 Acta Cryst. B41 27 - 32
Matsui Y & Horiuchi S 1981 Acta Cryst. A37 51 - 61

# Dynamical diffraction from modulated structures

D M Bird[1] and R L Withers[2]

1. School of Physics, University of Bath, Claverton Down, Bath BA2 7AY
2. H.H. Wills Physics Laboratory, University of Bristol, Tyndall Avenue, Bristol BS8 1TL

## 1. Introduction

Modulated structures are found in a wide variety of materials. They are characterised by having satellite reflections which decorate all the basic lattice reflections in a diffraction pattern. The modulation can usually be described as being predominantly compositional or displacive in origin, in the latter case it also has a polarisation (e.g. transverse, longitudinal) with respect to the displacement wavevector. Our aim in this paper is to analyse some aspects of electron diffraction from such structures and to show how information about the nature of the modulation may be deduced from large angle convergent beam patterns.

To describe the modulation the full crystal potential, $U(\underline{r})$, is split into two parts $U_0(\underline{r}) + \delta U(\underline{r})$. $U_0$ is the potential in the absence of any modulation and it is assumed that this is known, along with the Bloch states

$$\psi^j(\underline{r}) = \sum_{\underline{g}} C^j_{\underline{g}} \exp(i(\underline{k}^j + \underline{g}) \cdot \underline{r}) \tag{1}$$

which describe dynamical diffraction in $U_0$. $\delta U$ is the change of potential caused by the modulation and, like $U_0$, can be expressed in terms of its Fourier components $\delta U(\underline{g}+\underline{q})$ which act as structure factors of the modulation

$$\delta U(\underline{r}) = \sum_{\underline{g}} \sum_{\underline{q}} \delta U(\underline{g}+\underline{q}) \exp(i(\underline{g}+\underline{q}) \cdot \underline{r}). \tag{2}$$

The $\underline{q}$'s here lie in the first Brillouin zone of the unmodulated lattice. We assume that they form a discrete set of wavevectors which may be commensurate or incommensurate with respect to the underlying lattice. Because $\delta U$ is real $\delta U(\underline{g}+\underline{q}) = \delta U^*(-\underline{g}-\underline{q})$, so $\underline{q}$ and $-\underline{q}$ are always linked. The form of the modulation for each $\underline{q}$ is fully determined by the set of coefficients $\delta U(\underline{g}+\underline{q})$ for all $\underline{g}$. Therefore it is these that we aim to obtain in any diffraction experiment. In the kinematic approximation this is relatively straightforward because the amplitude diffracted to satellite reflection $(\underline{g} \pm \underline{q})$ is proportional to $\delta U(\underline{g} \pm \underline{q})$. Here we investigate whether any similar relation exists in a dynamical diffraction situation.

## 2. Theory

We assume that the modulation is weak in the sense that only a single scattering occurs in $\delta U$, whereas dynamical diffraction in $U_0$ is treated exactly. In this case, each $\underline{q}$ component of the modulation scatters independently. That is, the set of Fourier coefficients $\delta U(\underline{g} \pm \underline{q})$ diffract only to the set of satellites $(\underline{g}' \pm \underline{q})$. The dynamical diffraction in $U_0$ plays the crucial role of mixing the g-vectors between these sets of coefficients and satellites. In this paper we concentrate on systematic diffraction where all the relevant $\underline{g}$'s and $\underline{q}$'s lie in a line. In the symmetric Laue case the diffraction amplitude for satellite reflection $(\underline{g}_0 + \underline{q})$ from a crystal of thickness t is given by (Bird and Withers 1985)

$$A(\underline{g}_0 + \underline{q}) = \frac{t}{2k} \sum_{jj'} C_0^{j'}(\underline{K}')C_0^{j*}(\underline{K}) \sum_{\underline{gg}'} C_{\underline{g}'}^{j'*}(\underline{K}')\delta U(\underline{g}_0 + \underline{g}' - \underline{g} + \underline{q})C_{\underline{g}}^{j}(\underline{K}) \times \exp[i(k_z^j(\underline{K}) + k_z^{j'}(\underline{K}'))t/2] \frac{\sin[(k_z^j(\underline{K}) - k_z^{j'}(\underline{K}'))t/2]}{(k_z^j(\underline{K}) - k_z^{j'}(\underline{K}'))t/2} \tag{3}$$

where $\underline{K}$ and $\underline{K}' = \underline{K} + \underline{g}_0 + \underline{q}$ describe the incident and final electron orientations respectively. This can be expressed in the form

$$A(\underline{g}_0 + \underline{q}) = \frac{t}{2k} \sum_{\underline{g}} \delta U(\underline{g}_0 + \underline{g} + \underline{q}) a_{\underline{g}}(\underline{K}, t) \tag{4}$$

where the weighting factors $a_{\underline{g}}$ depend only on diffraction in $U_0$. Thus (3) can be used to compute the contribution each coefficient $\delta U(\underline{g} + \underline{q})$ makes to the diffraction amplitude. In practice however this would require extensive computation for all orientations and thicknesses. We will therefore look at certain limiting cases to obtain a qualitative feel for which orientations and coefficients make the significant contributions to (3).

The $\sin(\Delta k_z t/2)/(\Delta k_z t/2)$ part of (3) is equivalent to a kinematic shape factor. In the limit of large thickness it is strongly peaked when

$$\Delta k_z = k_z^j(\underline{K}) - k_z^{j'}(\underline{K} + \underline{g}_0 + \underline{q}) = 0. \tag{5}$$

(Single scattering in $\delta U$ is still a valid approximation provided $\delta U$ is sufficiently small that $\delta U(\underline{g} + \underline{q})t/2k \ll 1$ for all $\underline{g}$.) Using the symmetry properties $k_z^j(\underline{K}) = k_z^j(-\underline{K})$ and $k_z^j(\underline{K}) = k_z^j(\underline{K} + \underline{g})$, it can be shown that (5) is satisfied for $j = j'$ at the points $\underline{K} = -(\underline{g}_0 + \underline{q})/2 + \underline{g}/2$ (for all $\underline{g}$) and along lines running through these points at right angles to the systematic row (Fig. 1). These lines are the only orientations where (5) is satisfied, and it should be noted that it is satisfied for every branch along each line. They map out where we should look for strong diffraction into satellite $(\underline{g}_0 + \underline{q})$. However, whether there is significant intensity along any given line also depends on the Bloch wave parts of (3).

To examine this we look at the points $\underline{K}_1 = -(\underline{g}_0 + \underline{q})/2 + \underline{g}_1/2$, so that $\underline{g}_1 = 0$ describes diffraction at the Bragg condition. The amplitude along the line running through each point is the same as at the point itself. At $\underline{K}_1$ the largest terms in (3) are for $j = j'$ when it reduces to

$$A(\underline{g}_0 + \underline{q}) = \frac{t}{2k} \sum_j \exp(ik_z^j t)C_0^{j*}(\underline{K}_1)C_{-\underline{g}_1}^{j*}(\underline{K}_1) \times \sum_{\underline{gg}'} C_{\underline{g}'}^{j}(\underline{K}_1)\, \delta U(\underline{g}_0 - \underline{g} - \underline{g}' - \underline{g}_1 + \underline{q})\, C_{\underline{g}}^{j}(\underline{K}_1). \tag{6}$$

This again can be expressed in the form (4), where the coefficients $a_{\underline{g}}$ depend only on $\underline{g}_1$, and t, and their computation is now relatively straightforward. To pick out the significant features of (6) we examine it in the limit of weak diffraction in $U_0$. In this case, at all orientations away from the zone centre and zone boundaries, only one Bloch wave coefficient $C_{\underline{g}}$ contributes strongly to each branch. For a general $\underline{q}$, $\underline{K}_1$ will be such an orientation. It follows that the most important contribution to (6) will be when $\underline{g} = \underline{g}' = \underline{g}_1 = 0$. This corresponds to diffraction at the Bragg condition, and $\delta U(\underline{g}_0 + \underline{q})$ provides the major scattering. However, there will also be smaller contributions for non-zero $\underline{g}_1$ when $\underline{g} = \underline{g}' = 0$ and $\underline{g} = \underline{g}' = -\underline{g}_1$, provided both $C_0$ and $C_{\underline{g}_1}$ have some weight for the two branches concerned. This will be satisfied when both $\underline{g}_0$ and $\underline{g}_1$ are fairly small. $\delta U(\underline{g}_0 - \underline{g}_1 + \underline{q})$ and $\delta U(\underline{g}_0 + \underline{g}_1 + \underline{q})$ provide the dominant scattering mechanism at these orientations.

## 3. Discussion and Application to NiGeP

The limits of large thickness and weak diffraction discussed above lead to a picture in which there is strong diffraction into satellite reflections along lines of incident orientation and where certain components of $\delta U$ are dominant along each line (Fig. 1). Large angle convergent beam patterns (e.g. Tanaka 1980) provide a practical technique for observing the strength of diffraction into a given reflection as a function of incident orientation. In many diffraction situations the above limits are not unrealistic and so this picture should provide an initial, qualitative interpretation of the most significant features of such patterns taken using satellite reflections.

As an example of this we examine the diffraction from a modulated phase, NiGeP, based on the NiAs structure and found in Au/Ni/Ge contacts to InP devices. (See Vincent and Pretty (1985) for more details of this phase and experimental techniques and results.) A particular wavevector was found which has a systematic diffraction geometry with respect to a [hh0] row of the underlying reciprocal lattice. Large angle convergent beam patterns for satellite reflections $\underline{q}$, $(\underline{g} - \underline{q})$ and $(\underline{g} + \underline{q})$ are shown in Fig. 2. It can be seen that they do indeed consist of a series of lines along which there is strong diffraction into the chosen satellite. Some non-systematic features are also visible, but these are not considered here. For $\underline{q}$, there is very little intensity at the Bragg condition and two rather strong side bands corresponding to $\underline{g}_1 = \pm \underline{g}$. In the $(\underline{g} - \underline{q})$ pattern the side bands are weaker and there is now also diffraction at the Bragg condition. For $(\underline{g} + \underline{q})$ nearly all the scattering is centred about the Bragg position. These results are consistent with the picture of diffraction discussed above. Following the analysis of the weak diffraction limit, only low order side bands are observed as these involve a small $\underline{g}_1$, and all side bands are reduced as $\underline{g}_0$ increases. We can also obtain rough estimates of the size of the components $\delta U(\underline{g} + \underline{q})$. The intensity at the Bragg condition would indicate that $\delta U(\underline{q})$ is small, while $\delta U(\underline{g} - \underline{q})$ and $\delta U(\underline{g} + \underline{q})$ are both significant. The side band intensities are also consistent with this.

In many cases, such estimates should provide an indication of the nature of the modulation. Using this, detailed models can be constructed and refined by calculating the diffraction from them and comparing with experiment. In the case of NiGeP we have used a longitudinal displacive modulation to fit the data of Fig. 2 (Bird and withers 1985). The results shown in Fig. 3 have been computed using the full expression (3). The amplitude and phasing of the displacement waves have been varied to obtain the best fit; the final values of the components of $\delta U$ are in the ratio $\delta U(\underline{q}) : \delta U(\underline{g} - \underline{q}) : \delta U(\underline{g} + \underline{q}) \equiv 0.12 : 1 : 4.6$.

References

Bird, D.M. and Withers, R.L. 1985, to be published.
Tanaka, M. 1980, J. Electron Microsc. 29, 408.
Vincent, R. and Pretty, S.F. 1985, submitted to Phil. Mag.

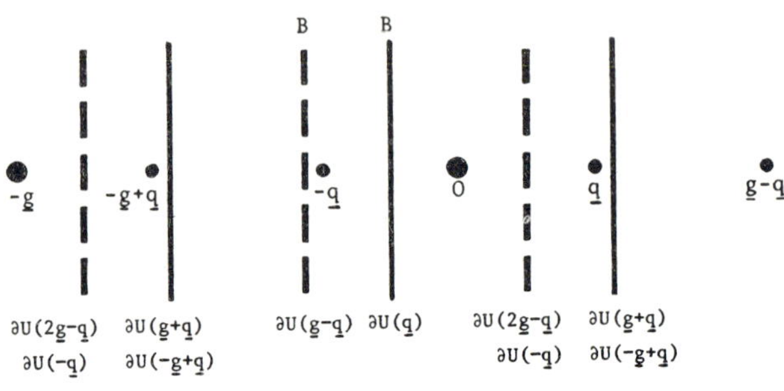


Fig 1. The systematic diffraction geometry, showing the lowest order lines along which one expects significant diffraction into satellites q (full lines) and g-q (broken lines). Lines at the Bragg condition are marked B. The major contributions in the weak diffraction limit to each line are also shown.

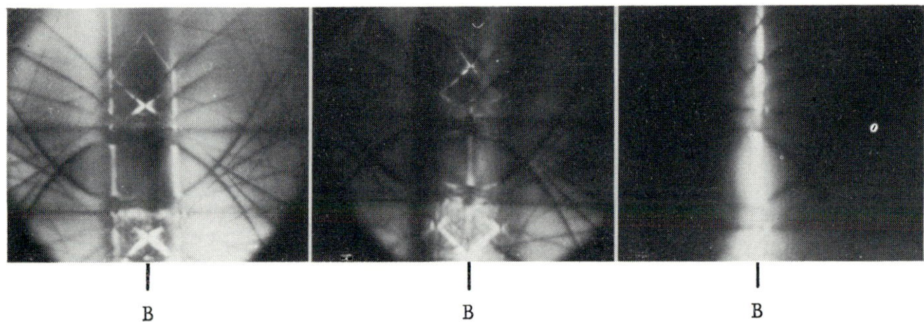


Fig. 2. Large angle convergent beam patterns taken using satellite reflections q, g - q and g + q (from left to right). In each case the Bragg position is marked B. Thickness increases from bottom to top.

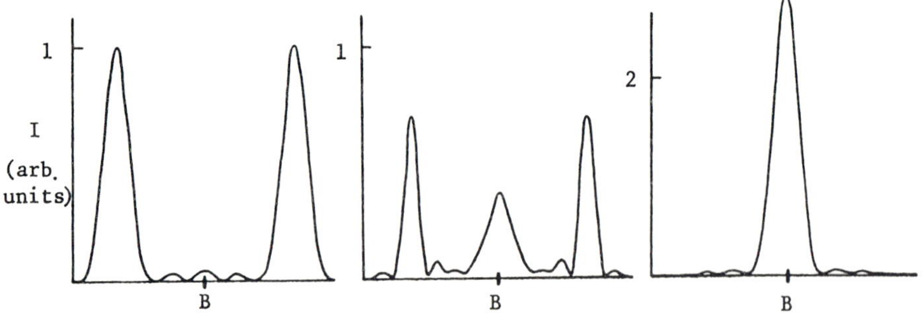


Fig. 3. Computed rocking curves, showing the intensity I diffracted into the satellites as in Fig. 2. Crystal thickness is 1000 Å .

# Interface structure determination by CB diffraction from epitaxial crystals

DJ Eaglesham* and CJ Kiely

H.H. Wills Physics Laboratory, University of Bristol, Bristol BS8 1TL
* Now at Dept of Metallurgy and Materials Science, Liverpool

## 1. Introduction

There has recently been widespread interest in studies of crystal interfaces. They are technologically important in a number of fields ranging from metallurgy to crystal growth, and in particular metal-semiconductor contacts. The interface structure is known to exert an influence on the electrical properties of the contact (Tung, 1984), and a large number of studies have involved attempts to determine interface structures in these systems. The established TEM technique for finding interface structures is by lattice imaging on a cross-section (see Cherns et al 1984). However, specimen preparation and image analysis present such formidable difficulties that very few interface structures have been unambiguously determined. Even for these systems, steps at the interface, relative twists of the crystals, and specimen-wedging may introduce artefacts. In addition, recent results show that interface strains may affect the observations (Gibson et al 1985). The purpose of this study is to show how diffraction effects from a plan-view specimen containing the interface can be used to obtain information on the interface structure.

## 2. Symmetry Effects

The symmetry of a bicrystal depends not only on the symmetry of each crystal separately, but also on their relative shift at the interface. This has been the subject of a number of theoretical studies (Buxton et al 1983, Pond and Vlachavas 1983). We have made calculations to show that the symmetry of convergent beam patterns (CBPs) should be sensitive to displacements at the interface as small as a few per cent (see Eaglesham et al 1985); hence they can be used to give information on the form of the interface. This has been used to study a known ($NiSi_2$/Si) and an unknown (Al/GaAs) interface structure.

Fig. 1 shows a [001] CBP from a $NiSi_2$/Si(001) plan-view specimen. Although both crystals have a four-fold in projection at this axis, the observation of a four-fold in Fig. 1 shows that the two crystals must have a high-symmetry displacement at the interface. $NiSi_2$ columns must lie either exactly on top of, or exactly in between, Si columns (Fig. 2). Identical experiments on Al/GaAs [001) again show four-fold CBPs (Fig. 3); this interface is also restricted to two possible high-symmetry displacements.

## 3. HOLZ Effects

In order to go beyond symmetry information in studying these interfaces,

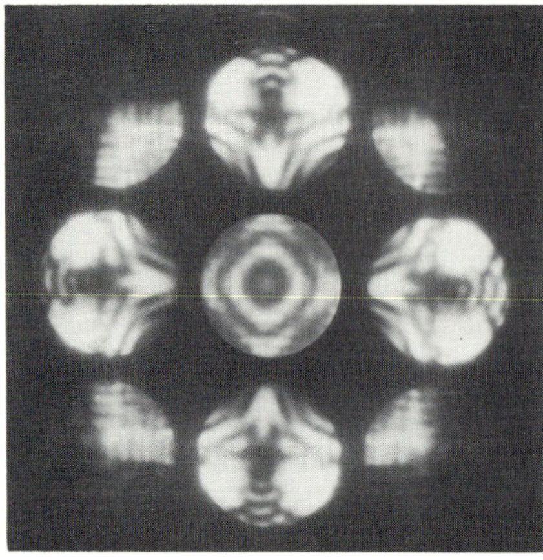

Fig. 1. An [001] CBP from $NiSi_2/Si$, showing four-fold symmetry in the zero layer; This implies a high-symmetry interface.

Fig. 2. Possible high-symmetry projections of $NiSi_2$ onto Si, consistent with Fig. 1.

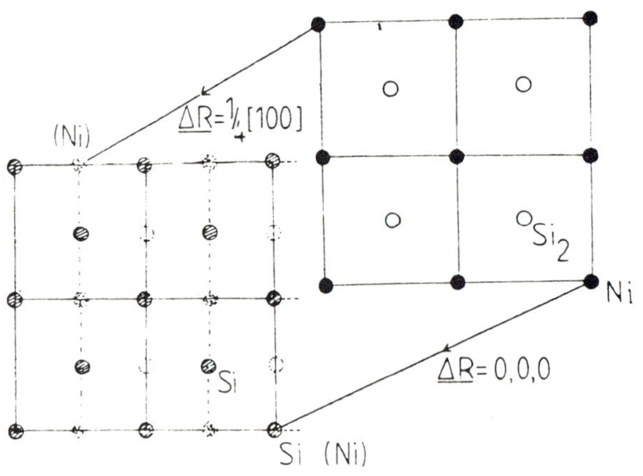


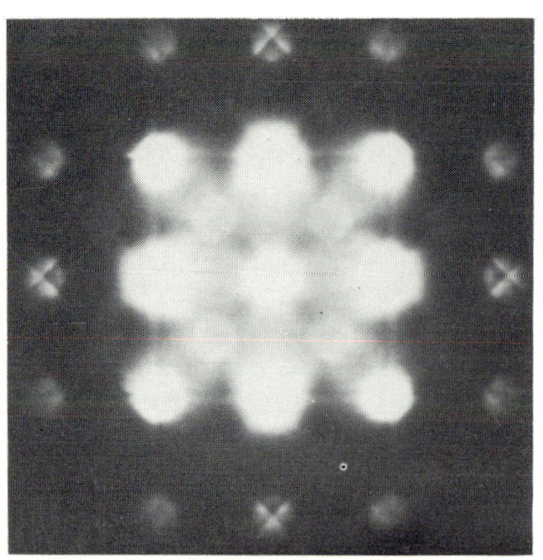

Fig. 3. [001] CP from Al/GaAs. Again, four-fold symmetry implies one of two high-symmetry projections.

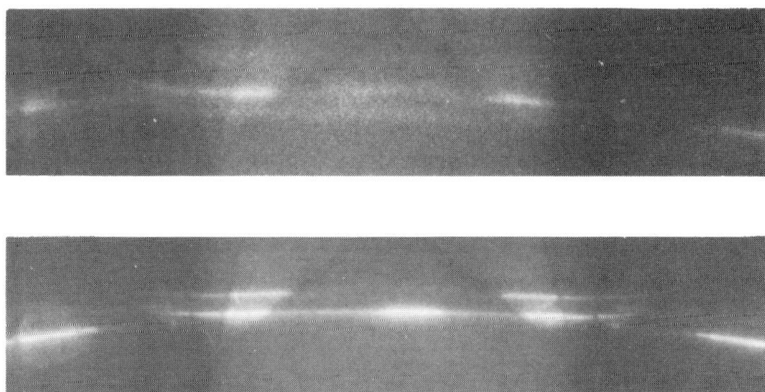

Fig. 4. Small regions of the HOLZ showing the Si branch structure for (a) $NiSi_2$/Si and (b) Si. The broadening of HOLZ for $NiSi_2$/Si may be attributable to strain. Note that, although the excitations are different, the branches are the same.

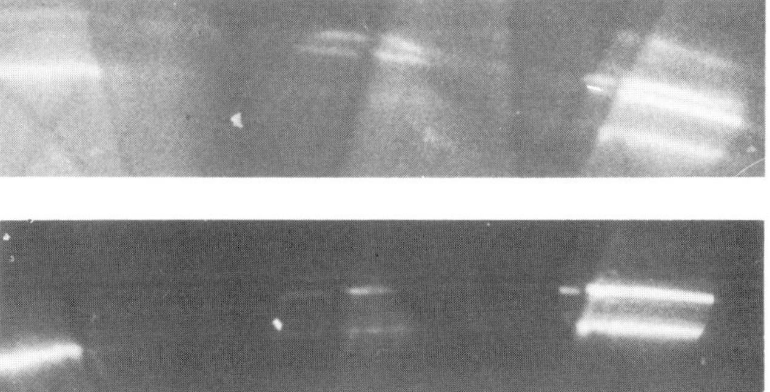

Fig. 5. Small regions of the HOLZ showing the GaAs branch structure for (a) Al/GaAs and (b) GaAs. Note the appearance of a new state in Al/GaAs. This state lies in the gaps in the GaAs structure, and its appearance allows us to deduce the displacement.

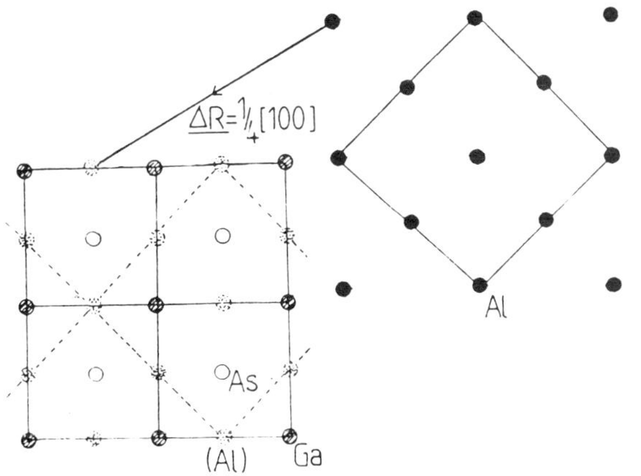


Fig. 6. The projected Al/GaAs interface displacement, as deduced from a combination of symmetry and HOLZ excitations.

we must consider bicrystal diffraction in detail. Although the top crystal must diffract in the normal way, diffraction from the bottom crystal is altered. Instead of an incident plane wave of electrons, it has an incident set of Bloch states, generally localised on the atomic strings of the top crystal. The diffraction in the bottom crystal is described in terms of its usual Bloch states, but the excited states are now those which lie underneath the atomic strings in the top crystal. More precisely the excitation of the $j^{th}$ Bloch state in the bottom crystal is

$$ {}^{B}\alpha^{(j)} = \sum_{i} {}^{T}C_{0}^{(i)*} \int {}^{T}\tau^{(i)}(\underline{R}) \,.\, {}^{B}\tau^{(j)}(\underline{R})^{*} \, d^2\underline{R} \times (\text{phase factor}) $$

in terms of the overlap between Bloch states $\tau^{(j)}$ in the Bottom and Top crystals, for a given displacement between the two crystals. (see Eaglesham et al 1985). This means that excited states are dependent on the shift between the crystals. Now since the "branch structure" in the higher order Laue zones (HOLZ) represents the excited Bloch states (Buxton 1976), comparison of observed excitations with predicted ones can be used to find the relative position of the two crystals, projected down this axis.

Examination of the HOLZ allows us to distinguish the two four-fold symmetry cases. For $NiSi_2$/Si the Si HOLZ shows the usual branch structure for Si, with slight changes in excitation (Fig. 4 a,b). This is only predicted if atomic strings in $NiSi_2$ lie above those in Si, so that the displacement has now been determined. This observation is in agreement with the known structure (Cherns et al 1984). However, for Al/GaAs the GaAs HOLZ shows that a normally unexcited state has become excited. (Fig.5 a,b). Comparison with calculations show that this arises only if Al strings sit exactly in the middle of gaps in GaAs (Fig. 6 ). The displacement has been determined, and the two high-symmetry cases are easily distinguished. Measurements of misfit dislocation spacing indicate a structure in agreement with that determined from CBPs.

## 4. Conclusions

CBPs can be used to obtain information on the interface displacement in a bicrystal. Symmetry information restricts the possibilities, and HOLZ features can be used to distinguish these. $NiSi_2$/Si (001) and Al/GaAs both have high-symmetry interfaces; but HOLZ studies show that (as expected) $NiSi_2$ strings sit above those in Si, whereas the unknown Al/GaAs interface has Al strings displaced from both Ga and As columns.

## References

Buxton BF, 1976, Proc.Roy.Soc.Lond. A350 335
Buxton BF, Forghany SKE and Schapink FW, Proc.EMAG 83, p.59
Cherns D, Hetherington CJD and Humphreys CJ 1984, Phil.Mag A49, 165
Eaglesham DJ, Kiely CJ, Cherns D and Missous M, submitted to Phil.Mag.
Gibson JM, Hull R, Bean JC and Treacy MMJ, 1985, Appl.Phys.Lett. 46 6049
Pond RC and Vlachavas DS, 1983, Proc.Roy.Soc.Lond. A386, 95
Tung RT, 1984, Phys, Rev. Lett. 52, 461

*Inst. Phys. Conf. Ser. No 78: Chapter 2*
*Paper presented at EMAG '85, Newcastle upon Tyne, 2–5 September 1985*



# Convergent beam diffraction study of the distorted structure of 1T-$VSe_2$

D M Bird[*], D J Eaglesham[**] and R L Withers[***]

[*] Dept of Physics, University of Bath, Bath
[**] Now at Dept of Metallurgy and Materials Science, Liverpool University
[***] H H Wills Physics Laboratory, University of Bristol, Bristol BS8 1TL

## 1. Introduction

This paper presents a study of the symmetry and atomic structure of a charge density wave (CDW) modulated phase of 1T-$VSe_2$. CDWs normally occur in "low dimensional" materials and involve a weak sinusoidal modulation of the undistorted atomic positions. This leads to weak extra "satellite" spots in diffraction patterns.

## 2. The Microstructure of $VSe_2$

The microstructure of $VSe_2$ has been investigated by a combination of satellite dark-field imaging and convergent beam diffraction patterns (CBPs). The material exists in two CDW phases. The higher temperature phase contains satellites in three trigonally equivalent directions and CBPs retain the three-fold symmetry of the basic crystal. Below 80K, however, dark-field images show that the structure has broken up into domains (Fig. 1). CBPs now show broken three-fold symmetry - each domain has only two of the three equivalent CDW wavevectors (Fig. 2). The existence of such a "2q" state is unique amongst CDW compounds (Eaglesham *et al* 1985).

## 3. Structure determination using CBPs

These features make $VSe_2$ worthy of a more detailed structural investigation. CBPs provide the only diffraction technique to avoid domain averaging. Recent studies have shown that higher order Laue zone (HOLZ) intensities in CBPs can be related to kinematic structure factors. This has been used to determine both crystal structures (Vincent *et al* 1984) and the form of commensurate CDW modulations (Bird *et al* 1985). Preliminary studies of HOLZ rings in $VSe_2$ indicate that at the 80K phase transition both the absolute intensity and the *relative* intensities of allowed spots in the HOLZ change. By the kinematic HOLZ argument, this, if correct, implies that the two phases may involve different forms of structural distortion.

As a first step towards investigating these structural changes we have begun a determination of the form of the in-layer displacement pattern in a single domain of the lower temperature 2q phase. This uses a sequence of CBPs, like that shown in Fig. 2. The theory differs from previous studies because the CDW in $VSe_2$ is incommensurate. We assume that the modulation can be described by longitudinal, sinusoidal displacement waves. The in-layer displacements in each unit cell $\underline{l}$ of the basic lattice are then written

$$\underline{\delta}_V = \hat{\underline{q}}^{\parallel} a_V \sin(\underline{q}.\underline{l} + \theta_V) \quad (1a) \qquad \underline{\delta}_{Se} = \hat{\underline{q}}^{\parallel} a_{Se} \sin(\underline{q}.\underline{l} + \theta_{Se}) \quad (1b)$$

where $\underline{q}$ is a CDW wavevector and a and θ are the amplitudes and phases of the two species. For an incommensurate modulation the displacement pattern depends only on $a_{Se}/a_V$ and $(\theta_{Se}-\theta_V)$. Using (1a) and (1b) expressions for satellite structure factors can be derived and fitted against the observed intensities. Independent values of $a_{Se}/a_V$ and $(\theta_{Se}-\theta_V)$ can be obtained from both satellite HOLZ rings in Fig. 2. For both rings, two solutions give equally good fits: the results are

| First ring | Second ring |
|---|---|
| $\frac{a_{Se}}{a_V} = 0.12\pm0.08, (\theta_{Se}-\theta_V) = 57°\pm30°$ | $\frac{a_{Se}}{a_V} = 0.16\pm0.04, (\theta_{Se}-\theta_V) = 42°\pm30°$ |
| $\frac{a_{Se}}{a_V} = 0.6\pm0.2, (\theta_{Se}-\theta_V) = 59°\pm10°$ | $\frac{a_{Se}}{a_V} = 0.55\pm0.15, (\theta_{Se}-\theta_V) = 41°\pm25°$ |

It can be seen that there is satisfactory agreement between the results from the two rings. A comparison of these results with an equivalent study of the higher temperature trigonal CDW state should indicate whether there is a significant structural difference between the two phases. This study, along with a refinement of the data for the 2q state, is currently being undertaken.

## References

Bird D M, McKernan S and Steeds J W, 1985 J.Phys.C 18 499

Eaglesham D J, Withers R L and Bird D M 1985 J.Phys.C in press

Vincent R, Bird D M and Steeds J W 1984 Phil.Mag.A50 764

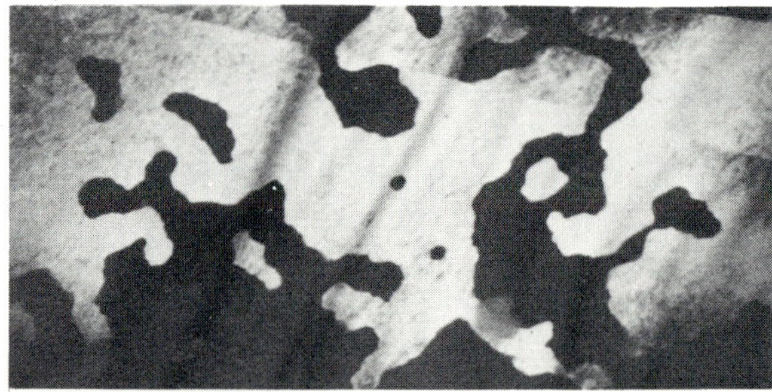

Fig. 1. Satellite dark-field of the low-temperature state, showing a network of 2q domains.

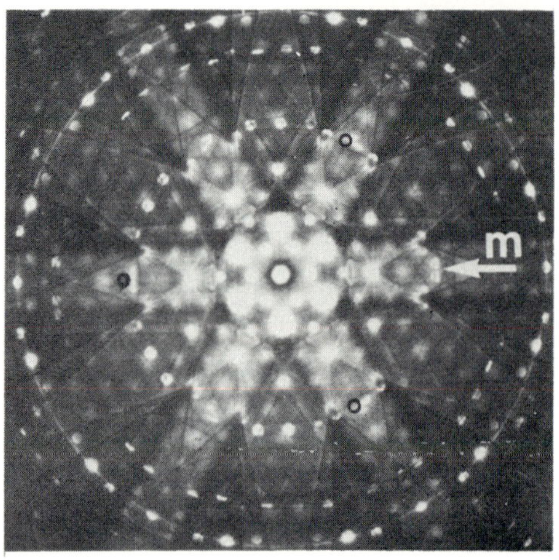


Fig. 2. [001] CBP from this state showing broken three-fold symmetry in the incommensurate FOLZ and SOLZ. Only two of the 3 possible qs appear.

*Inst. Phys. Conf. Ser. No 78: Chapter 2*
*Paper presented at EMAG '85, Newcastle upon Tyne, 2–5 September 1985*



# Measurement of strain in convergent beam electron diffraction patterns

A R Preston and D Cherns

H H Wills Physics Laboratory, University of Bristol, Bristol BS8 1TL

The use of convergent beam electron diffraction (CBED) to measure lattice parameters in perfect single crystals is now well established (eg Jones, Rackham and Steeds (1976)). However, comparatively little work has been carried out on the CBED measurement of lattice parameters in imperfect crystals. In this paper we investigate the use of CBED for this purpose. Studies were carried out both at 120kV using a Philips EM400 and at between 100 and 300 kV using a Philips EM430. The crystals studied include diamond and silicon with low densities of grown-in or deformation-induced dislocations, and FeNiCr alloys containing high densities of dislocations and voids introduced by ion-irradiation at elevated temperatures ( 600°C). These latter materials show some evidence for non-equilibrium radiation induced segregation of Ni to within a few hundred Å of defects, notably grain boundaries leading to changes in lattice parameter <0.1% in some alloys (ref. Marwick et al (1983, 1978)).

In a convergent beam zone axis pattern (ZAP) a change in lattice parameter can be detected in the diameter of the higher order Laue zone (HOLZ) given by d = 2Kh ; where K is the wave-vector, and h the reciprocal lattice spacing along the beam direction. This change can be more conveniently observed by its effect upon the intersections of the HOLZ deficiency lines in the central disc. Fig. 1 shows a set of such deficiency lines in the central disc from a [114] ZAP taken from a Ni-26Fe-19Cr alloy irradiated at 600°C to a dose of 5 dpa by 46.5MeV Ni ions, and containing a high density of dislocations. Fig. 2 shows an example of an irradiated FeNiCr alloy containing both dislocations and voids. The ZAP was taken at 300kV at 70K with a probe size of 100Å. In contrast with observations on unirradiated samples, deficiency lines in Fig. 1 each corresponding to diffraction to a single HOLZ reflection show a series of satellite lines disposed predominantly to one side of the main peaks and with a maximum peak-to-peak separation in the direction indicated. Moreover, in further contrast with perfect crystals the mirror symmetry about the ($2\bar{2}0$) plane expected for fcc FeNiCr alloys is no longer present. Although various factors might give rise to HOLZ satellite lines the magnitude of the peak separations ( .01°) suggests that the presence of strain fields due to defects is the most likely cause. The presence of rocking curve satellites due to the finite extent of the crystals, whose asymmetry from highly tilted crystals has previously been reported (Bird et al (1983)), offers an alternative explanation. We have investigated these two possibilities.

The effect of dislocations on ZAPs has been considered by Carpenter and Spence (1982) who demonstrated the splitting of HOLZ and Kikuchi lines into doublets in the case where the probe is symmetrically disposed about the dislocation core. In our work we have examined series of ZAPs from a 100Å probe taken along a line crossing individual 60° and screw dislocations in silicon and diamond samples. Fig. 3 shows one such pattern where the probe is wholly on one side of the dislocation. A splitting of the HOLZ and Kikuchi lines is observed into a main peak and a weaker satellite; this splitting being a maximum along the direction indicated which corresponds to the Burgers vector direction (determined by standard BF DF technique) and zero in the perpendicular direction. As the probe is brought closer to the dislocation core the line splittings increase until further peaks appear with the probe centred on the core itself. On tracking the probe to the opposite side of the dislocation the lines become doubly split once more, but with the positions of the main and subsidiary peaks reversed.

The features observed in ZAPs from near dislocations can be explained simply using a kinematical approach where the amplitude profile for a HOLZ reflection g, is given by: $\Phi_g = \frac{\pi i}{\xi_g} \int_0^t e^{2\pi i(sz + \underline{g}.\underline{R})} dz$

where, s is the deviation parameter, z the depth in the crystal, $\xi g$ the extinction length for the HOLZ reflection $\underline{g}$ and $\underline{R}$ is the displacement field of the dislocation. Fig. 4 shows the result of such a calculation for a typical case in which the 250 kV probe is 100Å across, passing 450Å from an edge dislocation with $\underline{b} = ^a/2[100]$, $\underline{b}//\underline{g}$, in the centre of an Si foil 1000Å thick. The result shows the appearance of a broad peak, in addition to the main peak whose position is shifted relative to that expected from a perfect crystal (Fig. 4c). The presence of the subsidiary peak can be simply interpreted as corresponding to the crystal rotation at which $d\underline{R}/dz$ passes a point of inflexion. A similar argument is used to explain the location of weak beam dislocation images (eg Cockayne (1973)). As the probe is moved closer to the core the peak splitting should increase: as observed. For probes at distances > 300Å from the core the observed peak splittings are in approximate agreement with the simple model. Studies on lightly irradiated FeNiCr samples have shown similar ZAPs to those in Fig. 3 when the probe is placed close to an isolated dislocation. Some preliminary studies have also been carried out for probes placed close to and on voids 200Å in size (cf Fig. 2). The ZAPs obtained in these experiments when the probe is on the void itself, show HOLZ lines to be multiply split and interference fringes are visible in the low order reflections similar to those observed from overlapping crystals. (Rackham et al (1978)). Work on the detailed nature of these diffraction and refraction effects is still in progress.

Bird et al (1983) reported asymmetries in the dark field rocking curves from diamond crystals under two-beam diffracting conditions and with large specimen tilts ( 45° from the horizontal). We have observed similar effects in FeNiCr samples, both unirradiated and irradiated, and also silicon and diamond samples tilted through large angles. We have investigated this phenomenon in some detail and have found that a number of additional factors are involved. These experiments will be described in detail in a subsequent publication (Preston and Cherns, to be published). The results indicate that change in sample orientation and film thickness over the probe area are important, such that the effect is reduced for smaller probe sizes. Fig. 5 illustrates rocking curve fringes from a tapered Si crystal and shows the marked reduction in asymmetry as the probe size decreases. Studies on undistorted MgO platelet samples suggest that sample tilt in the absence of other factors has only a minor effect on symmetry.

## Conclusions

Although the possible factors affecting the deficiency lines in Fig. 1 have been somewhat arbitrarily separated in our discussion we may reasonably ascribe the satellite lines as due to defects in the FeNiCr. The most intense satellite peaks can be reproduced by assuming the probe to pass near a dislocation,thus displacing the main deficiency lines in the direction indicated. Ascribing the most intense deficiency lines to regions away from defects it is clear from the broken symmetry that the crystal also supports long range strains. From the line width we may estimate the accuracy to which such strains may be measured in this image (300kV) as ±0·1 %.A full analysis of strains and compositional variations in FeNiCr alloys requires a detailed study of ZAPs near individual defects. However, our preliminary results suggest that the technique can be applied to quantitatively study defect strain fields and possible segregation effects.

Acknowledgements We would like to thank D J Eaglesham and Dr D M Bird for useful discussions. ARP acknowledges support from the SERC and AERE Harwell via a CASE award.

References

Bird D M, Walmsley J C and Vincent R 1983 Inst.Phys.Conf.Ser.68 41
Carpenter R W and Spence J C H 1982 Acta Cryst A38 55
Cockayne D J H 1973 J.Microsc.98 116
Jones P M, Rackham G M and Steeds J W 1977 Proc.R.Soc A354 197-222
Marwick A D, Kennedy W A D, Mazey D J and Hudson J A 1978 Scr.Met.12 1015
Marwick A D, Piller R C and Norton M E 1983 AERE report R10895
Rackham G M, Loveluck J E and Steeds J W 1978 Inst.Phys.Conf.Ser.41 435

Fig(1) (220),(000) and (220) discs of a [114] ZAP taken at 300kV.Each FOLZ line has an associated set of faint fringes.Second order Laue zone lines are also visible.

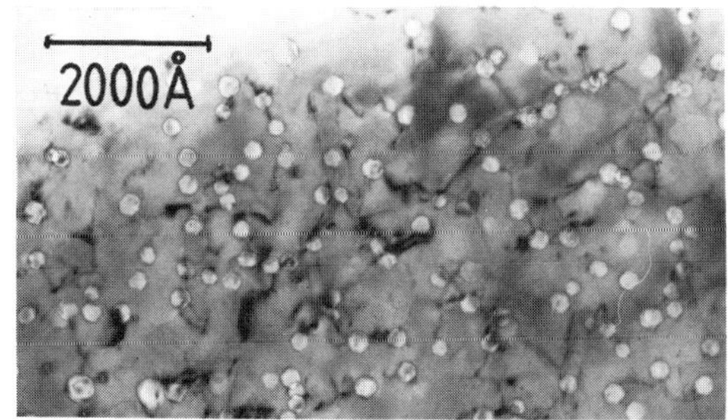


Fig(2) Radiation produced dislocations and voids in an Fe-15Ni-15Cr sample irradiated at 600°C to 10dpa.

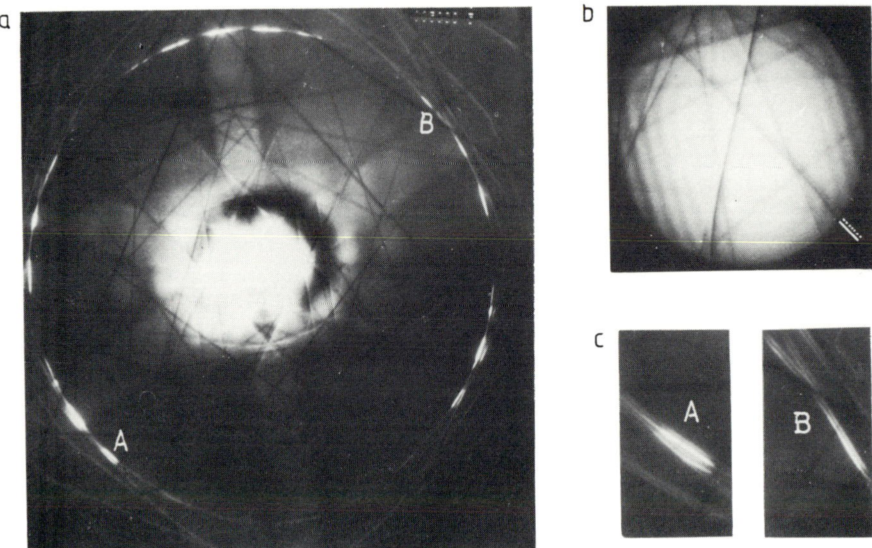


Fig(3) [114] ZAP Silicon,216kV,500A from dislocation.
a)Whole pattern:Split HOLZ and Kikuchi lines.
b)Bright field disc:Direction of maximum splitting shown.
c)Details of HOLZ from points in(a):faint fringes always to right of strong ones.

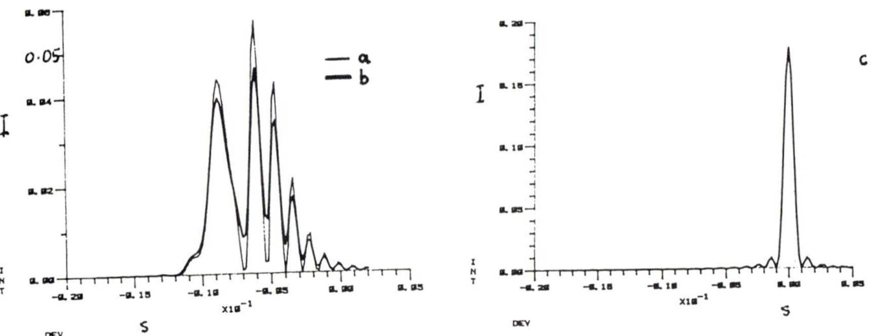


Fig(4) Kinematic calculation of HOLZ rocking curves,Intensity I versus deviation parameter s.
a) Single column 450Å from core.
b) Probe 100A across,450Å from core,weighted average of five columns.
c) Rocking curve for perfect crystal; note scale change between b and c.

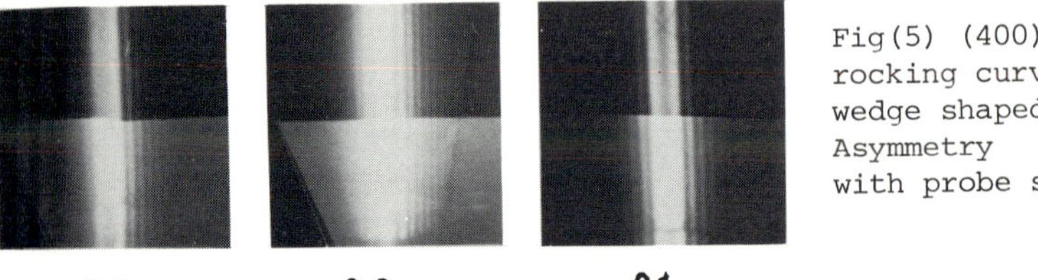


Fig(5) (400) Darkfield rocking curves from Si wedge shaped sample. Asymmetry decreases with probe size.

*Inst. Phys. Conf. Ser. No 78: Chapter 2*
*Paper presented at EMAG '85, Newcastle upon Tyne, 2–5 September 1985*


# The applicability of CBED methods for the determination of local alloy compositions in GaXAs heterostructures

E G Britton and W M Stobbs
University of Cambridge, Department of Metallurgy and Materials Science
Pembroke Street, Cambridge CB2 3QZ

The use of the symmetry properties of zone axis patterns for the determination of the space group of a crystal structure is well documented (Buxton et al. 1976, Steeds et al. 1984) and rarely questioned (Self et al. 1983, Shelton 1985), while it has also been demonstrated that dynamical aspects of the contrast are sufficiently well understood (Buxton 1976) to enable crystal structure refinement (Vincent et al. 1984). HOLZ line positions can be used both to measure locally uniform strains (Ecob et al. 1981) and to determine alloy compositions in situations in which Vegard's Law applies; the method has complementary application to the use of lattice fringe spacings (Self et al. 1981). In this respect it is the HOLZ line approach which should be the better method for alloy composition determinations in fine $Ga_xX_{1-x}As$ or $GaY_{1-y}As_y$ heterostructures, as viewed edge-on, because a fine layer spacing prevents the accurate use of relative lattice fringe spacing methods.

While simple dynamical corrections can be made (Jones et al. 1977) it is generally assumed that the movement of HOLZ lines is due only to changes in the lattice parameter. Hence composition determinations are normally made using crossovers showing only weak dynamical interactions. Changes in the splitting of HOLZ crossovers have however been reported as a function of thickness for $Ga_{1-x}Al_xAs$ (Lilienthal and Ishizuka 1982) and in this light we compare the relative applicability of the method for the different GaXAs alloys before going on to consider whether or not the strengths of these dynamical interactions might themselves have potential application for composition measurements when X is Al.

Lilienthal and Ishizuka (1982) obtained their results on thickness effects for $Ga_xAl_{1-x}As$ at the orientation shown in figure 1, the crossover point of interest (X) being indexed as $7\bar{3}\bar{1}/93\bar{1}$ at $(0\bar{1}\bar{3})$. The patterns are shown for both GaAs and AlAs as well as for an intermediate alloy, the two thicknesses in each case being multiples of $\xi_{400}$ for that particular alloy rather than the same absolute thicknesses. While it may be seen that locally at X crossover splitting is observed it is only of a simple form over a limited thickness range and becomes more complex for greater thicknesses and higher Al contents. Nevertheless and of more importance for the application of the method generally, comparison of these and other patterns demonstrates no gross thickness dependent line shift at positions distant from crossovers. In general, provided that the change in lattice parameter produces a change in HOLZ line position comparable to the thickness of the line, and that simple crossovers are chosen, it would appear that errors due to the thickness dependence of dynamical effects should be small. An example of an appropriate crossover region in a diffraction pattern of GaAs at (105) is shown in figure 2, together with computed kinematic plots of the most prominent lines, for a range of alloys of technological application where the relationships between lattice parameter and composition are known (Gratton and Woolley 1973, Ito 1985, Woolley and Smith 1958). Analysis of such plots demonstrates that for In or Sb alloys the InAs or GaSb molar fraction should be readily measurable to about 3-5% given line widths typically of the order of $0.01\text{Å}^{-1}$.

The substitution of Al for Ga is associated with only small changes in lattice parameter (a(GaAs) = 5.6538Å; a(AlAs) = 5.662Å) so that kinematically the line movements are unlikely to yield composition measurements of any useful accuracy. However the presence of Al has a strong effect on certain components of the lattice potential (e.g. $V_{200}$) so that it is of interest to see whether this results in significant systematic variations in appropriately chosen dynamical HOLZ line interactions. Some variation in the splitting behaviour with Al content is apparent in figure 1 for the lines related by 16 0 0 but the systematic trends are more clearly observed at the orientation shown in figure 3 (close to (102)) for which the major crossing lines are related by n 0 0. Qualitatively it seems that in this case increasing the Al content has a similar effect on the cumulative interactions to decreasing the thickness. In principle it would appear that the matching of such patterns from unknown alloys with standards could well provide, via the form of the dynamical behaviour exhibited, a more reliable measure of Al content than the application of spacing measurements in the standard manner.

In summary, compositions of $Ga_xIn_{1-x}As$ and $GaSb_{1-x}As_x$ should be measurable to an accuracy of about 3-5% using only a kinematical model for the formation of HOLZ patterns. Variations of such patterns with thickness need not be a significant source of error provided both that the crossovers chosen are not related by reflections of high $V_{g_1-g_2}$ and that crossover positions are extrapolated from relatively distant line positions. By contrast the substitution of Al for Ga has a stronger effect on dynamical aspects of the pattern than it does on its simple kinematic form. This in itself might provide a method for the measurement of the composition of $Ga_xAl_{1-x}As$ alloys.

Acknowledgements

Thanks are due to the SERC and GEC for financial support and also to the latter for the supply of materials. We are grateful to Prof. D. Hull for provision of laboratory facilities.

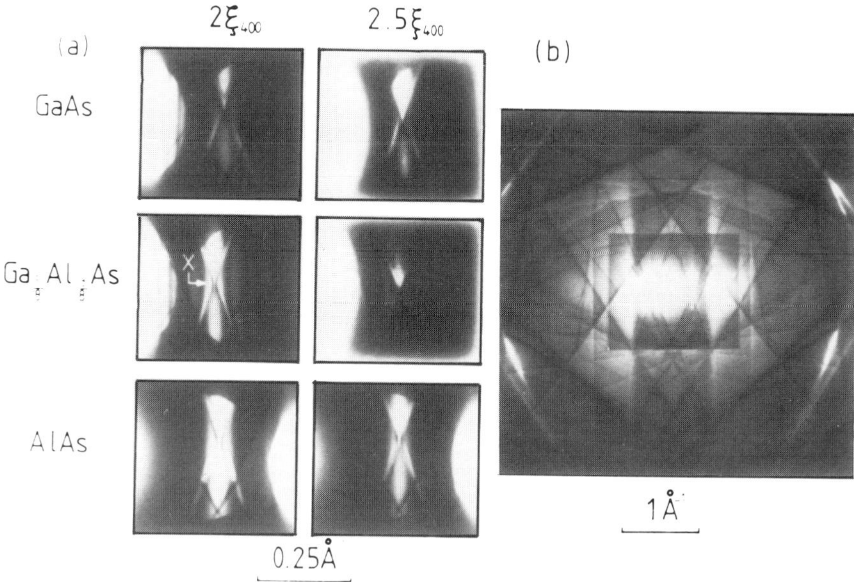


Fig.1 $\overline{7}3\overline{1}/93\overline{1}$ HOLZ crossover at 100kV

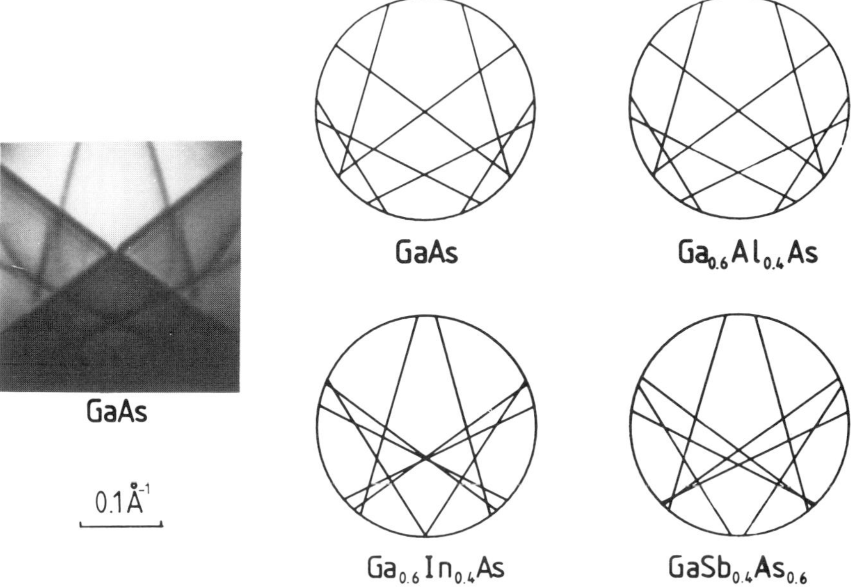


Fig.2 Variation of a HOLZ pattern near (105) as a function of X in GaXAs (120kV)

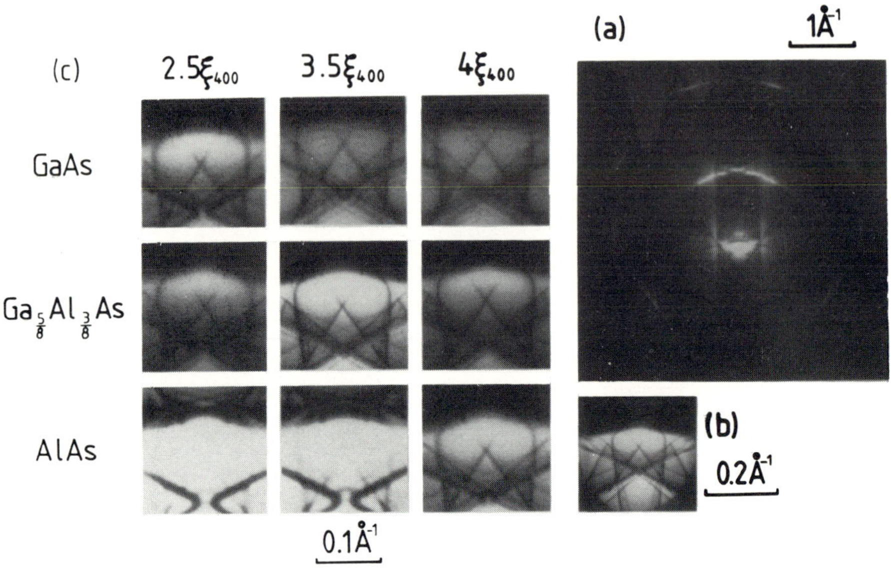


Fig.3 Example of a HOLZ pattern, at 100kV, showing dynamical variations with Al content of $Ga_xAl_{1-x}As$. (b) and (c) are successive enlargements of the central disc in (a)

## References

Buxton B F 1976 Proc. Roy. Soc. Lond. A350 335

Buxton B F, Eades J A, Steeds J W and Rackham G M 1976 Phil. Trans. Roy. Soc. Lond. 281 171

Ecob R C, Shaw M P, Porter A J and Ralph B 1981 Phil. Mag. A44 1117

Gratton M F and Woolley J C 1973 J. Electron. Mater. 2 455

Ito T 1985 Phys. Stat. Sol. (b) 129 559

Jones P M, Rackham G M and Steeds J W 1977 Proc. Roy. Soc. A354 197

Lilienthal Z and Ishizuka K 1982 Proc. 40th Annual Meeting EMSA ed. G W Bailey (San Francisco Press) p448

Self P G, Bhadeshia H K D H and Stobbs W M 1981 Ultramicroscopy 6 29

Self P G, O'Keefe M A and Stobbs W M 1983 Acta Cryst. B 39 197

Shelton C G 1985 Ph.D thesis, University of Cambridge

Steeds J W (the Bristol Group) 1984 Convergent Beam Electron Diffraction of Alloy Phases (Bristol: Adam Hilger Ltd.)

Vincent R, Bird D M and Steeds J W 1984 Phil. Mag. A50 765

Woolley J C and Smith B C 1958 Proc. Phys. Soc. Lond. 72 241

# Convergent beam electron diffraction and imaging of strained-layer superlattices

D M Maher*, H L Fraser**, C J Humphreys***, R V Knoell*, R D Field**, J B Woodhouse** and J C Bean*

* AT&T Bell Laboratories, Murray Hill, NJ 07974 USA
** Dept. of Metallurgy and Mining Engineering, Univ. of Illinois, Urbana, IL 61801 USA
*** Dept. of Metallurgy and Materials Science, Univ. of Liverpool, Liverpool L69 3BX, England

## 1. Introduction

In this paper we report the results of convergent-beam electron diffraction (CBED) experiments and analyses which are designed to measure local distortions, while taking into account surface relaxation effects in thin crystals. We also report the results of a new technique which will be referred to as convergent-beam imaging (CBIM). With the CBIM technique, crystal distortions are revealed and mapped with high angular resolution as displaced (or curved) high-order Laue zone (HOLZ) lines in images. The CBED and CBIM techniques have been used to study the distortions in a model system, namely a dislocation free, long-wavelength, dilute, $Si/Ge_xSi_{1-x}$, strained-layer superlattice (SLS).

The layer structure used for this work was grown by molecular-beam epitaxy (MBE) on a (100) silicon substrate (see Fig. 1 and Fraser et al. (1985) for details). The CBED patterns and CBIM images were recorded with a probe of diameter approximately 40 Å, and with the specimen cooled to ~ 85 K. All diffraction patterns were obtained using a nominal accelerating voltage of 120 kV. For brevity we limit the discussion to the Si-4.5%Ge alloy layer.

## 2. Results and Discussion

The HOLZ line pattern from the Si sample was used to determine an accurate value of the electron wavelength λ using a temperature corrected (Touloukian et al (1975)) Si lattice parameter (Dismukes et al (1964)). A comparison between the experimentally obtained [001] Si pattern and the computed pattern of HOLZ lines which yields λ is shown in Fig. 2.

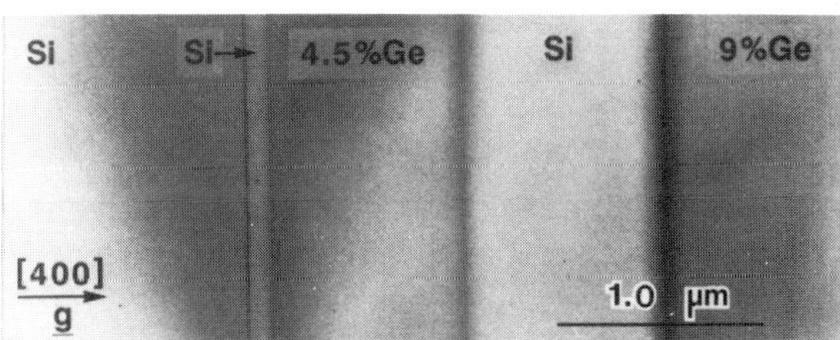


Fig. 1 Image of a cross-section of a Si/ $Si_xGe_{1-x}$ SLS showing the various layers, as indicated.

One of the most appealing properties of [001] zone-axis patterns in a study of this type is that a tetragonal distortion changes the symmetry of the bright-field disk from 4mm to 2mm. Therefore, the presence of such a distortion is easily "detected" by visual inspection, as can be seen by comparing Fig. 2a (Si) with Fig. 3a (4.5at%Ge). Simulations of Fig. 3a, including surface relaxations, give a range of combinations of $a_{100}$ and $a_{001}$ values which match the experimental pattern and these are given in the table. The importance of surface relaxation effects may be seen by comparing the "best fit" simulations obtained by either permitting $a_{001}$ to vary (relaxations), Fig. 3b, or restricting $a_{001} = a_o$ (no relaxations), Fig. 3c. It is clear that the pattern (Fig. 3c) corresponding to the unrelaxed condition does not match exactly that obtained in experiment (Fig. 3a), whereas excellent agreement is achieved for the simulation which includes surface relaxations (Fig. 3b). Values of the tetragonal distortions $(a_{100}-a_{010})/a_{010})$ which are obtained from these two analyses are given in the table and compared to those calculated from isotropic elasticity theory (Gibson, et al (1985)). The agreement between the values corresponding to Fig. 3b with relaxation with theory is excellent. Surface relaxation will vary according to both the distance from interfaces and the specimen thickness, and hence CBED patterns are sensitive to this, which we have observed.

The CBIM technique has been described in detail elsewhere (Humphreys et al (1985), Maher et al (1985)). In the present study, CBIM has been used to investigate the distortions that exist across the interface between a Si buffer region and an adjacent Si-5%Ge alloy layer. The CBIM image, recorded with the electron beam parallel to [013], is shown in Fig. 4; the position of the interface is marked by vertical arrows. As can be seen, the image exhibits a series of dark lines; these are the CBIM-HOLZ lines. The lines which are running almost perpendicular to the interface suffer large bends as they cross the interface, such as at the point marked R in Fig. 4. The patterns of lines on either side of the interface may be compared with those which form in the bright-field disks of CBED patterns recorded from these two layers; these are shown in Fig. 5(a,b), respectively. While the CBIM image clearly reveals the change in lattice parameters across the interface (by analogy with the CBED-HOLZ lines), additional important information concerning the relationship between the Si buffer and Si/Ge alloy layers may be obtained. Thus, the bending of the lines across the interface is thought to be due to a mutual rotation of the two crystal layers. Such a rotation results from strains produced during epitaxial growth, which are partially relaxed by rotations when the specimen is thinned. The magnitude of the lattice rotation may be estimated from CBED-HOLZ patterns as being approximately 1.0 mrad. These effects are not observed in thicker foils where surface relaxations are minimized.

## 3. Conclusions

We have obtained and analyzed (001) CBED patterns and also CBIM images from (001) $Si/Ge_xSi_{1-x}$ strained layers. The alloy layers distort to become orthorhombic due to a tetragonal distortion in the [100] growth direction and a surface relaxation along [001]. This surface relaxation is similar in magnitude to that in the growth direction. The measured distortions agree very well with elasticity theory provided surface relaxation effects are included.

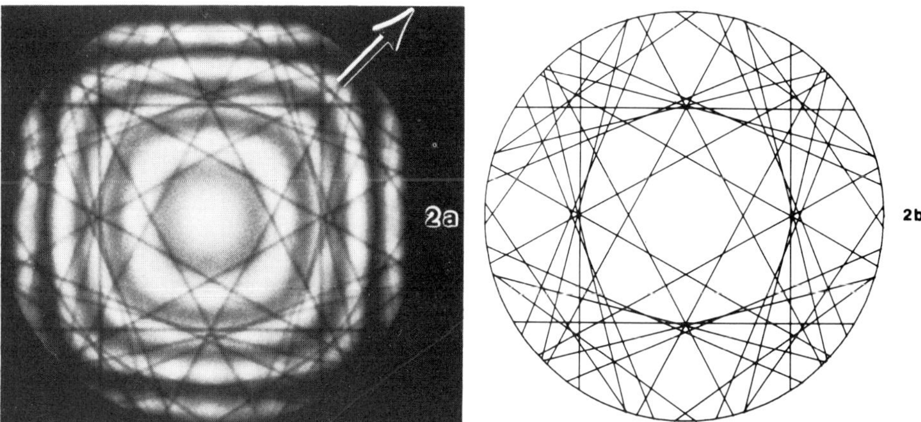


Fig. 2 Bright-field disks of [001] CBED patterns: a) Si substrate and b) simulation of the Si substrate pattern with $a_o$ = 5.4294 Å, $\lambda$ = 0.003365 Å and $2\alpha$ = 0.0053 rad. The [100] growth direction is indicated by the arrow.

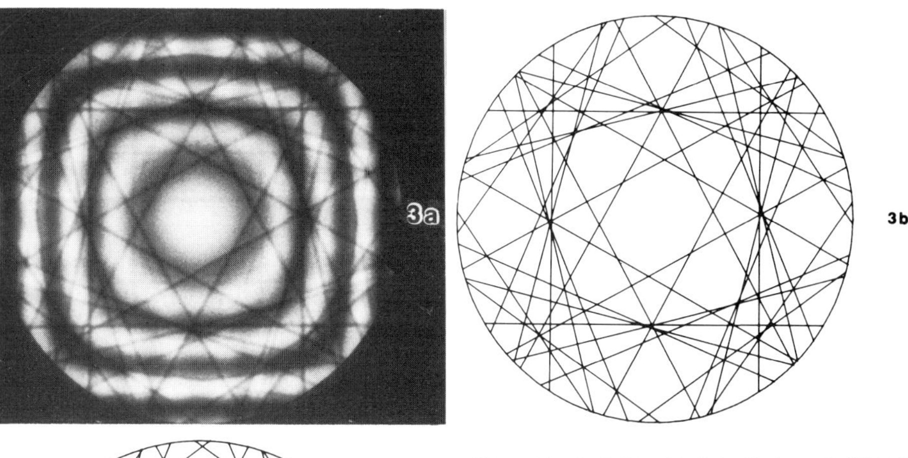


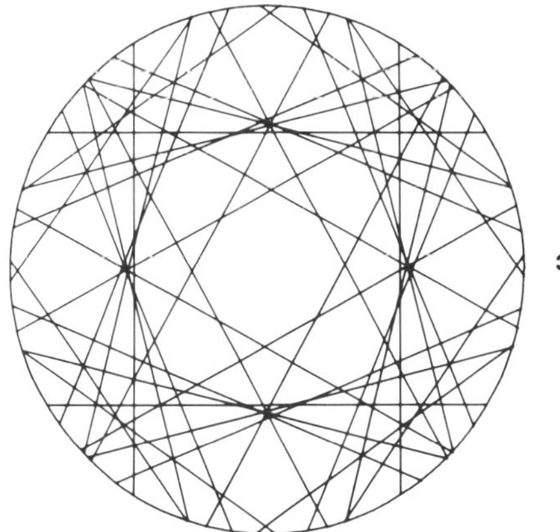


Fig. 3 Bright-field disks of [001] CBED patterns: a) 4.5at%Ge Si alloy; b) simulated pattern with $a_{100}$ = 5.441 Å, $a_{010}$ = 5.4294 and $a_{001}$ = 5.441 Å; and c) simulated pattern with $a_{100}$ = 5.4335 Å, $a_{010}$ = $a_{001}$ = 5.4294 Å.

Calculated distortions for (001)Si-4.5Ge growth direction [100]

| Origin | $a_{100}$ | $a_{010}$ | $a_{001}$ | $\varepsilon_T=(a_{100}-a_{010})/a_{010}$ |
|---|---|---|---|---|
| unrelaxed | 5.4335±0.0025 | 5.4294 | 5.4294 | 0.00076±0.00046 |
| relaxed | 5.4435±0.0025 | 5.4294 | 5.442±0.0018 | 0.00260±0.00046 |
| Bulk Crystal (elasticity theory) | | | | 0.0028 |
| Relaxed Crystal (elasticity theory) | | | | 0.0022 |

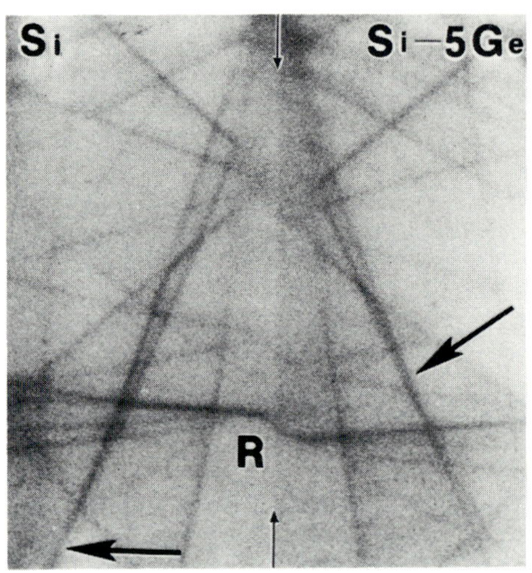


Fig. 4 CBIM image of an interface between Si and Si-5Ge (nominal composition).

Fig. 5 Portions of the bright-field disks of [013] CBED patterns of (a) Si and (b) Si-5Ge. A comparison may be made between these patterns of lines with those on the CBIM image of Fig. 4.

References

Dismukes J P, Ekstrom L and Paff R J 1964 J.Phys.Chem. 68 3021

Fraser H L, Maher D M, Humphreys C J, Hetherington C J D, Knoell R V, and Bean J C 1985 Microscopy of Semiconducting Materials (Inst.Phys. Conf.Ser. 76) p 307

Gibson A M, Hull R and Bean J C 1985 Appl.Phys.Lett. 46 649

Humphreys C J, Maher D M and Fraser H L to be published in Ultramicroscopy

Maher D M, Humphreys C J and Fraser H L, to be published in Materials Research Letters

Touloukian Y S, Kirby R K, Taylor R E and Desia P D 1975 Thermal Physical Properties of Matter, Vol. 13, Non-metallic Solids (New York: Plenum) p 154

*Inst. Phys. Conf. Ser. No 78: Chapter 2*
*Paper presented at EMAG '85, Newcastle upon Tyne, 2–5 September 1985*


# The symmetry of zone-axis patterns in reflection high-energy electron diffraction

M D Shannon*, J A Eades and B F Buxton+

Centre for Microanalysis of Materials, MRL, University of Illinois, Urbana, Illinois 61801
* Permanent address: ICI PLC, Mond Division, PO Box 11, Runcorn WA7 4QE
+ Hirst Research Centre, GEC, Wembley HA9 7AP

In reflection, high-energy, electron diffraction (RHEED), as in the transmission experiment, there is a complex variation in intensity of the diffracted beams as a function of the direction of the incident beam with respect to the lattice. Several experimental techniques (eg convergent beam, Tanaka, double rocking) have been reported in recent years to record the angular variation in the case of transmission. Recently these have been extended and developed in (S)TEM instruments to enable the equivalent experiments in a RHEED geometry (Shannon et al 1984, 1985). Against this background the authors became aware that the symmetries of zone-axis patterns in RHEED had not been discussed as they have for LEED (Holland and Woodruff 1973) and THEED (Buxton et al 1976).

By methods analagous to those used in the transmission case by Buxton et al we have shown that there are five possible "RHEED diffraction groups". However, the symmetries that are involved are not always the "metric" symmetries seen in transmission experiments, but may be "topological" as we will explain.

Such symmetries are a consequence of the geometry of the diffraction process, which is more complex in RHEED than in the transmission case. The periodicity of the surface imposes the condition $\underline{K}^1 = \underline{K} + \underline{G}$ on the diffracted beams, where $\underline{K}$ is the component of the incident beam, $\underline{k}$, parallel to the surface, $\underline{K}^1$ is the parallel component of the diffracted beam and $\underline{G}$ is a reciprocal lattice vector of the surface net. The perpendicular component of the diffracted beam, $\Gamma_{\underline{G}}$, is determined from $|\underline{k}^1| = |\underline{k}|$ for elastic diffraction. If we consider incidence near a zone-axis, $\underline{u}$, of the surface net and write $\underline{K}$ and $\underline{G}$ as $(K_x, K_y)$ and $(G_x, G_y)$ where the x direction is perpendicular to $\underline{u}$ and y parallel to it, our resulting diffraction pattern is a map of $(K_x + G_x, \Gamma_{\underline{G}})$ for each $\underline{G}$. We can easily show that:

$$(K_x + G_x)^2 + \Gamma_{\underline{G}}^2 = \Gamma_0^2 + K_x^2 - 2K_yG_y - G_y^2 \qquad (1)$$

Since $\underline{u}$ is a zone axis, $\underline{u}.\underline{G}$ is a mulitple of $2\pi$ , and the diffracted beams may be considered in layers corresponding to each multiple. Therefore, for a given layer and incident beam direction, $(K_x, -\Gamma_0)$, equation 1 shows that the diffracted beams for each layer lie on the circumference of a circle centred at the zone-axis. This is essentially the result of an Ewald sphere construction.

There are further consequences of this geometry. Firstly, if a convergent beam is incident on the surface, the specularly reflected beam is equal in size and shape, but the diffracted beams are considerably distorted in both aspects (Shannon et al 1985). Secondly, when looking for an internal dark field mirror in one of the zero layer reflections to establish the presence of a mirror line perpendicular to $\underline{u}$, a true mirror will not be found. Instead the symmetry related positions lie on the arc of a circle centred at the zone-axis (Fig 1). The $K_x^1$ components of the "mirror related" positions are still equidistant from the Bragg line but they are not true mirror images. This is what we refer to as a "topological symmetry".

By considering reciprocity and the point symmetry operators of the surface net, and including this "topological mirror" we have derived five RHEED diffraction groups. (A full account will be published elsewhere). These are listed in Table 1 using the nomenclature introduced by Buxton et al (1976). Mirror lines parallel or perpendicular to the zone-axis, a diad normal to the surface and all or none of these are the surface point group operators that may be detected. Other columns in the table give sufficient information to design experiments to distinguish the diffraction groups.

In principle, the ability to distinguish five RHEED diffraction groups considerably extends the usefulness of RHEED in surface structure determination. This used in conjunction with high-resolution reflection electron microscopy and surface EELS would offer a useful battery of techniques to study surfaces in electron microscope columns. However, calculation (Maksym and Beeby 1981) and our experiments show that non zero layer effects are very weak at least for the small unit cells of MgO and Ag. If these cannot be detected we can only distinguish whether or not a mirror line parallel to the zone-axis $\underline{u}$ is present.

REFERENCES


Buxton B F, Eades J A, Steeds J W, Rackham G M - 1976, Phil Trans Roy Soc London A281, 171

Holland B W, Woodruff D P - 1973, Surf Sci 36 488

Maksym P A, Beeby J L - 1981, Surf Sci 110 423

Shannon M D, Eades J A, Meichle M E, Turner P S, Buxton B F - 1984, Phys Rev Lett 53 2125

Shannon M D, Eades J A, Meichle M E, Turner P S - 1985, Ultramicroscopy 16 175


| DG | BF | WP | $(G_x,0)$ | $(0,G_y)$ | $\pm(G_x,0)$ |
|---|---|---|---|---|---|
| 1 | 1 | 1 | 1 | – | |
| $m_R$ | m | 1 | "m" | 1 | |
| $2_R$ | 1 | 1 | 1 | – | "$2_R$" |
| m | m | m | 1 | m | |
| $2_Rmm_R$ | m | m | "m" | m | |

Table 1 RHEED diffraction groups, DG
WP = whole pattern
BF = bright field

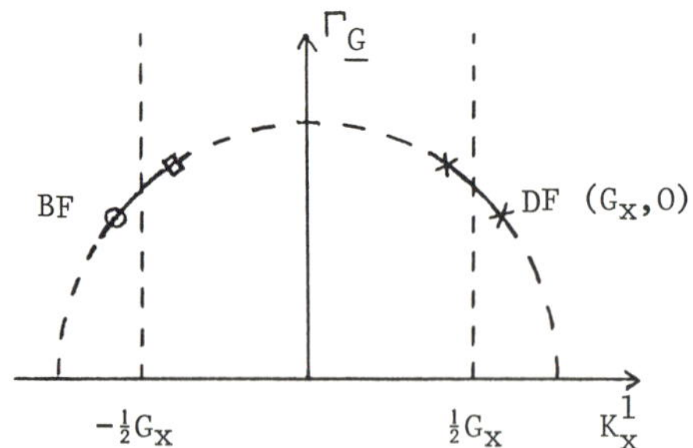


Fig 1 "Mirror related directions in $(G_x,0)$, marked x. 0 and □ are corresponding bright field directions.

# On line analysis of electron diffraction information

P M Budd and P J Goodhew

Department of Materials Science and Engineering, University of Surrey, Guildford GU2 5XH.

## 1. Introduction

The aim of this paper is to discuss the on-line collection and analysis of a line scan through an electron diffraction pattern of a material whilst it is being examined in the TEM. A computer program is being developed to allow interpretation of the resulting variation in intensity across the pattern for two immediate applications: the analysis of ring patterns from amorphous materials and sample thickness determinations from convergent beam diffraction (CBD) patterns. An on-line value for the thickness of the sample is of great value for x-ray absorption corrections and for consideration of spatial resolution and it is to this particular application that the paper will be mainly aimed.

## 2. Experimental procedure

The work has been carried out on a Philips EM400T/STEM with a VG ELS80 Spectrometer and a LINK Systems computer interfaced to the STEM. The ring or CBD pattern is set up in normal TEM mode and is scanned across either the STEM detector or EELS entrance aperture by post-specimen beam deflection coils, controlled by the hybrid diffraction unit when in SCID mode. By this means the diffraction pattern is transferred to the STEM monitor and the LINK Systems computer is used to control the positioning of the cursor for the line scan to measure the intensity variation. For the CBD pattern it was found to be most convenient to use the scan coils control to rotate the pattern until the fringes within the diffracted disc are vertical (this can be confirmed using the cross-wires) so that a horizontal scan taken through the centre of the disc yields the information required. For ring patterns the undiffracted spot is centred on the STEM monitor then a line scan through the central spot will allow the measurement of the ring diameters. The line scan on the videoscope monitor can be optimised to give the best signal to noise ratio using the brightness and contrast controls if in STEM mode or by altering the slit width and gain if using the EELS to image the diffraction pattern.

## 3. Aims of the computer program

The program allows the collection of a scan along any straight line in the diffraction pattern and the writing and reading of the data to and from disc. For the ring patterns a peak search routine is being developed to allow the ring diameters and their ratios to be determined. However further processing routines need to be developed to allow stripping of peaks to enable shoulders on the peaks to be examined and to determine peak widths which may be of interest for amorphous materials.

In the CBD analysis to determine the thickness then the parameters required are the disc spacing and the minima spacing within the diffracted disc. To find the disc edges differences from a running mean were used, however, particularly when using the STEM detector the edges are not clearly defined and it is not always easy to distinguish the precise position of each disc edge. Also the presence of Kikuchi lines running close to a disc may give rise to the determination of an incorrect position. Therefore a further facility was added to allow the user to move the cursor on the STEM monitor to mark the positions of the disc edge. A peak search routine using the change in gradient enables the minima to be determined and at each determination the user confirms the position of the minimum or may move the cursor to the appropriate position.

From this data, the deviation parameter, $S_i$, can then be calculated using the equation (Kelly et al, 1975),

$$S_i = \frac{\lambda}{d_{hkl}^2} \frac{(\text{minimum spacing})_i}{\text{disc spacing}} \quad \text{where } i=1, \text{ no of fringes}$$

If a graph of $(S_i/n_i)^2$ is plotted against $(1/n_i)^2$ then a straight line should be obtained where gradient = $(1/\text{extinction distance } (\xi_g))^2$ and intercept = $(1/\text{thickness})^2$.
The program uses a least squares fit to fit the data to a straight line each time for varying assignments to the value of $n_i$. If the appropriate $\xi_g$ is known therefore the fitted line which gives the closest value to $\xi_g$ will give the correct thickness. The program then allows the user to plot and examine the various fitted lines obtained, on the screen and prints out a table of results.

## 4. Results

Figures 1 and 2 give the plots collected for a ring pattern and CBD pattern respectively, each are comparing the traces obtained using the STEM and EELS detectors.

It was found for both types of pattern that to obtain sufficient resolution with the STEM detector that the highest camera length on the microscope was required. However even then there still always appears to be much better peak/background when using the EELS compared to the STEM detector. This is probably due to the fact that the diffraction pattern is being energy filtered to some extent even though the spectra were collected with the slit fairly wide open. Therefore processing of the data to find eg the minima in the CBD will be more accurate with the EELS due to the much better definition of the peaks.

With the ring pattern, because of the high intensity of the central spot relative to the diffracted rings the central region usually has to be saturated out on the line scan in order to see any of the peaks from the rings. This means that the position of the centre spot cannot be determined accurately and also leads to some problems using the EELS. Because the central spot is very bright there is an afterglow on the scintillator giving rise to tailing on one side of the central peak, increasing the background under the peaks to the right of the pattern.

For the CBD pattern, contamination is sometimes a problem causing the pattern to broaden and lose contrast the longer the spot remains on the same area. However once the microscope and STEM/EELS have been set up the collection of the line scan is very quick (≈ 30 secs.) and therefore this should not prove to be a problem. Also in CBD patterns Kikuchi lines may be present which could interfere with a minimum position in the diffracted disc. This may be eliminated by collecting two line scans at slightly different positions and comparing the two, using only the minima which occur in both scans.

Finally Figure 3 gives a table of results calculated using the program for the EELS pattern of Figure 2 along with a plot of the fitted line for the correct value of $\xi_g$. The accuracy of this method is ±3% for measurements using the EELS which is comparable to the conventional TEM method but using the STEM there is reduced resolution. This is therefore an accurate way of determining thickness quickly on-line, the procedure only taking a few minutes once the conditions have been set up correctly on the TEM.

Acknowledgements

Thanks are due to Dr. Steve Vale of LINK Systems for help with the computer software and to Ray Cox and Dawn Chescoe of the University of Surrey for useful suggestions and advice.

Reference

Kelly, P.M., Jostsons A., Blake, R.G., and Napier, J.G. (1975) Phys. Stat. Sol. A31, 771.

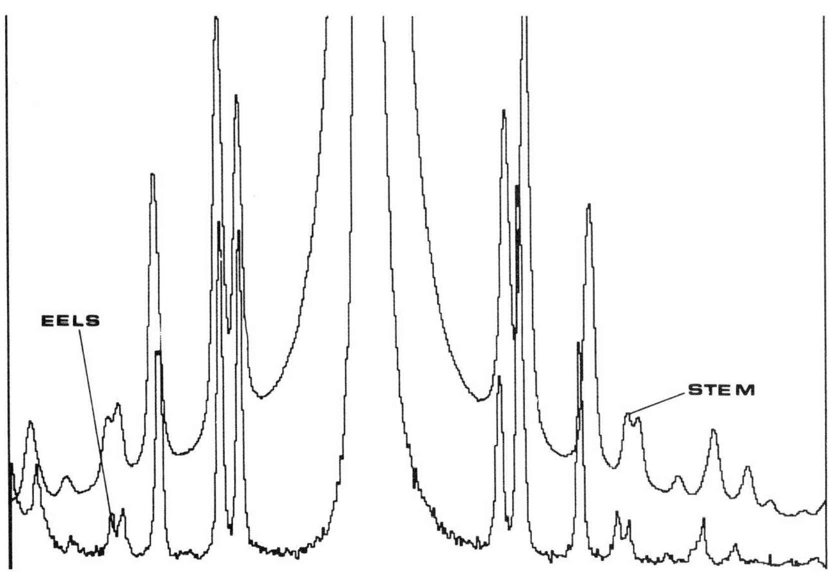


Fig. 1. Comparison of a ring pattern collected by the EELS and STEM detector.

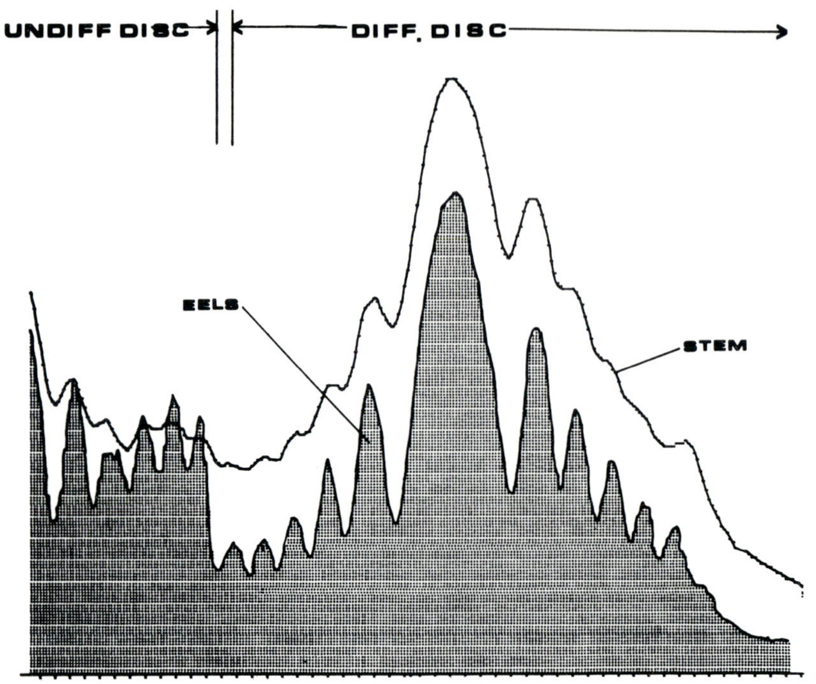


Fig. 2.Comparison of a CBD pattern collected by the EELS and STEM detector

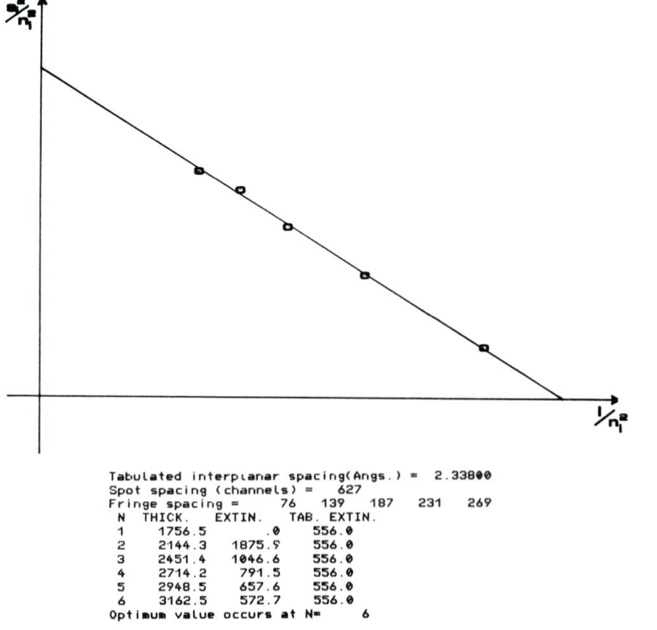


Tabulated interplanar spacing(Angs.) = 2.33800
Spot spacing (channels) = 627
Fringe spacing = 76 139 187 231 269

| N | THICK. | EXTIN. | TAB. EXTIN. |
|---|---|---|---|
| 1 | 1756.5 | .0 | 556.0 |
| 2 | 2144.3 | 1875.9 | 556.0 |
| 3 | 2451.4 | 1046.6 | 556.0 |
| 4 | 2714.2 | 791.5 | 556.0 |
| 5 | 2948.5 | 657.6 | 556.0 |
| 6 | 3162.5 | 572.7 | 556.0 |

Optimum value occurs at N= 6

Fig. 3. Table of results obtained for a thickness measurement with a plot of the best fit.

# Some aspects of the determination of crystal boundary geometry

Valerie Randle and Brian Ralph
Department of Metallurgy and Materials Science,
University College Cardiff.

## 1. Introduction

The geometrical relationship between two adjoining crystals can be conveniently described by an angular rotation, $\Theta$, about an axis of misorientation, $\ell$, (e.g. Mykura, 1980). Certain special rotations give rise to a coincident site lattice (CSL) which refers to the periodic array of coinciding lattice points which are obtained when the two crystals are notionally superimposed (e.g. Gleiter, 1982). A scheme which allows large numbers of boundaries to be analysed in terms of the CSL model has been described in detail by Randle and Ralph (1985). The particular aspect of this scheme which forms the main topic here relates to the twenty-four equivalent solutions which can be used to describe the same CSL in the cubic system. These symmetry-related solutions arise because there exists a choice of several sets of axes for the second crystal after those in the first crystal have been fixed.

## 2. Symmetry Related Solutions

Although an axis/angle pair is usually expressed in terms of the smallest angular rotation, any of the other twenty-three solutions are equally valid representations of the misorientation between two grains. This is illustrated with respect to a grain boundary in a Ni-Fe-Cr based alloy which has a face-centred cubic structure. An axis/angle pair of 0.990, 0.143, 0.035/50.5° is obtained by stereographic construction from the microdiffraction data shown in figure 1.

By inspection, this axis/angle pair is seen to be close to a $\Sigma = 5$, 100/53.13° CSL. In order to test if the experimental boundary lies within the limits for a $\Sigma$ = CSL description, it is necessary to express each axis/angle pair in terms of a 3x3 rotation matrix where the columns of the matrix are the base vectors of the first grain referred to the crystal axes of the second grain. The relevant matrices are:-

$$R\ (CSL) = \begin{matrix} 1 & 0 & 0 \\ 0 & 0.6 & -0.8 \\ 0 & 0.8 & 0.6 \end{matrix} \qquad R\ (expt) = \begin{matrix} 0.992 & 0.024 & 0.123 \\ 0.078 & 0.644 & -0.761 \\ 0.098 & 0.765 & 0.637 \end{matrix}$$

The angular differences between the column vectors are 7.2°, 3.5° and 7.7°. The overall variation of this boundary from a $\Sigma$ = CSL is thus 6.1°.

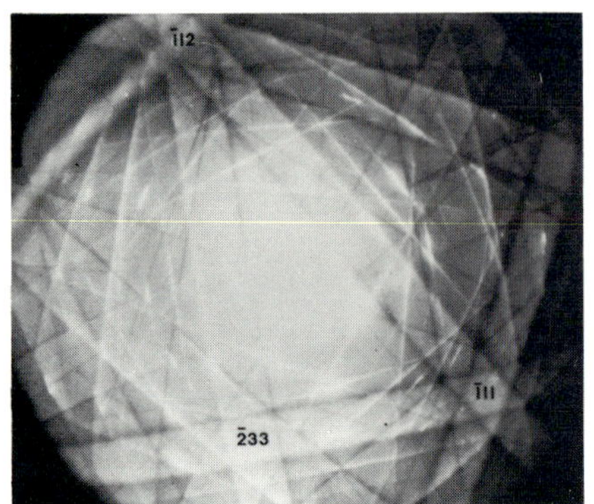


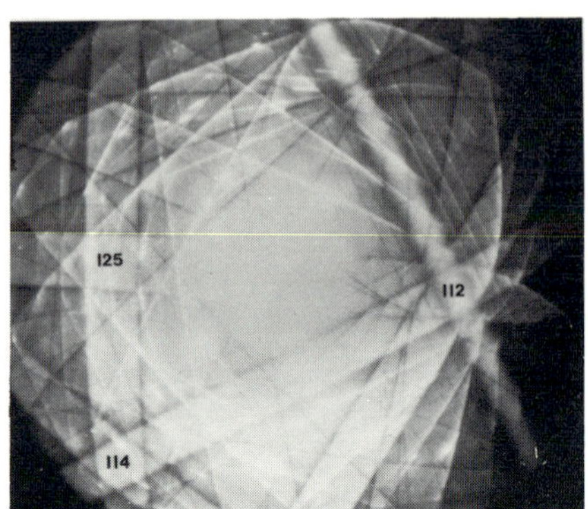


Figure 1. Low camera length microdiffraction patterns taken from either side of a grain boundary. Some major directions are labelled.

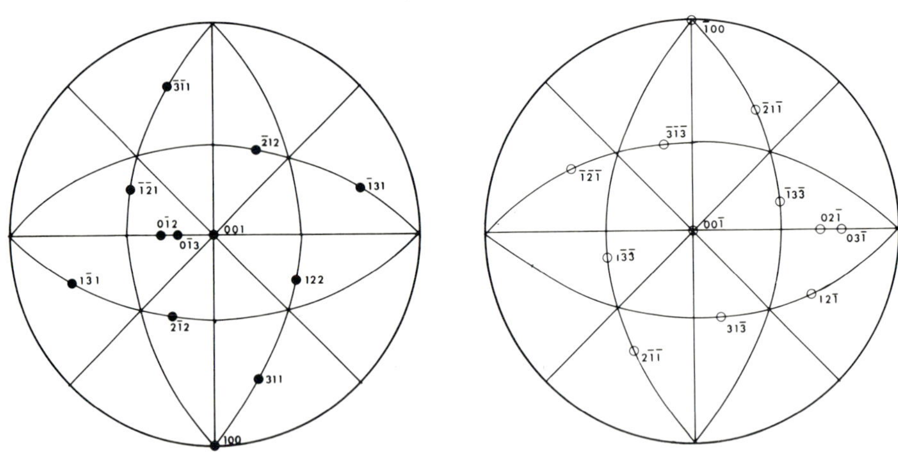


Figure 2. The distribution of misorientation axes for the $\Sigma = 5$ CSL. Poles on the upper and lower hemispheres are plotted separately for clarity.

Table 1

| Axis/angle pair for $\Sigma=5$ CSL. $\ell/\Theta$ | Axis/angle pair for experimental grain boundary. $\ell/\Theta$ | Deviation from exact CSL. $\Delta\Theta$ |
|---|---|---|
| 001/36.87° | 0.122,0.201,1.284/40.6° | 6.1° |
| $\bar{1}\bar{2}\bar{1}$/101.54° | $\overline{0.859}$,$\overline{1.636}$,$\overline{0.642}$/102.0° | 6.4° |
| $02\bar{1}$/180.00° | 0.004,0.054,$\overline{0.025}$/178.2° | 7.0° |
| $\bar{3}\bar{1}\bar{3}$/154.16° | $\overline{0.559}$,$\overline{0.227}$,$\overline{0.521}$/156.6° | 6.4° |
| $1\bar{3}1$/95.74° | 0.742,$\overline{1.753}$,0.613/91.0° | 6.4° |
| $\bar{1}3\bar{3}$/154.16° | $\overline{0.227}$,0.767,$\overline{0.715}$/147.5° | 6.2° |
| $\bar{1}31$/95.74° | $\overline{0.661}$,1.753,0.546/102.6° | 6.1° |
| $1\bar{3}\bar{3}$/154.16° | 0.231,$\overline{0.613}$,$\overline{0.546}$/154.9° | 6.2° |
| 311/95.74° | 1.757,0.767,0.559/93.4° | 6.1° |
| $12\bar{1}$/101.54° | 0.715,1.757,$\overline{0.521}$/100.4° | 6.4° |
| $31\bar{3}$/154.16° | 0.742,0.231,$\overline{0.661}$/149.3° | 6.4° |
| 122/143.13° | 0.348,0.888,0.859/140.1° | 6.2° |
| $0\bar{1}3$/180.00° | 0.007,$\overline{0.074}$,0.201/173.7° | 6.4° |
| $\bar{1}\bar{2}2$/143.13° | $\overline{0.355}$,$\overline{0.843}$,0.737/144.0° | 6.4° |
| $\bar{2}12$/143.13° | $\overline{0.642}$,0.348,0.663/150.4° | 6.1° |
| $03\bar{1}$/180.00° | 0.007,0.122,$\overline{0.045}$/176.4° | 6.5° |
| $2\bar{1}2$/143.13° | 0.785,$\overline{0.355}$,0.687/146.6° | 6.1° |
| $1\bar{3}1$/95.74° | 0.843,$\overline{1.629}$,0.785/93.9° | 6.4° |
| $\bar{1}00$/53.13° | $\overline{0.990}$,0.143,0.035/50.5° | 6.1° |
| $\bar{2}1\bar{1}$/101.54° | $\overline{1.629}$,0.843,$\overline{0.785}$/93.9° | 6.4° |
| $0\bar{1}2$/180.00° | 0.004,$\overline{0.102}$,0.221/173.0° | 6.4° |
| $2\bar{1}\bar{1}$/101.54° | 1.629,$\overline{0.737}$,$\overline{0.687}$/106.8° | 6.1° |
| $00\bar{1}$/143.13° | 0.074,0.045,$\overline{1.281}$/140.1° | 6.4° |
| 100/126.87° | 1.526,0.102,0.025/130.1° | 6.3° |

Small deviations, $\Delta\Theta$, from a precise CSL configuration can be accommodated by grain boundary dislocations, with the actual amount of mismatch which can be tolerated decreasing as $\Sigma$ increases. Brandon (1966) specifies the CSL unit ($\Delta\Theta$) thus:-

$$\Delta\Theta = V_o\Sigma^{-\frac{1}{2}}$$

Where $V_o$ is the low angle boundary limit ($\simeq 15°$). For a $\Sigma = 5$ boundary, the acceptable deviation is about $6.7°$. It is therefore justifiable to describe the boundary with an axis/angle of 0.990, 0.143, 0.035/50.5° as a $\Sigma = 5$ CSL. This axis/angle pair does not represent the lowest angle solution of the $\Sigma = 5$ CSL case. Generally the smallest angle solution is picked out by inspection when a complete set of twenty-four symmetry related rotation matrices are generated by computer. However, for the purpose of this example, all twenty-four variations for both the $\Sigma = 5$ CSL and the experimental boundary have been generated and an angular deviation from the exact CSL situation calculated for each case (table 1).

It is apparent that, apart from the 0.004, 0.054, 0.025/178.2° solution where precision is reduced the calculated deviation is almost constant for each solution. The distribution of misorientation axes for the $\Sigma = 5$ case is represented stereographically in figure 2. As mentioned previously there are twenty-four equivalent descriptions of an axis/angle pair for the cubic system and in the general case the twenty-four axes turn up in each of either the right handed or left handed unit stereographic triangles. Such is the case for some of the less densely packed CSL cases, such as $\Sigma = 39b$ (e.g. Mykura). However, when the CSL is more densely packed, as in the case considered here, the number of distinct forms is reduced since some or all of the rotation axes lie on the edges of unit triangles. Thus for the case where $\Sigma = 5$ the axes of misorientation are of seven distinct forms <100>, <311>, <211>, <221>, <331>, <210> and <310>.

From table I, the lowest angular description of the misorientation parameters (a frequently adopted method of describing $\ell/\Theta$) is .122, .201, 1.281 / 40.6° and this is seen to deviate from the exact $\Sigma = 5$ description by 6.1°. Using the Brandon criterion (equation 1) for allowable deviations from exact CSL which retain some element of CSL periodicity would make this a special boundary since on a $\Sigma^{-\frac{1}{2}}$ criterion the allowed $\Delta\Theta$ for $\Sigma = 5$ is 6.7°.

References

Brandon D G 1966 Acta Met. 14 1479
Gleiter H 1982 Mat. Sci. Eng. 52 91
Ishida T and McLean M 1973 Phil. Mag. 27 1125
Mykura H 1980 Grain-Boundary Structure and Kinetics (Ohio:ASM) pp 445-456
Randle V and Ralph B submitted to J. Mat. Sci.

*Inst. Phys. Conf. Ser. No 78: Chapter 2*
*Paper presented at EMAG '85, Newcastle upon Tyne, 2–5 September 1985*



# Accurate structure factor determination of binary cubic solid solutions

Alan G Fox and Robert M Fisher*

School of Engineering, The Polytechnic, Wulfruna Street, Wolverhampton WV1, 1LY, UK

*Center for Advanced Materials, Lawrence Berkeley Laboratory, University of California, Berkeley, California 94720, USA

## 1. Introduction

The Critical Voltage technique ($V_C$) in HEED can measure low angle x-ray structure factors of crystalline materials to within 0.1% provided accurate higher angle structure factors and Debye-Waller factors are available (for reviews of the $V_C$ method see e.g., Lally et al., 1972; Thomas et al., 1974). This is far more accurate than simply using structure factors obtained from atomic scattering factors (form factors) determined by Band Structure calculations or x-ray measurements, and by adopting this approach Smart and Humphreys (1978, 1980) have produced complete sets of accurate structure factors for many cubic elements, which they Fourier-analysed to produce finely detailed electron density maps.

Shirley and Fisher (1979) developed a model for the Debye-Waller factors, B, of binary cubic solid solutions. They correlated the variation of B with composition, temperature and short-range order (sro) by the use of two parameters $\tau$ and $\gamma$ which they found could be determined by $V_C$ measurements. Fox (1984) extended this approach and found that both sro and electron charge distribution changes due to alloying in Cu-25 at % Au could be detected by the $V_C$ technique. This method involves obtaining a set of accurate structure factors for the alloy by curve fitting to best pure element form factors and interpolation, and analysing the alloy $V_C$ measurements with an accurate alloy Debye-Waller factor to look for structure factor changes arising from possible electron charge redistribution. This method is an improvement over previous work on this topic which has always considered the form factors of the atoms in the alloy to be essentially the same as those in the pure elements. The extension of this technique to a range of alloys in the systems CuAu, CuAl and FeCr will now be presented.

## 2. Results and Discussion

CuAu. This solid solution comprises two monovalent metals with a large atomic size difference ($\sim$15%), and which shows an ordering tendency. There has been a great deal of $V_C$ work carried out on this alloy, and the best room temperature B and $V_C$ values necessary for a full analysis are shown in Table 1. Kuroda, et al., (1981) have also investigated the effects of changing sro on the 400 $V_C$ in Cu-15 at % Au, and their results are shown in Table 2.

Table 1 Room temperature B and $V_C$ values for Cu, Au, and their quenched solid solution alloys

| B($Å^2$) | Atomic % Au | $V_C^{222}$(kV) | $V_C^{400}$(kV) | $V_C^{440}$(kV) |
|---|---|---|---|---|
| 0.54[1] | 0.0 | 310 ± 3 [1] | 605 ± 3 [1] | 1750 ± 50[2] |
| 0.655[5] | 4.9 | 273 ± 10[3] | 532 ± 5 [3] | -- |
| 0.729[5] | 14.9 | <258 [3] | 433 ± 10[3] | -- |
| 074[6] 0.765[4] | 25.0 | 165 ± 5 [4] | 360 ± 10[4] | 1230 ± 60[4] |
| 0.51 | 100.0 | -- | 108 ± 2 [1] | 726 ± 5[1] |

Notes on Table 1: (1) from Thomas et al., (1974); (2) from Rocher and Jouffrey (1972); (3) from Kuroda et al., (1981); (4) from Fox (1984); (5) from present work; (6) from Borie (1957).

In order to analyse the results shown in Tables 1 and 2, the best pure element form factors of Wakoh and Yamashita (1971) for Cu and Doyle and Turner (1968) for Au were used. These were shown to be very accurate by the pure element $V_C$ measurements. The only accurate x-ray Debye-Waller factor available for this system appears to be that of Borie (1957) for a 25 at % Au alloy. This gives $\tau$ = 1.06 and $\gamma$ = 1.12 in good agreement with theory and with the sro results of Table 2.

It should be noted that sro does not affect the $V_C$'s significantly for the alloy compositions studied in the present work, and has very little effect on the charge redistribution studies if it is ignored. From the $V_C$ results of Table 1 it appears that all the alloy structure factors having (hkl) $\geq$ (200) are the same as those interpolated from the best pure element form factors (within experimental error). The (111) alloy structure factors on the other hand are significantly increased over those interpolated from the best pure element form factors and nearer to the free atom values of Doyle and Turner (1968), as shown in Figure 1. This increase suggests that the electron charge distribution of Cu is made more spherical by gold additions as is its Fermi surface (Coleridge et al., 1984).

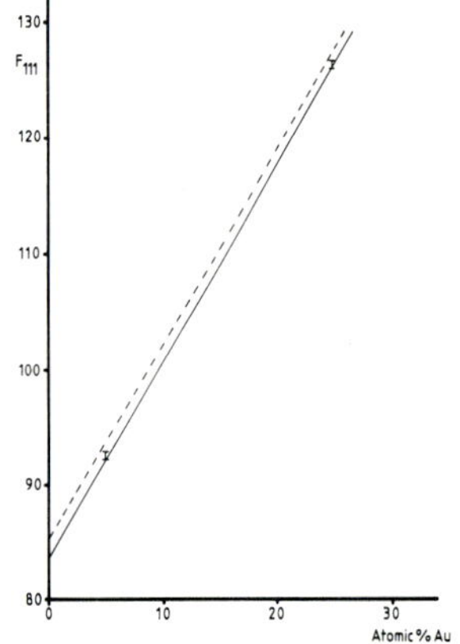


Figure 1. Graph of 111 structure factor, $F_{111}$, vs. composition for CuAu alloys. Dotted line - Doyle and Turner (1968). Solid line - interpolated between best pure element values. The bars show the experimental results.

Table 2 Changes in room temperature $V_C^{400}$ (for Cu-15 at % Au due to changing sro)

| Heat Treatment | sro parameters $\alpha_1$ | $\alpha_2$ | $\alpha_3$ | $V_C^{400}$(kv) |
|---|---|---|---|---|
| Furnace cooled from 1073K | -- | -- | -- | 438 ± 5 |
| Quenched from 673K | -0.106 | 0.156 | -0.007 | 433 ± 10 |
| Quenched from 1073K | -0.075 | 0.135 | -0.007 | 426 ± 10 |

CuAl In this case monovalent Cu is alloyed with trivalent Al, and the atomic radius disparity is ~6%. These alloys show sro, but it is too small to detect by the $V_C$ method unless the measurement accuracy can be improved to the order of that obtained by Sellar et al (1980). There is only sufficient electron diffraction information to perform a full analysis on one alloy of composition 14.8 at % Al. The best room temperature B and $V_C$ values required are shown in Table 3.

Table 3 Room temperature B and $V_C$ values for Cu, Al, and quenched Cu-14.8 at % Al

| B(Å$^2$) | At % Al | $V_C^{222}$ (kV) | $V_C^{400}$ (kV) |
|---|---|---|---|
| 0.54(1) | 0.0 | 310 ± 3(1) | 605 ± 3(1) |
| 0.611(2) | 14.8 | 332 ± 10(3) | 641 ± 10(3) |
| 0.85(1) | 100.0 | 425(1) | 918 ± 5(1) |

Notes on Table 3 (1) from Thomas et al., (1974); (2) from Houska and Averbach (1959), (3) from Kuroda et al (1981).

To analyse these results the Cu form factors used were as for CuAu, and for Al the form factors of Inkinen et al., (1970) were found to give best agreement with the $V_C$ measurements. The Debye-Waller factor of Houska and Averbach (1959) appears to be the most accurate available and gives $\tau$ = 0.97 and $\gamma$ = 1.4. An analysis of these results showed that the alloy low angle structure factors were significantly less than those found by interpolation between best pure element form factors as shown in Table 4. This suggests that not all the Al valence electrons are contributing to the conduction band of the alloy as suggested by Mott (1952) and recently confirmed by Coleridge et al., (1984).

Table 4. Room temperature low-angle x-ray structure factors, $F_{hkl}$, for Cu-14.8 at % Al

| hkl | $F_{hkl}$ (free atom) | $F_{hkl}$ (pure element) | $F_{hkl}$ (experimental) |
|---|---|---|---|
| 111 | 77.89 | 76.74 | 76.63 ± 0.15 |
| 200 | 72.07 | 71.46 | 71.27 ± 0.15 |

FeCr This alloy comprises two b.c.c. transition metals of variable valency with an atomic radius difference of 0.6%. These alloys have a segregating tendency, but this is too small to detect by $V_C$ measurements. The best room temperature B and $V_C$ values for the pure elements and three representative alloys are shown in Table 5.

Table 5 Room temperature B and $V_C$ values for Fe, Cr, and three solid solution alloys.

| B(Å$^2$) | At% Cr | $V_C^{220}$ (kV) | $V_C^{400}$ (kV) |
|---|---|---|---|
| 0.35 (1) | 0.0 | 305 ± 3(2) | 1278 ± 4 |
| 0.323(3) | 31.9 | 285 ± 3(3) | 1275 ± 15(3) |
| 0.31 (3) | 46.0 | 280 ± 5(3) | 1275 ± 20(3) |
| 0.277(3) | 75.0 | 270 ± 3(3) | 1280 ± 20(3) |
| 0.24 (1) | 100.0 | 265 ± 3(2) | 1285 ± 31(1) |

Notes on Table 5 (1) from Terasaki et al., (1975); from Shirley et al., (1975); (3) from present work.

It can be seen that the $V_C^{400}$ results all agree within experimental error and this immediately suggests that the 200 structure factor is unchanged by alloying. This gives the Debye-Waller factors shown in Table 5 and $\tau$ = 0.87. $\gamma$ could not be determined as the atomic radius disparity is very small. The 110 structure factors are markedly less than those obtained by interpolation between the best pure element form factors of Wakoh and Yamashita (1971) which were used in the analysis as shown in Figure 2. This is to be expected as it is well known that there are significant electronic changes associated with the alloying of Fe and Cr (see e.g., Starke et al., 1962).

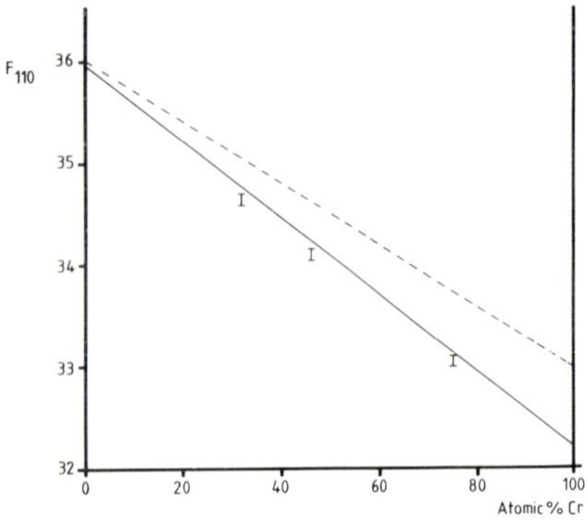

Figure 2. Graph of 110 structure factor, $F_{110}$, vs. composition for FeCr alloys. Dotted line - Doyle and Turner (1968). Solid line - Interpolated between best pure element values. The bars show the experimental results.

## 3. Acknowledgments

The authors would like to thank Dr. K.H. Westmacott, National Center for Electron Microscopy, Lawrence Berkeley Laboratory, Berkeley, California for the use of the 1.5 MeV HVEM. This work was supported in part by the U.S. Department of Energy under contract No. DE-AC03-76SF00098.

## References

Borie B S 1957 Acta. Cryst. 10 89

Coleridge P T, Templeton I M, and Vasek P 1984 J. Phys. F. 14 2963

Doyle P A and Turner P S 1968 Acta. Cryst. A24 390

Fox A G 1984 Phil Mag B. Vol50 No4 477

Fox A G and Fisher R M 1985 submitted to Phil. Mag.

Houska C R and Averbach B L 1959 J. Appl. Phys. 30 1525

Inkinen O, Pesonen A, and Paakkari T 1970 Ann. Acad. Sci. Fenn. AVI 344

Kuroda K, Tomokiyo Y, and Eguchi T 1981 Trans. Jap. Inst. Met. Vol22 No8 535

Lally J S, Humphreys C J, Metherell, A J F and Fisher, R M 1972 Phil. Mag. 25 321.

Mott, N F 1952 Prog. Met. Phys. 3 76

Rocher A M and Jouffrey B 1972 in Electron Microscopy 1972 (Inst. Phys. Conf. Ser. 14) 528

Sellar J R, Imeson D and Humphreys C J 1980 Acta. Cryst. A36 686

Shirley C G and Fisher R M 1979 Phil Mag A Vol39 No1 91

Shirley C G, Thomas L E, Lally J S and Fisher R M 1975 Acta Cryst. A31 174

Smart D J and Humphreys C J 1978 in Electron Diffraction 1927-1977 (Inst. Phys. Conf. Ser. No41) Chap3 145

Smart D J and Humphreys C J 1980 in Electron Microscopy and Analysis 1979 (Inst. Phys. Conf. Ser. No52) Chap4 211

Starke E A Jr, Cheng C H and Beck P A 1962 Phys. Rev. 126 1746

Terasaki O, Uchida Y and Watanabe D 1975 J. Phys. Soc. Japan 39 1277

Thomas L E, Shirley C G, Lally J S and Fisher R M 1974 in High Voltage Electron Microscopy (London and New York:Academic Press) 52

Wakoh S and Yamashita J 1971 J. Phys. Soc. Japan 30 422

# Supplements completing the Bloch-wave treatment of electron transmission

G Kästner

Institute of Solid State Physics and Electron Microscopy,
Academy of Sciences of the German Democr. Republ., 401 Halle, PSF 250

To take advantage of the Bloch-wave formulation of electron diffraction one has to solve the eigenvalue equation $C^{-1}A\,C = \gamma$ to a suitable approximation. Simplified solutions can often be refined by perturbation corrections as known from introducing anomalous absorption, but also reflections from non-zero order Laue zones as well as inclinations of the crystal surface (general Laue case) up to $\lesssim 85^{\circ}$ can be treated as usually weak corrections (Kästner 1985).

## 1. Eigenvalue Equation in Case of Systematic Reflections

A lasting interest in refining analytical solutions of the eigenvalue equation in case of systematics is stimulated by diffraction contrast investigations at higher accelerating voltages or of more complex crystal structures. To allow an analysis superior to the 2-beam case, the dynamical matrix A will be scaled to dimensionless elements and related to a single Brillouin zone (either Fig. 1a or 1b):

(a)

| | | | | | | |
|---|---|---|---|---|---|---|
| $-9$ $-3h$ | $u_1^*$ | $u_2^*$ | $u_3^*$ | $u_4^*$ | $u_5^*$ | $u_6^*$ |
| $u_1$ | $-4$ $-2h$ | $u_1^*$ | $u_2^*$ | $u_3^*$ | $u_4^*$ | $u_5^*$ |
| $u_2$ | $u_1$ | $-1$ $-h$ | $u_1^*$ | $u_2^*$ | $u_3^*$ | $u_4^*$ |
| $u_3$ | $u_2$ | $u_1$ | $0$ | $u_1^*$ | $u_2^*$ | $u_3^*$ |
| $u_4$ | $u_3$ | $u_2$ | $u_1$ | $-1$ $+h$ | $u_1^*$ | $u_2^*$ |
| $u_5$ | $u_4$ | $u_3$ | $u_2$ | $u_1$ | $-4$ $+2h$ | $u_1^*$ |
| $u_6$ | $u_5$ | $u_4$ | $u_3$ | $u_2$ | $u_1$ | $-9$ $+3h$ |

(b)

| | | | | | |
|---|---|---|---|---|---|
| $-6$ $-h5/2$ | $u_1^*$ | $u_2^*$ | $u_3^*$ | $u_4^*$ | $u_5^*$ |
| $u_1$ | $-2$ $-h3/2$ | $u_1^*$ | $u_2^*$ | $u_3^*$ | $u_4^*$ |
| $u_2$ | $u_1$ | $0$ $-h/2$ | $u_1^*$ | $u_2^*$ | $u_3^*$ |
| $u_3$ | $u_2$ | $u_1$ | $0$ $+h/2$ | $u_1^*$ | $u_2^*$ |
| $u_4$ | $u_3$ | $u_2$ | $u_1$ | $-2$ $+h3/2$ | $u_1^*$ |
| $u_5$ | $u_4$ | $u_3$ | $u_2$ | $u_1$ | $-6$ $+h5/2$ |

Fig. 1 a,b Scaled dynamical matrix A where $h = -2k_x d_{hkl}$ = deviation from any Bragg position of even order (1a) or odd order (1b)

The off-diagonal elements have been scaled to

$$U_{g-g'}d^2_{hkl} \equiv u_n \equiv v_n\, m_r/m_e \tag{1}$$

These quantities are known to characterise the degree of many-beam interaction. Examples can be taken from $v_n$ and mass ratios of Table 1. The diagonal elements agree with the sequence of the reflections. They are integers except of $|h|\leq 0.5$. For larger $|h|$ Figs. 1a and 1b interchange.

### 1.1 "Generalised 2-beam case" ( $|u_n| \lesssim 0.3$)

Only the diagonal and the cross-diagonal of A is taken into account,all further elements are neglected. This allows analytical diagonalisation similar to the 2-beam case. For any $n \neq 0$ a pair of eigenvalues results forming a pair of hyperbolae separated by the extinction number

$$1/\xi_n = |u_n|\sqrt{1+w_n^2} \,/\, Kd_{hkl}^2 = |v_n|\sqrt{1+w_n^2}\,\lambda_c \,/\, (v/c)d_{hkl}^2 \qquad (2)$$

where $w_n = n\ h/2|u_n| \equiv \mathrm{cotg}\, 2\sigma_n$ is the generalised deviation parameter, $\lambda_c$ = Compton wavelength, v = electron velocity. The eigenvector matrix C consists of a diagonal and a cross-diagonal:

$$C = \begin{bmatrix} \exp -i\pi_n \cos\sigma_n & \exp -i\pi_n \sin\sigma_n \\ -\exp i\pi_n \sin\sigma_n & \exp i\pi_n \cos\sigma_n \end{bmatrix} = \frac{1}{\sqrt{2}} \begin{bmatrix} \exp -i\pi_n & 0 \\ 0 & \exp i\pi_n \end{bmatrix} \begin{bmatrix} \sqrt{1+w_{rn}} & \sqrt{1-w_{rn}} \\ -\sqrt{1-w_{rn}} & \sqrt{1+w_{rn}} \end{bmatrix}$$

where $w_{rn} \equiv w_n/\sqrt{1+w_n^2}$ . Non-centrosymmetry yields $\mathrm{tg}\, 2\pi_n = \mathrm{Im}\, u_n/\mathrm{Re}\, u_n$.

### 1.2 Correction of the generalised 2-beam case ( $|u_n| \lesssim 0.5$ )

The elements neglected above are treated as a perturbation matrix. It corrects the diagonal elements (real) to become approximately

$$0 \rightarrow 2\left[|u_1|^2/(1-h^2) + |u_2|^2/(4-h^2) + |u_3|^2/9 + \mathrm{Re}(u_1^2 u_2^*)/(1-h^2)\right]$$

$$-1+h \rightarrow -1 - |u_1|^2 2/3 + |u_2|^2/8 + |u_3|^2 2/5 - \mathrm{Re}(u_1^2 u_2^*) + h(1 - 0.89|u_1|^2 - |u_3|^2/3)$$

$$h/2 \rightarrow |u_1|^2/2 + |u_2|^2 2/3 + |u_3|^2/4 - \mathrm{Re}\left[u_1(u_1 u_2^* - u_2 u_3^*)\right]/2 + h(1 + |u_1|^2/2 - |u_2|^2)/2$$

$$-2+3h/2 \rightarrow -2 - |u_1|^2/4 - |u_2|^2 2/5 + h\ 3/2$$

as well as the cross-diagonals (hermitian) to become

$$u_1 \rightarrow u_1 + u_1^* u_2 + u_2^* u_3/3 + u_3^* u_4/6 + \left[u_1^{*2} u_3 + u_2^2 u_3^* - u_1(|u_1|^2 + |u_2|^2)\right]/4 + h^2 0.62 u_1^* u_2$$

$$u_3 \rightarrow u_3 - u_1 u_2 + u_1^* u_4/2 + u_2^* u_5/5 + \left[u_1^3 + u_1^* u_2^2 - u_3(|u_1|^2 + |u_2|^2)\right]/4 - h^2 0.62 u_1 u_2$$

$$u_2 \rightarrow u_2 - u_1^2 + 2(u_1^* u_3/3 + u_2^* u_4/8) - |u_1|^2 u_2 - h^2(u_1^2 + 2|u_1|^2 u_2)$$

$$u_4 \rightarrow u_4 - u_2^2/4 + 2\ (-u_1 u_3/3 + u_1^* u_5/5) - h^2\ 0.37 u_1 u_3$$

It can be shown that for h = 0 the terms quadratic in u correcting $u_n$ agree with Bethe's known correction of the Fourier amplitudes.

### 1.3 Correction of the 3-beam or of the 4-beam case ( $|u_n| \lesssim 1.5$ or 2 resp)

The elements of the central 3x3 or 4x4 matrix of Fig.1a,b resp. can be corrected similarly as above. This improves analytical solutions known for Bragg positions (Howie 1966, Steeds 1970) and critical voltages (Lally et al. 1972). Corrected elements of the 3x3 matrix, Fig. 1a:

$$u_1 \rightarrow u_1 + 0.292 u_1^* u_2 + 0.410 u_2^* u_3 + 0.183 u_3^* u_4 + \ldots$$

$$u_2 \rightarrow u_2 + 2\ (\ u_1^* u_3/3 + u_2^* u_4/8 + u_3^* u_5/15 + \ldots)$$

$$0 \rightarrow 2\ (|u_2|^2/4 + |u_3|^2/9 + |u_4|^2/16 + \ldots)$$

$$-1 \rightarrow -1 + |u_1|^2/3 + |u_2|^2/8 + |u_3|^2 2/5 + |u_4|^2/6 + \ldots$$

## 2. Kinematical Approximation to Diffraction Contrast

This approximation treats as known the off-diagonals of A,C and related matrices as perturbation thus allowing analytical calculation of image amplitudes. It can be shown that the bright-field amplitude results from second-order perturbation on evaluating C = exp k as

$$C = I + k + k^2/2, \quad C^{-1} = I - k + k^2/2, \quad C^{-1}C = I + k^4/4 \approx I \tag{3}$$

where k is a diagonal-free matrix of known $k_{gg'} = A_{gg'}/(s_{g'} - s_g)$.
To realise kinematical conditions, the Ewald sphere has to cut in the region (at $h \approx 0.5$) between a Bragg position of order $H_k$ and $H_k+1$. For calculating the lowest necessary $H_k$ the contrast-determining $k_{g0}$ can be evaluated in terms of Fig.1 as $|k_{g0}| = |u_1| / |H_k - 0.5| = |1/2s_g\xi_g|$. Arbitrary restriction to $|k_{g0}|^2 \leq 0.1$ suggested by eq.(4), i.e. to $|s_g\xi_g| = |w_g| \geq 1.58$ yields the $H_k$ listed in Tab.1 for cases covered in practice. Comparison can be made to $H_{wb}$ obtained from the weak-beam criterion for beam -g : $|s_{-g}| = |H_{wb}+1.5| / 2Kd^2_{hkl} \geq 0.2\text{nm}^{-1}$. It requires large $H_{wb}$ for large $Kd^2_{hkl}$.

| lattice planes | | scaled Fourier amplitudes | | | | accel. voltage / relativistic mass ratio 100 kV 1.1957 | | 200 kV 1.3914 | | 500 kV 1.9785 | | 1000 kV 2.9569 | | 2000 kV 4.9139 | |
|---|---|---|---|---|---|---|---|---|---|---|---|---|---|---|---|
| | hkl | $v_1$ | $v_2$ | $v_3$ | $v_4$ | $H_k$ | $H_{wb}$ | $H_k$ | $H_{wb}$ | $H_k$ | $H_{wb}$ | $H_k$ | $H_{wb}$ | $H_k$ | $H_{wb}$ |
| Au | 111 | .6865 | .3515 | .192 | .111 | 4 | 4 | 4 | 8 | 5 | 15 | 7 | 24 | 11 | 43 |
| | 200 | .4632 | .2173 | .111 | .060 | 3 | 3 | 3 | 6 | 4 | 11 | 5 | 18 | 8 | 32 |
| | 220 | .1673 | .0619 | .026 | .011 | 2 | 1 | 2 | 2 | 2 | 4 | 3 | 9 | 4 | 15 |
| InP | 111r | .5549 | 0 | -.163 | -.142 | 3 | 11 | 4 | 17 | 5 | 31 | 6 | 52 | 10 | 90 |
| | 111i | .1386 | .1531 | .076 | 0 | | | | | | | | | | |
| | 220 | .1900 | .0757 | .036 | .018 | 2 | 3 | 2 | 6 | 2 | 11 | 3 | 19 | 4 | 33 |
| | 200 | .1402 | .2503 | .066 | .081 | 2 | 8 | 2 | 13 | 2 | 23 | 2 | 39 | 3 | 67 |
| GaAs | 111r | .4885 | 0 | -.156 | -.141 | 3 | 11 | 3 | 16 | 4 | 29 | 6 | 48 | 9 | 83 |
| | 111i | -.0261 | -.0137 | -.002 | 0 | | | | | | | | | | |
| | 220 | .1701 | .0732 | .036 | .018 | 2 | 3 | 2 | 5 | 2 | 10 | 3 | 17 | 4 | 31 |
| Ge | 111 | .4933 | 0 | -.157 | -.144 | 3 | 11 | 3 | 16 | 4 | 29 | 6 | 48 | 9 | 84 |
| | 220 | .1704 | .0741 | .037 | .019 | 2 | 3 | 2 | 5 | 2 | 10 | 3 | 17 | 4 | 31 |
| Si | 111 | .3559 | 0 | -.079 | -.069 | 2 | 10 | 3 | 15 | 3 | 27 | 4 | 44 | 7 | 77 |
| | 220 | .1059 | .0363 | .018 | .009 | 1 | 2 | 1 | 4 | 2 | 9 | 2 | 16 | 3 | 28 |
| Mo | 110 | .4758 | .2149 | .113 | .067 | 3 | 4 | 3 | 7 | 4 | 13 | 5 | 22 | 8 | 38 |
| | 200 | .1670 | .0624 | .030 | .016 | 2 | 1 | 2 | 2 | 2 | 6 | 3 | 10 | 4 | 19 |
| α-Fe | 110 | .3214 | .1439 | .072 | .038 | 2 | 3 | 2 | 6 | 3 | 11 | 4 | 18 | 6 | 32 |
| | 200 | .1122 | .0402 | .016 | .007 | 1 | 1 | 1 | 2 | 2 | 4 | 2 | 8 | 3 | 15 |
| γ-Fe | 111 | .3463 | .1540 | .075 | .038 | 2 | 3 | 3 | 6 | 3 | 11 | 4 | 19 | 6 | 34 |
| | 200 | .2253 | .0920 | .041 | .019 | 2 | 2 | 2 | 4 | 2 | 8 | 3 | 14 | 4 | 25 |
| Cu | 111 | .3279 | .1595 | .080 | .040 | 2 | 3 | 2 | 6 | 3 | 11 | 4 | 19 | 6 | 34 |
| | 200 | .2183 | .0964 | .043 | .020 | 2 | 2 | 2 | 4 | 2 | 8 | 3 | 14 | 4 | 25 |
| Al | 111 | .2130 | .0767 | .037 | .019 | 2 | 4 | 2 | 8 | 2 | 14 | 3 | 24 | 4 | 42 |
| | 200 | .1314 | .0453 | .020 | .009 | 1 | 3 | 2 | 6 | 2 | 11 | 2 | 18 | 3 | 32 |

Table 1 Bragg position of order $H_k$ and $H_{wb}$ to be exceeded for obtaining kinematical and weak-beam condition, resp.. $v_n$ by courtesy R. Hillebrand

Anomalous absorption is weak in the kinematical case but not neglible in thicker specimens. Particular attention must be paid to diffuse scattering arising from the strong primary beam. It spreads "bright-field information" to the neighbouring beams where it compensates dark-field contrast. Fig.2 demonstrates scattering of information.

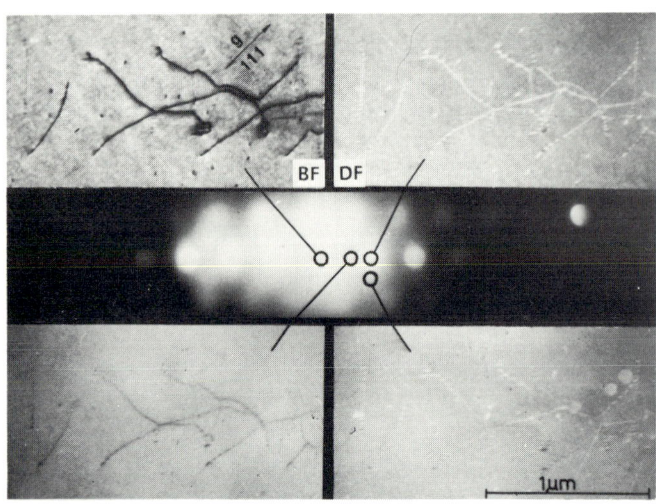


Fig.2 Dislocations in stainless steel imaged at $w_g = 1.2$, accel. voltage 1 MV. Lower micrographs taken on using diffuse scattering.

## 3. Absorption-Correction of the Eigenvectors

This principally known but usually neglected correction maximises at moderate deviations $0 < |w| \lesssim 1$ from Bragg conditions. It can affect oscillating and uniform diffraction contrast. Example: Fig.3. Uniform contrast of both BF and DF is affected by the same amplitude factor

$$1 - a'w \sin\alpha \Big/ \sqrt{1+w^2}\,(1+\cos\alpha + 2w^2) \gtrless 1 \qquad \text{if} \quad w\alpha \lessgtr 0 \qquad (4)$$

(2 -beam approximation where $a' = V_g'/V_g$ = absorption ratio $\approx 0.1$).

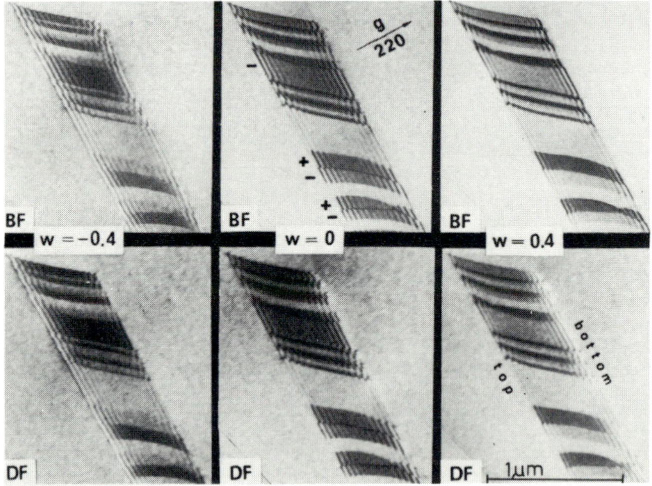


Fig.3 Sequence of $\alpha = 2\pi/3$ stacking faults in GaAs. Accel. voltage 1 MV. Uniform contrast depends on the sign of $w\alpha$.(sgn$\alpha$ indicated by + and -). DF behaves like BF. Explanation by absorption-correction of the eigenvectors, eq.(4). Also confirmed at reversed vector g.

## References

Howie A 1966 Phil. Mag. 14 223
Kästner G Ultramicroscopy submitted
Lally J S et al. 1972 Phil. Mag. 25 321
Steeds J W 1970 phys.stat.sol. 38 203

# Diffraction studies of solid inert gases

R J Cox, P J Goodhew and J H Evans[+]

Dept. of Materials Sci. & Eng. University of Surrey, Guildford GU2 5XH.
+ Materials Development Division, AERE Harwell, Didcot, OX11 ORA

## 1. Introduction

The development of materials for controlled fusion reactors has stimulated interest in the behaviour of high concentrations of inert gases in metals. In materials close to a fusion plasma high concentrations of helium gas will be both implanted and generated by transmutation reactions. It is known that high concentrations of inert gases in metals cluster to form bubbles which are able to grow thermally by vacancy collection and coalescence or by emitting self interstitial atoms. Until recently it has been difficult to measure the gas density in bubbles and so it has not been possible to test the various theoretical models proposed eg. athermal bubble growth by Greenwood et al (1959), gas driven blister growth by Evans (1976). Rife et al, using EELS (1981), found helium densities within bubbles sufficient for the gas to become solid. These and other results indicated that if other inert gases behaved in a similar way to helium they would become solid over a much wider range of pressures. This has been shown to be the case and a number of papers have now been published showing solid Ar, Kr and Xe bubbles in several metals, (Templier et al, 1984; Vom Velde et al, 1984; Evans and Mazey 1985).

Diffraction studies of solid inert gases not only provide evidence for the extremely high pressures which can exist within bubbles but also allow accurate measurement of the gas density from the lattice parameter, $l_p$, of the solid inert gases. This work examines the behaviour of argon bubbles in aluminium. The lattice parameter for the solid Ar is measured from diffraction patterns of the solid gas and from moiré fringe patterns formed between the epitaxial solid argon and the aluminium matrix.

## 2. Diffraction from solid argon

A sample of Al was implanted to give 1.5at% Ar. This was then annealed in vacuum for 10hrs at 673K. This treatment gave rise to a population of bubbles of radius 20-30Å. The thermal vacancy concentration in Al becomes appreciable above about 450K and so, provided sufficient vacancy sources exist and they are able to sink at the bubble surfaces, the bubble population should be in equilibrium at 673K, the anneal temperature.
The equilibrium pressure of bubbles in the size range 20-30Å radius is insufficient to cause solid Ar at room temperature (RT). However it is sufficient to cause an elevation in the normal freezing point, as would be expected from P-V-T behaviour of gases. To determine freezing point and $l_p$ data for this sample experimentally, a cryogenic stage was used on the Philips EM400T. The sample was cooled to 78 K and epitaxial

diffraction spots from solid Ar could be clearly seen (Fig.1). This is the first reported occurance of solid Ar forming epitaxially with the host metal: A diffuse ring from solid Ar was reported by Vom Feld et al (1985). An area of the foil was selected and a line trace through the Al and Ar (111) diffracted spots was collected using a hybrid diffraction unit and our own software (Budd and Goodhew, 1985). Keeping the same area of the foil, the temperature was increased and further line traces collected at 10K intervals up to 150K. The variation of $l_p$ with temperature is plotted in Fig.2. The Al (111) spot, corrected for thermal contraction, was used as a calibration for the argon $l_p$ measurements. A selection of the line traces can be seen in Fig.3, which shows a decrease in the average argon lattice parameter during heating.

TABLE 1

| Temp./K | Measured $l_p$ /Å | Gas density at Temp/ atoms/cm$^3$ | (a) Gas density at 673K atoms/cm$^3$ | (b) Pressure at 673K /GPa | (c) Equiv. bubble rad/Å | p=2γ/r for measured bubs at 673K/GPa |
|---|---|---|---|---|---|---|
| 87 | 5.43 | $2.49 \times 10^{22}$ | $2.40 \times 10^{22}$ | 1.18 | (17) | 0.83 |
| 150 | 5.33 | $2.64 \times 10^{22}$ | $2.54 \times 10^{22}$ | 1.40 | (14) | 1.00 |

(a) Using coeff. of linear expansion for Al (b) Using equation of state by Ronchi (1981) (c) Using $P=2\gamma/r$ ($\gamma = 1\,Jm^{-2}$).

Calculations shown in Table 1 indicate that the reason for the apparent decrease in $l_p$ on heating is that larger bubbles are melting, reducing the average bubble size. Because bubble pressure is inversely proportional to bubble radius, $p=2\gamma/r$, the smaller bubbles are at higher pressures and hence remain solid to higher temperatures. This is illustrated in Fig.4 which shows the P-T curve from the modified Simon equation (Hardy et al, 1971). Superimposed on this curve are P-T traces for different bubble radii cooled from 673K, corrected for thermal contraction of Al, showing the effect of bubble radius on freezing point.

It is possible using these results to test the assumption of equilibrium bubbles after annealing for 10hrs at 673K. The measured $l_p$ and hence gas density can be extrapolated up to 673K because a bubble is effectively a sealed pressure vessel. After taking account of the thermal expansion of the Al matrix (not a large effect) use can be made of the accurate equation of state work by Ronchi (1981) to obtain the pressure at any temperature, given the density. This has been done in Table 1 and the results indicate pressures at 673K which would arise from equilibrium bubbles of average radius 17Å from those solid at 87K and 14Å from those solid at 150K. From the measured bubble population the mean bubble radius was 24Å with 93% of the population between 20 and 30Å radius.

## 3. Moire fringes from solid argon

Diffraction from solid inert gas bubbles will, when an epitaxial relationships with the Al matrix exists, give rise to parallel moiré fringes. The fringes are perpendicular to the g-vector and will have a characteristic spacing given by:-

$$D = d_1 d_2/(d_1 - d_2)$$

where $D$ = fringe spacing

$d_1, d_2$ = (111) planar spacings of Ar and Al

Using this equation moire fringe spacings can be used to find the solid argon $l_p$ at any temperature. Two samples were examined for this effect at RT. Sample A93 was irradiated to give 6at% Ar at RT and given no further treatment. Sample A97 was irradiated to give 1at% Ar at RT and then vacuum annealed for 10hrs at 623K. Both samples were photographed at RT using a Philips EM400T with g=111. Moiré fringes were found on bubbles from both samples and fringe spacing measurements were made. The results are shown in Table 2. Sample A93, which was not annealed, gave a solid argon $l_p$ of 4.92Å at 300K. Sample A97 which was annealed at 623K gave a solid argon $l_p$ of 5.09Å at 300K.

TABLE 2

| Sample | Anneal Temp. /K | Fringe Spacing /Å | $l_p$ of Ar at 300K/Å | Gas density at 300K atoms/cm$^3$ | (a) Pressure at anneal temp. /GPa | (b) p=2γ/r for measured bubbles at 623 /GPa |
|---|---|---|---|---|---|---|
| A93 | 300K | 13.2 | 4.93 | $3.34\times10^{22}$ | 2.7 | - |
| A97 | 623K | 11.5 | 5.09 | $3.03\times10^{22}$ | 2.4 | 1.54 |

(a) Equation of state from Ronchi (1981) (b) Using $p=2\gamma/r$, with $\gamma=1\,\mathrm{Jm^{-2}}$.

It is reasonable to expect the bubbles in A93 to be above equilibrium pressure as the thermal vacancy concentration in Al only becomes appreciable above about 450K. If the growth mechanism for bubbles in this sample was by athermal processes e.g. dislocation loop punching, as suggested by Greenwood et al (1959) then the bubble pressure may aproach the minimum value required for loop punching. The average measured bubble radius was 17Å, the minimum pressure for loop punching from this size bubble is given by:

$$p \simeq 2\gamma/r + \mu b/r = 4.7\mathrm{GPa} \qquad \text{where } \mu = 26\ \mathrm{GPa},\ \gamma = 1\mathrm{Jm^{-2}},\ b = 2.3\times10^{-10}\ \mathrm{m}$$

The bubble population from sample A97, annealed for 10hrs at 623K should be at thermal equilibrium; from previous arguments. Calculations of the bubble pressure from the measured bubble radii are shown in Table 2 to be significantly less than those measured.

## 4. Conclusions

Sample A93 which was in the as irradiated condition contained solid Ar epitaxial with the host lattice at room temperature. The $l_p$ from moire fringe spacing measurements indicated a pressure of 2.7GPa and the measured average bubble radius was 17Å. This pressure is somewhat lower than the minimum pressure required for growth by loop punching of 4.7GPa.

Samples A78 and A97 were both annealed after irradiation and would be expected to be at thermal equilibrium. Diffraction spot spacing measurements from A78 indicates over-pressurisation by a factor of 1.4 at the annealing temperature of 673K. Moire fringe spacing measurements on A97 indicate the bubbles to be over-pressure by a factor of 1.6 at the annealing temperature of 623K. Two ways to account for this behaviour are either the value chosen for the surface tension γ of $1\mathrm{Jm^{-2}}$ was too low or insufficient thermal vacancies have been able to sink at the bubble surface. The values of surface tension γ which would need to be used are for A78 - $\gamma = 1.4\mathrm{Jm^{-2}}$ and for A97 $\gamma = 1.6\mathrm{Jm^{-2}}$. For the bubbles not to have reached thermal equilibrium a barrier to vacancies sinking at the bubble surfaces could be a possible cause. This might occur where bubbles have oxides on internal surfaces as discussed elsewhere (Cox et al, 1984).

References

Budd and Goodhew (1985) Inst.Phys.Conf.Ser. 78 (Bristol: Adam Hilger)
Cox R.J., P.J.Goodhew & J.H.Evans (1984) J. Nucl. Mater. 126 117.
Evans J.H (1976) J. Nucl. Mat. 61 1.
Evans J.H. and Mazey D.J. (1985) Scripta. Met. 19 621.
Greenwood, Foreman A.J.E and Rimmer D.E. (1959) J. Nucl. Mat. 4 305.
Hardy W.H., Crawford R.K. and Daniels W.B. (1971) J. Chem. Phys. 54 1005.
Rife J.C., Donelly S.E. and Lucas A.A. (1981) Phys. Rev. Lett. 46 1220.
Ronchi C. (1981) J. Nucl. Mat. 96 314.
Templier C., Jaouen C. et al (1984), C.R. Acad. Sci. Paris 299 613.
Vom Velde, A., Fink J., Muller T. et al (1984) Phys. Rev. Lett. 53 922.

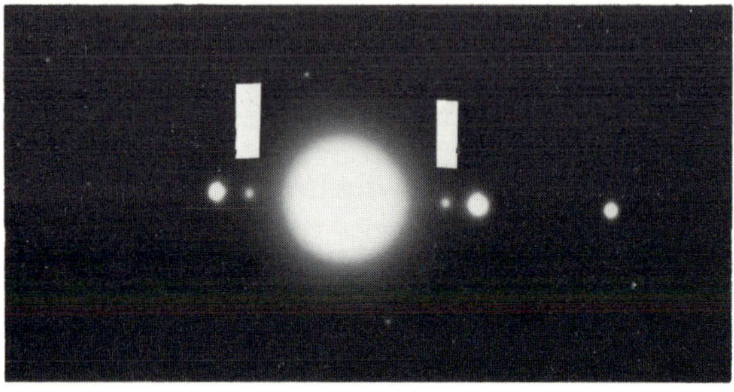

Fig.1 Diffraction pattern showing Ar(111) spots

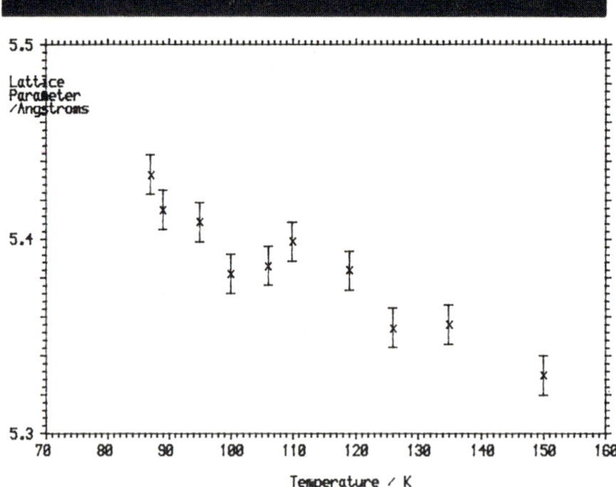


Fig.2 Sample A78 solid argon lattice parameter vs temperature

Fig.3 Modified Simon equation for Ar showing freezing point vs pressure

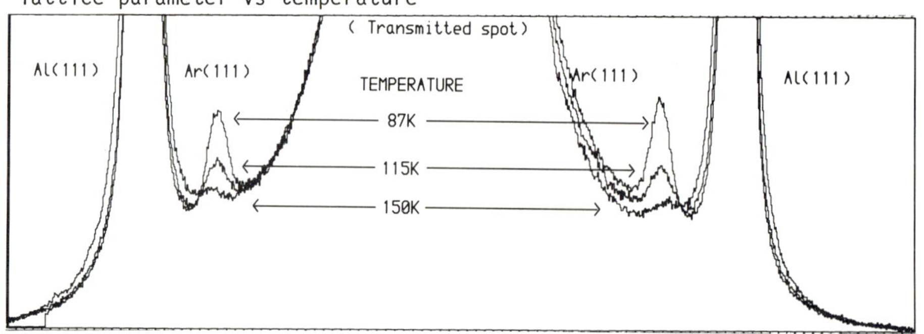


Fig. 4 The shift in the Ar(111) spot spacing during heating from 80 K

*Inst. Phys. Conf. Ser. No 78: Chapter 3*
*Paper presented at EMAG '85, Newcastle upon Tyne, 2–5 September 1985*


# Dark field images as paired complements to bright

W O Saxton, K M Knowles and W M Stobbs

Dept. of Metallurgy and Materials Science, Pembroke St.,
Cambridge, CB2 3QZ.

## 1. Introduction

The recording of dark field (DF) images using axial illumination and a central beam stop, rather than the more usual tilted illumination conditions, is attractive theoretically for a number of reasons, particularly if the images are recorded in complementary pairs with bright field (BF) images with otherwise identical imaging conditions.

We have reported previously (Stobbs 1984; Saxton and Stobbs 1984) some images of zirconia using a JEM 200CX and an aperture from a thin tungsten wire mounted across a normal circular aperture. We have now now been able to use apertures produced to order by Agar Aids Ltd, with a small stop supported from one side only, in the Cambridge University 600kV HREM, and have examined several materials with the aim of establishing the expertise necessary for the reliable recording of complementary BF/DF pairs.

## 2. Theoretical advantages

The potential benefits, in theory, can be summarised as follows:

(i) the images, as for other DF modes, are more sensitive to atomic scattering factor variations than BF images, since the image *intensity,* and not simply the contrast, is dependent on the scattering factors, and the dependence is moreover on the second rather than first power.

(ii) image interpretability is improved in that local beam excitation conditions may be deduced from a comparison of the two images, whereas diffraction patterns cannot be obtained from local areas except in the STEM.

(iii) nonlinear contributions to the BF image, such as the '(g,−g)' interferences responsible for the notorious dumbell images in silicon, are directly observed in the DF image, allowing their relative influence on the BF image to be assessed immediately; moreover, with the aid of digital image processing, these contributions can be subtracted from the latter yielding entirely *linear* images from which the wavefunction can be recovered by such methods as series filtration (Saxton 1980).

## 3. Results and discussion

Fig. 1 shows a small area of graphitised carbon in the two modes. The most striking feature is that although the BF image is extensively traversed by 0.34nm fringes in all directions, few areas at all show fringes in the DF image, though the presence of 0.17nm fringes in some areas (and in fact in many orientations) indicates clearly that the imaging conditions are sufficiently stable to reproduce this periodicity. It is probable that over much of the specimen area the graphite is oriented so as to cause largely *two*-beam rather than three-beam excitation, leaving the diffracted beam nothing to interfere with once the primary beam had been removed; this view is consistent with the way in which 0.17nm fringes, when encountered in BF images, are rarely found over more that small areas, and with the marked contrast changes commonly observed at bends in the graphite ribbons.

Figs. 2 and 3 show images of small gold particles on a carbon substrate. In fig. 2, the loss of the low spatial frequencies in the DF images is particularly evident, depriving many particles of clear outlines altogether. Many particles however show fringes in DF: in fig. 3 clear crossed 0.144nm fringes are seen, arising from the mutual interference of the four (200)-type reflections, which is masked in BF by their (stronger) interference with the primary beam to form crossed 0.204nm fringes rotated by 45°.

Fig. 4 shows images of a GaAs/AlAs multiple quantum well (MQW) structure produced by Philips Research Laboratories, Redhill. The low difference in scattering factor between gallium and arsenic means that high resolution BF images normally show little contrast between the layers, though Hetherington et al. (1985) have identified two conditions in which contrast can be obtained, from the (002) reflection that depends only the scattering factor difference. While no samples prepared in the (100) orientation were available to us, we have investigated their other configuration, namely tilting the specimen a few degrees off the (110) axis so as to excite the (002) beam preferentially, and find that the DF images show a markedly higher contrast between the layers than do the BF images, though the same is not true for images exactly down the zone axis. No second order (004) fringes are however discernible in the DF images, which in this case continue to show mainly first order 0.28nm fringes.

In seeking to explain the persistence of these first order fringes, we are driven to the conclusion that a significant proportion of the primary beam is bypassing the beam stop (10μm diameter in this case), allowing some first order interference to persist in the image: the mutual interference of the (002) and (004) reflections would be eliminated by the level of focus spread present in the microscope. The DF images in fact show a surprisingly high intensity relative to the bright, and suggest that as many as a third of the electrons transmitted by the specimen may be scattered around the aperture. Inelastically scattered electrons will be particularly likely to evade interception of course, and a very substantial part of the DF image intensity may in fact result from the elastic scattering of initially inelastically scattered electrons. This notion is further supported by the finding that the DF images are commonly improved markedly when the defocus is increased by perhaps a Scherzer unit relative to the BF conditions; it may also help account for why the apparent loss of low spatial frequency content from the images can be so severe - a mere high pass filter would not render particle boundaries difficult to recognise.

These considerations, amongst others, emphasise the need for careful evaluation of all the possible factors affecting this powerful new technique if the full theoretical benefits are to be realised, and interest in energy filtered imaging may be intensified for yet another reason.

Acknowledgments

We are very grateful to Dr. J. Gowers of Philips Research Laboratories for permission to make use of the MQW specimen. We acknowledge the support of the SERC and the Royal Society for this work, and thank Prof. D. Hull for the provision of laboratory facilities.

## References

Hetherington C J D, Barry J C, Bi J M, Humphreys C J, Grange J and Wood C 1985 Mat. Res. Soc. Symp. Proc. Vol 37 41

Saxton W O 1980 J. Microsc. Spectrosc. Electron. 5 661

Saxton W O and Stobbs W M 1984 Proc. 8th Eur. Cong, EM Budapest, 287

Stobbs W M 1984. J. Microsc. 142 137

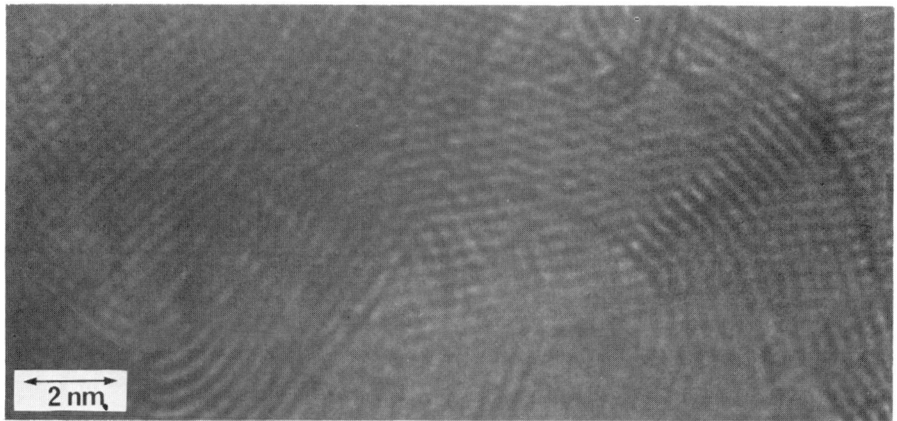


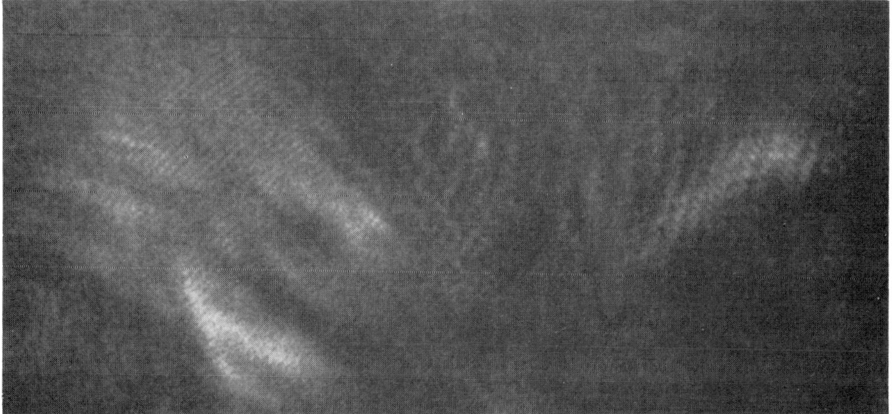

Fig. 1: BF (above) and DF (below) images of graphitised carbon.

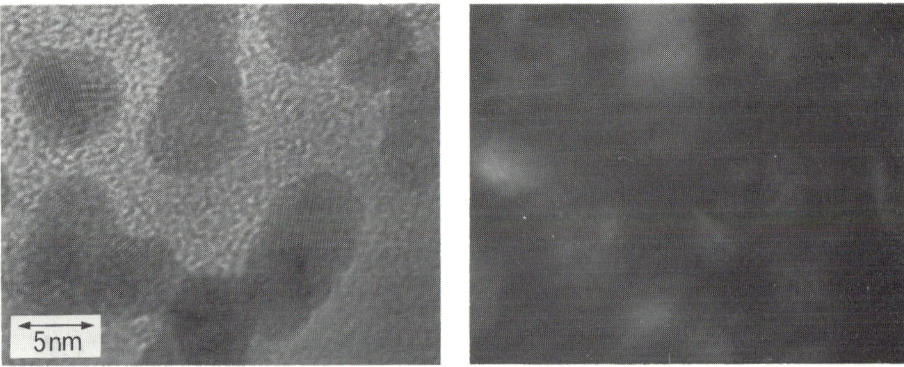


Fig. 2: BF (left) and DF (right) images of gold particles on carbon.

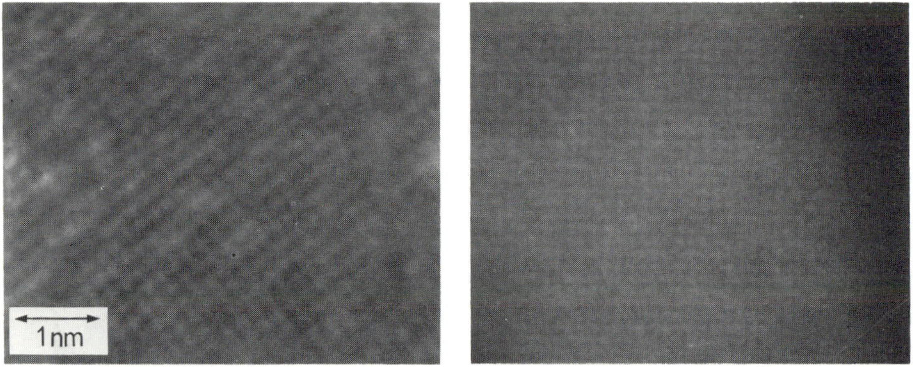


Fig. 3: Magnified image of single gold particle: BF (left) and DF (right).

Fig. 4: BF (left) and DF (right) images of GaAs/AlAs MQW structure.

*Inst. Phys. Conf. Ser. No 78: Chapter 3*
*Paper presented at EMAG '85, Newcastle upon Tyne, 2–5 September 1985* 

# Computer aided analysis of electron channelling patterns

S H Vale
Link Systems Ltd., Halifax Road, High Wycombe, Bucks. HP12 3SE

## 1. Introduction

Electron Channelling Patterns (ECP's) are a well established technique for obtaining crystallographic information from samples in the scanning electron microscope (Joy et al 1982). On many SEM's it is a simple matter to set up the experimental conditions to acquire an ECP. However, the identification of the poles represented in the ECP in unfamiliar materials is not always straightforward. Moreover, although the techniques for converting the crystallographic information represented on the ECP to orientations on a stereographic projection are well known (Ayers and Joy 1972) it is a tedious procedure when repeated on many different patterns. Other than the work of Newbury and Joy (1971) little work has been performed on computer aided analysis of ECP's. This paper describes a program which presents a unified approach to ECP collection and processing of patterns and display of the results.

## 2. Hardware

The program was written for a Link Systems AN10000, a computer normally used for X-ray analysis. This has a 16 bit word length, 20 MHz CPU, 128 kbyte of memory, 256 kbyte of extended memory for intermediate image storage and processing and a 512 x 512 pixel display. Digital control of the position of the electron beam is provided through two 12 bit DAC's and the signal intensity is read through a 5 MHz VFC. The ECP's were acquired on a Jeol 840 SEM. With the microscope set up to acquire an ECP the computer takes control of the electron beam incident angle and rasters about the chosen point on the sample. The intensity of the signal was stored at 512 x 512 points to 8 bit precision in extended memory and simultaneously displayed on the computer monitor.

## 3. Software

The basic function of the software is to identify unknown poles on an ECP and to store the crystallographic orientation of the specimen so defined. The system is initially calibrated using a known standard and a recognisable pole. Figure 1 shows a photograph of the computer monitor after acquiring an ECP of the (111) pole from a sample of electro-polished silicon using back scattered electron detection at 25 kV. A cursor can be moved over the pattern to locate points along a band edge. In figure 2 several points have been identified along both edges of a (220) band and straight lines fitted to them. The d-spacing represented by this band is typed into the program which calculates the camera length (i.e. the relation between scanned angle and linear distance on the screen) for this pattern. The program is now ready to analyse an unknown pattern. The standard is replaced by the sample and an ECP acquired from it. The positions of at least two bands from the same pole on the pattern are measured as described above. The program can calculate the d-spacing and interplanar angles represented by these bands from these measurements and the calculated camera length. The crystal structure for the sample can be entered from a file stored on disk. The program proceeds by comparing the measured d-spacings and angles with those calculated for the chosen crystal structure until consistent sets are found which produce fits within preset errors in the measurement and camera length. A list of possible poles is produced in descending order of fit and the ECP's generated for each case can be overlaid on the original pattern for comparison (figure 3). The crystallographic orientation of the sample with respect to the microscope screen edges and the electron beam direction (assumed to be the centre of the ECP) is calculated using the method of Young, Steele and Lytton (1973) and stored as a matrix on disk. A cursor can be moved around the ECP to identify the crystallographic direction represented by any point on the pattern.

## 4. Example

ECP's were collected at 25 kV using back scattered electron detection from a sample of electropolished 99.997% pure aluminium mounted normal to the beam. The material had been 90% cold rolled and annealed until the structure was completely recrystallised (figure 4). ECP's were taken from 10 grains in the sample and each was analysed as described above and the orientation stored. The time for the analysis of each pattern was approximately 2 minutes which includes acquisition of the pattern and identifying the band edges. Solving the pattern and plotting the theoretical solution takes a matter of seconds. Figure 5 shows a typical ECP from the aluminium sample overlaid by the calculated solution. The results of these analyses were plotted in figure 6 which shows the (001) poles calculated for each ECP plotted on a stereographic projection with the rolling direction horizontal. The development of the cube texture associated with recrystallised fcc materials is apparent after a few measurements.

## 5. Discussion

Making measurements of ECP band width and position directly on the computer monitor involves errors of comparable magnitude to those when taken from photographs (Dingley 1981). Routine use of the software gives errors in orientation of about 1°. Errors in measurement of the width of ECP bands vary from about 1% for the diffuse low order bands to about 0.2% for the sharper high order bands. This does not necessarily favour using high order bands for measurement since there tend to be more planes with spacings conforming within a given range for high order than for low order bands. Use of high order band edges could therefore result in spurious solutions, particularly for low symmetry materials. The computer generated scan is set up as far as possible in hardware to be square although this is difficult to better than 2%. To improve calibration, measurements can be taken of the width of the same ECP band oriented both horizontally and vertically with respect to the microscope screen so that the software can correct for the aspect ratio in subsequent measurements.

Acknowledgement

I wish to acknowledge the co-operation of Jarle Hjelen, SINTEF, Trondheim, Norway in providing the aluminium sample.

References

Ayers J D and Joy D C 1972 Acta Metall. 20 1371

Dingley D J 1981 Scanning Electron Microscopy ed. O Johari (Chicago SEM) 273-286

Joy D C, Newbury D E and Davidson D L 1982 J.appl. Phys. 53 R81

Newbury D E and Joy D C 1971 Proc. 25th EMAG 306-309

Young C T, Steele J H and Lytton J L 1973 Met.Trans. 4 2081

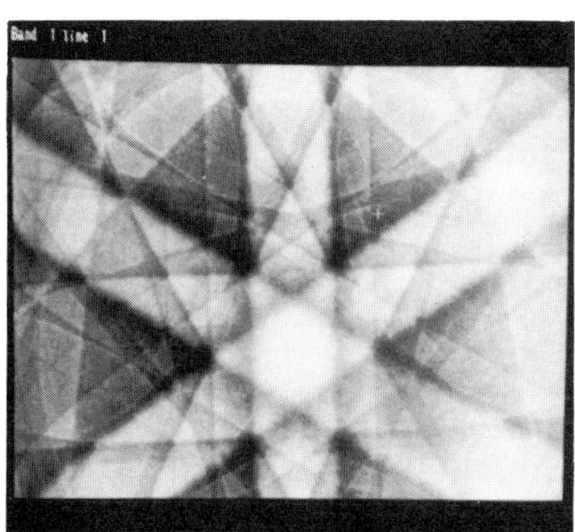


Fig. 1 ECP collected from silicon sample photographed from computer monitor.

Fig. 2 Identifying band edges of ECP on computer monitor.

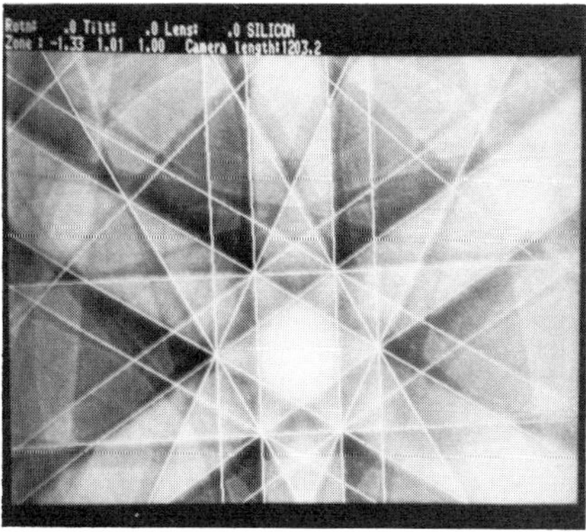


Fig. 3 (111) pole ECP overlaid on original pattern.

Fig. 4 Montage of ECP's and backscattered electron image of polycrystalline aluminium photographed from computer monitor.

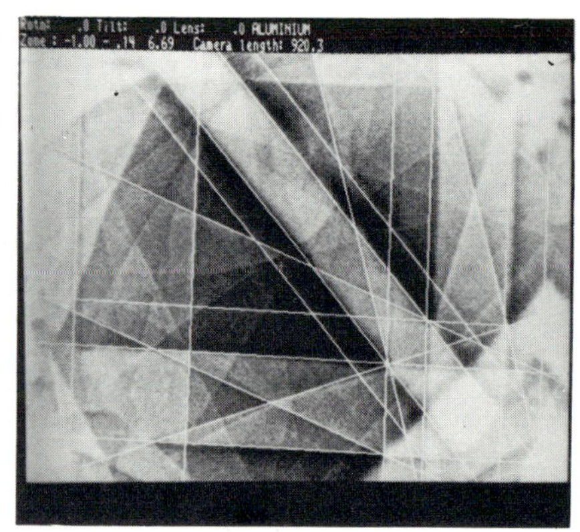

Fig. 5 ECP from aluminium sample overlaid by calculated solution.

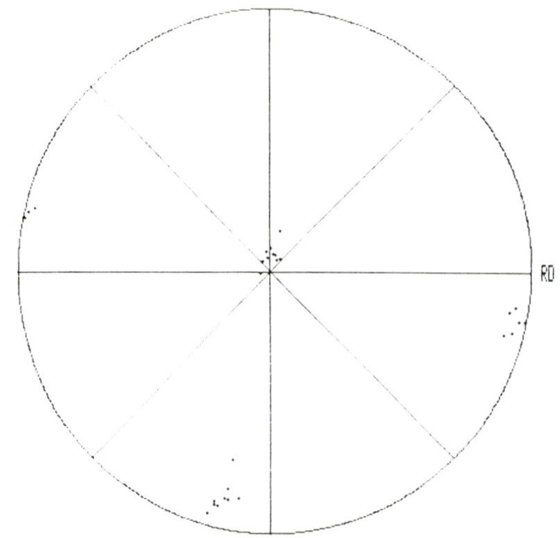


Fig. 6 Computer generated stereographic projection of (001) poles from 10 grains of aluminium sample calculated from their ECP's. Rolling direction (RD) horizontal.

*Inst. Phys. Conf. Ser. No 78: Chapter 3*
*Paper presented at EMAG '85, Newcastle upon Tyne, 2–5 September 1985* 

# Fourier smoothing of scanning Auger microscope images using an optimal Wiener filter technique

C G H Walker

Department of Physics, University of York, Heslington, York, YO1 5DD, U.K.

## 1. Introduction

Images from a Scanning Auger Microscope (SAM) are typically very noisy as compared to ordinary SEM images. This is due to the small number of electrons collected for each image pixel. Several methods for the smoothing of SAM images have been attempted in the past (Peacock and Prutton 1983). These smoothing techniques are intended to enhance the contrast in the images by reducing the noise content.

An optimal Wiener filtering technique (Kosarev and Pantos 1983) has been employed for smoothing spectra at York with results of higher quality than those from other spectrum smoothing methods. It was therefore decided to adapt the technique to two dimensions for the smoothing of SAM images.

This technique can be employed to smooth any image to the required degree and gives better results than several other successful smoothing algorithms that were developed at York (Peacock et al 1983).

## 2. The Wiener Filter

This filter is applied to the Fourier transform of the data to be processed as described by Kosarev and Pantos 1983; for two dimensions it is given by the equation:-

$$W_f = \frac{C^2}{C^2+N^2} \qquad (1)$$

Where C are the two dimensional Fourier components of the power spectrum intensity of the signal and N is the noise power intensity. Since the true image is not known, C is calculated from the Fourier transform of the data. The noise level of the image is calculated from the average of those points of the power spectrum whose frequency exceeds half the maximum frequency. The power spectrum is then circularly averaged and a frequency $f_1$ chosen such that the nearest point to the zero frequency which is below the noise level is chosen as frequency $f_1$. If the value of $f_1$ is the same in all directions (i.e. isotropic in Fourier space), then this is termed circular filtering. However the noise power may not be isotropic due to a combination of 1/f noise in the beam current and the scan rate in the x direction being greater than that in the y direction. In this case elliptical filtering may be required. If elliptical filtering is required then the frequency $f_1$ varies elliptically with the major and minor axes along either x or y axes. The frequencies $f_{1x}$ and $f_{1y}$ in the x and y directions are also chosen as the nearest point to the zero frequency which are below the noise level. As an option the frequency $f_1$ can be chosen manually if desired.

For points which have frequencies less than $f_1$, the Wiener filter is calculated assuming the true data is the collected data. However for points beyond $f_1$ the Wiener filter is calculated assuming the power spectrum data continues to fall with frequency according to the equation

$$\mathrm{Ln}(C^2) = A.f+B \qquad (A<0) \qquad (2)$$

where A and B are constants and f is the frequency. Fig 1 shows a cross section of the power spectrum of an image. Notice the linearity of the signal below $f_1$ and the white noise above $f_1$. The constants A and B are calculated by fitting equation (2) by least squares to the power spectrum below frequency $f_1$. A frequency $f_2$ is chosen such that the Wiener filter is zero above $f_2$. Frequency $f_2$ is determined by finding when equation (2) falls to 20 dB below the noise level (Signal power/Noise power = 0.01).

3. Subtracting the Least Square Plane

In order to reduce Gibbs oscillations at the edge of the image, a plane, which is a least squares fit to all the edge points of the image is subtracted prior to the smoothing operation. This plane is added back to the image after the image has been smoothed. On some images the subtraction of the least squares plane results in low frequencies being substantially reduced, such that equation (2) is no longer valid in the region of the power spectrum below $f_1$. This has been determined by visual examination of the power spectra of each image during processing. In such cases the least square fit to equation (2) is made between the largest value in the power spectrum and $f_1$.

4. Outlier Removal

It may be expected that some images will have outliers (possibly caused by mains transients), or points that vary greatly from the neighbouring average of points. These outlier points can be removed (optionally) from the image by comparing smoothed and unsmoothed images. If the difference between smoothed and unsmoothed isolated pixels is greater than a certain threshold then the original point is replaced by the smoothed point. The smoothing operation is then recommenced. The threshold is chosen such that 1% of all random noise points will be rejected. The programe will smooth iteratively until the number of outliers on the present iteration is the same as the number of outliers on the previous iteration.

5. Assessment of the Technique

In order to ascertain how the technique can remove noise, it was decided to study artificial images with varying amounts of added noise . Obtaining the noise images required setting up the SAM (Prutton et al 1982) to collect images in an unusual manner. The electron beam was arranged to scan a wide fixed area of the specimen at high speed while the computer acquired the images as though the beam were being scanned in the slow digital form that Auger images are usually collected. The noise thus measured should be representative of the fluctuations due to the column, measuring system and electron statistics and not dependent on the specimen or its drift. The signal image used was a vertical bar seen in Fig 2a and the noise image in Fig 2b. The noise was then scaled and added to the signal image ready to be smoothed.

It is important to analyse the Fourier transform of the noise in order to establish whether there is a low frequency component in the noise. If there is, then heavily smoothed images may reflect the low frequency noise and not the low frequency variations in the signal. For images normalised using the scheme (N1-N2)/(N1+N2) where N1 is the number of electrons counted on an Auger peak and N2 is the counts on the background above the peak, the noise appears to have a slightly stronger low frequency component than the higher frequency white noise. It is uncertain where this low frequency noise is generated as any beam fluctuations should be accounted for in the above normalisation scheme (Prutton et al 1981).

Artificial images before and after smoothing are shown in Figs 3a and 3b respectively. The amount of smoothing is heavy because of the low signal to noise count ratio (0.01). On smoothing less noisy images the frequency $f_1$ will be higher resulting in a lighter smooth. Fig 4 shows the same image smoothed by a 3x3 salt and pepper technique, which was previously considered the best method (Peacock et al 1983). Although the vertical bar is apparent, the Wiener filtered image shows the original image more clearly.

## 6. Conclusions

a. The Optimal Wiener filter technique enables smoothing of images of either good or poor signal to noise ratio.

b. The technique offers outlier removal, better smoothing than all previous techniques and is an objective procedure.

c. The technique takes 20 seconds of CPU time per iteration for a 128x128 image on a DEC system-10 computer.

Acknowledgements

The author would like to thank P Heesterman who wrote some of the software, M Prutton for helpful discussion of the problem and to the SERC for the funding of this work.

## References

Kosarev E L, Pantos E, J Phys. E: Sci. Instrum. 1983 16 p537-543.
Peacock D C, Prutton M, Inst. Phys. Conf. Ser. No. 68 (1983) p231.
Prutton M, Peacock D C, El Gomati M M, Larson L A, Poppa H, Inst. Phys. Conf. Ser. No. 61 (1981) p443.
Prutton M et al, Vacuum TAIP, 1982, 32, 351.

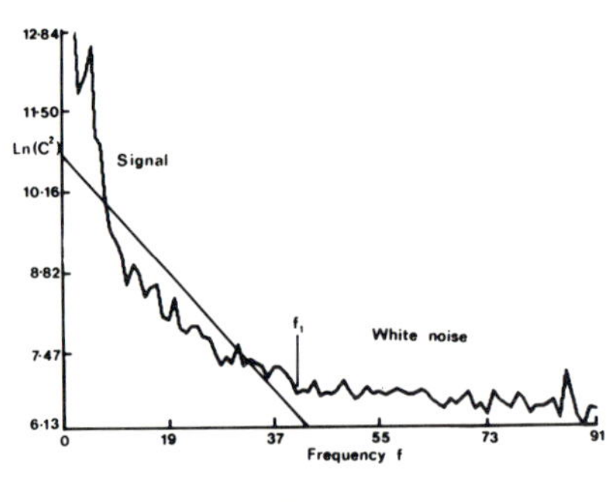


Fig 1. A plot of log (power spectrum$^2$) versus frequency showing regions containing signal and white noise. The straight line is a least squares fit to all the points below $f_1$.

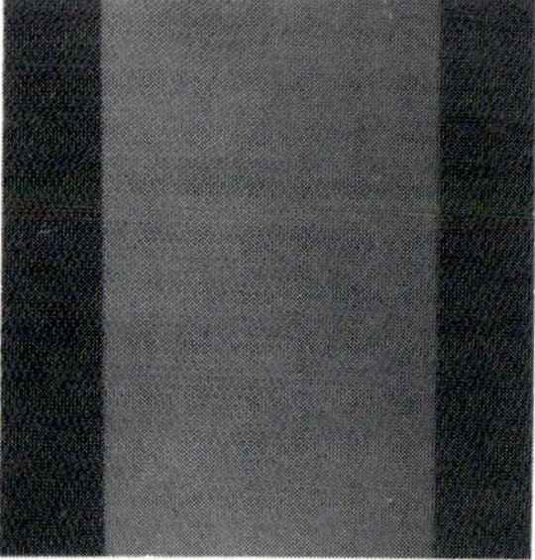

Fig 2(a) Signal image

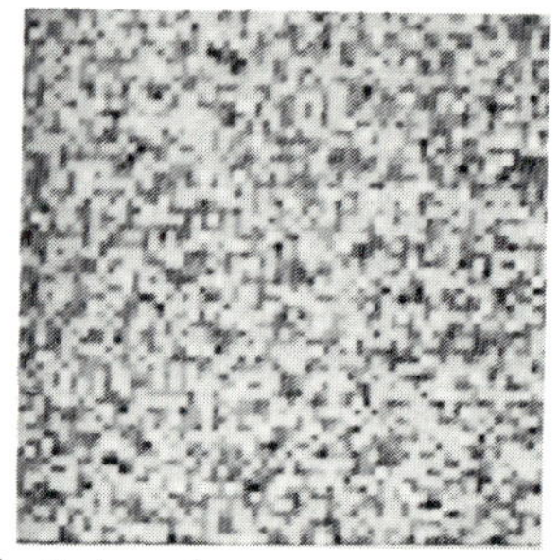

(b) Noise image

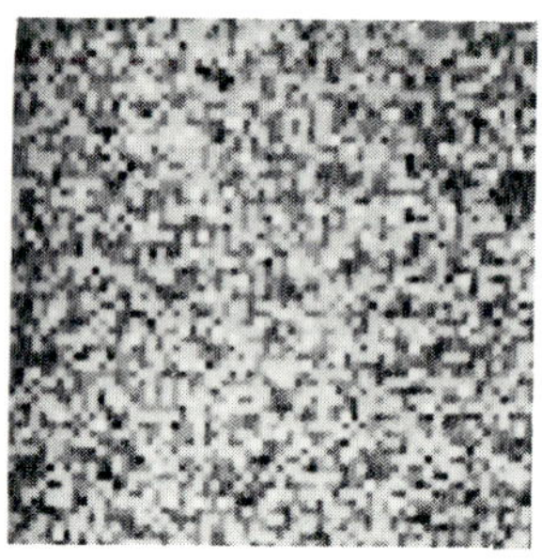

Fig 3(a) Signal image & scaled noise image

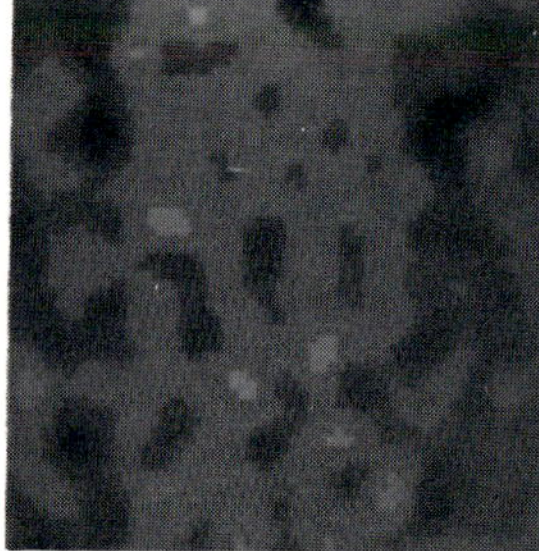

(b) Fig 3a After Fourier smoothing $f_1$=7,152 outliers & smoothed in 14 iterations

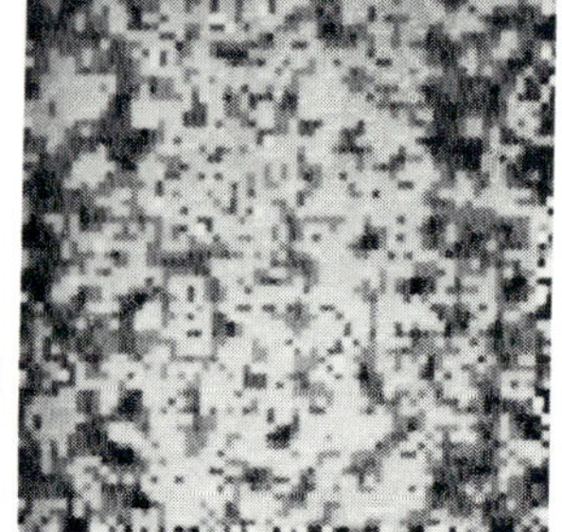

Fig 4. Fig 3a After 3 x 3 Salt and Pepper smooth

# Theory of image formation by plasmon loss electrons in small spheres

H Kohl and C Colliex, Laboratoire de Physique des Solides, Université Paris-Sud, Bâtiment 510, 91405 Orsay (France)

## 1. Introduction.

For the study of very small inclusions either of denser materials (precipitates, Guinier-Preston zones) or of gas bubbles or voids it is vital to determine the position of boundaries with high spatial resolution. Currently used methods are based on diffraction contrast, weak beam cases, phase contrast and Z-contrast. Each of these methods has its limitations. In the last years new types of electron microscopes have become commercially available, which yield excellent information on inelastic processes with high spatial resolution. The Scanning Transmission Electron Microscope (STEM) with an energy loss spectrometer or the Fixed Beam Electron Microscope (FBEM) equipped with an imaging energy filter offer the possibility of obtaining both energy loss spectra and inelastic images of small objects. Using inner-shell losses one can thereby determine the projected local chemical composition of the specimen. For a small object embedded in a large matrix this method suffers limitations due to the poor signal to noise ratio. It would be preferable to use a strong signal well localized at the interface. If the two materials have a difference in their electron densities, the surface plasmon modes offer such a strong signal. But how localized is it ?

## 2. Theory.

As a model system we study the surface plasmon of small spheres (Fujimoto and Komaki, 1968 ; Hénoc and Henry, 1970 ; Schmeits, 1981 ; Batson, 1982a,b ; Manzke et al., 1983 ; Barberan and Bausells, 1985 ; Achèche, 1985). In the following we shall discuss image formation in a STEM. The image taken in a FBEM with axial illumination is equivalent to a STEM image recorded with a small axial detector.

For the calculation of the probability of plasmon excitation as a function of the impact parameter $\vec{r}$ two theoretical approaches are available. In the semi-classical model one considers the influence of the external potential, caused by the passing of a classical charge, on a quantum object (Schmeits, 1981). Thus the quantum properties of the incident electrons are neglected. As a result the excitation probability $P(\vec{r},\omega)$ can be written as the sum of three terms (Schmeits, 1981)

$$P(\vec{r},\omega) = P_o\ \delta(\omega) + P_B(\vec{r})\ \delta(\omega - \omega_p) + \sum_\ell Q_\ell(\vec{r})\ \delta(\omega - \omega_\ell) \qquad (1)$$

associated respectively to the zero-loss peak $P_o\ \delta(\omega)$, to the bulk plasmon mode $P_B(\vec{r})\ \delta(\omega - \omega_p)$, and to all surface plasmon modes $Q_\ell(\vec{r})\ \delta(\omega - \omega_\ell)$, where $\omega_\ell$ and $\omega_p$ are the surface and bulk plasmon frequencies.

The quantum-mechanical approach treats the total system (electron + object) as one single entity and yields the transition probability from an initial to a final state. One then computes the current per unit energy in the detector as a function of the probe position (Rose, 1976 ; Kohl, 1983)

$$\frac{dI}{dE}(\vec{r}) = \frac{I_o}{\pi^3 \alpha_o^2 \hbar} \frac{k k_o^3 E_H}{E_o}$$

$$\int A(\vec{\alpha}) A(\vec{\alpha}') D(\vec{\beta})\, e^{i\{\gamma_o(\vec{\alpha}) - \gamma_o(\vec{\alpha}')\}} e^{i k_o \vec{r}(\vec{\alpha} - \vec{\alpha}')} \frac{S(\vec{K}, \vec{K}', \omega)}{K^2\, K'^2}\, d^2\vec{\alpha}\, d^2\vec{\alpha}'\, d^2\vec{\beta} \qquad (2)$$

where $I_o$, $\alpha_o$, $E_o$, $k_o$ and k are the incident beam current, the objective aperture angle, the energy and the wave numbers of the incident ($k_o$) and the scattered (k) electron respectively. The functions

$$A(\vec{\alpha}) = \begin{cases} 1 & \text{for } |\vec{\alpha}| < \alpha_o \\ 0 & \text{otherwise} \end{cases}$$

and

$$D(\vec{\beta}) = \begin{cases} 1 & \text{for } |\vec{\beta}| < \beta_o \\ 0 & \text{otherwise} \end{cases}$$

denote the objective and the collecting aperture ; $\gamma_o(\vec{\alpha})$ denotes the phase shift due to the spherical aberration and defocus and $S(\vec{K}, \vec{K}', \omega)$ is the mixed dynamic form factor describing the object properties as a function of the two scattering vectors $\vec{K} = k_o[\theta_E \vec{e}_z + (\vec{\alpha} - \vec{\beta})]$ and $\vec{K}' = k_o\,[\theta_E\, \vec{e}_z + (\vec{\alpha}' - \vec{\beta})]$ (Rose, 1976 ; Kohl, 1983). $E_H$ is the ionisation energy of the hydrogen atom.

Ritchie (1981) has shown, that for an infinite objective aperture and a perfect lens ($\gamma_o = 0$) the quantum approach yields the classical result. Then the integrationsover $\alpha$ and $\alpha'$ yield two-dimensional $\delta$-functions at the position of the classical impact parameter.

In the limiting case of an infinitely large spectrometer acceptance angle ($\beta_o \to \infty$) the integration over $\beta$ can be performed independently from the $\alpha$ and $\alpha'$ integration (Rose, 1976) yielding

$$P(\vec{\theta}, \omega) = \frac{k k_o^3}{\pi^2 \hbar} \frac{E_H}{E_o} \int \frac{S(\vec{K}, \vec{K} - k_o\vec{\theta}, \omega)}{K^2 (\vec{K} - k_o\vec{\theta})^2}\, d^2\vec{\beta} \qquad (3)$$

This is the Fourier transform of the excitation probability $P(\vec{r}, \omega)$. Introducing the Fourier transform of the current density distribution $J(\vec{r})$ in the probe (Colliex and Mory, 1984)

$$J(\vec{\theta}) = \frac{I_o}{\pi\, \alpha_o^2} \int A(\vec{\alpha})\, A(\vec{\alpha} - \vec{\theta})\, e^{i[\gamma_o(\vec{\alpha}) - \gamma_o(\vec{\alpha} - \vec{\theta})]}\, d^2\, \vec{\alpha} \qquad (4)$$

we can rewrite eq.(2) as

$$\frac{dI}{dE}(\vec{r}) = \int P(\vec{\theta}, \omega)\, J(\vec{\theta})\, e^{i k_o \vec{\theta}\vec{r}} d^2\vec{\theta} = P(\vec{r}, \omega) * J(\vec{r}) \qquad (5)$$

In the limit $\beta_o \to \infty$ the intensity distribution in the image is given by a convolution of the current density in the probe with the inelastic excitation probability. This property is very similar to incoherent dark-field imaging with elastically scattered electrons (Colliex and Mory, 1984).

## 3. Calculated Images.

To demonstrate the limitations of the semi-classical model we present a comparison of inelastic images of a dipole surface plasmon calculated with both approaches as they should be visualized in a typical FBEM situation. As a guide for our calculations we introduce the following values, which constitute a rough estimate of the exact parameters used by Hénoc and Henry (1970) in their experiments concerning the detection of gas bubbles : $E_0 = 75$ keV, $\Delta E = 11$ eV, $\alpha_0 = 1$ mrad, and vary the radius of the sphere from $R_s = 10$ Å (fig. 1a,d) to 20 Å (fig. 1b,e) and 50 Å (fig. 1c,f). The resolution limit $d = 0.6\ \lambda/\alpha_0$ is then equal to about 27 Å. We have neglected aberrations ($\gamma_0 = 0$) and assumed axial illumination in a FBEM $[D(\vec{\beta}) = \delta(\vec{\beta})]$. Whereas the semi-classical description is a good approximation for the large spheres (c,f) its failure for small spheres is evident. One should note that for the 10 Å spheres the convolution of the excitation probability with the point-spread function would yield essentially a slightly broadened Airy-disc, whereas the exact quantum-mechanical calculation shows a ring-shaped structure. For sphere radii smaller than the resolution limit the position of this ring is determined by instrumental parameters and is independent of the size of the sphere. To increase the degree of localization of the signal one can increase the objective aperture angle. The small value of $\alpha_0$ used for the calculation has been considered because it contains most of the surface plasmon scattering. If one increases its size, it also introduces a large amount of other excitations thus decreasing the signal to noise ratio for the signal of interest.

We conclude that in a typical STEM instrument with a large spectrometer acceptance angle the image formation process can satisfactorily be described by a convolution of the probe shape with a semi-classical excitation probability. In a FBEM with axial illumination (or in a STEM with a small acceptance angle) and a finite resolution the quantum nature of the incident electrons must be taken into account when interpreting inelastic images.

*Acknowledgements.*

*We wish to thank Drs. M. Achèche and A. Nourtier for fruitful discussions. Financial support by a bursary Ko 885/1-1 under the DFG-CNRS exchange program is gratefully acknowledged.*

## References

Achèche M. (1985), Thèse 3ème cycle, Orsay.
Barberan N. and Bauscells J. (1985), Phys. Rev. B31, 6354.
Batson P.E. (1982a), Ultramicroscopy, 9, 277.
Batson P.E. (1982b), Phys. Rev. Lett. 49, 936.
Colliex C. and Mory C. (1984), in "Quantitative Electron Microscopy", 149, , Eds. J.N. Chapman and A.J. Craven, Scottish Universities Summer School in Physics, 25.
Fujimoto F. and Komaki K. (1968), J. Phys. Soc. Jap. 25, 1679.
Hénoc P. and Henry L. (1970), J. Physique, 31, C1, 55.
Kohl H. (1983), Ultramicroscopy, 11, 53.
Manzke R., Crecelius G. and Fink J. (1983), Phys. Rev. Lett. 51, 1095.
Ritchie R.H. (1981), Phil. Mag. A44, 931.
Rose H. (1976), Optik, 45, 139 and 187.
Schmeits M. (1981), J. Phys. C14, 1203.

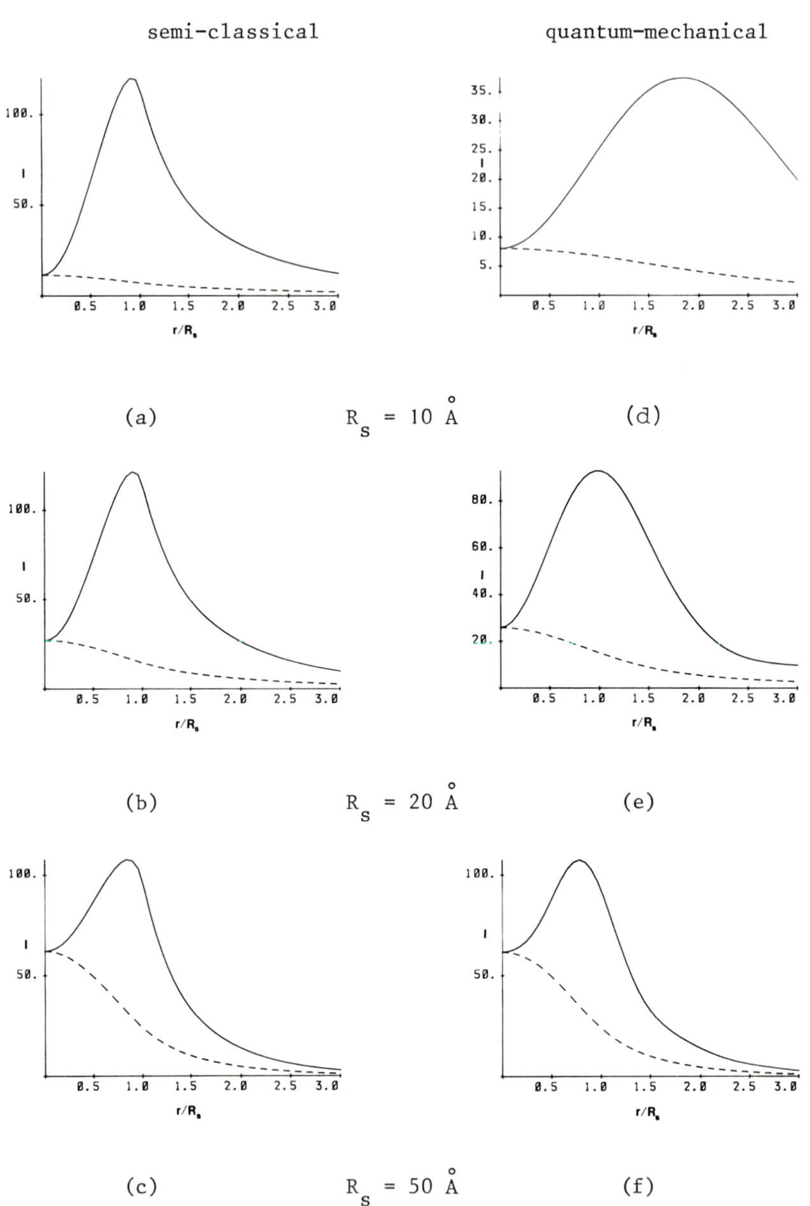


Fig. 1 - Calculated intensity distribution in the image of a small metallic sphere, recorded in a FBEM by using only those electrons, which have excited a dipole surface plasmon. In the left column (a-c) we have displayed the semi-classical results ; the fully quantum mechanical curves are drawn on the right (d-f). The dashed curves show the intensity due to the excitations parallel to the optic axis, whereas the full curves indicate the total intensity. For our calculation we assumed a primary energy of 75 keV, an energy loss of 11eV and an objective aperture angle of 1 mrad.

Paper presented at EMAG '85, Newcastle upon Tyne, 2–5 September 1985

# Dynamical calculations of quasi-periodic structures

M Cornier, R Portier and D Gratias
C.E.C.M./C.N.R.S. 15, rue G. Urbain 94400-Vitry/France

## Introduction.

Dynamical simulations of HR electron microscopy images of an icosahedral quasi-lattice are presented and compared to the experimental images obtained on a rapidly solidified Aluminum-Manganese Alloy (Shechtman et al 1984,1985, Portier et al 1985). Although the computed model uses a simple average spherical motif corresponding to 4 atoms per quasi-lattice node, the simulated images are in perfect agreement with the experimental ones : in metallic close or relatively close packed structures the HR images are mostly governed by the topology of the lattice (or the quasi-lattice); the contributions of the actual nature and locations of the atoms are strongly dominated by the intrinsic form factor of the quasi-lattice. More quantitative diffraction techniques (X-rays and/or Neutrons) are necessary for collecting precise informations about the actual local atomic arrangement.

## Formalism.

Quasi-periodic structures are aperiodic long range ordered structures with a discrete Fourier spectrum. The usual diffraction theories (Cowley 1975, Gratias et al 1983) requiring only an enumerable set of diffracted beams, can still be used for simulating quasi-periodic objects. The elastic scattering of fast electrons in the small angle approximation can be described with the diffraction pseudo-Hamiltonian:

$$H = \sum_{q} |q\rangle q^2 \langle q| + \sum_{q}\sum_{q'} |q'\rangle\langle q'|\bar{V}|q\rangle\langle q| \qquad (1)$$

where the off-diagonal elements represent the Fourier components of the projected potential and the diagonal ones the obliquity factors corresponding to the excitation errors of the beams.

The Cut and Projection Method (C.P.M.) independently derived by Duneau and Katz (1985), Elser (to appear) and Kalugin et al. (1985) leads to an unambiguous labelling of the diffraction peaks together with an intrinsic form factor $\Phi(q)$ due to quasi-periodicity. Since no clear informations are yet available concerning the actual positions of the atoms, the atomic form factors have been approximated by an average value $\langle f \rangle$ adjusted to match with the experimental density (Kelton et al 1985) of the Al6Mn icosahedral phase:

$$\langle q|\bar{V}|0\rangle = \langle f \rangle \, \Phi(q) \qquad (2)$$

Quasi-periodicity: a dense reciprocal space.

The primary difference between crystals and quasi-crystals is that, although they both have a discrete Fourier spectrum, quasi-crystals are built on Z-moduli which make the spectrum dense. The numerical calculations obviously require a finite number of terms. This is easily achieved for the case of crystals by limiting the allowed diffracting beams inside a certain cone of diffraction around the central beam. It may be demonstrated that enlarging the cone leads to convergent solutions of the eigenstate problem. This limitation is not sufficient for quasi-crystals since any finite volume in the reciprocal space contains an infinite number of allowed reflections: as a direct result of the C.P.M., the reciprocal spectrum is an irrational cut of a dense N-dim spectrum, the support of which is of zero measure onto the cutting subspace (this condition is sufficient for insuring quasi-periodicity); the N-dim dense spectrum results in the convolution of the Fourier transform of a basic cut function - used in the direct space for selecting the quasi-lattice nodes - with a N-dim lattice of Dirac peaks. The particular structure of the Fourier spectrum is such that any cut-off based on the magnitude of the intrinsic form factor leads to a relatively dense set of surviving peaks with no accumulation points: there exists a minimum finite distance between nearest neighbouring peaks. This procedure avoids the "small divisor" instability usually encountered in quasi-periodic Hamiltonians (Belissard 1982, Kolmogorov 1957). Reconstructions of the projected potential are seen on Fig.1 as an function of the cut-off.

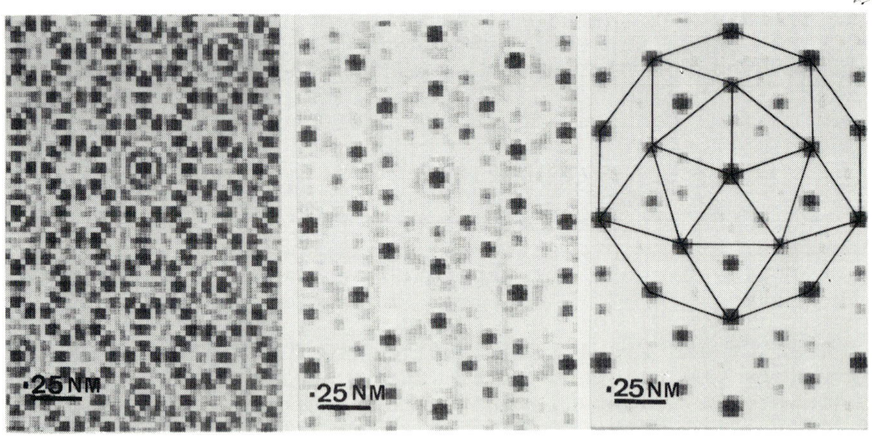


Figure 1: Reconstruction of the projected potential along a 5-fold orientation with decreasing cut-off; 1-a cut-off=.8; 1-b cut-off=.5; 1-c cut-off=.05.

An increasing number of beams (small cut-off) weaken the background noise. In fact, the cut-off in the reciprocal space makes the lattice nodes of the 6-dim direct space spread perpendicularly to the projection 3-dim subspace; hence the nodes close from the cutting slice perturb the reconstructed potential. The decrease of the cut-off narrows the tails of the lattice nodes and cleans up the potential image.

Simulations.

The HR images (Figs.2 and 3) have been calculated by standard diagonalization of (1) for a number of beams corresponding to an optimum cut-off of .05. The specific optical and diffraction parameters are those of the experimental images presented in reference (Portier et al,this issue).

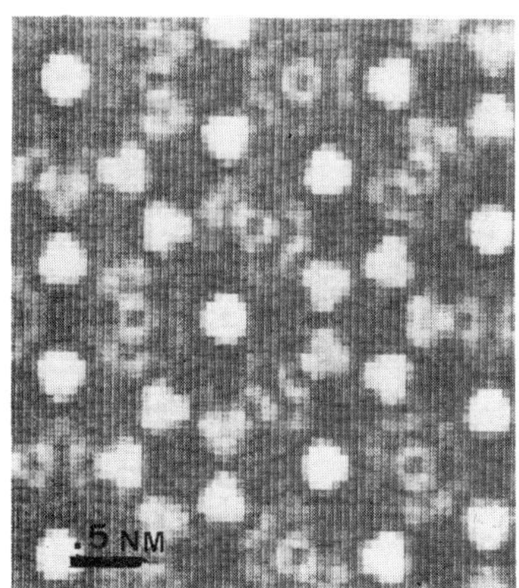


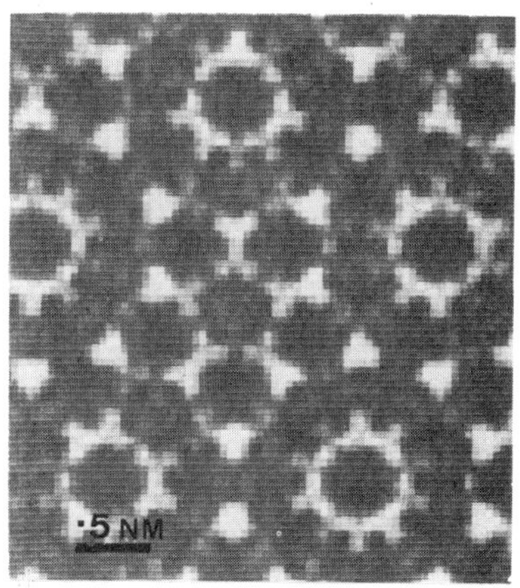


Figure 2: Simulated HR images along a 5-fold orientation. 2-a and 2-b correspond to Fig. 1-a and 1-b of Portier et al (this issue).

A remarkable agreement is observed between experimental and simulated images although the model is based on an average spherical motif! The most sensitive effect is obtained by varying the cut-off: a small number of allowed beams leads to noisy images with spurious oscillations, quite different from the actual ones. The non-harmonicity of the Fourier components of the image allows no pseudo-repetitivity of the images as a function of the defocus since the magnitudes of the diffracting vectors are in an irrational ratio.

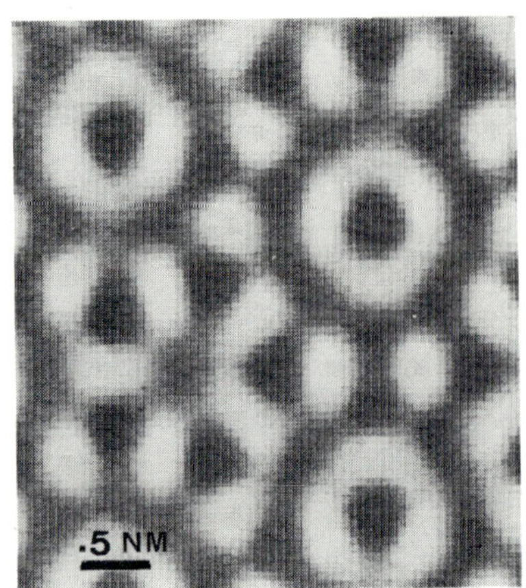


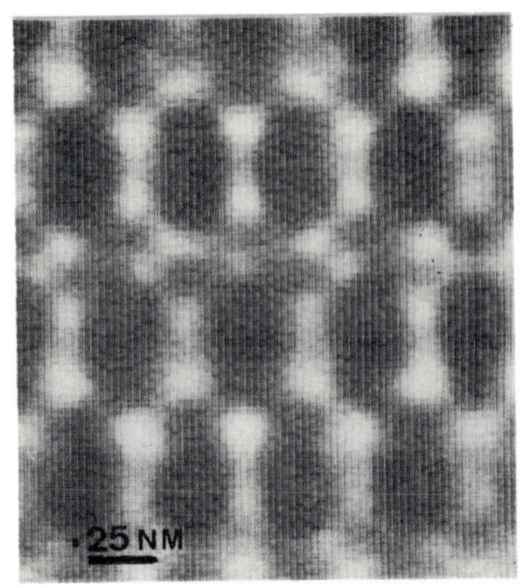


Figure 2-c: HR image along a 5-fold orientation with a small objective aperture as in Fig.1-c of Portier et al (this issue). Notice the similarity with the density distribution of the order parameter in Mermin et al (1985).

Figure 3: HR image along the 2-fold orientation corresponding to Fig.2 in Portier et al (this issue)

References.


Belissard J. Lectures Notes in Phys. 153, 356 (1982)

Cowley J.M. Diffraction Physics, North Holland Amsterdam 1975

Duneau M. and Katz A. Phys. Rev. Lett. 54, 2688 (1985)

Elser V. Preprint to appear in Acta Cryst. A

Gratias D. and Portier R. Acta Cryst. A39, 576 (1983)

Kalugin P.A., Kitaev A.Y. and Levitov L.S. JETP Lett., 41, 145 (1985)

Kelton K.F and Wu J.W. Appl. Phys. Lett. 46, 1059 (1985)

Kolmogorov A.N. Theorie generale des systemes dynamiques en mecanique classique Proc. Int. Conf. Math., Amsterdam (1957)

Mermin N.D. and Troian S.M. Phys. Rev. Lett. 54, 1524 (1985)

Portier R., Shechtman D., Gratias D. and Cahn J.W. J. Micr. Spect. Elec. 10, 107 (1985)

Portier R., Shechtman D., Gratias D. and Cahn J.W., this issue

Shechtman D., Blech I., Gratias D. and Cahn J.W Phys. Rev. Lett. 53, 1951 (1984)

Shechtman D., Gratias D. and Cahn J.W. C.R.A.S. 300 Serie II, 909 (1985)

# Dynamic focusing for reflection microscopy in STEM

R H Milne and D McMullan

Cavendish Laboratory, Madingley Road, Cambridge CB3 0HE

## 1. Introduction

Recently there has been a revival of interest in imaging surfaces by using electron beams at a glancing angle of a few degrees as was first tried by von Borries (1940). Further work was done in the 1950s by Menter (1952), Fert and Saporte (1952), and Haine and Hirst (1953), but since then it appears to have been little used until Osakabe et al. (1980, 1981) showed that with flat, clean specimens, features on an atomic scale such as monatomic steps and emerging dislocations can be imaged successfully. Most of this work has been done in conventional transmission electron microscopes (CTEMs) and the basic criterion for obtaining high resolution images is that the surface is flat, clean and crystalline enough to give clear reflection diffraction patterns.

Some preliminary work using a scanning transmission electron microscope (STEM) has been done by Cowley (1982, 1983) but the main advantages over CTEM will come when the various signals available in the STEM such as electron energy loss spectroscopy (EELS) and energy dispersive X-ray spectroscopy are used to analyse different areas of an inhomogeneous surface. The highest resolution in reflection electron microscopy (REM) is obtained by imaging with diffracted electrons but when more interesting surfaces are examined e.g. those with epitaxial growth, oxidation or corrosion, the various imaging techniques in STEM such as using secondary electrons or annular dark field detectors could be important. Because of the low angle of incidence of the electron beam (a few degrees) there is a problem in that the depth of focus is such that only a central strip of the image is in focus. With flat clean surfaces imaged in TEM (or the STEM spectrometer) this is partially overcome because of the focussing effect due to the small divergence of the diffraction-collimated crystal reflection (Cowley and Nielsen, 1975). However, when using other imaging modes or when examining rough or less clean surfaces this effect is lost. Because of the raster method of imaging in the STEM it is possible to adjust the objective lens current while the specimen is being scanned so as to keep the whole field of view in focus, an option not possible in CTEM.

## 2. Dynamic Focussing

Dynamic focusing in SEM for keeping an inclined specimen in focus was proposed many years ago by one of us (McMullan, 1953) and it was shown that whether or not focus modulation is needed depends on the beam angle, the number of lines in the display monitor or recorder, and of course the angle of inclination of the specimen. Briefly, if the raster in the microscope is of side d, and the specimen is inclined at an angle $\theta$ to the beam, the edges of the scanned area will be $(d/2)\cot\theta$ above and below the plane of best focus and if the beam semi-angle is $\alpha$ the out-of-focus spot diameter will be $\alpha d\cot\theta$ assuming that the beam is in focus at the centre of the scanned area. The out-of-focus spot diameter ought to be no greater than the width of the scanning lines in the microscope, d/N, so that focus modulation will be desirable if N is greater than $(\alpha\cot\theta)^{-1}$. In the SEM $\theta$ may be say 70 deg and so with $\alpha$ = 10 mrad N should not exceed ~ 270 lines; three or four times that number of lines are commonly used and focus modulation is often a standard feature of SEMs.

With reflection electron microscopy in the STEM when $\theta$ is typically 5 deg, N is only 10 if $\alpha$ = 10 mrad and focus modulation is essential even if the depth of focus is larger due to diffraction collimation, as mentioned above. In practice even smaller values of $\theta$ may be used.

We have made a quite simple unit for producing focus modulation in the VG Microscopes HB501 STEM at the Cavendish. Because of the inductance of the objective lens winding it is desirable to use the frame scan waveform rather than the line scan, the specimen being correctly oriented with the scan rotate control. When adding facilities to the STEM it is our practice to make the minimum possible modification to the instrument, particularly when critical parts of the electronics are involved, in this case the objective lens current stabiliser. The compensation waveform, which is derived from the frame time base, is fed into the stabiliser through the existing wobble circuit; no alteration therefore has had to be made to the stabiliser. Unfortunately, the design of the stabiliser is such that only a limited excursion of the lens current can be produced through the wobbler input but magnifications down to 5 x $10^4$ times can be used with full compensation. The amplitude of the frame scan compensating waveform can be adjusted manually while observing the image to produce the correct amplitude of focus modulation for the specimen angle being used; it is also changed automatically with alterations in the magnification setting.

The frequency response of the compensation is limited but a one-second scan can be used for visual observation: this is fast enough for setting the ramp and the microscope focus by eye provided there are clear features near the top and bottom of the picture. A small adjustment has to be made when switching to a slower scan for recording.

## 3. Results

Preliminary investigations of the surfaces of copper and GaAs have been performed. The single-crystal copper specimens were mechanically polished and electropolished in 50% orthophosphoric acid. This gave a fairly flat surface but one that was covered with oxide and possibly remnants of the polishing solution. Consequently, the diffraction pattern consisted mainly of diffuse scattering and the images obtained were not helped by the focussing effect of imaging with a strong diffraction spot. The reflection cartridge used to hold the copper could be put into a side chamber of the microscope, tilted towards an argon gun and bombarded to clean up the surface, but at present there are no facilities for giving the specimen the required annealing. The resulting very rough surface again gave a very poor diffraction pattern. In both these cases the use of dynamic focussing greatly improved the images that could be obtained, as can be seen in Fig. 1 which shows the argon bombarded surface with and without correction.

GaAs crystals were cleaved along their (110) plane, held in a normal tilting cartridge by a folding grid and put into the microscope as quickly as possible (a few minutes after being cleaved). These surfaces were more suited for examination by REM, giving clear diffraction patterns, but even here there is a marked difference between the images obtained with the focus modulation, as shown in Fig. 2.

In some cases it might be desirable or necessary to use one of the other forms of imaging in the STEM so that diffraction effects give no help to the focussing. The secondary electron signal can show details of fairly small structures on the surface although the image widths of small steps is broadened compared to REM taken with diffracted electrons (Fig. 3).

## 4. Conclusions

Although the highest resolution for reflection microscopy is obtained by using a diffracted beam there will be occasions where imaging with such detectors as annular dark field or secondary emission is desirable because of greater signal intensities. There will also be a future trend towards examining less perfect, but more interesting, surfaces; in these cases the ability to have the whole field in focus will be very useful. In this respect the STEM, due to its raster type of imaging, has an advantage over the CTEM for surface imaging which, coupled to its potential for analysis using various signals, suggests that it could be used to an increasing extent for surface studies in the future.

## Acknowledgements

The authors would like to thank the SERC for provision of equipment and for a research assistantship (RHM).

## References

Cowley, J.M. and Nielsen, P.E.H. (1975), Ultramicroscopy, 1, 145.
Cowley, J.M. (1982), Surface Sci., 114, 587.
Cowley, J.M. (1983), J. Microsc., 129, 253.
Fert, C. and Saporte, R. (1952), C. R. Acad. Sci., 235, 1490.
Haine, M. and Hirst, W. (1953), Br. J. Appl. Phys., 4, 239.
McMullan, D. (1953), Proc. Inst. Elect. Eng., 100 II, 245.
Menter, J.W. (1952), J. Inst. Metals, 81, 163.
Osakabe, N., Tanishiro, Y., Yagi, K. and Honjo, G. (1980), Surface Sci., 97, 393.
Osakabe, N., Tanishiro, Y., Yagi, K. and Honjo, G. (1981), Surface Sci., 102, 424.
von Borries, B. (1940), Z. Phys., 116, 366.

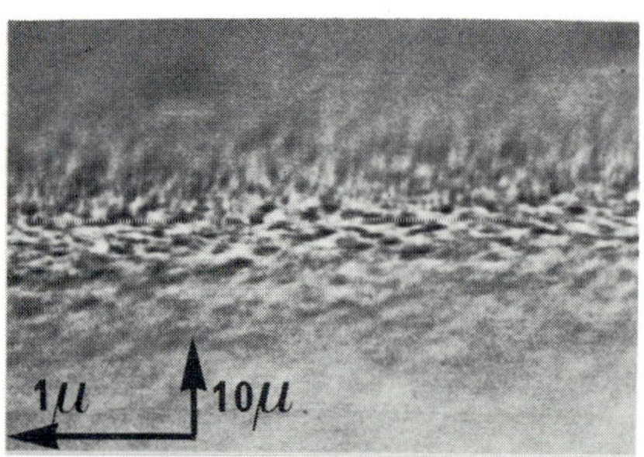


Fig. 1a An argon bombarded Cu surface imaged with the STEM spectrometer without focus modulation.

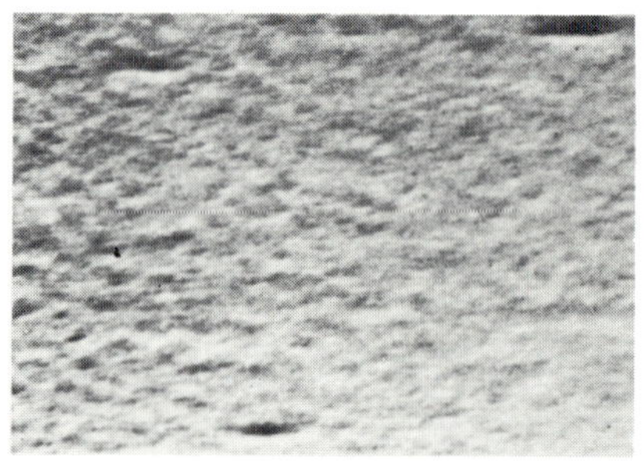

Fig. 1b The same area of Cu as 1a but with focus modulation.

Fig. 2 GaAs (110) surface showing surface steps in focus throughout the picture.

Fig. 3a Secondary electron image of GaAs surface without focus modulation.

Fig. 3b Secondary electron image taken with focus modulation.

*Inst. Phys. Conf. Ser. No 78: Chapter 4*
*Paper presented at EMAG '85, Newcastle upon Tyne, 2–5 September 1985*



# High resolution secondary electron imaging in a VG HB501 STEM

S D Berger, D Imeson, R H Milne and D McMullan

Cavendish Laboratory, Madingley Road, Cambridge CB3 OHE

## 1. Introduction

High quality secondary electron images can be produced in the VG HB501 STEM with a minor modification to the Everhart-Thornley secondary electron detector. A very low background signal and good signal to noise ratio give the images good contrast, and the small probe size (~ 1 nm) gives high spatial resolution in transmission images of thin specimens. These images can be obtained simultaneously with the more conventional annular dark field (ADF) and 'bright field' energy filtered images and can supplement the information obtainable by adding topographical detail. In particular, unique information on the microtopography of specimens such as high surface area supports for catalysts can be obtained, (see Imeson et al 1985 and the references therein).

In addition, the UHV environment, the high spatial resolution and the use of thin specimens allows investigation of a number of basic features of secondary electron emission. The small probe of primary electrons allows investigation of the generation processes and of escape depths in various materials. In such studies, the possibility of specimen charging must also be taken into account. In this paper we discuss some preliminary results.

## 2. The Detection System

The Cavendish VG HB-501 microscope has been fitted with an Everhart-Thornley secondary electron detector situated above and to one side of the objective lens (i.e. on the electron exit side). As a modification the voltage on the input mesh of the system has been raised to a maximum of 2 kV to allow efficient collection of secondary electrons from specimens immersed within the objective lens. Collection is possible with this arrangement because a high excitation objective is used. In this operating mode a strong post-specimen field is produced causing an angular compression of the transmitted electrons. This also forces the secondary electrons to spiral up the bore of the cartridge without impinging on its sides and into the detector under the influence of the high input mesh potential. A typical low magnification image taken with this detector is shown in figure 1 and compared with the annular dark field (ADF) image. The specimen is a mixture of small gold particles and graphitized carbon on a holey carbon film. The most striking difference between the images is that only part of the graphitized carbon can be seen in the secondary picture, confirming that the secondaries originate at the specimen. Since thc majority of the secondary electrons have energies below 50 eV only those electrons leaving the specimen exit surface with an unobstructed flight path will be imaged, the rest being absorbed in the specimen. In certain circumstances it is however possible to image one object underneath another due to the complexities of the system (Imeson et al 1985). The graphitized carbon SE image clearly shows topographic detail as distinct from the 2D ADF image. At higher magnification such an image can be used to reveal fine details of the morphology of samples with a resolution comparable to the probe size.

Although only secondary electrons are detected by this system not all secondaries originate from the specimen/beam interaction. High energy electrons scattered through large angles may strike various parts of the microscope, generating secondaries and causing a spurious signal which would mimic the ADF signal. Close examination of images such as that in fig.1 shows that in normal operation such a spurious signal is very small.

## 3. Origin of Secondary Electrons

The topographic detail seen in SE images results from the small escape depth of these electrons. Only secondary electrons generated within a certain distance of the specimen surface, commensurate with the electron's energy and the specimen's work function, will escape and be detected. Figure 2 shows a number of MgO smoke cubes on a holey carbon film and it is clear that those cubes with their edges parallel to the incident beam appear brighter than those that are more randomly oriented. The increase in signal is due to the increased path length of the incident beam within the SE escape depth of the material. However secondary electrons are generated not only by the incident beam but also by other secondaries which have sufficient kinetic energy. Thus even though the detected secondary electrons evolved near the specimen surface the original point of energy loss may have been well within the specimen. This fact has ramifications for both the ultimate spatial resolution of the technique and the observed contrast.

Figure 3 shows ADF and SE line traces across the edge of a 250 nm MgO cube oriented on a 100 pole. The ADF signal (curve a) is a plot of the object shape convoluted with the incident probe and by comparison shows that the SE signal (curve b) rises ~ 2 nm before the electron beam strikes the edge of the cube. Similar line traces using electron energy loss images have shown that the excitation of surface plasmons occurs as the probe approaches the specimen (see for example Howie and Milne 1985) and it is thought that their decay is responsible for the aloof production of secondaries. This argument suggests that the SE signal should rise ~ 8 nm before the cube edge rather than the 2 nm observed. However since MgO is an insulator the specimen surface tends to charge up under electron beam bombardment greatly reducing the number of SEs which can escape. This effect explains the results obtained and also the observation that some smoke cubes (whatever their orientation) can appear very much brighter than others.

Notice also that the edge effect in the SE signal is asymetric, decaying slowly within the cube. This effect is caused not only by direct excitations within the escape depth but also by energy loss processes within the bulk of the material. For example bulk plasmons excited outside the escape depth may travel into it before decaying to a SE. The plasmon decay route to SE production has a large impact parameter (~ 5 nm) and is a limiting factor as far as spatial resolution is concerned.

SEs are generated both by plasmon decay and by single electron excitation. The latter type of excitation may result in the excited electron having enough kinetic energy to move through the material and generate further secondaries which may then be detected. Consideration of the uncertainty principle shows that the impact parameter of an inelastic process is inversely proportional to the energy transferred. Therefore, unlike plasmon excitations, we would expect original energy loss processes of sufficient energy to lead to channelling effects in the secondary electron signal. In the ADF image of figure 4 a bend contour in a nickel based superalloy is imaged by underfocusing the probe. The contour is also visible, with much less contrast, in the corresponding SE image. The low contrast can be explained qualitatively since only energy loss processes of the order of a few hundred eV and above are expected to show channelling effects in this material. Since the differential cross section for single electron excitation falls off as approximately $1/\Delta\varepsilon^4$ only a small proportion of the total secondary electron signal is produced in this way.

A system is being constructed which will extend the present capabilities by allowing energy analysis of the secondaries and biasing the specimen potential, but this will be reported at a later date. Even without this facility, it is clear that the technique as presented here has considerable potential.

References

Howie, A. and Milne, R.H. Ultramicroscopy (in press) (1985).

Imeson, D., Milne, R.H., Berger, S.D. and McMullan, D. Ultramicroscopy (in press) (1985).

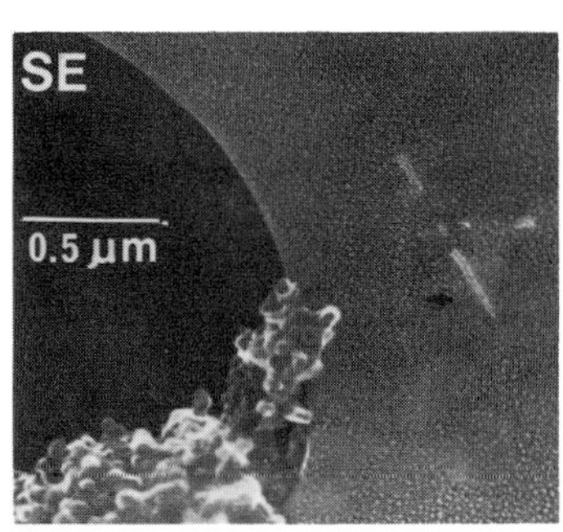


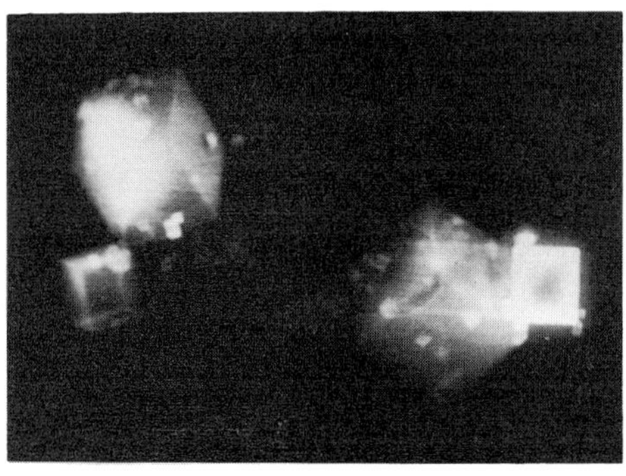

Fig. 2 SE image of MgO smoke cubes showing large differences in secondary yield.

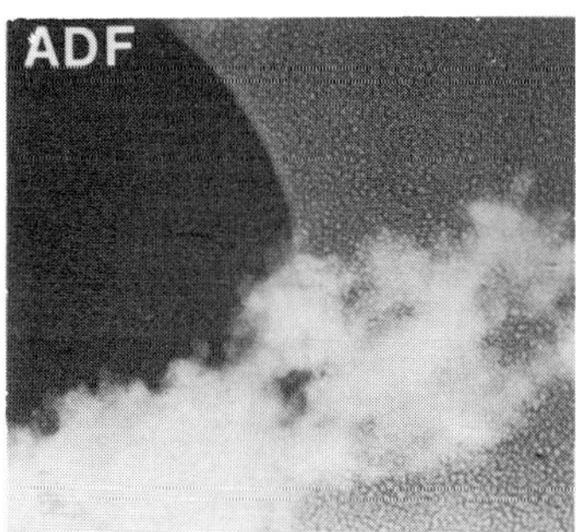


Fig. 1 Secondary electron (SE) and annular dark field (ADF) images of graphitized carbon and small gold particles on holey carbon film.

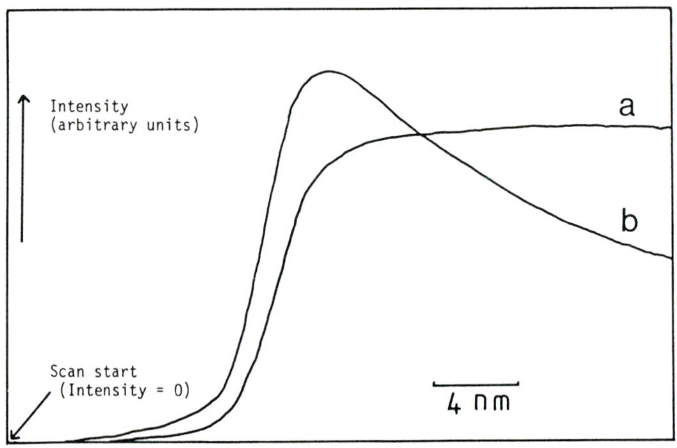


Fig. 3 ADF (a) and SE (b) line traces across an MgO cube edge.

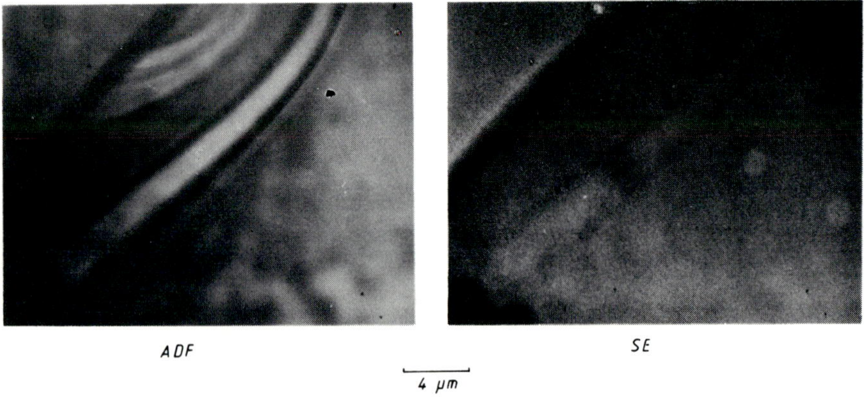


Fig. 4 Bend contour in a Ni based superalloy image in ADF and SE mode.

Support from BP (DI) and SERC (RHM) is gratefully acknowledged.

# A STEM technique for quantifying angular correlations in amorphous specimens

J M Rodenburg

Cavendish Laboratory, Madingley Road, Cambridge. CB3 OHE

## 1. Introduction

Recent calculations (Howie et al, 1985) suggest that various disparate models of the amorphous state resulting in very similar RDFs can be differentiated by observing cross-correlations in intensity at two points in reciprocal space lying on a diffuse ring maximum. These assume that an isolated cluster of, say, 500 atoms is illuminated by a plane wave and the intensity cross-correlation is measured as a function of azimuthal separation, $\phi$, and averaged over all possible orientations of the cluster. Though a useful method for quantifying different models, direct experimental measurements are difficult to perform. A coherent convergent probe allows a small area of a thin specimen to be illuminated but the resulting diffraction plane is confused by random interference effects (speckle) because each point in reciprocal space is convoluted in amplitude with the angular distribution function of the probe. For examples of such experiments on colloidal suspensions using a laser probe see Clark et al (1983). Corresponding patterns recorded at high angular resolution from a STEM probe illuminating ~ 500 atoms of a-Ge display similar characteristics (Rodenburg, 1985). This paper describes a new techique for measuring cross-correlations in the microdiffraction plane of a STEM.

## 2. Technical

A separate scanning unit has been attached to an HB501 STEM to allow the Grigson coils to scan the collector aperture around diffuse ring maxima at constant radius, $2\theta$, in the microdiffraction plane. The sine and cosine outputs are also superimposed upon the x, y scan coils via variable gain amplifiers to compensate for movement of the probe due to stray field from the Grigson coils to within 0.1 nm. A further signal increments the computer hardware which is normally used to collect an energy-loss spectrum by an integral number channels per scan. The spectrometer drift tube is earthed. After every second complete scan, the probe is stepped to a new position on the specimen.

When the probe is defocused, there remains a background intensity modulation because the plastic scintillator on the spectrometer, which is partially imaged in this mode, becomes unevenly damaged when subjected to the full bright field intensity. After collecting a "spectrum" of data (i.e. 1024 channels), it is therefore necessary to collect another set of data with the probe defocused, before returning to the bright field image to provide the instantaneous instrument function. Each spectrum contains 2M circular scans from M probe positions. M is between 1 and 64, depending on the resolution in $\phi$. The averaged instrument function, obtained by adding up 2M scans in the defocused condition, is used to divide each scan in the focused probe data. An autocorrelation is then performed on the central section of each pair of scans from the same probe position and scaled to unity at $\phi = 0$. The resulting curve is given by

$$F(\phi) = \left\langle \frac{\sum_{i=0}^{N} I(\Phi_i).I(\Phi_i+\phi)}{\sum_{j=0}^{N} I^2(\Phi_j)} \right\rangle$$

where $I(\Phi)$ is the intensity at a channel corresponding to an azimuthal angle $\Phi$. The average is taken over a total of M probe positions, which is equivalent to all cluster orientations in the theoretical calculation.

## 3. Results

It has become apparent that with a self-supporting a-Ge specimen of sufficient thinness (5 nm) to allow only 500 atoms to be illuminated (probe diameter ~ 0.8 nm), temporal correlations in the microdiffraction plane are surprisingly short - ~ 15 ms - which presumably results from specimen and instrumental instability. Coherence in the probe results in large changes in amplitude at the collector aperture from drift distances of the order of the atomic spacings. A scan cycle time of 25 ms was therefore used to ensure correlations up to $\phi = 90°$ could safely be measured. At such rates, the number of counts/channel becomes low (~ 100 for a resolution in $\phi$ of $11^{o}$: collector aperture 2 mrad, objective aperture 8 mrad). Statistical noise must be overcome by accumulating many scans. Certainty in the observed correlations also increases with number of specimen volumes sampled. However, with the present system, handling the quantity of data generated is difficult. The memory of the on-line computer is too small and so all data files (i.e. all "spectra") must be loaded into a mainframe IBM for processing.

The summed correlation data from 310 scans from a-Ge (evaporated onto a room temperature NaCl substrate, 6nm thick) has not shown any significant correlation peaks over the measured noise level of 0.2%. Decreasing noise to a more satisfactory level (0.02%) would require 100 times more information which could not practicably be handled by the present hardware. Optical correlation experiments (for example, Rarity and Randle) are often performed with extremely low count rates (1 - 5 counts/channel) and short sample times but the data is reduced on-line by a hard-wired correlator. This is presumably the direction in which to develop the experimental technique described here.

The author acknowledges financial support from the SERC and VG Scientific Ltd.

## References

Clark, A., Ackerson, B.J. and Hurd, A.J. (1983) Phys. Rev. Lett. 50, 1459.
Howie, A., McGill, C.A. and Rodenburg, J.M. (1985) J. de Phys. Conf. Ser. (in press).
Rarity, J.G. and Randle, K.J. (1984) Optica Acta 31, 629.
Rodenburg, J.M. (1985) J. de Phys. Conf. Ser. (in press).

*Inst. Phys. Conf. Ser. No 78: Chapter 4*
*Paper presented at EMAG '85, Newcastle upon Tyne, 2–5 September 1985*


# The effect of systematic variation of the data in the Munro programs on the predicted optical properties of a high excitation STEM objective lens

A J Craven and C P Scott

Department of Natural Philosophy, University of Glasgow, Glasgow G12 8QQ.

## 1. Introduction

The Munro suite of computer programs is a very useful tool for calculating the optical properties of magnetic and electrostatic lenses (Munro 1973). Previous calculations of the post-specimen optical properties of an extended VG HB5 scanning transmission electron microscope (STEM) have been carried out at Glasgow using these programs (Craven and Buggy 1981). Recent discussions with other groups have shown discrepancies in the optical properties calculated for the high excitation objective lens using these same programs. Fig 1 shows the post-field magnification, M, as a function of excitation parameter, $NI/\sqrt{V_r}$, assuming a source in the selected area diffraction (SAD) aperture plane. One was calculated by the Glasgow group and the other by the Orsay group, illustrating the magnitude of the discrepancy. In both cases the axial fields were obtained using the M12 subroutine with identical amp-turns but small differences in the mesh size, mesh distribution, iron circuit and coil position. In an attempt to reconcile such discrepancies, the data used to calculate the axial field of the lens was varied systematically.

## 2. Variations in the Number of Mesh Lines

A reference mesh which consisted of 25 radial lines by 50 axial lines was chosen. (Following Munro, radial lines are defined as lines along the optic axis and axial lines as lines across the optic axis). The number of lines was systematically varied while maintaining the relative number in each area of the lens. Subroutine M12 was used to calculate the axial fields at an excitation, NI, of 7310.03 amp-turns. Fig 2 shows M as a function of $NI/\sqrt{V_r}$. The effect of changing from a mesh of 17 x 31 lines to one of 49 x 99 lines is to give a change of ∿6% in the value of $NI/\sqrt{V_r}$ for M = −10 (virtual image). Following a suggestion by Mulvey (private communication ), Ampere's Law was applied to the axial field data by calculating the line integral of B along the optic axis using a standard Numerical Analysis Group (NAG) routine. (B is identically zero around the perimeter of the mesh). NI inferred from Ampere's Law was compared to that given as data to the program, revealing discrepancies which varied with the data used to calculate the axial field. These discrepancies can be expressed in terms of a fractional error, Δ, equal to (Ampere's Law NI-Program NI)/Ampere's Law NI. Δ for the reference mesh was found to be 0.85% . The values of NI for each axial field were altered to give the same Δ as the field from the reference mesh and the optical properties were recalculated. The results are shown in fig 3 indicating that the spread in $NI/\sqrt{V_r}$ for M = −10 is reduced to ∿1.5%. If further adjustments are made to NI to overlay the curves, the shape is found to be identical. Thus, over this range of $NI/\sqrt{V_r}$, the effect can be considered a scaling error in $NI/\sqrt{V_r}$. Since the difference in the theoretical and experimentally observed $NI/\sqrt{V_r}$ is greater than this spread (Craven and Buggy 1981), it is more sensible to compare the optical properties as a function of M rather than $NI/\sqrt{V_r}$ so that direct comparison can be made with experiment. Figs. 4, 5 and 6 plot the post-field image position, $Z_i$, spherical aberration coefficient, $C_s$, and chromatic aberration coefficient, $C_c$, in this way. The spread due to change in number of mesh lines is ∿3% in $Z_i$, ∿6% in $C_s$, and ∿3.5% in $C_c$. For a given value of M, $C_s$ and $C_c$ are found to converge to an asymptotic value with increasing number of mesh lines. The above procedure was repeated using the M13 subroutine for saturated magnetic lenses and no significant differences were observed.

3. Variations in the Mesh Line Spacing

Two distributions of the same number of mesh lines on an identical lens were compared. Initially the difference in $NI/\sqrt{V_r}$ to obtain M = -10 was ∿1.5%. The difference between $C_s$ and $C_c$ for a given M was ∿3% and ∿0.5% respectively and there was no difference in $Z_i$. By changing from one mesh to the other in small steps, it was noted that the biggest change in optical properties came with adjustments to the radial line spacing and density in the air-gap resulting in a smoother change of spacing of mesh lines when travelling out from the optic axis. (The line spacing travelling along the axis was already reasonably smooth). Meshes with smoothly varying radial lines in the air gap were found to give the lowest Δ.

4. Change of Origin

During the above investigation, a dependance on the origin of the coordinate system used to define the mesh was observed. Over a 120mm range of origin positions, no discrepancies were found in the axial fields calculated either by M12 and M13. However some variation was found in the optical properties as calculated by M21. A ∿1% spread in $NI/\sqrt{V_r}$ at M = -10 was noted, along with ∿1% spread in $C_s$, and ∿0.5% spread in $C_c$ for a given M. $Z_i$ relative to the lens was independent of origin.

5. Modification of Iron Circuit

A detailed mesh was constructed to represent accurately the iron circuit of the lens. The results from this were compared to those from the same mesh with the ironwork simplified as far as possible (see fig 7a and 7b for the type of simplification). The spread in $NI/\sqrt{V_r}$ at M = -10 was ∿0.5% and spread in $C_s$ was ∿1% for a given M. $Z_i$ and $C_c$ were independent of change in ironwork.

6. Change of Coil Shape and Position

Fig 8 shows the various coil positions examined using the reference mesh. Coils were placed in areas ( 1 + 3 + 4 + 5), (1 + 4), (3 + 5), (1), (3) and (1+ 2 + 3 + 4 + 5). In each case the current density was altered to give NI = 7310.03 amp-turns. The resultant spread in $NI/\sqrt{V_r}$ for M = -10 was ∿1%. For a given M, $Z_i$, $C_s$ and $C_c$ were independent of coil position.

7. Change of B - H Data

Axial fields were calculated from the reference mesh using the M13 subroutine with four different B - H curves. One to represent the VG pole-pieces and the other three obtained from Munro (1975). The resultant spread in NI /$\sqrt{V_r}$ at M = -10 was ∿1%. Variations in $C_s$ and $C_c$ were both ∿1% for a given M while $Z_i$ was insensitive to changes in B - H data.

## 8. Conclusions

Near the telefocus condition of this high excitation objective lens, large discrepancies in optical properties are caused by small variations in the axial field calculated by the programs. Such variations might be caused by errors inherent in such numerical routines. Assuming no abrupt discontinuities in mesh spacing the largest variations occur as the number of mesh lines is changed. The resultant spread in NI/$\sqrt{V_r}$ for a given M can be reduced to ∿1.5% by replacing the value of NI in the axial field file returned by the program with that calculated from Ampere's Law. This, however, still leaves considerable discrepancy in the optical properties for a given NI/$\sqrt{V_r}$, since they are very sensitive in the telefocus region. Much better agreement is observed if the other optical properties are plotted as a function of M rather than NI/$\sqrt{V_r}$. $Z_i$ for a given M varies by ∿3% for the mesh size changes considered here but is not affected by other changes. $C_s$ changes by ∿6% and $C_c$ by ∿3.5% over the range of mesh sizes used but, as the mesh size increases, they converge asymptotically to fixed values. Other changes in parameters cause changes in $C_s \lessapprox$ 3% and in $C_c \lessapprox$ 1% for a given magnification. Thus the calculations referred to one optical parameter are quite consistant but great care must be exercised when comparing properties at a given NI/$\sqrt{V_r}$, particularly near a telefocus condition. If the data from fig 1 are replotted in this way, the discrepancy in $Z_i$ is ∿1.5%, $C_s$ ∿2.5% and $C_c$ ∿1% showing close agreement between Glasgow and Orsay calculations.

## References

Craven A J and Buggy T W, 1981 Ultramicroscopy 7 27.
Munro E, 1973 Image Processing and Computer Aided Design in Electron Optics ed. P W Hawkes pp 234-323 (London: Academic Press).
Munro E 1975 Cambridge Engineering Department Report.

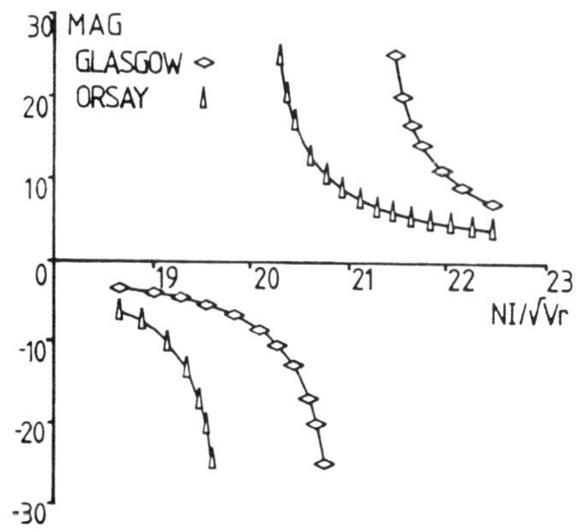


Fig. 1. Comparison of Glasgow and Orsay results.

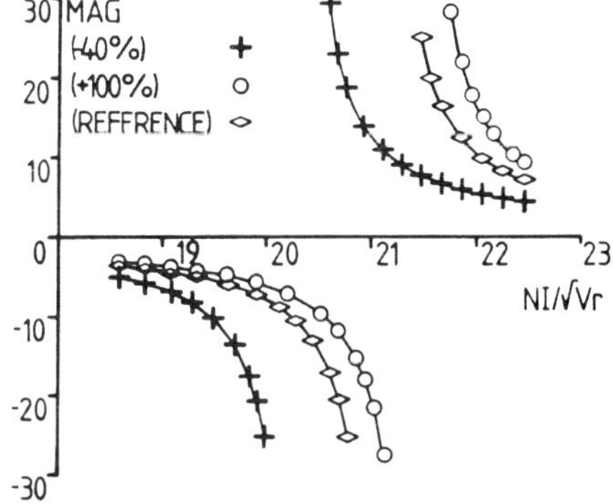


Fig. 2. Effect of mesh size changes.

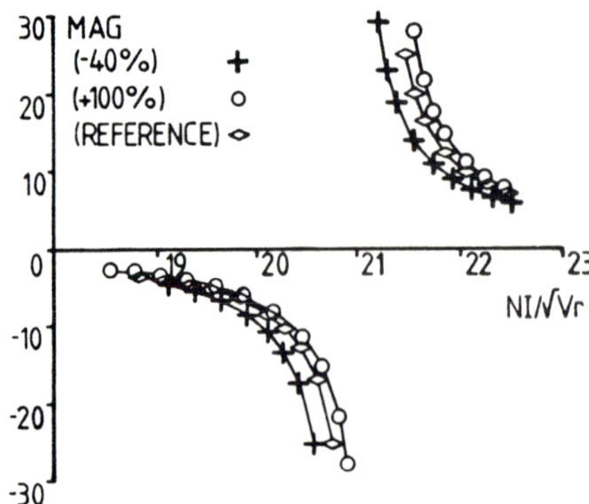


Fig. 3. Effect of scaling axial fields to Δ of reference mesh.

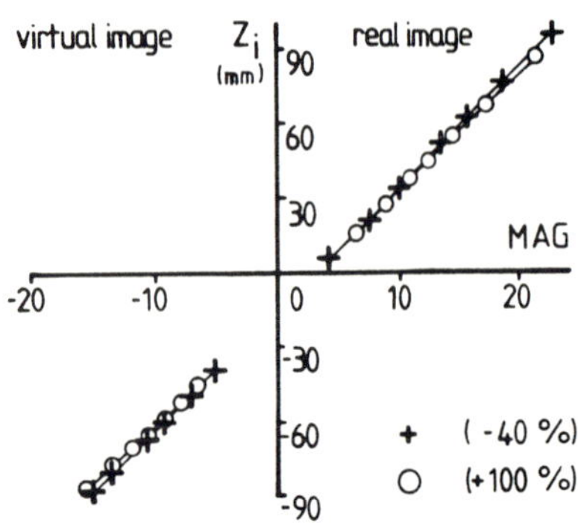


Fig. 4. Post field position, $Z_i$, versus magnification, M, as a function of mesh size change.

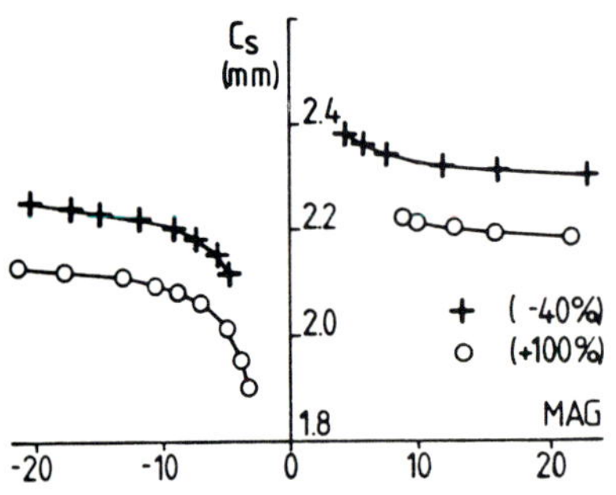


Fig. 5. Spherical aberration coefficient, $C_s$, versus magnification, M, as a function of mesh size change.

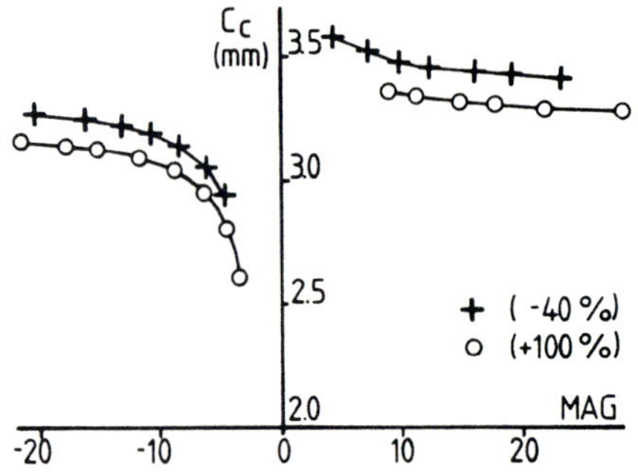


Fig. 6. Chromatic aberration coefficient $C_c$ versus magnification, M, as a function of mesh size change.

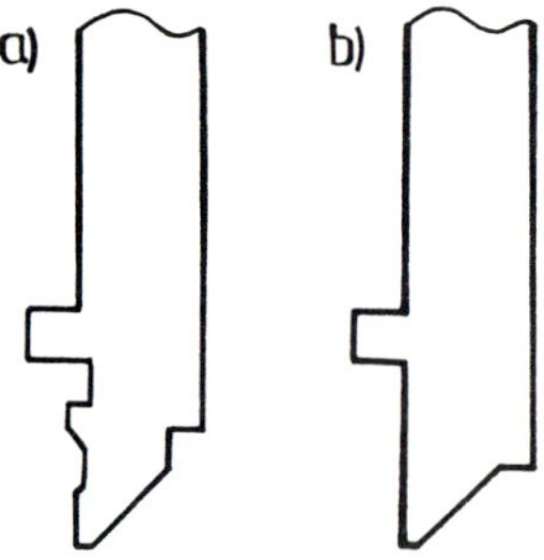


Fig. 7. a) VG polepiece
b) Simplified polepiece

Fig. 8. Diagram showing the various coil positions

# Small angle scattering of electrons for the study of amorphous silicon films

A J Craven, A M Patterson and A R Long

Department of Natural Philosophy, University of Glasgow, Glasgow G12 8QQ

## 1. Introduction

The size and nature of the voids and defects present in amorphous semiconductors influence their electrical, thermal and optical properties. Small angle scattering (SAS) in the diffraction pattern provides information about them. The SAS in an electron diffraction pattern is obscured by inelastic scattering which can be removed by energy filtering. Energy filtered line traces through the diffraction pattern can be recorded in the VG Microscopes HB5 scanning transmission electron microscope (STEM) so that this technique together with a range of other analytical techniques is available to study a specimen in the STEM (Patterson et al 1984).

## 2. Specimen Preparation

Thin films of hydrogenated amorphous silicon were grown on KCl substrates using either RF sputtering (RFS) of a silicon target with an argon:hydrogen gas mixture or RF glow discharge (GD) decomposition of silane. The films could be doped by suitable additions to the gas mixture and the substrate temperature could be set in the range 283 to 523K. By appropriate masking of the substrate, several regions of differing thickness were obtained for each set of preparation conditions. The films were floated onto distilled water and mounted on specimen grids for study in the STEM.

## 3. The Technique

To obtain the line traces, the rocking beam selected area diffraction (SAD) mode is used with the electron spectrometer and the data acquisition system for electron energy loss spectroscopy (EELS). Fig. 1 illustrates this schematically. A uniformly illuminated SAD aperture is imaged onto the specimen by the objective lens. The image of the source is focussed on the distant collector aperture so that the beam divergence at a point on the specimen is very small. By rocking the beam direction about a point in the plane of the SAD aperture, the specimen illumination rocks about a point in the specimen plane, leaving the illuminated area unchanged for small angles of rocking. Only scattered electrons close to the axis pass through the collector aperture and the inelastic scattering is then filtered out using the spectrometer. By the reciprocity theorem, this mode is equivalent to the SAD mode in the CTEM. The angular diameter of the centre spot is determined by the angle subtended at the specimen by the collector aperture, $2\beta$. This can be varied either by changing the physical size of the aperture or by using post-specimen optics. The camera length is controlled by the angular range of the rocking.

Since inelastic scattering obscures the SAS, it is important that the fraction of inelastic scattering left after energy filtering is negligible. Fig. 2 shows part of a series of EELS spectra taken at different mean angles of scattering with $\beta \sim 0.3$ mrad. The angle of scattering, $2\theta$, is expressed as a scattering vector $s = (4\pi \sin \theta)/\lambda$. If the mean angle of scattering is zero, the elastic signal is much greater than the inelastic signal. However, for mean angles of scattering of $\sim 1$ and $\sim 2$ mrad, the reverse is true, A 12 eV energy window centred on the elastic peak leaves only a small inelastic contribution to the signal even in the worst case. Note the additional loss signal between 0 and 16 eV when the mean angle of scattering is zero. This is attributed to Cerenkov losses due to the high refractive index of the specimen (e.g. Raether 1980).

Fig. 3a shows an energy filtered diffraction trace from an undoped GD film taken with $\beta \sim 0.3$ mrad and fig. 3b shows a trace from the same specimen with $\beta \sim 0.05$ mrad. Since the top of the centre peak represents the focussed image of the source passing through the collector aperture, its height changes (for a given region of specimen) only if the current incident on the specimen changes or the source image becomes larger than the collector aperture. The rest of the signal is relatively slowly varying and so is proportional to $\beta^2$. Hence the much reduced ratio of the signal height to the centre peak height in fig. 3b. The fact that this ratio has dropped by less than $\sim 36$ for a change $\sim 6$ in $\beta$ may indicate that the size of the image of the source is approaching that of the collector aperture when $\beta \sim 0.05$ mrad and hence that this is the best angular resolution available. Because the signal varies as $\beta^2$, the signal to noise ratio varies as $\beta$ assuming that shot noise is the limiting factor.

## 4. Processing the Data

Guinier and Fournet (1955) showed that a plot of ln I against $s^2$ gives information about the size of the voids or defects. Such a plot is known as a Guinier plot and I is the signal intensity at a scattering vector s. If such a plot is a straight line, the radius of gyration of the voids, $R_o$, can be obtained from the slope using the approximate Guinier relationship $I = I_o \exp(-s^2R_o^2/3)$. $I_o$ gives information on the number of voids present in the film. Even if the plot is not straight, the slope gives an estimate of the scale of the structure responsible.

Before such plots are made, two corrections are necessary. The first is for the stray scattering which appears at a particular value of s because of the instrument response in the absence of the specimen scattering. This can be estimated by recording a trace under identical conditions but through a hole in the specimen. Such stray traces, normalised so that they have the same area under the centre peaks as the specimen traces, are shown in figs. 3a and 3b. The second correction is for the background signal between the first diffraction ring and the centre spot. Some of this is the tail of the first diffraction ring but the ratio of the height of the maximum intensity in the first ring to the minimum intensity between it and the centre spot tends to decrease as films of increasing thickness are examined, indicating that some additional scattering may be involved. If the specimen is tilted so that the projected thickness increases, this ratio remains constant. Thus, in the same specimen, the peak and the minimum scale in the same way with thickness (Craven et al 1985). The values of this ratio observed in the electron SAS measurements reported here and elsewhere in the literature are higher than those in corresponding x-ray experiments (e.g. D'Antonio and Konnert 1979). It is not clear exactly what the form of this background correction should be and, as a first approximation, a background signal $I_B$ equal to the minimum intensity between the centre spot and the first diffracted ring is subtracted from the intensity at all values of s. Fig. 4 shows the effect of such corrections on the data of fig. 3a. All the curves are normalised by $I_D$, the intensity of the peak of the first diffraction ring minus $I_B$. Curve 1 is the signal. Curve 2 is the signal corrected for the stray which makes little difference for a film of this thickness (330A). Curve 3 is curve 2 after correction for $I_B$. This correction is most important for larger s. Curve 4 shows the stray scattering on the same scale. The traces are expected to be symmetric. The error bars indicate the spread of values from the left and right side of the traces. Fig. 3b shows deviations from symmetry but the results give a very similar average curve as shown on fig. 5a. The corresponding error bars are much larger of course.

## 5. The Results

Fig. 5a shows the results obtained from GD films. Curves 1 and 2 are the results from the data of figs. 3a and 3b respectively and are for an undoped film, showing good agreement despite the difference in angular resolution. Curve 3 is for a similar film which has been doped with phosphorus, showing some change of the SAS. Curve 4 is plot of the x-ray data of D'Antonio and Konnert (1979) normalised to the first diffraction peak, showing reasonable agreement for similar material. The slopes of the curves indicate structure on the scale of 10A. Fig. 5b shows some results for RFS material. The much straighter Guinier plots give $R_o$ values of 4.0A and 6.0A for lines 1 and 2 respectively. Line 1 was obtained from a specimen prepared at 283K and line 2 from one prepared at 513K. Both had an average thickness ∿240A determined from the plasmon intensity. These results are typical of those obtained from RFS films at normal incidence. If the specimen is tilted about an axis in the rocking plane, the SAS is unchanged except for an increase in intensity caused by the increased projected thickness. This is expected since no component of position parallel to the scattering vector changes. However, if the tilt axis is perpendicular to the rocking plane, this is no longer the case and the SAS is observed to change indicating that the voids have a long dimension normal to the film surface (Craven et al 1985). This is consistent with the columnar voids postulated to exist in this material.

## Acknowledgment

The authors would like to thank Dr. J.I.B. Wilson for the GD films.

## References

Craven A J, Patterson A M, Long A R and Wilson J I B 1985 Proc 11th Int Conf on Amorphous and Liquid Semiconductors, Rome (to be published in J. Non-Cryst. Solids).

D'Antonio P and Konnert J H 1979 Phys. Rev. Lett. 43, 1161.

Guinier A and Fournet G 1955 Small Angle Scattering of X-rays (New York: Wiley).

Patterson A M, Craven A J, Chapman J N and Long A R. 1984 Electron Microscopy and Analysis 1983 ed P Doig (London: IOP conf. ser. 68) p 429.

Raether H 1980 Excitation of Plasmons and Interband Transitions. Springer Tracts in Modern Physics 88 (Berlin: Springer Verlag).

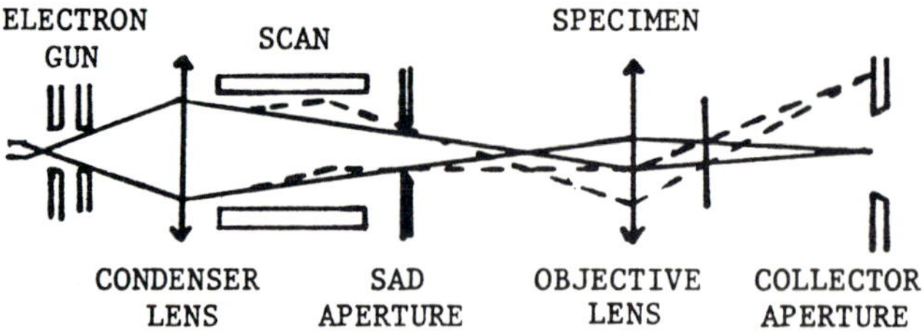

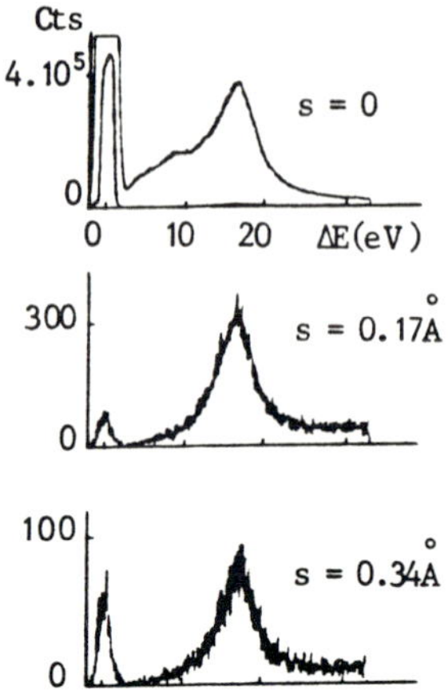

Fig. 1 (above) Schematic ray diagram of the selected area diffraction mode.

Fig. 2 (right) EELS spectra with $\beta \sim 0.3$ mrad and mean scattering angles 0, $\sim 1$ and $\sim 2$ mrad.

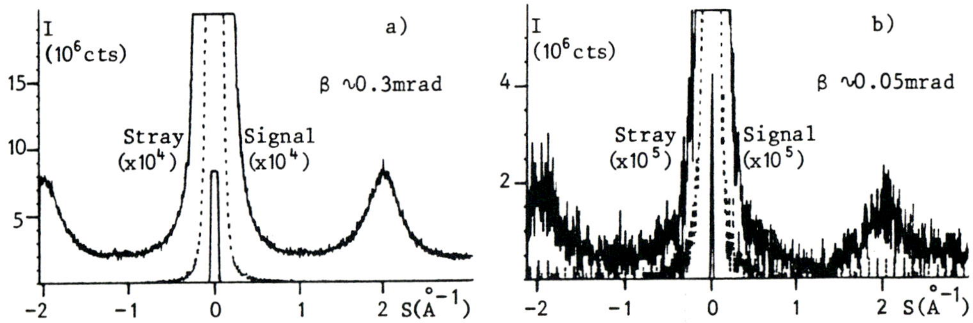

Fig. 3 Energy filtered diffraction traces from undoped GD material 330 Å thick. a) $\beta \sim 0.3$ mrad. b) $\beta \sim 0.05$ mrad.

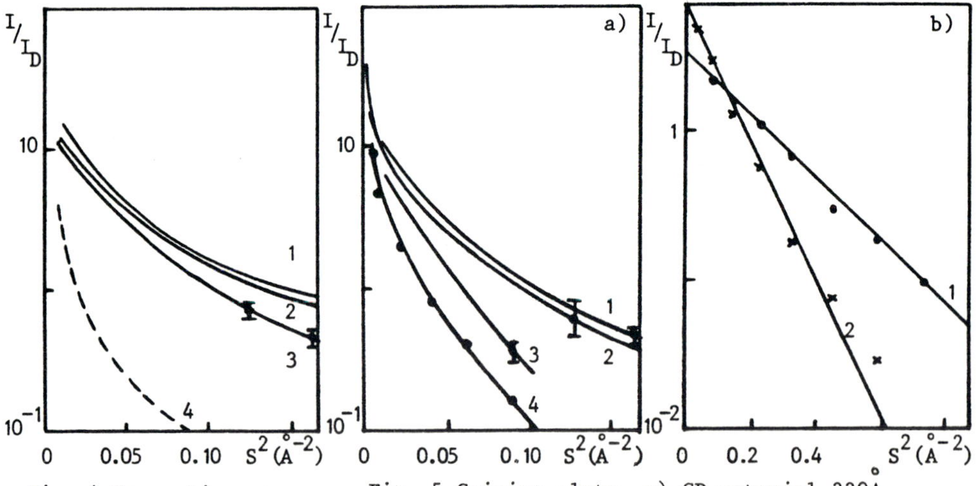


Fig. 4 Corrections to data. 1) Recorded data. 2) 1 - Stray. 3) 2 - Background. 4) Stray.

Fig. 5 Guinier plots. a) GD material 330Å thick. 1)undoped $\beta \sim 0.3$ mrad. 2) undoped $\beta \sim 0.05$ mrad. 3) P doped $\beta \sim 0.05$ mrad. 4) X-ray data (D'Antonio and Konnert 1979). b) Undoped RFS material 240 A thick. 1) 283 K. 2) 513 K.

*Inst. Phys. Conf. Ser. No 78: Chapter 4*
*Paper presented at EMAG '85, Newcastle upon Tyne, 2–5 September 1985*



# Fourier analysis of focused electron probes in STEM

G Cliff and P B Kenway
Department of Metallurgy and Materials Science, University of Manchester/UMIST, Grosvenor Street, Manchester M1 7HS, U.K.

## 1. Introduction

It is often tacitly assumed that focussed electron probes in STEM operating modes are gaussian, that the image resolution is dependent on the probe size and that the smallest probe (yielding the highest image resolution) is at the plane of the disc of minimum confusion. Using previously reported expressions (based upon geometrical optics) which allow numerical computations of electron density distributions in any plane of a probe-forming lens (Kenway and Cliff 1983) computed distributions have been produced for two planes of major interest, the gaussian image plane (GIP) and the disc of minimum confusion (DOMC). These have been subjected to fourier analysis using a commercial machine code fast fourier transform (FFT) designed for the BBC B microcomputer (the FFT is available as an EPROM from Structured Software, 23 Redcar Drive, Eastham, Wirral, L62 8HE).

## 2. Results and discussion

To assess the characteristics of gaussian electron density distributions the FFT power spectra of four gaussian distributions as a function of full width half maximum (FWHM) are shown in figure 1. It is seen that the power spectra of gaussian distributions decay smoothly with increasing spatial frequency intercepting the abscissae at higher spatial frequencies as the FWHM of the distribution decreases. A reconstruction of each of the original distribution is shown above each spectrum.

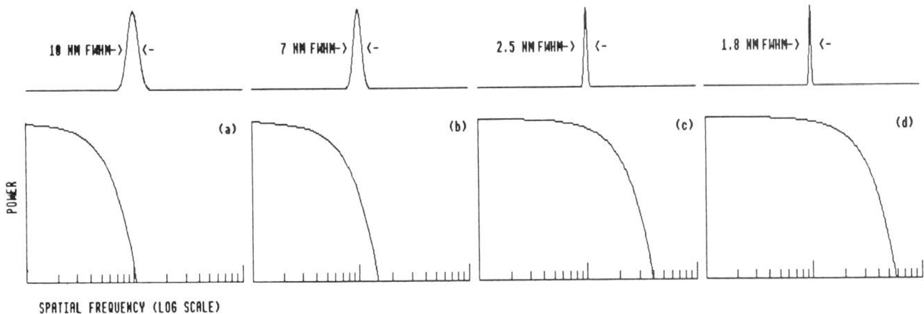


Figure 1. FFT power spectra of gaussian electron density distribution cross-sections. Above each spectrum is the original gaussian re-constructed from the transform.

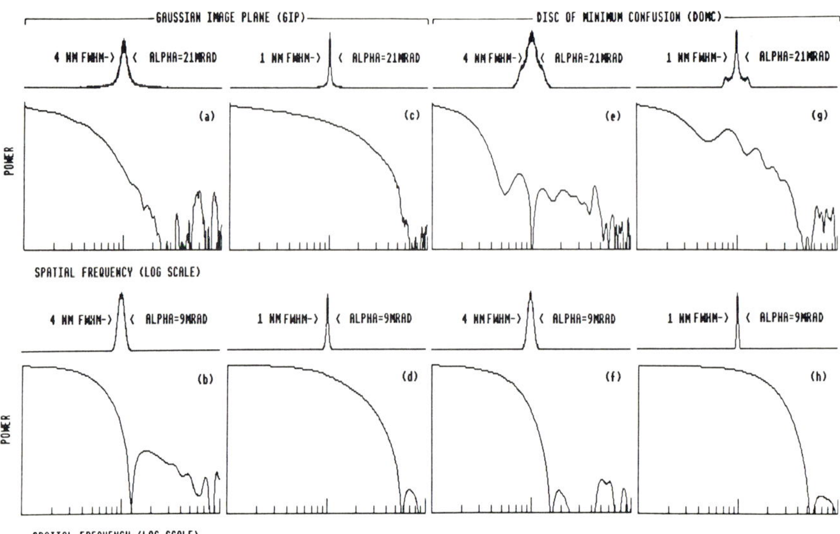


Figure 2. FFT power spectra of electron density distributions at the gaussian image plane (GIP) and the plane of the disc of minimum confusion (DOMC) as a function of FWHM and illumination half angle. Above each spectrum is the original distribution re-constructed from the transform.

The power spectra of computed distributions of probes formed under conditions likely to prevail during x-ray microanalysis (small FWHM and large illumination angle) are shown in figure 2. These distributions were computed for the GIP and the DOMC of a probe-forming lens of spherical aberration coefficient Cs=2.8mm for two probe sizes of 4nm. and 1nm. (FWHM) and two aperture semi-angles ( $\alpha$ ) of 21mrad and 9mrad at an accelerating voltage of 120kV.
It is seen from the distributions and spectra, figure 2 (a)(c)(e)(g), that small probes formed with large apertures are non-gaussian and cannot be made gaussian by defocussing to the DOMC. When the aperture is reduced the effects of spherical aberration are reduced, figure 2 (b)(d)(f)(h), there is little difference between the distributions and spectra in the GIP and DOMC and they closely resemble the gaussian spectra of figure 1.
As in the case of the gaussian distributions the higher spatial frequencies in the computed probe increase as the probe size is reduced, for any given aperture. (Note:- the high frequency contribution in all the spectra beyond the abscissae intercept is caused by noise in the original numerical description of the electron distribution).
Generally for any given probe size and aperture the highest spatial frequencies in the probe occur in the GIP. As the image detail seen on the CRT of a STEM system is the result of the convolution of the spectral frequencies in the probe and the specimen detail being imaged, it follows that best focus STEM images are produced when the specimen occupies the GIP as observed experimentally by Venables and Janssen (1980) and Cliff and Kenway (1982), this is despite the fact that

the overall probe diameter in the GIP will be four times the diameter of the probe in the DOMC (under conditions where spherical aberration dominates).

| Probe size FWHM nm. | Aperture mrad | GIP nm. | | DOMC nm. | |
|---|---|---|---|---|---|
| | | 50% | 90% | 50% | 90% |
| 4 | 21 | 22.4 | 54.6 | 10.8 | 17.6 |
| 4 | 9 | 4.2 | 8.0 | 3.8 | 7.2 |
| 1 | 21 | 22.8 | 54.6 | 10.8 | 15.8 |
| 1 | 9 | 1.7 | 4.5 | 1.0 | 2.1 |

Table 1. Probe diameters containing 50% and 90% of the incident electrons in the GIP and DOMC for FWHM of 4nm and 1nm, at 9mrad and 21mrad illumination semi-angles.

Since the probe current is constant for any given probe size and aperture it is seen from Table 1 that for x-ray microanalysis it is desirable to defocus the probe to bring the DOMC coincident with the specimen plane to maximise the probe current to size ratio. The required under focus is the axial distance between the GIP and the DOMC (0.75 Cs $\alpha^2$). Some STEM image resolution is lost with this defocus, but the x-ray spatial resolution can be improved by as much as a factor of four. This improvement is illustrated in figure 3 which shows a focussed STEM image figure 3(a) of an Nb particle in a Cr/Ni stainless steel foil and the x-ray spectrum figure 3(b) obtained when the probe is placed centrally on the Nb particle. The improved x-ray spectrum figure 3(c), with a reduced matrix contribution, was obtained when the probe was underfocussed to place the DOMC in the specimen plane on the Nb particle. The slight reduction in STEM image detail under these conditions can be seen in figure 3(d).

## 3. References

1 Kenway P B and Cliff G (1983) Inst. Phys. Conf. Ser. No. 68, 83

2 Venables J A and Janssen A P (1980) Ultramicroscopy 5, 297

3 Cliff G and Kenway P B (1982) Microbeam Analysis-1982, ed. K F J Heinrich, San Francisco Press, 107

4 Takaaki H and Hibino M (1984) Jap. J. of Microsc., (1984) Vol. 33, No.2, 116

## 4. Note added in preparation.

During the preparation of results reported in this paper the authors' attention was drawn to a recent paper by Takaaki and Hibino (1984) which described computed electron density distributions in probe-forming lenses in a 50kV STEM. Their paper additionally allowed for phase and showed that a small defocus, not to the DOMC, was needed to compensate for phase. This defocus maximised the central intensity of the electron distribution at the GIP. They also noted that maximum image resolution occurred at the GIP and not at the DOMC. Their paper concluded that geometrical optics gave a good description of their computed results.

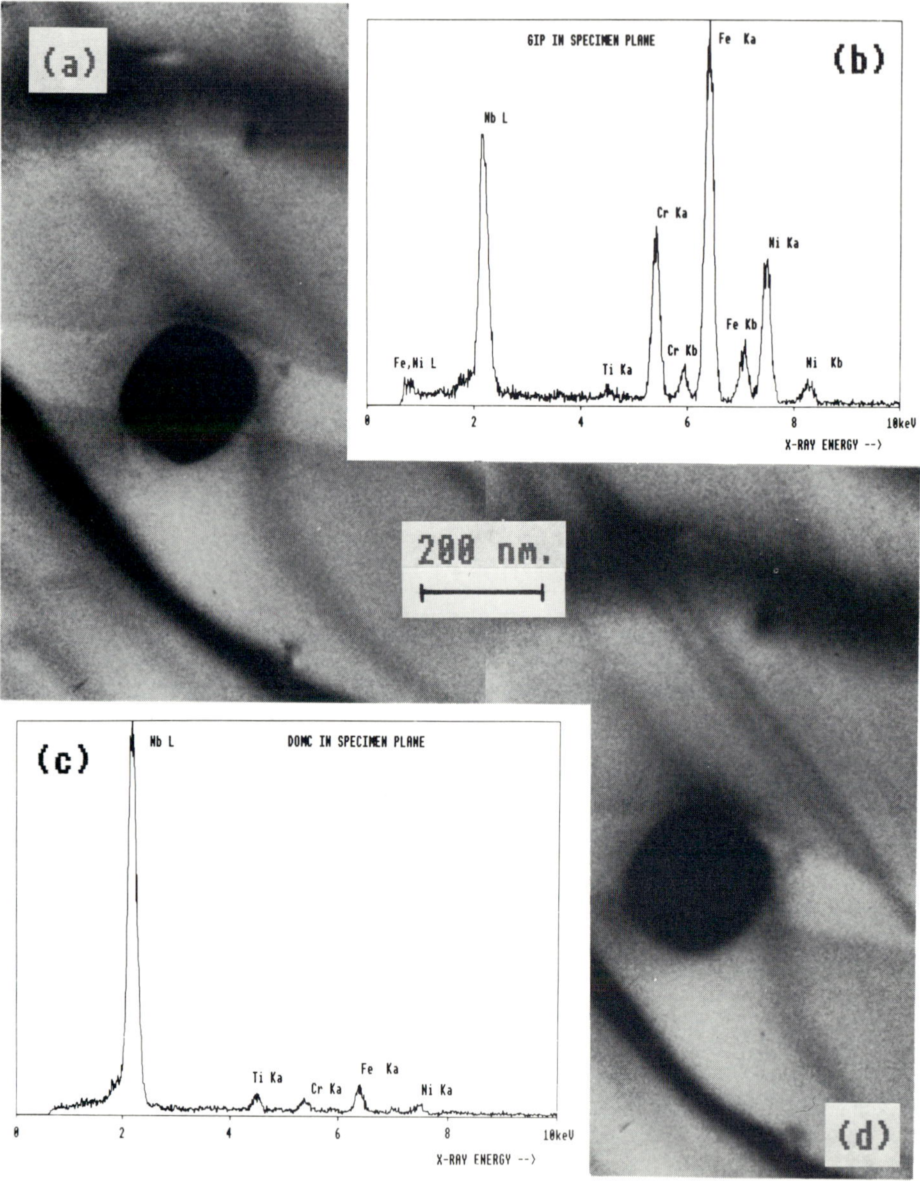


Figure 3. In focus STEM image of an Nb particle in a Cr/Ni stainless steel foil when the GIP of the electron probe is in the specimen plane (a), and the x-ray spectrum (b) obtained with the probe stationary on the particle. An improved x-ray spatial resolution , as shown by the reduced matrix contribution, is seen in (c) where the probe DOMC is in the specimen plane although the image is now underfocussed (d).

*Inst. Phys. Conf. Ser. No 78: Chapter 4*
*Paper presented at EMAG '85, Newcastle upon Tyne, 2–5 September 1985*



# Dielectric excitations at surfaces and interfaces

A Howie, R H Milne and M G Walls

Cavendish Laboratory, Madingley Road, Cambridge. CB3 OHE

## 1. Introduction

Conventional electron microscopy currently provides the most high resolution information about surfaces or interfaces when sufficiently thin, well aligned samples can be produced. In the somewhat lower resolution range of 0.5 nm to 2.0 nm however, the STEM yields a greater variety of signals providing local information about structure, composition and electronic properties. Moreover this can be achieved not only with thin specimens when the beam is aligned parallel to the interface but also with more easily produced surfaces on bulk samples by operating in glancing angle mode. Here we consider the valence excitation region of the electron energy loss spectra obtained in such situations. Previous work on interfaces (see Howie and Milne, 1984, 1985 with references therein) has indicated that the observations can be qualitatively explained by classical dielectric theory. More recent studies by Milne and Echenique (1985) of the dependence of the loss spectra on the distance of the STEM probe from the surface of well-aligned MgO particles demonstrated good agreement with observations, provided relativistic corrections were included in the theory. In Milne's latest experiments where the spectra were calibrated directly against the incident beam intensity, the relativistic excitation theory was confirmed to an accuracy of about 20%, a discrepancy which can be produced by an error of only 0.5 nm in positioning the probe. Here we concentrate on the glancing angle situation, which apart from its practical significance, offers the possibility of checking the data against incidence angle $\theta$ - a potentially well defined parameter. Multiple scattering effects are known to be important, particularly at small values of $\theta$, but if these can be successfully computed it may be possible to interpret the data in terms of penetration distances into the sample (Krivanek et al 1983).

## 2. Theory

In the relativistic dielectric excitation theory, (see for example Garcia-Molina et al 1985) a point electron travelling in medium 2 at distance x parallel to an interface with medium 1 has a differential probability per unit distance for energy loss $\Delta E = \hbar\omega$ which is given by

$$\frac{dP}{d\hbar\omega} = \left(\frac{e^2}{2\pi^2\varepsilon_o\hbar^2 v^2}\right)\int_0^{K_y^m} dK_y \; \mathrm{Im}\,\{F - (1 - \beta^2\varepsilon_2)/\varepsilon_2\alpha_2\} \qquad (1)$$

where $\beta = v/c$, $\alpha_j{}^2 = K_y{}^2 + \omega^2/v^2 + \varepsilon_j\omega^2/c^2$ and $\varepsilon_j(\omega)$, $j = 1, 2$ is the complex dielectric constant of the medium j. The integral covers various momentum transfers $\hbar K_y$ parallel to the surface and normal to the beam direction with an upper limit set either by some specimen-dependent cut-off or by the instrument aperture. The quantity F in eqn. (1) is given by the expression

$$F = \left(\frac{2\alpha_2^2(\varepsilon_1 - \varepsilon_2)}{\varepsilon_2\alpha_1 + \varepsilon_1\alpha_2} + (\alpha_1 - \alpha_2)(1 - \varepsilon_2\beta^2)\right) \frac{\exp(-2\alpha_2|x|)}{\varepsilon_2\alpha_2(\alpha_2 + \alpha_1)} \qquad (2)$$

It may be verified, most simply in the non-relativistic limit, that this produces both interface excitations and a correction to the bulk excitation typically significant for impact parameters $|x| < 5$ nm.

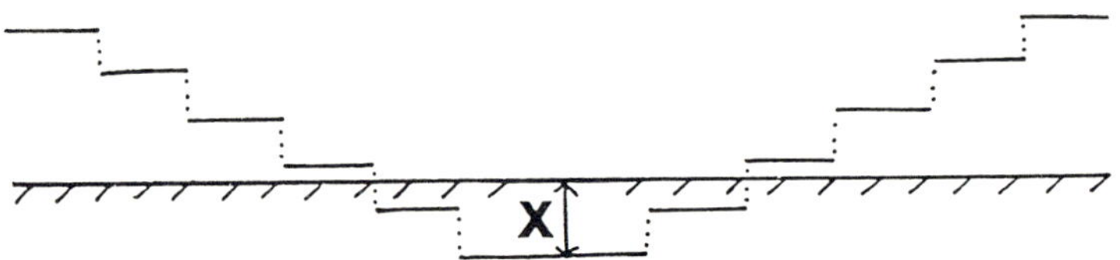


Fig.1. Path of reflected electrons assumed in the theory.

These relations can be adapted to the glancing angle case by treating the path as a series of segments at different distances from the interface (see fig. 1). This approximation can be shown to work well at small incidence angles $\theta$. Since medium 2 is a vacuum $\varepsilon_2 = 1$. Multiple inelastic scattering effects have been included by convoluting the results for each segment and including a factor 2 to allow for the incoming and outgoing parts of the track. An upper limit $K_y^m$ corresponding to the effective aperture was taken, ignoring therefore some multiple scattering effects into or out of the aperture. Including dielectric data up to 40 eV and covering the loss spectrum up to 80 eV with 1 eV channels, we find that sufficient accuracy is generally achieved by starting 20 nm out from the surface and moving into the surface with 1 nm steps. By interchanging subscripts 1 and 2 and allowing for the refraction effect on $\theta$, further steps can then be taken allowing for a certain path length in and out of the medium 1. Comparison with experimental data at various angles of incidence then gives some indication of penetration distances.

## 3. Results

A crystal of GaAs with a (110) surface produced by cleaving was transferred quickly to the HB501 and examined in the glancing angle mode. The scanning images obtained revealed substantial areas free from surface steps or other artefacts and therefore suitable for electron spectroscopy. The crystal tilt relative to the incident probe (i.e. instrument axis) and the deflection angle of the post-specimen Grigson coils must be adjusted to bring a given diffraction spot into the spectrometer under known diffraction conditions. These adjustments were facilitated by running the incident probe initially in an angular scan mode, thus displaying a diffraction pattern. A series of spectra using the specular beam with the crystal successively set to the 440, 660 and 10,10,0 Bragg positions were thus obtained. Fig. 2 shows some of the results for the 660 condition where first and weak second surface plasmons are visible together a bulk plasmon at 16 eV. The theoretical curves shown have been relatively scaled to equalise approximately the total areas under the curves. Improved agreement with experiment could obviously be achieved by modifying slightly the dielectric data to move all these peaks to slightly higher energies.

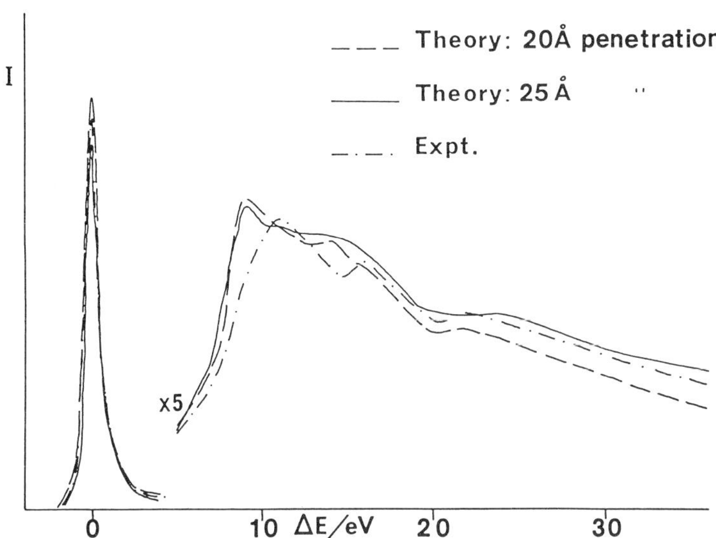


Fig.2. GaAs(660) reflection E.E.L.S.

Penetration depths for the 440, 660 and 10,10,0 cases are determined approximately from such fits to the data to be 1, 2.5 and 3.5nm respectively corresponding to a total effective beam path in the specimen of about 150nm. This can be compared with a figure of ~ 120 nm obtained by comparing the areas under the characteristic Ga edge at 1116 eV in a transmission spectrum and in the 660 glancing angle loss spectrum. In making this latter estimate, the zero and valence loss regions of both spectra were used to correct the relative intensities of the characteristic losses and to deduce the thickness of the transmission sample.

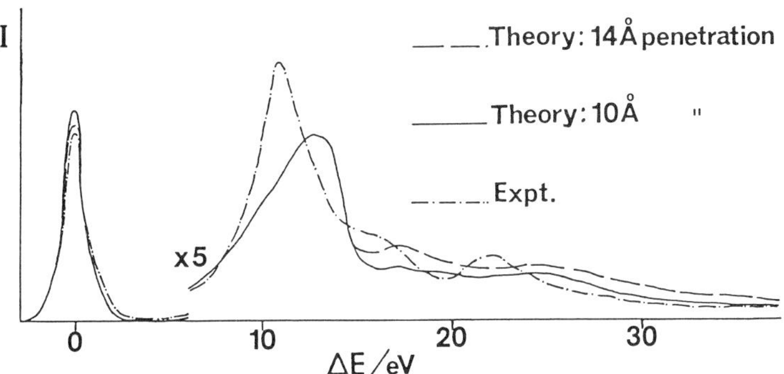


Fig.3. Si(666) reflection E.E.L.S.

A somewhat more detailed check on the theory is provided by Si where both the surface and bulk plasmon peaks are considerably sharper and more intense than in GaAs and thus exhibit rather well defined multiple loss peaks. Figs. 3 and 4 show experimental data obtained by Krivanek et al 1983 for an Si (111) surface at the 666 and 222 Bragg conditions respectively. In these experiments a rather small effective aperture in the

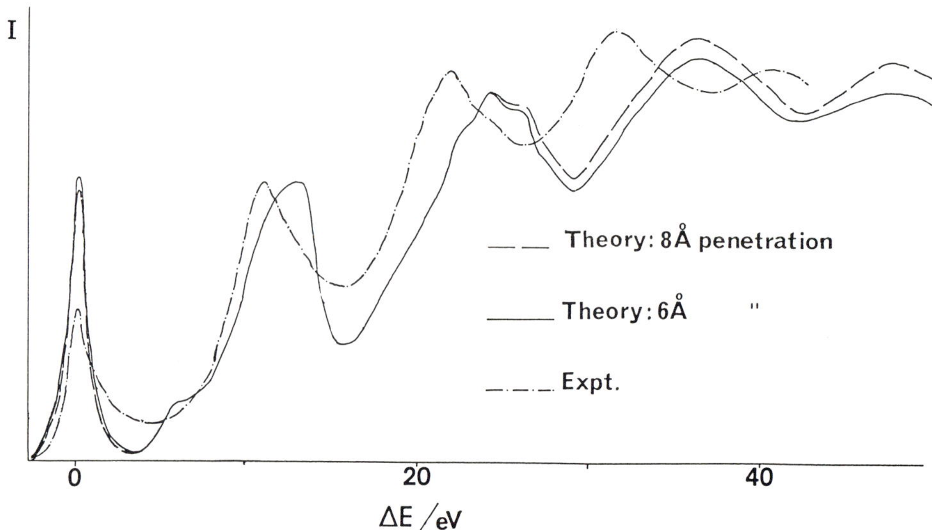


Fig.4. Si(222) reflection E.E.L.S.

range $10^{-3}$ to $10^{-4}$ rad was employed, tending to suppress the bulk peaks relative to the surface plasmon peaks. The results have been compared with the theory for a number of different aperture sizes and penetration depths. In Si somewhat smaller beam path lengths of 10nm or less seem to occur inside the specimen but the estimate is rather unreliable because of the lack of information about the precise aperture used. Clearly there is also considerable scope here for improving the fit to the observations by modifying the dielectric data. In particular the surface plasmon peak should be lowered and sharpened.

## 4. Discussion

It appears that the classical dielectric theory is capable of giving quite a good quantitative fit to glancing angle loss spectra particularly if good dielectric data are available. Indeed this mode of spectroscopy may be quite a sensitive test of such data. Investigation of the results shows that even at 100 keV, relativistic effects can be quite important particularly in the low loss region where Cherenkov losses can occur. The theory used here which simplifies the relation between multiple scattering and scattering angle may not be completely adequate but does seem to reproduce several of the observed effects. The evidence suggests that in the glancing angle mode the effective beam path in the specimen may be 100nm or more and thus adequate to generate useful characteristic losses for microanalysis.

We are grateful to the SERC for financial support.

## 5. References

Garcia-Molina, R., Gras-Marti, A., Howie, A. and Ritchie, R.H. 1985 J. Phys. C 18 (in press).

Howie, A. and Milne, R.H. 1984 J. Microsc. 136, 279.

Howie, A. and Milne, R.H. 1985 Ultramicroscopy (in press).

Krivanek, O.L., Tanishiro, Y., Takayanagi, K. and Yagi, K. 1983 Ultramicroscopy 11, 215.

Milne, R.H. and Echenique, P.M. 1985 Solid State Comm. 55, 909.

# Computation of lens shape with minimum spherical aberration coefficient in TEM

K Tsuno* and K C A Smith

University Engineering Department, Trumpington Street, Cambridge CB2 1PZ.
* Adress from March 1986: JEOL Ltd., Akishima, Tokyo 196, Japan.

## 1. Introduction

The shape and optical properties of an objective lens determine the main characteristics of the electron microscope. Specimen holder, goniometer, and in-situ observation equipment limits the shape of the pole-pieces and therefore influences the resolution of the microscope.

To design objective lenses with a variety of requirements, computer-aided design methods are very useful. The finite element method (FEM) is a familiar computer-aided method in mechanical and electrical engineering and was first introduced in electron optics by Munro (1971). The use of the computer-aided design method is wide spread, but the method for generating the mesh for FEM requires a lot of experience and it is difficult to find the optimum shape of lens because FEM and numerical electron trajectory calculation only give field distributions and aberrations for the given shape of lens and do not suggest which shape is the optimum.

In this paper, an improved program is described which can automatically change the pole-piece shape and search for the optimum shape of pole-piece for lenses with various requirements.

## 2. Lens parameters influencing optical properties

There are 8 important parameters in the pole-piece of a magnetic-electron lens: gap length (S1+S2), specimen position measured from the lower pole face (S2), taper angle of upper and lower pole-pieces (G1, G2), top face radius of both pole-pieces (D1, D2) and bore radius of both pole-pieces (B1, B2). Of course, there are other parameters which sometimes strongly influence the optical properties of the lens, such as inner or outer yoke radius, pole-piece radius, but these parameters are mainly concerned with magnet technology rather than electron optics.

## 3. Optimizing the pole-piece shape

Table I shows a diagram of the computation. First, two data tables are read: one is the numerical table of parameters of pole-piece geometry (Table II). In this example, S1+S2 is constant in all of the calculation and the specimen is located at the center of the gap (S2 is constant). The second data table is the initial coarse mesh of the objective lens and magnetization curves of materials used in the lens. The second step is to create a new coarse mesh according to Table I. Fig. 1 shows 5 coarse meshes in a pole-piece region where B2 is changed from 0.75 mm to 9 mm. The ratio D2/B2 has a constant value of 2 in this example. In every case, only the pole-piece region of the mesh is changed.

The 3rd step is to see if the new lens shape is already calculated or not. All data of optical properties and pole-piece parameters which previously calculated are stored in the file. If the same lens shape has already been calculated, calculation is skipped. The 4th step is to generate the fine mesh which is used for the FEM calculation. The spacings between lines of fine mesh are logarithmically increased (Hill and Smith 1982) from the gap region of the pole-piece. The creation of coarse and fine meshes is made so as to avoid a sudden change of mesh spacing. The 5th step is to calculate the axial field distribution using FEM, and from this, the coefficients Cs, Cc and focal length f0 are calculated.

At the 6th step, the specimen position is checked. If the Z0 is greater than 0 (Z0=0 is the pre-set specimen position), excitation is increased. If Z0<0 for the first calculation, excitation is decreased and FEM calculation is repeated. If Z0 becomes lower than 0, optical properties at Z0=0 is interporated. This process corresponds to the focussing of the actual microscope by changing the excitation of the objective lens. Optical properties at Z0=0 are stored in a file of the computer and then one of the parameters is slightly changed according to Table II. The above process is repeated, and then obtained value of Cs is compared with the Cs of the previous calculation. If the Cs of the second calculation is lower than that of the previous calculation, the value of the parameter is changed again. If Cs has greater value than the previous Cs, the previous coarse mesh is replaced with the first coarse mesh, and then the value of another parameter is changed.

If all the parameters are changed, the resulting shape is an optimum for the given initial conditions. Figs. 2 and 3 show values of pole-piece parameters changed according to Table II, and values of Cs and Cc for corresponding pole-piece shapes. First, B2 is changed without changing D2/B2. The Cs has a minimum at B2=1.2 mm. So that, for the further calculation, B2 is constant at b2=1.2 mm and D2 is increased. The Cs then decreased with increasing of D2. Then the maximum value of D2=7 mm is chosen for further steps, and G2, B1, D1 and G1 are also changed. Finally, B1=0.6, B2=1.2, D1=0.9, D2=8.4 mm, G1=60 and G2=55 deg. are determined. As a result, Cs=0.58 mm and Cc=1.02 mm are obtained at 200 kV.

For finding another minimum of Cs, the calculation was repeated using the first optimum shape obtained by the first cycle. If the minimum value of Cs obtained by the second cycle is nearly the same as the first cycle, the calculation stops. In this example, repeated calculation gave a minimum Cs value of 0.57 mm, therefore, the calculation order given in Table II was reasonable. But, when the calculation was started from B1, D1, G1 to B2, D2, G2, the minimum Cs value obtained was 0.6 mm. Thus, the minimum value of Cs obtained is influenced by the order which parameter is first changed.

References

Hill R and Smith K C A 1982 Inst. Phys. Conf Ser. 61 71-74
Munro E 1971 Ph. D. Dissertation, Cambridge University

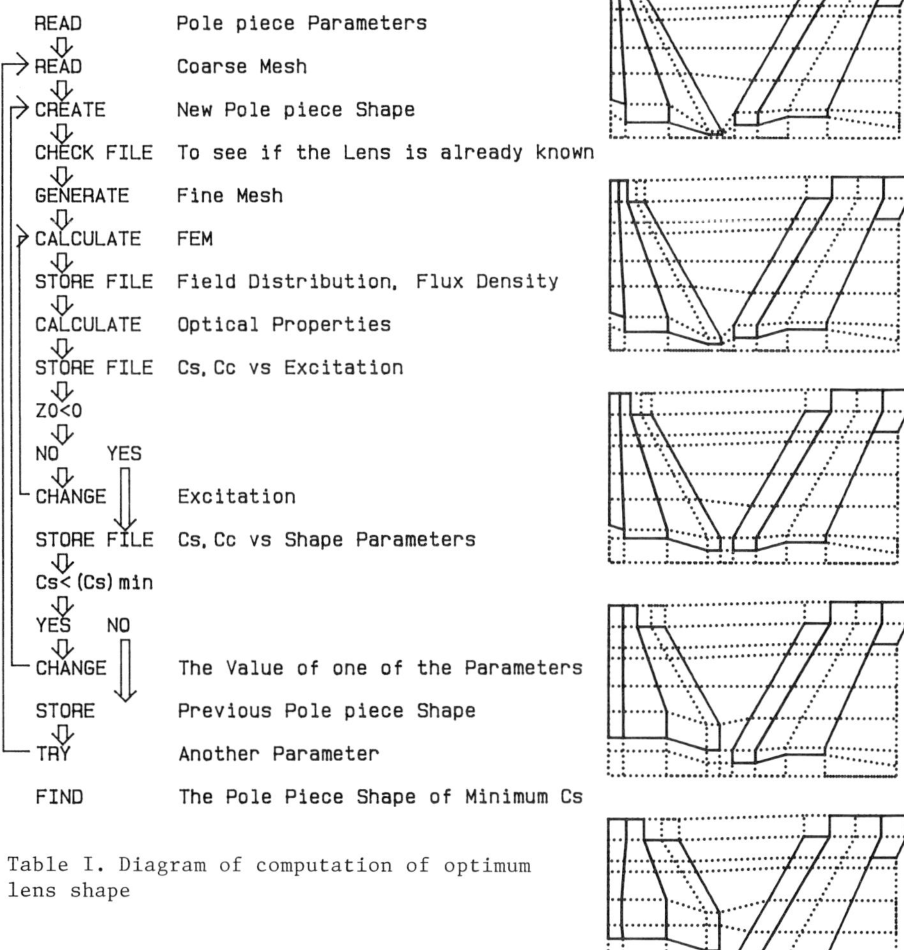


Table I. Diagram of computation of optimum lens shape

Fig. 1. Automatic change of lower bore radius B2

| | | 1 | 2 | 3 | 4 | 5 | |
|---|---|---|---|---|---|---|---|
| 1 | B2 | 0.6 | 1.2 | 1.8 | 2.4 | 3.0 | mm |
| 2 | D2 | 1.5B2 | 2B2 | 3B2 | 5B2 | 7.5B2 | |
| 3 | G2 | 70 | 60 | 55 | 50 | 45 | deg. |
| 4 | B1 | 0.6 | 1.2 | 1.8 | 2.4 | 3.0 | mm |
| 5 | D1 | 1.5B1 | 2B1 | 3B1 | 5B1 | 7.5B1 | |
| 6 | G1 | 70 | 60 | 55 | 50 | 45 | deg. |
| 7 | S1 | 1.5 | 1.5 | 1.5 | 1.5 | 1.5 | mm |
| 8 | S2 | 1.5 | 1.5 | 1.5 | 1.5 | 1.5 | mm |

Table II An example of data table for pole-piece parameters

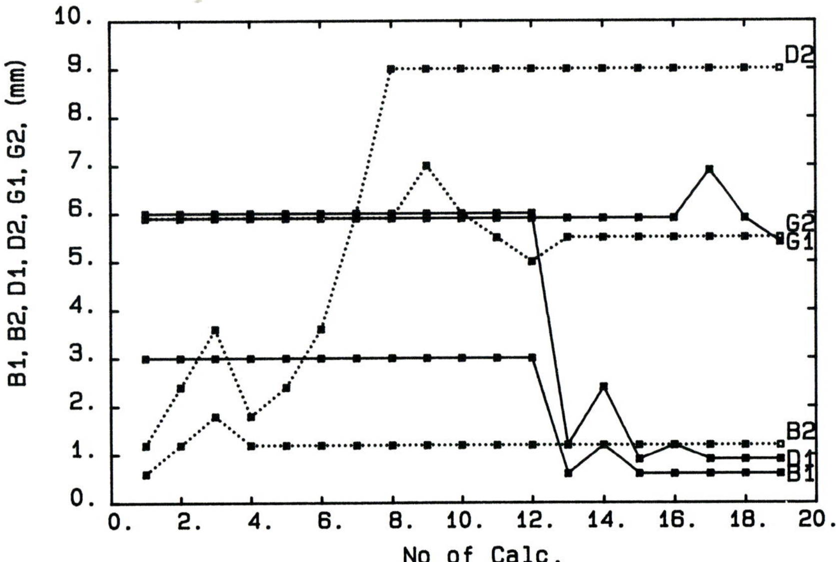


Fig. 2 Values of pole-piece parameters changed according to Table II

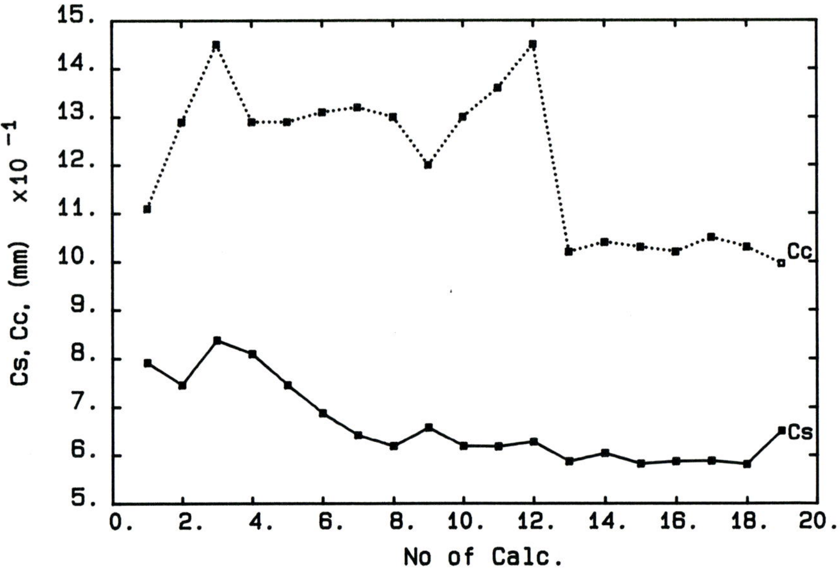


Fig. 3 Calculated values of Cs and Cc of pole-pieces with parameters shown in Fig. 2

*Inst. Phys. Conf. Ser. No 78: Chapter 5*
*Paper presented at EMAG '85, Newcastle upon Tyne, 2–5 September 1985* 

# Computations on the electron-optical parameters of saturated objective lenses

I S Al-Nakeshli and S M Juma*

Department of Mathematics and Physics, Aston University, The Triangle, Birmingham B4 7ET, UK
*Department of Physics, College of Science, University of Baghdad, Jadiriyah, Baghdad, Iraq

Abstract. Computations have been carried out on the effect of the energizing coil on the magnetization of saturated magnetic electron lenses. The objective focal properties have been computed as function of the current density, which is the limiting factor. Designs of single and double polepiece lenses of low aberration coefficients have been suggested for high resolution and high voltage electron microscopes.

## 1. Introduction

The design of objective electron lenses of low aberration coefficients is of considerable interest in various fields of electron optics. In order to reduce aberrations, the magnetic electron lens should produce the highest possible flux density peak consistent with the smallest halfwidth of the axial field distribution. Computations of Al-Nakeshli et al (1984a,b) have shown that a favourable pole shape in single and double polepiece lenses is in the form of a hemisphere or a spherical end cap joined to a conical section. With such pole shapes of soft iron, various coils have been investigated for magnetizing the lenses in order to achieve the parameters which satisfy the requirements for a high resolution objective lens. The single and double polepiece lenses whose magnetization is investigated in this paper are assumed to have the same outer diameter and polepiece dimensions. The pole face is rounded off in the form of a hemisphere.

## 2. Polepiece Magnetization

At any point on the lens axis the magnetic field $B_z$ is the sum of the flux density $B_c$ due to the energizing coil plus the flux density $B_{Fe}$ due to the iron magnetization. The maximum contribution of polepiece magnetization to $B_z$ is 2 tesla (saturation flux density of soft iron). Figure 1 shows the magnetization curves of such a double polepiece lens as energized by four different coils. Here only the upper half of the symmetrical lens is shown. In calculating the magnetization, the polepiece gap was 8 mm and the bore diameter was negligibly small. Ideally, the pole tip should reach magnetic saturation for the lowest value of field strength H. It is seen that dB/dH is highest when the lens is energized by two identical coils surrounding the polepieces and placed in the conventional position (see lens 1). Therefore, such a coil shape and position is favourable as an excellent magnetizer. The main disadvantage of this arrangement, however, is that it leads to premature saturation in the conical part of the polepiece that

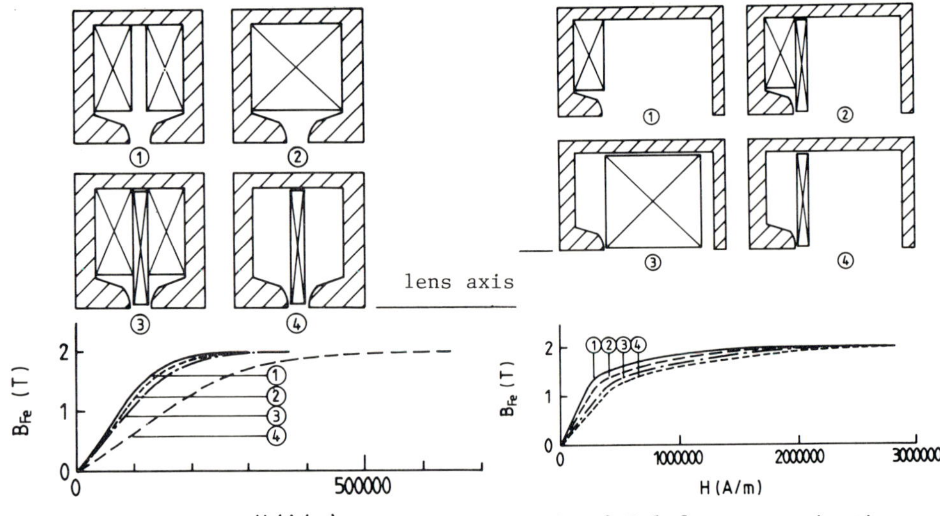


Fig. 1 Poleface magnetization curves of a double polepiece lens energized by four different coils.

Fig. 2 Poleface magnetization curves of a single polepiece lens energized by four different coils.

lies in the region of the centre plane of the coil well before the pole tip. In addition, the peak of the axial field is not a maximum at the centre of the air gap but two peaks appear near the tips of the two poles. Although a thin coil of large mean diameter is not a good magnetizer (lens 4), it has the advantage, however, of producing the highest maximum value of flux density at the centre of the air gap and the smallest halfwidth of the axial field $B_z$ compared with those of other coils. In order to combine the advantages of the two types of magnetizers mentioned above, the coils have been arranged to energize the double polepiece lens (lens 3). This combination produces a magnetization better than that of lens 4 and a smaller halfwidth and higher flux density at the air gap centre than those of lenses 1 and 2.

For single polepiece lenses of zero bore (figure 2), the above argument also holds. The favourable coil arrangement is the combination shown in lens 2. The field halfwidth is smaller and its peak, situated at the pole face, is higher than those of lenses 1 and 3.

## 3. Objective Focal Properties

The choice of a favourable lens design in an electron microscope depends on its objective focal properties. Once the polepiece is saturated, any further increase in the flux density must come from the coil. The limiting factor to the total flux density is therefore the current density $\sigma$ that can be supported by the energizing coil. The value of $\sigma$ will determine whether the windings must be conventional or superconducting (Mulvey 1982). The objective focal properties of the lenses shown in figures 1 and 2 have, therefore, been computed in terms of $\sigma$. The computations have been carried out for an electron beam entering the lens field parallel to the optical axis with an energy of 1 MeV (i.e. relativistic voltage $V_r$ = 2 MV). The beam intersects the optical axis at the centre of the double polepiece lens (telescopic mode of operation) or at the poleface of the single

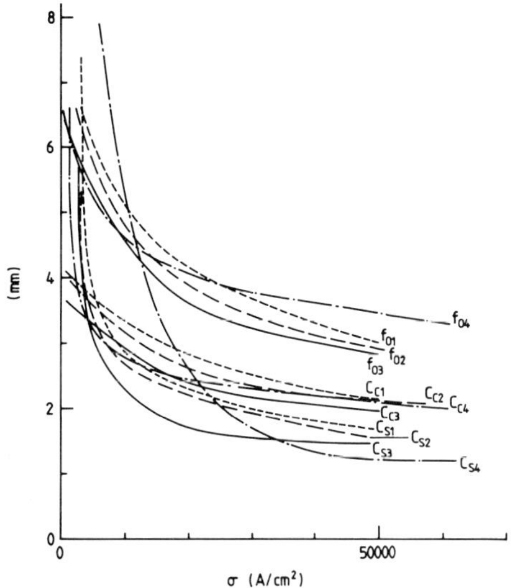


Fig. 3 Objective focal properties of 1 MV double polepiece lenses as function of current density σ .

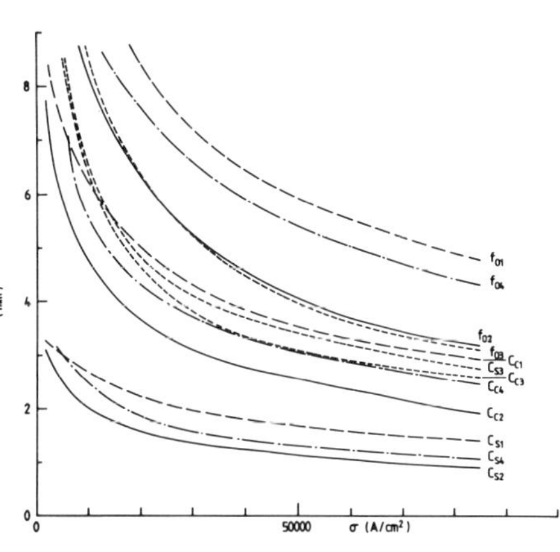


Fig. 4 Objective focal properties of 1 MV single polepiece lenses as function of current density σ .

polepiece lens. Figure 3 shows the objective focal properties of the 1 MV double polepiece lenses as function of σ . Of the four lenses, lens 3 of figure 1 has the shortest objective focal length $f_o$ and the smallest chromatic aberration coefficient $C_c$. Lens 3 has also the smallest spherical aberration coefficient $C_s$ provided $\sigma < 35000$ A/cm$^2$. Thus the favourable coil combination of lens 3 can be operated by superconducting windings to achieve low focal properties. For higher values of σ , the field of the thin superconducting coil in lens 4 is dominant over $B_{Fe}$ and hence provides smaller values of $C_s$ than those of lenses 1, 2 and 3.

Of the four 1 MV single polepiece lenses, lens 2 in figure 2, energized by a combination of two coils, has the lowest values of $C_c$ and $C_s$ as shown in figure 4. The advantages of both coils have, therefore, combined to produce low aberrations under the conditions of iron saturation. For a constant σ , the maximum flux density $B_m$ of lens 3 in figure 1 at the centre of its air gap is higher than that of lens 2 in figure 2 at its pole tip. However, the halfwidth of the axial field of the single polepiece lens is smaller than that of the double polepiece lens. Due to the characteristics of their axial fields, $C_s$ of the single polepiece lens is smaller than that of the double polepiece lens but $C_c$ of the double polepiece lens is smaller than that of the single polepiece lens. Both lenses are therefore capable of yielding low aberrations and the choice of a favourable lens will depend on the operational requirements; for example, in biological microscopes $C_c$ is very important. It should be noted that in designing a double polepiece lens there are many variables need to be considered such as gap width and bore size which would require further investigations.

## 4. Objective Lens Design

Considering the favourable designs of single and double polepiece lenses, their computed focal parameters can be scaled to achieve a lens suitable for operation at other beam voltages. For an accelerating voltage of 1 MV and keeping $\sigma$ constant at 20000 A/cm$^2$, the dimensions and the focal parameters of lenses 3 and 2 in figures 1 and 2 respectively have been scaled and produced the lenses of figure 5. They can be fabricated in practice and their small size provides many advantages. The parameters of the single polepiece lens excited at 25800 ampere-turns are: $f_o$ = 6.3 mm, $C_c$ = 3.66 mm, $C_s$ = 1.57 mm and $B_m$ = 3.2 T while those of the double polepiece lens at 32100 A-t are: $f_o$ = 3.6 mm, $C_c$ = 2.4 mm, $C_s$ = 1.7 mm and $B_m$ = 4.6 T. In this particular case, the single polepiece lens is optimized for $C_s$ and the double polepiece lens is optimized for $C_c$. However, $C_s$ of the double polepiece lens can be improved at the expense of $C_c$. The resolution parameter $\delta$($= 0.7(C_s\ \lambda^3)^{1/4}$, $\lambda$ being the electron wavelength) of the single polepiece lens is 0.125 nm which is very slightly better than that of the double polepiece lens (0.128 nm). For the same $\sigma$ , a scaled down version of the lenses shown in figure 5 indicates that it is possible in practice to construct them for 500 kV accelerating voltage and preferably energized by superconducting coils.

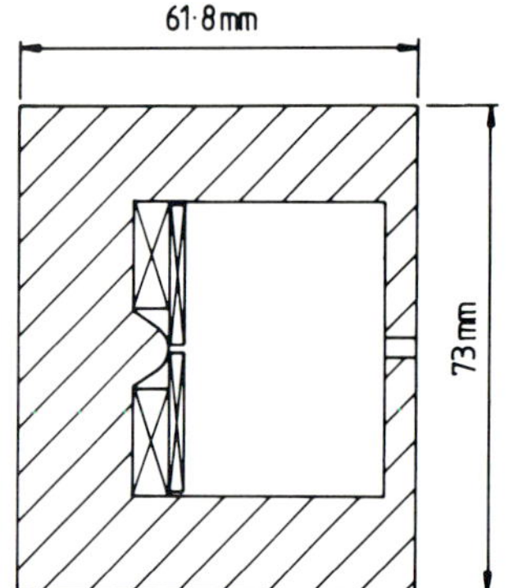


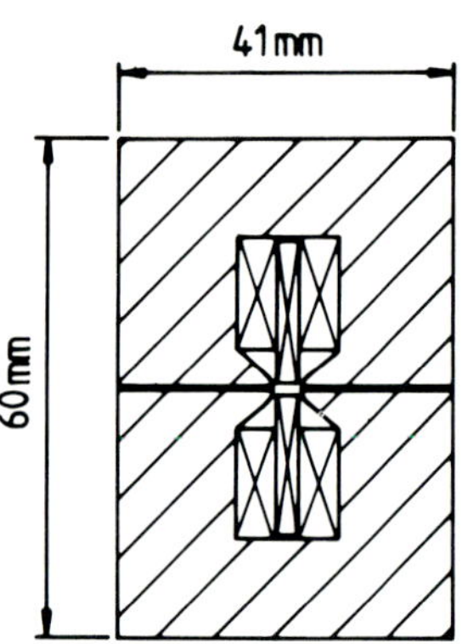


Fig. 5 Suggested designs of saturated 1 MV single and double polepiece objective lenses operated at a current density of 20000 A/cm$^2$. The lenses have thick iron shrouds to eliminate flux leakage.

## 5. Conclusions

It appears that the design of the magnetizing coil is important in saturated magnetic electron lenses. A combination of the optimum designs of both polepiece and coil suggests that high flux density objective lenses can lead to better resolution in present high voltage electron microscopes.

## 6. Acknowledgment

The authors are very grateful to Professor T. Mulvey of Aston University for reading the manuscript and for the helpful and stimulating discussions.

## 7. References

Al-Nakeshli I S, Juma S M and Mulvey T 1984a Electron Microscopy and Analysis 1983 ed P Doig (Inst. Phys. Conf. Ser. No. 68) pp 475-8

Al-Nakeshli I S, Juma S M and Mulvey T 1984b Electron Microscopy 1984 vol 1 eds Á Csanády, P Röhlich and D Szabó (Budapest: Programme Committee of the 8th Eur. Cong. on Elec. Micros.) pp 21-2

Mulvey T 1982 Magnetic Electron Lenses ed P W Hawkes (Berlin: Springer) pp 359-412

*Inst. Phys. Conf. Ser. No 78: Chapter 5*
*Paper presented at EMAG '85, Newcastle upon Tyne, 2–5 September 1985*


# Matrix derivation of aberration shift relations

P W HAWKES

Laboratoire d'Optique Electronique du CNRS, BP 4347 - F-31055 TOULOUSE CEDEX.

## I. Introduction.

The computer-aided design of complex electron optical systems proceeds in several stages : typically, the optical properties of the various elements of the system are established and then those of the entire column are obtained by suitable addition. The characteristics of the individual elements are then varied until the column possesses the desired properties and certain practical constraints are satisfied - the individual settings must not be too critical, for example.

The variations of the characteristics of each element are of two kinds. The geometry may be altered, in which case the calculation of the new optical properties may be complicated, requiring a field calculation and trajectory tracing unless a preliminary survey based on field or potential models is in question. Alternatively, the working conditions of the element may be altered : the object position, the position of the aperture, the distance from neighbouring elements and the operating strength when the relation between this and the optical characteristics is simple. Here we are particularly concerned with variations of the second kind though in a sophisticated design program, provision must of course be made for both kinds.

Provided that neighbouring elements are sufficiently far apart for their fields to overap only negligibly, it is convenient to represent each member of the sequence by a matrix showing the transformation effected by the element on incoming rays. The choice of matrix is not unique and we now consider the various possibilities. We should however stress that the limitation to non-overlapping fields should not be an unduly severe restriction. It certainly permits us to establish the behaviour of a complex system and the trends associated with the various principal configurations; furthermore, if the overlap is severe or indeed an intrinsic part of the design, as in the case of superimposed deflection and round lens fields, then a single matrix characterizes the multiple functions of the element, though at the cost of much recalculation when the parts are altered (when the relative position of the deflectors and polepieces is altered, in the case mentioned above).

## II. Choice of matrices.

The transfer matrix of an optical element - lens, prism, deflector - characterizes the coupling between object and image space for that element and should include the effect of paraxial focusing and primary aberrations. There are numerous matrices that encode this information, differing in the way in which the aberrations are represented and in the planes in object and image space between which the coupling is analysed. The most important difference in pratice is the first; the choice of planes is largely a matter of convenience. Aberrations may be expressed in terms of position and gradient in some object plane, in which case the transfer matrix characterizes the relation between these quantities and the powers and multiples of them that occur in the aberrations in object space and in image space. These quantities are organized as a vector with paraxial properties first, giving the transfer matrix a typical block form :

| | | | | |
|---|---|---|---|---|
| x<br>x'<br>y<br>y' | = | paraxial transfer matrix | aberration matrix for position and gradient | x<br>x'<br>y<br>y' |
| aberrations | | null | addition matrix coding rules for adding aberrations, generated by the paraxial transfer matrix | aberrations |

The column vector on the left represents quantities in image space, that on the right in object space.

Alternatively, aberrations may be expressed in terms of object and aperture coordinates, an important point when the aperture position is to be regarded as a free parameter. The paraxial zone of the object and image vectors is now occupied by two pairs of position coordinates, those in the object plane and those in the entrance pupil plane for the former and those in the image plane and those in the exit pupil plane for the latter. The block structure illustrated above is maintained but the paraxial transfer matrix now simplifies and in most cases of practical interest, becomes diagonal. There is a corresponding simplification, frequently a diagonalization, of the addition matrix.

So far as the choice of planes is concerned, the natural choice for an imaging system in which the conjugacy between various pairs of planes is of interest, is a pair of conjugates or, when aperture coordinates are used, two pairs of conjugates, the obvious choice being those mentioned earlier. The question that thus arises is how the matrix elements vary with these choices of planes : in particular, how do the elements of the aberration matrix vary with the positions of the object and image planes and of the entrance and exit pupils ? There are two quite different ways of answering this question.

## III. Shift relations for the elements of aberration matrices.

We shall not go into the details of the matrix calculations here, these are dealt with fully for round lenses in two companion papers (Hawkes, 1984 a, 1985). Our purpose is to bring out the underlying structure of the calculation to show that the technique can be straightforwardly applied to most optical elements once the basic relations have been established : quadrupoles, deflection units, prisms and indeed any optical element for which the coupling between object and image space is asymptotic.

The first step is to establish the vectors that characterize the paraxial relations between the two spaces and the associated aberrations. We have so far considered only the primary (third-order) aberrations of round lenses and quadrupoles (the latter not by the algebraic approach) ; this should enable the study of projector lens combinations to be put on a systematic footing, though it is not very likely that any results not already anticipated by Mulvey and Tsuno will emerge (Mulvey, 1982, 1984 ; Tsuno 1981, 1982, 1984). The next task with important practical application concerns the optical elements with overlapping deflection fields and focusing fields used in microlithography machines, where a very complex sequence is required.

The algebraic principle is most easily explained by considering round lenses alone ; we assume that the aberrations are expressed in terms of position in the object and entrance pupil planes. The input and output vectors in complex notation then have two paraxial elements, $x + iy$ and $p + iq$, where $p$ and $q$ are cartesian coordinates in the pupil planes and $x$ and $y$ are cartesian coordinates in the object and image planes (with appropriate suffices to distinguish these).

Suppose now that the object and hence the image plane is shifted, altering the magnification from $M$ to $\bar{M}$ and that the entrance pupil and hence the image pupil is also shifted, so that the pupil magnification is changed from $P$ to $\bar{P}$. A shift matrix $S_o$ will relate the unshifted $w_o = x_o + iy_o$ and $t_o = p_o + iq_o$ to their shifted counterparts $\bar{w}_o$ and $\bar{t}_o$. This shift matrix is extended to include the aberrations and hence has the block form already illustrated. It does not, however, introduce any new aberrations and the aberration sub-matrix (upper right) is thus null :

$$S_o = \left[\begin{array}{c|c} S_1^{(o)} \text{ (paraxial)} & S_2^{(o)} \text{ (aberrations)} \\ \hline S_3^{(o)} & S_4^{(o)} \text{ (addition formulae)} \end{array}\right]$$

with $S_2 = S_3 = 0$. The important elements are those of $S_1$, from which the matrix $S_4$ may be derived. We find

$$(\bar{p} - \bar{m})\, S_1^{(o)} = \begin{pmatrix} p - m & m - \bar{m} \\ \bar{p} \quad p & \bar{p} \quad \bar{m} \end{pmatrix}$$

and similarly,

$$(P - M)\, S_1^{(i)} = \begin{pmatrix} P - \bar{M} & \bar{M} - M \\ P - \bar{P} & \bar{P} - M \end{pmatrix}$$

We now use these object and image shift matrices to establish the aberration coefficients for the new values of the two magnifications. It is easy to show that the aberration matrix after shifting the object and entrance pupil planes is given by

$$S_1^{(i)}\, M_2\, S_4^{(o)}$$

where $M_2$ is the aberration sub-matrix of the lens system. The shift matrices do not of course relate conjugate planes and are therefore less sparse than the lens matrices but the necessary matrix multiplication is nevertheless not very complicated. Explicit formulae for the results are given in the papers already cited.

IV. Manipulation of aberration integrals.

The alternative to this algebraic method is to examine the aberration integrals directly. The integrals traditionally include field or potential functions of the paraxial equations. By choosing these linearly independent solutions in such a way that they are independent of the operating conditions, we bring out the aberration structure, the magnification(s) appearing explicitly, so that the coefficients can be evaluated straightforwardly for any situation (Hawkes, 1968, Ade, 1982, Hawkes, 1984 b). Such solutions are typically those that enter and emerge from the lens parallel to the optic axis. This approach has been explored in great detail and we shall say no more about it here. We simply confirm that the two methods do of course lead to identical results.


. Ade G 1982 Optik 63 43

. Hawkes P W 1968 Optik 27 287

. Hawkes P W 1984 a Ultramicroscopy 15 227

. Hawkes P W 1984 b Optik 66 379

. Hawkes P W 1985 Optik (to be published)

. Mulvey T 1982 In Magnetic Electron Lenses (P.W. Hawkes, ed.) 359 (Berlin : Springer).

. Mulvey T, (1984), In Electron Optical Systems, (J.J. Hren, F.A. Lenz, E. Munro and P.B. Sewell, eds) 15 (AMF O'Hare : SEM).

. Tsuno K 1984 J. Phys. E : Sci. Instrum. 17, 1038.

. Tsuno K and Harada Y 1981 , J. Phys. E : Sci. Instrum. 14, 955.

. Tsuno K and Harada Y 1982, Proc. 10th Int. Conf. Electron Microscopy Hamburg 1, 301.

# A miniature variable energy electron probe for nanometric analysis

M M El Gomati and M Prutton

Department of Physics, University of York, Heslington, York, YO1 5DD, U.K.

## 1. Introduction

Field emitters have been successfully used over the last 10-15 years as either the electron or ion source of a microprobe (El Gomati et al 1985, Cleaver et al 1983). The drawback to their use, however, has been their relatively high flicker noise as electron emitters in addition to a high energy spread that contributes unfavourably to the chromatic aberration component in the focussed beam. For electrons, the energy spread of thermally assisted field emitters can approach a value of 1-3 eV (Cumming and Smith 1979), while typical values for the liquid metal ion sources can be of the order of 6-10 eV (Swanson et al 1979). In conventionally designed electron probe lenses, the interest is in minimisation of the spherical aberration contribution to the focussed beam. However, when the same lens configurations are adopted in the case of field emitter sources, larger beam diameters usually arise because of the chromatic aberration. Therefore it helps greatly to attempt lens designs that take into account such a disadvantage of field emitters in order to fully exploit their many other advantages. With this philosophy in mind, a new miniature variable energy electron probe for nanometric analysis has been designed and built at York. The system is based on a similar five element configuration to that realised by Veneklasen et al (1979). Electron optical properties of this system are reported here. It is found that an electron beam current of about 10 nA could be focussed into a spot diameter of less than 20 nm at 15 kV using a thermally assisted ZrO/W(100) field emitter and a total emission current of 100 uA. Such a spot diameter is largely limited by the intrinsic source size of the emitter assumed to be of the order of 10 nm (Swanson 1980). Higher beam currents in a slightly larger spot diameter are also feasible.

## 2. Calculation

Since the description by Butler in 1966 of a first electron lens to be used with field emitters, several authors have reported other lens shapes and configurations specifically suited to the field emitter (Kuroda et al 1974, Riddle 1978). The majority of these studies have focussed attention on reducing the spherical aberration in comparison to the chromatic aberration. These lenses could also be adopted for ion probes, but even a greater contribution of chromatic effects to the probe diameter will result.

In the present system, which consists of two lenses, a gun lens and an

objective lens, the aim was to reduce the chromatic aberrations of both lenses. This task was made easier by referring the aberration coefficients into object space (i.e. the source of electrons). In this case, the diameter of the circle of least confusion due to the chromatic aberration would be expressed as a function of the extract voltage of the emitter assembly and the accelarating voltage ratios between the final beam energy and that of the emitted electrons (El Gomati 1985). A number of triode and tetrode lenses exhibitng small chromatic aberrations were considered. A further criterion for these lenses was that the beam current should be of the order of 10-100nA. Higher beam currents at the same total electron emission will increase the spherical aberration effects. All estimates of beam currents and diameter are based on the use of either a ZrO/W(100) or a built-up W(100) emitter both operated in a thermally assisted mode with total emission current of 100 μA.
A tetrode type lens was chosen for the gun and a triode type for the objective. The first of these two is used to accelerate and collimate the electrons entering the objective lens. It has a variable focal length that can be changed via a control electrode. The objective lens function is to focus the beam at a distance $S_i$ away from the last element of the column, i.e. the column working distance. A value of $S_i$=10 mm is selected in the present design because of other constraints in the system. It should be noted that the chromatic and spherical aberrations of the objective lens will change for a different $S_i$ value. The optical properties of the lenses were calculated by means of the finite element method (Munro 1975). Details of the method of addition of the contributions of the different aberrations to the focussed electron beam size are given elsewhere (El Gomati et al 1985).

## 3. Results

Fig (1) is a schematic of the column depicting its various components. A block diagram of the electrical supplies is also shown. This is a purpose built supply consisting of four 20 kV high stability supplies (better than 1 part in $10^4$) driven in a master/slave fashion, and a heater supply isolated to 20 KV for the operation of the field emitter. A second supply provides the deflection and astigmatism correction voltages, a raster generator and an SEM head-amplifier. The whole column is 10 cms long and 5 cms in diameter and can be easily attached to any existing UHV system. Details of assembly and alignment will be published elsewhere (El Gomati and Prutton 1985). In Fig. 2 a, b, are shown the electron optical properties of the Gun and objective lens respectively. In the case of the Gun lens, the aberration coefficients are referred to object space. As can be seen from the figure, the choice of the working distance in the case of the objective lens or the emitter to emission electrode distance in the case of the Gun lens determines the values of the aberration coefficients. These are low for small values of $S_i$ and $S_o$ and increase for larger values. However, this increase is slow for the chromatic aberration which is the dominant contributor to the spot diameter. In Fig. 3 an estimate of the beam diameter versus the elctron energy is given for the present geometry in the case of a constant beam current of 10 nA. For a low electron beam energy, the lenses are weak and the focussed spot will have a large diameter dominated by the chromatic aberration. For electron energies above ca 5 keV, a beam diameter of less than 20 nm should be obtained under the above conditions. This latter region of operation has the emitter's source size (10 nm) as the largest single contributing factor to the focussed electron beam diameter.

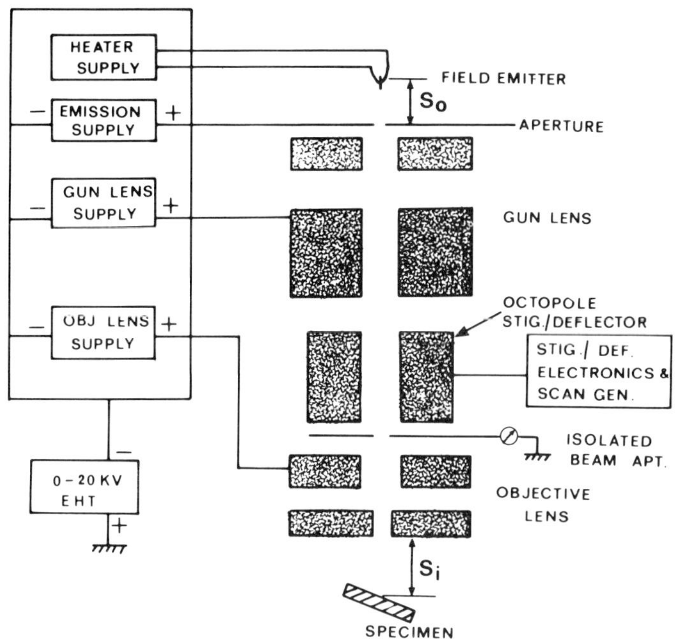


Fig 1. Schematic of the mini field emission column.

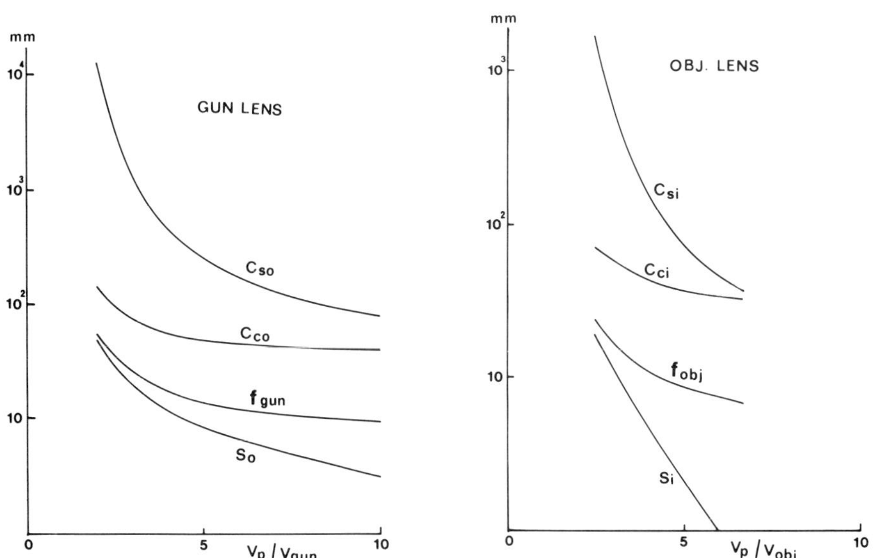


Fig 2. The electron optical properties of the column for:
(a) the gun lens as a function of $V_p/V_{gun}$.
(b) the objective lens as a function of $V_p/V_{obj}$, where $V_p$ is the final voltage, $V_{gun}$, $V_{obj}$ are the control voltages of the gun and objective electrodes and $f_{gun}$ and $f_{obj}$ are their respective focal lengths.

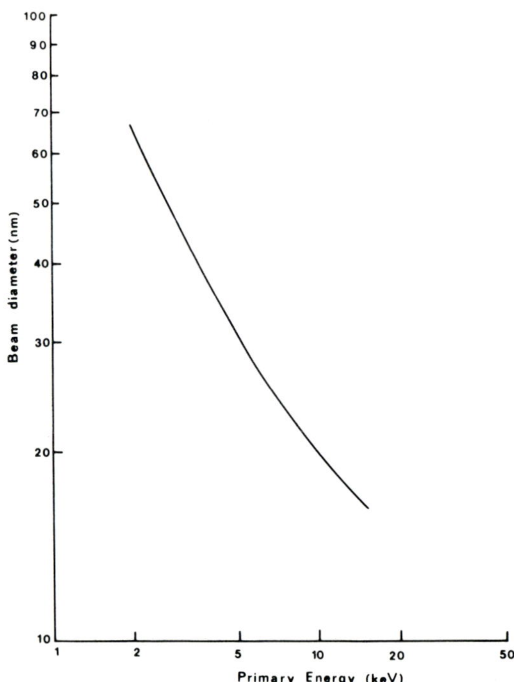


Fig.3 The relation between the probe diameter and primary energy of the mini field emission column for $S_i$ = 10 mm. The calculation is for an intrinsic source size of about 10 nm, and an energy spread of 0.6 eV.

This column has now been built and assembled and preliminary tests completed. The results of beam current and diameter in a test chamber having a large working distance are encouraging and compare well with the above theoretical estimates. The voltages required to focus the column agreed well with the theoretical focal-loci. A fuller theory/experiment comparison will be published later.

Acknowledgements

We would like to thank the SERC for financial support of this project and Dr L H Veneklasen of Perkin-Elmer Corporaton, Hayword, California for discussion and convincing demonstration of his miniature field emission column.

References

Butler J W. 1966, Proc. 6 Int. Conf. Elect. Mic. Koyoto, Japan, p191.
Cleaver J R A and Ahmed H. 1985, J. Vac. Sci. and Tech. 3, (1) p144.
Cumming A D C and Smith K C A, 1979, Microcircuit Eng. Cambridge: CUP,
El Gomati M M, 1985, Optik. In Press.
El Gomati M M, Browning and Prutton M, 1985, J. Phy. E, 18, p32-38.
El Gomati M M and Prutton M, 1985, to be published.
Kuroda et al, 1974, J. App. Phys. 45, 5.
Munro E, 1975, Cambridge Univ. Report, CUEDIB-ELECT. TR45.
Riddle G H N, 1979, J. Vac. Sci. and Tech. 15, p857.
Swanson L. W et al, 1979, J. Vac. Sci. and Tech. 16, (6) p1864.
Veneklasen L H et al Proc 9th Int. Cong. Elect.Microsc. Toronto, p12.

# The spectrum of cathodoluminescence from thin films in STEM

S D Berger, D McMullan, J Yuan and L M Brown
Cavendish Laboratory, Madingley Road, Cambridge. CB3 OHE

## 1. Introduction

A variety of cathodoluminescence (CL) detectors have been built for use in scanning transmission electron microscopes (STEMs) and used to study radiative recombination at individual defects in semiconductors (Petroff et al 1978, Pennycook et al 1978, 1981, Roberts 1981). Apart from those due to Pennycook et al (1978, 1981) all the designs have been used in conventional TEMs, with scanning attachment, producing CL images having a spatial resolution of typically ~ 0.2 μm (Myhajlenko et al 1985) and high quality CL spectra. The resolution in the CL image is a function of incident probe size, specimen thickness and electron hole pair generation volume while the number of photons produced in a specimen is proportional to the incident current. In a dedicated STEM, as used by Pennycook et al, the incident probe size is ~ 1 nm and in suitably thin specimens a spatial resolution of ~ 50 nm can be achieved (Pennycook 1981). However, the use of a field emission source in the dedicated STEM limits the beam current at the specimen to ~ $10^{-10}$ A making the collection of good quality spectra in thin films difficult. In this paper we describe a simple CL detection and analysis system based on confocal optics which allows spectra to be collected from thin specimens (~ 500 Å) in a dedicated STEM with a 1 nm wavelength resolution.

## 2. The Detection System

Fig. 1 shows a schematic diagram of the CL collection system which has been described briefly elsewhere (Berger & Brown 1983). The system consists of a plain polished aluminium mirror inclined at 45° to the electron optic axis. The mirror is inserted beneath the specimen (i.e. the electron entrance side) and is pierced by a 0.5 mm diameter hole to allow the passage of the incident beam. An achromatic doublet lens (f/3.6), as supplied by Melles Griot, is placed one focal length (i.e. 25 mm) away from, and orthogonal to the specimen such that any CL produced is directed into the lens by the mirror and collimated. The parallel beam of light thus produced falls directly onto the diffraction grating of a modified Spex Minimate monochromator, and is then refocused onto the output slits to be detected by a photomultiplier. Spectra are produced by scanning the grating and recording the digitised signal of the photomultiplier on a Link Systems computer. By using this arrangement the source seen by the spectrometer is the source of the CL, genuinely a point source in thin films, rather than the input slits as is more commonly the case. By dispensing with the input optics of the spectrometer, one reduces greatly the number of components in the system and their associated losses. Another benefit of the confocal arrangement is that spurious CL sources due to scattered electrons are detected with much less efficiency since only photons collimated by the lens are focussed onto the output slits. CL images, both pan- and mono-chromatic, can be taken by directing the P.M.T. output into a Link Systems Digimap frame store.

If the photon emission is too low for an acceptable signal to noise ratio with the digital system a parallel detection system can be used. The PMT is then replaced by an image intensifier tube, and with the diffraction grating stationary a wavelength-dispersed image of the source is produced at the front face of the intensifier. The intensified image at the output can be photographed and a densitometer trace of the resulting negative yields a CL spectrum. This has been done for single dislocations in diamond in 100 nm thick specimens though calibration and correction of the spectra for instrumental response is difficult (Berger 1984). Calibration of the digital system is achieved by reference to standard sources.

## 3. Experiments and Results

The system has been used to observe the CL spectra of rare-earth-doped phosphors and natural diamond as a function of specimen thickness. The phosphor specimen, commercially available europium-doped yttrium-vanadium oxide ($YVO_4$:Eu), was prepared by crushing the powder and mounting it on a holey carbon film to observe crystallites with electron-thin edges. Figure 2 shows CL spectra from regions of varying thickness, this being estimated from the energy loss spectrum. Figure 2a is the spectrum from the bulk material (i.e. not electron transparent) and shows a number of sharp peaks (FWHM resolution ~ 1 nm) due to transitions between different Stark components within the 4f shell of the europium ions (Brecher et al 1967). As the specimen becomes thinner, figure 2b (t ~ 2000 Å) and figure 2c (t ~ 1000 Å), some of the higher energy peaks (those arrowed) lose intensity and eventually disappear below the noise level. A similar phenomenon is observed in diamond where the specimen used is a thin flake produced by oxidation. Figure 3, curve a shows the spectrum from a bulk crystal of natural diamond. The spectrum consists of two broad peaks centred at 440 nm and 510 nm. These are band A (440 nm) and probably H3 (510 nm) emissions (see, for example, Collins 1974). As the specimen thickness is reduced from 2000 Å (curve 3b) to ~ 200 Å (figure 3e) through intermediate thickness (curves c and d) two changes occur. The H3 peak becomes smaller and eventually disappears while the band A peak is shifted by ~ 20 nm to longer wavelengths.

It is found that similar results are obtained by measuring the CL spectra for constant specimen thickness as a function of incident beam current density. The results are shown in figures 4 and 5. In figure 4 the spectra are taken from the bulk $YVO_4$:Eu specimen with the specimen current being lowered by an order of magnitude from curve a, through to c. Figure 5 shows the results from the diamond specimen for a current change of 1.5 orders of magnitude, again the current falls from curve a to curve d. It can be seen that the changes in the spectra produced by a change in current density are qualitatively the same as those produced by reducing the specimen thickness.

## 4. Discussion

The results suggest that the excitation density of free carriers is a function of specimen thickness. The effect this has on the CL emission of $YVO_4$:Eu samples can be understood in terms of saturation effects. The peaks seen in the spectra are due to atomic transitions from $^5D_1$ and $^5D_0$ to $^7F_j$ levels of the europium ion. At high excitation densities the lowest excited state, $^5D_0$, becomes saturated, causing the higher energy states to become populated and higher energy transitions occur. As the excitation density decreases the relative probability of the high energy transitions also decreases. The band A system of diamond is generally accepted as being due to donor-acceptor pair recombination involving nitrogen and boron impurities. This type of recombination mechanism is expected to show a peak wavelength shift as a function of incident beam current density again as a result of saturation effects (see for example Shinoya 1968).

The volume in which the free carriers exist after excitation is their generation volume convoluted with the carrier diffusion length. If the carrier diffusion length is very much less than the incident electron range, as is the case for the highly-doped phosphors, then the density of free carriers will closely follow the generation density. It is well known that the generation density is peaked below the specimen surface, the peak increasing in depth as a function of incident electron energy. Thus the changes in the CL spectra from the phosphors is directly attributable to a change in excitation density as a result of a reduction in specimen thickness. For the diamond however, the bulk diffusion length is of the same order as the incident electron range, resulting in a carrier density more uniformly distributed throughout the specimen. To interpret the results on the above model requires that in thin films the carrier diffusion length must be greatly reduced from its bulk value. Such a reduction has been observed experimentally (Berger 1984) and will be discussed fully elsewhere. The main reason for the effect is a change in the population of the recombination centres (usually generated by ionization of the impurities under electron bombardment) by a surface depletion layer.

The gradual disappearance of the H3 centre as a function of current density/specimen thickness is more difficult to explain. A similar phenomenon has been observed in the band A emission from type IIb diamond which is associated with individual dislocations. Here no CL is seen in specimen thicknesses below ~ 100 nm (Berger and Brown 1984). We can understand this in terms of a surface dead layer, where no radiative recombination occurs because of a competitive recombination route produced by the large number of surface states.

Although these interpretations are far from complete it is clear that the nature of the dependence of the spectra on specimen thickness contains information as to the nature of the recombination centres.

References

Berger, S.D. Ph.D. Thesis, University of Cambridge (1984).
Berger, S.D. and Brown, L.M. Inst. Phys. Conf. Ser. No. 68, 115 (1983).
Brecher, C., Samelson, H., Lempicki, A., Riley, R. and Peters, T. Phys. Rev. 155, 2, 179 (1967).
Collins, A.T. Ind. Diam. Rev. (April) 131 (1974).
Myhajlenko, S., Hutchinson, H.J. and Steeds, J.W. Proc. R.M.S., 20, 2, MSCM10 (1985).
Pennycook, S.J. Inst. Phys. Conf. Ser. No. 61, 55 (1981).
Pennycook, S.J. and Brown, L.M. J. Lumin., 18/19, 905 (1978).
Petroff, P.M., Lang, D.V., Strudel, J.L. and Logan, R.A. SEM 1, 325 (1978), SEM Inc. Chicago.
Roberts, S.H., Inst. Phys. Conf. Ser. No. 61, 51 (1981).

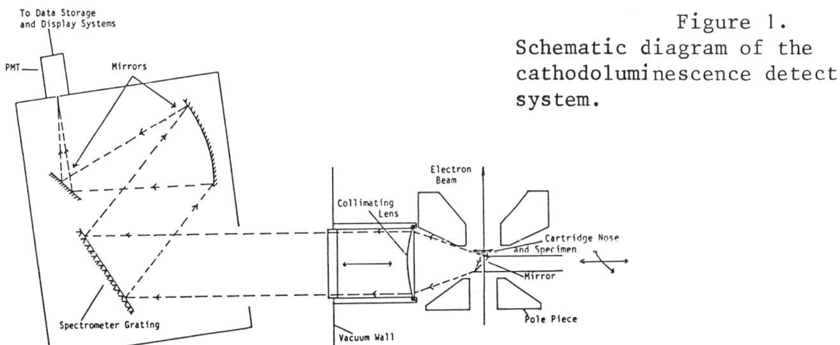


Figure 1.
Schematic diagram of the cathodoluminescence detection system.

For all spectra probe diam. ~ 20 nm, beam current ~ $10^{-11}$ A acquisition time ~ 5 mins.

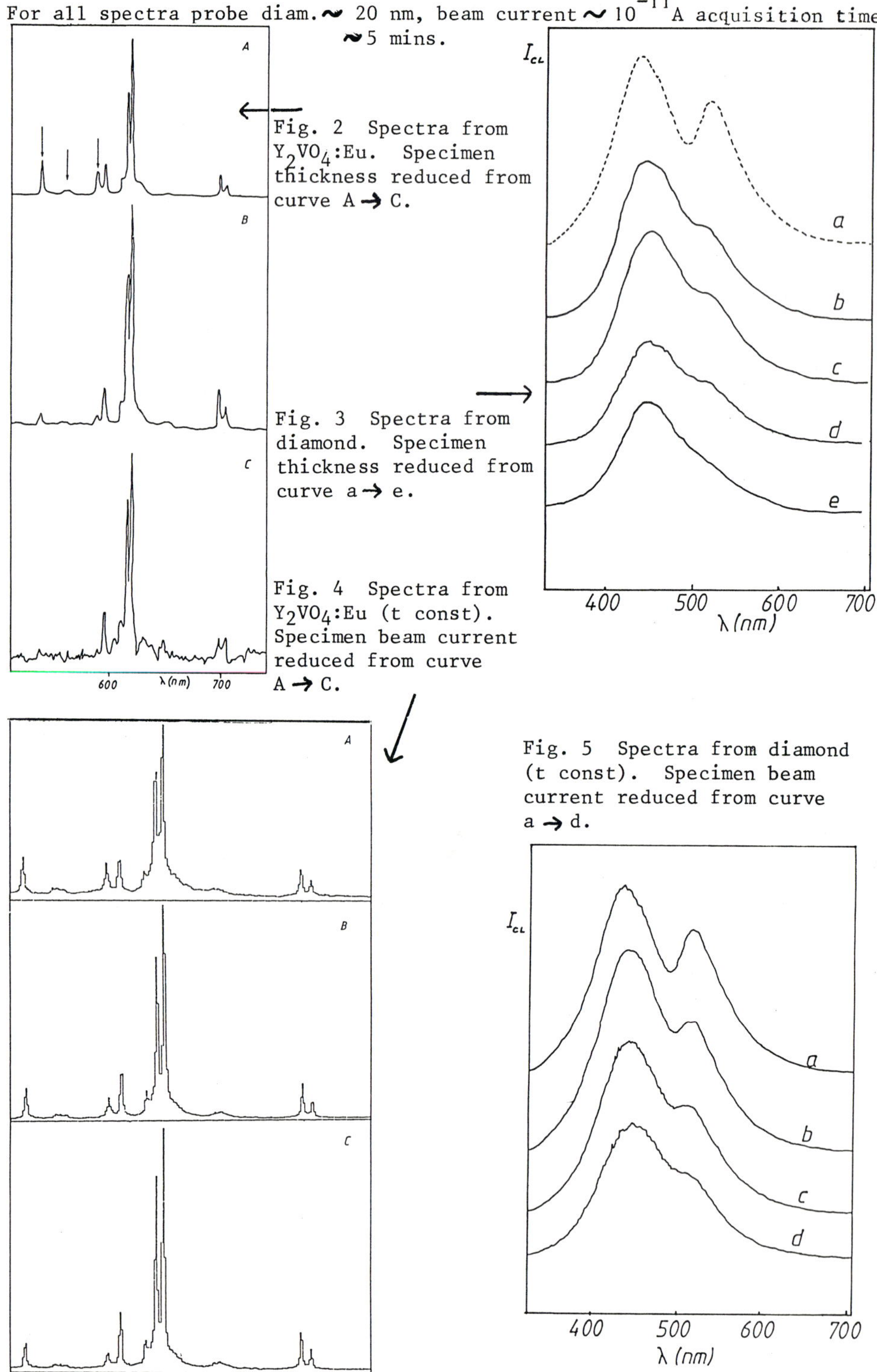


Fig. 2 Spectra from $Y_2VO_4$:Eu. Specimen thickness reduced from curve A → C.

Fig. 3 Spectra from diamond. Specimen thickness reduced from curve a → e.

Fig. 4 Spectra from $Y_2VO_4$:Eu (t const). Specimen beam current reduced from curve A → C.

Fig. 5 Spectra from diamond (t const). Specimen beam current reduced from curve a → d.

*Inst. Phys. Conf. Ser. No 78: Chapter 5*
*Paper presented at EMAG '85, Newcastle upon Tyne, 2–5 September 1985*



# A transfer device for studies of air-sensitive materials and surfaces

D Cherns and D A Carter

H H Wills Physics Laboratory, University of Bristol, Bristol BS8 1TL

and B K Ambrose

W A Technology Ltd, Chesterton Mills, French's Rd, Cambridge CB4 3NP

Transmission electron microscope studies in our laboratory of the battery materials $Li_xTiS_2$ and $Na_xTiS_2$ (0<x<1) (Cherns and Ngo 1983) have been complicated by the fact that these materials are $O_2$- and $H_2O$-sensitive. We have designed and built a transfer device to effect transfer of battery materials from a dry box atmosphere to the electron microscope without atmospheric contamination. The transfer device has been built to accept single tilt side entry holders for the Philips 300 and 400 series transmission electron microscopes. The design is also such as to allow for subsequent transfer of samples from separate vacuum plant for surface studies.

As illustrated in Fig. 1 the transfer device consists of a stainless steel chamber A into which the loaded specimen holder can be retracted along sliding rails as indicated and subsequently sealed using the valve B. On bringing the chamber A up to the microscope goniometer (on supporting rails) a second chamber C is formed by bolting to a flange which replaces a nut securing the goniometer entry tube.(This nut and the sample holder locating pin can be simply replaced when the transfer device is not required). The intermediate chamber is flushed for some minutes with a slight overpressure of inert gas (in our case usually $O_2$-free $N_2$), before opening the valve B. The sample holder is then inserted to a point where the clamping jaws (D) release, adjusted such that the holder O-ring forms a seal in the goniometer insert tube. The transfer device may then be unbolted and retracted allowing the sample holder to be fully inserted by hand. If required the procedure could be reversed to retain the sample uncontaminated.

In preliminary work the transfer device is being used to investigate for the first time the microstructure of the rare earth halide $LaI_2$ and the transition metal halides $TiCl_3$ and $TiBr_3$. These materials, which degrade readily in air giving various oxidised and hydrolysed derivatives, are of interest for their low temperature electrical and magnetic properties e.g. Wilson (1972). Samples were in all cases synthesised directly from the elements in quartz tubes at elevated temperatures and these tubes were broken in a Barretts dry box with specified $O_2$ and $H_2O$ concentrations below 5ppm and 15ppm respectively. In studies of $TiBr_3$, single crystals showing some light transmission were selected and mounted between Cu folding grids and loaded into the transfer device. On loading into the electron microscope thin flakes of good crystal quality were achieved (c.f. Fig. 2(a)), subsequent exposure to the atmosphere for a few seconds being sufficient to totally degrade these regions.

The experience gained in these observations has proved useful in further studies of $TiBr_3$ using a He-cooled low temperature holder not compatible with our transfer device at present. In this case, transfer was effected by protecting the sample holder tip with a cover tube which was removed prior to inserting the holder into the microscope inside a protective polyethylene cover strapped to the goniometer and flushed with $O_2$-free $N_2$. Although this method resulted in contamination of some samples (<50%) low temperature studies of samples showed a transformation around 180K leading to domain formation (Fig. 2(b)(c)). The details of this transformation, observed also in optical absorption studies, will be reported shortly (Maule,Wilson,Carter and Cherns: to be published).

Preliminary studies with our transfer device have therefore demonstrated its suitability for work on bulk structures of a range of air-sensitive materials. Further work is planned to assess the usefulness of the device for reflection studies of surfaces, in particular for cleaved semiconductors and thin metal deposits on semiconductors.

We are grateful to Mr J Burrow for growing the $LaI_2$, $TiCl_3$ and $TiBr_3$ crystals used in this work. The financial assistance of the SERC is gratefully acknowledged.

References

Cherns D and Ngo G P 1983 J.Solid State Chem. 50 7
Wilson J A 1972 Adv. in Phys. 21 143

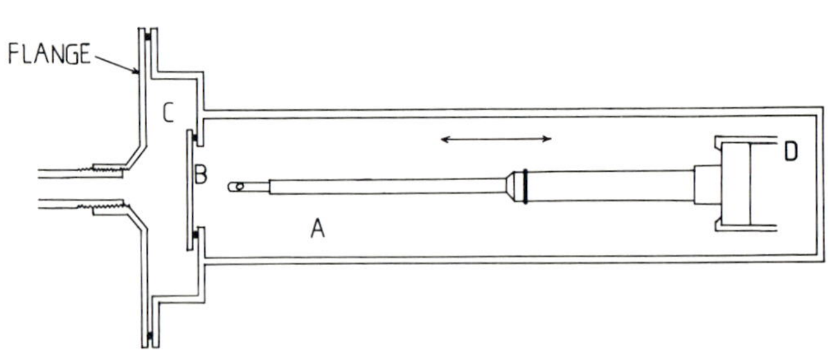


Fig. 1

Schematic cross-section of the transfer device.

See text for details

a

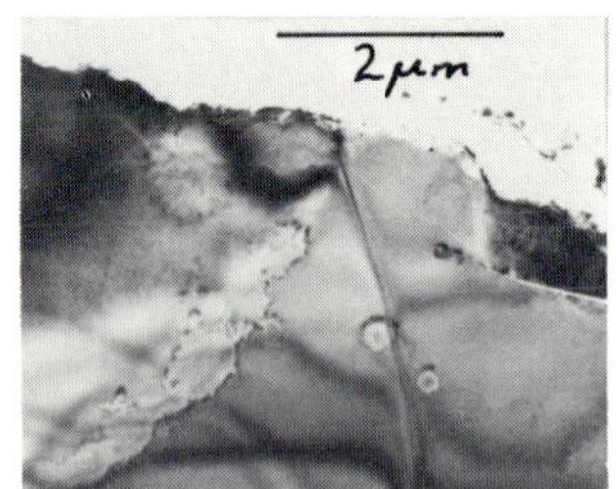


b

c

Fig. 2 a) SAD pattern from a $TiBr_3$ sample (below 180K)

b) Corresponding area of $TiBr_3$ at room temperature

c) Same area below 180K

# TEM specimen preparation for semiconductors using iodine ion milling

N G Chew and A G Cullis

Royal Signals & Radar Establishment, St Andrews Road, Malvern, Worcs WR143PS

## 1. Introduction

The use of ion beams for final thinning during the preparation of transmission electron microscope (TEM) samples from semiconductors has gained widespread acceptance, particularly when cross-sectional specimens are employed. However, the use of conventional inert gas ion or atom milling can lead to the formation of artefactual structure in specimens prepared from a range of compound semiconductors. This undesirable ion-beam induced structure may, as in the case of indium-containing III-V semiconductor compounds, be in the form of surface precipitation arising from gross elemental disproportionation during the ion-milling process (Cullis and Farrow 1979). Alternatively it may take the form of small localised lattice defects present in the near surface regions of final specimens, as in the case of the II-VI semiconductor compounds CdTe, ZnS and ZnSe (Cullis et al 1985).

A reactive ion-milling process employing $I^+$ ions has been developed which substantially reduces, and in some cases totally eliminates, the artefactual structures described above (Chew and Cullis 1984, Cullis et al 1985). The present paper describes the apparatus used for this $I^+$ ion milling process and compares, for a number of semiconductors, the properties of samples prepared by reactive $I^+$ ion milling and conventional $Ar^+$ inert gas ion milling.

## 2. Experimental Details

The compound semiconductor specimens used in this work were prepared in both plan-view and cross-sectional configuration. Cross-sectional specimens (Pettit and Booker 1971) were prepared by bonding pairs of samples together face-to-face, using "quick-setting" epoxy resin. Bonded samples were then mechanically polished normal to the joint and from both sides to a final thickness of $\sim$40µm, the final polishing process employing 3µm diamond paste on a soft lap. Plan-view samples were prepared by double-sided polishing, the bonding stage of course being omitted. Sample discs were next cut from the polished specimens using an ultrasonic drill. The $\sim$40µm thick sample discs thus obtained were then thinned to electron transparency by ion-milling from both sides using either $Ar^+$ or $I^+$ ions as described below.

The ion-milling apparatus employed for these studies was an Ion Tech unit which was modified as shown in Fig. 1. In addition to the normal needle-valve controlled argon gas inlet lines (labelled Ar) to the ion guns, there were inlet lines (labelled I) which carried iodine vapour from a solid

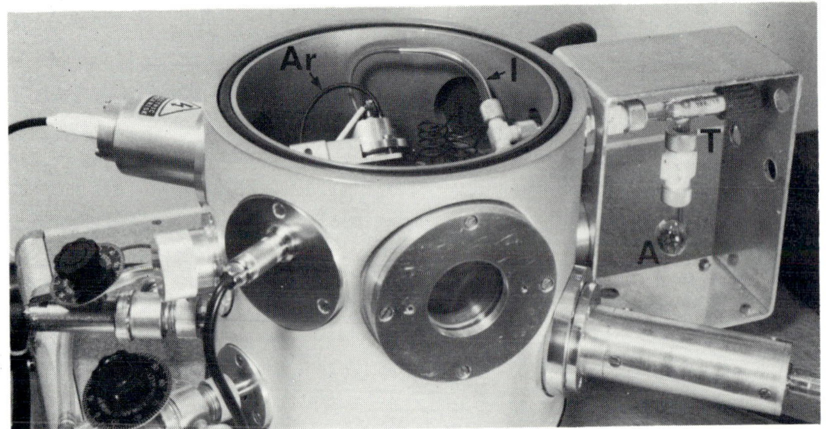


Fig. 1 Modified apparatus used for $I^+$ ion-milling of semiconductors

source of elemental iodine contained in an ampoule (A) situated outside the vacuum chamber. The flow of iodine vapour was controlled by a large aperture glass/PTFE tap (T) positioned above the ampoule. The only other significant alteration to the commercial ion-milling apparatus was the substitution of stainless-steel ion-gun cathodes in place of the aluminium cathodes normally employed. In this modified apparatus beams of either $Ar^+$ or $I^+$ ions could be obtained simply by admitting the appropriate gas. The operating conditions of the ion guns using either argon or iodine have proven to be very similar (Chew and Cullis 1986) allowing stable beams of either species to be produced.

During ion-milling, specimens were cooled using liquid nitrogen and constantly rotated. The beam incidence angle was maintained at $\sim 15^o$. Beam energies employed were typically 6keV with measured beam currents in the range 20-40μA, the beam diameter being ~3mm. Thinning was terminated with reduced beam energies of 2-3keV (and reduced beam current) in order to minimise the ion damage remaining in the surface regions of the final specimens.

Thinned specimens were examined, using diffraction contrast, in a JEM 120C TEM operated at 120keV. High resolution imaging of specimens in [110] projection was carried out using either the Oxford University JEM 200CX TEM (resolution 0.24nm at Scherzer defocus) or a JEM 4000EX TEM (resolution 0.18nm at Scherzer defocus).

## 3. Results and Discussion

Thin specimens of CdTe prepared using $Ar^+$ ion-milling for the last stage of the thinning process invariably were found to contain a very high number of small contrast features when examined under strong-beam diffraction conditions in the TEM. This is evident in Fig. 2a which shows a cross-sectional micrograph from a CdTe layer deposited onto an InSb substrate using molecular beam epitaxy. The strain contrast features present in the CdTe region dominate the image and make analysis of other crystallographic defects in the layer extremely difficult. The nature of the small defects, which had a projected number density of $\sim 10^{11} cm^{-2}$, was investigated using high resolution lattice imaging which revealed (Fig. 3a) that the defects were small (~5-10nm) faulted dislocation loops (see also Sinclair et al

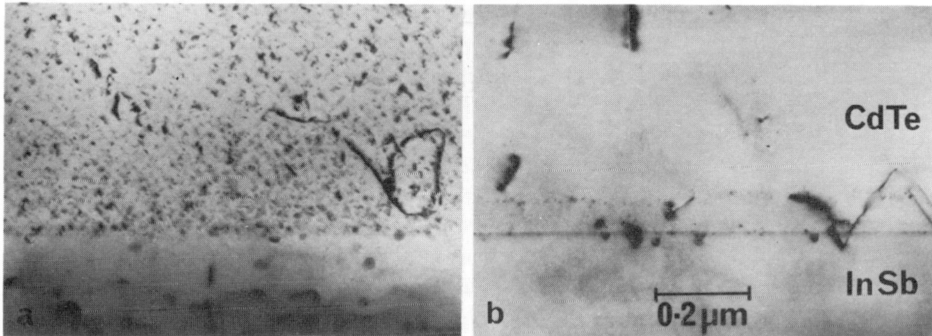


Fig. 2 Cross-sectional transmission electron micrographs (strong-beam, g=220) of heteroepitaxial CdTe/InSb specimens prepared by (a) $Ar^+$ ion-milling and (b) $I^+$ ion-milling

1983). These lay on vertically aligned {111} planes and, therefore, appear edge-on in the lattice image. For comparison with this $Ar^+$ ion milled material Fig. 2b shows a strong-beam micrograph from the sample of Fig. 2a but taken after it had been subjected to a further ~2mins of milling using $I^+$ ions. It is evident that the small dislocation loops present in the CdTe after $Ar^+$ ion-milling are now absent! The lack of these defects allows the extended defects present within the grown layer to be clearly revealed and facilitates their analysis. The absence of the small faulted loops in this material is confirmed by the high resolution micrograph of Fig. 3b in which no defect structures are visible. Specimens thinned using $I^+$ ions throughout the ion-milling process were similarly free from the small dislocation loops and it is concluded that these loops are an artefact arising from the $Ar^+$ ion-milling procedure and that their formation may be avoided by using reactive $I^+$ ion milling.

In addition to this, the micrographs of Fig. 2 also confirm that the deleterious indium droplets which are normally present on indium-containing III-V semiconductors after $Ar^+$ ion-milling are not formed during the reactive $I^+$ ion-milling process (Chew and Cullis 1984).

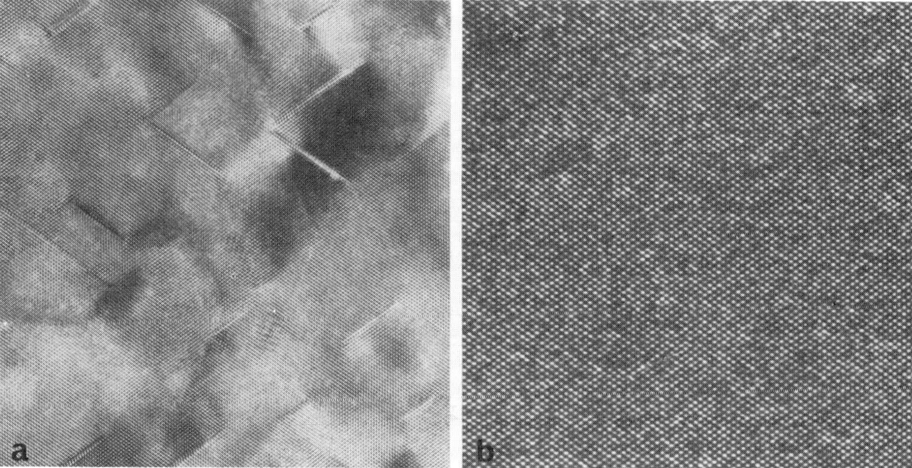


Fig. 3 High resolution electron micrographs of CdTe after (a) $Ar^+$ ion-milling and (b) $I^+$ ion-milling

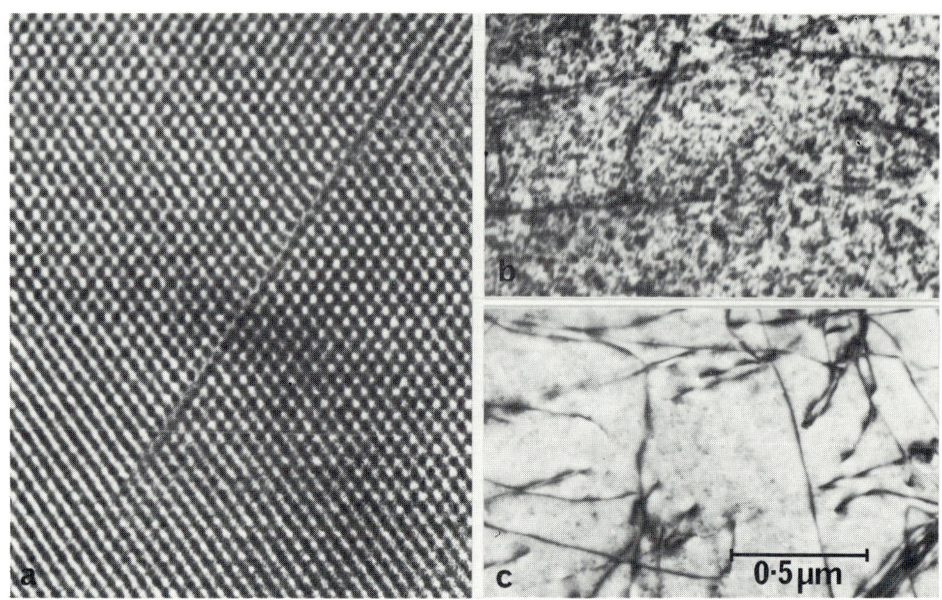


Fig. 4 (a) High resolution electron micrograph of $Ar^+$ ion-milled ZnS. (b) Strong-beam image of $Ar^+$ ion-milled ZnS ($\underline{g}$=220). (c) Strong-beam image of $I^+$ ion-milled ZnS ($\underline{g}$=220)

Specimens of ZnS prepared for TEM analysis using $Ar^+$ ion-milling were also found to contain a high projected number density of small faulted dislocation loops, and an example is shown in the high resolution lattice image of Fig. 4a (see also Ponce et al 1983). These defects, as in the case of CdTe, dominated the contrast in images obtained under strong-beam diffraction conditions - see Fig. 4b. The fact that these defects are not inherent to the bulk ZnS sample but are an artefact of the $Ar^+$ ion-milling is established by comparing this image with a corresponding image taken from the same material after $I^+$ ion-milling - see Fig. 4c. This micrograph shows that there is a substantial reduction in the number of small loops remaining after the ion-milling process, although they are not totally eliminated as in the case of CdTe. Results similar to those obtained for ZnS also have been obtained for ZnSe (Cullis et al 1985) which further demonstrates the widespread application of this new $I^+$ ion-milling process.

Acknowledgement

The authors would like to thank Dr J L Hutchison (Oxford University) for collaboration in the high resolution electron microscopy.



References

Chew N G and Cullis A G 1984 Appl. Phys. Lett. 44 142
Chew N G and Cullis A G 1986 - to be published
Cullis A G, Chew N G and Hutchison J L 1985 Ultramicroscopy 17 in the press
Cullis A G and Farrow R F C 1979 Thin Solid Films 58 197
Pettit H R and Booker G R, 1971. Inst. Phys. Conf. Ser. 10 290
Ponce F A, Stutius W and Werthen J G 1983 Thin Solid Films 104 122
Sinclair R, Ponce F A, Yamashita Y and Smith D J 1983 Inst. Phys. Conf. Ser. 67 103

# Some observations of Pd/C specimens supported on various TEM grids

B C Smith
Department of Metallurgy and Science of Materials,
University of Oxford, Parks Road, Oxford, OX1 3PH, U.K.

## 1. Introduction

During a series of in-situ experiments on model catalyst materials (some of which are reported elsewhere in these proceedings) results were occasionally found to be seriously affected as a consequence of using certain types of electron microscope grid to support the specimens. Some of the phenomena observed while using copper and gold grids are described here along with a more detailed analysis of how arsenic impurities affect palladium specimens. Uncontaminated results using molybdenum grids are included for comparison. The specimens were prepared by evaporating thin (50Å) films of metal on to amorphous carbon substrates and then supporting them on various types of TEM grid which were chosen to give the least interference with microanalysis spectra. The in-situ work was done in a gas reaction cell fitted to an AEI EM7 HVEM. The microanalysis was performed on a Philips FEG EM400T at I.C.I., Runcorn and the high resolution pictures were taken using a JEOL 200CX.

## 2. Palladium on Gold grids

Extra rings were first noticed in diffraction patterns (while doing in-situ heating and reduction experiments in the HVEM) as the evaporated palladium film broke up to form particles at temperatures up to 400°C. The actual ring patterns obtained depended on the environment used but they could not be identified as hydrides carbides or carbonyls which might have formed. Microanalysis later showed the presence of arsenic in the specimens (Fig. 1). After eliminating other sources of contamination in the apparatus and confirming the purity of the bulk Pd used in the preparation, a gold TEM grid from the same batch as that used to support the specimens was examined by wavelength dispersive X-ray analysis in a Camebax Cameca Microprobe. About 0.5% As was found on the surface of the grids. The As desorbs from the grid on heating and is deposited on the Pd specimens, compounds of $Pd_xAs$ being formed.

## 3. Palladium Arsenides

The bulk phase diagram for Pd-As is very complex and it is not well characterised for high Pd content. However the majority of rings seen in the diffraction patterns following reduction in 10%CO/He (Fig. 2) could be assigned to Pd metal or $Pd_5As$. After reduction in 10%$H_2$/He a doubling of some Pd lattice spacings was observed. A similar effect was seen after heating in vacuo, but most of the rings remain unassigned. A doubling of the lattice spacings can be obtained by changing one in eight of the Pd atoms for As atoms (as shown in Fig. 3) to form $Pd_7As$. This is isostructural with $Pd_7Ce$ described by Smith et al (1982). Fig. 4 shows

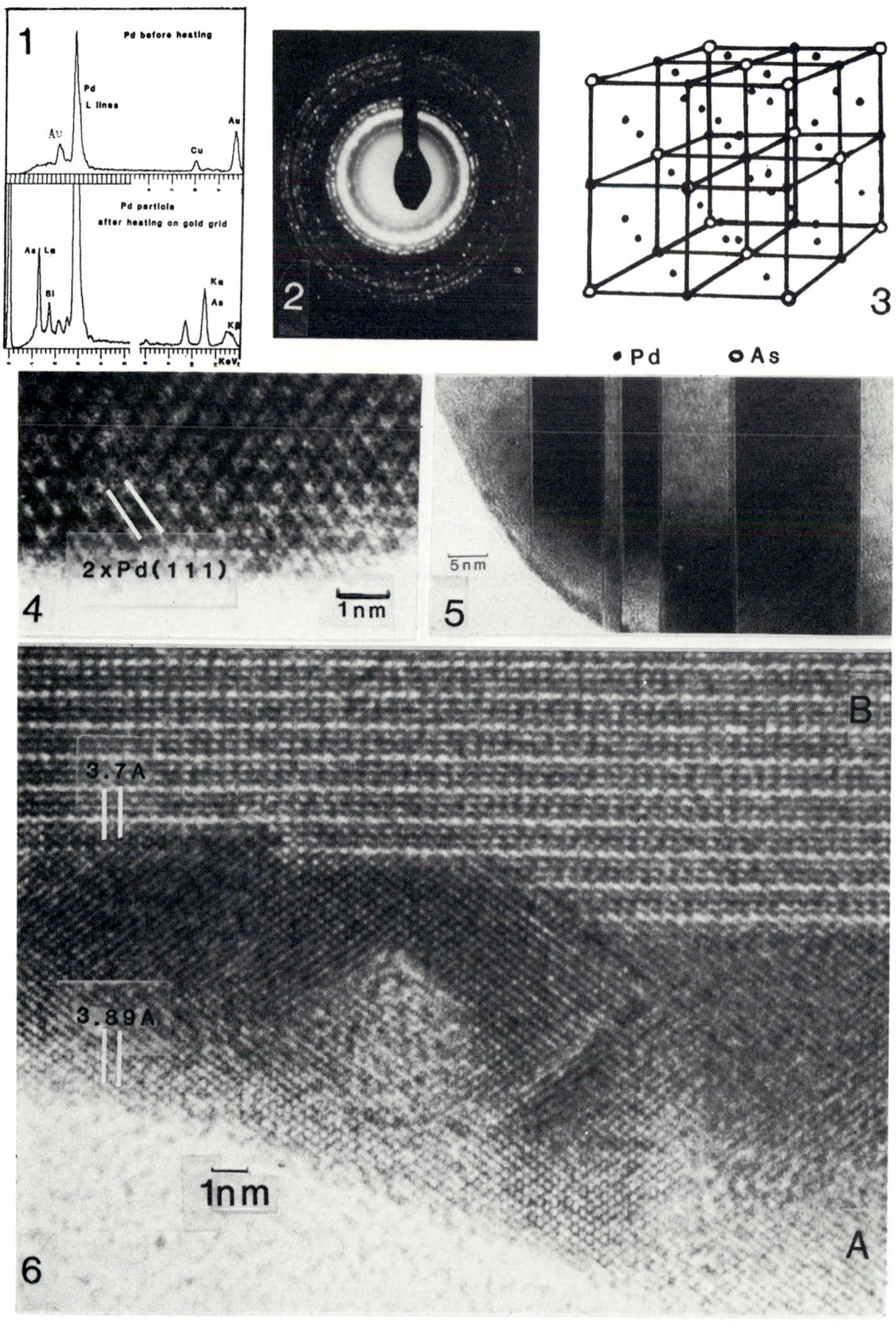


Figs. 1) Microanalysis 2) DP after reduction in 10%CO/He on Au grid
3) $Pd_7As$ structure 4) HREM $Pd_7As$ 5) Twins in $Pd_7As$
6) HREM FCC phase + superlattice

a high resolution image of the new phase with the double (111) spacing marked. Fig. 5 shows twin boundaries in the same structure which are very similar in appearance to twins in Pd. Occasionally phase boundaries were observed, between $Pd_7As$ and $Pd_5As$ as well as between known and unidentified phases. Fig. 6 shows two phases coexisting in a particle following heat treatment in vacuo. The lower half of the image (A) has the FCC structure of palladium with a (200) spacing of 1.95Å and a (110) spacing of 2.77Å. The upper half (B) shows a new superlattice phase with a large spacing of 7.98Å which is three times the secondary spacing of 2.66Å. The horizontal spacing of 3.70Å is slightly less than the (100) FCC spacing of 3.89Å. This mismatch is taken up by the introduction of a stacking fault tetrahedron in the FCC phase. Stacking fault tetrahedra were not a general feature of the FCC lattice and were not observed in any of the pure Pd specimens.

## 4. Palladium on Copper grids

As may be expected when heating copper in oxygen gas the surface of the grids oxidised with a consequent blackening of the previously shiny grid surfaces. Heating Pd specimens supported on Cu grids in vacuo or in a reducing gas was not expected to drastically affect the appearance of the grids. However, while heating specimens to about 400°C in vacuo ($10^{-4}$ torr) the grids crystallised (Fig. 7). Each crystallite appears to have (100) faces and edges. In areas of the grid covered by the Pd/C specimen, needles originating from the grid were observed (Fig. 8) consisting of a head containing Pd followed by a corrugated stem which could be made of carbon from the support. It is not known what proportion of the copper peaks seen in microanalysis traces could be attributed to the needles, as there was a high background count from the support. The needles are similar in appearance to carbon filaments reported by Audier et al (1983) which have iron carbide heads.

## 5. Uncontaminated Palladium

Fig. 9 shows a Pd sample being reduced at 350°C in 10%CO/He on a molybdenum grid. The uncomtaminated sample in Fig. 10 was heated to a similar temperature in 10%$H_2$/He. The particles are clearly much smaller than those contaminated with arsenic described above. They are well faceted and often defective with twins on (111) planes in some particles. They are slightly smaller than the copper-palladium particles described in the paper on bimetallic systems.

## 6. Conclusions

When carrying out in-situ heating experiments with model systems, supported on TEM grids, it is important to consider carefully the choice of grid used. Grids are often treated to make one side appear shiny and the other dull. Arsenic impurities, introduced during the manufacture of gold TEM grids, easily contaminate reactive metals like palladium. Copper grids are usable at low temperatures (up to about 250°C) for reduction experiments. Molybdenum grids can be used at higher temperatures successfully.

## Acknowledgements

I would like to thank Dr. G. Owen and Dr. M. D. Shannon (I.C.I., The Heath, Runcorn) and the SERC for supporting this work.

## References

Audier M., Bowen P. and Jones W., 1983, J. Cryst. Growth, 63, 125

Smith D. A., Jones I. P. and Harris I. R., 1982, J. Mat. Sci. Lett. 1 463

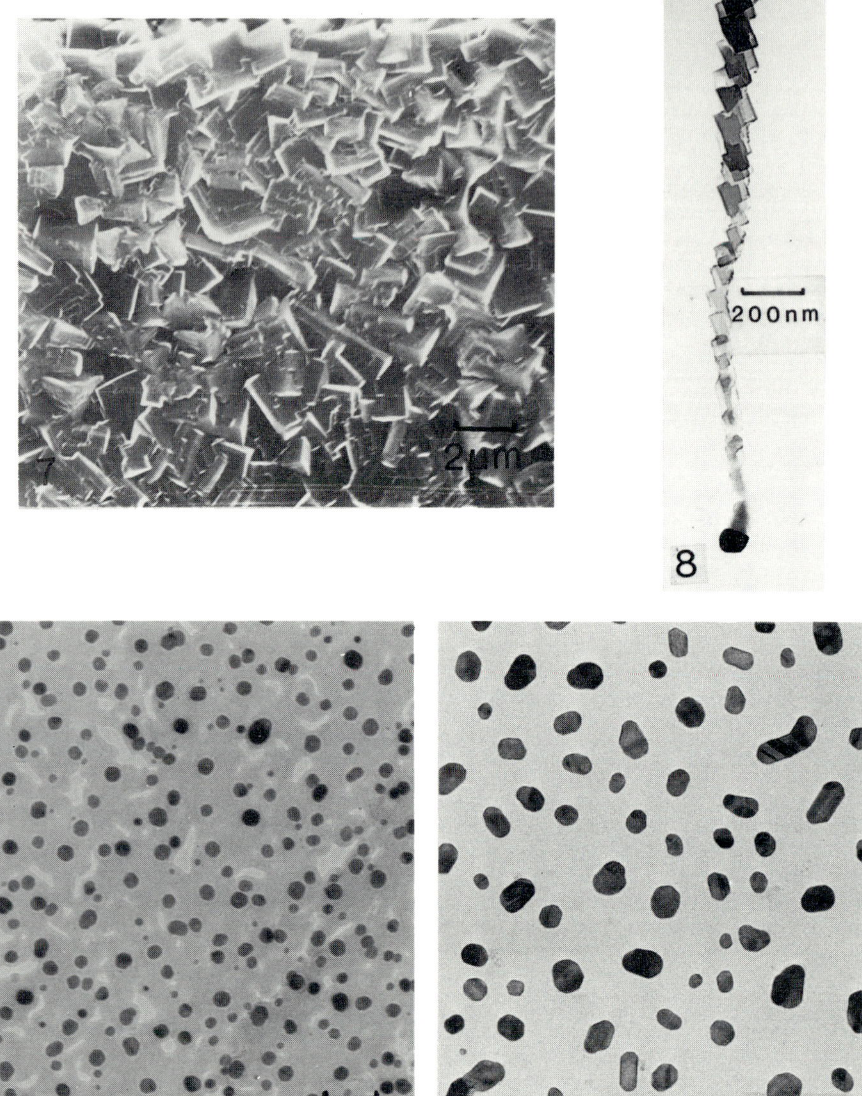


Figs. 7) Cu grid after heating in vacuo 8) Pd/C (Cu grid) needle
9) Pd in 10%CO/He 350°C 10) Pd after heating in 10%$H_2$/He 350°C

# Carbon coating of heat sensitive materials

D S Finch and D Vesely

Department of Materials Technology, Brunel University, Uxbridge, Middlesex.

## 1. Introduction.

The electron microscopy of non-conducting materials is usually affected by charging as a result of electron bombardment of the specimen. It is for this reason that the surface of a thin section has to be coated with or supported by an electrically conductive layer such as metal or more often, carbon, the latter interfering less with the observed structure. This process requires vacuum evaporation of the carbon or metal from a heated source and therefore there is a danger of permanent heat damage to the specimen. Previous work by D Vesely and S Woodisse (1982) established a reproducible method for carbon coating, however, the heating effect during coating had not been studied. It has been found that under normal coating conditions the temperature of the specimen can reach the degradation point and the structure can be significantly altered [figure 1]. The purpose of this work therefore is to correlate the temperature rise of the substrate with the thickness of carbon deposited in order to determine the optimum conditions for prevention of heat damage. The minimum thickness of carbon required to prevent charging of the specimen is evaluated by measurement of the change in resistivity as a function of the thickness of carbon deposited.

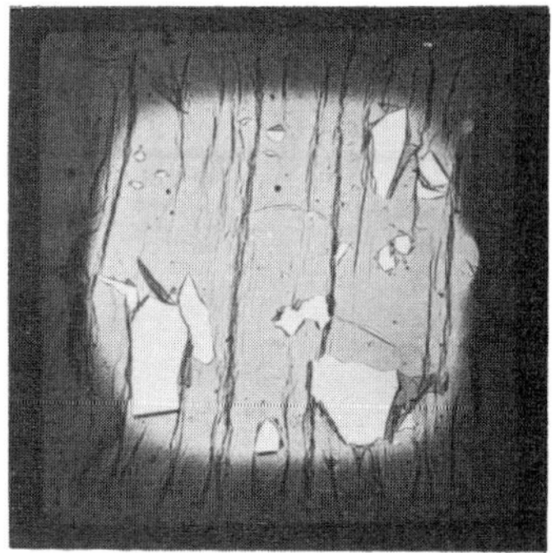

Figure 1.shows a polymer blend damaged by heat.

## 2. Experimental.

All of the evaporated carbon films were produced using an Edwards 306A coater with a vacuum measured at better than $5\times10^{-4}$ Torr. The carbon was evaporated either from a carbon rod source or from a carbon fibre filament (obtained from Balzers). Carbon films were evaporated onto glass slides and evaluated using a microdensitometer comparing with standards previously measured with an interferometer. The substrate temperature has been measured during evaporation using a thermocouple attached to a chart recorder and confirmed by calibrated, commercially available temperature sensitive substrates which change colour at their characteristic temperature. The temperature of the substrate could thus be measured to an accuracy of ±1 percent. The temperature of the evaporating source has been evaluated as follows. Both current and voltage were monitored during coating, and assuming a resistance which is only temperature dependent (Arnold 1965),a graph of filament current versus power input has been produced. The power input has been related to the temperature of the filament by calibrating it with metals of known melting points.

The thickness of the carbon layer is dependent also upon the time of evaporation and therefore on the lifetime of the filament. At high voltages (30 Volts) the average lifetime of the carbon filament is short (0.5 seconds) and measurement of the time for evaporation was only possible by using the electrical interference recorded directly on the chart recorder via the thermocouple. Finally, the resistance of evaporated films was measured directly on the glass slide with evaporated carbon film and aluminium electrodes, using a Schlumberg 7065 microprocessor voltmeter, and related to the thickness of the deposited layer.

## 3. Discussion.

The rate of carbon deposition, G, on the specimen is dependent primarily on the carbon vapour pressure P,

$$G = P\sqrt{M/2\pi RT} \qquad [1]$$

where M is the molecular weight and R the gas constant. The vapour pressure is dependent exponentially on the temperature of the source as shown in figure 2, demonstrating the need for source temperatures above 3000K to achieve the required vapour pressure in excess of $10^{-2}$ Torr.

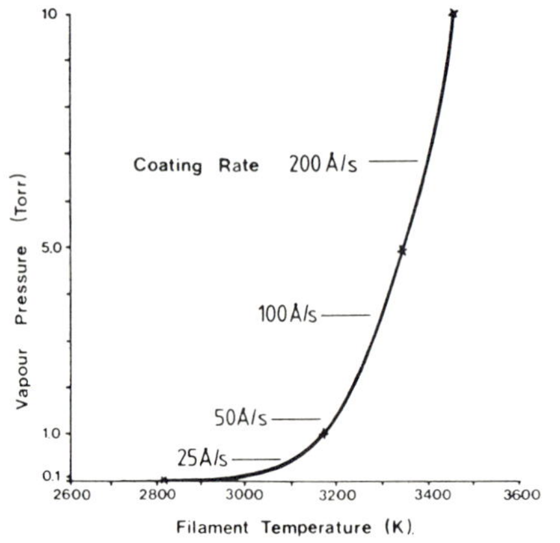


Figure 2. Temperature of the filament versus vapour pressure.

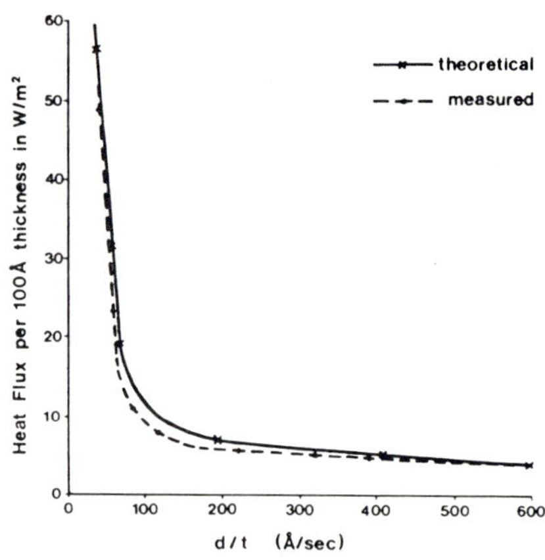


Figure3. Heat flux versus carbon deposition rate.

It is apparent that during coating the substrate will be heated by the filament. The hotter the filament, or the closer the substrate to the filament, the greater the heat flux, but also the greater the deposition rate. Figure 3 shows a plot of the deposition rate (equivalent to the evaporation rate assuming all carbon molecules remain attached to the substrate) against the heat flux from the source. This demonstrates clearly that, for a given thickness of carbon to be deposited, the higher the rate of evaporation, the lower the heating effect. The temperature of the source must therefore be maximised. Figure 3 shows that there is a close correlation between both theoretical and experimental data. The theoretical thickness has been calculated using a Langmuir treatment of evaporation from a free surface based on equation [1], using the geometric parameters for the evaporation from a cylindrical source (Hottel 1967). The heat flux in Watts per square meter emitted from the source is given by

$$E = \sigma \alpha T^4 \qquad [2]$$

where $\sigma$ is the Stefan-Boltzman constant and $\alpha$ is the degree to which the emitter behaves as a black body ( $\alpha$ =1). The variation in $\alpha$ with temperature has been reported by Hottel (1967). It follows from figure 3 that with a working distance of 50mm, an evaporation rate in excess of 200Å per second is required for a significant reduction in heating. From the measurement of the filament current, the filament temperature is found to be in excess of 3400K and the corresponding vapour pressure is greater than $5\times10^{-1}$Torr.

The heat deposited on the substrate will increase its temperature. This temperature has been measured and is plotted in figure 4 versus the deposition rate. A similar trend to that of figure 3 which considers the heat flux from the source versus the deposition rate is clearly shown. This again demonstrates the necessity to use high coating rates in order to reduce specimen heating. Some specimen heating however cannot be avoided even at very high deposition rates. The coating rate and the substrate temperature are also dependent on the working distance. The variation in minimum substrate temperature with working distance is given in figure 5 for a fixed coating time of 0.5 seconds.

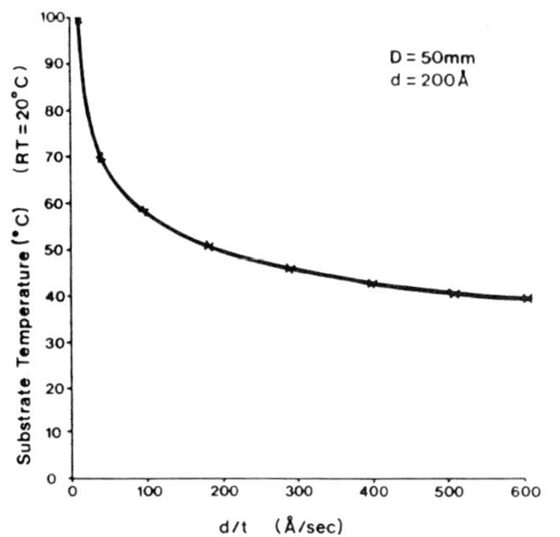

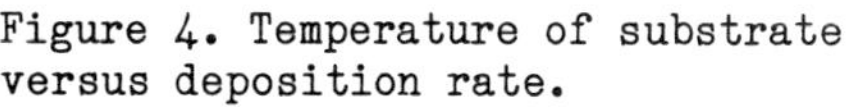

Figure 4. Temperature of substrate versus deposition rate.

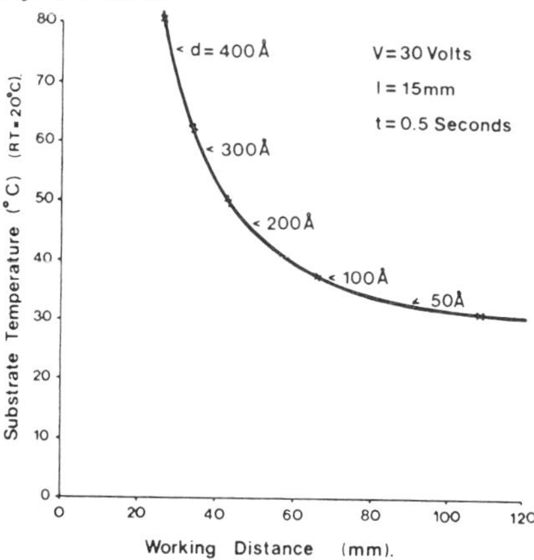


Figure 5. Variation in substrate temperature with working distance.

It might be useful at this point to compare the two different types of carbon source, ie, the carbon rod and the carbon filament. The temperature of the source is dependent on the electrical energy dissipated on its resistance. This can be calculated from the voltage applied and current passing the carbon source. The carbon filament has an obvious advantage in its homogeneity, as compared to a pointed carbon rod, where the resistivity is inhomogeneous and unknown at the tip. Also, the changes in resistivity with time are more predictable for the carbon filament. The high evaporation rate and thus low heating of the substrate can only be achieved with the carbon filament. However, its lifetime is often unpredictable and reliable coating can only be achieved by using an electronically controlled shutter. In order to obtain high thicknesses of carbon, ie greater than 200Å, without substantial heating of the specimen, a multiple coating process could be used.

As the temperature of the specimen is directly proportional to the thickness of carbon deposited, it is important to determine the minimum thickness required to protect the specimen from the build up of charge during analysis, and from this, to evaluate the maximum degree of specimen heating. Figure 6 shows a plot of specific resistivity against thickness of carbon deposited on to a glass slide. It shows clearly an exponential increase in resistivity with a reduction in the thickness of the carbon layer. A model which can explain this behaviour was described originally by Fuchs (1938) and Sondheimer (1950) for thin film conductivity and is in agreement with the results found. The thickness of carbon coating required for a particular application, will be dependent upon the conductivity and the thickness of the specimen; the thicker the specimen the more conductive a coating it requires. To overcome the charging of thin specimens used in electron microscopy, only a very thin carbon coating layer is required. Carbon films 50Å thick are sufficiently strong for conductive coating of most polymeric specimens, thicknesses up to 200Å however are required for replication work. In order to produce films of the order of 50Å a carbon filament with D = 50mm at a voltage of 30V would give minimum specimen heating (28 to 30 degrees Celsius at a working distance of about 90mm and a burning time of 0.5 seconds. For 200Å thickness however, the temperature of the specimen cannot be kept below 45 degrees Celsius unless a multiple coating procedure is used.

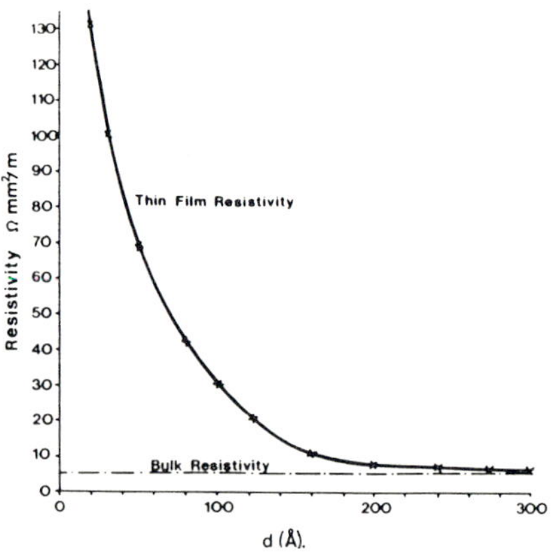


Figure 6. variation in resistivity with thickness carbon deposited.

## 4. Conclusion.

It has been shown that the temperature of the substrate can rise substantially during the carbon coating process. Temperatures approaching 200 degrees celsius can be achieved under normal coating conditions. The heating can be minimised by using a high source temperature as the vapour pressure is increasing exponentially with the temperature of the source. The advantages of a carbon filament over a carbon rod are sufficiently significant to make its use for heat sensitive specimens a necessity.

## 5. References.


Arnold R 1965 Z. Angew. Phys. 7 453

Fuchs K 1938 Proc. Camb. Phil. Soc. 34 100

Hottel H and Sarofim A 1967 Radiative Transfer. Mc.Graw Hill

Sondheimer E 1950 Phys. Rev. 80 401

Vesely D and Woodisse S 1982 Proc. R. Mic. Soc. 17 137

*Inst. Phys. Conf. Ser. No 78: Chapter 6*
*Paper presented at EMAG '85, Newcastle upon Tyne, 2–5 September 1985*



# Voltage dependence in electron energy loss spectroscopy

SEVELY J

Laboratoire d'Optique Electronique du C.N.R.S.*, B.P. 4347, 31055 TOULOUSE CEDEX, France.

Abstract. The use of high voltage for the development of electron energy loss analysis is considered. Through the results of experiments mainly performed at 1 MV and spectra calculated by the procedure devised by Burge and Misell, extended to K shell excitations, the visibility of the inner shell excitation profiles is studied in the case of boron, carbon and boron nitride. Its variation as a function of voltage and specimen thickness is determined. The advantage of using high accelerating voltages for the development of elemental and EXELFS analysis by EELS is discussed.

## 1. Introduction

There has been considerable efforts devoted to the determination of elemental composition at high spatial resolution in the electron microscope using the gross features of electron energy loss spectra of core losses (Colliex, 1984). Though electron energy loss spectroscopy (EELS) with the new dedicated scanning transmission electron microscopes (STEM) produces energy loss spectra of valence shell excitations with a wealth of precise information regarding the local chemical and electronic environment in the interface region (Scheinfein et al. 1985), core electron excitation studies remain of considerable interest for the chemical analysis of the sample. Not least since the associated electron energy losses bring the chemical information directly and because these excitations are more highly localized.

In the electron energy loss spectra, the visibility and the use of core shell ionization edges is greatly dependent on various experimental parameters, among which primary electron energy is certainly less frequently considered. Practically, all electron energy loss experiments are still performed at 100 ± 20 keV. The development of new electron microscopes of medium energy 300-400 keV, may give a practical interest in this parameter.

Through experimental results got at 1 MV, the influence of the incident energy on the electron energy loss spectrum profile and on the visibility of the core shell excitation distributions will be studied. After a description of particular experimental problems associated with the use of high incident energy electrons, the various kinds of interactions : elastic scattering, valence shell and core shell excitations, which occur in the specimen and which are shown by the spectra will be considered. The

* *Associated with the University Toulouse III.*

determination of their contribution to the spectrum formation allows the determination of their characteristic parameters and discussion of their variation as a function of the different experimental parameters, particularly the primary electron energy.

## 2. Experimental problems associated with the use of high voltages

Essentially, they concern the source, the spectrometer and the detection.

### 2.1. The source

The adaptation of the field emission gun to high voltage microscopy is not yet achieved. So the energy dispersion will always be larger than 1 eV. In spite of a good accelerating voltage stability, about $10^{-6}$, one cannot expect an energy resolution of the analysis better than 1 eV at high voltage.

### 2.2. The spectrometer

It is now well established that the magnetic prism is the suitable solution for electron energy analysis in the microscope. This is more true at high voltage where otherwise high electrostatic fields would be necessary. The variation of the prism dispersive power D coefficient as a function of the incident voltage is $\gamma/\gamma^2-1$, where $\gamma = (1-v^2/c^2)^{-1/2}$ is the usual relativistic factor (Perez et al., 1975). Increasing the accelerating voltage has to be compensated by larger values of the prism radius of curvature a. On the 1 MV microscope where a = 20 cm, at 1000 kV, D varies between 3 and 0.5 μm/eV and is equivalent to the typical values of D got at 100 kV.
So as at low voltage, the energy resolution R of the dispersive system is dependent on the detector resolution power S. R = S/D. In the parallel detection mode, S>10 μm and the spectrum has to be enlarged before its detection (Sevely et al, 1973).

### 2.3. The detection

For the present, the detection of electrons is certainly the main experimental problem in the field of EELS, particularly for quantitative purposes. This problem is enhanced at high voltages where existing imaging techniques decrease in effectiveness (Fotino, 1982). This fact is the consequence of features of the electron interaction with matter at high energy : reduced energy loss - dE/dx, increased penetrating power and range and increased X-ray yield. They result in cross talk of information and production of a parasitic background due to X-rays and stray electrons (Kihn et al.,1980). These problems can be partially solved : the broad cross-talk is solved by sequential acquisition which unfortunately results in a considerable waste of information ; protection against X-rays is efficiently realised by locating the electron detector out-of the microscope axis and bending the electron beam by a magnetic field (Fotino, 1982, Sirvin, 1984). These precautions are necessary for quantitation an high voltage transmission electron microscopy and spectroscopy (Kihn et al., 1980). They are adaptable to low voltage with surely the same benefits (Craven and Buggy, 1984).

## 3. Valence and core shell electron excitations

They produce the inelastic scattering processes of the incident electrons in the sample which are clearly observed in the spectra. Their excitation probability can be deduced from the spectra and is measured by an interaction mean free path $\lambda$.

3.1. Valence electron excitations

Generally, they result in collective excitations of free electrons : "plasmons". The plasmon mean free path expression $\lambda_p$ can be deduced from the theory developed by Ashley and Ritchie (1970), (Sevely et al., 1974).

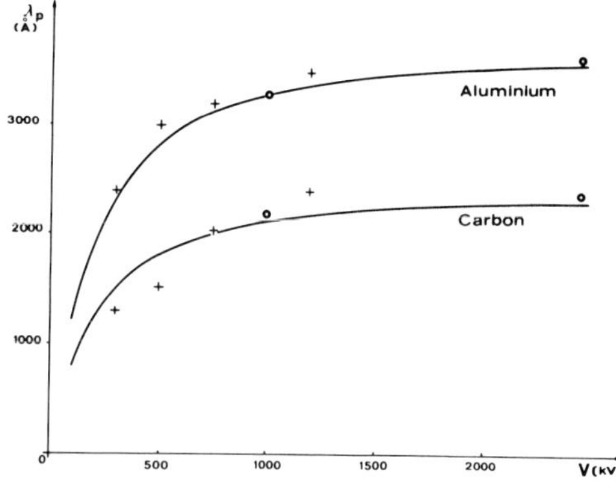


Fig. 1 - Variation of $\lambda_p$ as a function of the electron accelerating voltage in the case of carbon and aluminium. Some experimental results got up to 2.5 MV are compared to the calculated values.

Figure 1,shows the decrease of the plasmon excitation probability when the voltage increases. $\lambda_p$ changes significantly up to 500 kV then it saturates due to relativistic effects. Between 100 and 500 kV, $\lambda_p$ is multiplied by about 2.25 and by 1.16 between 500 and 1000 kV. It remains practically unchanged above 1 MV.

3.2. Core-shell excitation

The variation of the mean free path as a function of the accelerating voltage V can be deduced from the Bethe theory (Inokuti, 1971, Perez et al., 1974). Figure 2a shows the variation of $\lambda_K$ (K shell excitation mean free path in the case of aluminium). $\lambda_K$ has a maximum value at 400 kV then it decreases slowly. In all the range of the accelerating votages used in electron microscopy $\lambda_K$ variation is not important. Figure 2b allows to compare experimental values with the calculated values in the case of carbon.

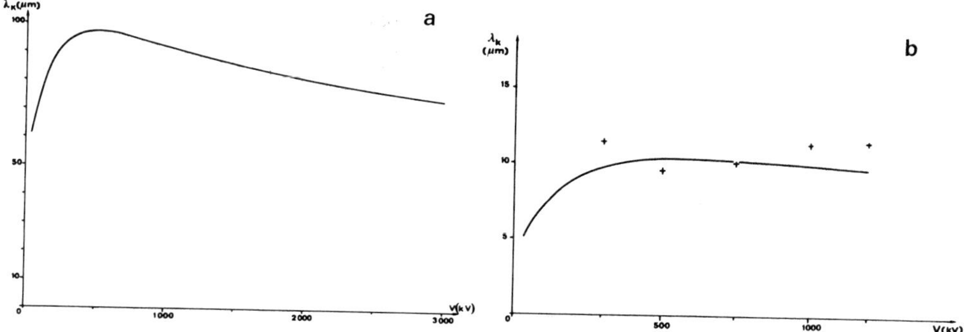


Fig. 2 - Calculated mean free path values for K shell excitation as a function of accelerating voltage V. a) for aluminium ; b) for carbon,dots show experimental values.

## 4. Contribution of the basic scattering processes to the spectrum formation

Electron energy loss spectra can be considered as the result of convolution effects of the basic scattering processes in the sample. In the case of experimentally well characterized and "simple" energy loss spectra got with carbon, boron and boron nitride samples, the procedure devised by Burge and Misell (1968) for the prediction of the low energy loss part of the spectrum was extended to K shell excitation (Zanchi et al.,1983). A good agreement, within a deviation of a few percent as the experimental uncertainty, is found between the spectra got by experiment and the spectra calculated by using this procedure for energy losses less than 600 eV in the case of samples of different thicknesses (10 nm to 1 μm) and different accelerating voltages (80 to 1000 kV).

By fitting the experimental and calculated spectra in a wide range of thicknesses and incident energies, it is possible to determine the characteristic parameters of each kind of excitation processes.
- the parameter $t/\lambda_p$ is the fundamental parameter for the low energy loss part of the spectrum, before the first core electron excitation edge. The profile of the spectrum depends on this parameter only, whatever the thickness or $\lambda_p$, that is to say the microscope accelerating voltage V, may be.
- the preedge background is due to a contribution from plural scattering due to valence electron excitations.
- the method allows to determine the thickness t of the sample when $\lambda_p$ is known and reciprocally.
- it provides a basis for predicting the spectrum profile variation as a function of multiple scattering effects in the sample, particularly when the accelerating voltage changes.

## 5. K edge visibility as a function of the accelerating voltage

A point of interest for the development of EELS analysis on a very small scale is the visibility of the core excitation distribution over the background in the spectra. This visibility may be characterized by the jump ratio J.R.

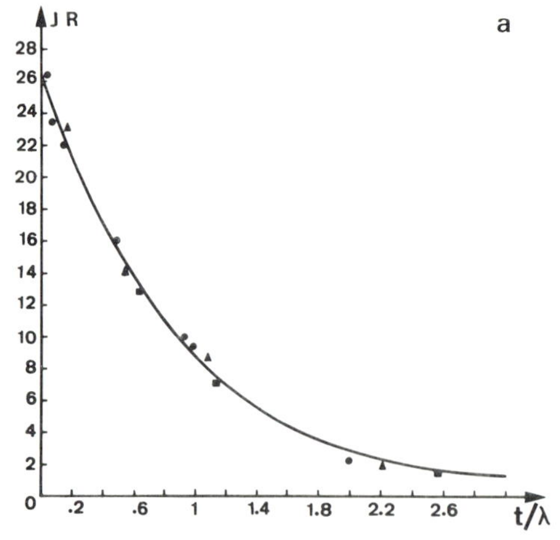


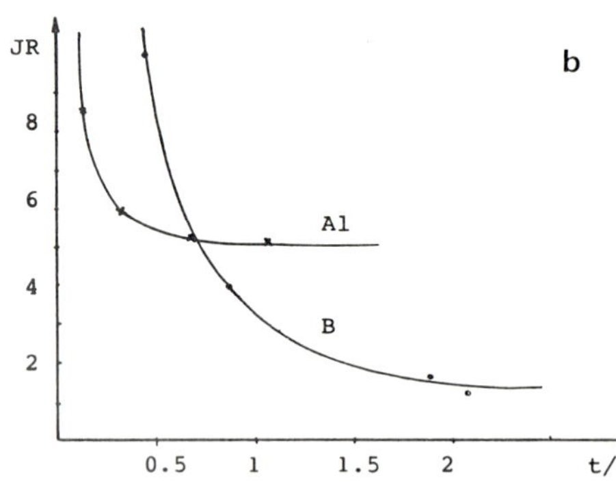


Fig. 3 - Jump ratio at the K edge as a function of $t/\lambda_p$. a) case of carbon; points are experimental values got at 1000 kV ●, 500 kV ▲, 300 kV ■, full curves correspond to the calculated results. b) case of boron and aluminium ; experimental results got at 1000 kV with different thicknesses.

at the edge as defined by Egerton (1978). The fitting procedure of the spectra in the case of "simple" energy loss spectra as obtained with carbon or boron allows the determination of the jump ratio variation as a function of the parameter $t/\lambda_p$. It is to say as a function of thickness t and accelerating voltage. The variation is given figure 3a for carbon. Some experimental values deduced from aluminium and boron spectra, figure 3b, confirm the model validity.

The conclusion is that J.R. decreases when $t/\lambda_p$ increases. It confirms previous results on thickness influence by Egerton (1978) or Sarikaya and Rez (1983). As concerned the accelerating voltage dependence, it is introduced through the volume plasmon mean free path $\lambda_p$. The K edge visibility is improved at high voltage, this improvement follows the $\lambda_p$ variation as a function of V.

. For the same thickness, the J.R. is multiplied by a factor about 3 between 100 and 1000 kV.

3 times thicker samples can be analysed at 1000 kV as compared to 100 kV. This is experimentally confirmed if one observes the limit thickness for the analysis of biological samples. It is 50-80 nm at 100 kV while its typical value is 200-250 nm at 1000 kV (Berry et al., 1984).

. The quantitative analysis formulae are only valid when $t/\lambda_p < \sim 1$ (Egerton 1980, Zaluzec, 1980). The quantitative potentialities of the method are enlarged at high voltage.

- The simplification in the core electron excitation spectral shape at high voltage enlarges the energy loss range which is currently observed. At 1 MV, it is typically 3000 eV. That involves an improvement of the sensitivity of the method for the elemental characterization (Berry et al., 1984, Fourdeux et al., 1984) and for the EXELFS studies (Garg et al., 1984).

. In this case, at 1000 kV, it was possible to observe the EXELFS modulation on the copper (fig. 4a), and zinc $L_{23}$ distribution (at 930 and 1050 eV respectively) while reported experiments at 100 kV by Leapman et al. (1981) show they were only observed up to the chromium $L_{23}$ edge at 575 eV.

For the K shell, EXELFS modulations have been observed and studied on the aluminium and silicon spectra. About 8 modulations have been detected on an energy window about 500 eV (fig. 4b). The aluminium EXELFS K edge profiles at 100 kV (Leapman and Cosslett, 1976, Johnson et al., 1981) clearly show 4 modulations in an about 150 eV range only.

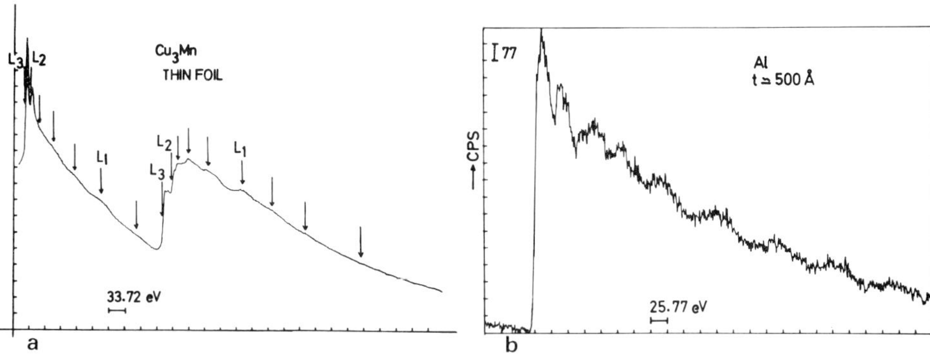


Fig. 4 - EXELFS spectra got at 1 MV, collection angle $\theta_D$ = 3 mrad -
a) with a $Cu_3Mn$ thin film after the Mn $L_{23}$ excitation edge at 640 eV and the Cu $L_{23}$ edge at 930 eV, recorded photographically - exposure time 1 mn.
b) with an aluminium 50 nm thick sample, after the K edge at 1560 eV, counting time 0.6 sec./channel and channel steps 0.98 eV x 3. T = 25° K.

The thickness of the sample which produces the best visibility of the modulation is estimated to be so that $t/\lambda_p \sim 0{\cdot}2$,that means $t \sim 500$ Å at 1000 kV and $\sim 150$ Å at 100 kV.

## 6. Discussion

The experiments and predictions of the electron energy loss spectra show that for all the points which does not need an energy resolution below 1eV, using the highest incident accelerating voltage improves energy loss analysis. That is mainly due to the fact that multiple scattering effects are reduced. The main scattering process in the sample (volume plasmon) mean free path saturates significantly above 500 kV due to relativistic effects. So, one must not expect considerable improvement between 500 and 1000 kV. This is confirmed, for instance, by comparing aluminium K EXELFS modulation got with a 50 nm thick sample at 500 kV and 1 MV (Sevely et al., 1985). Due to the variation of $\lambda_p$, the improvement in electron energy analysis results is significant up to 500 kV. In this respect, the new medium energy microscopes seem to correspond to very good conditions for the development and use of the EELS technique.

## References


Berry JP, Galle P, Kihn Y, Zanchi G, Sevely J, Jouffrey B 1984 J. de Phys. 45 C2 581.

Burge RE and Misell DL 1968 Phil. Mag. 18 251.

Colliex C 1984 Advances in Optical and Electron Microscopy (Academic Press) 9 65.

Craven AJ and Buggy TW 1984 J. of Microscopy 136 227.

Egerton RF 1978 SEM/1978 1 133.

Egerton RF 1980 SEM/1980 1 41.

Fotino M 1982 J. of Microscopy 125 265.

Fourdeux A, Sevely J, Zanchi G and Kihn Y 1984 J. de Phys. 45 C2 691.

Garg RK, Sevely J, Kihn Y, Zanchi G, Jouffrey B 1984 in Electron Microscopy 1984 Proc 8th Eur Cong on E.M. 1 437.

Johnson DE, Csillag S and Stern EA 1981 SEM/1981 1 105.

Inokuti M 1971 Rev Mod Phys 43 297.

Kihn Y, Perez JP, Sevely J, Zanchi G and Jouffrey B 1980 in Electron Microscopy 1980 (7th European Congress The Hague) 4 42.

Leapman RD, Cosslett VE 1976 J. Phys.D. 9 L29.

Leapman RD, Grunes LA, Fejes PL, Silcox J 1981 in EXAFS Spectroscopy (Plenum) 217.

Perez JP, Sevely J and Jouffrey B 1974 8th Int Cong on E.M. Canberra 1 380.

Perez JP, Zanchi G, Sevely J and Jouffrey B 1975 Optik 48 487.

Sarikaya M and Rez R 1982 40th Ann Proc EMSA 486.

Scheinfein M, Muray A and Isaacson M 1985 Ultramicroscopy 16 233.

Sevely J, Perez JP et Jouffrey B 1973 C.R. Acad. Sci. (Paris) 276 B 515.

Sevely J, Perez JP and Jouffrey B 1974 High Voltage Electron Microscopy (Academic Press) 32

Sevely J, Garg RK, Zanchi G and Jouffrey B 1985 Proc 25th Meeting French Electron Microscopy Society 33.

Sirvin R 1985 CNAM Thesis Toulouse.

Zanchi G, Kihn Y, Sevely J and Jouffrey B 1983 Proc. 7th Int Cong on HVEM Berkeley 85.

Zaluzec NJ 1980 38th Ann Proc EMSA 112.

# A dual parallel and serial detection spectrometer for EELS

A J Bourdillon[+], W M Stobbs[+], K Page[+], R Home[+], C Wilson*, B Ambrose*, L J Turner* and G P Tebby[o]

[+]University of Cambridge, Department of Metallurgy and Materials Science, Pembroke Street, Cambridge, CB2 3QZ

*W.A. Technology Ltd., Chesterton Mill, French's Road, Cambridge, CB4 3NP

[o]University of Cambridge, Cavendish Laboratory, Madingley Road, Cambridge, CB3 OHE

Parallel detection of electron energy-loss spectra (EELS) has numerous advantages. The high detection efficiency can be used (1) to increase sensitivity for elemental analysis of low concentration elements; (2) to allow measurements on beam-sensitive specimens; (3) to improve fine structure and Compton scattering data; (4) to overcome problems of specimen drift and contamination growth which distort the quantification of serially recorded spectra; and (5) for fast elemental mapping.

The need for parallel recording spectrometers has been appreciated by a number of authors who have attempted to use solid-state array devices (e.g. Egerton, 1984; Shuman 1981 and Johnson et al. 1981), but our design differs from theirs in several important respects, among which are that ours is dual to allow for both serial and parallel recording, our spectrometer has a large dispersion, and we have avoided the difficulties imposed by optical coupling. The reasons for choosing these features will emerge after a brief discussion of the needs of EELS and the features of solid state arrays.

At a high probe size and high convergence the beam current can be of the order of $10^{12}$e/s but at the lowest probe sizes and a low convergence the current can be as low as $10^7$ e/s. A typical electron probe at 120 kV, 40 nm diameter and 6 $10^{-3}$ rad convergence contains $\sim 10^9$ e/s. At a specimen thickness of an extinction distance, typical count rates in a spectrum of e.g. aluminium, will contain after background subtraction $\sim 10^7$ e/s/eV in the plasmon loss, $\sim 10^6$e/s/eV in the L loss above 73eV, and as few as $10^3$e/s/eV in the K loss above 1.5 kV.

A major requirement of a suitable spectrometer is thus that it should have a very high dynamic range, i.e. able to measure signals at count rates from $10^{12}$ electrons/s at the zero loss, to 1e/s at 2.5 kV loss. Since solid-state array devices do not have this range, and since they damage under high doses, it is in fact necessary to use a dual detection system, including a fast serial scan for the low energy region and an array device for the remainder of the spectrum. This dual system can be arranged by deflecting the beam emerging from the spectrometer between the array and a conventional slit, phosphor and photomultiplier. On deflection the beam can be automatically refocussed.

Likewise it is important that the array device should have as wide a dynamic range as possible. Some control over this is possible by ranging between fast readout speeds and slow speeds with long dwell times, but the inherent limit lies in the saturation levels of individual pixels. Normally, it is also important, however, that the array should be sensitive, i.e. able to detect individual incident electrons. If energy-analysed electrons are made to excite a phosphor (Shuman 1971), then the required sensitivity is only achieved by signal amplification with a vidicon tube, and poor spatial resolution at the phosphor also requires subsequent magnification beyond the spectrometer image plane, with accompanying reduction in spectral range. These undesirable features (which involve a long chain of signal conversions from electrons to photons, to electrons to photons and back to electrons again) can be circumvented by employing the natural amplification of signal achieved (without photon conversion) by the scattering in the device of the fast incident electrons. For example, a Reticon S-series self-scanned linear diode array containing the normal 3 μm thick silica surface is reported (Tull, Choisser and Snow 1975) to provide a gain of about $2 \times 10^4$ electrons which is above the switching noise of about $10^3$ electrons. This source of noise can be reduced by slow clocking and thereby extending dwell times (up to ∿10s) till diodes approach saturation, but then a second source of noise, thermal noise of $\sim 10^3$ electrons/s tends to dominate it unless the array is cooled. Thus individual 100 kV incident electrons can be detected up to a saturation limit of 14 pico coulombs in each pixel, corresponding to $5 \times 10^3$ incident 100 kV electrons. The dynamic range is limited - assuming the data is to be digitized for computation - by the speed of analogue to digital conversion with devices of sufficient resolution (14 bit). Stepping times of faster than 30 μs are possible with increasing difficulty, but this corresponds to a speed which can be easily linked to a microprocessor-supported data store and regular display. Then the dynamic range lies between 0 and $10^5$ for a dwell time of 15ms (512 diodes x 30 μs).

Resulting from these considerations our detection system includes a slit and photomultiplier capable of both pulse and analogue counting over the range of $0\text{-}10^{12}$e/s, and an S-series Reticon 512 line array which is normally cooled to -50°C. The array is clocked by a microprocessor interface to a microcomputer for easy interactive data processing, which includes dark noise subtraction and normalization for individual pixel response.

The energy resolution achieved by this system depends on the characteristics of the spectrometer which is accessory to a Philips EM 400T electron microscope. The spectrometer is of the usual magnetic prism design with a large bending radius of 25 cm for large dispersion and pole face geometry (Table 1) calculated using Shuman's program (1980). Focussing is achieved with a double magnetic quadrupole and alignment with tilt and deflector coils at the magnet entrance, while coils at the magnet exit deflect for serial scanning. The dispersion of about 5 μm/eV provides a resolution of 5 eV at the linear diode array, which is sufficient for most applications, while serial recording over narrow spectral regions provides much higher resolution for near-edge structure studies. This is limited by stray fields which are screened, except at the viewing chamber, by mu metal or soft iron. The magnet windings, however, provide compensation for stray fields, and also a de-gausser, while scanning can be either magnetic or electronic at the isolated drift tube.

Figure 1 displays counts in individual pixels from the energy-loss spectrum for graphite. These data match the design parameters of the spectrometer. They were acquired in a time of 0.1 s, though acquisition in one tenth of this time is easily achieved. We also obtained spectra using a simple optical coupling and found that cross-talk reduced the resolution from ∿5 to ∿20 eV. (Figure 2) The phosphur used was JEOL P22 (see Camps and Coslett 1975) with 6 μm grain size and fibre diameter of 6 μm. We assume that the cross-talk arises from the angular distribution of emitted light. In this case the signal intensity (due to equivalent probes) was reduced by orders of magnitude owing to the loss of energy and amplication in the signal which results from photon conversion.

It is known (Egerton 1984) that diode arrays used in the way we have described eventually suffer some damage. We avoid the worst case by deflecting away the relatively intense low loss region to the slit system. Nevertheless we have noticed some increase in dark current after moderate exposure, though this increase can be corrected by annealing to 300°C (Egerton 1984). When large signal currents are to be used however, as in EXELFS, it may in future prove advantageous to protect the linear array by using photon conversion and optical fibre coupling. This can be done without loss of information provided the signal level is well above noise, and can be achieved by increasing the dispersive power by a factor of 3 or 5. A lens beyond the spectrometer image plane will not only serve this purpose but will facilitate energy filtered imaging with a 2-dimensional array (Shuman and Somlyo 1981), and adapt the diode array to near edge studies if used without optical coupling. It is moreover interesting to note in this case, that a similar but asymmetric spectrometer, with entrance at the viewing screen instead of the projector cross-over, would have a larger dispersion of up to 10 μm / eV at the image plane; though it would be far less practical to use.

Finally, our detection chamber is designed to take advantage of future developments in array devices. These are likely to include position sensitive arrays (McMullen, Williams and Sparrow, 1985) or low (switching) noise charge coupled linear arrays when commercially available. The last, however, are in principle more difficult to use than linear diodes as the incident beam must be switched off during readout.

Quantification procedures for EELS are constantly improving (e.g.Bourdillon and Stobbs, 1985), so on the experimental side our spectrometer opens the way to a number of exciting applications, spectra now being obtainable with greater reliability and greater sensitivity even on damage-sensitive materials.

## Acknowledgements

We are grateful to the Royal Society Paul Fund for the grant which enabled this research and to Prof. D. Hull for laboratory facilities.

## References

Bourdillon A J and Stobbs W.M, 1985. Elastic scattering in EELS-fundamental corrections to quantification. To be published in Ultramicroscopy.

Camps R A and Cosslett V E 1975, Mic.elec.a haute tension IV Int. Cong. Toulose.

Egerton R F 1984, J. El. Micr. Tech. 1 37

Johnson D E, Monson K L, Csillag S and Stern E A (1981), Analytical Electron Microscopy, Ed. R H Geiss, San Francisco Press
McMcullan D, Williams B G and Sparrow T, Parallel detection for EELS. EMAG 1985.
Shuman H, 1980, Ultramicroscopy 5 45.
Shuman H, 1981, Ultramicroscopy 6 163.
Shuman H and Somlyo A P 1981, Analytical Electron Microscopy, Ed. R H Geiss. San Francisco Press p.202.
Tull R G, Choisser J P and Snow E H, 1975, The self-scanned digicon: a digital image tube for astronomical spectroscopy. Applied Optics 14 May 1975.

Table 1 Spectrometer parameters

| | | | |
|---|---|---|---|
| Bending radius | 25 cm | Pole separation | 2.0 cm |
| Sector angle | 80° | Fringe field coefficient | 0.174 |
| Entrance tilt | 13.6° | Object distance | 66 cm |
| Exit tilt | 32.5° | Image distance | 52 cm |
| Entrance radius | 15.6 cm | Dispersion | 4.96 μm/eV |
| Exit radius | -15.3 | | |

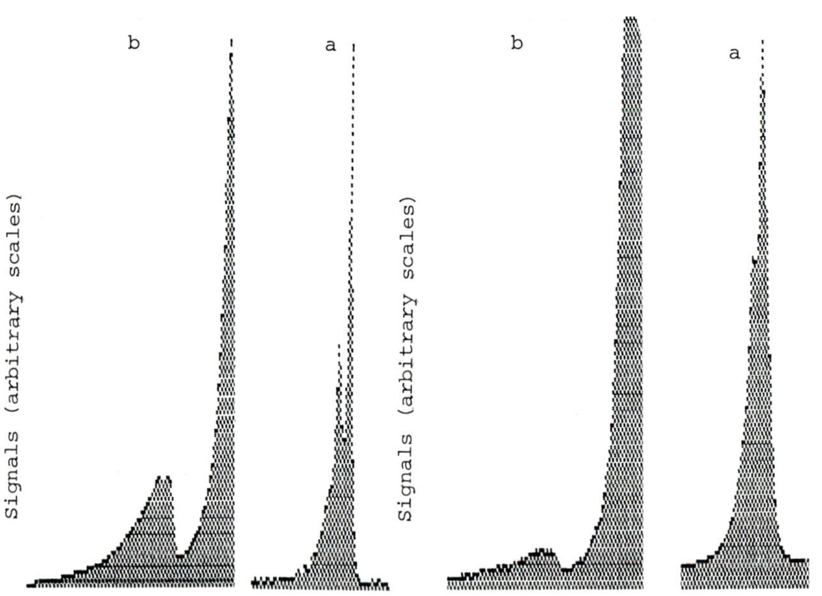


Figure 1. Zero loss and plasmon spectrum (a) and carbon K edge spectrum (b) acquired with self-scanned linear diode array having normal 3μm silica surface, and 100 kV incident electrons, 1 vertical line per diode.

Figure 2. Low energy (a) and carbon K edge (b) spectra acquired as in figure 1 but with optical coupling via phosphor and light guide, and using a larger probe at the specimen.

# Energy filtered imaging by the scanning CTEM image technique

Y Kondo, T Yoshioka, T Oikawa and Y Kokubo

JEOL Ltd., 1418 Nakagami, Akishima, Tokyo 196, Japan

Abstract The energy filtered imaging technique by the combination of the sector type energy analyser and the scanning CTEM image technique, which is provided by the deflectors between the objective and 1st intermediate lenses, is shown. On MgO smoke particles, the zero loss, surface plasmon loss and bulk plasmon loss images were observed with the equal thickness fringes which were the evidence that the specimen was illuminated by parallel beam.

## 1. Introduction

The energy filtered imaging technique has so far been carried out in a scanning transmission electron microscope (STEM) fitted with a sector type energy analyser. (Marks 1981, Oikawa 1982). For phase contrast, the STEM requires a small collection angle which means the detected signal is small. In the conventional transmission electron microscope (CTEM) mode a small angle of illumination is required but all pixels are available simultaneously giving good phase and diffraction contrast which can be observed directly. The technique to obtain energy filtered CTEM images has thus far been carried out by a Castaing-Henry type filter or an Ω type filter. (Castaing and Henry 1964, Zanchi et al 1977) However, at present time, these filters have some disadvantages, those are, lower energy resolution than the conventional sector type energy analyser, that large modification of image forming lens system required, and difficulty to obtain energy loss spectra. This paper reports how to obtain energy filtered CTEM images using scanning CTEM image technique and high energy resolution sector type energy analyser.

## 2. Instrument

Figure 1 shows the ray path of scanning CTEM image technique. In the normal TEM mode, a fixed image is projected onto the fluorescence screen, while in the scanning CTEM image technique, the electron beam is deflected by deflectors located between the objective lens and 1st intermediate lens, and so that the entrance aperture of energy analyser scans the whole CTEM image. In our system, the double deflection system was used in order to obtain stationarydiffraction pattern on the back focal plane of projector lens. The electrons passing through the entrance aperture are filtered by sector type energy analyser ASEA10. The filtered CTEM image signal is fed into frame memory (256x256 or 1024x1024 pixels) of EDS computer (TN-5500), or directly recorded on photo film through the CRT of scanning image device ASID20.

## 3. Experiment and results

In order to show this technique, the MgO smoke particles were selected as a well defined simple specimen. Figure 2 shows the non-filtered image of MgO smoke particles at 200KV, because of the high parallelity of the electron beam, the equal thickness fringes are clearly observed. The spacial resolution (d) of scanning CTEM image technique is defined as following equation.

$$d=D/M$$

D: diameter of entrance aperture
M: magnification of CTEM image

In fig. 2, the entrance aperture of 1 mm diameter and $10^6$ magnification were used, therefore the spacial resolution was 1 nm.
Figure 3(a) and (b) show the electron energy loss (EELS) spectra of MgO on the crystal and at the face of the crystal respectively. In fig. 3(a), large peak which corresponds to bulk plasmon loss is observed at 22 eV. In fig. 3(b), four large peaks are observed, they are good agreement with the result of L. D. Marks who carried out by STEM. From the fig. 3(a) and (b), the energy windows were decided to 0 eV, 10 eV and 22 eV, which were corresponding to zero loss, surface plasmon loss, and bulk plasmon loss, respectively. The reason we chose 10 eV as surface plasmon loss was that the ratio of surface and bulk plasmon loss signals was maximum at 10 eV.
Figure 4(a), (b) and (c) show energy filtered images obtained by the scanning CTEM image technique and sector type energy analyser. Figure 4(a) shows energy filtered image at 0 eV. Because the image is formed by only elastic scattered electrons, the image is clearer than non-filtered image. And the equal thickness fringes also observed. Figure 4(b) shows 10 eV loss image. The area near the (100) face of MgO crystal is bright, because the surface plasmons are excited efficiently at the face. And also, there is slight bright intensity at the edge of the crystal, which might be caused by the tail of bulk plasmon peak. Figure 4(c) shows 22 eV loss image. The areas near the edges of crystals are bright, because those areas have the thickness to excite single bulk plasmon efficiently. The equal thickness fringes are not observed on a large center crystal because the crystal was tilted, however, the equal thickness fringes are observed in left upper small crystal.
In order to show the

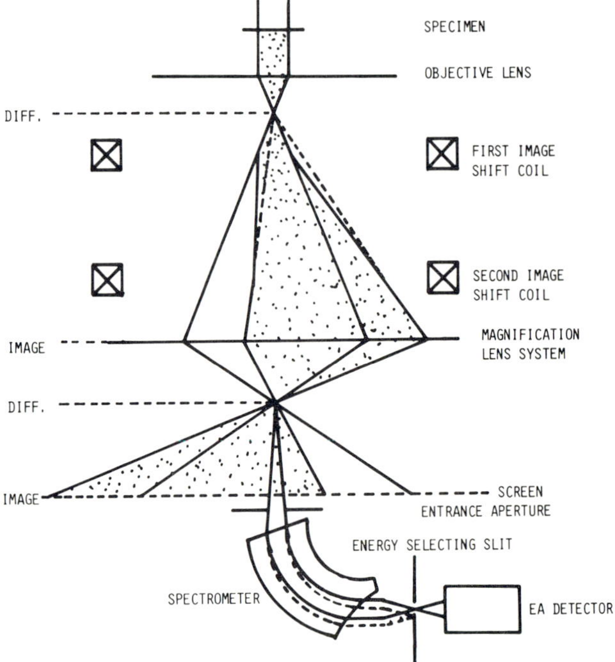


Fig. 1 Ray path of scanning CTEM image technique

Fig. 2 Non-filtered image taken by scanning CTEM image technique.

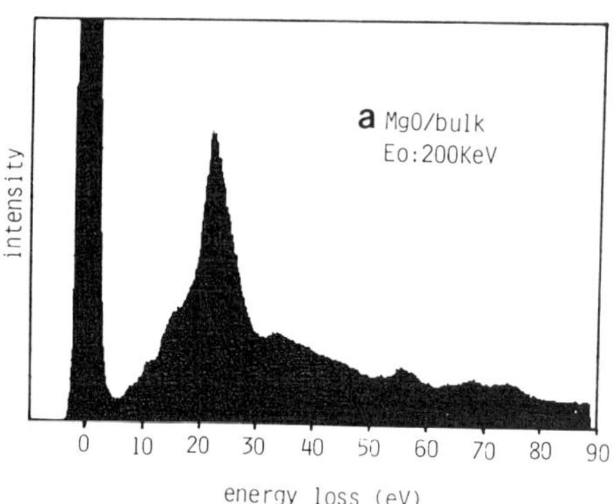


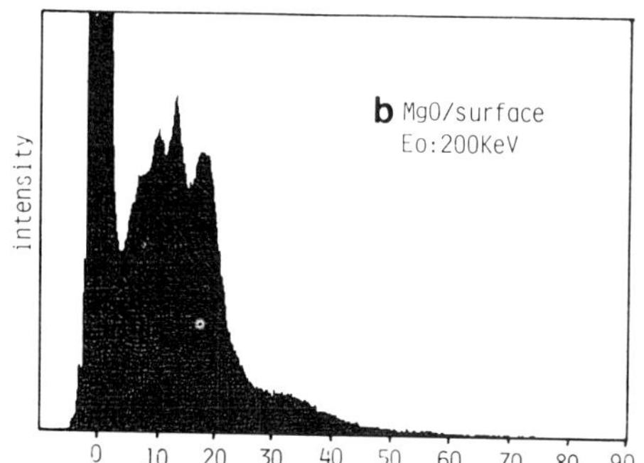


Fig. 3 EELS spectra from (a) area on the MgO crystal and (b) area at the face of crystal

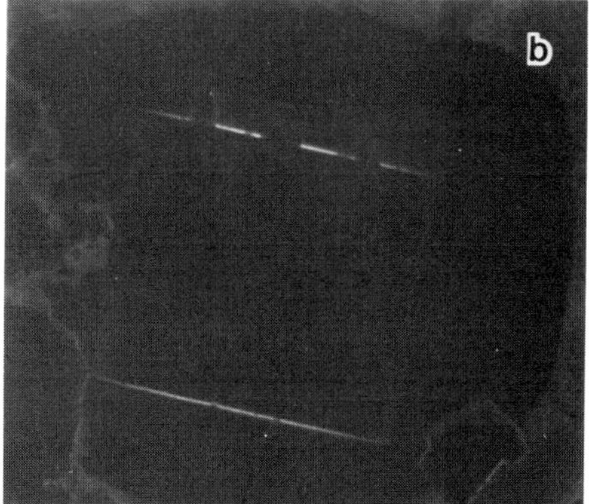


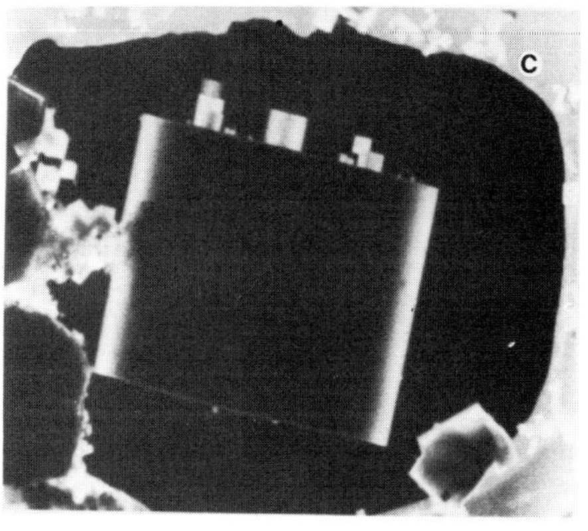


Fig. 4 Energy filtered images of MgO crystal at (a) 0eV, (b) 10 eV, and (c) 22 eV

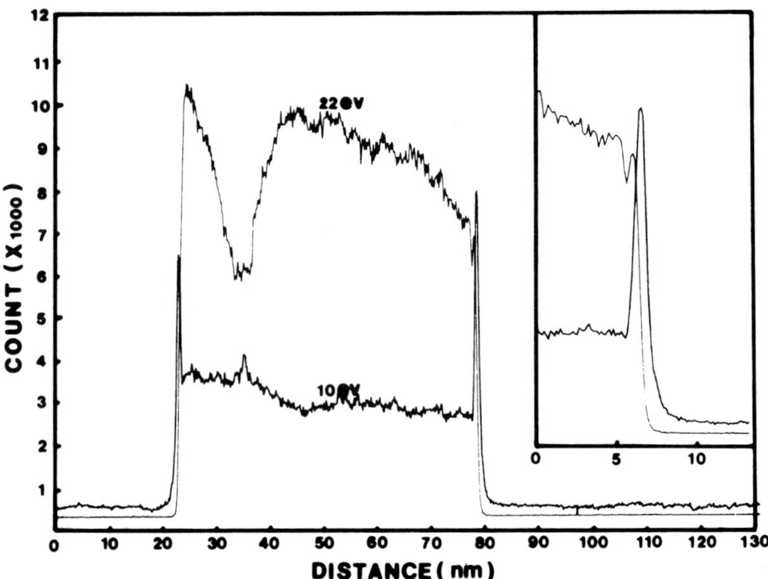


Fig. 5 Line trace of Mgo crystal using 10 eV loss signal and 22 eV loss signal

distribution of surface and bulk plasmon excitation, the line trace of MgO crystal which is perpendicular to (100) face is showed in fig. 5. In order to correct the specimen drift, the both (100) faces of MgO crystal were traversed.The right figure shows the expanded line trace at MgO face. From fig. 5, it is clear that the bulk plasmon was excited inside the crystal, while the surface plasmon was excited outside the crystal. The full width of half maximum of the distribution of surface plasmon excitation was observed 2~3 nm when the crystal thickness was about 10 nm.

## 4. Summary

The advantages of the energy filtered imaging technique using the combination of the scanning CTEM image technique and the sector type energy analyser are showed as follows:

1) The energy filtered image with high energy resolution can be obtained,
2) The specimen can be illuminated by parallel beam (the equal thickness fringes are observed),
3) It is easy to attach to standard analytical electron microscope.

## References

Castaing R and Henry L 1964 J. Microscopie 3 pp 133
Marks L D 1981 Inst. Phys. Conf. Ser. 61 pp 259
Oikawa T et al. 1982 Proc. of EMSA meeting pp 736
Zanchi G, Severy J and Jouffrey B 1977 Optik 48 pp 173

# Parallel detection for EELS

D McMullan

Cavendish Laboratory, Madingley Road, Cambridge CB3 OHE

B G Williams and T Sparrow

Department of Physical Chemistry, Lensfield Road, Cambridge CB2 1EP

## 1. Introduction

Over recent years there have been a number of proposals for the parallel recording of EELS spectra. There have been two main approaches: either to detect the electrons from the spectrometer directly or else to interpose a scintillator and detect the photons (for a review see Egerton 1984). Apart from the difficulties arising from the wide dynamic range, the resolution of the diode arrays used for direct detection and that of the optically coupled detectors are too coarse for the image scale of the compact spectrometers generally used (typically 2 μm/eV). Both types of detector have maximum resolutions of about 20 μm set either by the diode spacing or by the scintillator and the transfer optics, and the spectrometer image scale must therefore be increased either by interposing a magnifying electron lens or by using a larger spectrometer with a correspondingly larger image scale. With all detectors there is the problem of the wide dynamic range that has to be coped with: if the zero-loss peak is included and the spectrum extends to 2 keV the ratio of the signals is of the order of $10^7$ which implies that the detector must be able to operate in both analogue and pulse counting modes.

We have been investigating the possibility of making direct electron detecting arrays having a good enough resolution to work with compact spectrometers without magnifying electron optics and able to cover the wide dynamic range. Our proposal is that the detector should be divided into two completely different sections: one working in analogue or high count-rate modes but accepting only a fraction of that part of the spectrum it covers at a time, say 10%, so that it is equivalent to a multi-slit spectrometer, while the other part provides true parallel pulse-counting detection with a maximum rate per channel of perhaps 5 kHz.

## 2. Position-Sensitive Detector (PSD)

The type of detector we are proposing for the low count rate part of the spectrum is one that has been used for X-ray and optical detection in astronomy, and in particular for charged particle detection in high-energy nuclear physics (e.g. Laegsgaard 1979, England et al 1982). The principle is illustrated in Fig. 1: the charged particle, in the present case an electron, enters a silicon p-n junction on the surface of which is a resistive layer provided with electrodes at each end. The holes and

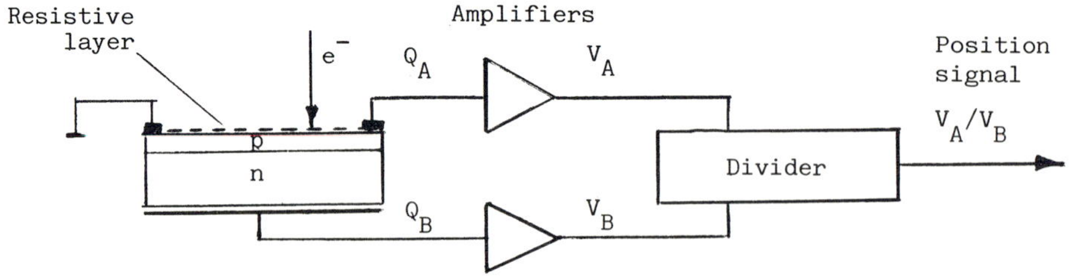


Fig. 1 Position-sensitive detector.

electrons produced by the electron (~ 2.5 x $10^4$ for a 100 keV electron) are collected respectively by the n-bulk contact and by the resistive layer which shares the charge so deposited between the contacts at its ends. One contact is grounded and the other is connected to a charge-sensitive amplifier, the output of which, normalised with respect to the total charge (from a charge-sensitive amplifier connected to the n-bulk), is proportional to the distance from the grounded contact to the point where the high energy electron was incident. The positional accuracy is limited by the signal-to-noise ratio and this depends critically on the Johnson noise produced by the resistive layer and the amplifier noise. Briefly, if the resistance of the layer is R and the time-constant of the amplifier is $\tau$, it can be shown that the root mean square noise charge is given by

$$q = (4kT\tau/R)^{\frac{1}{2}}$$

where k is the Boltzmann constant and T the absolute temperature. If $\tau = 5 \times 10^{-7}$ s and R = 4 MΩ, q is ~ 300 electrons rms and if the amplifier noise is also 300 electrons rms (a fairly easily achieved level) the total noise is ~ 425 electrons giving with 100 keV electrons a signal-to-rms noise ratio of ~ 60. So that if the length of the array is 100 μm the rms positional noise will be 1.67 μm or 4 μm FWHM and there will be effectively 25 channels per 100 μm section.

To cover a 1000 eV length of spectrum 20 sections will be required with 21 amplifiers and associated signal processing channels, each of which will consist of three discriminators, two sample and holds and an analogue divider. Using integrated circuits the cost and complexity will not be high; a typical circuit is shown in a paper by McWhirter et al (1982). In addition there will have to be a computer interface with buffer storage.

Since only one electron event can be handled at a time by any one channel the maximum counting rate is limited by the processing time of the electronics. With the simple electronics envisaged this is ~ 5 μs giving a maximum count rate per section of 200 kHz or 8 kHz per channel if the count rate in all channels were the same. Incidentally this type of PSD does not produce non-linearity due to pulse pile-up because if the pulse rate is too high the proportion of counts lost will be the same in all channels; even so, to avoid problems with normalisation of the different sections the counting rate will probably have to be kept to about half the above figures.

## 3. Multi-Slit Detector

The above count rates are obviously too low for the lower energy loss part of the spectrum and for this we are proposing a simple array of diodes on the same chip. The pitch of the array will be 20 μm but the individual diodes will be masked so that only a 2 μm width of each will be exposed. It appears to be feasible to deposit a 4 μm layer of gold, with 2 μm slots, on the surface of an array (Ahmed 1985); although such a thickness of gold will not stop a 100 keV electron, pulses produced by electrons passing through the gold are expected to produce pulses about $\frac{1}{2}$ the amplitude of those entering the silicon directly. Discrimination between the two sets of pulses will be easy although some false counts will be produced by the coincidence of two or more electrons passing through the mask. Since only one tenth of the spectrum will be recorded it will be necessary to scan the spectrum by an amount greater than the diode pitch to fill up the gaps. Each diode will be connected to its own pulse amplifier and buffer counter but although there will be some 50 channels this again is not considered to be a problem; a similar detector used in astronomy has 212 separate diodes and associated channels (Beaver et al 1976). The maximum count rate per channel will be ~ 2 MHz but this will be insufficient for the zero loss peak. For this it is proposed that analogue detection will be used with the signal collected by a narrow conductor on the surface of the chip. There will remain the low energy loss region where the signal is too high for pulse counting: here the diodes could be used in the analogue mode but discrimination against electrons passing through the gold will not be possible, and some alternative method will have to be found unless the thickness of the gold mask can be increased.

Although this part of the combined detector will not give true parallel detection the gain over a single slit spectrometer will be ~ 50 times for the range 0 - 1 keV. With the PSD section there will also be dead areas at the ends of each section (amounting to ~ 20% in all) but the gain over single slit detection for the range 1 - 2keV will be ~ 200 times.

## 4. Practical Tests

The silicon detector we expect to use in the final system is being developed by Dr. J.B.A. England of the Physics Department of Birmingham University, but some preliminary work has been done using a PSD chip (S1543) manufactured by Hamamatsu for optical detection. The sensitive area is 3 mm long and 1 mm wide and the resistive element is formed by the p-layer itself which is thin and has low doping. The resistance is 100 kΩ and the capacitance is rather high (8 pF). Tests were made with 40 keV electrons in a SEM and by setting the discriminators at a level to accept only double electron coincidences the effect of 80 keV electrons could be simulated. The signal-to-rms noise ratio at the output of the divider circuit was 12 which was rather better than that calculated (10) but even so the resolution was, as expected, too low to be of use. The short detectors mentioned above will have a much lower capacitance and higher resistance and therefore better signal-to-noise.

It was found that under electron bombardment the resistance of the p-layer increased because of radiation damage, probably to the silicon dioxide layer on the surface of the chip. The type of diode being made at Birmingham has a resistive layer of germanium and is radiation resistant.

## 5. Channel Plate Detector

Pending the development of the system described earlier in the paper an attempt is being made to produce a low resolution detector for use in the analysis of Compton scattered electrons (Williams and Thomas 1983). The requirement is for the parallel detection of a very low flux of electrons (producing a spectrum with a single slit spectrometer can take as much as an hour and there is also the problem of radiation damage to the specimen). Only low spectral resolution (~ 30 eV, i.e. ~ 60 μm) is required and it appears possible to obtain this performance from a single PSD 3 mm long if a channel plate with a gain of ~ 1000 precedes the silicon detector.

Again, the initial tests have been made in a SEM at 40 keV and although the channel plate has a poor response to electrons of this energy the beam could be turned up to compensate for this. The even lower response of the plate to 100 keV electrons has not been overlooked, but it is believed that adequate response will be attained with the plate at an angle of ~ $45^{o}$ to the beam. Measurements have been published by Oms et al (1979) on the single electron response of channel plates placed at this angle for energies up to 200 keV. They found that with a plate having 25 μm diameter channels the response fell to a minimum of ~ 5% at an incident energy of 80 keV and then rose to ~ 85% at 180 keV. The fall is due to the low secondary emission ratio at high electron energies and the subsequent rise is because electrons penetrating the channel walls lose energy and produce secondary electrons with a reasonable yield in adjacent channels. If instead of a plate with 25μm channels one with 12.5 μm channels is used the penetration will occur at a lower energy and the curve given by Oms et al can be rescaled accordingly showing that the minimum is now at ~ 50 keV and the response at 100 keV is ~ 70%.

Unfortunately the radiation damage susceptibility of the Hamamatsu device precludes its use but as an alternative we are trying an array of 16 individual diodes, also 3 mm long. The diodes are connected up by resistors to provide the charge division and the charge-sensitive amplifiers are connected to the ends of the resistor chain. The potential of the output surface of the channel plate is held at - 6 kV and the EBIC gain is ~ 1000 so that a single electron entering the channel plate produces ~ $10^{6}$ electron hole pairs in the silicon. The gap between the channel plate and the diode array must be large enough so that each burst of electrons covers at least two diodes; preliminary tests indicate that ~ 10 mm separation is necessary. The measured width of the line profile is ~ 40 μm FWHM equivalent to a resolution of ~ 20 eV.

## References

Ahmed, H. 1985 Private communication.

Beaver, E.A., Harms, R.J. and Schmidt, G.W. 1976 Adv. Elect. & Electron Phys. 40B, 745.

Egerton, R.F. 1984 J. Elect. Microsc. Technique 1, 37.

England, J.B.A., Hyams, B.D., Hubbeling, L., Vermeulen, J.C. and Weilhammer, P. 1982 Nucl. Inst. & Methods 196, 149.

Laegsgaard, E. 1979 Nucl. Inst. & Methods 162, 93.

McWhirter, I., Rees, D. and Greenaway, A.H. 1982 J. Phys. E: Sci. Instrum. 15, 145.

Oms, J., Pouthas, J., Landois, G. and Girard, J. 1979 Nucl. Inst. & Methods 165, 549.

Williams, B.G. and Thomas, J.M. 1983 Intl. Rev. Phys. Chem. 3, 39.

# Digital filters and their limitations in data analysis for electron energy loss spectroscopy

Nestor J Zaluzec and John F Mansfield
Argonne National Laboratory, EM Center for Materials Research
Materials Science and Technology Division, Argonne, Illinois 60439, USA

## 1. Introduction

Background subtraction and removal in electron energy loss spectroscopy (EELS) is one of the more operator intensive and judgemental aspects of EEL microanalysis. Conventional data analysis procedures for EELS generally rely on the fitting of a predetermined function to an operator defined "background" window followed by an extrapolation of the resulting curve underneath subsequent core-loss edges. For the case of high edge/background ratios this method works both quickly and reliably. In the low loss regime (50 - 300 eV), where the edge/background ratios are usually poor, the technique works less successfully and in many cases fails due to the rapidly changing background signal which is not amenable to simple power law modeling. In the worst cases, the failures have entirely precluded the identification of the presence or absence of edges. It has been shown elsewhere (Zaluzec, 1985) that the application of digital filters to EELS data can in many cases remove ambiguities in the interpretation of EELS data by removing the judgemental nature of the background modeling step, particuliarly in situations of poor edge/background ratios. Using this technique to identify the presence or absence of an edge, independent of a background model, allows the analyst to employ without serious reservation more elaborate, yet operator-defined, background modeling routines to recover the relevant characteristic edge profiles. This paper will briefly discuss the advantages and limitations of using these types of filters to spectral analysis of electron loss data in practical data analysis situations.

## 2. Results and Discussion

In this work, symmetric zero-area digital filters of the top-hat (fig. 1a) and second-order top-hat (fig. 1b) type were applied to EEL spectral data to remove the background intensity. The application of this type of filter to EELS data is straightforward. For example, for the case of the top-hat filter, the algorithm is to sequentially pass a step function filter through the spectrum replacing the contents of the centroid channel of the filter with the value obtained by summing the contents of the central positive lobe (C channels wide fig. 1a) multiplying the result by its scaler amplitude $A_C$, and then subtracting the sum of the two lower negative lobes (L channels wide fig. 1a) multiplied by their scaler amplitude $A_L$. If the amplitudes of the positive and negative lobes are chosen such that the areas sum to zero (i.e. $A_C = [-2*A_L*L/C]$, then the application of this type of filter to any linear signal will result in a processed spectrum of zero slope and amplitude. After filtering, the processed data resembles a differentiated spectra similiar to that seen in Auger Electron Spectroscopy (AES). The second-order top-hat filter is a straightforward extension of the first-order top-hat and is useful in cases of extreme background slope. Generally, experience has shown that the central lobe width (C) should be approximately equal to the energy resolution of the EELS data measured at

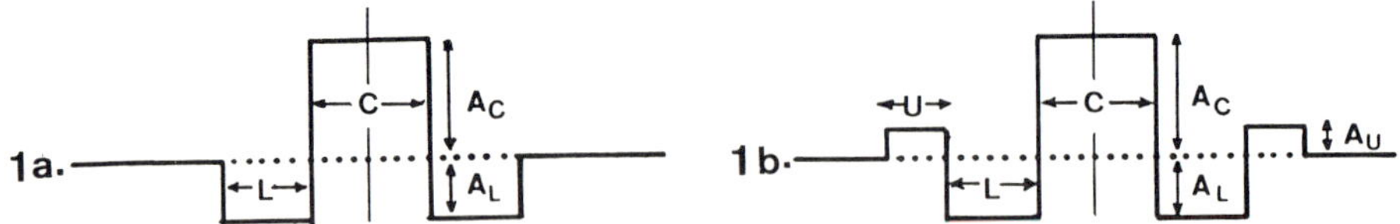


Figure 1. Spectral weighting functions for top-hat (1a) and second-order top-hat (1b) digital filters. The parameters C, L, and U correspond to the filter widths in channels, while $A_C$, $A_L$, and $A_U$ to their amplitudes.

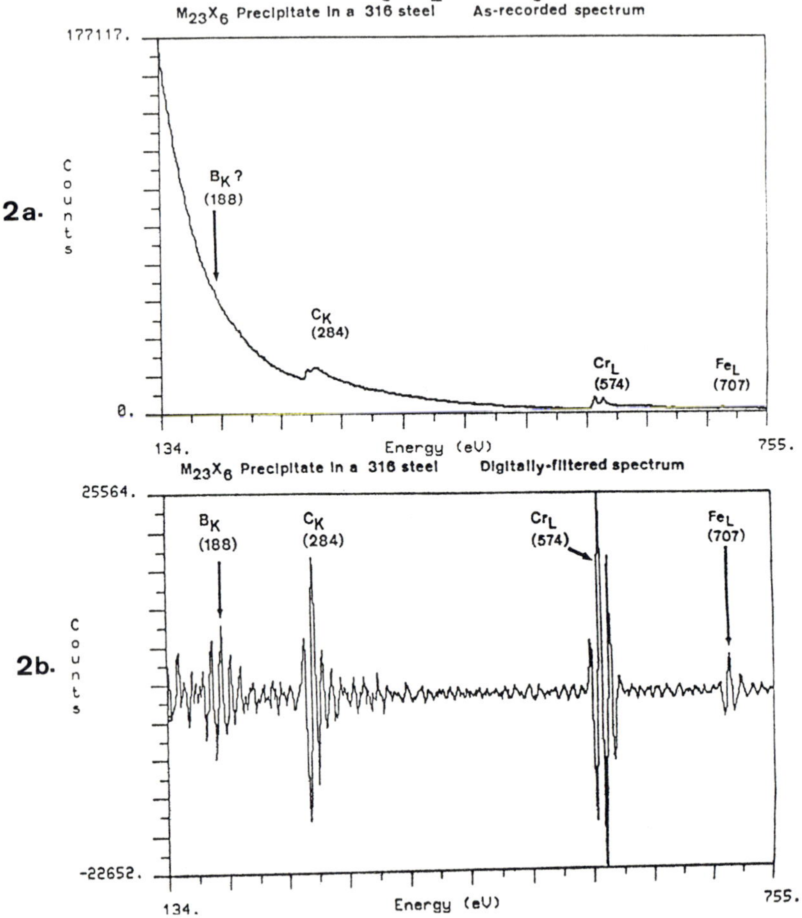

$M_{23}X_6$ Precipitate in a 316 steel As-recorded spectrum

2a.

177117. Counts 0. Energy (eV) 134. 755.

$B_K$? (188) $C_K$ (284) $Cr_L$ (574) $Fe_L$ (707)

$M_{23}X_6$ Precipitate in a 316 steel Digitally-filtered spectrum

2b.

25564. Counts -22652. Energy (eV) 134. 755.

$B_K$ (188) $C_K$ (284) $Cr_L$ (574) $Fe_L$ (707)

Figure 2. a) Partial EEL spectrum from a $M_{23}X_6$ precipitate in 316 SS, location of $B_K$, $C_K$, $Cr_L$, and $Fe_L$ edges is indicated. Note owing to the poor edge/background ratio Boron cannot be identified in the spectrum. b) Digital filter [3,6,2] of spectrum in fig. 2a. $B_K$, $C_K$, $Cr_L$, and $Fe_L$ edges are now clearly identifiable.

Figure 3. a) Partial EEL spectrum from NBS multi-element glass, location of $O_K$, $Fe_L$, $Mg_K$, and $Si_K$ edges is indicated. b) Digital filter [3,6,0] of spectrum in 3a, note problems associated with noise at the $Mg_K$ edge. c) Digital filter [3,60,0] of spectrum in 3a, note the enhanced edge visibility of Mg and Si relative to fig. 3b.

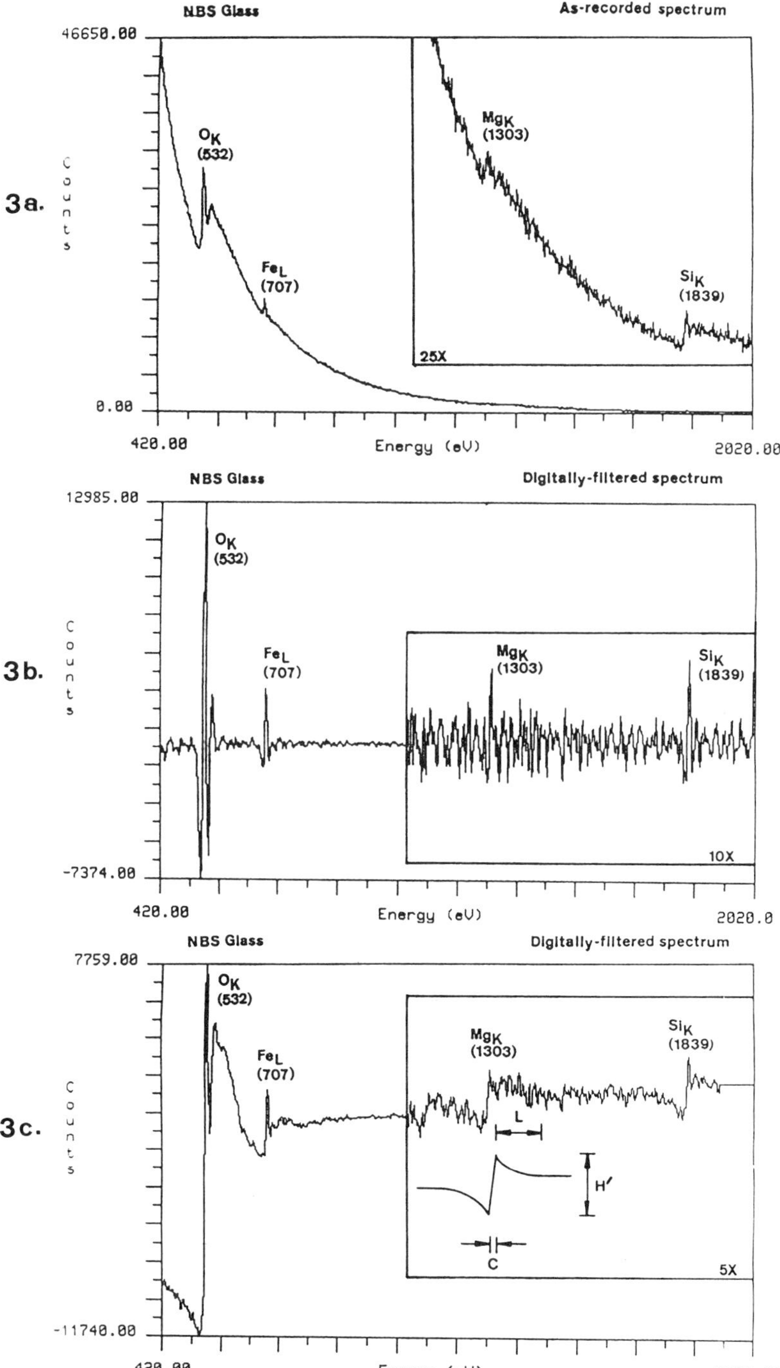
NBS Glass
As-recorded spectrum
46650.00
0.00
Counts
3a.
$O_K$ (532)
$Fe_L$ (707)
$Mg_K$ (1303)
$Si_K$ (1839)
25X
420.00
Energy (eV)
2020.00
NBS Glass
Digitally-filtered spectrum
12985.00
-7374.00
3b.
Counts
$O_K$ (532)
$Fe_L$ (707)
$Mg_K$ (1303)
$Si_K$ (1839)
10X
420.00
Energy (eV)
2020.0
NBS Glass
Digitally-filtered spectrum
7759.00
-11740.00
3c.
Counts
$O_K$ (532)
$Fe_L$ (707)
$Mg_K$ (1303)
$Si_K$ (1839)
L
H′
C
5X
420.00
Energy (eV)
2020.00

a characteristic edge onset (typically 3-4 eV in these studies) and the negative lobe (L) twice that value. Three zero-area filters have been found to be the most useful and will be referred to by the widths in channels of the C, L and U lobes they are: [3,6,0], [3,6,2] and [3,60,0].

As an example of the method, figure 2 shows an experimental EEL spectrum recorded from a grain boundary $M_{23}X_6$ precipitate in a 316 steel specimen. In this phase, M is a transition metal (Cr, Fe, Ni & Mo) while X is a low atomic number element, principly Carbon with possible additions of Boron and/or Nitrogen. Boron autoradiography and SIMS analysis of this alloy has identified the presence of Boron at the grain boundaries of the material. However, neither technique has sufficient spatial resolution to localize the species (Mansfield, 1983). Although the observation of grain boundary precipitates is common in this system, to date no one has been able to unambiguously identify the presence of Boron within the precipitates. The limitation here is the steeply sloping background just before the $B_K$ edge. This signal results from low loss structure (plasmon, inter and intraband transitions, etc.) and M-shell tails of Cr, Fe and Ni edges at lower energies. This precludes the application of power law models for background fitting. Figure 2b shows the result of a [3,6,2] filter operating on the data of figure 2a. Immediate identification of the presence of Boron in the $M_{23}X_6$ phase is possible. It should be stressed here that, unlike conventional background fitting, no operator intervention was required as the filter parameters were determined independent of this analysis.

Previously (Zaluzec, 1985), it was pointed out that the ultimate limitation of this technique is not the edge/background ratios, but rather the signal/noise ratio. An example of this is given in figure 3, which illustrates a spectra obtained from an experimental NBS K411 (Si, Mg, Fe, O, Ca) glass. As noted, for the case of good signal to noise (O, Fe, Si) a standard [3,6,0] filter (fig. 3b) readily identifies the major elements. For Mg, the signal to noise is ~2/1 (fig. 3a) and as one can see in figure 3b this is about the practical limit of detectability using a [3,6,0] filter. This can be improved upon by using a [3,60,0] filter as shown in figure 3c, where the detectability of the Mg and Si edges is substantially improved. This improvement as a function of the width of the negative lobe is not suprising. One can show that the operation of a zero-area (C,L,0) digital filter upon a feature having an original step height of H results in a new spectral shape having an amplitude $H'=2*H*L*A_L$ and a peak to valley width of C (fig. 3c inset). The pre and post edge decay in the filtered spectrum has a characteristic decay width of L (fig. 3c inset), which also serves to enhance the visibility of small edges in poor signal/noise conditions. The disadvantage of the [3,60,0] filter is that one must have a sufficient number of data points between edges to insure that neighboring edges do not interfere with the filter by overlap. This can be overcome by recording spectra at lower eV/channel values rather than the typical 1 to 2 eV/channel that many experimentalists use routinely.

Additional research is in progress to determine appropriate methodology of further noise reduction and quantification of digitally filtered EELS data. This work was supported by the U.S. Department of Energy at ANL.

3. References

Mansfield J F Ph.D. Thesis, Univerity of Bristol, Bristol U.K. 1983

Zaluzec N J Proc. of the Arizonia State University HREM Symposium, Jan. 1985, Tempe, AZ. (in press, Ultramicroscopy) 1985.

*Inst. Phys. Conf. Ser. No 78: Chapter 6*
*Paper presented at EMAG '85, Newcastle upon Tyne, 2–5 September 1985*



# The separation of characteristic signals from complex EELS spectra

J N Chapman, J D Steele, J H Paterson and J M Titchmarsh*

Department of Natural Philosophy, University of Glasgow, Glasgow G12 8QQ.
* Materials Physics & Metallurgy Division, AERE Harwell, Oxon.

## 1. Analysis Techniques

The quantitative analysis of electron energy loss spectra is normally carried out in two stages (e.g. Egerton 1984). Firstly, the characteristic signals are separated from the rapidly varying and generally uninformative background; then, using theoretically predicted cross-sections, the characteristic signals are converted to elemental concentrations. The commonest method of accomplishing the former is to assume that the background varies as $AE^{-r}$ where A and r are chosen to best fit the regions preceding the edges of interest. Unfortunately in many instances the proximity of edges means that the pre-edge regions available for this procedure are limited in extent so that appreciable errors can occur when extrapolation beneath the edges is undertaken. Such errors are particularly serious for weak signals and the resulting poor estimate of the magnitude of characteristic signals frequently limits the accuracy of the analysis.

In this paper we discuss an alternative single-stage technique in which pre- and post-edge spectral regions are fitted to a function of the form $AE^{-r} + k\sigma(E)$ (Steele et al. 1985). Here $\sigma(E)$ is the appropriate cross-section (which is only used in the second stage of the standard procedure) and k is proportional to the number of atoms of the element present. Use of the new technique ensures that values assigned to A and r are compatible with the form of the spectrum beyond as well as in front of the edge and further constrains the characteristic signal to a shape consistent with the known variation of $\sigma$ with E.

To compare the two techniques, spectra from second phase carbides extracted from a ferritic steel are analysed. The energy region of interest is restricted to lie between 500 and 800 eV where the O K, Cr L, Mn L and Fe L edges occur. A spectrum from a typical particle is shown in Fig. 1 and it can be seen that the background in front of each of the transition metal edges is perturbed and that only limited energy windows are available for determination of the characteristic signals. In section 2 we show, using each technique, how the magnitude of these signals varies as the pre-edge fitting region is changed. Thereafter, in section 3, we comment on how determination of the transition metal ratios is affected by the choice of theoretical cross-section model. Finally, as in previous papers (Crozier et al. 1983 and 1984) where the standard EELS analysis scheme was used, we compare the results of the EELS analyses with those from simultaneously recorded x-ray spectra. Details of all experimental conditions can be found in the two earlier papers.

## 2. Extraction of Characteristic Cr, Mn and Fe Signals

In selecting pre-edge fitting regions for the L23 edges of Cr, Mn and Fe a compromise has to be reached between using a relatively wide energy range, which is desirable whenever extrapolation is to be undertaken, and a short energy range which avoids as much as possible perturbation to the background due to preceding edges. The problem is further exacerbated in the case of the Fe L23 edge by the presence of the Cr L1 edge which necessitates all background fitting being performed in front of the latter. To determine to what extent variation of the background fitting region affected the magnitude of the characteristic signals extracted, the number of counts in windows typically 50 eV wide (denoted W1) beginning at the onsets of the edges of interest were calculated as the pre-edge fitting region was extended from 15 to 30 eV in 5 eV steps. The use of larger windows was precluded in the cases of Cr and Mn by the presence of a following edge.

This procedure was carried out using both the techniques described in section 1, the post edge fitting range required in the single-stage technique being held constant. These fitting ranges are shown in Fig. 1 and were typically 25 eV windows (denoted W2) displaced beyond the edge onsets by similar amounts. Such regions were selected to avoid problems associated with transitions to bound states (the 'white' lines) which dominate the spectra near the edge onsets and which are not included in the available cross-section models. Because of this it is also of interest to note the number of counts in the characteristic signals in windows W2 as well as over the maximum energy range available (windows W1).

Table 1: Number of counts in windows W1 and W2 determined using the two techniques

| | W1 x $10^{-5}$ | | W2 x $10^{-4}$ | |
|---|---|---|---|---|
| | Extrapolation | Single-stage | Extrapolation | Single-stage |
| Cr | 18.1 ± 0.7 | 19.3 ± 0.8 | 61 ± 4 | 68 ± 4 |
| Mn | 1.25 ± 0.21 | 1.09 ± 0.12 | 4.2 ± 1.1 | 3.3 ± 0.7 |
| Fe | 15.3 ± 0.5 | 16.4 ± 0.9 | 81 ± 5 | 88 ± 9 |

Table 1 shows the mean number of counts in windows W1 (50-70 eV) and W2 (20-30 eV) obtained for each element as the pre-edge fitting region was varied as described. Also shown is the standard deviation in the means, from which it can be seen that for the larger signals the variations obtained using either technique were <10% for Cr and Fe but considerably greater for the small Mn signal. In the latter case the spread obtained using the constrained fitting of the single-stage technique was significantly smaller than when the simple extrapolation method was used. More disconcerting, however, was the difference in the mean values obtained using the two methods and it is clear that appreciably different estimates of, for example, the Mn/Cr ratio would occur dependent on the background fitting technique employed. We return to this point following a discussion of the different cross-section models in use.

## 3. Use of Cross-Sections in Analysis

As noted in section 1, partial cross-sections are required in the conventional analysis technique to convert ratios of characteristic counts to ratios of atoms present and are required from the outset in the single-stage technique. The cross-sections available for this purpose are the Hartree-Slater (HS) model of Leapman et al. (1980) and the semi-analytic modified hydrogenic (H) model of Egerton (1984). Both are based on atomic models and as such neglect solid state effects, and neither considers transitions to the unfilled d-band which dominate the region near the edge onset for Cr, Mn and Fe. The cross-sections used in section 2 followed the HS model with account being taken of the angular acceptance and resolution of the spectrometer as well as plural scattering within the specimen.

The edge shapes calculated according to both HS and H models and taking account the factors listed above are shown in Fig. 2. It is clear that they differ significantly in shape and magnitude, although the differences become less marked as energies well above the edge onset are attained. To emphasise the differences between experiment (irrespective of the technique used to analyse the data) and either theory, Table 2 shows the ratio of the counts in window W1 to those in window W2. The much higher values of this ratio as determined experimentally emphasise how serious is the neglect of the 'white' lines in the theories and the dangers of using the counts in the full window in conjunction with either theoretical cross-section model if accurate quantitative analyses are required.

Table 2: Ratio of counts in window W1 to those in window W2

| | HS | H | Extrapolation | Single-stage |
|---|---|---|---|---|
| Cr | 2.39 | 2.17 | 2.95 | 2.85 |
| Mn | 2.07 | 2.11 | 2.97 | 3.26 |
| Fe | 1.51 | 1.55 | 1.94 | 1.91 |

Although the most significant differences between the HS and H edge shapes occur within the first 30 eV, their shapes within the fitting regions shown in Figs. 1 and 2, also differ. Thus different results are obtained when H cross-sections rather than those predicted by HS are used in the single-stage technique. To date, we have found no evidence (e.g. goodness of fit as measured by a $\chi^2$ test or agreement with the results of background extrapolation using simpler spectra) to suggest that one model is superior to the other for a range of elements. To give an indication of the reliability of the single-stage technique for analysing the second phase particles of interest, Fig. 3 shows a graph of the Cr/Fe ratio obtained by EELS versus that obtained when simultaneously recorded energy dispersive x-ray spectra were analysed. The latter values were believed to be accurate to 4% (Crozier 1985). EELS results using HS and H models are shown and whilst differing from each other as expected, both show fair agreement with the x-ray results.

We conclude from the results presented here that considerable difficulties exist when trying to analyse EELS spectra containing a range of adjacent transition metals. This is partly attributable to the inadequacy of the theoretical cross-sections when only small energy windows close to the edge onsets are available and affects both analysis techniques. Despite this, the single-stage technique offers an internally consistent approach, appears to offer some advantages when dealing with small signals and provides results in every way comparable to those described previously by Crozier et al. (1983, 1984) and Crozier (1985).

Crozier P A 1985 Ph.D. Thesis, University of Glasgow

Crozier P A, Chapman J N, Craven A J and Titchmarsh J M, Proc EMAG 1983 (ed. P Doig, IOP) 107 and Analytical Electron Microscopy 1984 (eds. D B Williams and D C Joy, San Francisco Press) 79.

Egerton R F in: Quantitative Electron Microscopy (eds. J N Chapman and

A J Craven, Edinburgh University Press) 1984 ch. 7.
Egerton R F 1984 Scanning Electron Microscopy 1984 (ed. O Johari) 505
Leapman R D, Rez P and Mayers D F 1980 J. Chem. Phys. 72(2) 1232.
Steele J D, Titchmarsh J M, Chapman J N and Paterson J H 1985 Ultramicroscopy - accepted for publication.

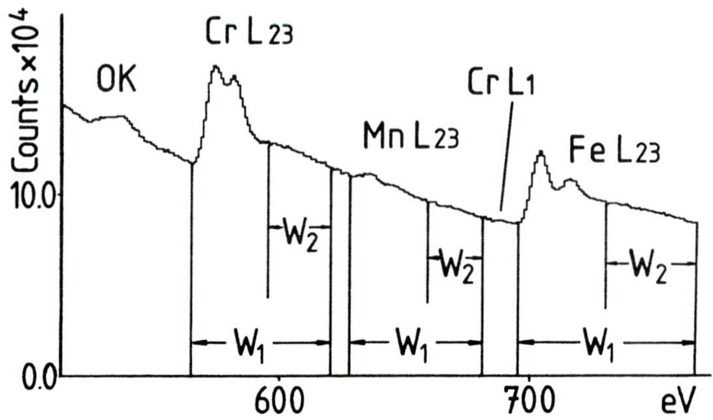


Fig. 1. EELS spectrum taken from a second phase particle extracted from a ferritic steel.

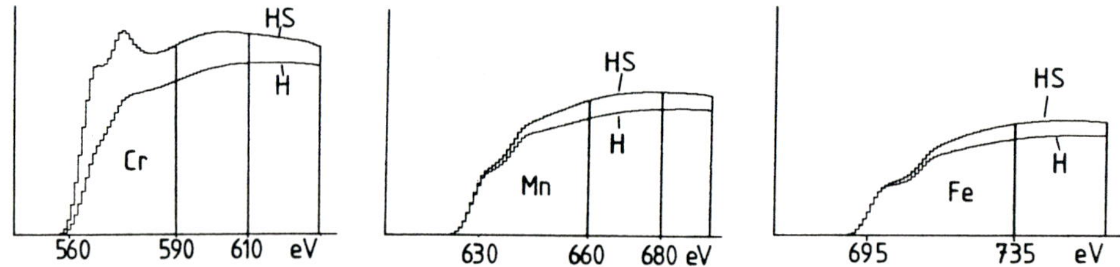


Fig. 2. Calculated L23 edge shapes taking account of the conditions under which fig. 1 was recorded.
H.S. - Hartree-Slater (Courtesy Dr. P. Rez)
H - Hydrogenic.

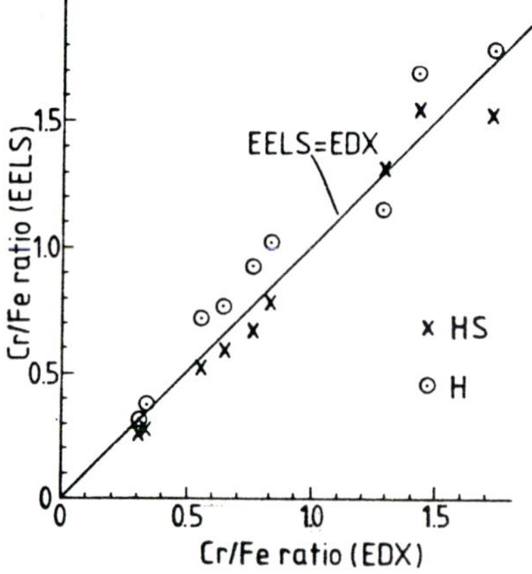


Fig. 3. Cr/Fe ratios determined using the single-stage EELS analyses technique vs. those obtained from simultaneously recorded X-ray spectra.

*Inst. Phys. Conf. Ser. No 78: Chapter 6*
*Paper presented at EMAG '85, Newcastle upon Tyne, 2–5 September 1985*



# A 'k-factor' approach to EELS analysis

T Malis* and J M Titchmarsh+

* PMRL/CANMET, 568 Booth St., Ottawa, Canada K1A 0G1
\+ A.E.R.E. Harwell, Didcot, Oxon, U.K.

Accurate chemical analysis in the AEM using EDX normally is performed by the use of experimentally determined sensitivity factors for the required elements. However, EELS quantification has been achieved only by using calculated partial differential cross-sections. In view of the known discrepancies between the experimental and theoretical edge shapes, it is surprising that no serious attempts have been reported to determine, from standards, experimental cross-section ratios ('k-factors') for EELS analysis. Such an attempt is summarized in this paper.

Experiments were performed using a Gatan 607 spectrometer in the pulse counting mode at 120 keV on a Philips EM400T, operating in normal imaging mode. Convergence and acceptance angles were held constant for all spectra (2.0 and 5.6 mr, respectively). Fifteen standard samples either of known homogeneous composition or accepted to be stoichiometric were analysed to obtain 'k-ratios' involving 11 elements. Several spectra were processed for each sample, with $t/\lambda<1$ for all spectra. Both $Ae^{-R}$ and log-polynomial fitting were employed to remove background signals, before deriving ratios of counts for energy windows, $\Delta$, of 20 to 100 eV. The averaged count ratios for each $\Delta$ were converted to partial cross-section ratios and compared with relativistic theoretical values from SIGMAK and SIGMAL (Egerton, 1984). Examples are shown in Fig.1. Close agreement was found for ratios of K-edges, e.g. Fig.1a, however ratios of L/K and L/L edges often showed differences for the whole or part of the energy window range, e.g. Fig.1b.

It was then assumed that SIGMAK accurately predicts the value for the B-K ($\Delta$=80 eV) cross-section. The absolute values of the cross-sections for the other 10 elements were derived from combinations of the experimental ratios and/or this value for B-K. These values are plotted in Fig.2 together with the SIGMAK and SIGMAL values. Whereas agreement for K-edges is within about 10% (Fig.2a), there appears to be a systematic departure of experimental L-edge results from SIGMAL values for the transition elements (Fig.2b), e.g. Zn is a factor of two different. This is significantly greater than the differences ($<$10%) between SIGMAL and the more rigorous Hartree-Slater calculations (Egerton, 1984). Other published values of cross-section ratios, for various collection conditions, have been used to derive L-edge values assuming the K-edge value is accurately predicted by SIGMAK. These are also shown in Fig.2 and are consistent with the trend revealed by the present work. Therefore this work suggests that any quantitative analysis involving L-edges will be more accurate using experimentally determined 'k-factors' than using existing theoretical values for L-shell partial cross-sections.

## References

Allison C.,Williams W.S. and Hoffman M.P.,1984,Ultramicroscopy,13, p.253.

Egerton,R.F. ,1984, SEM/1984/II (Chicago, SEM Inc.) p.505
Grande M. and Ahn C.C.,1984, Inst.Conf.Ser.No.68 (Bristol, Inst.Phys.)p.123

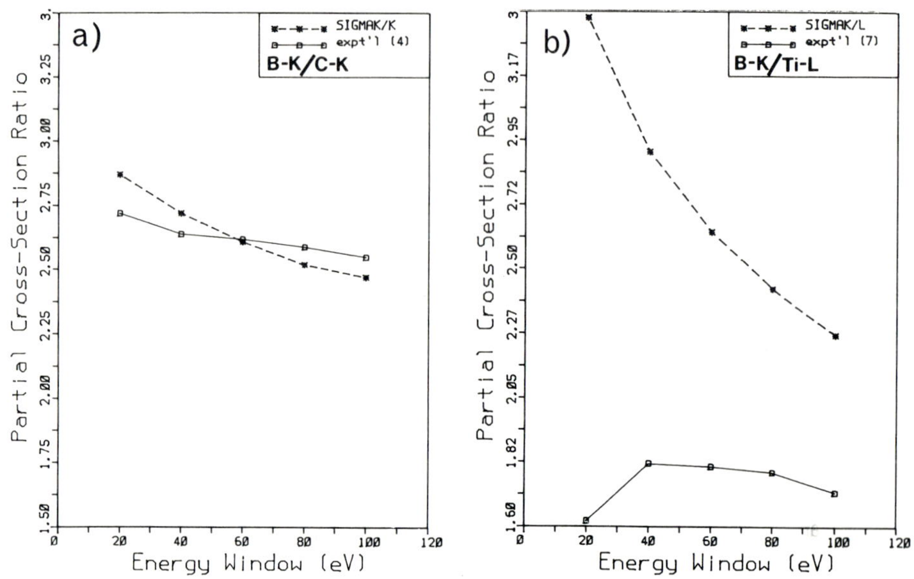


Figure 1 Partial cross-section ratios for a) $B_4C$ and b) $TiB_2$ averaged over 4 and 7 spectra, respectively.

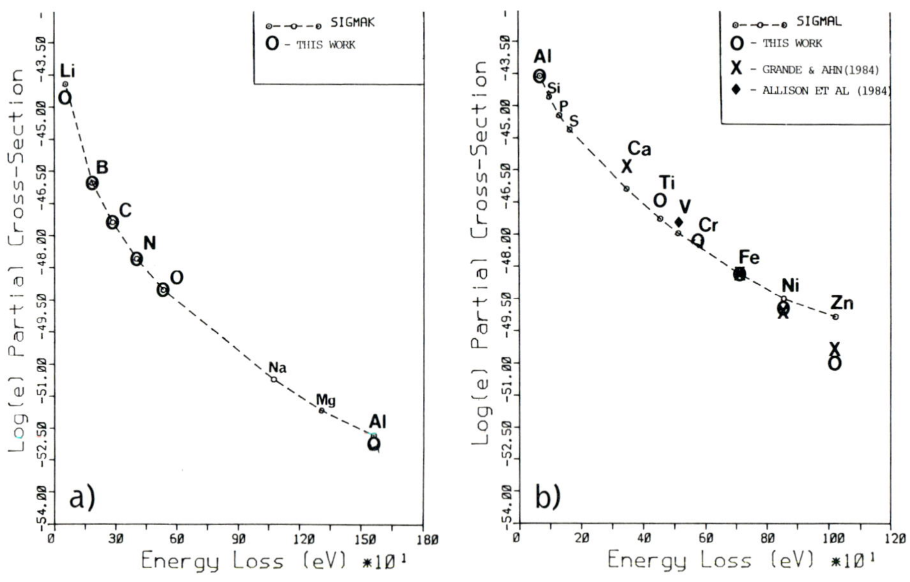


Figure 2 Comparisons of experimental partial cross-sections with calculated cross-sections from a) SIGMAK (Z=3 to 13) and b) SIGMAL (Z=13 to 30) assuming B-K ($\Delta$=80eV)-SIGMAK is accurate.

*Inst. Phys. Conf. Ser. No 78: Chapter 7*
*Paper presented at EMAG '85, Newcastle upon Tyne, 2–5 September 1985*


# Light element analysis in the TEM using windowless x-ray detectors

P J Goodhew

Department of Materials Science and Engineering, University of Surrey, Guildford GU2 5XH.

## 1. Introduction

X-ray microanalysis of thin specimens in the TEM is now commonplace, and techniques for the quantitative analysis of elements in the range sodium to uranium are fairly well established (eg 1). In the last five years Si(Li) detectors with ultra-thin polymeric windows (UTW) or with removable beryllium windows ("windowless") have been in use on SEMs, where constraints on space are relatively minor and where ample X-ray signals are generally available. X-ray peaks with energies down to 185eV (boron K) can then be detected. More recently a significant number of TEMs around the world have been fitted with windowless detectors and it is the purpose of this paper to report on the success of this approach to the analysis of light elements. It was less clear a few years ago that windowless EDS would be the best way forward, since EELS appeared to offer an excellent technique for the quantitative analysis of light elements. I hope to show that both techniques are necessary, but that in most cases windowless EDS offers a simpler, faster route to useful compositional information on a fine scale.

## 2. The detector

Since the absence of a window exposes the cooled detector crystal to the microscope environment it is necessary either to withdraw the detector behind a vacuum valve or to cover it with a beryllium window whenever it is not in use. The majority of designs currently available use a rotating turret to provide two or three alternative configurations - a conventional vacuum-tight beryllium window, an ultra-thin window (usually an aluminium-coated organic film) or no window. The active area of the detector is typically 10mm$^2$ and because of the inevitable bulk of the turret mechanism it can rarely be mounted closer to the specimen than 30mm in a conventional TEM/STEM. Consequently the detector collection efficiency (manifested as count rate) tends to be a factor three to ten lower than for "conventional" 30mm$^2$ detectors which can be mounted a little closer.

Even in its "windowless" mode the detector is still not perfectly efficient for soft x-rays. There remains a gold contact layer, of thickness typically 20nm, and what used to be called the "silicon dead layer" at the surface of the crystal. The gold layer is responsible for some absorption, so that some x-rays hit the detector but are not counted and therefore contribute to no part of the spectrum. This effect is in principle calculable but is complicated by the fact that the layer may not be of uniform thickness and absorption effects are not linear with thickness. Consequently a patchy layer does not absorb as if it were a uniform layer of the same mean thickness (eg Statham 1981). The effect of the silicon "dead layer" is even more complex, since it is likely to be responsible for incomplete charge collection (eg Nicholson et al 1984). This will cause some x-rays to be registered as having a lower energy than they actually possess and spectral peaks distorted towards lower energies will result. This is a serious effect in the soft x-ray region since there are potentially so many peaks which could appear (see below).

## 3. The potential for analysis

A windowless detector is in principle capable of analysing all elements apart from H, He and Li. However the low energy noise generally extends to at least 100eV and it is unlikely to be practicable to analyse for Be with a K peak at 110eV. One of the key parameters in determining whether analysis is likely to be successful must be the x-ray production efficiency, given by the product of the ionization cross-section and the x-ray fluorescence yield divided by the atomic weight. Thomas (1984) has shown that, for K x-rays excited by 100keV electrons, this term is about the same for B as for Cu, rising to a peak about 2.5 times this value at O. In terms of the number of x-rays produced it should therefore be no more difficult to analyse for boron than for copper. The other factor which makes it in practice more difficult to analyse for boron is absorption, in both specimen and detector dead layers. An example which illustrates this problem is considered below (section 5).

A second problem arises not from the efficiency of the detector or from the lack of x-rays to interpret, but from the analytical inconvenience that most specimens worth analysing contain two or more elements. The K peaks from the light elements are quite close together. Detector resolution improves at low energies so that peaks from adjacent light elements can usually be resolved, unless the intensity of one is very different from that of the other. However the problem arises because the L lines of the transition elements also lie in the region below 1keV, as shown in Figure 1. In a modern detector one can expect a FWHM resolution of about 100eV for peaks below 1keV so separating oxygen from vanadium, chromium, titanium or even manganese is difficult. (This author has the ultimate analytical misfortune of wishing to analyse for neon in nickel. All other workers should be encouraged that their problems are unlikely to be more intractable.) M lines tend to be less of an inconvenience only because few analytical problems feature the elements xenon to samarium.

Apart from peak overlap problems the use of a windowless detector for qualitative analysis is no different from the familiar situation with a conventional detector. The detection limit is controlled by count rate, which can be selected by the usual trade-off between analysed volume and the beam current.

## 4. Quantitative analysis

The usual k-factor approach to quantitative analysis can be applied to spectra collected with a windowless detector and in principle better results should be obtained in many cases because it is no longer necessary to ignore carbon, oxygen or nitrogen. However additional problems arise, among which are:

a. absorption corrections will almost always be significant,
b. detector noise thresholds at the low energy end of the spectrum must be set carefully, particularly if boron is sought,
c. suitable standards are needed, which show light element peaks well separated from other peaks,
d. k-factors may need calibrating much more frequently than with a conventional detector, because of changes in detector performance with time.

There is little point in tabulating k-factors in this paper since, even more than for a conventional detector, they depend critically on the hardware. $K_{OFe}$ values ranging from 1.08 to 3.6 have been reported in the literature (Garratt-Reed et al 1984, Maher et al 1984, Thomas et al 1984) and we have measured, on our single system, values from 1.5 to 4.25. The reasons for this vast range are not entirely clear but it is easy to identify several contributory factors. Any windowless detector will tend to ice up over a long period of use. The development of an ice layer results not only in the absorption of soft x-rays but also in the fluorescence of an additional oxygen K signal. These factors operate on the measured k ratio in opposite directions, and the fluorescence of oxygen depends on what else is in the standard. This is a very likely reason for the variation of k factors from standard to standard. A further reason for variability lies in the choice of standard. It is not easy to select an ideal standard for oxygen. Mineral standards tend to be fairly thick, as usually prepared, and hence need thickness measurements and large absorption corrections. They may also contain an unknown number of hydroxyl groups. Sputtered oxide glass standards of uniform, small thickness would be an improvement but the NBS glasses intended for this purpose (Steel et al 1981) have not yet been fully characterised and released.

Since absorption corrections are likely to be important in most windowless analyses it is of interest to examine how robust they are to errors in the experimental parameters. The principle of absorption correction has been described many times (eg Goldstein 1979). The parameters which it is necessary to know are the local specimen thickness and density, the take-off angle (toa) to the detector and the mass absorption coefficients for all the lines used in all elements in the specimen. Specimen thickness is notoriously difficult and tedious to determine rapidly with any accuracy and in few analytical experiments would the workers claim an experimental error better than ±10%. Similarly the local density of the specimen cannot be measured and must be assumed to carry a similar error. Take-off angle may be known to within a degree if the sample is lying perfectly flat in the holder. This is rather unlikely for most specimens and an uncertainty of ±5 degrees is probably optimistic. Mass absorption coefficients (macs) have been tabulated and paramaterized by numerous workers and there is a reasonable measure of agreement even for soft x-rays in many elemental absorbers. The tabulation by Henke and Ebisu (1973) is probably the most widely used for light elements. However the greatest area of disagreement between authors is for light element x-rays in absorbers of medium atomic number, and it is possible to find differences of more than a factor of two between (already large) macs in this vital area. For the purposes of the present analysis we assume an uncertainty of only ±10% on all macs. Using these uncertainties the data in Table 1 have been calculated for a variety of simple analytical situations. Shown in the table are the "ideal" and two "worst case" values of the absorption correction which would need to be applied to a thin-film k-factor.

Table 1

| Sample | thickness (nm) | toa (deg) | density (g/cc) | correction factors for light element/heavy low | ideal | high | % variation |
|---|---|---|---|---|---|---|---|
| $SiO_2$ | 100 | 37 | 1.73 | 1.039 | 1.060 | 1.094 | + 2/- 2 |
| $Cr_{23}C_6$ | 100 | 37 | 6.7 | 1.407 | 1.660 | 2.059 | +24/-15 |
| $Cr_{23}C_6$** | 100 | 37 | 6.7 | 4.881 | 7.425 | 11.109 | +50/-34 |
| WC | 50 | 37 | 15.8 | 1.813 | 2.358 | 3.250 | +38/-23 |
| WC | 20 | 37 | 15.8 | 1.298 | 1.482 | 1.785 | +20/-12 |
| WC ** | 20 | 37 | 15.8 | 6.006 | 9.088 | 13.634 | +50/-34 |

** = macs from Bracewell & Viegele (1971)
other macs from Heinrich (1966) and Henke & Ebisu (1973)

It is clear that if a light element is to be analysed in a medium to heavy matrix not only are the values of the macs crucial, but also the uncertainty in determining the appropriate absorption-corrected k-factor may be very large. What is needed is a set of well-characterised standards of different uniform thicknesses, which could be used to determine k-ratios and to check macs. There is no reason why such a set of sputter-deposited oxide glasses should not be prepared, but an alternative approach may be needed for carbon.

## 5. Examples of windowless analysis

Two recent analytical problems from our laboratory illustrate some of the strengths and pitfalls of windowless analysis.

Oxygen in glassy CuZr: A spectrum from this material is shown in Figure 2. It is clear that a substantial amount of oxygen is present and we need to distinguish between a surface oxide film and a uniform oxygen-containing glass. Conventionally one would do this by collecting spectra at a variety of specimen thicknesses and observing the oxygen to copper ratio, which would drop for a film but remain constant for a homogeneous sample. However, for oxygen the analyses need to be corrected substantially for absorption as one would expect the raw O/Cu ratio to drop even if the composition is uniform. The corrected analyses from such an experiment are shown in Table 2. The straightforward interpretation is that since the corrected concentration is virtually constant the oxygen must be uniformly distributed throughout the foil. However when the raw data are studied more carefully it becomes clear that this is not the correct interpretation. If the oxygen is homogeneously distributed then the O K intensity should increase with thickness, reaching a constant level when the oxygen signal from the bottom of the sample is totally absorbed (at about 500 nm thickness). However the raw (background subtracted) oxygen peak intensity actually falls steadily with increasing thickness (Figure 3), which is only consistent with a surface oxide film. This example illustrates how careful it is necessary to be to avoid misinterpretation of "corrected" data.

Boron-containing glassy phase: This phase contains boron and oxygen as well as a number of heavier elements (Gregg et al 1985). Although the composition appears to be variable the average oxygen content implied by the windowless spectra (ignoring boron) is 62wt%. However EELS analysis of the same samples indicates a ratio of boron to oxygen atoms of about 20. Thus the rather small boron peak shown in Figure 5 of Gregg et al apparently comes from a boron concentration of 90wt%. On this basis the detection limit for boron with this detector is as high as 8.5wt%. The lesson from this exercise is that the combination of EELS and windowless x-ray detector is needed for this type of problem.

## 6. Acknowledgements

I would like to thank LINK Systems for their cooperation and Pam Budd, Dawn Chescoe, Ray Cox, Nick Gregg and Robin Payne for their results and advice.

## References

Bracewell B.L. and W.J. Viegele 1971 in Dev. in Applied Spectroscopy Eds E.L. Grove and A.J. Perkins 9, 357-400.

Garratt-Reed A.J., T. Thorvaldsson & J.B.Van der Sande 1984, in Analytical Electron Microscopy - 1984 Eds D.B. Williams and D.C. Joy, San Francisco Press, 345.

Goldstein J.I. 1979, in Introduction to Analytical Microscopy, Eds J.J.Hren, J.I. Goldstein and D.C. Joy, Plenum 83.

Gregg N.R., P.M. Budd and P.J. Goodhew 1985, this volume

Heinrich K.F.J. 1966, in The Electron Microprobe, Ed T.D. McKinley, 296.

Henke B.L. and E.S. Ebisu 1973 Adv. in X-ray Analysis 17, 150-213.

Maher D.M., D.C. Joy and G. Cliff 1984, in Analytical Electron Microscopy-1984 Eds D.B.Williams and D.C.Joy, San Francisco Press, 341.

Nicholson W.A.P., P.F. Adam, A.J. Craven and J.B. Steel 1984, in Analytical Electron Microscopy - 1984 Eds D.B. Williams and D.C. Joy, San Francisco Press, 258.

Statham P.J., 1981, NBS Special Technical Publication 604, 127.

Steel E., D. Newbury & P. Pella 1981, in Analytical Electron Microscopy - 1981, Ed R.H. Geiss, San Francisco Press, 65.

Thomas L.E. 1984 in Analytical Electron Microscopy - 1984 Eds D.B. Williams and D.C. Joy, San Francisco Press, 358.

Thomas L.E. et al. 1984, in Analytical Electron Microscopy - 1984 Eds D.B. Williams and D.C. Joy, San Francisco Press, 333.

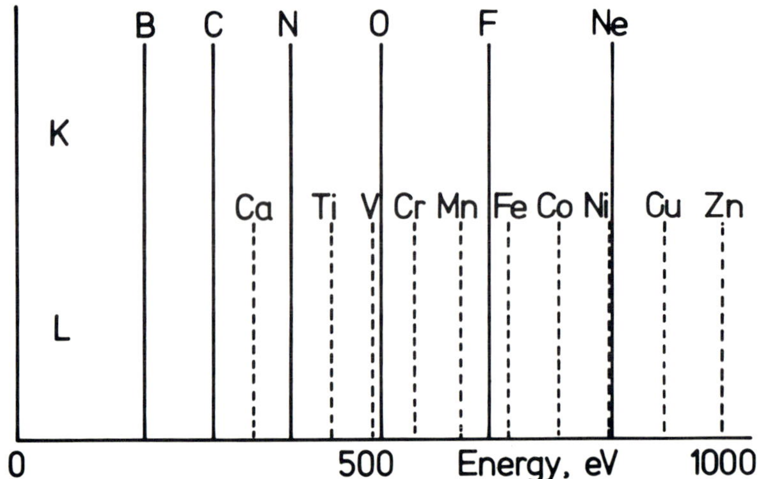


Figure 1. The K and L line overlaps at energies below 1 keV.

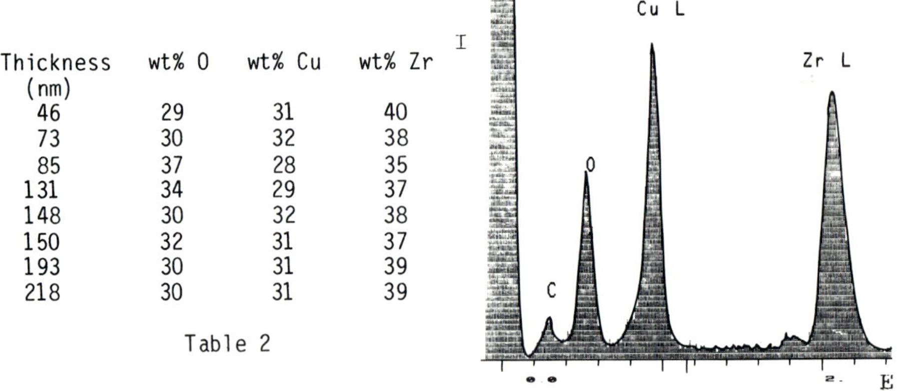

| Thickness (nm) | wt% O | wt% Cu | wt% Zr |
|---|---|---|---|
| 46 | 29 | 31 | 40 |
| 73 | 30 | 32 | 38 |
| 85 | 37 | 28 | 35 |
| 131 | 34 | 29 | 37 |
| 148 | 30 | 32 | 38 |
| 150 | 32 | 31 | 37 |
| 193 | 30 | 31 | 39 |
| 218 | 30 | 31 | 39 |

Table 2



Figure 2. A spectrum from CuZr glass

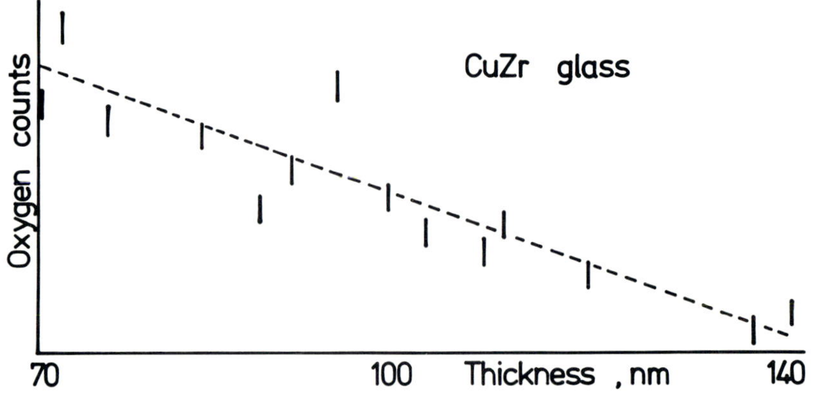


Figure 3. The oxygen K peak area versus thickness for the oxidised CuZr glass.

*Inst. Phys. Conf. Ser. No 78: Chapter 7*
*Paper presented at EMAG '85, Newcastle upon Tyne, 2–5 September 1985*


# Incomplete charge collection and x-ray microanalysis

A J Craven, P F Adam and R Howe

Department of Natural Philosophy, University of Glasgow, Glasgow G12 8QQ

## 1. Introduction

The response of a Si(Li) energy dispersive x-ray detector to monoenergetic photons is normally idealised as a Gaussian peak whose mean position is proportional to the photon energy. The full width at half maximum height is determined by noise in the measuring circuitry for low photon energies and statistical fluctuations in the number of carriers generated for high photon energies. There are many deviations from this ideal behaviour (e.g. Statham 1981). For a fraction of the incident photons, not all the energy is deposited in the crystal or not all the carriers are separated into the external circuit. Counts then appear on the low energy side of the main peak. An example of the former is the escape peak caused by the escape from the crystal of a Si K x-ray produced in the cascade of processes resulting from the initial absorption. The latter occurs if there are bulk or surface recombination traps in the crystal. It also occurs if there is a layer at the surface of the crystal which is outside the depletion layer of the pn junction but from which the minority carriers can diffuse into the depletion layer. Such mechanisms give rise to a continuous tail on the low energy side of the peak. This effect is known as incomplete charge collection (ICC). Previous work studying the x-ray performance of the VG Microscopes HB5 scanning transmission electron microscope has shown the significance of ICC (Craven et al 1984) and its effect on the analysis biological systems (Steele et al 1984) and a proposed self absorption correction (Craven and Adam 1984).

## 2. Experimental Measurement of ICC

A source producing characteristic x-rays without bremsstrahlung can be made using a strong $Fe^{55}$ radioactive source. The source can be used directly to give Mn $K_\alpha$ and $K_\beta$ x-rays or used to produce x-rays of lower energy by fluorescence. In addition to the fluoresced x-rays, there is always a fraction of backscattered Mn K x-rays present but their contribution can be subtracted using a suitably scaled spectrum from them alone. With the exception of Ar x-rays, the experiments with the higher energy x-rays were carried out in a $N_2$ atmosphere to avoid fluorescing the Ar present in air while, for the lower energy x-rays, a He atmosphere was used to minimise absorption. For Ar x-rays, an Ar atmosphere was used as the target.

The results are plotted in fig. 1 with a logarithmic intensity scale. The tail due to ICC is clearly seen as are the escape peaks. The ratio of the intensity in one channel of the tail to that in the highest channel of the $K_\alpha$ peak is $\sim$0.001 for Mn x-rays, rising with decreasing photon energy to

∿0.03 for P x-rays and then dropping again for Al x-rays. The sharp drop of signal below ∿0.4 keV is not caused by the Be window of the detector, since no photons of this energy are incident on the detector, but is the result of the decreasing efficiency of pulse recognition in the electronic circuitry as the low energy threshold is approached (Statham 1981). Thus it is reasonable to assume that the actual pulse height distribution continues down to zero energy.

The sudden drop of the ICC as the x-ray energy drops below the silicon absorption edge suggests that ICC is related to the number of x-rays absorbed close to the detector surface and hence their mass absorption coefficient, $\mu$, in silicon. By fitting ideal Gaussians to the $K_\alpha$ and $K_\beta$ peaks, the number of counts outside this idealised shape (ICC counts) can be measured. The ratio of the total ICC counts to the total counts in the spectrum is plotted against $\mu$ in fig. 2. The error bars represent the range of acceptable Gaussian fits. Since the low energy photons lose a larger fraction of their ICC counts due to the low energy cut-off, an estimated correction has been made for this loss in the points marked with •. The surface related nature of the phenomenon is confirmed although the non-zero intercept may indicate some bulk effect. The magnitude is very large for x-rays with energies just above the silicon absorption edge. Fig. 3 compares two detectors from the same manufacturer, showing almost identical ICC. Since the effect depends on the surface treatment, it is likely to vary from manufacturer to manufacturer and even with time for a given manufacturer. ICC may also vary with time for a given detector.

## 3. Modelling the Shape of the ICC

To understand the effect that ICC has on an electron generated spectrum, which has bremsstrahlung as well as characteristic x-rays, it is necessary to have a model of the ICC shape as a function of incident energy. Previous calculations by Llacer et al (1977) suggest that, if the spectra are replotted on an energy scale normalised to the incident x-ray energy, they will have the same shape and the number of counts per unit normalised energy will be proportional to $\mu$. This assumes only a small fraction of the incident x-rays is absorbed in the surface layer. Study of fig. 1 suggests an exponential decay from the peak with decreasing energy in the spectrum. A function of the form:-

$$\frac{I_{ICC}}{I_T} = A\left(\frac{\mu}{E_p}\right)\exp\left(\frac{\alpha E}{E_p}\right) + B\left(\frac{\mu}{E_p}\right)\exp\left(\frac{\beta E}{E_p}\right) + \frac{C}{E_p}\left(1 - \frac{\gamma E}{E_p}\right).H(E)$$

was found to fit the ICC shape for incident photon energies from 1.2 to 5.9 keV. $E_p$ is the photon energy, E is the energy in the spectrum, $I_{ICC}$ is magnitude of the ICC in a fixed energy interval at E and $I_T$ is the total counts in the spectrum. $\alpha,\beta,\gamma$ and C are constants. $A(\mu/E_p)$ and $B(\mu/E_p)$ vary almost linearly with $\mu/E_p$ and can be expressed as polynomials in $\mu/E_p$ with a few terms. H(E) is 1 for $E < E_p/\gamma$ and 0 otherwise. The first two terms maintain the shape of the ICC with normalised energy and scale very nearly with $\mu/E_p$ as suggested above. The dependence on $\mu/E_p$ rather than $\mu$ comes from the fixed channel width used for displaying the spectra and the third term contains $C/E_p$ for the same reason. However, this term does not have the same behaviour. It is independent of $\mu$ and represents a fixed fraction of counts independent of $E_p$. These counts decrease approximately linearly from E = 0 to $E = E_p/\gamma$ above which they make no contribution. The success of the parameterisation can be seen in figs. 4 and 5. These compare the experimental data with that calculated from the parameterised fit. A large number of counts in the correct ratio are placed in the centre channels of the $K_\alpha$ and $K_\beta$ peaks. These are then redistributed to give the distribution of counts caused by ICC. This distribution is convolved with the detector resolution and then multiplied by a suitable Gaussian error function to take account of the loss of electronic detection efficiency at low E. As shown, the fit copes well with the large change in magnitude going from P to Si x-rays and is good for photon energies in the quoted range.

## 4. The Effect of ICC on the Bremsstrahlung Shape

The effect of ICC on the bremsstrahlung shape in an electron generated spectrum can be seen by redistributing the intensity in each channel of the ideal shape reaching the detector surface according to the above equation. The contribution to the ICC of x-rays with $E_p > 6$ keV can be neglected. The ideal shape can be calculated using a parameterisation of the modified Bethe Heitler equation (Chapman et al 1984) corrected, as necessary, for absorption in the specimen of interest and the various absorbing layers in the detector e.g. the Be window and Au contact layer. Fig. 6 shows such an ideal bremsstrahlung and the effect of ICC. Shown dashed is the effect of a dead layer which absorbs x-rays but allows all the carriers to recombine. The thickness of $\sim$0.5 μm is chosen to give the same step height at the silicon absorption edge. ICC redistributes the counts whereas the dead layer removes them. ICC raises the signal above the ideal shape in some places and lowers it in others, whereas a dead layer always lowers it. Just below the absorption edge, ICC gives a small peak which could be confused with a small silicon peak. Fig. 7 shows a portion of the ICC tail from Mn x-rays. If a dead layer exists, then Si K photons should escape into the depletion layer to give a silicon peak in the spectrum. Its size can be estimated in the same way as that of the Si escape peak (Reed and Ware 1972). The Si peak shown dashed is that estimated for a 0.1 μm dead layer indicating that any dead layer is $\ll$0.1 μm in thickness. These calculations confirm the qualitative predictions of Statham (1981).

Fig. 8 shows an experimental spectrum from a thin copper film together with the theoretical bremsstrahlung shape corrected for self absorption, the Be window, the Au layer and ICC. While the agreement is not perfect, the residual no longer has a step at the absorption edge. A similar residual is also found in a spectrum from carbon. With so many imprecisely known experimental parameters, it may be possible to improve the agreement by varying them. However, for consistency, one set of detector parameters is required to give agreement for any spectrum.

## References

Chapman J N, Nicholson W A P and Crozier P A 1984 J Microscopy 136 179.

Craven A J, Adam P F, Nicholson W A P, Chapman J N and Ferrier R P 1984 J de Physique 45 C2-437.

Craven A J and Adam P F 1984 Electron Microscopy 1984 ed A Csanady, P. Rolich and D Szabo pp 371-2 (Budapest: Organising Committee of 8th European Congress on Electron Microscopy).

Llacer J, Haller E E and Cordi R C 1977 IEEE Trans Nucl Sci NS-24 53

Reed S J B and Ware N G 1972 J Phys E 5 582

Statham P J 1981 J Microscopy 123 1

Steele J D, Chapman J N and Adam P F 1984 Electron Microscopy 1984 ed A Csanady, P Rolich and D Szabo pp 373-4 (Budapest: Organising Committee of 8th European Congress on Electron Microscopy).

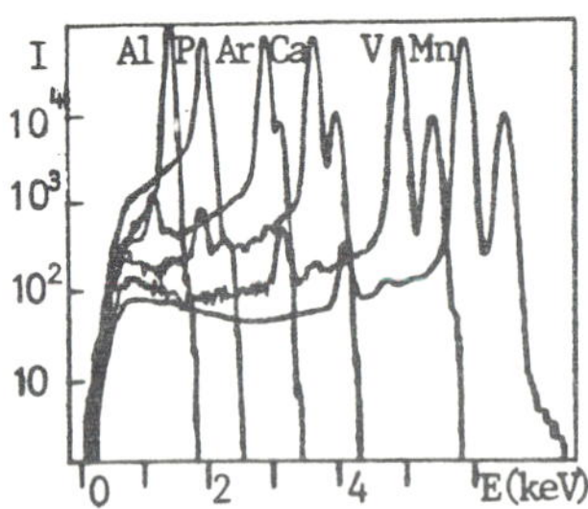


Fig. 1 Spectra recorded from various characteristic x-rays. Logarithmic intensity scale.

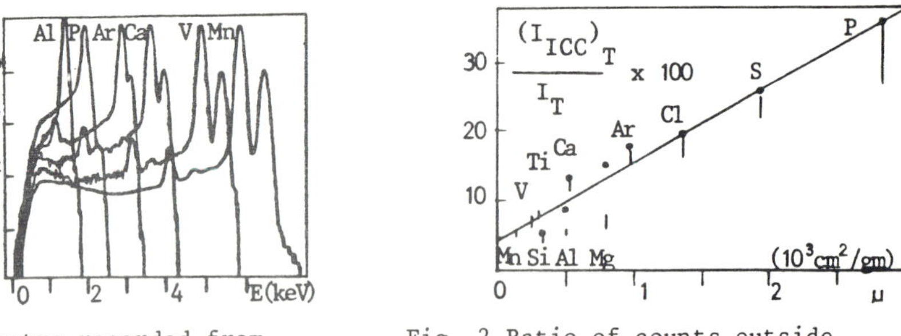


Fig. 2 Ratio of counts outside Gaussian peaks to total counts vs. mass absorption coefficient.

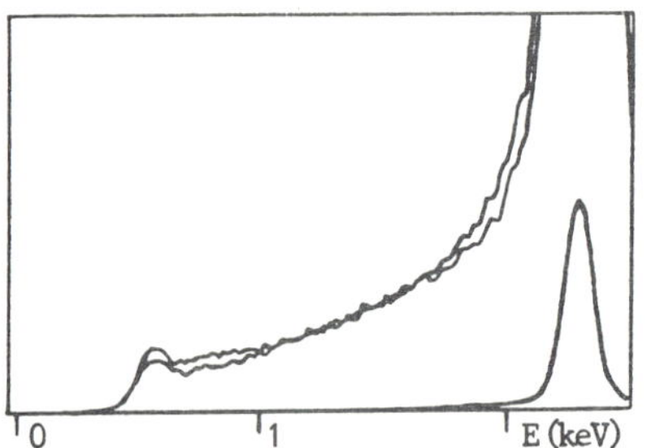


Fig. 3 Comparison of the ICC for sulphur in two detectors, scaled to equal numbers of counts.

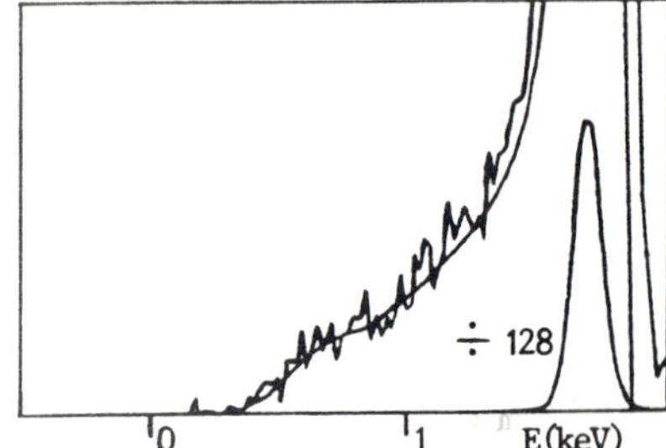


Fig. 4 Experimental and fitted ICC from Silicon.

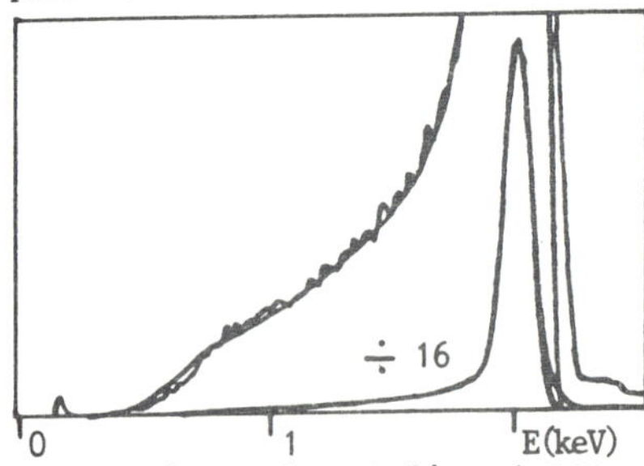


Fig. 5 Experimental and fitted ICC from phosphorus.

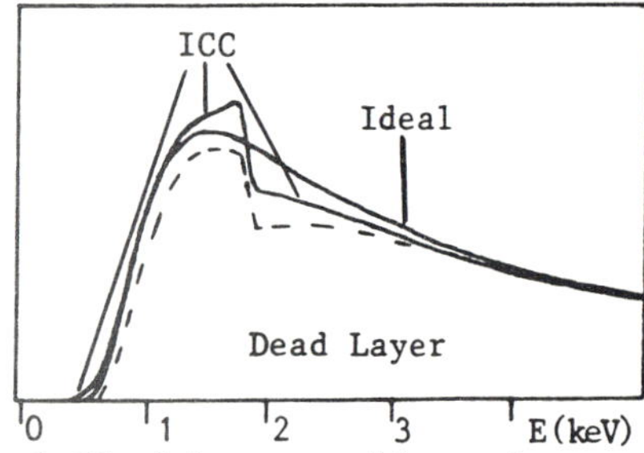


Fig. 6 Ideal bremsstrahlung shape reaching the detector and the effect of ICC or a dead layer (dashed).

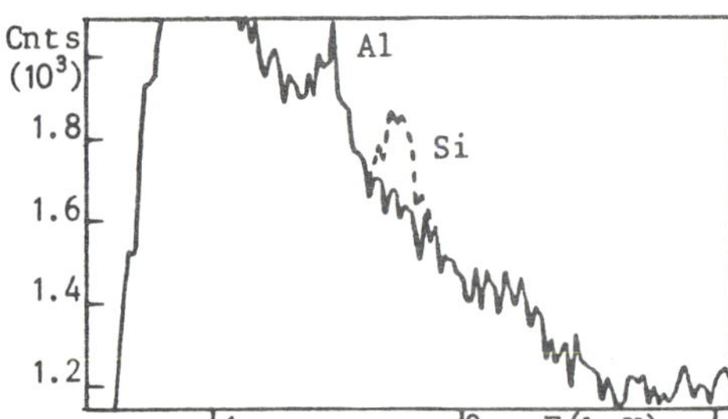


Fig. 7 Portion of ICC tail from Mn x-rays. Fluoresced Si peak expected from 0.1μm dead layer shown dashed.

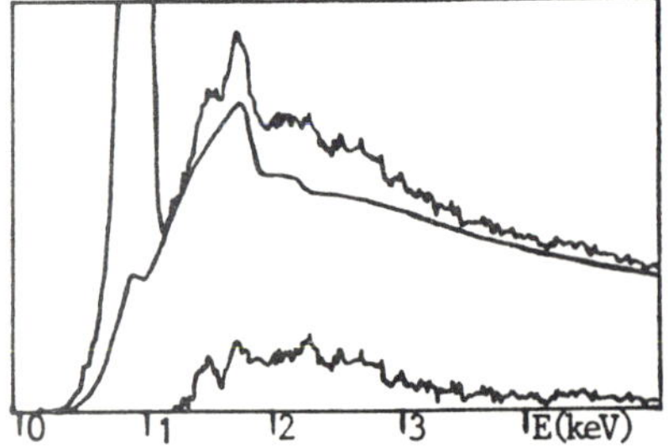


Fig. 8 Copper spectrum and background corrected for ICC. The difference is also shown with $Cu_L$ peak suppressed.

*Inst. Phys. Conf. Ser. No 78: Chapter 7*
*Paper presented at EMAG '85, Newcastle upon Tyne, 2–5 September 1985*


# Accuracy and sensitivity of EDS for light element microanalysis

D J Bloomfield, G Love and V D Scott

School of Materials Science, University of Bath, Bath, BA2 7AY, UK.

## 1. Introduction

Although 'windowless' energy-dispersive (ED) analysis is increasingly used for locating very light elements (5 < Z < 11) in specimens, until recently little systematic work has been carried out to assess its potential for quantitative investigations. Here we describe a method of spectrum processing for an EDAX ECON detector and 711 analyser, and evaluate its effectiveness for quantitative light element analysis by comparison with data obtained by wavelength-dispersive (WD) spectrometry. Finally, some comments are made concerning the sensitivity of the ED system for the elements carbon, nitrogen and oxygen.

## 2. Data Acquisition

In order to obtain reliable ED data when recording soft x-ray spectra, the following experimentation has been adopted:

i) a thin aluminised plastic detector window to prevent activation of the detector by light photons (e.g. from cathodoluminescent specimens and to avoid contamination of its surface by organic vapours from the electron column;

ii) a low energy electron beam (<15keV) to maximise efficiency of soft x-ray emission;

iii) count rates kept below 2000cps because of ineffective pulse pile-up rejection for x-ray energies <2keV;

iv) manual measurement of the analysis period because the automatic Barnhardt system often underestimates dead time (Bloomfield et al 1983).

## 3. Spectrum Processing

Firstly, the electronic noise peak must be removed since otherwise low concentrations of boron or carbon may be obscured, but this is a relatively straightforward operation. The electron beam is switched off and the noise spectrum recorded. This is then scaled for the counting period used when recording the spectrum of interest, and subtracted from it to leave a noise-free background. We have found, however, that the residual background still shows an artefact, a small rise in background at low energies which we have termed the 'spur'. (Note that low energy radiation will be readily absorbed in specimen and plastic detector window and so the 'true' continuum background tends to zero at low energies). This artefact is not in any way associated with residual noise, its magnitude being unaffected by counting time, but depends upon the total number of counts in the spectrum. Despite the fact that the origin of the 'spur' has not been clearly established, its size and shape may be modelled mathematically and then subtracted, as shown in Fig.1.

Next the 'true' continuum must be removed and we have developed a method based upon that of Smith et al (1975), where x-radiation from a reference standard is used to predict the continuum background for the spectrum of interest. An advantage of this method is that knowledge of the variation in detector efficiency as a function of x-ray energy is unnecessary. The modifications introduced to extend Smith's method to low energy spectra involved, essentially, (a) using the absorption correction of Love and Scott (1978) rather than that of Philibert (1963), (b) reverting to the original form of Kramers' law (Kramers 1932) and (c) selecting a reference standard of mean atomic number similar to that of the analysed specimen. The result of fitting the background in this manner and then adding the contributions from noise and spur is illustrated in Fig.2. for a spectrum from silicon carbide.

Finally, overlapping peaks must be dealt with, a particularly difficult problem in much soft x-ray analysis because of the proliferation of L and M lines from heavier elements which occur in this region of the spectrum. We have chosen to carry out deconvolution by least-squares fitting of stored spectra from elemental standards. This approach was preferred to the alternative of generating peaks using mathematical functions because low-energy peaks are markedly assymmetrical and difficult to model accurately.

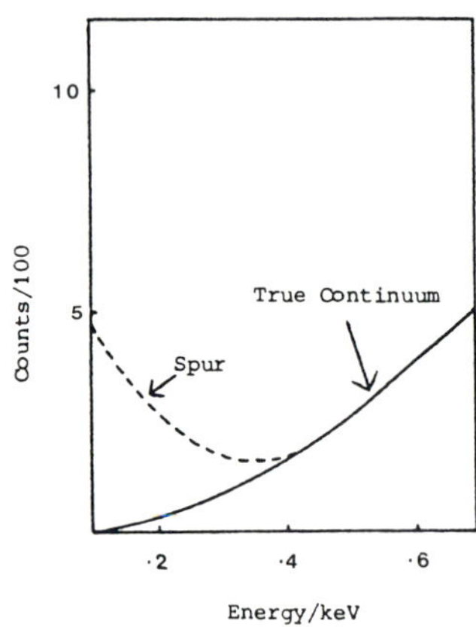


Fig.1. Low energy spectrum from copper

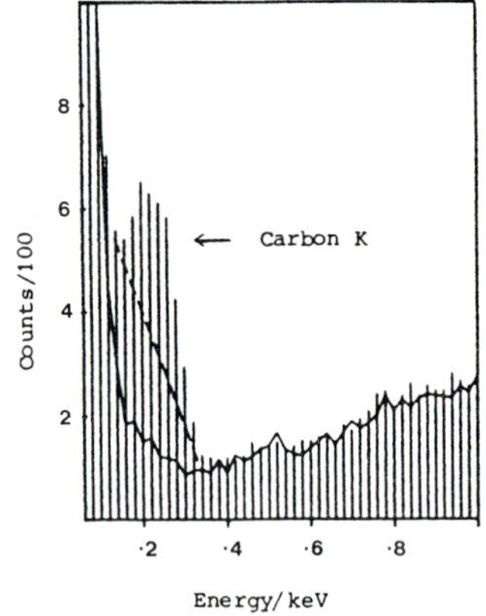


Fig.2. Spectra from SiC; --- background by linear interpolation, —— background using silicon reference standard.

4. Comparison of ED and WD Data

The effectiveness of the ED background subtraction and deconvolution routines were separately tested using two sets of specimens, the first having well separated soft x-ray lines and the second giving closely spaced lines with severe overlap problems. Measurements were carried out at 7kV and 15kV and the data are given as x-ray intensity ratios k, ($k = I_{spec}/I_{stnd}$) in Tables 1 and 2. Included in the Tables are the corresponding WD data, acquired under identical experimental conditions (electron probe voltage and x-ray take-off angle). The similarity between the ED and WD values indicates that our method of spectrum processing is working satisfactorily and that it can cope with x-ray line overlap problems.

Table 1 Comparison of k ratios obtained by WD and ED analysis: light element peaks clearly resolved.

| X-ray line | Specimen | Standard | k ratios 7kV WD | ED | k Ratios 15kV WD | ED |
|---|---|---|---|---|---|---|
| $F_K$ | $PbF_2$ | $MgF_2$ | 0.196 | 0.208 | 0.105 | 0.110 |
| $O_K$ | NiO | $Al_2O_3$ | 0.444 | 0.454 | 0.458 | 0.463 |
| $C_K$ | SiC | Graphite | 0.0951 | 0.0933 | 0.0374 | 0.0366 |
| $C_K$ | $Fe_3C$ | Graphite | 0.0497 | 0.0504 | 0.0191 | 0.0198 |

Table 2 Comparison of k ratios obtained by WD and ED analysis: light element peaks overlapped by those from heavier elements

| X-ray line | Specimen | Standard | k ratios 7kV WD | ED | k ratios 15kV WD | ED |
|---|---|---|---|---|---|---|
| $F_K$ | $MnF_2$ | $MgF_2$ | 0.339 | 0.363 | 0.144 | 0.148 |
| $O_K$ | $Fe_2TiO_5$ | $Al_2O_3$ | 0.593 | 0.589 | 0.472 | 0.460 |
| $O_K$ | $MgCr_2O_4$ | $Al_2O_3$ | 0.826 | 0.837 | 1.060 | 1.100 |
| $N_K$ | $Fe_4N$ | $Si_3N_4$ | 0.258 | 0.252 | 0.451 | 0.284 |

5. Minimum Detection Limits

Minimum detectable levels for carbon, nitrogen and oxygen are given in Table 3. The figures were obtained for samples of silicon carbide (SiC), silicon nitride ($Si_3N_4$) and alumina ($Al_2O_3$), although it must be borne in mind that the values would be different for different specimens and would be affected also by electron probe voltage; in addition, ED figures would be particularly sensitive to the presence of overlapping x-ray lines. It is, nevertheless, evident that the ED system can, in many cases, measure concentrations of less than 1% of these elements. The WD system shows results an order of magnitude better, a situation similar to that found when analysing heavier elements.

Table 3 Minimum detection levels in weight percent; 200 seconds counting time; 7kV; beam currents WD 200nA, ED 10nA.

| Element | WD | ED |
|---|---|---|
| Oxygen | 0.015 | 0.12 |
| Nitrogen | 0.070 | 0.41 |
| Carbon | 0.024 | 0.33 |

## 6. Conclusions

We have shown that, provided appropriate spectrum processing methods are adopted, the acquisition of quantitative light element data using a 'windowless' ED detector is a practicable proposition. Although data processing is more complex than for WD analysis, the quality of the final data is remarkably similar. Nonetheless, WD spectrometry remains the better method for carrying out quantitative light element studies since it is a simpler and more sensitive technique.

## 7. Acknowledgements

To the SERC for support.

## 8. References

Bloomfield DJ, Love G and Scott VD 1983 X-ray Spectrom 12 2.
Bloomfield DJ and Love G. 1985 X-ray Spectrom 14 8.
Kramers HA 1932 Phil. Mag. 46 836.
Love G and Scott VD 1978 J. Phys.D.: Appl. Phys. 11 1369.
Philibert J. 1963 X-ray Optics and X-ray Microanalysis eds. HH Pattee VE Cosslett and A. Engstrom (New York:Academic Press) p.379.
Smith, DGW Gold CM and Tomlinson DA 1975 X-ray Spectrom 4 149.

*Inst. Phys. Conf. Ser. No 78: Chapter 7*
*Paper presented at EMAG '85, Newcastle upon Tyne, 2–5 September 1985*



# Simulations of characteristic fluorescence behaviour in wedge shaped targets

E Van Cappellen and J Van Landuyt

Universiteit Antwerpen, RUCA, Groenenborgerlaan 171, B-2020 Antwerpen, Belgium

Abstract. An evaluation is made of the characteristic fluorescence emission in a wedge shaped (S)TEM specimen of a binary compound. In the proposed model the primary X-ray emission is perpendicularly projected onto the optical axis and subsequently integrated along it. As a result the formula for the characteristic fluorescence yields a four dimensional integral which was solved numerically for a wide range of practical situations, in particular for different wedge angles $\alpha$ which turns out to be a very important parameter.

## 1. Model

A binary compound $A_xB_{1-x}$ is considered in which the element "A" can be fluoresced by X-radiation of the element "B" (fig. 1). "$C_A$" and "$C_B$" are the mass concentrations while "T" is the mass thickness along the optical axis. The primary emission of "B" in a point "P" situated at a mass depth "t" can be written as follows :

$$dI_B^o = C_B \cdot \phi(t) \cdot dt \quad (1)$$

The mass depth distribution $\phi(t)$ proposed by Philibert et al. (1975) will be used :

$$\phi(t) = \exp(-\sigma t) \quad (2)$$

where $\sigma$ is an electron absorption coefficient. The fraction of the isotropically emitted radiation leaving in the solid angle $d\Omega$ and reaching the element of volume dV reads :

$$dI_B = \frac{1}{4\pi} \cdot C_B \cdot \phi(t) \cdot \exp(-\mu_B|\bar{r}|).d\Omega.dt \quad (3)$$

where "$\mu_B$" is the mass absorption coefficient of X-radiation from element "B" in the specimen. A fraction of this intensity will be absorbed by the element "A" present in the volume dV. If the mass absorption coefficient of X-radiation from element "B" in element "A" is denoted $\mu_B^A$ the absorbed B-intensity may be expressed as :

$$dI_B^A = dI_B \cdot C_A \cdot \mu_B^A \cdot d|\bar{r}| \quad (4)$$

This intensity multiplied by "$(r_A-1)/r_A$", (Philibert et al. 1975; Nockolds et al. 1980) the ionisation yield and "$\omega_A$", the fluorescence yield of the considered shell yields the amount of secondary A emission "$dI_A^F$" generated by primary B emission originating from "P", (fig. 1). To account for absorption in the specimen of this secondary emission $dI_A^F$ must be multiplied by $\exp(-\mu_A.q(\bar{r},t))$ where "$\mu_A$" by analogy is the mass absorption coefficient of X-radiation from element A in the specimen. "$r_A$" is the absorption edge jump ratio and "$z_A$" is the weight of the spectral line under consideration.

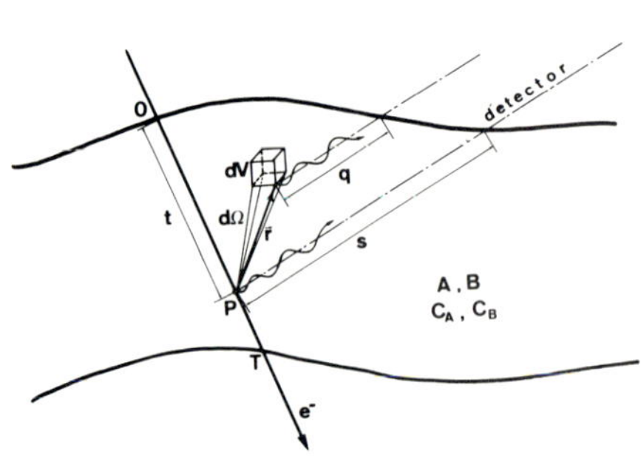


*Fig. 1 The primary X-radiation of element B isotropically emitted in point P generates secondary A emission in the element of volume dV.*

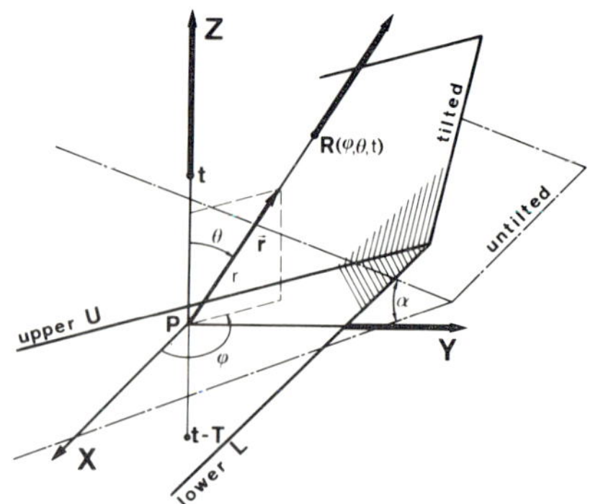


*Fig. 2 The Z-axis is the optical axis of the microscope and for an untilted specimen the X-axis is chosen parallel to the specimen edge.*

$$dI_A^F = \frac{r_A - 1}{r_A} \cdot z_A \cdot \omega_A \cdot dI_B^A \cdot \exp(-\mu_A . q(\bar{r},t)) \tag{5}$$

Integration over the specimen volume V and subsequently over the mass thickness T along the optical axis yields the total secondary emission of A due to characteristic fluorescence.

$$I_A^F = \int_0^T \int_{\text{spec.}} dI_A^F = \epsilon \int_0^T \int_{V(t)} \phi(t) . \frac{\exp(-\mu_B . |\bar{r}|)}{|\bar{r}|^2} . \exp[-\mu_A . q(\bar{r},t)] dV dt \tag{6}$$

where all the constants are condensed in the symbol $\epsilon$.

## 2. A wedge-shaped specimen

To examine the behaviour of characteristic fluorescence in a typical (S)TEM specimen, equation (6) will be integrated for a wedge-shaped specimen. The influence of different parameters such as the wedge angle, detector position and the different mass absorption coefficients will be examined. Spherical coordinates proved to be the most convenient for solving the problem of the integration boundaries.

### 2.1. Coordinate system

A reference frame has to be chosen to calculate the upper boundary $R(\phi,\theta,t)$ of the integral over the variable r and the distance q. It turned out to be favourable to take the origin in "P" and the Z-axis along the optical axis of the electron microscope. For an untilted specimen (symmetry plane of the wedge perpendicular to the optical axis) the X-axis is chosen parallel to the edge of the wedge, consequently the Y-axis is perpendicular to it (fig. 2). To allow for an arbitrary electron impact direction the specimen is first rotated around the X-axis over an angle "a" and then around the Y-axis over an angle "b". The limit $R(\phi,\theta,t)$ is the intersection of the position vector $\bar{r}$ with the wedge, therefore a mathematical description of the wedge is needed. The upper (U) and lower (L) planes of the wedge will be described by their normal vectors $\bar{n}_+$ and $\bar{n}_-$.

2.2. The upper boundary $R(\phi,\theta,t)$

The intersection "$R_+$" with U and "$R_-$" with L are obtained from straightforward calculations (Van Cappellen et al. 1984).

$$R_+ = \left|\frac{t \cdot \cos b}{\bar{r} \cdot \bar{n}_+}\right| \quad ; \quad R_- = \left|\frac{(t-T) \cdot \cos b}{\bar{r} \cdot \bar{n}_-}\right| \qquad (7a,b)$$

Clearly the boundary $R(\phi,\theta,t)$ can be either $R_+$, $R_-$ or infinite depending on the direction of the position vector $\bar{r}$. A convenient way to determine the correct $R(\phi,\theta,t)$ is to check the signs of the scalar products. The integral (6) becomes :

$$I_A^F = C \int_0^T dt \left\{ \int_0^{\pi} d\phi \left( \int_0^{\theta^\star(\phi,t)} d\theta \left( \int_0^{R_+(\phi,\theta,t)} dr.I \right) + \int_{\theta^\star(\phi,t)}^{\pi} \left( \int_0^{R_-(\phi,\theta,t)} dr.I \right) \right) \right. \qquad (8)$$

$$\left. + \int_{\pi}^{2\pi} d\phi \left( \int_0^{\theta_+(\phi)} d\theta \left( \int_0^{R_+(\theta,\phi,t)} dr.I \right) + \int_{\theta_+(\phi)}^{\theta_-(\phi)} d\theta \left( \int_0^{\infty} dr.I \right) + \int_{\theta_-(\phi)}^{\pi} d\theta . \left( \int_0^{R_-(\phi,\theta,t)} dr.I \right) \right) \right\}$$

whereby :

$$I = \phi(t)\exp(-\mu_B.r).\exp(-\mu_A.q(r,\theta,\phi,t)).\sin\theta \qquad (9)$$

Analytic expressions for the integration boundaries $\theta^\star$, $\theta_+$, $\theta_-$ and for the distance q can be deduced from geometrical considerations (Van Cappellen et al. 1984).

## 3. Results

All graphs shown below represent curves of characteristic fluorescence F in arbitrary units as a function of the mass thickness T along the optical axis. All calculations were carried out for the Cu-50 at % Cr alloy. For this material the density is approximately 8 g/cm$^3$ and the mass absorption coefficients are :

$$\mu_A = \mu_{Cr} = 120 \text{ cm}^2\text{g and } \mu_B = \mu_{Cu} = 152 \text{ cm}^2/\text{g} \qquad (10)$$

The first set of curves (graph 1) shows the influence of the absorption of secondary emission and of the deceleration of the electrons in the material. For the dashed curve both absorption phenomena were neglected whereas for the dash-dot curve only the electron absorption was omitted. The full curve is the result when both phenomena are accounted for. Graph 2 shows the influence of the accelerating voltage. In graphs 3 and 4 the parameter is the wedge angle $\alpha$. Careful inspection of graph 4 reveals the enormous importance of the shape of the specimen. At a realistic thickness of 50 nm a wedge with a small angle (10°) emits twice as much secondary emission than a perfect plane parallel foil and for a wedge angle of 40° the secondary emission is already six times as intense. Interesting to note is that the linear term of the curves is proportional with the wedge angle. As a consequence the curve for $\alpha$ = 0° starts as a pure parabola which is in good agreement with the formula of Philibert et al. (1975). Further results (not printed here) reveal that the tilting conditions of the specimen and the position of the detector have only a slight influence on the fluorecence emission at least up to thicknesses of 500 nm.

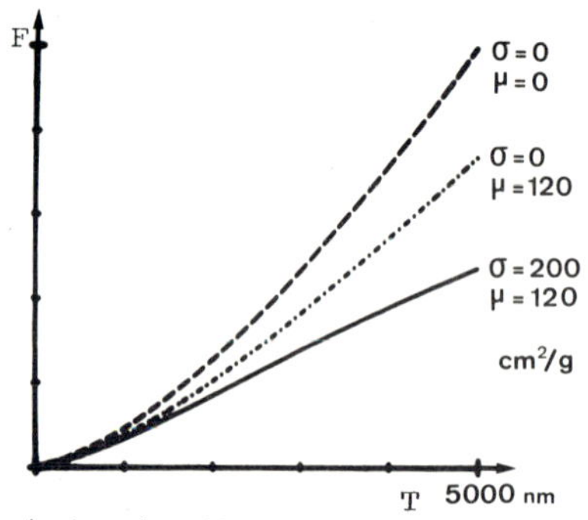


Graph 1 The fluorescence emission for a fixed wedge angle of 30°. σ=200 corresponds with an accelerating voltage of 100kV.

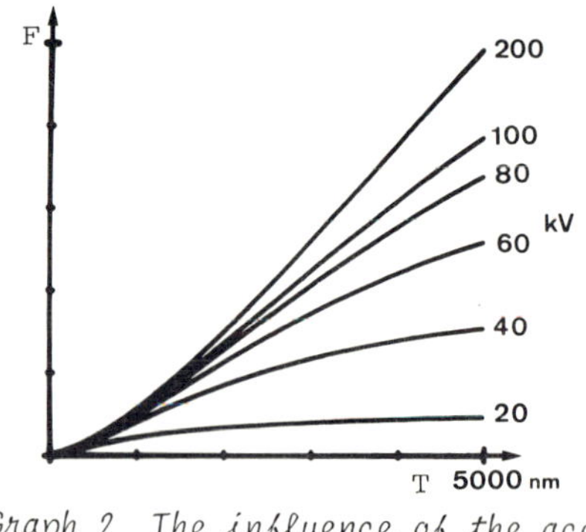


Graph 2 The influence of the accelerating voltage on the fluorescence emission for a fixed wedge angle of 30° (a=-30°, b=0°, Φ=90°, Θ=90°).

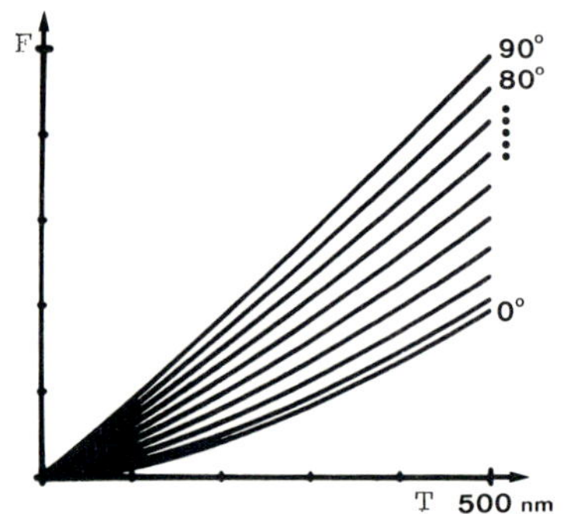


Graph 3 The influence of the wedge angle α. The operating voltage is 100 kV. (a=-30°, b=0°, Φ=90°, Θ=90°).

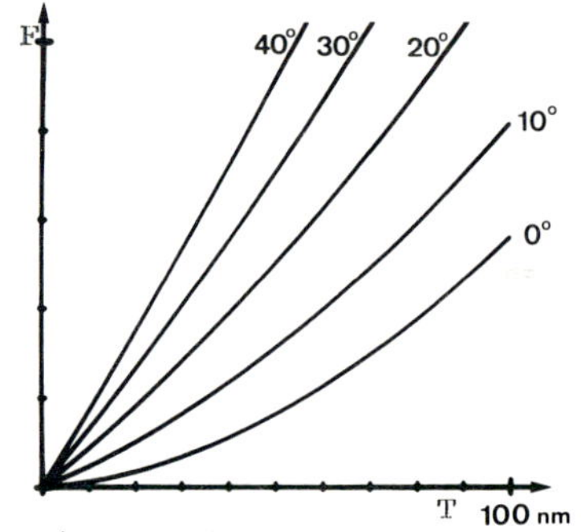


Graph 4 A detail of graph 3.

## 4. Conclusions

The new parameter α turned out to be very important since it influences the fluorescence yield dramatically. It is therefore concluded that no fluorescence correction can be entirely satisfactory unless the exact shape of the specimen is introduced. Developing such a correction formula might be feasible but it remains doubtful whether the exact shape of an analysed specimen is easy to determine.
In contrast with the absorption correction of primary X-radiation in E.D.X.-microanalysis, fluorescence emission is not too sensitive to the orientation of the specimen and or position of the detector.

## 5. References

Nockolds C, Nasir M J, Cliff G and Lorimer G W 1980, Inst. Phys. Conf. Ser. N°52, pp.417-9.

Philibert J and Tixier R 1975 Physical Aspects of Electron Microscopy and Microbeam Analysis (New York : Wiley & Sons) pp. 333-21.

Van Cappellen E, Deblieck R, Van Landuyt J and Adams F 1984, J. Trace and Microprobe Techniques 2(2) 139.

*Inst. Phys. Conf. Ser. No 78: Chapter 7*
*Paper presented at EMAG '85, Newcastle upon Tyne, 2–5 September 1985*


# On the atomic number dependence of the x-ray fluorescence analysis in the electron microscope

I Pozsgai

Research Institute for Technical Physics of the Hungarian Academy of Sciences,
1325 Budapest,Ujpest 1. P.O.B.76

## 1. Introduction

In the electron microscope electron excitation is used for chemical analysis apart from some special cases when X-rays are excitation source (Middleman and Geller 1976,Linneman and Reimer 1978,Pozsgai 1982, Eckert 1982). The usage of X-ray excitation is justified by a lower background radiation and as a consequence by lower detection limits compared to electron excitation. Roughly speaking 1000 ppm detection limits can be achieved by electron excitation and energy dispersive detection of X-rays. The X-ray fluorescence analysis (EDXRF) makes it possible that trace element analysis can be carried out in a scanning electron microscope but for the gain in concentration sensitivity we have to pay by the loss of the good lateral resolution of the electron beam microanalysis. Not only the improved detection limits but other advantages like the possibility of analysing non-conductive specimens and of beam sensitive materials have to be mentioned.
The aim of this paper is to demonstrate the possibilities of improving the detection limits (DL) in a scanning electron microscope and to show the atomic number dependence of the DL of the X-ray fluorescence analysis.

## 2. Experimental

The details of how the optical column of an electron microscope can be converted into a X-ray tube has been published elsewhere (Pozsgai 1982). The scheme of the experimental set-up can be seen in Fig.1. An energy dispersive X-ray spectrometer of type EEDS-II attached to a scanning electron microscope of type JSM 35 was used. As the detector of the spectrometer is retractable microanalysis can be performed after retracting the detector with the fluorescence attachment. The measurements from which the detection limits were calculated have been carried out on glass standards of the National Bureau of Standards, on No.612 and K1727 respectively. The concentration of the relevant elements in the standards,denoted by $c_o$,can be found in the Tables 1-4. Mo,GaAs,Ti and Al thin targets were applied as X-ray sourcesand for each target we tried to find an optimum thickness and an optimum acceleration voltage of the microscope.Thinner targets and lower accelerating voltages of the microscope result in broadband excitation while thicker targets and higher accelerating voltages are required for monochromatic excitation. The diameter of the collimator and the spot size of x-rays on the specimen surface were 3 and 6 mm respectively. The applied electron currents varied in the range of 1-20 μA. The count rates were about 5000 count/s.

## 3. Results

The blind spectra taken on a graphite block were fairly clean they contained only a negligible amount of iron spectral contamination. Detection limits referring to 1000s livetime measurements (the total time was about 1300 s) were calculated according to equ.(1)(Jenkins 1981)

$$c_{min} = \frac{2.33\sqrt{B}}{P-B}\, c_o \qquad (1)$$

where P stands for peak height, B - background, $c_o$ - concentration of element in the standard. Equ.1. can be used in cases when the width of ROI (region of interest) is the same for peak and background.

For the excitation of elements in different ranges of atomic numbers, different target foils were applied .

Mo target (Table 1. and Fig.2.) 100 um in thickness provided DL between 1 and 7 weight ppm in the range Z=26-38 at 35kV accelerating voltage of the microscope.

GaAs target (Table 2. and Fig.3.) 70 um in thickness irradiated at 25 kV ensured DL from 1 to 12 ppm in the range Z=22-29.

Ti foil (Table 3. and Fig.4.) 4 um in thickness proved to be effective at 15 kV for Ca and K analysis (DL= 11.4 and 17.7 ppm respectively).

Finally, 7 um Al foil was the X-ray source for Na and Mg analysis at 12kV (Table 4. and Fig.5.). The DL obtained were 29 and 14 ppm respectively.

It has to be noted that although the measurements were carried out with a lateral resolution of 6 mm, the technique described above allows easy analysis even with a lateral resolution of 0.5 mm and the detection limits cited here can be maintained as long as the count rate is not much less than 5000 cps.

Summarizing the results it can be said that by means of X-ray fluorescence analysis the detection limits can be improved in the scanning electron microscope by 1 to 3 orders of magnitude depending on the atomic numbers of elements.

## Acknowledgements

The author wishes to thank Drs.K.F.J.Heinrich,D.E.Newbury and R.Eckert for the N.B.S. glass standards.

Table 1.
Detection limits measured on NBS 612 glass standard /Mo $K_\alpha$ excitation/

| | $c_o$ /ppm/ | $c_{min}$ /ppm/ |
|---|---|---|
| Rb$K_\beta$ | 41.4 | 1.0 |
| Sr$K_\alpha$ | 78.4 | 1.3 |
| Pb$L_\alpha$ | 38.5 | 1.0 |
| Cu$K_\alpha$ | 37.7 | 2.2 |
| Ni$K_\alpha$ | 38.8 | 4.2 |
| Co$K_\alpha$ | 35.5 | 6.8 |
| Fe$K_\alpha$ | 51.0 | 3.3 |

Table 2.
Detection limits measured on NBS 612 glass standard /GaAs excitation/

| | $c_o$ /ppm/ | $c_{min}$ /ppm/ |
|---|---|---|
| Cu$K_\alpha$ | 37.7 | 0.9 |
| Ni$K_\alpha$ | 38.8 | 1.0 |
| Co$K_\alpha$ | 35.5 | 1.5 |
| Fe$K_\alpha$ | 51.0 | 1.4 |
| Mn$K_\alpha$ | 39.6 | 2.2 |
| Cr$K_\alpha$ | 37.8 | 3.1 |
| Ti$K_\alpha$ | 50.1 | 12.3 |

Table 3.
Detection limits measured on K1727 glass standard /Ti $K_\alpha$ excitation/

| | $c_o$ /%/ | $c_{min}$/ppm/ |
|---|---|---|
| Ca$K_\alpha$ | 3.57 | 11.4 |
| K $K_\alpha$ | 2.49 | 17.7 |

Table 4.
Detection limits measured on K1727 glass standard /Al $K_\alpha$ excitation/

| | $c_o$ /%/ | $c_{min}$/ppm/ |
|---|---|---|
| Mg$K_\alpha$ | 3.00 | 14.0 |
| Na$K_\alpha$ | 2.96 | 29.0 |

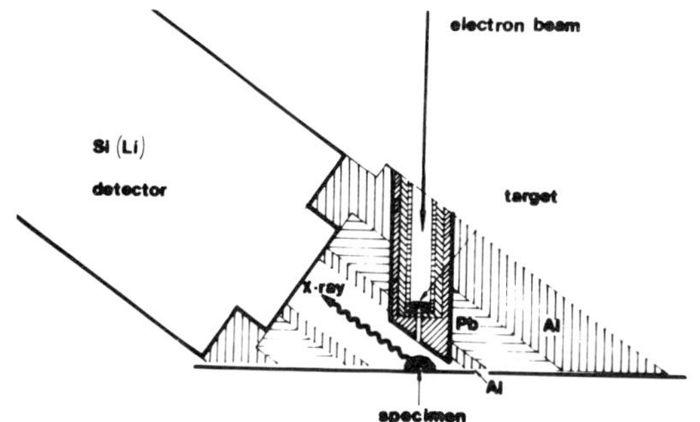


Fig.1.The scheme of attachment for X-ray fluorescence analysis in the scanning electron microscope

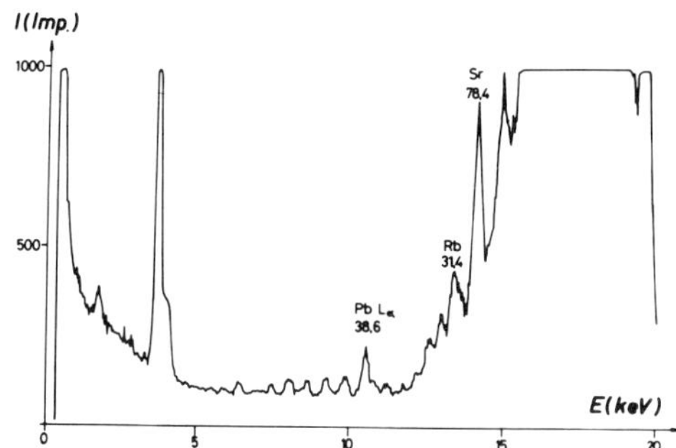


Fig.2.EDXRF spectrum of NBS 612 glass standard /Mo$K_\alpha$ excitation/

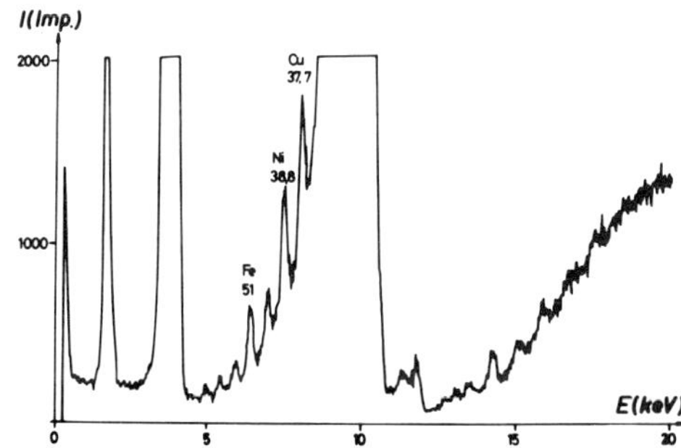


Fig.3.EDXRF spectrum of NBS 612 glass standard /GaAs excitation/

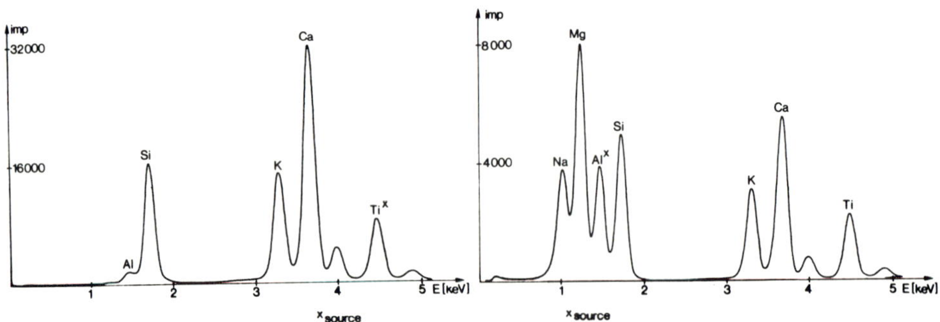


Fig.4.EDXRF spectrum of K1727 stanadard /Ti$K_\alpha$ excitation/

Fig.5.EDXRF spectrum of K1727 standard /Al$K_\alpha$ excitation/

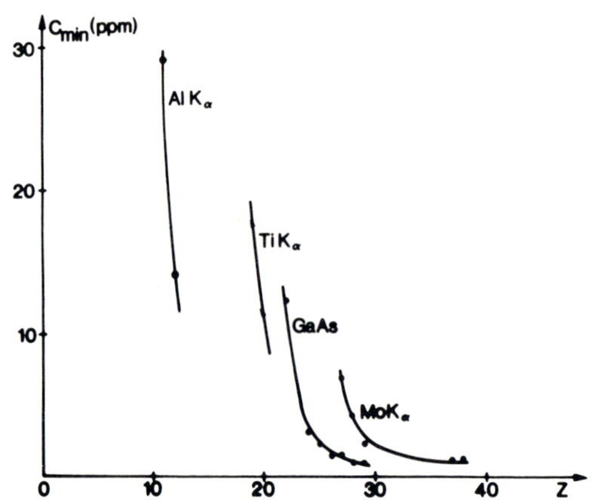


Fig.6.Atomic number dependence of the detection limits of EDXRF in the Z=11-40 range

References

Eckert R 1982 10th International Congress on Electron Microscopy, Hamburg, Vol.1.677-678

Jenkins R,Gould R W ,Gedcke D 1981 Quantitative X-ray Spectrometry ,Marcell Decker Inc. N.Y. 516.

Linneman B and Reimer L 1978 Scanning Vol.1. 109-117

Middleman L M and Geller J D 1976 Scanning Electron Microscopy, Vol.1. 171-177

Pozsgai I 1982 10th International Congress on Electron Microscopy, Hamburg, Vol.1. 681-682.

*Inst. Phys. Conf. Ser. No 78: Chapter 7*
*Paper presented at EMAG '85, Newcastle upon Tyne, 2–5 September 1985*


# Coincidence counting techniques in analytical electron microscopy

A W Nicholls, I P Jones and M H Loretto

Department of Metallurgy and Materials, University of Birmingham, P.O. Box 363, Birmingham B15 2TT, U.K.

## 1. Introduction

Analytical Electron Microscopy, which relies on the interaction of incident fast electrons with atomic inner shell electrons, has proved to be an invaluable technique for material characterisation as a result of its high spatial resolution. In particular energy dispersive X-ray analysis (EDX) and electron energy loss spectroscopy (EELS) are capable of providing quantitative analysis with reasonable accuracies and detection efficiencies which can vary from 0.1% to 20%. These limits are generally set by peak to background ratios and tend to be worse for EELS than EDX. Obviously, any technique that is capable of increasing peak to background ratio is worth investigating.

In principle the background could be reduced by employing a coincidence counting technique where an X-ray photon would be accepted only if the loss electron which gave rise to the specific X-ray photon was detected at the appropriate instant in time. Coincident counting techniques, widely used in nuclear physics, have previously been applied to electron microscopy by Voreades and Crewe (1974, 1975) who used the technique for measuring the number of secondary electrons generated directly from primary electron energy loss events. It has been suggested that electron spectroscopy could be improved using coincident counting (Wittry 1976) but little experimental work has been carried out.

The aim of the current work has been to assess the applicability of coincident counting using a conventional analytical electron microscope and standard EDX and EELS detectors. From this it was hoped to gain a better understanding of the factors that affect the signal and thus develop a more suitable experimental configuration.

## 2. Factors Affecting the Measurement of Coincident Events

The measured coincident signal is made up of both genuine coincidences between X-ray photons and their loss electrons and false coincidences which are random chance coincidences between unconnected events. As the latter are random events their count rate is given by the product of three terms

$$N_f = N_x \cdot N_e \cdot \Delta t \qquad - 1$$

where $N_x$ and $N_e$ are the count rates for X-ray and energy loss respectively and $\Delta t$ is the resolving time (that is the time after the first signal is detected that a pulse on the other channel will be counted as coincident). In order to reduce the number of false coincidences the X-ray and energy loss count rates could be reduced, or the resolving time must be reduced.

The number of genuine coincidences depends strongly on the efficiency with which the signals can be detected. This proves to be the major limitation where conventional EDX and EELS detectors are concerned. For example if we consider the detection of the products of the aluminium K shell ionisations we find that from $N_i$ ionisations a conventional Si(Li) X-ray detector would detect

$$N_x = N_i \cdot \omega \cdot \gamma \cdot \phi(E) \qquad - 2$$

events, where $\omega$ is the fluorescent yield ($\sim$ 0.03 for AlK), $\gamma$ the proportion of the characteristic signal that reaches the detector ($\sim$ 0.01) and $\phi(E)$ the detection efficiency at an energy E ($\sim$ 0.7). Hence from 1000000 ionisations 210 AlK X-rays would be detected. For the EELS spectrometer, from $N_i$ ionisations we detect

$$N_e = N_i \cdot \mu_\alpha \cdot \mu_\Delta$$

events, where $\mu_\alpha$ and $\mu_\Delta$ are efficiency terms for angular detection and energy detection ($\mu_\alpha \sim 0.13$ and $\mu_\Delta \sim 0.1$ for a 70 eV window set on the AlK edge). Hence of the same 1000000 ionisations 13000 are detected. The typical genuine coincidence signal would be expected to be given by

$$N_t = N_e \cdot \omega \cdot \gamma \cdot \phi(E) = N_x \cdot \mu_\alpha \cdot \mu_\Delta$$

which gives 3 coincidence events detected from 1000000 ionisations. This is too large a loss in signal to make the technique worthwhile with conventional detectors.

## 3. Experimental Results

The results of a preliminary investigation into the ability to measure coincidence signals in a conventional analytical electron microscope have been presented previously (Nicholls et al. 1984). The aim of this paper is not to repeat the details of the results but to discuss more fully the implications. However it is useful to review the results.

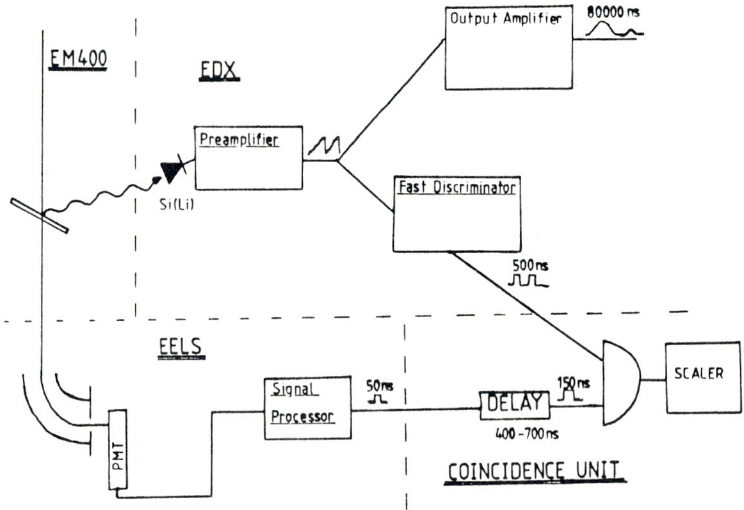


Fig. 1. Schematic of experimental equipment.

The experimental set up is shown in figure 1. A delay is necessary on the energy loss signal as the X-ray photon takes a finite time to be detected. The standard Gaussian amplifier used to measure photon energy is slow (80 μsec/photon) and not suitable for coincident counting and so the fast discriminator signal, which is not energy filtered, was used instead. Figure 2 shows that when no delay is used the coincident count rate can be calculated using equation 1, i.e. false coincidences only.

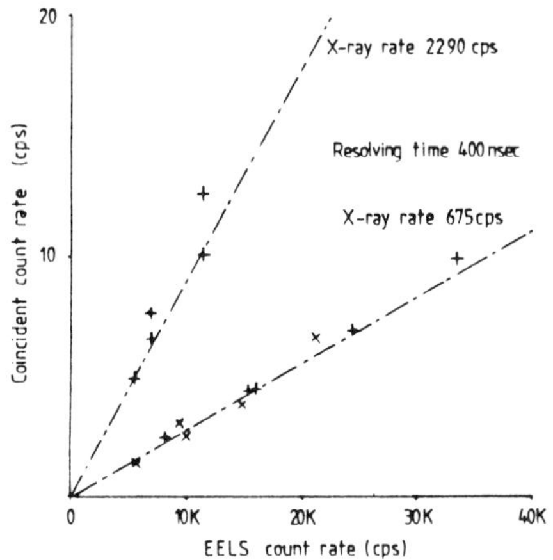


Fig. 2. Comparison of experimental false coincidence rate with theory.

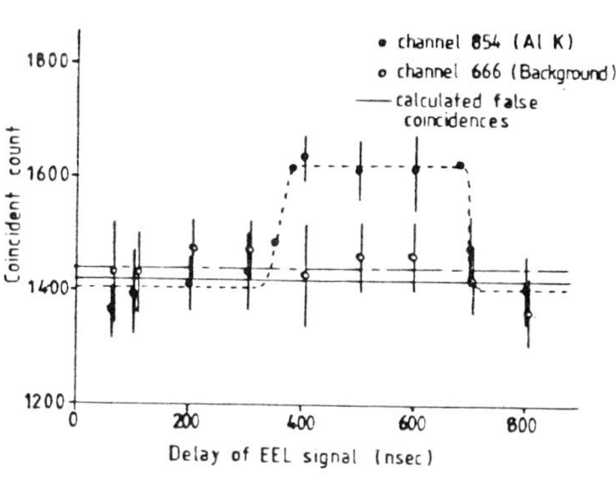


Fig. 3. Result of delaying EELS signal on apparent no of coincidences in two channels (counting time 60 s).

The effect of delay on the coincidence signal is shown in figure 3. A small but significant increase in coincidence count can be seen for delays between 380 and 680 nsec for a channel placed just above the AlK edge, when compared with a channel on the EELS spectrum with the same count rate (fig. 4).

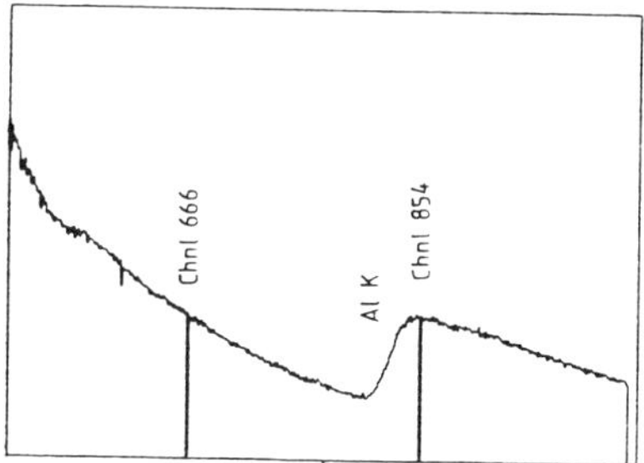


Fig. 4. EELS spectrum showing location of channels used.

4. Discussion

It is clear from the results presented here that coincident counting using conventional detectors is feasible but their detection efficiencies are not good enough. Hence, if the technique is to become useful it is necessary to increase the detection efficiencies. This is a reasonably easy task in the case of the electron energy loss spectrometer. For a serial detector the pmt slit width could be increased in size (allowing the $\mu_\Delta$ term to be increased to $\sim$ 0.5 for 400 eV at the AlK edge) or, perhaps, a parallel detector could be used although at present the necessary electronics would be slow. Increasing detection efficiencies would increase the count rates and hence increase the false coincidence signal. This is not a serious problem as experimental results have shown that false coincidences can be mathematically modelled (eqn. 1) and hence subtracted. It would be advantageous to reduce the resolving time: the effect of such a reduction on the ratio of true to false coincidences for fixed input count rates is shown in figure 5. Resolving times of 50 nsec should be possible.

It is not obvious, however, how the detection efficiency of a conventional Si(Li) X-ray detector could be improved. The efficiency is limited by the size of available crystal and the need to keep the detector at cryogenic temperatures. The standard energy filtering electronics are too slow and some other form with much faster response even at poorer energy resolution is needed. Some success has been reported using a timing filter amplifier (Kruit et al 1984) which can reduce X-ray pulses to 0.25 μsec but with much poorer energy resolution. The best hope of increasing detection efficiency is that a room temperature solid state detector could be used. Experiments have shown (Dabrowski 1982) that $HgI_2$ is the most suitable alternative detector. With such a detector angular collection efficiencies of > 40% should be possible compared with the current 1%.

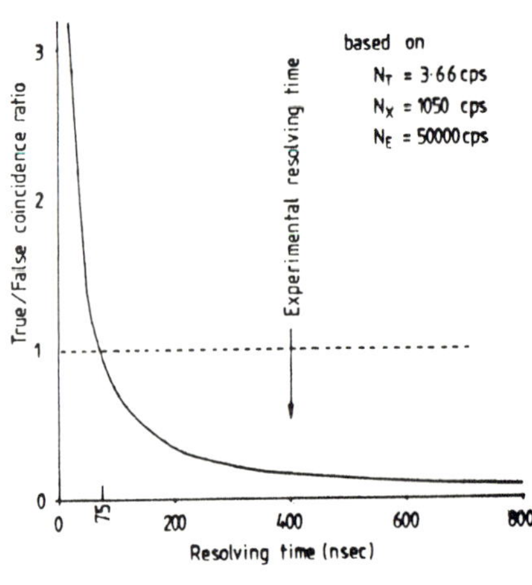


Fig. 5. Calculated effect of reducing resolving time on true/false coincidence ratio.

Another possibility is to use some other result of the ionisation. An obvious choice would be Auger electrons. Problems associated with this would include the position of the specimen in the centre of the objective lens magnetic field and again poor detection efficiencies.

## 5. Summary

This investigation has shown that it is possible to measure X-ray photons and energy loss electrons in time coincidence but the technique is limited by the inefficiencies of currently available detection systems. Improvements in both room-temperature X-ray detectors and energy filtering electronics will be needed before coincidence counting can reach a stage where both detectability limits would be improved and spatial resolution accurately defined.

## References

Dabrowski A J 1982 Adv. in X-ray anal. 25 1-21.
Kruit P, Shuman H and Somlyo A P 1984 submitted to Ultramicros.
Nicholls A W, Colton G J, Jones I P and Loretto M H 1984 Proc. of 3rd AEM Workshop in press.
Voreades D and Crewe A V 1974 Proc. 32nd EMSA meeting pp 428/9.
Voreades D and Crewe A V 1975 Proc. 33rd EMSA meeting pp 100/1.
Wittry D B 1976 Ultramicros 1 297-300.

*Inst. Phys. Conf. Ser. No 78: Chapter 7*
*Paper presented at EMAG '85, Newcastle upon Tyne, 2–5 September 1985*


# A set of standards for calibration of energy-dispersive x-ray detectors

A J Bourdillon

Department of Metallurgy and Materials Science, University of Cambridge, Pembroke Street, Cambridge, CB2 3QZ, England.

There is a need for a set of stable mineral standards to be used for calibrating energy-dispersive X-ray detectors. Calibration is especially important at the low energy part of the spectrum when account must also be taken of specimen self-absorption.

While thin film microanalysis by energy dispersive X-ray spectroscopy (EDS) is simpler than bulk analysis (see e.g. Reed 1975), a completely standardless approach would imply not only that approximations used in calculating ionization and emission probabilities (Green and Cosslett 1961; Manson and Kennedy, 1974; Schofield, 1974; Keski-Rahkonen and Krause 1974 and Zaluzec 1979) were sufficiently accurate but also that the detector response was accurately known. This assumes knowledge of beryllium window thickness if assumed uniform and also of electrodes, dead layers etc. Supposing that secondary fluorescence - generated by scattered electrons of X-rays whether inside or outside the probed volume - can be reduced to negligible proportions, the measurement of relative elemental concentration/emission intensity ratios, or k-factors (Cliff and Lorimer, 1975), require careful correction for specimen self-absorption in standard specimens. This is particularly true for the lighter elements or the L and M emissions of heavier ones. Besides, published values are not complete and are not general e.g. sometimes not all lines $\alpha,\beta$ etc. of an emission series are ratioed.

Self-absorption can be easily calculated if enough is known (as below) about the foil thickness, density and orientation with respect to the incident probe and the EDS detector. k-factors are usually measured relative to that for silicon because of the variety of stable minerals containing this element (cf.Wood, Williams and Goldstein 1983 who use iron). We chose the following set which does not have emission lines overlapping with the Si K: Kyanite ($(Al_2O_3)SiO_2$), Jadeite ($NaAl(SiO_2)_2$), Olivine ($(MgFe)_2SiO_4$), Rhodonite ($MnSiO_3$), Wollastonite ($CaSiO_3$), Willemite ($Zn_2SiO_4$), Zircon ($ZrSiO_4$), Terbium silicide ($TbSi_2$), Celsian ($BaAl_2Si_2O_8$) and Eulytine ($Bi_4Si_3O_{12}$). These minerals were ground, and picked out of suspension on holey carbon films.

The foil thickness for each specimen was measured internally by comparison of the SiK emission rate in the standard with the emission rate of a foil of Si of known thickness (measured by thickness fringes of known extinction distance) at constant operating conditions (probe intensity). We thus measured for each standard foil an effective silicon thickness, $t^{Si}$, but ignored second order self-absorption corrections in this measurement of relative thickness.

The self-absorption correction, $a_i$, for any emission line of wavelength $\lambda_i$, depends on an elementary calculation of the proportion of fluorescent emission transmitted through the specimen in path to the detector, and is given by

$$a_i = \frac{\alpha(\lambda_i)}{1 - \exp(-\alpha(\lambda_i))} \tag{1}$$

where the absorption coefficient

$$\alpha(\lambda_i) = \sum_j \left(\frac{\mu}{\rho}(\lambda)\right)_j \rho_j \frac{t^{Si}}{\sin\theta} c_j \tag{2}$$

summed over all elements, j, present in the standard with mass absorption coefficients $(\mu/\rho)_j$ (Leroux, 1961), density $\rho_j$ and concentrations $c_j$, while $\theta$ is the take-off angle, in our case simply 45 degrees. At small t, $a_i \simeq (1 - \alpha/2)^{-1}$, and this last term provides the self-absorption correction relating relative concentrations of two elements, $c_i$, to their emission intensities, $I_i$, in the normal way:

$$\frac{c_1}{c_2} = \frac{I_1}{I_2}\frac{k_1}{k_2}\frac{a_1}{a_2} \tag{3}$$

For the standards listed above values of $\alpha$ have been calculated for foils of 100 nm effective Si thickness (i.e. giving the same Si emission rate as a pure Si foil 100 nm thick, $I_i^{\text{pure Si}}(100)$). From these values the specimen self-absorption from foils of various thicknesses or take-off angles can be easily calculated (i.e. by writing $\alpha(\lambda_i, t^{Si}) = \alpha(\lambda_i)\, I_i/I_i^{\text{pure Si}}(100)$ and substituting in equation 1). For light elements 20 per cent of the emitted radiation can easily be self-absorbed. The fact that all of the standards contain silicon simplifies the application of this correction.

With self-absorption corrections, the k-factors for various elements were calculated from measurements taken in a Philips EM 400T with Edax detector, and accelerating voltage of 100 kV. Where necessary substitutional corrections were made for detected impurities, and resulting values plotted in figure 1. The k-factors shown are atomic, and these are easily converted to mass concentration factors by products with corresponding atomic weights. The results, though of course specific to our detector, are in reasonable agreement, where they overlap, with previously published results (Cliff and Lorimer, 1975, Wood, Williams and Goldstein 1983, and the Bristol group, 1984), which are in turn particular to various operating conditions including accelerating voltage. Our k-factors however, include all emission lines: $K_\alpha + K_\beta$, or $L_\alpha + L_\beta + L_\gamma$ etc.

An assumption made is that the specimens are uniformly thick over the probe area; though this is not critical since self-absorption corrections can always be kept below 20% of signal by measurement of sufficiently thin areas. However the use of the curves shown in figure 1 imply more regularity than is the case. Koster-Kronig transitions, for example, produce discontinuities with increasing Z in the X-ray emission rates. For the most accurate work there is at present no substitute for careful comparison of spectra with those from known and similar standards, with systematic accounting for thickness dependent specimen self-absorption. Other factors, such as crystallographic orientation (Bourdillon, Self and Stobbs, 1981) must also be considered. However for routine work the set of standards described is commercially available and can be used more generally to calibrate a detector.

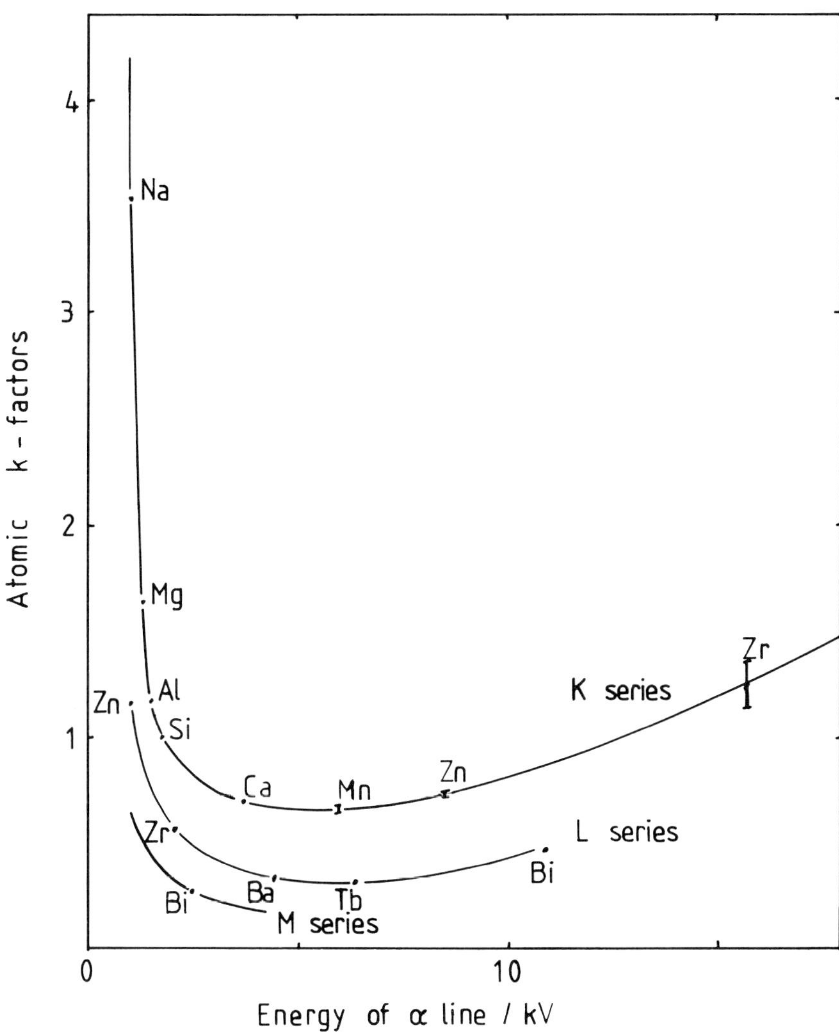


Figure 1: Atomic $k_{Si}$ factors measured with the standards described in the text, for K, L and M X-ray emission series.

Acknowledgements

We are grateful to Micro-analysis Consultants Ltd. of St. Ives and to G.A. Chinner and S.B. Reed for giving us standards, and to Prof. D. Hull for the provision of laboratory facilities.

References

Bourdillon A J, Self P G and Stobbs W M (1981), Phil. Mag. A44 135.

Bristol group (1984) Convergent beam electron diffraction of alloy phases. Hilger.

Cliff G and Lorimer G.W. 1975, J. Micr. 103 203-207.

Green M and Coslett V E 1961, Proc. Phys. Soc. 78, 1206-1214.

Keski-Rahkonen O and Krause M O 1974, At. da. and Nucl. da. Tables 14 139-159.

Leroux J 1961, Adv. in X-ray Analysis, Vol. 5 p.153-160, Ed. W. Muelle Plenum.

Manson S T and Kennedy D J 1974, At. Da. and Nucl. Da. Ta. 14 111-120.

Reed S J B, Electron Microprobe Analysis C U P 1975.

Schofield J H 1974, At. Da. and Nucl. Da. Ta. 14 121-137.

Wood J E, Williams D B and Goldstein J I, 1983, J. Micr. 133 255-274.

Zaluzec N 1979, in Introduction to Analytical Electron Microscopy, ed. Hren J J, Goldstein J I and Joy D C, Plenum Press p.121.

*Inst. Phys. Conf. Ser. No 78: Chapter 7*
*Paper presented at EMAG '85, Newcastle upon Tyne, 2–5 September 1985*


# The influence of contamination on EDS detector performance and on the measurement of phosphorus in the presence of molybdenum

P Doig and P E J Flewitt

Central Electricity Generating Board, South Eastern Region, Scientific Services Department, Gravesend, Kent, DA12 2RS, U.K.

## 1. Introduction

High and low temperature mechanical properties of ferritic steels can be related to the distribution of minor alloying and impurity elements within the microstructure such as the segregation of phosphorus to molybdenum rich carbides (Stevens and Flewitt 1985) or silicon rich inclusions (Hippsley and Druce 1983). Any correlation of mechanical properties with the microstructure therefore requires measurement of such segregations which can be effected using STEM-EDS X-ray microanalysis.

Quantitative X-ray microanalysis of thin foil samples uses the approximation that the X-ray intensity recorded from a particular element, $I_x$, is related to its concentration, $C_x$, by a calibration factor, k, derived from the X-ray yield and the detector efficiency (Cliff and Lorimer 1975; Goldstein 1979):

$$C_A/C_B = k_{AB} (I_A/I_B)$$

This is independent of the sample thickness and composition since it does not include any contribution from absorption or fluorescence. These k-factors are measured experimentally (Cliff and Lorimer 1975; Wood, et al., 1981) or may be calculated (Goldstein 1979; Zaluzec 1979) and evaluated with reference to a standard element, e.g. B = silicon. The calculated values are given by the product of the generated X-ray intensity and the detector efficiency. The first of these terms is derived from fundamental physical parameters and the second is determined by the geometry and construction of the detector (Goldstein 1979; Zaluzec 1979). Although there is some debate concerning the precise values of the fundamental physical parameters (Goldstein, 1979; Zaluzec 1979), they are constants that are independent of the particular experimental system. The detector efficiency, however, is recognised to vary from system to system and for that reason an experimental evaluation of the relevant k-factors for each detector is recommended and a number of determinations have been reported (Cliff and Lorimer 1975; Goldstein 1979; Wood et al., 1981; Metcalfe and Broomfield 1983).

This paper examines the implications of contamination on these k-factors and its effect on the measurement of small phosphorus concentrations in the presence of molybdenum.

## 2. Evaluation of k-factors

The theoretical evaluation of k-factors has been described (Goldstein 1979). In this work we use the fluorescence yield given by Wentzel (1927) and ionisation cross section due to Powell (1976) with values for L lines derived using the relationship given by Reed (1975). The detector efficiency is determined by the absorption of X-rays in the beryllium window, the silicon dead layer and the gold layer and also the transmission of X-rays through the detector. Absorption coefficients for these components are given by Heinrich (1966). The range of evaluated k-factors varies with the assumed detector parameters for transmission and absorption of X-rays. This is shown in Figure 1 for the range of beryllium window thickness = 5-20μm, silicon dead layer thickness = 0.1-0.3μm, gold film thickness = 0.01-0.05μm and a silicon detector thickness of 3mm for (a) K lines and (b) L lines.

Regardless of the particular detector parameters and the absolute value for the ionisation cross sections and fluorescence yields, these plots give a measure of the k-factor variation to be found between different detector systems. For a given detector, the sensitivity across the X-ray energy range will vary depending on its particular parameters. Because of the uncertainties in the theoretical descriptions of the ionisation cross section and fluorescence yields, however, unique evaluation of the detector parameters from méasured k-factors is not straightforward. The variability in k-factors is also revealed by the range of the ratio of k-factors for K and L lines shown in Figure 2 for elements with atomic number $\geqslant 31$ derived from Figure 1. Here the range of values for molybdenum (atomic number = 42) is 1.43-2.28. This ratio is conveniently measured from $MoO_3$ crystal samples and is routinely evaluated by the authors to monitor the performance of the STEM-EDS X-ray detector system. The ratio is observed to change progressively with time to lower values indicating a variation in the detector efficiency. Similar variations have been identified on a number of instruments in other laboratories. Thus, k-factors do not remain constant. These changes in the k-factor can result from the deposition of contamination on the specimen or the cooled detector window derived from the specimen preparation or the microscope vacuum system. Contamination produces an additional absorption layer for the X-rays generated within the specimen thereby reducing the detector efficiency by an amount that depends on the particular X-ray energy and the nature of the contamination. We assume the contamination to be either carbon or silicon based. The influence of carbon and silicon contamination on the measured k-factor ratio for molybdenum K and L lines is shown in Figure 3. Clearly, very significant changes occur for relatively thin layers of contamination.

## 3. Influence of Contamination on the Measurement of Phosphorus in the Presence of Molydenum.

Silicon or carbon based contamination changes the detector efficiency non-linearly because of the wavelength dependence of X-ray absorption, particularly in the region of absorption edges. The influence of carbon and silicon contamination on the k-factors for phosphorus K and molybdenum L lines is shown in Figure 4 where the ratio k(P(K))/k(Mo(L)) increases with contamination thickness. Since the phosphorus K line is almost coincident with the molybdenum Ll line, estimation of phosphorus in the presence of large concentrations of molybdenum is effected by examining the residual X-ray intensity after stripping of the molybdenum Ll line. The characteristic intensity of this line with respect to the molybdenum Lα line is reduced by contamination such that any additional intensity at this energy derived from phosphorus Kα will be interpreted as molybdenum and a reduced phosphorus concentration will result, Figure 5.

## 4. Discussion

When phosphorus is measured in the presence of molybdenum, the use of k-factors established under conditions of no contamination on the sample or detector window will always result in an underestimate of the true phosphorus concentration. This effect is not confined to phosphorus and molybdenum but will apply for all analyses where the true k-factor is changed. In this particular case, where overlapping peaks are concerned, the effect is enhanced by the overestimate of the molybdenum Ll intensity thereby further decreasing the remaining X-ray intensity ascribed to the phosphorus. As discussed above, the development of contamination is readily monitored by recording the relative intensities of the molybdenum L and K lines using thin $MoO_3$ crystal calibration specimens.

## 5. Conclusions

1. Contamination on either the detector window or the thin foil specimen will change the k-factors for X-ray microanalysis.
2. Phosphorus concentrations in the presence of a large concentration of molybdenum will always be underestimated when contamination occurs on the specimen or detector window.

## 6. References

Cliff G and Lorimer G W 1975 J. Microsc. 103 203.
Goldstein J I 1979 Introduction to Analytical Electron Microscopy, ed. J J Hren, J I Goldstein and D C Joy, New York, Plenum Press.
Heinrich K F J 1966 The Electron Microprobe, ed. T D McKinley, K F J Heinrich and D B Wittry, Wiley, New York.
Hippsley C A and Druce S G 1983 Acta Metall. 31 1861.
Metcalfe E and Broomfield J 1983 Electron Microscopy and Analysis 1983, ed. P Doig, Institute of Physics Conf. Series 68, Lond.
Powell C J 1976 NBS Special Publication 460 ed. K F J Heinrich, D E Newbury and H Yakowitz.
Reed S J B 1975 Electron Probe Microanalysis, Cambridge University Press.
Stevens R A and Flewitt P E J 1985 Acta Metall.(in press).
Wentzel G 1927 Zeit Phys. 43 524.
Wood J Williams D B and Goldstein J I 1981 Quantitative Microanalysis with High Spatial Resolution, ed. G W Lorimer, M H Jacobs and P Doig, Met. Soc. Lond.
Zaluzec N J 1979 Introduction to Analytical Electron Microscopy, ed. J J Hren, J I Goldstein and D C Joy, New York, Plenum Press.

Acknowledgement: This paper is published by permission of the Executive Director, South Eastern Region, Central Electricity Generating Board.

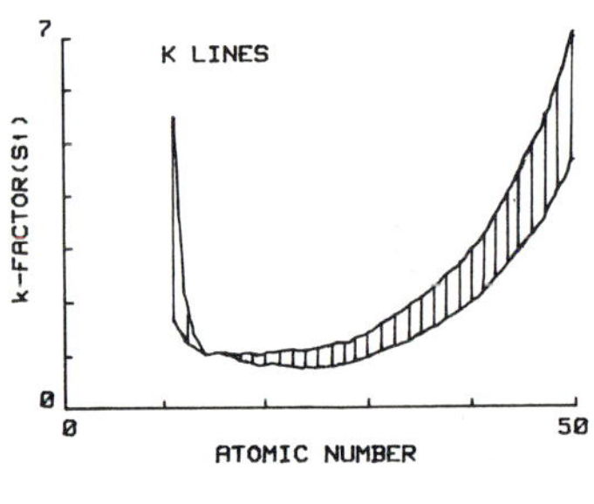


(a)

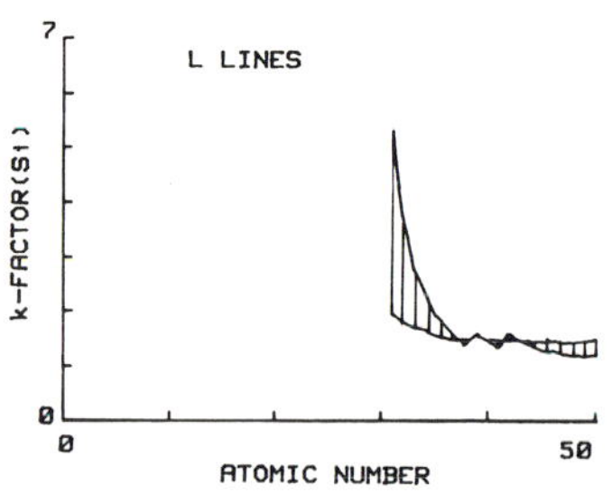


(b)

Figure 1. Calculated range of k-factors for a) K lines and b) L lines using the range of detector parameters : Si dead layer = 0.1-0.3μm, Be window thickness = 5-20μm, Au film thickness = 0.01-0.05μm and a silicon detector crystal thickness = 3mm.

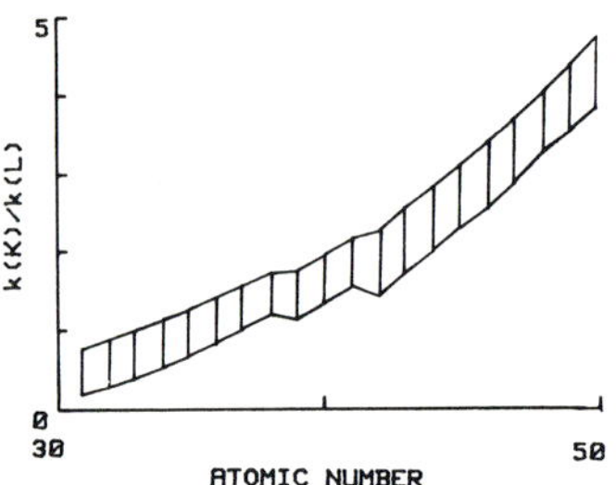


Figure 2. Calculated range of the ratio of k-factors for K and L lines for elements with atomic number ⩾ 31.

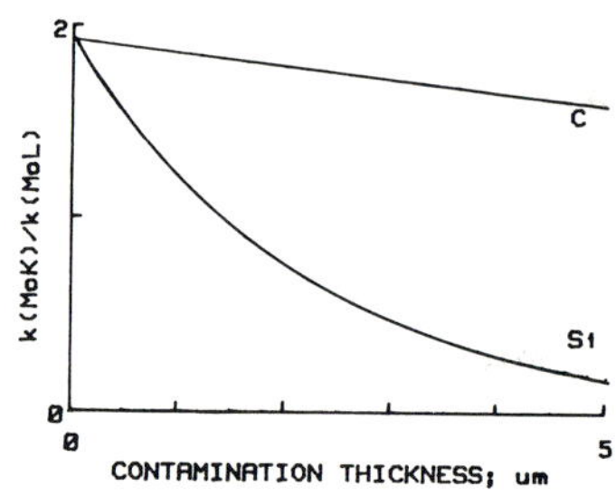


Figure 3. Calculated influence of C and Si contamination on the L and K line k-factor ratio for Mo.

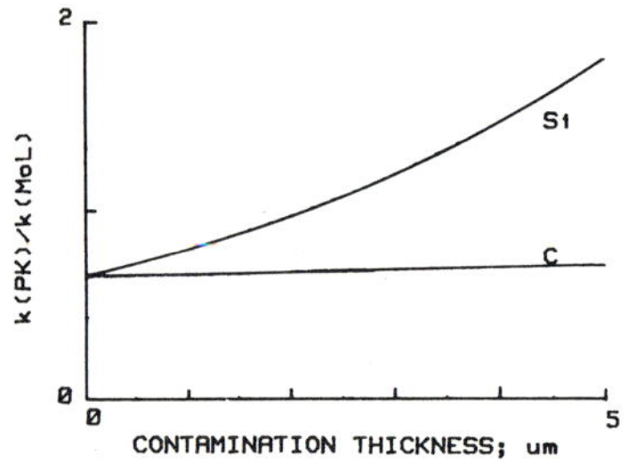


Figure 4. Calculated influence of C and Si contamination on the ratio of k-factors for P(K) and Mo(L) lines.

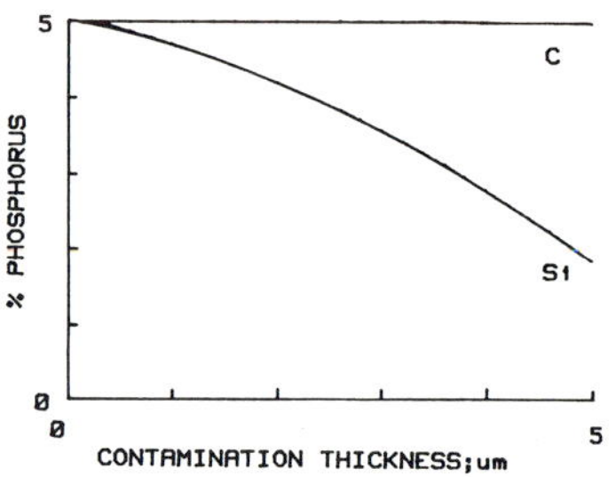


Figure 5. Calculated influence of C and Si contamination on the measured P concentration in the presence of 95% Mo.

# The use of channelled electron induced x-ray emission for the detection of atomic displacements

F Glas and P Henoc

C.N.E.T. Laboratoire de Bagneux, 196 rue de Paris, 92220 BAGNEUX, France

We consider here the possibility of detecting atomic displacements from the average planes of a crystal using the X-ray emission induced by a channelled electron beam, and apply this technique to the semiconducting alloy $In_xGa_{1-x}As$.

$In_xGa_{1-x}As$ crystals have a cubic lattice whose parameter follows Vegard's law: $a(x) = x\ a(1) + (1-x)\ a(0)$ ; the parameters a(1), a(0) of the sphalerite structures InAs and GaAs differ by 7%. However, EXAFS experiments show that the distances between III and V neighbour atoms do not correspond to the Virtual Crystal Approximation (sphalerite of parameter a(x)) but nearly retain the values of the corresponding binary (e.g. Ga-As distances in any alloy and in GaAs are close). Thus some atoms at least are considerably displaced from their virtual sublattice. EXAFS suggests that As (but not III) atoms are displaced in polycrystalline samples (Mikkelsen et al 1983), but is inconclusive for $In_{.53}\ Ga_{.47}\ As$ layers epitaxied on (001) InP (Bellessa et al 1983). In order to assess displacements in such layers, we remark that if :

(a) the angle of incidence of a parallel electron beam impinging on a flat thin crystal is varied around the Bragg angle relative to a family of lattice planes $\underline{g}$

(b) some B atoms happen to be displaced from these average lattice planes and other A atoms are not (or less)

then not only the count rates of A and B characteristic X-rays will vary with the angle (a) (Cherns et al 1973), but also their ratio (b), provided the ionization inducing the emission is sufficiently localized. More precisely, the A/B count rates ratio should be larger for negative deviation parameters s (when the beam density is maximum on the lattice planes) than for positive s (when the beam has minima on them).

A VG HB5 STEM delivers, in SAD mode, a parallel beam on a μm size unbent uniformly thinned area. The angle of incidence (at a fixed excitation of the rocking coils) is measured on the diffraction screen and the X-rays analyzed by EDS.

To estimate quantitatively the effect expected, we suppose the average positions of all A atoms exactly in the lattice planes and those of all B atoms displaced out of these by $\pm\ \delta$, without correlations. Thus we consider that the electron wave is the same as in the undistorted crystal. If t is the sample thickness and $\xi_g$ the extinction distance, we calculate, in a 2-beam approximation and following Cherns et al (1973), the ratio of A/B count rates :

$$R_{A/B}(s) = R_{A/B}(s=0)\ x\ (1 - P_g^A\ f(t)\ \sin\beta\ \cos\beta)\ /\ (1 - P_g^B\ f(t)\ \sin\beta\ \cos\beta)$$

with : $f(t) = 1 - \sin(2\pi\Delta k\, t) / 2\pi\Delta k\, t$

$$\Delta k = (s^2 + \xi_g^{-2})^{1/2} \quad , \quad \cot\beta = s\,\xi_g$$

$$P_g^C = \int_0^{1/g} P^C(u)\cos 2\pi gu \; du \qquad (C = A \text{ or } B)$$

if $P^C(u)$ is the density of probability of C characteristic X-ray emission by a unit charge at a distance u from the lattice planes. $P^C(u)$ is a Dirac distribution broadened by phonons and the range of the ionizing interaction; we suppose that: $P_g^C = \exp(-\alpha g^2)$ with : $\alpha = D + (B/2)^2$

(D : Debye - Waller parameter, B : impact parameter associated to the ionization of C (Bourdillon 1984)). Moreover : $P_g^B = \cos(2\pi g\delta)\, P_g^A$ .

In $In_{.53}\,Ga_{.47}\,As$, we consider the (220) planes , containing all the crystal atoms . Fig. 1 gives the expected variation of $R_{Ga\,K\alpha/As\,K\alpha}$ if Ga atoms are in these planes and As 0.01 nm away from them.

The variation with s of Ga Kα/As Kα in an $In_{.53}\,Ga_{.47}\,As$ specimen (Fig. 2a) diplays a minimum around s=0 (expected for s > 0) and a maximum around the symmetric 220,$\bar{2}\bar{2}0$ position, where channelling effects on 220 and $\bar{2}\bar{2}0$ reinforce (s < 0 for both). Channelling thus produces qualitatively the effect expected of a localization of Ga closer to (220) planes than As. We only performed a perfect crystal 2-beam calculation, neglecting absorption: this explains the relative discrepancy between calculated and experimental curves, and also makes difficult the quantitative estimation of the displacements, but the observation of the effect suggests that they are at least of the order of the EXAFS determined ones. Fig. 2b demonstrates the strong variation with s of the Ga and As L to K count rates ratios, illustrating the large difference in localization of the ionizations of L (B ∿.06 nm (Bourdillon 1984) ) and K (B ∿.006 nm) levels; this also confirms that the specimen thickness and flatness conditions required to observe strong channelling are met.

Bellessa J, Gors C, Launois P, Quillec M and Launois H 1983 Inst. Phys. Conf. Ser. No 65 (London) 529
Bourdillon A J 1984 Phil. Mag. A50 839
Cherns D, Howie A and Jacobs M H 1973 Z. Naturforsch. 28a 565
Mikkelsen Jr J C and Boyce J B 1983 Phys. Rev. B28 7130

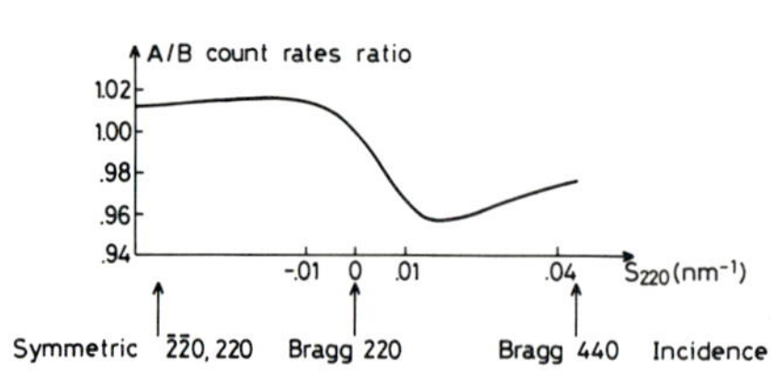


Fig. 1 Calculated variation around 220 Bragg angle of the ratio of A to B X-ray count rates, if A atoms are in 220 planes and B .01 nm away from them, supposing the interaction perfectly localized, $g_{220} = 4.8\ \text{nm}^{-1}$ $\xi_{220} = 56$ nm (normalized scale). (sample thickness = 80 nm)

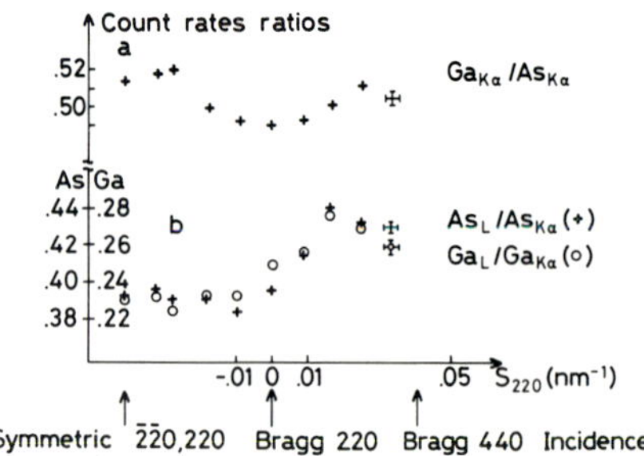


Fig. 2 Measurement in an $In_{.53}Ga_{.47}As$ specimen, around (220) Bragg incidence of the ratios of Ga Kα/As Kα (a) and As L / As Kα and Ga L / Ga Kα (b) count rates. (estimated sample thickness = 85 nm)

Paper presented at EMAG '85, Newcastle upon Tyne, 2–5 September 1985

# Quantitative compositional mapping on an electron microprobe

R L Myklebust, R B Marinenko, D E Newbury and D S Bright

## 1. Introduction

Since the first demonstration of x-ray area scans obtained with the scanning electron microprobe of Cosslett and Duncumb (1956), the use of the electron microprobe to produce two-dimensional elemental distribution maps has become a popular adjunct to conventional fixed-beam microanalysis. In the past, these maps have been generated almost exclusively by analog signal processing. X-ray signals detected with either wavelength-dispersive spectrometry (WDS) or energy-dispersive spectrometry (EDS) are used to modulate the intensity of a cathode ray tube (CRT) scanned in synchronism with the electron beam on a specimen, and the resulting image is recorded photographically. Very little quantitative information could be obtained from these maps. The technique of recording allowed only 1-bit dynamic intensity range (on-off), and gray scale information could be obtained only in special circumstances. Figure 1 is an example of an analog map near the practical lower limit of concentration (1 wt%). It shows the zinc distribution at the grain boundaries in polycrystalline copper after diffusion-induced grain boundary migration (DIGM) (Butrymowicz et al 1984 ). The map required about 6 hours to accumulate.

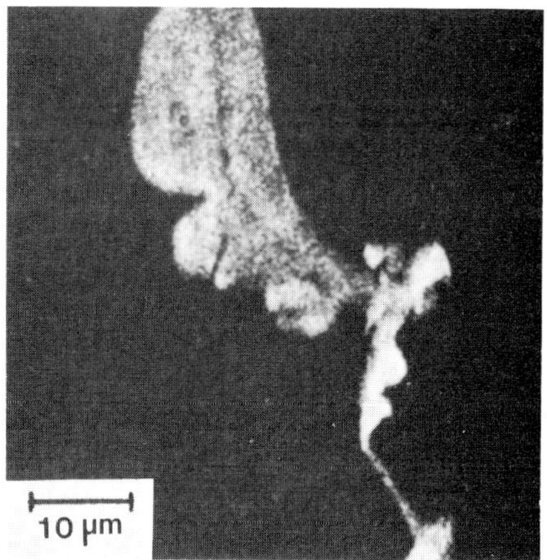

Fig. 1 Analog zinc x-ray map of DIGM in copper.

The analog methods for efficiently using x-ray signals for mapping have been reviewed by Heinrich (1975). It is difficult to form x-ray images with gray-scale information similar to conventional images produced by electron signals since the efficiency of x-ray production is so poor. Methods of obtaining gray-scale concentration maps by setting thresholds on the x-ray signals have been described by Heinrich (1962, 1963). The principal limitation to analog mapping is the use of photographic film as the data storage device. The analyst is unable to modify the x-ray information after recording in order to subtract background, change the contrast to enhance features, or perform quantitative computations.

The advent of laboratory computers dedicated to controlling electron microprobes has made possible the quantitative digital compositional map. A new generation of multichannel analyzers is now capable of processing both EDS and WDS signals as well as controlling the beam position on the specimen. In the digital method, the beam is stepped in a raster pattern over the surface of the specimen, pausing at each step to integrate x-rays for each element or to accumulate an EDS spectrum. The EDS spectrum may be processed to yield data on several elements or one WDS signal from each available spectrometer can be recorded. An x-y grid of counts for each x-ray signal is stored and may be displayed as an image or processed into quantitative compositional maps or maps that display enhanced features.

## 2. Instrumental Considerations

The electron microprobe used in this work was equipped with three wavelength-dispersive spectrometers and an energy-dispersive spectrometer. A multichannel analyzer and a computer automation system were used to control the electron probe and accumulate the data. The original program for generating x-ray maps did little more than produce elemental dot maps. A threshold value was set and if the x-ray count was above the threshold, the element was assumed to be present; otherwise, the element was assumed to be absent. This software was modified to store the accumulated x-ray counts at each of the analysis points (pixels) for each element analyzed. The software can produce maps of 32x32, 64x64, or 128x128 pixels for both the WDS and EDS systems. We have recorded eight elements with the EDS system; however, the time required to acquire EDS maps was large due to the lower counting rate per element of the EDS system compared to the WDS system. High beam currents are used when possible to increase the counting rate. In the case of the EDS system care must be taken not to saturate the detector since it accumulates all x-rays emitted by the specimen not just those from the elements of interest. Accumulation times of a few tenths of a second per pixel for high concentration elements to several seconds for low concentration elements may be required for WDS maps.

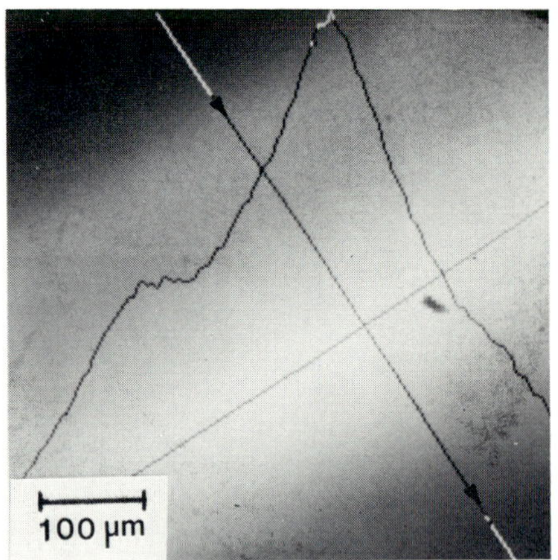


Fig. 2 Digital WDS x-ray map of pure chromium with spectrometer intensity profile taken along diagonal line.

Low magnification (200x - 800x) WDS maps were observed to exhibit errors due to spectrometer defocussing effects that may be greater than 50% relative. A digital x-ray map of chromium is shown in Figure 2 superimposed on the intensity profile taken along a line perpendicular to the ridge of maximum intensity. The intensity profile is actually the peak profile of chromium $K\alpha 1$ and $K\alpha 2$ as a function of wavelength. In these figures, black is the minimum intensity and white is the maximum intensity in each map.A more effective way of looking at these images is to present the data as a contour map. The intensities are divided into ten levels, each level representing a 10% band of intensities. The intensities at each pixel are sorted into these levels which can then be displayed as contours on an image as is shown for the silver image in Figure 3. The angle of the contours represent the angle of the spectrometer with respect to the orientation of the rastered area on the specimen. Since the orientation of the raster may be arbitrarily set on the microprobe, care must be taken to not rotate the rastered area during an analysis.

3. Quantitative Computations

The highest level of sophistication is the determination of a quantitative digital compositional map. The first step is to correct all of the WDS data for dead-time loss and in the case of EDS, system live-times must be used. This is particularly important since high counting rates are used to minimize the data collection time. The next step is to remove the background. In the case of WDS, the wavelength profile of the continuum background near the peak for each element is constant across the entire image for a homogeneous standard; therefore, only one background measurement per element is required. The background for EDS measurements must be computed from the spectrum accumulated at each pixel. For multiphase unknowns, one spectrometer can be used to measure a background map, and by compensating for differences in spectrometer efficiency, this map can be used to correct for background on other spectrometers. Digital maps are acquired for each standard as well as for each unknown. The intensity ratios (k-ratios) for each element are then computed at each pixel in the array by dividing the intensity from the unknown by the intensity from the standard at the corresponding pixel in the standard array. A matrix correction procedure is then used to compute the concentrations for each element at each pixel (Myklebust and Thorne 1984).

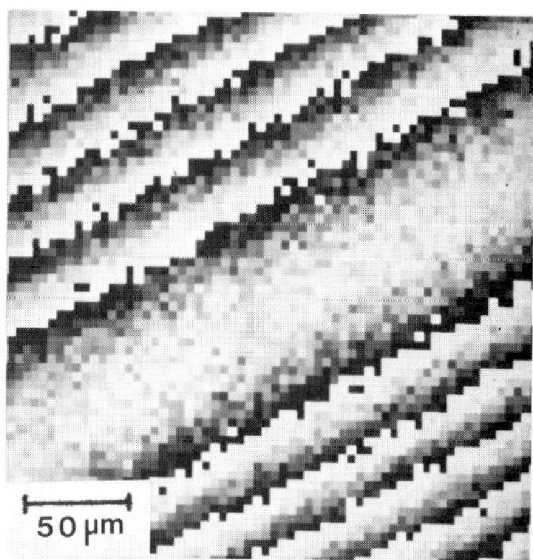


Fig. 3 Contour map of silver where each level is equivalent to a 10% step in intensity.

The resulting concentration maps are then displayed through our image analysis system. Figure 4 is a compositional map of the same alloy as Figure 1. Comparison of these two maps shows the superior contrast of the compositional map which allows discrimination of distinct compositional changes within the zinc-enriched region.

To test the validity of this approach, standard reference materials known to be homogeneous on a micrometer scale were mapped and the data from the maps were analyzed using statistical procedures. A compositional map of silver in a 40%Ag-60%Au alloy is shown in Figure 5 where the minimum value for silver (36.8 wt%) is black and the maximum value for silver (43.2 wt%) is white. The mean value for silver

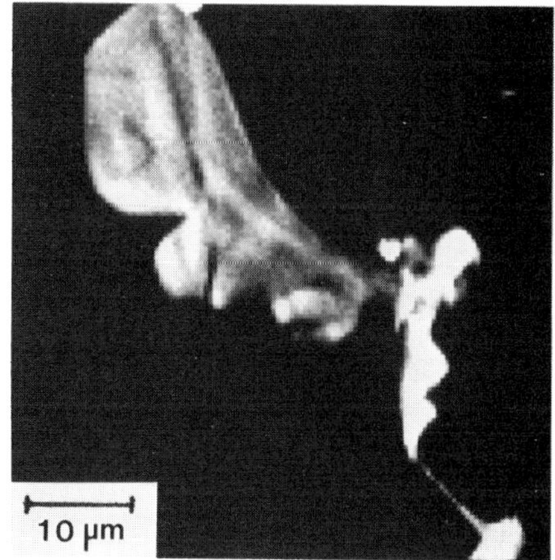


Fig. 4 Quantitative digital x-ray map for zinc in DIGM in copper. Same area as in Fig. 1.

obtained from the 4096 pixels in the map is 40.0 wt% (certified 39.9 wt%) with a relative standard deviation of 2.2 wt%. It is evident from Figure 5 that the silver composition is higher on the right side than on the left. From counting statistics on a single pixel, the relative standard deviation is 1.7 wt% which seems to indicate either a tilted standard or specimen, or a small error in the quantitative imaging procedure.

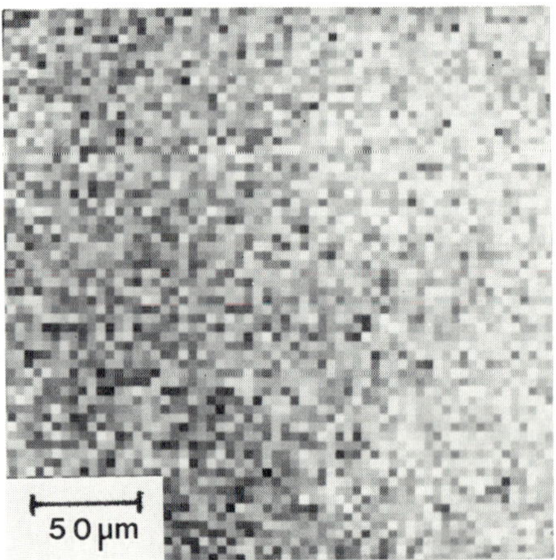


Fig. 5 Quantitative digital x-ray map for silver in 40wt%Ag-60wt%Au standard.

## 4. Summary

For the purposes of mapping, EDS and WDS systems offer different advantages. The EDS map is practically free from defocussing effects, even at low magnifications, whereas defocussing is significant in WDS maps even at high magnifications. WDS images can be used to display much lower values of concentration than EDS images because of the higher peak/background ratio for WDS. Any number of elements are mapped simultaneously with the EDS system; however, the WDS system limits the number of elements to the number of spectrometers on the instrument. It generally takes less time to acquire a WDS map since much higher counting rates per element are possible with WDS than with EDS.

We have demonstrated only one method for correcting the spectrometer defocussing. There is no defocussing if the beam is stationary and the specimen is moved; however, this requires a specimen stage that is capable of reproducing movements of less than one micrometer. It is also possible to move the spectrometer crystal so that it is always focussed on the peak regardless of beam position. This requires very high precision spectrometers, but is otherwise a very good method. We are studying a method for modeling the intensities across an image; thereby eliminating the need for a precision stage, or high precision spectrometers, or of measuring an entire array on each standard.

## 5. References


Butrymowicz D B, Newbury D E, Turnbull D and Cahn J W 1984, Scripta Met.18 1005

Cosslett V E and Duncumb P 1956, Nature 177 1172

Heinrich K F J 1975, in Advanced Optical and Electron Microscopy ed Barer and Cosslett (London:Academic) 6 p 275

Heinrich K F J 1962, in Advances in X-ray Analysis ed Mueller and Fay (New York:Plenum) 6 pp 291-300

Heinrich K F J 1963, in Advances in X-ray Analysis ed Mueller, Mallet and Fay (New York:Plenum) 7 pp 382-394

Myklebust R L and Thorne B B 1984, NBS Tech. Note 1200

*Inst. Phys. Conf. Ser. No 78: Chapter 7*
*Paper presented at EMAG '85, Newcastle upon Tyne, 2–5 September 1985*


# Development of a universal correction procedure for EPMA

G Love, V D Scott and D A Sewell

School of Materials Science, University of Bath, Bath, BA2 7AY, UK.

## 1. Introduction

In quantitative electron-probe microanalysis (EPMA), x-ray intensities from the specimen are compared with those from a reference standard of known composition, and the intensity ratio then corrected for atomic number, absorption and fluorescence effects (the ZAF procedure) to give elemental mass concentrations. Whilst, however, current correction programmes often provide satisfactory results, problems arise when dealing with light elements ($Z < 11$) and with specimens inclined to the electron beam. In this paper we describe a new, more flexible, correction method, assess its accuracy, and comment briefly an how it may be extended to the analysis of thin films on substrates.

## 2. The model

The principal problem is to develop an improved absorption correction, since existing atomic number and fluorescence corrections give sufficient accuracy in quantitative analysis (see Love et al 1975). In particular, formulae have to be produced which can describe the distribution of x-rays as a function of mass depth in the sample for a wide range of analysis conditions. This may be achieved directly from theoretical considerations of the physics of electron-induced x-ray generation or, alternatively, an empirical approach can be adopted which involves curve fitting. We have chosen the latter method and have used a combination of experimental measurements on tracer-type specimens (Sewell et al 1985a) and calculations via Monte Carlo simulations of electron trajectories (Love et al 1977) in order to produce the necessary x-ray depth distributions, or $\Phi(\rho z)$ curves.

A typical $\Phi(\rho z)$ curve is illustrated in Fig.1 and although its scale and detailed shape depends upon mean atomic number, incident electron energy and x-ray energy, the same general trends are evident for almost all experimental circumstances. We have adopted a simplified approach to $\Phi(\rho z)$ curve fitting, choosing a quadrilateral profile defined by the points, A, B, and C (Fig.1) to represent the x-ray depth distribution. This approach simplifies the mathematics involved but, before proceding, its validity needs to be assessed.

Geometrical considerations of the quadrilateral profile show that the absorption correction factor, $f(x)$ is given by

$$2\left[(\rho z_n - \rho z_m)(\rho z_m + h\rho z_n)\chi^2\right]^{-1}\left\{\left[-\exp(-\chi\rho z_m) + h.\exp(-\chi\rho z_n) + \chi(\rho z_n - \rho z_m) - h + 1\right] + \left[\frac{\exp(-\chi\rho z_m)(\rho z_n - h.\rho z_n) + h.\rho z_n - \rho z_n}{\rho z_m}\right]\right\}$$

where $\chi = \mu/\rho.\mathrm{cosec}\psi$, $\mu/\rho$ is the mass absorption coefficient and $\psi$ the x-ray take-off angle; the terms $\Phi(0)$, $\Phi(\rho z_m)$, $\rho z_m$ and $\rho z_n$ refer to the coordinates in Fig.1 and $h = \Phi(\rho z_m)/\Phi(0)$.

Values of $f(\chi)$ calculated from this equation, $f(\chi)$quad, are compared with Monte Carlo calculations, $f(\chi)$mc, in Fig.2. It may be seen that agreement is good, within a few percent for $f(\chi)$ values down to ≈0.2 (equivalent to 80% absorption).

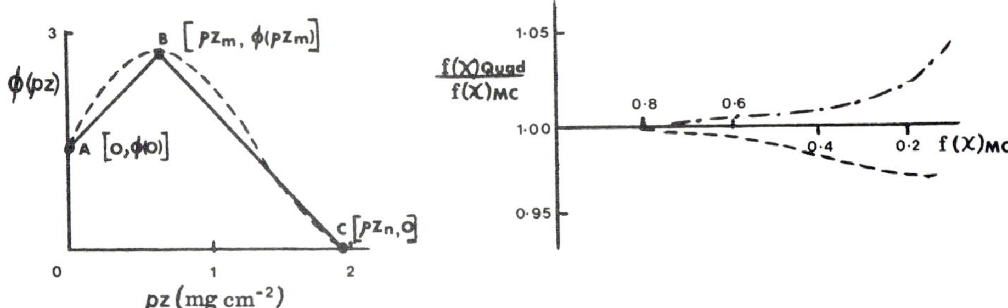


Fig.1. Dashed line Monte Carlo $\phi(\rho z)$ distribution for aluminium at 10kV, solid line quadrilateral representation.

Fig.2. Plots of $f(\chi)$quad/$f(\chi)$mc for 10kV probe voltage: dashed line aluminium matrix; dash-dot line carbon matrix.

Now equations for $\Phi(0)$, $\Phi(\rho z_m)$, $\rho z_m$ and $\rho z_n$ can be constructed which describe their variation with atomic number, electron energy, etc. These have been produced using our tracer measurements and Monte Carlo calculations and are given in Sewell et al (1985b).

## 3. Test of model using microanalysis data

Two sets of microanalysis data have been used for evaluating the new absorption correction, the first a collation of 554 measurements on binary alloys and the second our own measurements on well characterised oxides and fluorides (Sewell et al 1985b). We have adopted the atomic number correction of Love et al (1978) and the characteristic fluorescence correction of Reed (1965), whilst mass absorption coefficient formulae of Springer and Nolan (1976) for radiation above 2keV and data from Henke et al (1982) for less energetic radiation have been used.

Measured intensity ratios from specimen and standard, k, have been compared with intensity ratios predicted by the correction model, k', from prior knowledge of specimen composition. Hence when the model corrects properly, $k'/_k = 1$, whereas $k'/_k < 1$ indicates over correction and vice versa. The RMS error for the 554 binary alloys (heavier element data) is 2.9%, a figure close to the estimated precision of the 'raw' microanalysis measurements (≈2%). For the light element data (oxides and fluorides), the RMS error is 4.8%. These values are compared with RMS errors given by an established ZAF method incorporating the absorption correction of Philbert (1963) in Table 1, and the superiority of our new model is self evident.

Table 1. Percentage RMS errors

| | New Model | Established Model |
|---|---|---|
| Binary alloy data | 2.9 | 4.8 |
| Oxide and fluoride data | 4.8 | 16.5 |

4. Samples inclined to the electron beam

First, $\Phi(\rho z)$ data were produced using tracer specimens positioned at a variety of tilt angles and the coordinates for A, B, and C measured. The tilt-angle parameter was then incorporated into the quadrilateral model such that the equations for $\Phi(0)$, etc, matched those above. Further details will be available later (Sewell et al, to be published). Next, aluminium radiation was measured in samples of $NiAl_3$ and $CuAl_2$ for tilt angles of $30^o$, $40^o$ and $50^o$. The ratios $k'/_k$ are plotted in Fig.3 as a function of tilt angle and it may be seen that in no case does the error in the correction model exceed 2.5%.

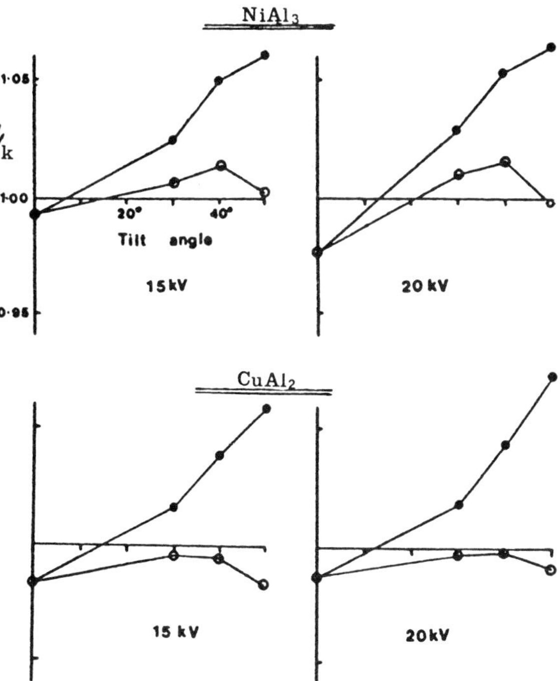


Fig.3. Correction for tilted sample:

o our method of incorporating the tilt factor into the equation $\phi(0)$ etc.

● old approach replacing $\chi$ with $\chi \sin \tau$.

For comparison purposes we show also results of using the conventional method of dealing with tilted samples, where the value for $\chi$ in $f(\chi)$ equation is replaced by $\chi.\sin\tau$, $\tau$ being the angle between the specimen surface and the electron beam direction. Clearly this method is far less satisfactory since errors increase with tilt angle, up to 7 or 8% at a beam energy of 20keV.

## 5. Concluding remarks

A new correction procedure for quantitative EPMA has been developed which we consider can be regarded as universal in its range of application. It works well for all systems, including the light elements where conventional methods break down, and can also cope with tilted samples.

Because the procedure is based upon modelling a $\Phi(\rho z)$ curve, it has the potential to deal with non-standard boundary conditions such as the case of a thin ($\approx 1\mu m$) coating on a substrate. Work is, therefore, currently under way to accumulate the necessary experimental data showing how the nature of the substrate influences the shape of the x-ray generation profile in the coating.

## 6. Acknowledgements

To the SERC and AERE, Harwell, for support.

## 7. References

Henke BL, Lee P Tanaka TJ Shimabukuro RL and Fujikawa BK 1982 Atomic Data and Nuclear Data Tables 27 1.
Love G, Cox MGC and Scott VD 1975 J Phys D:Appl Phys 8 1686.
Love G, Cox MGC and Scott VD 1977 J Phys D:Appl Phys 10 7.
Love G, Cox MGC and Scott VD 1978 J Phys D:Appl Phys 11 7.
Philibert J 1963 X-ray Optics and Microanalysis eds H H Pattee VE Cosslett and A Engstrom (New York: Academic Press) p.379.
Reed SJB 1965 Brit J Appl Phys 16 913.
Sewell DA, Love G and Scott VD 1985a J Phys D: Appl Phys 18 1233.
Sewell DA, Love G and Scott VD 1985b J Phys D: Appl Phys 18 1245.
Springer G and Nolan B 1976 Canad J Spectrosc 21 134.

*Inst. Phys. Conf. Ser. No 78: Chapter 7*
*Paper presented at EMAG '85, Newcastle upon Tyne, 2–5 September 1985* 

# Theory of crystalline effects in x-ray and Auger electron microanalysis

J F Bullock, C J Humphreys, J M Titchmarsh* and H E Bishop*

Department of Metallurgy and Science of Materials,
University of Oxford, Parks Rd., Oxford OX1 3PH.

*AERE Harwell, Didcot, Oxon., OX11 ORA.

## 1. Introduction

The intensities of characteristic X-rays and of Auger electrons produced from a crystalline specimen by an incident electron beam are strongly influenced by the orientation of the specimen relative to that beam. This effect was first pointed out by Hirsch et al.(1962). Some theoretical work (Cherns et al.(1973), Andersen and Howie(1975), Bishop et al.(1984), Bullock et al.(1985)) has been done for both the X-ray and the Auger cases in order to explain fully this effect, although little detailed comparison with experiment has been made. In microanalysis this effect is detrimental and errors are usually avoided by orienting the crystal away from a channelling condition. The variation in intensities can be as much as three fold as the crystal is tilted from the symmetry position. In the ALCHEMI technique of Spence and Taftø(1983), however, channelling is exploited to determine atomic site location. Theoretical calculations on ALCHEMI have been limited to crystals with alternate planes of atoms of different types (Krishnan et al.(1983), though no account was made of the absorption of the incident beam, or of the contribution to the characteristic X-ray intensity of X-rays produced by backscattered electrons.

## 2. Calculation of X-ray and Auger Intensities

The theories of the orientation effect for X-rays and Auger electrons are very similar, as both arise from the core level ionisation of the atom by the incident beam.

For a simple centrosymmetric crystal consisting of only one kind of atom the X-ray intensity produced by the incident beam can be calculated following Cherns et al.(1983), and using standard dynamical theory notation:

$$I^{(x)} = \sum_{j}\sum_{l} C_o^{(j)} C_o^{(l)} P^{jl} \int_o^t \exp(2\pi i(k_z^{(j)} - k_z^{(l)})z)\exp(-2\pi(q^{(j)} + q^{(l)})z)\,dz$$

Where $$P^{jl} = \sum_{g}\sum_{l} C_g^{(j)} P_h C_{g-h}^{(l)}$$

This assumes an interaction potential $P(\underline{r})$ for x-ray production, localised at the atomic positions, which thus has the periodicity of the lattice, so

$$P(\underline{r}) = \sum_h P_h \exp(2\pi \underline{h}.\underline{r})$$

In a compound we consider the production of X-rays from atoms of type m, which have an interaction potential $P^m(\underline{r})$, and a summation must be made of X-ray production from all atoms of type m in the unit cell. $P^{jl}$ is now written as

$$\sum_{gh}\sum (\sum_m \exp(2\pi i(\underline{g}-\underline{h}).\underline{r}_m) C_g^{(j)} P_h C_{g-h}^{(1)}$$

and used in equation 1 to give the characteristic X-ray intensity, $I_m^{(x)}$, of atom m. This can then be repeated for the other elements present in the unit cell, and for a series of orientations. Thus although absolute intensities are difficult to calculate, a measure of the variation of the characteristic X-ray intensity of one element relative to another can be evaluated.
In the case of Auger electron production in compounds the equation of Bullock et al.(1985) can be used, with the interaction potential $A_n$ replaced by a similar sum over atoms of type m.

## 3. Calculation of the Orientation effect

Calculations were performed for CsCl using 20 beams in a systematic (001) row, tilting though the g=(001) reflection. Figure 1 shows the calculated variation in the ratio $I_{Cl}^{(x)}/I_{Cs}^{(x)}$. $k_g/g$=0.5 corresponds to the (001) Bragg reflection position. The thickness was taken to be 1000Å, and incident beam voltage was 100kV.

## 4. Discussion

Calculations show a large variation in the ratio of the two characteristic X-ray intensities, with peaks appearing after the $k_g/g$=1.0,2.0,3.0 and 4.0 positions. It is to be expected that these calculations will overestimate the observed variations as no account has been made of the contribution of backscattered and inelastically scattered electrons, which would be expected to add a relatively flat background to the signal produced by the primary beam (Cherns et al.(1973)). However the orientation effect should be large enough to observe experimentally.
The calculated variation is best explained by examination of the various Bloch waves excited in the crystal. Figure 2 shows the the variation of the excitation amplitudes $C_0^{(j)}$ of the first 8 bloch waves with orientation. Bloch wave 1 is preferentially excited at the symmetry position, wave 2 at between the first and the second order Bragg reflection positions, and so on. The Bloch wave intensities are plotted in figure 3 as a function of distance x. x=0.0 and x= 1.0 are the Cs atom positions, x=0.5 is the Cl atom position. In figure 3(i) at $k_g/g$=0 ,wave 1 peaks strongly at Cs, and less so at Cl, wave 2 peaks between the atom positions. At $k_g/g$=1.0 (fig 3(ii)) waves 2 and 3 are the most

strongly excited, wave 3 is peaked at the Cl atom sites, thus the ratio of X-ray intensities increases (fig 1). Figure 3(iii) shows that at $k_g/g=2$ the wave most strongly excited is wave 5, which peaks at the Cl atom sites, so the ratio $I_{Cl}^{(x)}/I_{Cs}^{(x)}$ is much larger (fig. 1).

## 5. References


Andersen S K and Howie A 1975 Surf. Sci. 50 197

Bishop H E, Chornik B, Le Gressus C and Le Moel A 1984 Surf. Interface Anal. 6 116

Bullock J F, Humphreys C J, Mace A J W, Bishop H E and Titchmarsh J M 1985 Proc. 4th Oxford Conf. Micros. Semiconducting Materials (in press)

Cherns D, Howie A and Jacobs M H 1973 Z.Naturforsch. 28a 565

Hirsch P B, Howie A, and Whelan M J 1962 Phil. Mag. 7 2095

Krishnan K M, Rez P and Thomas G 1983 7th Internat. HVEM Conf. Berkeley CA p365

Spence J C H and Taftø J 1983 J.Microsc. 130 147


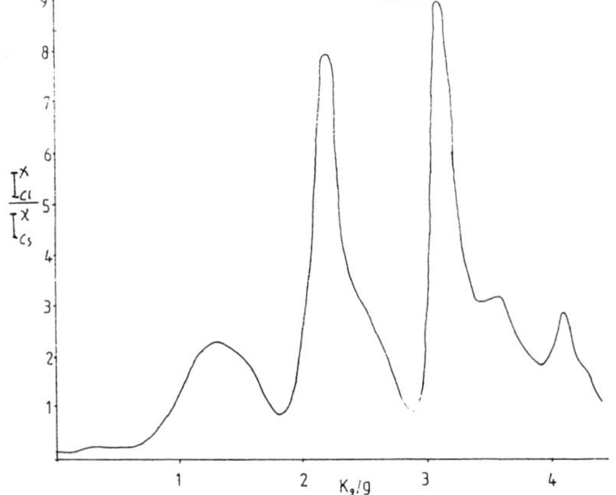


Figure 1. Variation of the ratio $I_{Cl}^{(x)}/I_{Cs}^{(x)}$ with orientation.

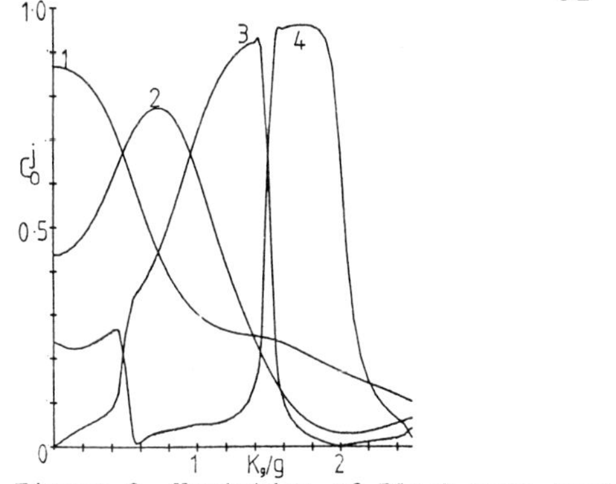


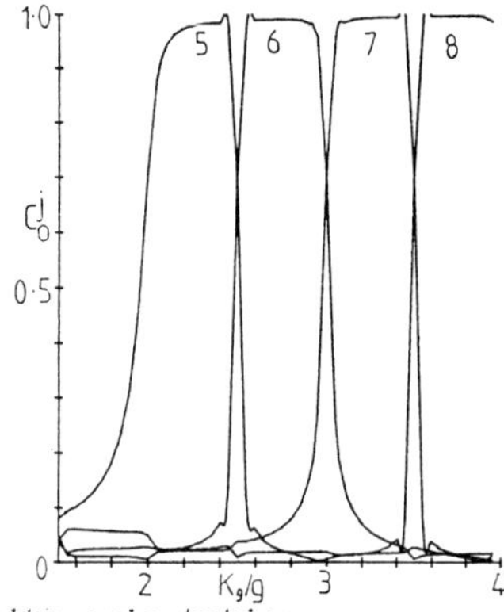


Figure 2. Variation of Bloch wave amplitude with orientation.

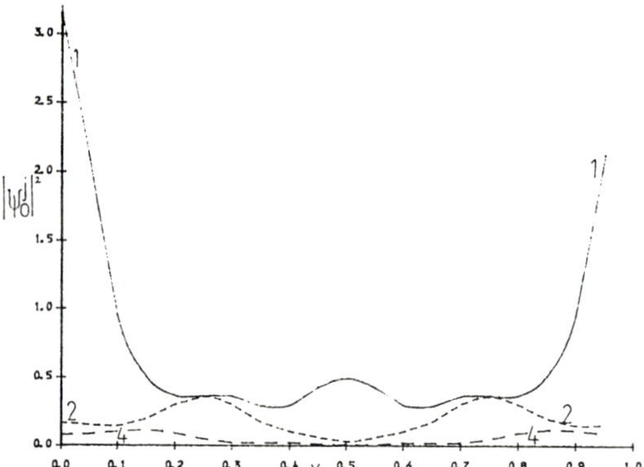

Figure 3(i). Variation of Bloch wave intensity with x at $\kappa_g/g=0$.

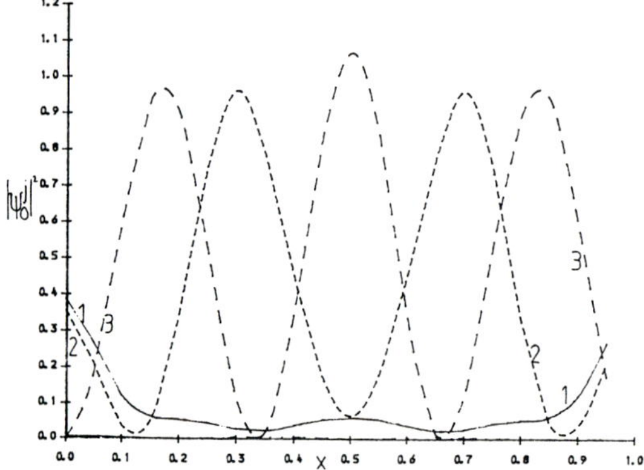

Figure 3(ii). As figure 3(i) at $\kappa_g/g=1$.

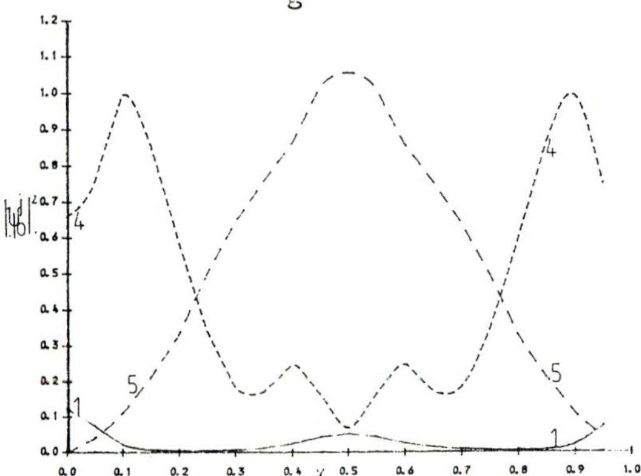

Figure 3(iii). As figure 3(ii) at $\kappa_g/g=2$.

*Inst. Phys. Conf. Ser. No 78: Chapter 7*
*Paper presented at EMAG '85, Newcastle upon Tyne, 2–5 September 1985* 

# Multispectral Auger imaging

R Browning

Stanford/NASA Joint Institute for Surface and Microstructure Research, NASA-Ames Research Center, MS 230/3, Moffett Field, Ca 94035.

## Introduction.

In the analysis of materials by Auger microscopy most of the useful experimental information is normally obtained by using small area Auger spectroscopy or Auger linescans. In a typical analysis, comparatively little additional information would be garnered from Auger imaging, and in publications Auger micrographs are usually only included for clarification and illustration. The low information content of Auger images is despite the considerable time taken for Auger image collection and is not related to the Auger measurement's signal-to-noise ratio either in a single picture element (pixel) or in the image as a whole. The low information content is caused by a lack of knowledge about the meaning, the statistics, and the reliability of the signal information. In Auger microscopy, the lack of information about the multiple contrast mechanisms, the low signal to noise ratios, and the presence of instrumental drifts and errors are the deciding factors in the interpretation of images.

Multispectral scanning Auger imaging is a technique that allows the recovery of a considerable part of the Auger image information.

## The Multispectral Auger Imaging Technique.

The multispectral Auger image technique is a nearly simultaneous collection of several Auger images using a computer controlled scanning Auger microscope. An electron beam is stepped across a specimen surface, and at each point in the image raster several energies in the secondary electron spectrum are sampled. Usually the energies sampled are at the major Auger transition peak positions and the adjacent high energy backgrounds found from spectra. The Auger peak heights are then the difference of the peak and background measurements. Several Auger peaks are measured in this way and it would be normal to divide these Auger peak heights by the adjacent backgrounds to give some normalization against topographic contrast. The resultant image is a multi-elemental Auger image with all the elemental information in spatial correspondence. The spatial mismatch between the separate images will only depend on the short-term image position noise and not on long-term mechanical stability or electronic drifts. In the absence of a large short-term image noise, the multi-elementalinformation from each pixel will have a spatial resolution determined largely by the electron probe size, rather than by the image step size or magnification.

The recovery of information from the spatially correlated multi-elemental image will depend partly on the amount of noise in the image and partly on the complexity of the image. The image noise can come from several sources. The fundamental limit to measurement is the quantum shot noise determined solely by the number of electrons randomly arriving at the detector during a measurement period. A good estimate of this noise can always be made and compared with the total noise found. For many specimens the most important additional noise source is the topography. Very often the surface of a sample is microscopically rough, and at high spatial resolution the roughness gives a random spread in signal. There are several other contrast mechanisms that may appear to produce noise ie. on polycrystalline samples there may be diffraction contrast. Although contrast noise is actually information, when interpreting an image it may appear as noise. The recovery of information from the multiple image is therefore a statistical problem and multivariate statistical techniques are required. The Auger measurement of the concentration of each atomic element is a variable in an N-dimensional image, and it is useful to think of the measured signal at each pixel as an N dimensional vector.

There is a very large body of literature on multivariate statistics and the reduction of information from multispectral images. A common approach to the classification or partitioning of multispectral images is to initially use the technique known as the "principal component transform". This technique is fully explained in the literature and has the following important properties for imaging (Ready and Wintz 1973).

(i) The signal is compressed into a minimum number of dimensions by finding the principal directions of the signal in the N-dimensions used.

(ii) The signal, which can also be understood as the variance in a measurement, is a constant in the transformation, or alternatively the total signal-to-noise in the multiple image is conserved.

The effect of the transformation can be similar to rotating the axis of the multiple vectors in N-dimensional space to find the largest sum of the signals in the image. The reason why this is particularly important to Auger imaging is that the random noise content of a measurement appears as an N-dimensional sphere of confusion in an N-dimensional histogram. The sphere of confusion is invariant under rotation whereas the N-dimensional signal is not. Since the Auger signal-to-noise ratio is only partly dependent on counting statistics and can be largely dependent on contrast noise, multiple images will always outperform single images. Improving the counting statistics may do little to improve the information content, but there will always be a principal direction in the multiple image. The effects of topography and other contrast mechanisms can thus be alleviated or, at least, better results can be obtained by multispectral Auger imaging.

With the major components of the image signal determined either by principal component analysis or similar transformations,the signal can be divided up into regions that represent different contrast levels or image features. In a one-dimensional signal, this would mean assigning multiple thresholds for imaging from a histogram of the signal strengths over the image. In a two-dimensional signal, the contrast levels appear as areas in a scatter diagram, and in higher dimensions as volumes in N-dimensional space. In practical terms, for Auger imaging the aim of the data transformation would be to maximize the signal into two dimensions.

One problem in partitioning a single elemental image into discrete contrast levels is determining what part of an overlapping distribution is noise and what compositional variation. In multiple elemental imaging there can in some cases be a dramatic improvement in separating levels. An example of the usefulness of two-dimensional scatter diagrams in partitioning signal is presented below.

Two types of transforms using Auger image data have been published. Ratioed scatter diagrams (Browning 1984) and simply rotated histograms (Browning 1985). The ratioed scatter diagram has the advantage of reducing three dimensional data to two, while the simple rotation of the histograms means less distortion of the results and the retention of the Gaussian noise distributions.

## Illustrative results.

To illustrate some of the above comments on multispectral Auger imaging, results from the analysis of an Al2124 specimen are presented. Al2124 is a precipitate hardened aluminium alloy. The precipitates found were intermetallics of Cu, Mn, Fe and Al. The experimental details are published elsewhere (Browning 1985). The recording of an Auger image using the Cu and Mn LMM transitions was found to give a high discrimination between the aluminium matrix and the intermetallic precipitates. The combination of information from these two transitions illustrates the major features of multispectral imaging. Fig 1 shows an SEM image of the rough extruded alloy, and areas of the harder precipitates can be suspected from the very slightly rougher regions. A Cu-Mn image was taken over one of the larger precipitate regions using a limited raster of 32x32 pixels. The spatial resolution of the electron probe was 100 nm. The sampling time per energy was relatively long, 500 ms, and the shot noise was found not to be the dominant noise. Fig 2 shows the result of histogramming the Cu and Mn signals. The background-ratioed signals are negative because of the slope of the E.N(E) curve. The matrix signal distributions can be seen as the better defined peaks at the left hand side of the histograms and one can notice that the precipitate and the matrix levels are not well separated. However, if the results are plotted as a two-dimensional scatter diagram, a clear separation between the signals from the matrix and the precipitates can be achieved. In this scatter diagram the matrix signals are scattered into a circle of confusion that is largely contrast noise. The maximum separation between the matrix and the precipitate signal occurs at a direction of approximately 45°to the Cu axis. The histograms can be rotated 45°so that one histogram (fig 4b) coincides with this direction. The left hand peak is the matrix distribution and it is clearly separated from that of the precipitate. The precipitate signals form an extended distribution along a 135° (90+45) direction. Using fig 4b to set a theshold, the matrix points can be supressed and the precipitate distribution alone can be histogrammed (fig 4a). More than one precipitate composition can be inferred from this histogram. The data in fig 4 was used as an experimental input for collecting semiquantitative images of the precipitate distribution.

In this example with more than one precipitate composition there is a large gain in image interpretation over the single element image. But there is also a gain in the effective signal to noise because of the rotation into the maximum signal separation.

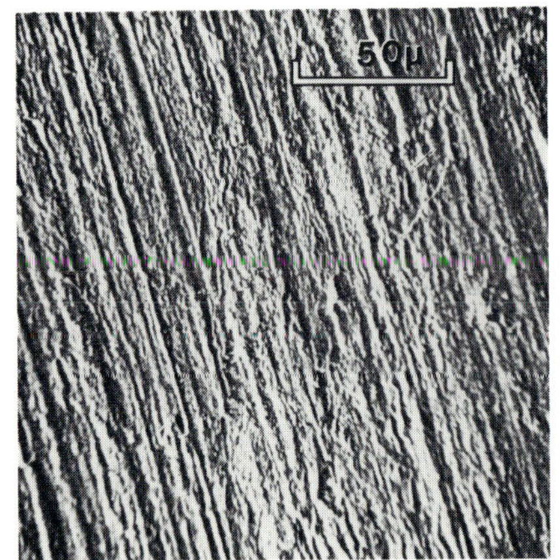


Fig 1. SEM of extruded Al2124 alloy.

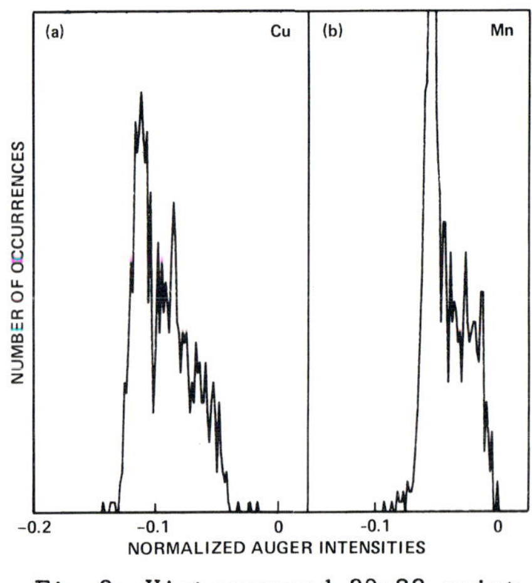


Fig 2. Histogrammed 32x32 point multispectral Auger image.

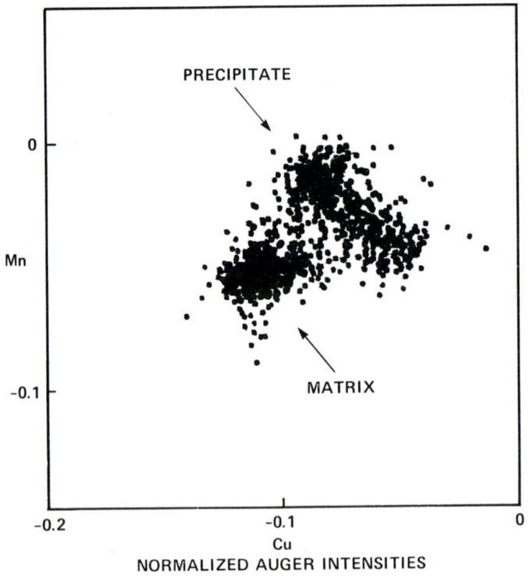


Fig 3. Scatter diagram from a 32x32 point multispectral Auger image.

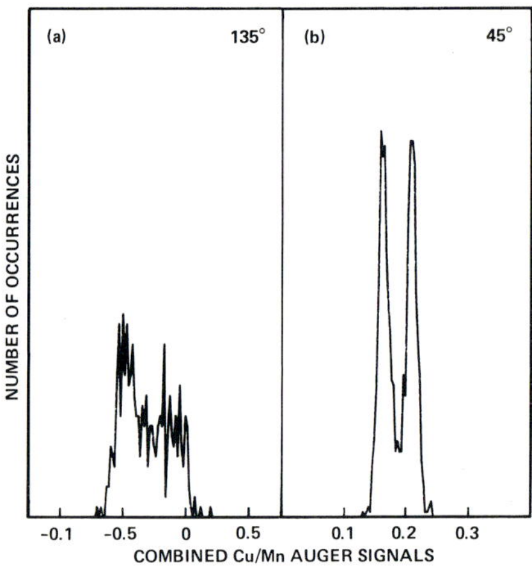


Fig 4. Rotated histograms from a 32x32 multispectral Auger image.

Although this is a very brief discussion of multispectral Auger imaging, it can be stated that there are some clear advantages of this technique over conventional single-element imaging. First, the spatial correspondence of the images allows the realization of the spatial resolution of the electron probe, second, the increased discrimination against contrast noise is unique, and third, there can be considerable advantages in partioning the signal. The degree to which these advantages are realized will depend on the specimen and the instrument.

## References

Ready P J and Wintz P A 1973 IEEE Trans. Com-21 1123
Browning R 1984 J. Vac. Sci. Technol. 2 1453
Browning R 1985 J. Vac. Sci. Technol. Sept/Oct.

*Inst. Phys. Conf. Ser. No 78: Chapter 7*
*Paper presented at EMAG '85, Newcastle upon Tyne, 2–5 September 1985*



# The effects of peak to background ratios in quantitative Auger analysis: Illusions in SAM images

M M El Gomati

Department of Physics, University of York, Heslington, York, YO1 5DD, U.K.

## 1. Introduction

In quantitative Auger electron spectroscopy (AES) and scanning Auger electron microscopy (SAM) one is normally faced with the task of estimating the surface atomic concentration of an element in a multi-variable system (Seah 1983).

This can presently be achieved with modest effort to a degree of accuracy of the order of 10-20 atomic per cent, but there is still a great need to improve on this figure. Therefore the strategies for data acquisition and processing in this field are under continuous development. A recent example of the efforts in this direction has been in the extension of a pragmatic technique successfully used to normalise against sample microtopography and beam current variations (Prutton et al 1981) to extract quantitative information from the collected Auger data. This technique involves the use of a combined ratio of the Auger peak height (P) to the height of its high energy background (B). In particular, Langeron et al (1984) and Venables et al (1985) have applied P/B ratios to Auger spectra to obtain quantitative AES information, while Bishop (1983) has used this ratio to normalise against atomic number variations in the substrate. The purpose of the present work is to assess critically the use of these ratio techniques in quantitative and qualitative AES and SAM. In addition, the effects of primary beam energy and the substrate atomic number Z on P/B are also studied.

## 2. Experiment

Homogeneous carbon (C) films of thickness 4 ± 1 nm and 25 ± 3 nm have been deposited on substrates of polycrystalline C, Al, Cr, Cu, Mo, Ta, W, Pt and Pb. A Cu electron microscope grid and a specially prepared sample consisting of Au islands on a C substrate were also covered with thin C films. Details of sample preparation have been described elsewhere (El Gomati et al 1985). Because of the fragility of the samples, no cleaning of their surfaces took place in situ. However, AES spectra has shown negligibly small Auger peaks of elements other than C. All data were collected under UHV conditions using a computer controlled high resolution SAM (Prutton et al 1982).

## 3. Results and Discussion

The height P, of an Auger peak is defined here as the difference between the counts on the Auger peak and the counts, B1 on a closely chosen point on its high energy side of the background at energy $E_{B1}$. In the case of the 260 eV Auger peak of C $E_{B1}$ is chosen as 290 eV. In addition, the background count B2 at 1990 eV was also recorded for comparison between P/B1 and P/B2 as a function of Z and Ep. A summary of this comparison is given in table 1. It is found that although the surface C concentration is the same for the supported and un-supported regions of the film, P/B ratios are always higher for the thin film in comparison to the thin film plus a substrate. Furthermore, the P/B ratio depends on the substrate atomic number and is high where the substrate atomic number is low (i.e. the Auger backscattering factor $r_A$ is low) and vice versa. This implies that peak to background ratioing tends to overcompensate for backscattering.

| Material | P/B1 0.02 | P/B2 | $r_A$ | $\eta$ |
|---|---|---|---|---|
| C film | 1.4 | 34 | 1. | |
| C/C | 1.2 | 10.7 | 1.24 | 0.066 |
| C/Cu | 1.08 | 10.4 | 1.68 | 0.325 |
| C/W | 0.98 | 9.0 | 1.85 | 0.492 |
| C/Pt | 0.99 | 8.2 | 2.1 | 0.507 |

Table 1

A comparison of P/B1 and B/B2 for some different Z materials. C/W denotes a C film on a W substrate. P measured at 260 eV, B1 = 290 eV and B2 = 1990 eV.

An illustrative example of the ill-effects of using these ratios in SAM images is shown in fig (1). A C film of about 6 nm thickness was deposited on a C substrate covered with Au islands of 300 nm thickness. Although the deposited C film has a uniform thickness over the two regions of the substrate, the collected Auger map implies a smaller effective C concentration on the Au containing region than on the simple C region. This misleading information is obtained in spite of the increased Auger peak height from areas on the Au islands due to backscattering because of the use in the image of the ratio P/(P+2B1) for brightness modulation. The reason for such a behaviour is that the counts on the background B1 and the Auger peak height P (which is proportional to the actual Auger current) depend on the substrate in different ways. Contributions to P comes from two sources: a direct ionisation component caused by the incident electron beam which is not substrate dependent, and a backscattering ionisation component which depends on the substrate. The background B1 on the other hand, which is proportional to the substrate backscattering coefficient $\eta$ for E> 50 eV, contains two environmental dependent contributions; backscattered primary electrons and true secondaries. This implies that B1 varies more strongly with environment than P.

Fig (2) shows the behaviour of P/B1 as a function of electron beam energy for a 25 nm C film deposited on different Z materials. These results clearly demonstrate the strong dependence of P/B on Z and Ep. In

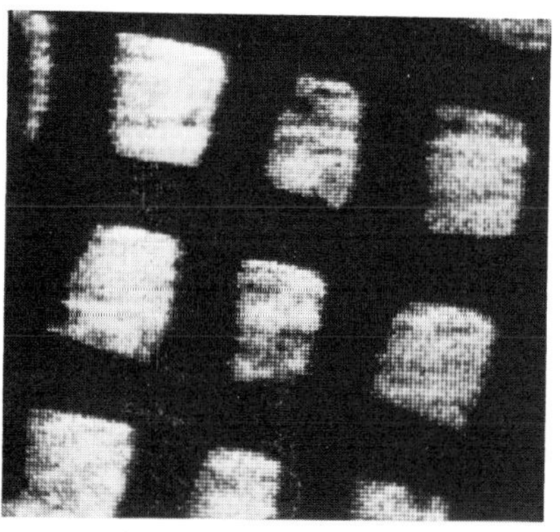

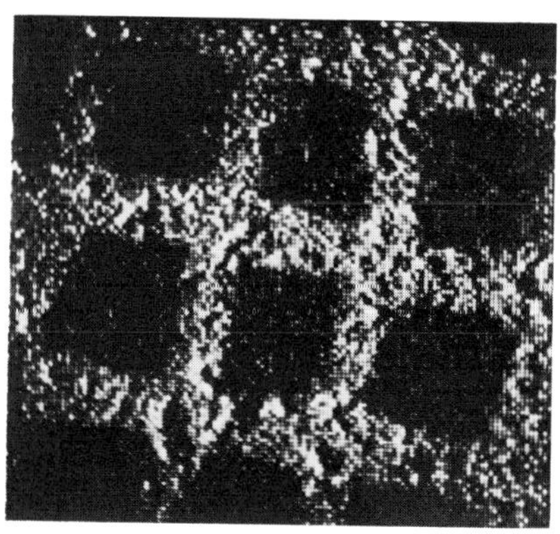


Fig 1. Illusion in SAM images due to the use of P/B ratio.
(a) SEM of a thin C film (6 nm thickness) deposited on a C substrate covered by Au islands (bright regions) of about 300 nm thickness.
(b) A 2 grey level C Auger image of the same area shown in (a) collected using the ratio P/(P+2B). Images are 128 x 128 pixels, 10 m sec/pixel, $I_b$ = 3nA, $\theta$=45°.

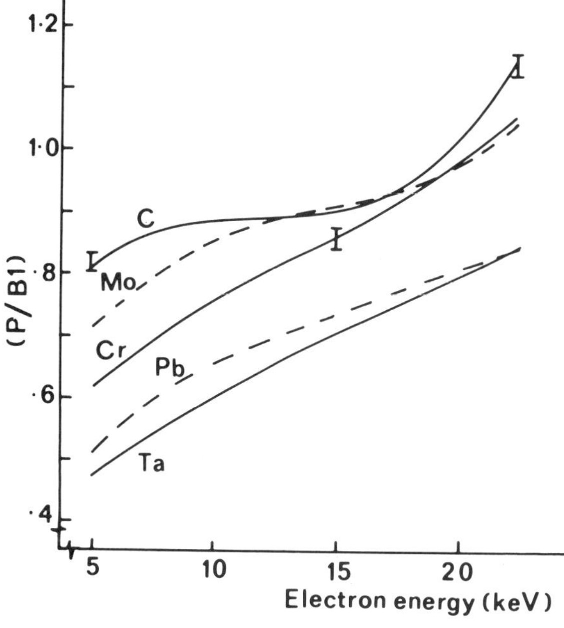


Fig 2. The effect of the substrate atomic number and the primary energy on P/B1.

particular, this ratio is a strong function of Z at low energies; it ranges from 0.47 for C/Ta (C/Ta means a C film on a Ta substrate) to 0.82 for C/C at 5 keV. The corresponding ratios at 22 keV are 0.84 and 1.09 respectively. The difference between Ta and Pb are comparable to the precision of the P/Bl. Furthermore, it is found that the increase in P/Bl with increasing primary energy Ep is faster for high Z materials than for low Z. The essence of these results is that although the Auger current depends through the ionisation cross-section upon Ep, this dependence is less pronounced than the dependence of the background on Ep and the substrate atomic number Z.

In conclusion, it has been demonstrated here that the use of a combined ratio of P and B in AES and SAM may give misleading indications of the surface concentrations of a certain element.

Accordingly, the use of P/B ratios in quantitative spectroscopy is fraught with difficulty, and should be avoided. In imaging, qualitative information can be obtained from P/B from which the effects of beam current fluctuations and surface roughness have been eliminated. However, the same constraints apply to qualitative and quantitative imaging as to spectroscopy. In the latter case it is suggested that either a signal directly proportional to the incident beam should be used (El Gomati and Prutton 1985) or the SEM signal. Auger maps should be collected and displayed using the actual Auger current. In this case the image will reflect the contribution of both the backscattering effects of the substrate and angle of incidence of the incoming beam on surface microtopography. The latter can be easily correlated with the corresponding SEM image and may require the collection of a set of images of differing angles of beam incidence. However, for the determination of absolute surface concentration one would need to separate the contribution of the backscattered electrons. This may be achieved by collecting from each point of the image a set of coarsely spaced energy counts at the background. This data can then be used to estimate the Auger backscattering factor at each point as described by Prutton and El Gomati (1985). An alternative method that requires a knowledge of the substrate's mean atomic number is outlined by Peacock and Walker (1985).

Acknowledgements

The author would like to thank M Prutton and J A D Matthew for useful discssion and the SERC for financial support through the ALVEY project VLSI-29.

References

Bishop H, 1983, SEM 83 (SEM Inc. A M F. O'Hare, IL 60666, U.S.A).
El Gomati M M, Matthew J A D and Prutton M, 1985, App. Sur. Sci. In Press.
El Gomati M M and Prutton M, 1985, EMAG 85, Inst. Phys. These proc.
Langeron et al, 1984, Surf Sci, 138, 610.
Peacock D C and Walker C H, EMAG 85, Inst. Phys. Conf. Ser. These proc.
Prutton M, Browning R, El Gomati M M and Peacock D C, 1982, Vacuum , 32, 351.
Prutton M and El Gomati M M, 1985, ECASIA 85, to be published.
Prutton M et al, 1981, EMAG 81, Inst. Phys. Conf. Ser. No 61, 443.
Seah M, 1983, SEM 83, (SEM Inc. A M F O'Hare, IL 60666, U.S.A) 521.
Venables J et al,1985,Phil. Trans. In Press.

*Inst. Phys. Conf. Ser. No 78: Chapter 7*
*Paper presented at EMAG '85, Newcastle upon Tyne, 2–5 September 1985*


# Quantitative multi-element imaging in the scanning Auger microscope

D C Peacock and C G H Walker

Department of Physics, University of York, Heslington, York, YO1 5DD, U.K.

## 1. Introduction

The calculation of quantitative elemental compositions of surfaces from Auger spectra has been frequently employed in the past (El Gomati et al 1983, Van Langeveld et al 1983, Minni 1983). However, quantification of images from a Scanning Auger Microscope has never been attempted before, to the authors knowledge. This paper outlines the method by which quantification of images may be achieved, as well as other aspects of multi-element imaging such as scatter diagrams, and "phase images". The method for collection of the images is also outlined. It is necessary to collect an Auger image for each element present on the surface, otherwise the quantified images will not have the correct compositions. A quantitative imaging technique has recently been applied to EPMA (Mayr et al), now quantitative imaging can be performed using SAM.

## 2. Multi-element Imaging on the SAM

If the Auger images are to be quantified then the signal for each pixel from each Auger image has to be collected from the same place on the sample. However if the images were collected one after the other, then specimen drift may cause considerable movement between each image. To overcome this problem each line of each image is collected in turn. Image "sets" are collected. These "sets" consist of the counts at the Auger energy, the counts on the background at the high energy side of the Auger peak and an SEM image collected simultaneously. To improve statistics on a particular image the dwell time may be increased while collecting data from the required Auger peak. The SAM used to collect the data is described elsewhere (Prutton et al 1982). 5 sets of 128x128 images takes approximately 10 hours to collect at 100 ms per pixel.

## 3. The Quantitative Algorithm

Prior to applying quantification, the raw Auger current (I) at each pixel is calculated using the scheme :-

$$I=(N1-N2)/N_{SEM} \qquad (1)$$

where N1 are the counts on the Auger peak, N2 are the counts on the background above the Auger peak and $N_{SEM}$ are the counts from the SEM detector (a scintillator). The raw images are not good Auger images because of the division by the counts from the scintillator. However this corrects for beam current and since each pixel should be from the same place on the specimen then topography and Z contrast in the images will be eliminated in the quantified images. An initial estimate , $C_j$, of the composition of element j for each pixel is given by the equation

$$C_j = (I_j / I_j') \left\{ \sum_{i=1}^{n} I_i / I_i' \right\}^{-1} \qquad (2)$$

Where $I_j$ is the Auger current from the specimen for the jth element, $I_j'$ is the Auger current from the elemental standard and n is the number of elements in the specimen. Correction for backscattering and electron mean free paths is performed iteratively using the equation :-

$$C_j = (I_j / I_j') \left\{ \sum_{i=1}^{n} F_{ij} \; I_i / I_i' \right\}^{-1} \qquad (3)$$

where:-

$$F_{ij} = \frac{(1+r_m(E_j))(1+r_i(E_i)) \; \lambda_m(E_j)\lambda_i(E_i)[a_j]^3}{(1+r_m(E_i))(1+r_j(E_j)) \; \lambda_m(E_i)\lambda_j(E_j)[a_i]^3} \qquad (4)$$

In equation (4) $r_k(E_k)$ is the Auger backscattering factor for elemental standard k, $r_m(E_k)$ is the Auger backscattering factor for element k in the specimen, $\lambda_m(E_k)$ is the specimen electron mean free path at energy $E_k$, $\lambda_k(E_k)$ is the electron mean free path in elemental standard k and $a_k$ is the atom size for element k. The derivation of $F_{ij}$ is outlined in greater detail in Seah 1983. The calculation continues until there is convergence or the maximum number of iterations has been reached. Notice that equation (2) should correct for topography and Z contrast introduced by the division of the SEM counts assuming that each pixel of each image is from the same place on the specimen.

## 4. Display and Analysis of Quantified Images

Quantitative 64x32 pixel images of a NiCrAl alloy are shown in Fig 1. The mean value of each image agrees closely with estimates of the surface composition using large area electron spectra collected from the surface. The data may also be displayed in the form of a scatter diagram for direct comparison with bulk phase diagrams (see Fig 2). This is an extension of the ratio histogram method (Browning et al 1983) and is the quantitative equivalent of the ratio scatter diagrams (R Browning 1985, R Browning et al 1985). Different phases of the specimen with different compositions will appear as separate clusters on the scatter diagram. By selecting a particular cluster of points on a spatial map a quantitative "phase image" is obtained. Such an image is shown for one phase on the NiCrAl alloy in Fig 3. Using scatter diagrams and "phase images" information can be revealed which previously would not have been available such as the number of distinct phases present and the proportions of the scanned area occupied by each, the spread of compositions within a phase and the average composition of dispersed regions comparable in size to the maximum SAM resolution. For instance, a precipitate scattered throughout a sample may be missed using ordinary Scanning Auger images, but may be picked up using the "phase image" technique.

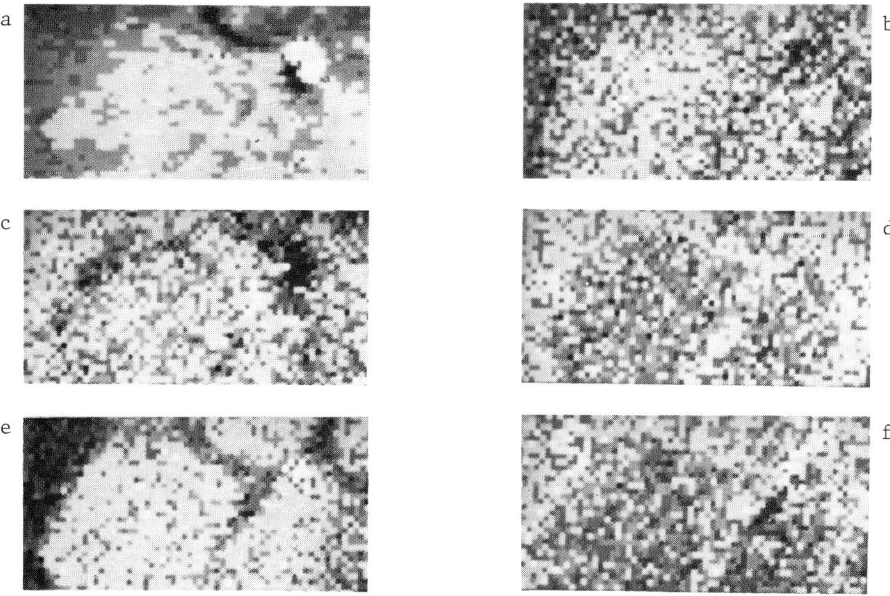


Fig 1 (a) SEM image of the surface of a NiCrAl-Zr alloy and Quantified Auger Images of this alloy. (b) Al 1388 eV. (c) Ni 843 eV. (d) Cr 526 eV. (e) O 5Ø5 eV. (f) Zr 147 eV.

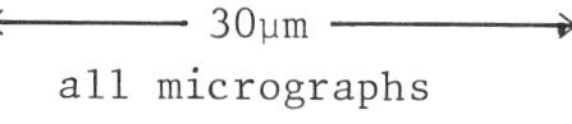


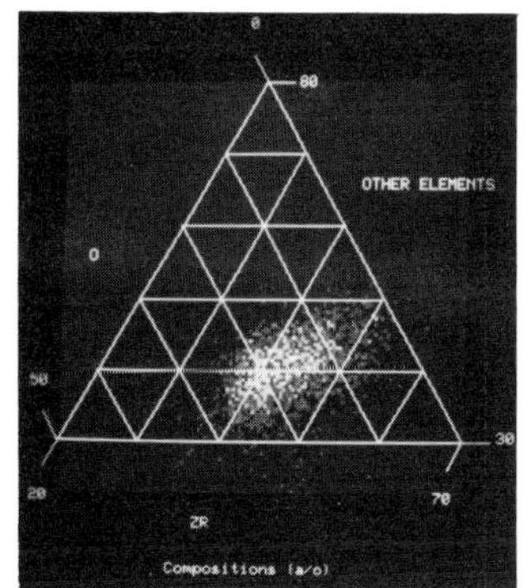


Fig 2 (above) Scatter diagram in the form of a phase diagram created from Fig 1(e) and 1(f).

Fig 3. Phase image created from Fig 2 by brightening all points within Ø-15 a/o O and 45-55 a/o Zr. If the points are distributed all over the image, then it could indicate a distinct phase but consisting of small particles eg. a precipitate.

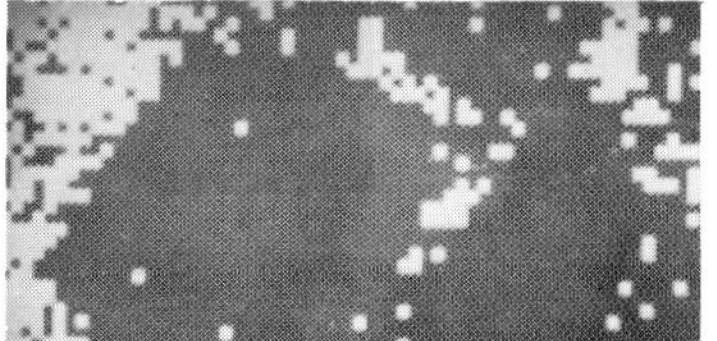

Multi-imaging can also give information on how the energy dependence of the contribution of the electron background to the energy distribution of emitted electrons varies across the specimen. For instance, experimental evidence suggests that this background obeys the equation:-

$$N(E) = AE^{-m} \qquad (4)$$

where A and m are constants and E is the electron energy (Sickafus 1977, Peacock et al 1984). Thus images showing the spatial variation of m and A may be constructed and may lead to images of the spatial variation of r, the Auger backscattering factor (Prutton and ElGomati 1985).

## 5. Summary

Multi-imaging allows:-

1. Quantification of Auger images.

2. Analysis of small phases within the sample surfaces by use of scatter diagrams and "phase images".

3. Images of the variation of the energy dependence of the electron background across the specimen.

## Acknowledgements

The authors would like to thank M Prutton and M M El Gomati for useful discussion and the SERC for the funding of this work.

## References

Browning R, Peacock D C, Prutton M, Walker C, 1984, Inst. Phys. Conf. Ser. No. 68: p127.
Browning R, 1985, J. Vac. Sci. and Tech. In Press.
Browning R, Peacock D C, Prutton M, 1985, App. Surf. Sci. 22/23 145-159.
El Gomati M M, Peacock D C, Prutton M, Roberts R, SEM '83 Inc. A M F O'Hare, (Il 60666 U.S.A) p1091-1099.
Mayr M , Angeli J, 1985, X-ray Spec. 14,89-98.
Minni E, 1983, App. Surf. Sci. 15, 293-306.
Peacock D C, Prutton M, Roberts R, 1984, Vacuum 34, 497-507.
Prutton M and El Gomati M M, ECASIA '85. To be published.
Prutton M, Browning R, El Gomati M M, Peacock D C, 1982, Vacuum TAIP 32,
Seah M P, SEM '83 SEM Inc. A M F O'Hare (Il 60666 U.S.A) p521.
Sickafus E N,1977,Phys. Rev.B,16 1436-1447.
Van Langeveld A D, Hendrickx H A C M, Nieuwenhuys B E, 1983,Thin Solid 351.Films 109, 179-192.

*Inst. Phys. Conf. Ser. No 78: Chapter 7*
*Paper presented at EMAG '85, Newcastle upon Tyne, 2–5 September 1985* 

# Sub-micron backscattered electron microanalysis in the SEM

E D Boyes, M D Hill, M J Goringe and C J Salter,
Department of Metallurgy and Science of Materials, University of Oxford, Parks Road, Oxford OX1 3PH.

The use of bulk specimens has many practical advantages. They should be more representative of the material; simpler to prepare and with fewer artefacts. Features which are more widely spaced or with a lower frequency of occurrence than can be readily handled in a typical transmission electron microscope (TEM) specimen with an open area which may be no more than a few micrometers in diameter, can be analysed quite simply with the wide field of view available in the scanning electron microscope (SEM). The specimens should be more robust and therefore suitable for ongoing experiments such as heating or the attachment of the connections needed to operate devices. However a major disadvantage of electron opaque specimens is the relatively poor micron sized spatial resolution and hence reduced sensitivity of chemical analysis by X-ray spectrometry; particularly when compared with the sub-nanometer, and in the most favourable cases sub-monolayer, capability demonstrated with EDX analysis of thin foils in the FEG STEM. In extreme cases, of relatively high energies and light matrices, the resolution may be many micrometers. Diffraction contrast analysis of defects is also less well developed. On the other hand, when fully quantitative corrections are applied to the data the chemical microanalysis of bulk specimen is inherently more accurate ; due principly to the better defined physics.

It is well established that :-
(a) there is a strong energy dependence to the range of electron beam interaction with a bulk specimen and a lower voltage means less spreading,
(b) for any given element there is a minimum energy below which the excitation of characteristic x-rays is inefficient,
(c) the range and hence the degree of lateral spreading is much reduced for those electrons which are backscattered with sufficient energy to be re-emitted through the entrance surface,
(d) the yield of backscattered electrons (BSE) from a polished sample at normal incidence is a monotonic function of atomic number, and simple rules of mixture govern the yield from alloys and materials of different density, including where there is porosity,
(e) the absolute yield of backscattered electrons is high - up to 0.5 - but the corresponding figures for x-rays are very much lower.

It is possible to construct highly efficient detectors for backscattered electrons, based on either a scintillator/photomultiplier or solid state diodes; with a very wide angle of acceptance of up to at least 1.7Sr, compared to 0.1Sr for a good EDX x-ray detector and much less for WDX. The efficiency of the method means that low beam currents of <0.5nA can be used for BSE analysis; minimising the probe size contribution to the overall resolution, and restricting the effects on beam sensitive specimens. Continuous digital analysis of the experimental data is used to control acquisition interactively (Fig.4) and to enhance the results, without corrupting the basic data which is securely stored digitally.

To highlight small differences, BSE microanalysis images are generally presented most usefully as false colour coded maps. Fig.2(c) is a monochrome reproduction of a high resolution BSE image, subjected to non-linear enhancement in which adjacent grey levels are assigned different colours. Under computer control it is possible to have multiple regions of interest. The data in the Fig.3 histogram were subsequently processed in this way. To do this effectively it is necessary to have extremely even illumination of the specimen and a high degree of uniformity in signal collection.

A number of minor perturbations in the Z dependence of the BSE signal have been reported, but within the restricted range of carefully prepared samples so far studied, reliable relative compositions have been consistently obtained : the data in Figs.1(d) & 2(a) are examples. Of course only in a binary material is the result unique, since in more complex compositions it is in principle possible to have more than one way of combining the constituents to attain any given mean Z, which forms the basis of this method of analysis. In practice there may be other beneficial constraints to the possible combinations, or as in Fig.2(b&c) one of the elements is constant.

The technique has the potential for accurately analysing and efficiently mapping the distribution of the major elements associated with future sub-micron microelectronic devices. Being proportional to the mean atomic number, it is insensitive to the minor elements. However it is particularly effective for materials, such as oxides, which contain the lighter elements; although for a given energy the spatial resolution is then more limited.

A DEC LSI-11/23 minicomputer has been used but the most useful functions of digital line traces and spot measurements of intensity, together with the necessary calibration and video expansion routines, could have been run almost as well on a low cost micro system. Chemically accurate images require a framestore with digitisation and storage to better than 8-bit precision. A high precision 12-bit ADC has been connected through the computer bus, bypassing the framestore's integral digitiser, and operating much more slowly. The earlier system required the analogue video signal to be controllably amplified in gain and offset to fit the restricted useful range of the 256 level 8-bit standard input digitiser. In practice this presented few problems with modern electronics, and the limiting stability is that of the SEM beam. This was addressed by using internal standards, as close to the target material as possible; and in all respects similar to the procedure used in EPMA, including EDX spot calibration for mean Z.

Some problems were experienced with attaining the necessary quality of surface finish to obtain a purely compositional BSE signal. The annular detector geometry does not entirely suppress the modification of the yield as a function of angle of incidence, although it copes well with any associated changes in angle of emission, which are successfully integrated out. We have used a JEOL 35X, which has been fitted with an original design of scintillator backscattered electron detector (to be published; M D Hill, Inst of Phys Conf Ser 76, 1985).

A lateral spatial resolution of 0.1μm and a chemical accuracy of 1% have been demonstrated for the analysis of mercury in CdHgTe using the calibrated and digitally processed integral backscattered electron signal from a regular SEM fitted with an efficient annular detector system. At Z = 60 the discrimination is better than 0.25AMU.

Fig.1 Alternate layers of GaAs, each 1.2μm wide, and 5.6μm GaAlAs with different proportions of Ga/Al in multilayer cross-section specimen. The compositional BSE image has better resolution, despite the higher kV, and signal to noise ratio (S/N), than the X-ray map.

(a) Quantitative EPMA x-ray data

| | | | | | |
|---|---|---|---|---|---|
| Ga | .09 | .19 | .30 | .39 | .50 |
| Al | .44 | .33 | .21 | .12 | – |
| As | .47 | .48 | .49 | .49 | .50 |

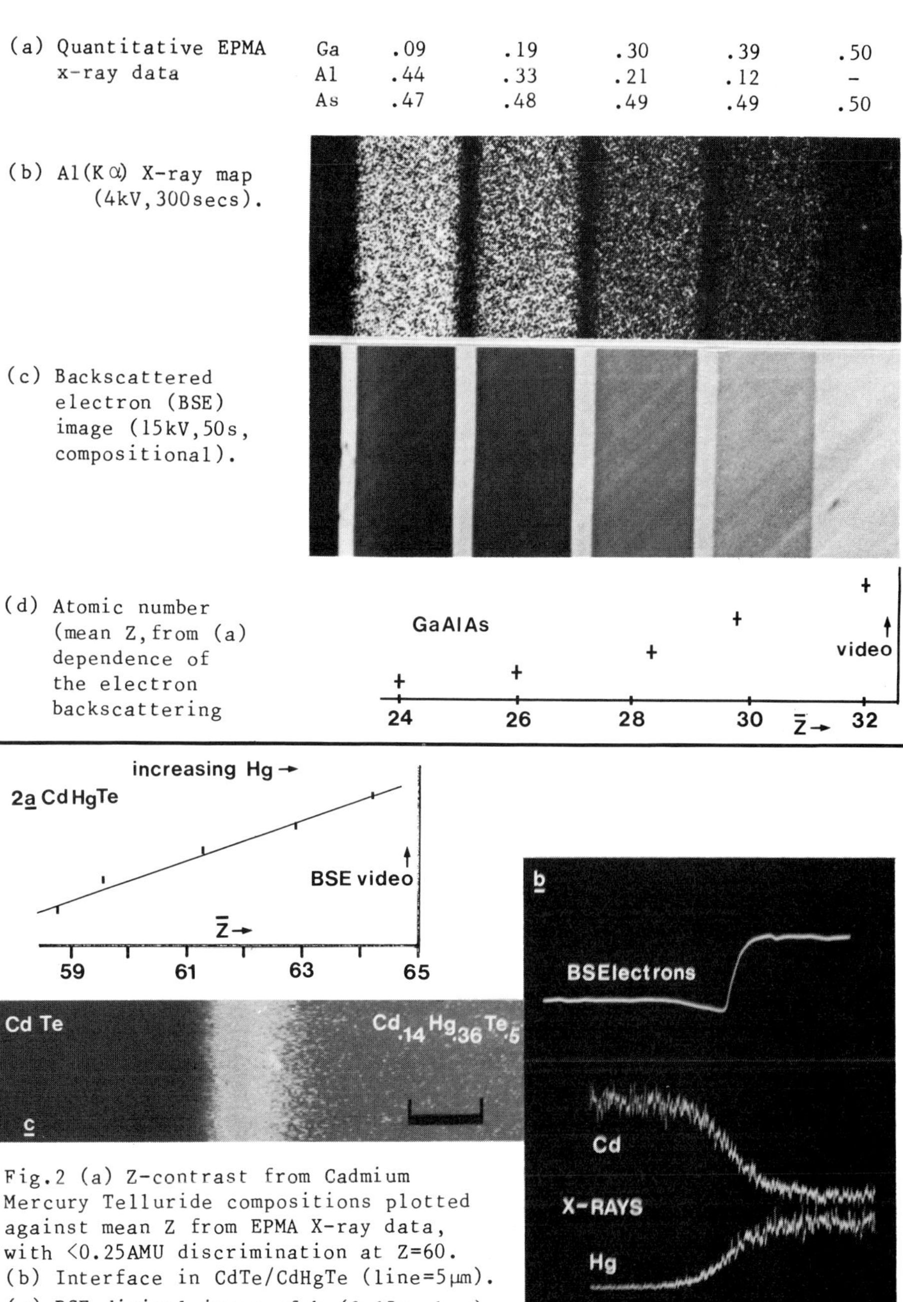


Fig.2 (a) Z-contrast from Cadmium Mercury Telluride compositions plotted against mean Z from EPMA X-ray data, with <0.25AMU discrimination at Z=60.
(b) Interface in CdTe/CdHgTe (line=5μm).
(c) BSE digital image of b (0.15μm bar) non-linear enhanced for colour.

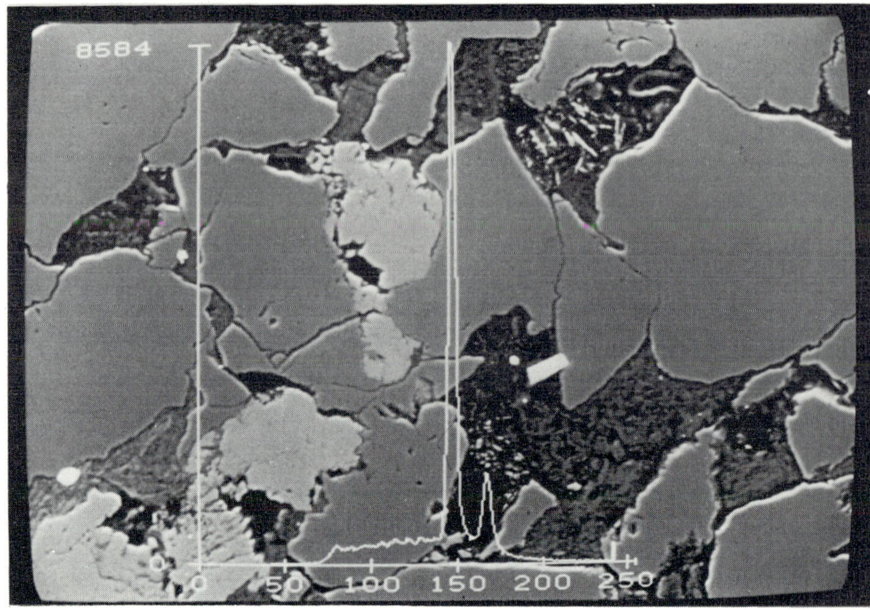


Fig.3 Digital backscattered electron image of artificial sandstone with superimposed luminance histogram as basis for x,y,Z and area fraction image analysis.

INTERACTIVE DIGITAL IMAGING

Fig.4

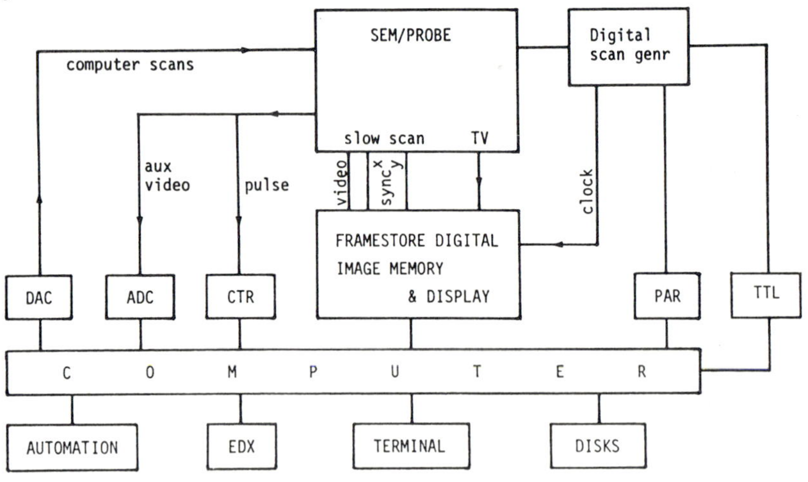


We thank the SERC and AERE Harwell for support. The semi-conductor specimens were provided by Dr G R Booker, M Lyster and Dr Astles (RSRE) and the sandstone by Dr S Graham of Shell Oil, to all of whom the authors are most grateful.

*Inst. Phys. Conf. Ser. No 78: Chapter 8*
*Paper presented at EMAG '85, Newcastle upon Tyne, 2–5 September 1985*



# Analytical electron microscopy of steels

J M Titchmarsh

Materials Physics and Metallurgy Division, AERE Harwell, Didcot, Oxon

## 1. Introduction

Steels are the most commonly used metallic alloys. They are the cheapest metals available which have the necessary strength, formability and durability required for many different components subjected to widely different service conditions. Microstructures in steels are rarely, if ever, in equilibrium. Those produced by fabrication processes at high temperatures are likely to be stable during service at ambient temperatures. However, exposure during service to elevated temperature and irradiation allow a gradual evolution of the microstructure. Those mechanical properties which are ultimately dependent upon crystallographic imperfections and chemical inhomogeneities change accordingly. Plastic deformation, creep, fatigue and corrosion can all be influenced by precipitation and segregation. Therefore, it is hardly surprising that analytical electron microscopy (AEM) is now an established technique for characterising the microstructure of steels because of its unique capability for combining high resolution imaging with crystallographic and chemical analysis.

AEM, in common with other sophisticated analysis methods, is costly and time consuming. Other techniques should always be considered at the onset of any investigation to determine if they can provide the required information more easily or accurately; additional, important information will almost certainly be obtained by using more than one technique. For example, Auger Electron Spectroscopy (AES) might be more useful than AEM for studies of fracture surface chemistry; X-ray diffraction of extracted particles could be an easier method for estimating the relative proportions of mixtures of precipitate phases (Lai and Galbraith, 1980). AEM, however, is unrivalled as a means of obtaining information from different, selected regions of the microstructure. In particular, variations in the composition close to boundaries and the differences in particle types at boundaries, compared with those within grains, are almost impossible to determine by other methods. Solid-state transformations involving several phases occur in some alloy steels and these are ideally suited to AEM analysis. An example (Figure 1) is the reaction $\delta \rightarrow \gamma + M_{23}C_6$+Laves, by which $\delta$-ferrite in stainless steel weld metal can transform during prolonged service at 625°C. No other method can reveal and identify phases on such a fine scale. It is towards such investigations that AEM is best directed when studying relationships between mechanical properties and microstructures and, coincidentally, many problems of embrittlement, cavitation and corrosion arise through changes in boundary chemistry and precipitation.

## 2. Problems of Specimen Preparation

Specimens are almost always prepared in the form of thinned discs, or as extraction replicas; the latter allow particle analysis without the presence of a surrounding matrix. Artefacts can arise during preparation of both forms. During electropolishing or etching the electrochemical potential of a particle in the acid solution will be different from that of the matrix (Kurosawa et al 1980). Particles can be dissolved preferentially or can fall out of thinned regions when the adjacent matrix is preferentially removed. Grain boundaries with composition different from that of the matrix can be preferentially etched. This can lead to difficulties in data interpretation. For example, many etching solutions for ferritic steels contain HCl. Gage (1984) has shown that this acid preferentially attacks $M_6C$ particles, the presence of which greatly affects embrittlement of $2^1/4$%Cr-1%Mo alloys by phosphorus. Previous investigators in this field had often failed to detect $M_6C$ during analysis. Commercial electropolishing equipment can be a source of copper contamination (James 1985) deposited as a thin film on ferritic alloys. Almost all electropolished discs will have oxidised surfaces (Figure 2), and chromium-rich particles yield EELS spectra which sometimes show evidence of oxidation during preparation. Other problems which arise from the preparation of replicas include incomplete removal of the matrix from around particles or the preferential deposition of elements such as silicon on particles. This latter effect has already led to errors in the interpretation of the role of silicon in the tempering of some steels (Barnard et al, 1981).

Fig.1 Decomposition of ferrite δ to austenite (A), $M_{23}C_6$(M) and Laves phase (L) in austenitic weld metal

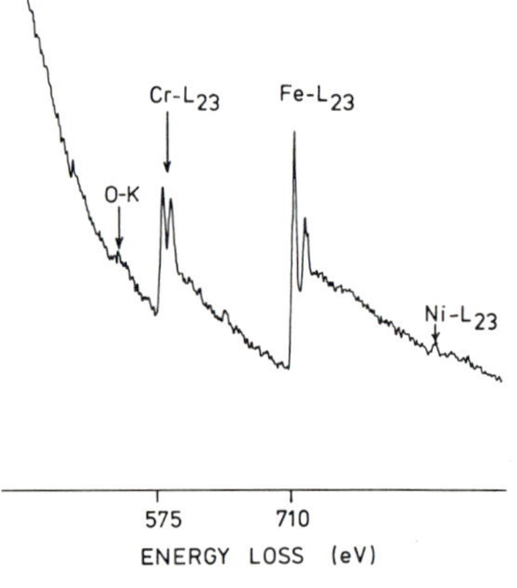


Fig.2 EELS spectrum revealing oxidation of electropolished disc of austenitic steel

## 3. The Analysis of Second Phase Particles

The identification and the measurement of the chemical composition of second phase particles is a major aim of many experiments. AEM is ideal for the study of the removal, by precipitation from solid solution, of many elements both harmful and beneficial to mechanical properties. The detection of $Ni_3Sn$ particles at boundaries in Ni-Cr steels, which can render harmless the potential embrittling element, tin (Titchmarsh et al, 1979) is a good example. Another is the identification of sensitised grain boundaries in many austenitic alloys by the formation of chromium-rich phases, which reduce the adjacent grain boundary chromium concentration to a critical, low level. AEM is particularly important for determining the microstructural stability under neutron irradiation especially of those austenitic alloys used for nuclear reactor core components (e.g. Lee et al, 1980; Williams et al, 1982). Silicon atoms are strongly bound to the irradiation-induced vacancies and are preferentially transported to vacancy sinks such as voids and grain boundaries (Lam and Okamoto, 1978). Other elements diffuse away from such sinks, leaving behind enhanced concentrations of slowly moving atoms such as nickel (Anthony, 1972). This leads to the formation of precipitates such as $Ni_3Si$ (Cawthorne and Brown, 1977) and G-phases (Williams et al, 1979) which are not usually found in unirradiated samples of these alloys. Potentially harmful effects resulting from the removal of these elements from solution are: a reduction in void swelling resistance; and the inhomogeneous transformation of the $\gamma$-matrix to the $\alpha$-phase.

The structures of precipitates commonly occurring in steels have been known for many years (Andrews et al, 1971), and it is unlikely that completely new phases will be discovered. Hence, the acquisition of an EDX spectrum and one or two microdiffraction patterns of low-index zone-axes are often sufficient to enable identification. Occasionally, however, more detailed diffraction analysis is necessary using the procedures described by the Bristol Group (1984). Such analysis may be impossible when particles are very small or completely embedded in the matrix, as is often the case with irradiated materials. When this is so, and the compositions of the likely particles have been pre-determined, for example from an extraction replica or from particles at the edge of the hole in the foil, subtraction of increasing proportions of the matrix component can leave a spectrum typical of the particle alone (Titchmarsh and Williams, 1981). Cliff et al (1983) have formalised this approch following the initial proposal by Feest (1980) for analysis of particles in bulk specimens by EPMA. As an example of the method, Figure 3 shows how embedded particles of Laves phase and $M_{23}C_6$ in a foil of 9%Cr-1%Mo steel are clearly distinguished, with the Fe vs Cr and Fe vs Mo contents from composite matrix/particle spectra defining two separate lines for each phase.

Many investigations require the comparison of a series of specimens; in such cases it may be necessary to examine hundreds of particles in order to reveal all the phases which are present in significant numbers. It is most convenient to use extraction replicas for such experiments. The first sample in any such series should be examined at length to confirm the identity of phases by microdiffraction and to determine the typical ranges of composition, using EDX, for each phase. This establishes an EDX "fingerprint" for each phase by which means individual particles can subsequently be identified within seconds. Particles in specimens from other samples in the series will often have similar fingerprints, and identification is then very fast. However, diffraction patterns should always be obtained for each new specimen to confirm the equivalence of similar fingerprints. The following results emphasise the need for this (Titchmarsh et al, 1984b). The spectra in Figure 4 were obtained from particles extracted from a cast of 9Cr-1Mo ferritic steel following ageing at 500–550°C for times up to 50,000 hours: a and b are from $M_{23}C_6$ and $M_7C_3$, c and d from $Cr_2N$ and CrN, and e and f from $M_6C$ and Laves. It can be seen that for each pair of spectra only slight differences occur in apparent peak-height ratios. Detailed diffraction information was required to establish the correct sequence of precipitate transformation during ageing. Such results act as a warning against the use of EDX alone for particle identification.

## 4. Choice of Instrumentation

The field emission electron gun allows the effective spatial resolution of AEM to be greatly improved, compared with the thermal gun. Such improvements reveal details previously obtainable only by using alternative techniques. For example, Garrett-Reed et al (1983) have shown peaks in the chromium concentration within individual cementite laths of a pearlitic steel, with a resolution of ≃2nm; such peaks had previously only been revealed by the atom probe. Diffusion profiles across grain boundaries in irradiated austenitic steels have also shown silicon segregation on a similar scale (Figure 5, collaborative work with the CEGB). Such narrow profiles reflect the true spatial distribution of the elements in question, convoluted with the electron probe intensity profile and any beam broadening. Thus, true segregation profiles of atomic width, such as are known to occur with impurities segregated to grain boundaries in embrittled ferritic steels, will always have an apparent width at least as broad as the probe. However, the greatly improved detectability of such segregation using the field emission gun (Titchmarsh et al, 1984a) promises to make AEM a possible alternative to AES for embrittlement studies.

The detection of the light elements B,N and C, which have low solubilities in steels and readily form precipitates, is now possible using EELS (e.g. Figure 2). Previously their presence in particles was based indirectly on diffraction information. The increasing use of "windowless" EDX detectors offers an alternative method for light element analysis. Thomas (1983) has shown that measurement of carbon in TiC may be more accurate in all but the thinnest specimens using EDX rather than EELS at 100keV. However, for many steels the close proximity of the transition element L-edges and the K-edges of carbon and nitrogen require the use of peak stripping to improve both peak visibility and accurate measurement of the light element content. Figure 6 illustrates the results of stripping the Cr-L and C-K peaks from a spectrum from a particle of $Cr_2N$; the presence of the V-L peak, masked in the original spectrum, is revealed on the high energy side of the N-K peak. In the general case, however, the accuracy of quantitative light element analysis by EDX will depend upon the accuracy of corrections for self-adsorption.

Increasing the beam accelerating voltage will improve EELS detection and measurement of the light element content of particles in steels by reducing the complication of multiple scattering. No such advantages apply to EDX; the absorption correction is independent of the accelerating voltage. Some improvement will occur in the detectability of segregated impurities at higher voltage, through a reduction in beam-broadening, and there will be great advantages to be gained from imaging thicker foils. The major advantage in using higher voltages for the microanalysis of steels, however, might well be the improved capability for correcting the strong astigmatism, which occurs in both images and in focused probes, when strongly-magnetic, tilted, ferritic foils are immersed in the objective lens field.

## 6. Conclusions

AEM is an ideal technique for the analysis of the complex microstructures which occur in many steels. It is possible to identify and measure the chemical composition of most of the phases found in steels and to determine compositional profile at important sites such as grain boundaries. The data enhances our knowledge of the influence of microstructural features on physical properties in fields as diverse as fracture, corrosion and void swelling. The trends towards higher voltage microscopes, high brightness sources and improvements in EDX and EELS analysis will extend the areas of application further, as spatial resolution and sensitivity limits improve.

## References


Andrews K W, Dyson D J and Keown S R (1971) The Interpretation of Electron Diffraction Patterns (London:Adam Hilger).

Anthony T R (1972) Proc Conf on Radiation-induced Voids in Metals, Albany, USAEC Symp Series 26, Conf 710601, 630.

Barnard S J, Smith G D W S, Garratt-Reed A J and Vander Sande J (1981) Proc Conf on "Advances in the Physical Metallurgy and Applications of Steels," Univ. of Liverpool (London:Metals Soc).

Bristol Group (1984) Convergent Beam Electron Diffraction of Alloy Phases (London:Adam Hilger).

Cawthorne C and Brown C (1977) J Nucl Mater 66, 201.

Cliff G, Powell D J, Pilkington R, Champness P E and Lorimer G W (1983) Inst Phys Conf Ser No 68, Ch3, 63.

Feest E A (1980) Proc Conf on "Solidification Techniques in the Foundry and Casthouse" (London:Metals Soc) 3211.

Gage G (1984) Proc 4th AEM Workshop, Lehigh Univ, Bethlehem, Pa. (San Francisco:San Francisco Press).

Garrett-Reed A J, Mottishaw T D, Worrall G M, and Smith G D W (1983) Inst Phys Conf Ser No 68, Ch9, 311.

James A W (1985) Unpublished work.

Kurosawa F, Taguchi I and Matsumoto R (1979) J Jap Inst Metals 43 1068.

Lai J K and Galbraith I F (1980) J Mater Sci 15, 1297.

Lam N Q and Okamoto P R (1978) J Nucl Mater 78, 408.

Lee E H, Maziasz P J and Rowcliffe A F (1980) Proc Symp of Met Soc AIME on "Phase Stability during Irradiation," Pittsburgh, USA, 191.

Thomas L E (1983) "Microbeam Analysis-1983," (San Francisco:San Francisco Press), 70.

Titchmarsh J M, Edwards B C and Eyre B L (1979) Nature 287, 38.

Titchmarsh J M, Kerr R T and Boyes E D (1984a) Proc 4th AEM Workshop, Lehigh Univ, Bethlehem, Pa (San Francisco:San Francisco Press), 64.

Titchmarsh J M, Wall M and Edwards B C (1984b) ibid, 247.

Titchmarsh J M, and Williams T M (1981) Proc Conf on "Quantitative Microanalysis with High Spatial Resolution," Metals Soc Book 277, 223.

Williams T M, Titchmarsh J M and Arkell D R (1979) J Nucl Mater 82 199.

Williams T M, Titchmarsh J M and Arkell D R (1982) J Nucl Mater 107 222.

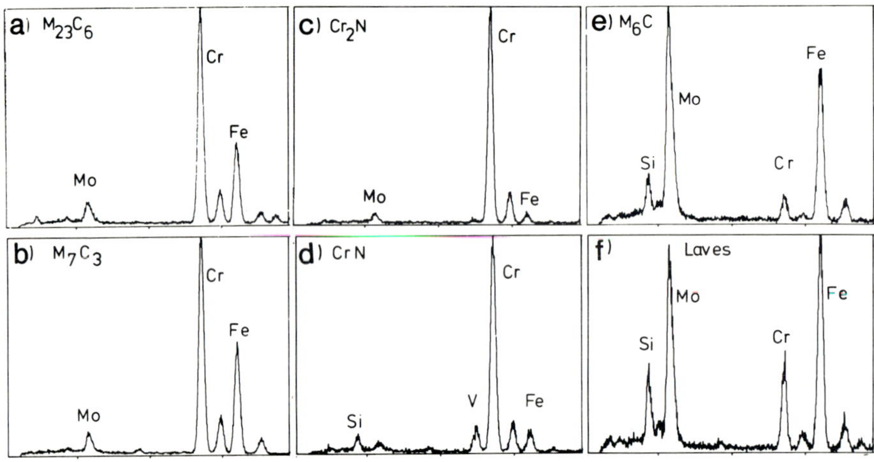


Fig.4 Similar pairs of EDX spectra from particles in 9Cr 1Mo steel

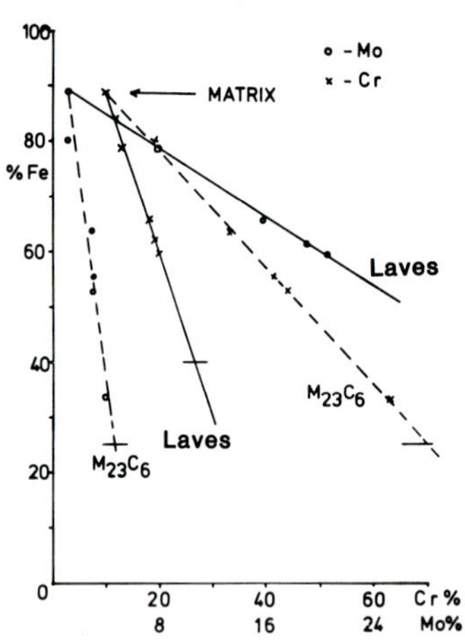


Fig.3 Analysis of embedded particles in 9Cr 1Mo steel by the method of Cliff et al (1983).

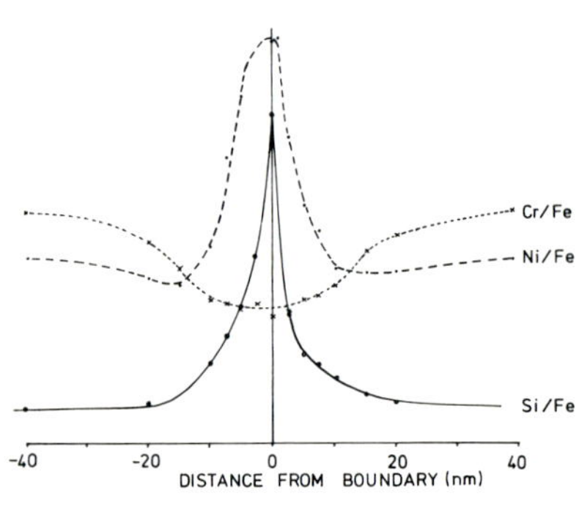


Fig.5 Compositional variations across a grain boundary in an irradiated stainless steel.

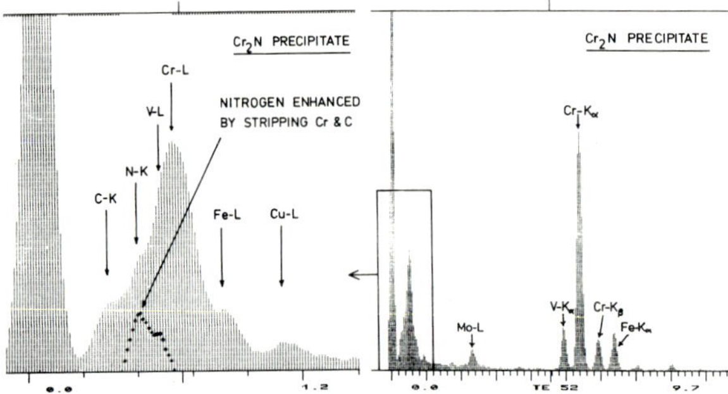


Fig.6 Use of "windowless" EDX detector to reveal presence of nitrogen in a $Cr_2N$ precipitate.

# Pre-precipitation phenomena associated with $\gamma'$ formation in stainless steels

R A Ricks

Department of Materials Science and Engineering, University of Surrey, Guildford GU2 5XH.

## 1. Introduction

Based on the composition Fe-20%Cr-25%Ni, austentic stainless steels are intrinsically suitable for use in corrosive environments or at elevated temperatures, (Weiss and Stickler, 1972). These relatively soft alloys may be mechanically improved by precipitation hardening reactions involving the formation of either carbides, based on the NaCl crystal structure, or by the formation of the metastable $LI_2$ $\gamma'$ phase; this latter reaction being directly analogous to the precipitation of $\gamma'$ in nickel-base superalloys. In view of the low volume fraction of carbides which may be formed, the hardening due to coherent $\gamma'$ formation is the favoured strengthening mechanism despite the metastable nature of this phase.

This paper is concerned with an investigation into the symmetry of the austenite matrix prior to the onset of $\gamma'$ nucleation in two alloys strengthened by either aluminium or titanium additions. Both convergent beam electron diffraction and two dimensional lattice-fringe imaging techniques have been used in this study and the results show that the cubic symmetry normally associated with the austenite is lost before the formation of $\gamma'$ precipitates. Such symmetry losses are also apparent in specimens aged above the $\gamma'$ solvus for the alloy, which contain directly-nucleated equilibrium precipitates. These results will be discussed and compared with those published recently concerning aluminium-base alloys.

## 2. Experimental procedure

The two alloys used in this investigation were prepared from high purity base elements to the following compositions:-

Fe-20%Cr-25%Ni-4%Al    Fe-20%Cr-25%Ni-4%Ti    (wt%)

The ingots were homogenised at 1200°C prior to rolling to strip form. Heat treatments were carried out on solution treated (1200°C for 30mins) specimens either in a tin-bath for short times (up to ≈ 5mins), or in electrical muffle furnaces. Specimens for electron microscopy were punched from the strip specimens and thinned to perforation using conventional techniques. Convergent beam electron diffraction was carried out on a Philips EM400T electron microscope whilst lattice-fringe imaging was performed on a JEOL 120CX. Analysis of the fringe images was done either by direct fringe counting using a travelling microscope or by optical (laser) diffraction.

## 3. Results

The two alloy compositions were chosen for their known sequence of precipitation, which involves the formation of γ' prior to the equilibrium precipitate in the temperature range 600-750°C. Above this the alloys age directly to the equilibrium precipitate phase associated with the alloy (Wilson and Pickering, 1966) - ie NiAl (ordered bcc) for the aluminium-bearing alloy and $Ni_3Ti$ (ordered hcp) for the alloy containing titanium. Examples of these precipitate types are illustrated in Figure 1.

Symmetry investigations in both alloys initiated with a study of the as quenched, solution treated alloys. These specimens were examined using convergent beam electron diffraction and in all cases diffraction patterns were obtained which were entirely consistent with the matrix having the expected cubic space group (Fm3m). Figure 2 shows an example of a <111> zone axis diffraction pattern with the higher order Lane zone (HOLZ) lines displaying 3m symmetry consistent with a cubic lattice (for a review of convergent beam electron diffraction see Steeds, 1979). This result was obtained irrespective of the angle of tilt of the specimen within the microscope, and indicated that possible thin-foil relaxation effects observed by Shelton et al (1983) in titanium alloys could be ignored in this investigation.

Aging both alloys at temperatures below the γ' solvus for times shorter than those required for the observation of γ' (either in the image or diffraction pattern) resulted in a loss of the 3m symmetry of <111> zone axis HOLZ pattern. In the majority of cases the pattern retained a single mirror plane although occasionally all symmetry was lost. Figure 3 shows an example of such a pattern taken from the 4%Ti alloy aged at 720°C for 2mins. The degree of symmetry loss was observed to increase with time until the onset of γ' formation, as seen by the superlattice reflections in the diffraction pattern. At this stage the HOLZ lines present in the convergent beam diffraction patterns became highly broadened due to the strain associated with the coherent γ' precipitates.

When aged at temperatures above the γ' solvus similar symmetry losses could be observed, again in both alloys. At this temperature the majority of equilibrium precipitation events took place on grain boundaries and twin boundaries leaving the centre of each grain relatively free from precipitation. Thus diffraction patterns could be obtained without interference from the particles and an example is shown in Figure 4, taken from the 4%Al alloy aged at 800°C for 1hr. Of particular interest were those HOLZ patterns obtained close to a growing equilibrium precipitate. Previous work (Ricks, 1984) has shown that the diffusion fields associated with an equilibrium precipitate could extend for up to 250nm ahead of the precipitate/matrix interface. Within these diffusion fields, aluminium or titanium levels were depleted and HOLZ patterns obtained from these regions displayed a return to the expected cubic symmetry (Figure 5).

Lattice-fringe imaging using 002/020 reflections in a tilted illumination configuration produced crossed fringes as shown in Figure 6a. Analysis of such images revealed that in all cases where symmetry losses had been detected by convergent beam similar symmetry losses could be detected by measurement of fringe spacings. This result was confirmed by the use of optical diffraction patterns obtained from the crossed lattice fringes, shown in Figure 6b. Measurements taken from such patterns showed the same degree of symmetry loss as that obtained by fringe spacing measurements.

## 4. Discussion and Conclusions

The observation of symmetry losses associated with aging the two alloys at temperatures below the γ' solvus prior to the onset of γ' formation indicate that atomic rearrangements take place within the lattice either to act as a precursor to γ' precipitation, or, as seems more likely, as an attempt to relieve some of the strain associated with the inclusion of the oversize solute atoms into the matrix. The simplest of such rearrangements would be the formation of single atomic layers of solute atoms throughout the matrix grains (somewhat analogous to GP zonal formation in Al-Cu alloys) to localise the strain in one dimension thus reducing the total strain associated with the solute atoms. The outcome of such a rearrangement of atoms would be the overall loss of cubic symmetry in favour of a tetragonal unit cell. It should be appreciated that such a diffusional process would be driven by the reduction in strain energy of the lattice in much the same way as the association of solute atoms with dislocations, if there is an appreciable size difference with the matrix.

The observation of similar losses of symmetry at aging temperatures above the γ' solvus would confirm that the postulated atomic movements are not associated directly with the formation of γ' and are more likely to be due to a strain energy reduction. The retention of cubic symmetry close to a growing equilibrium precipitate where solute levels are depleted gives further evidence that solute atoms are responsible for the observed symmetry losses.

It is of some interest to compare the behaviour of the alloys reported here with those which exhibit spinodal decomposition. Certainly if a compositionally modulated matrix results from the postulated atomic rearrangements then there is a great deal of similarity between the two transformations. Furthermore, recent studies have show that in some aluminium alloys containing lithium additions, which undergo δ' precipitation (isostructural precipitate and matrix to the alloys discussed here), phase diagram calculations indicate the possibility of a misability gap at the Al-rich end of the phase diagram (Sigli et al, 1985). If similar miscibility phenomena associated with strain energy reductions could be considered as a "legitimate" spinodal decomposition then the results presented in this paper would imply that such reactions may take place prior to the formation of many of the familiar metastable phases which themselves are precursors to the precipitation of the equilibrium intermetallics.

## References

Weiss, B. and Stickler, R. Met. Trans. 3 p851 (1972).
Wilson, F.G. and Pickering, F.B., J.I.S.I., p628 (1966).
Steeds J.W. (1979) in "Introduction to Analytical Electron Microscopy" eds J.J. Hren et al, Plenum Press, New York.
C.G. Shelton, A.J. Porter and B. Ralph, Inst. Phys. Conf. Ser.No. 68, p47 (1983).
R.A. Ricks, Acta Met. 32 p1105 (1984).
C. Sigli, J.M. Sanchez and J.M. Papazian, Paper presented at 3rd Int. Conf. on Al-Li alloys, Oxford 1985, to be published.

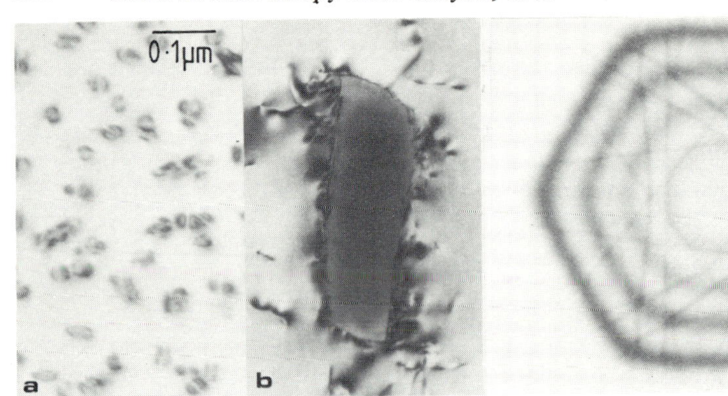


Figure 1. TEM images showing '(a) & equilibrium precipitate morphologies (b) in 4% Al alloy.

Figure 2. HOLZ pattern displaying 3m symmetry from solution treated 4%Ti alloy

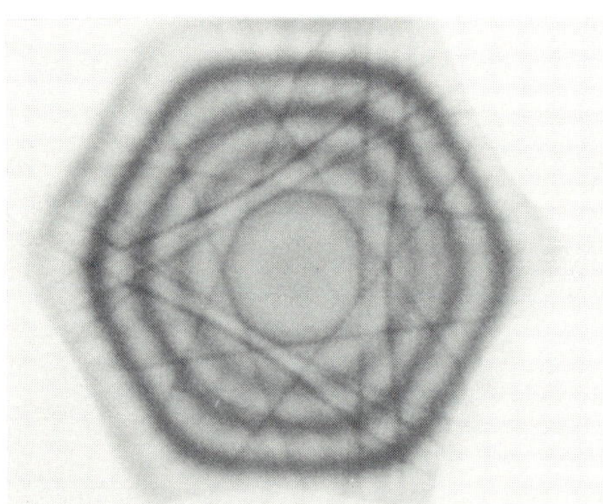

Figure 3. HOLZ pattern showing loss of cubic symmetry. 4%Ti alloy aged 2mins at 720°C.

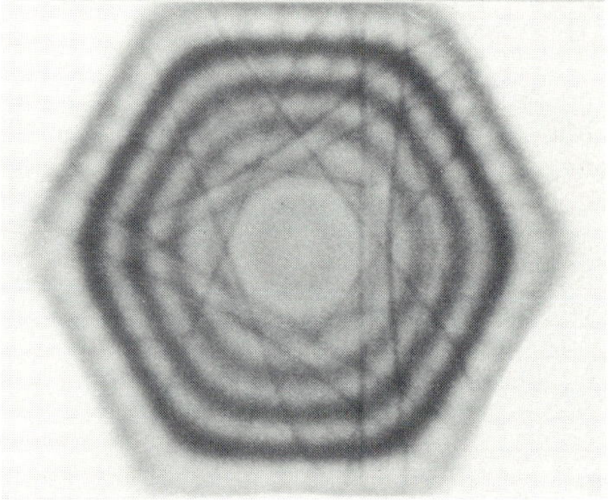

Figure 4. HOLZ pattern showing loss of cubic symmetry. 4%Al alloy aged 1 hr at 800°C.

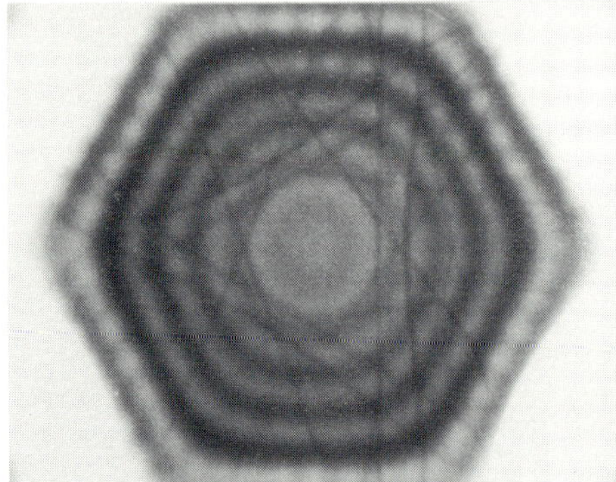

Figure 5. HOLZ pattern showing return to 3m symmetry in diffusion field of equilibrium precipitate. 4%Al alloy aged 1hr at 800°C

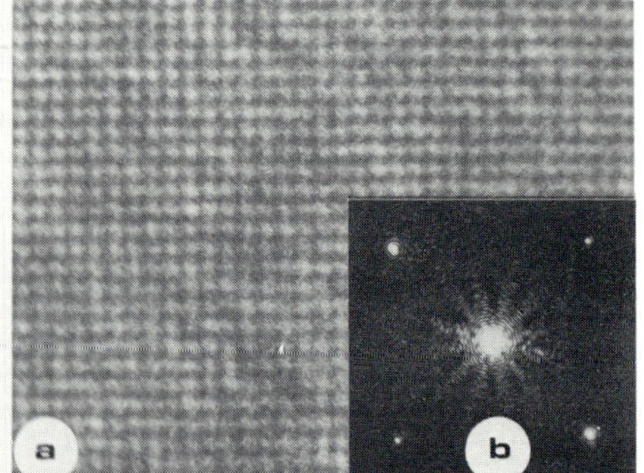


Figure 6. 2D 002/020 lattice image (a) and associated optical diffraction pattern (b) from 4%Al alloy. d(020)/d(002) ≠ 1. For discussion see text.

# Microstructural examination around weldments in AISI 316 stainless steel

GT Finlan, YP Lin, JW Steeds

H.H. Wills Physics Laboratory, University of Bristol, Bristol BS8 1TL

## 1. Introduction

Type AISI 316 austenitic stainless steels have found extensive application in nuclear power plant design. The creep properties of the steel can be improved by the addition of a small amount of boron. There remains, however, a notable lack of understanding as to the exact role of the boron addition.Invariably welds cannot be avoided in the design practices employed in large and complex components. Understanding of the complex microstructures of welds and their performance under service conditions over several decades are thus of singular importance to the design engineer. The work presented here represents initial investigations into the microstructures of industrial weldments of 316 plate and is an application of conventional TEM coupled with Convergent Beam Electron Diffraction (CBED) for phase identification, and EDS for routine chemical analysis. The results relate principally to the parent plate before and after the application of a post weld stress relieving heat treatment (800°C for 10 hrs)and to the weld metal itself.

The welding procedures followed model closely those used practically in industry, 316 plate sections butt welded manually in a multipass sequence. TEM foil samples were spark cut and trepanned prior to electropolishing in a Struers Tenupol using standard A8 solutions. Carbon extraction replicas were prepared from electrochemically etched surfaces in the usual manner.

## 2. Results

Boron autoradiographs, revealed significant segregation to the major microstructural feature of the plate material, the 'ferrite' stringers, (Marshall 1984) (Fig.1). As would be expected these structure impart a significant anisotropy to the mechanical properties, (Kimmins and Horton 1984). Schaeffler diagram calculations using the manufacturers composition show up to 5% δ-ferrite may exist, although this depends on solution treatment history. The δ-ferrite forms from the melt prior to austenite solidification, and subsequent hot rolling forms the characteristic sheets (Fig. 1). Normal solution treatments of the plate appear to have been unable to remove all the ferrite. TEM examination of the foils and extraction replicas revealed large quantities of $M_{23}C_6$ precipitation along the stringers,closely associated with the remaining δ-ferrite, (Fig. 2a); little or no precipitation was seen elsewhere. More detailed examination of the replicas revealed large numbers of other chromium rich precipitates though far less numerous than $M_{23}C_6$, (Fig. 2b). These have been identified using CBED and EDS and the results summarized in Tables 1 and 2. The presence of the boride $M_3B_2$ can be expected from previous work,(Evans 1980). However, other than a recent observation in MARMOO2,(Lin and Steeds 1985) $M_5B_3$ is a previously unreported boride phase in commercial alloys. The presence of both boride phases supports the boron autoradiographs. Fig. 3 shows typical convergent beam patterns from which the I4/mcm space group was determined. Fig.3a showing the 4-fold [001] zone axis from which the tetragonal structure was determined, a from the spot patterns and c estimated from the FOLZ radius. Fig. 3b shows the [301] zone axis the dynamic absences Gjønnes-Moodie dark lines along the (020) mirror revealing the c-glide and the space group. An EDS spectrum is shown in Fig. 4a. The $M_5B_3$ structure was identified by comparison of space group and lattice parameters with $Cr_5B_3$,(Bertaut and Blum 1953). However, the density of boride precipitation is insufficient to explain the degree of segregation observed in autoradiographs. As to the exact location of the remaining boron several theories exist, the most likely being its partial substitution for carbon in $M_{23}C_6$,(Williams and Talks 1972). It should be noted the advantages CBED and EDS together provide. However, given the range in compositions typically shown by $M_{23}C_6$ (eg. atomic % iron from 17% to 26%) EDS alone could not provide an absolute means of identification, since the spectra of several major phases in 316 stainless steel are often comparatively similar, eg. $M_{23}C_6$ and $M_5B_3$ (Figs. 4a,b). CBED is then required to distinguish space groups for a final identification. Optical and SEM examination of the stress relieved plate revealed relatively small changes; the stringers still dominate the microstructure. Boron autoradiographs still indicated segregation to the stringers. Extraction replicas and foils, however, revealed significant changes in precipitation distribution. Along the stringer the dominant phase is still $M_{23}C_6$ with fine 0.1µm Laves dispersions. However, intergranular $M_{23}C_6$ precipitation now occurs widely away from the stringers.

Analysis of the weld metal composition showsthat up to 8% ferrite can exist, providing the weld with its characteristic duplex structure. The disadvantages of long term phase instability inherent in such a microstructure are outweighed by the necessity of preventing hot cracking. Fig. 5 shows typical $\gamma$ matrix and $\delta$ with $M_{23}C_6$ precipitation at the $\gamma/\delta$ interface. The presence of $M_{23}C_6$ is probably due to the complex bead reheating cycles experienced within multipass weldments. Inclusions commonly seen in the weld metal show from EDS (Fig. 4c) high silicon and manganese contents. These were invariably found in close association with spinel, again manganese rich, their respective morphologies distinguishing the two. Si-Mn have a heavy globular appearance compared to the spinel's rectangular platelet form. These do not nucleate on $\gamma/\delta$ grain boundaries but show a random distribution. Due to their high silicon content they are usually termed 'slag' particles, where the silicon is thought to originate from flux coatings on welding electrodes. Although many Si-Mn inclusions appear amorphous or highly polycrystalline, diffraction work carried out from single crystals revealed the structure and chemical composition of a mineral Rhodonite, Tables 1, 2, commonly seen in stainless steels (Kiessling 1977).
The authors wish to thank the SERC and CEGB Berkeley for financial support and Dr. P. Marshall for his assistance.

References

Bertaut F and Blum P, 1953 Comptes Rendus 236 1055
Evans N 1980 Bristol University Ph.D Thesis
Kiessling R 1977 Non-Metallic Inclusions in Steel(London:Metals Soc)p 44
Kimmins ST and Horton CAP 1984 CEGB Rep.TPRD/L/MT0213/M84
Lin YP and Steeds JW 1985 This Proceedings
Marshall P 1984 Unpublished research
Williams TM and Talks MG 1972 J.Iron Steel Inst.London 210 870

Table 1. Listing of phases identified in this work

| Phase | Location | Space Group | Lattice constants (Å) |
|---|---|---|---|
| $M_{23}C_6$ | Plate, Weld | Fm3m | a = 10.67±0.03 |
| Spinel | Plate, Weld | Fd3m | a = 8.48±0.01 |
| $M_3B_2$ | Plate | P4/mbm | a = 5.72±0.05, c = 3.10±0.03 |
| $M_5B_3$ | Plate | I4/mcm | a = 5.56±0.05, c = 10.57±0.12 |
| σ | Plate | $P4_2$/mnm | a = 8.84, c= 4.66 |
| χ | Plate | $I\bar{4}3m$ | a = 8.83±0.10 |
| Si-Mn | Weld | $P\bar{1}$ | a = 7.25, b - 12.50, c = 6.67 |
| Laves | Plate | $P6_3$/mmc | a = 4.73, c = 7.72 |

Table 2. Typical atomic % compositions of phases identified

| Phase | Cr | Fe | Ni | Mo | Ti | Si | Mn | V | Al |
|---|---|---|---|---|---|---|---|---|---|
| $M_{23}C_6$ | 68.8 | 22.8 | 1.4 | 7.0 | - | - | - | - | - |
| Spinel | 55.4 | - | 1.0 | - | 7.8 | - | 30.7 | - | 5.1 |
| $M_3B_2$ | 39.9 | 12.0 | 1.6 | 48.5 | - | - | - | 1.0 | - |
| $M_5B_3$ | 66.2 | 9.1 | 1.2 | 22.5 | - | - | - | 1.0 | - |
| σ | 34.0 | 55.1 | 3.7 | 7.2 | - | - | - | - | - |
| χ | 28.2 | 50.4 | 3.6 | 14.6 | - | 3.2 | - | - | - |
| Si-Mn | 5.9 | - | - | - | 1.0 | 48.8 | 44.3 | - | - |
| Laves | 17.8 | 38.1 | 1.7 | 32.3 | - | 10.1 | - | - | - |

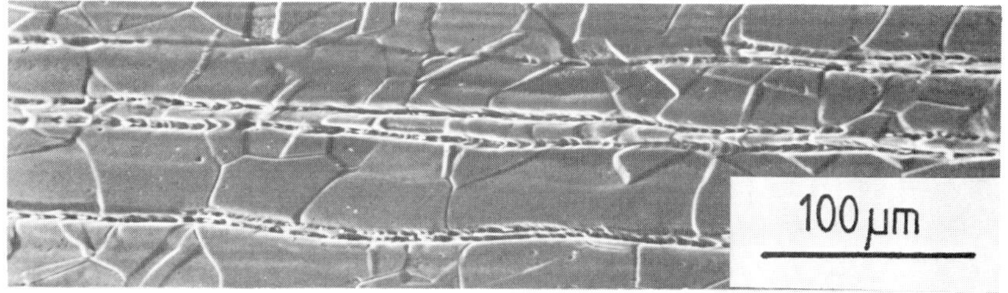


Fig. 1. Scanning electron micrograph showing stringers in the parent plate material.

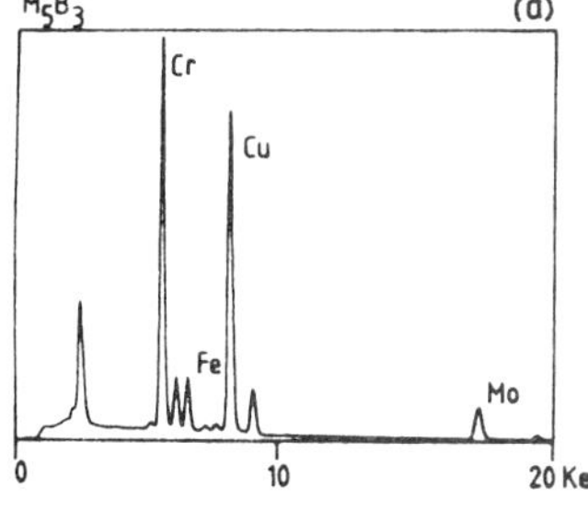


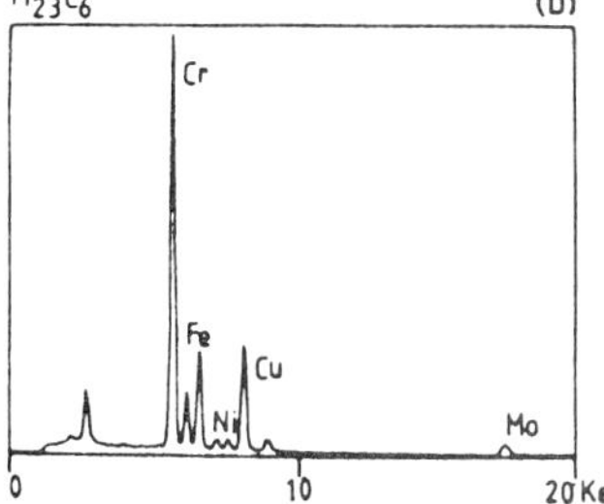


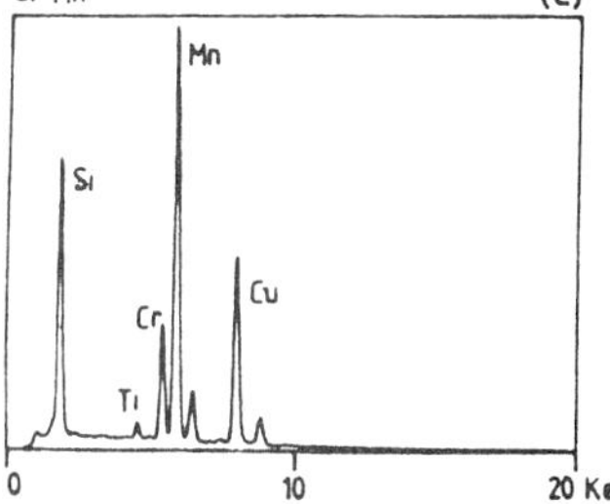


Fig. 4. Typical EDS spectra from some of the phases identified.

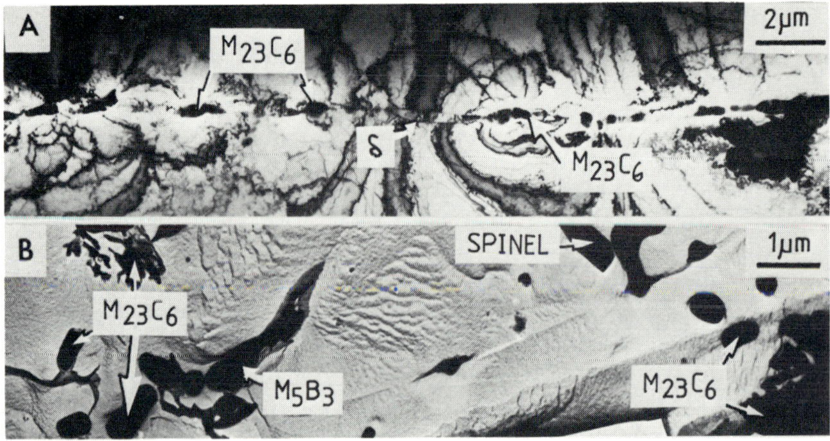


Fig. 2. Transmission electron micrographs of stringers in the parent plate, (a) foil (b) carbon extraction replicas.

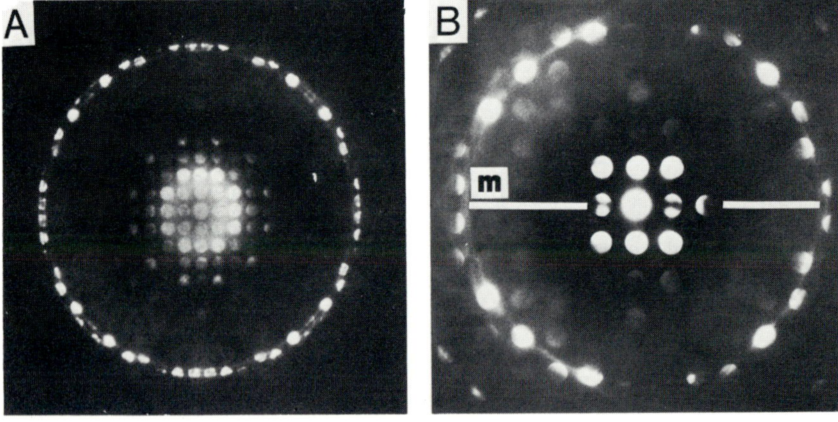


Fig. 3. Typical CBED patterns from $M_5B_3$ (a) [001] (b) [301]

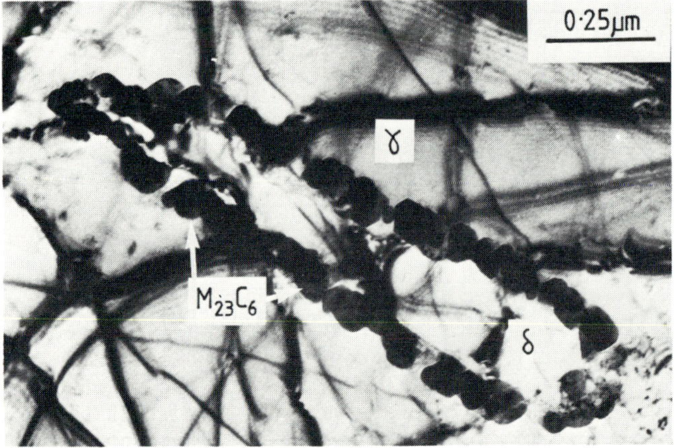


Fig. 5. TEM micrographs from the weld metal showing δ-ferrite lath in γ-matrix.

# Measurement of C/N ratio in (TiNb)(CN) precipitates using EELS

Z Chen and M H Loretto, Department of Metallurgy & Materials, University of Birmingham, P.O. Box 363, Birmingham B15 2TT U.K.
R Cochrane, British Steel Research Laboratories, Swinden House, Moorgate, Rotherham S60 3AR

## 1. Introduction

Recently electron energy loss spectroscopy (EELS) has been used as a microanalytical technique since it is useful for the analysis of thin precipitates which contain light elements and is potentially useful for analysis where high spatial resolution is required (Joy 1981). For example, studies of precipitates in Ti and V microalloyed steel (Loberg 1984 et al., Garratt-Reed 1981) and very small defects in silicon and germanium (Bourret and Colliex 1982) have been reported. In the present work the precipitates in a Ti-Nb-V high strength low alloy steel have been analysed using EELS in an attempt to determine the C/N ratio of the precipitates formed at different temperatures. In conjunction with energy dispersive X-ray analysis (EDX) the analysis should allow the chemical composition of all the precipitates to be determined completely but it has been found that there are major problems in the analysis of precipitates which contain more than about 40% niobium. The reasons for this problem are discussed together with the limitations of EELS.

## 2. Experimental

The overall composition of the steel is given in Table 1. Specimens 20 x 20 x 100 mm were annealed in vacuum for 4 h at 1050 and 1250°C. Both carbon and aluminium extraction replicas were prepared and preliminary examination of typical replicas showed that the morphology and size distribution of the extracted precipitates were similar on the two types of replica. Energy dispersive analysis and electron energy loss spectroscopy were carried out on a Philips EM400T fitted with a tungsten filament and interfaced to a STEM unit, a SiLi EDX detector and an EELS Gatan Model 607 spectrometer.

Table 1 Chemical composition (wt-%) of the Ti-Nb-V steel

| C | Si | Mn | Mo | Al | Cu | Ti | Nb | V | N |
|---|---|---|---|---|---|---|---|---|---|
| 0.082 | 0.34 | 1.32 | 0.22 | 0.030 | 0.30 | 0.019 | 0.039 | 0.057 | 0.0076 |

EDX analysis was carried out using live counting times up to 500 seconds in order to obtain analyses of reasonable accuracy. EELS analyses were carried out in the image mode for particles >100 nm and in the diffraction mode or in STEM for smaller particles. In image mode the optimum parameters such as objective aperture size, slit width, and spectrometer entrance aperture were obtained empirically. It was found that the optimum semi acceptance angle (i.e. scattering angle at the specimen) was

∿4 mrad with the other parameters adjusted to give a resolution of ∿3 eV. In the diffraction and STEM mode it was found that the optimum semi acceptance angle was about 7 mrad, i.e. larger than that in the image mode, because of the larger convergence angle of the incident beam used in the STEM mode. An advantage of the STEM mode was that the influence of the replica on the EELS spectrum was reduced very significantly; no oxygen edge was visible for example, on spectra from Al replicas obtained in the STEM mode whereas oxygen edges were clearly visible on spectra recorded in the image mode. Some analyses were carried out in the STEM mode using a slightly smaller convergence by switching on C2 and for some particles more obvious edges could be obtained in this way. The increase in probe size, associated with this mode of operation, occasionally led to an increased contribution to the spectrum from the Al replica and the results obtained in this mode are not reported in this short paper. The carbon contamination was negligible in both modes and the reproducibility of results was very good. Quantification of EDX data was carried out by stripping the background from the characteristic X-ray peak and then using the integrated intensities in the $TiK_\alpha$, $NbK_\alpha$ and $NbL_\alpha$ to calculate the ratio of Ti to Nb. No significant amount of V was observed in precipitates from samples which had been heated to >1050°C. All the precipitates were thin enough so that there was no need to apply an absorption correction as was apparent for the sensibly constant value of the ratio of the intensities of $NbK_\alpha$ and $NbL_\alpha$ X-rays. The programme used to quantify the observations has been calibrated against many standards. Quantification of the EELS data was done using a background fitting programme which extrapolated the background, fitted if possible on the high energy (i.e. low loss) side of an edge, under the observed edges, and extrapolating assuming that the background was of the form $AE^{-r}$.

## 3. Results

EDX and EELS spectra obtained from precipitates extracted from specimens annealed at 1250 and 1050°C are shown in Figs. 1 and 2 respectively. It is apparent from the EDX spectra (Figs. 1a and 2a) that the Ti content is much higher in the precipitate from the sample annealed at 1250°C (Fig. 1a) than in the sample annealed at 1050°C (Fig. 2a). More importantly it is apparent that the delayed Nb M45 edges (Siegbahn et al. 1967, Ahn and Krivanek 1983) at about 230 and 250 eV, which are visible in Fig. 2(b), make it impossible to fit a background prior to the C K edge. Figure 3 shows EDX and EELS data obtained from a small particle extracted from a sample which had been annealed at 1250°C. The EDX spectrum shows that the precipitate contains a smaller fraction of Nb than does the precipitate which yielded the spectrum shown in Fig. 2. In agreement with this the EELS data (Fig. 3b) shows no obvious Nb edges at 230 or 250 eV. Nevertheless it is again not possible to fit a background of the form $AE^{-r}$ prior to the C edge as is evident from Fig. 3(c) where attempts to fit such a background are shown.

## 4. Discussion

In principle it should be possible to analyse small (TiNb)(CN) precipitates using a combination of EDX and EELS. Thus the ratio of Ti to Nb can be obtained with an accuracy limited by counting statistics using a standard EDX analytical programme. Since the Ti L edge is available on the energy loss spectrum, together with the C K-edge the N K-edge and the Nb M45 edge the EELS data should allow complete analysis. The obvious

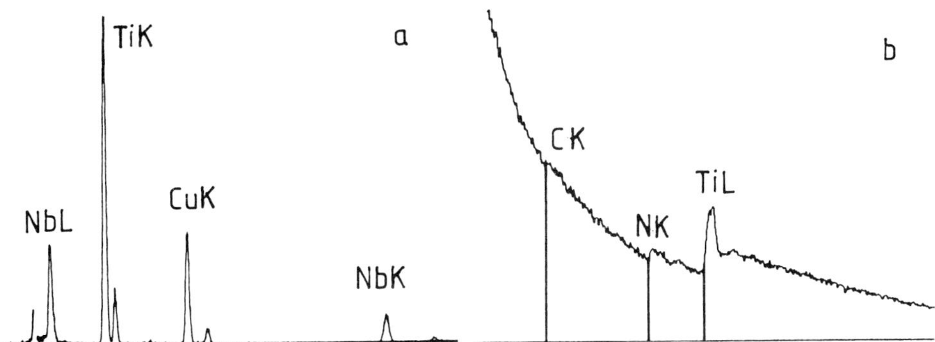


Fig. 1(a) EDX spectrum taken from a precipitate extracted onto an Al replica from a sample after heat treating at 1250°C for 4 h. Full scale 850.

(b) EELS spectrum taken from the same precipitate as used for the EDX spectrum in (a). Full scale 15,625. 100 A probe. Semi-collection angle 6.8 mrad. Dwell time 300 ms.

Table 2 Data of EELS and EDX from precipitates with various Nb/Ti ratios

| Nb/Ti | C/N | N/Ti | (C+N)/(Ti+Nb) |
|---|---|---|---|
| 0.28 | 1.31 | 0.53 | 0.96 |
| 0.35 | 0.73 | 0.72 | 0.92 |
| 0.38 | 0.71 | 0.71 | 0.88 |
| 0.50 | 0.81 | 1.19 | 1.43 |
| 0.70 | 2.44 | 0.75 | 1.52 |
| 1.28 | 4.27 | 0.92 | 2.04 |

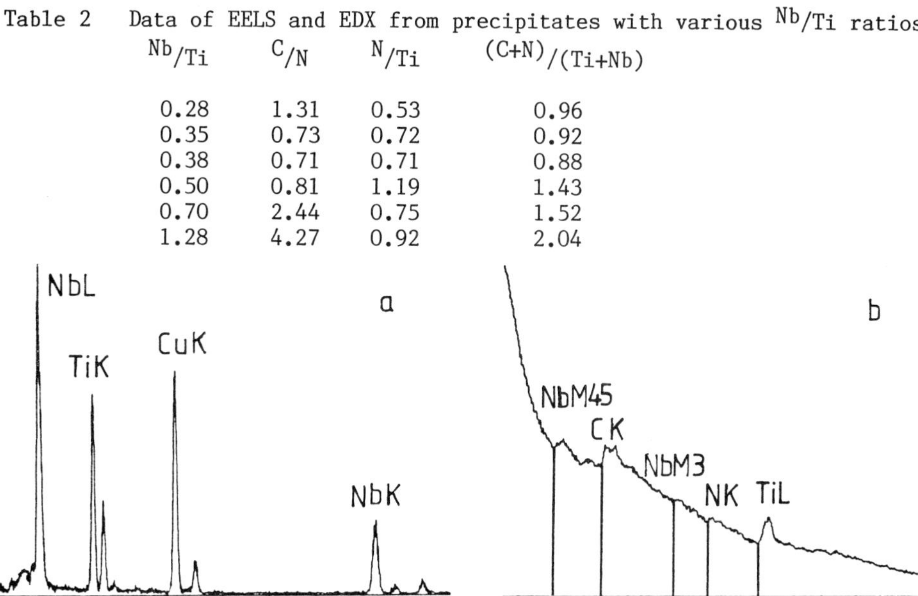


Fig. 2(a) EDX spectrum taken from a precipitate extracted onto an Al replica from a sample after heat treating at 1050°C for 4 h. Full scale 1190.

(b) EELS spectrum taken from the same precipitate as used for the EDX spectrum in (a). Full scale 7812. 100 A probe. Semi collection angle 6.8 mrad. Dwell time 300 ms.

problem in using EELS is that the cross section for the Nb M45 edge, is not known with any confidence but this can be calibrated from the EDX data. A more difficult problem arises from the fact that the Nb M45 edges lie just before the carbon edge so that, as pointed out earlier, it is not possible to fit a background to allow straightforward quantification of carbon. This problem is highlighted in the results in Table 2 which summarises the EDX

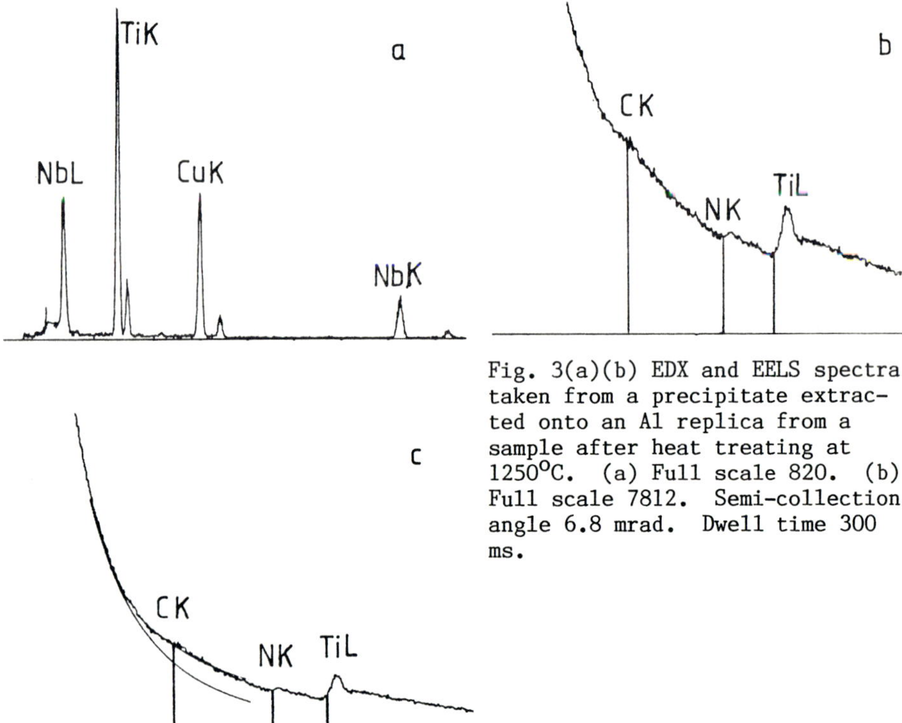


Fig. 3(a)(b) EDX and EELS spectra taken from a precipitate extracted onto an Al replica from a sample after heat treating at 1250$^{o}$C. (a) Full scale 820. (b) Full scale 7812. Semi-collection angle 6.8 mrad. Dwell time 300 ms.

(c) Background fitting to the EELS data shown in (b). The lower background was fitted over an energy range centred at 200 eV and clearly results in an overestimate of the C edge. The higher background fitted either side of 250 eV underestimates the C content. (Full scale 15625).

and EELS data for specimens with a range of Ti and Nb contents. These observations show that when the Nb/Ti ratio is greater than $\sim$0.4 the apparent ratio of (C+N)/(Ti+Nb) is very different from the value of 1. Since the general formula for such carbonitrides is (TiNb)(CN) it is apparent (cf. Fig. 3) that the analyses are incorrect for the high Nb contents. It is not obvious how this type of problem, which is likely to arise in EELS when heavy elements are present, can be overcome.

## 5. References

Ahn C C and Krivanek O L 1983 Atlas (ASU HREM and GATAN).

Bourret A and Colliex C 1982 Ultramicroscopy, 9, 183.

Garratt-Reed A J 1981 Quantitative Microanalysis with High Spatial Resolution, (Metals Society, London) p.165.

Joy D C 1981 Quantitative Microanalysis with High Spatial Resolution, (Metals Society, London) p.127.

Loberg B, Norden A, Strid J and Easterling K E 1984 Metal Trans. 15A p.33.

Siegbahn K, Nordling C, Fahlman A, Nordberg R, Hamrin K, Hedman J, Johansson G, Bergmark T, Karlsson S E, Lindgren J and Lindberg B 1967 Nova Acta Reg. Soc. Sc., Ups. Ser. IV 20 p.224.

# Analysis of borides in Mar M002

Y P Lin and J W Steeds

H.H. Wills Physics Laboratory, University of Bristol, Bristol BS8 1TL

## 1. Introduction

Phase identification on a microscopic scale is easily achieved in modern electron microscopes combining the powerful techniques of convergent beam electron diffraction (CBED) for space group and lattice parameter determination and energy dispersive X-ray spectroscopy (EDS) and/or electron energy loss spectroscopy (EELS) for composition determination. Once the phases have been fully identified using CBED, the X-ray spectra for the various phases are in most cases sufficiently distinct to enable subsequent identification based on EDS alone. However, in cases where phases with similar X-ray spectra are present, CBED remain the only method for unambiguous identification of the precipitates. We present below the identification of phases in Mar M002, an advanced Ni-based superalloy, after $10^4$ hours ageing at 850°C, in particular the analysis using CBED of some borides which can have similar X-ray spectra. The alloy has the following nominal composition (wt %):

9% Cr, 10% Co, 10% W, 1.5 % Ti, 2.5% Al, 2.5% Ta, 2.5% Hf, 0.15% C, 0.02% B, balance Ni.

Prior to ageing for $10^4$ hours at 850°C, the sample had received a five stage heat treatment/surface coating process, starting with solution treatment at 1190°C for 4 hours and ending with ageing at 870°C for 16 hours. For preparation of carbon extraction replicas, samples were etched in Kalling's solution prior to carbon evaporation and the replica released electrolytically in 10% HCl in methanol at 3V. The replicas were examined in either a Philips EM400 electron microscope equipped with a Link 860 EDS system, or a Philips EM430 microscope with a EDAX 9100 system.

## 2. Results and Discussion

The extraction technique used does not extract the major $\gamma'$ strengthening phase. On carbon replicas, large (up to ~20μm) blocky MC carbides containing variable proportions of Ti, Ta, Hf and W can be easily identified. Other phases identified include $M_{23}C_6$, $M_5B_3$ and two forms of $M_2B$, one tetragonal and one orthorhombic. The metal content of these precipitates based on EDS are given in Table 1 together with their space groups and lattice parameters. The observation of such a variety of borides rather than the $M_3B_2$ commonly found in Ni-based superalloys is unusual. Although their existence in the binary Cr-B system is known (Bertaut and Blum 1953 and Aronsson and Åselius 1958), only the orthorhombic $M_2B$ in stainless steels (Padilha and Schanz 1980 and Stoter 1979) and recently the Cr-rich $M_5B_3$ in Type 316 stainless steel (Finlan et al, this proceeding) have been reported. No observation of the tetragonal $M_2B$ in commercial alloys have been reported.

In the present alloy, the orthorhombic $M_2B$ is distinguishable by its lath-like appearance and by faults on the pseudohexagonal (100) planes, Fig. 1. The 2mm whole pattern symmetry of the [100] CBED pattern is most clearly shown by the higher order Laue zone (HOLZ) reflections, Fig. 2. For $M_5B_3$, two distinct morphologies were observed: as large (μm sized) Cr-rich blocky precipitates or more commonly as small (up to ~ 0.2μm) W-rich diamond shaped particles. The latter is easily distinguished from other phases present by its morphology, Fig. 3, and X-ray spectrum. However, the blocky $M_5B_3$ together with $M_{23}C_6$ and the two types of $M_2B$ are all Cr-rich, Table 1, and reliable distinction between them based on EDS alone may not always be possible. Although the orthorhombic $M_2B$ is distinguishable by its morphology, Fig. 1, and most $M_{23}C_6$ precipitates by their cube-like appearance, Fig. 3, blocky $M_{23}C_6$ are also present and easily confused with the blocky tetragonal $M_2B$ and $M_5B_3$. Distinction between these borides and $M_{23}C_6$ was thus based on analysis of CBED patterns. The $M_{23}C_6$ is easily recognised from its CBED patterns (The Bristol Group 1984). For $M_5B_3$ and the tetragonal $M_2B$, both with a space group of I4/mcm, inspection of the CBED patterns of the major axes reveal a body-centred tetragonal lattice and the measured lattice parameters are in accord with those of $Cr_5B_3$ and $Cr_2B$ (Bertaut and Blum 1953). The comparable a but different c for these two borides is illustrated by the different first order Laue zone (FOLZ) diameters in their respective [001] CBED patterns, Fig. 4. The I-lattice of these two borides can also be identified by inspection of the relative position of FOLZ and zero layer reflections in a [100] CBED pattern, as shown in Fig. 5 for $M_2B$. In this pattern, the reflections in the zero layer form a simple rectangular mesh, while the FOLZ reflections a face centred rectangular mesh. This indicates the presence of a c-glide on the y-z plane perpendicular to the zone axis (i.e. Okl type reflections with l odd are absent). Dynamic absences or Gjønnes-Moodie (1965) dark lines due to the c-glide on the x-z plane also occur in the FOLZ (arrowed). This pattern alone thus allowed the lattice parameters and space group of the crystal to be determined. Further confirmation of the c-glide can be obtained by tilting to the <101> or <301> axes, where dynamic absences would be observed, and Fig. 6 shows the expected dark lines in a [101] CBED pattern of $M_5B_3$.

To fully confirm the identification of the borides, detection of boron using EELS would be necessary. This is however hampered by the carbon support film, particularly at low accelerating voltages. Nevertheless, at 300kV, boron in the various borides has been detected and an example for the Cr-rich $M_5B_3$ is shown in Fig. 7, where the boron K edge is visible above the fast rising background.

Acknowledgement

The authors wish to thank the SERC for financial support, Rolls Royce Ltd., Derby and Dr W Plumbridge for provision of sample material.

References

Aronsson B and Åselius J 1958 Acta Chem. Scan. 12 1476
Bertaut F and Blum P 1953 Compt. rend. 236 1055
Bristol Group 1984 Convergent Beam Electron Diffraction of Alloy Phases compiled by Mansfield J F (Bristol and Boston: Adam Hilger)
Finlan G T, Lin Y P and Steeds J W, this proceeding
Gjønnes J and Moodie A F 1965 Acta Cryst. 19 65
Padilha A F and Schanz G 1980 J. Nucl. Mat. 95 229
Stoter L P 1979 Thesis University of Bristol.

Table 1 Typical metal content and crystallographic data for phases in MM002 (excluding $\gamma'$ and MC).

| phase | metal content (at %) Cr | W | Co | Ni | space group | lattice parameters, Å |
|---|---|---|---|---|---|---|
| $M_{23}C_6$ | 87.3 | 7.3 | 1.8 | 3.6 | Fm3m | a = 10.65 |
| $M_5B_3$ | 41.6 | 53.4 | 1.0 | 4.0 | I4/mcm | a = 5.63<br>c = 10.7 |
| $M_5B_3$ (blocky) | 77.3 | 21.0 | 0.6 | 1.1 | I4/mcm | a = 5.55<br>c = 10.6 |
| $M_2B$ | 92.2 | 6.4 | 1.1 | 0.3 | I4/mcm | a = 5.20<br>c = 4.33 |
| $M_2B$ | 93.4 | 4.4 | 1.1 | 1.1 | Fddd | a = 14.6<br>b = 4.25<br>c = 7.47 |

Fig. 1. Orthorhombic $M_2B$ containing planar defects adjacent to MC.

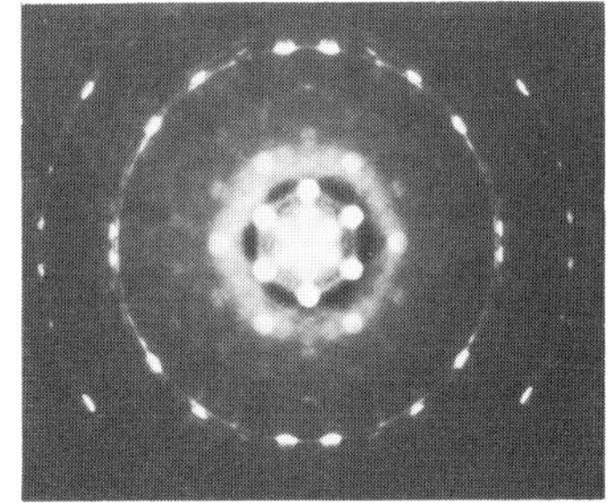

Fig. 2. [100] CBED pattern for the orthorhombic $M_2B$.

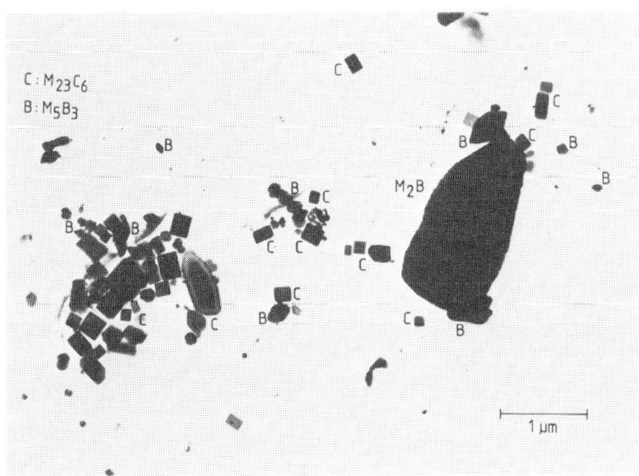


Fig. 3. Carbon extraction replica showing $M_{23}C_6$, W-rich $M_5B_3$ and tetragonal $M_2B$.

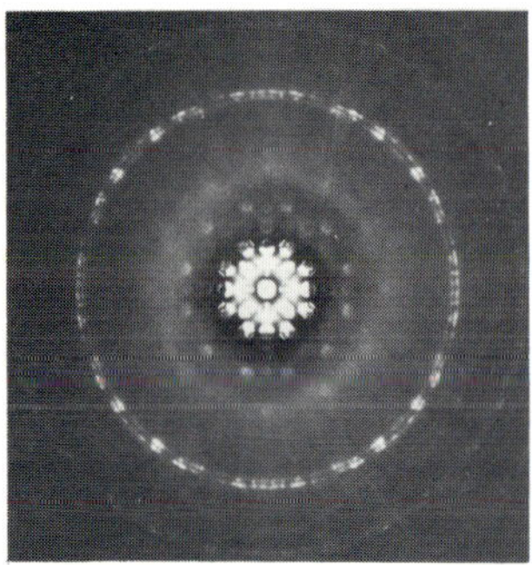

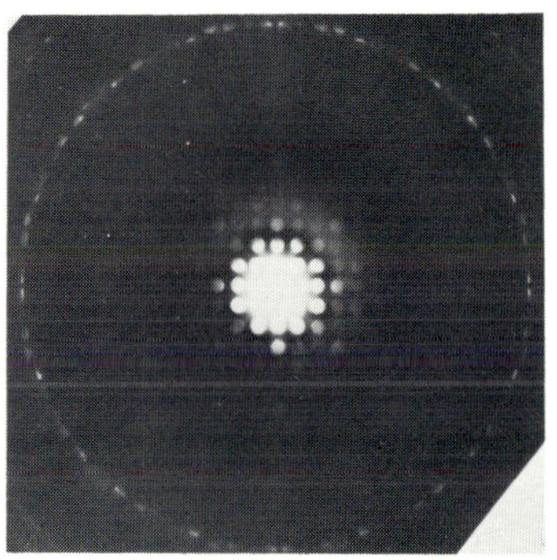

Fig. 4. [001] CBED pattern for $M_5B_3$ and the tetragonal $M_2B$(right).

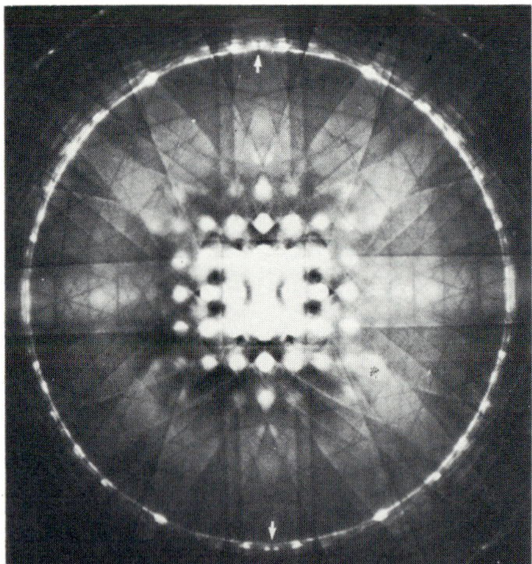

Fig. 5. [100] CBED pattern for the tetragonal $M_2B$.

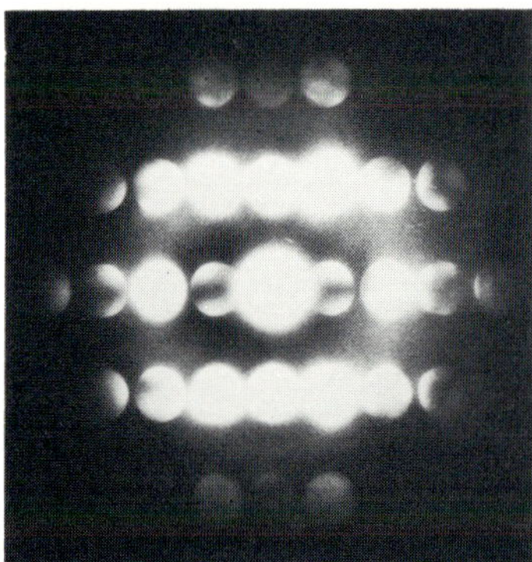

Fig. 6. [101] CBED pattern for $M_5B_3$ showing Gjønnes-Moodie dark lines in the $10\bar{1}$ discs.

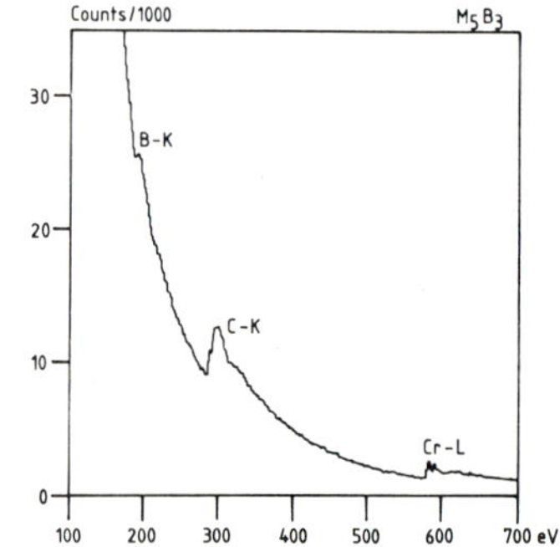


Fig. 7. EELS spectrum for the Cr-rich $M_5B_3$ on extraction replica recorded at 300kV.

*Inst. Phys. Conf. Ser. No 78: Chapter 8*
*Paper presented at EMAG '85, Newcastle upon Tyne, 2–5 September 1985*


# Phosphide phases in fast reactor irradiated 12%Cr ferritic–martensitic stainless steels

F A Little and L P Stoter*

Materials Development Division, AERE, Harwell.

* Pye-Unicam Ltd., Cambridge.

## 1. Introduction

Phosphides are rarely observed in ferritic steels containing normal commercial impurity levels of phosphorus (viz. < ∿200 ppm); rather, the typical behaviour of this element is to segregate to grain boundaries during thermal ageing, giving the loss of cohesive strength and intergranular fracture characteristic of temper embrittlement (e.g. Eyre et al., 1982).

In the present studies, two distinct phosphide types have been detected in 12%Cr ferritic-martensitic steels following elevated temperature neutron irradiation. Their identification serves to illustrate the combined application of convergent beam electron diffraction (CBED) an energy dispersive X-ray analysis (EDX) as a "fingerprint" technique for rapid precipitate characterisation.

The observations have possible relevance to temper embrittlement processes in both irradiated and unirradiated ferritic steels.

## 2. Experimental

The data were established from two commercial grades of 12%Cr steel, viz. (i) type A - an unalloyed variant with 13.3%Cr, 0.38%Ni, 0.046%C and 140 ppm P; and (ii) type B - a complex 12%CrMoVNb alloy with 10.7%Cr, 0.65%Ni, 0.6%Mo, 0.14%V, 0.26%Nb, 0.10%C and 200 ppm P, in standard heat treatment conditions (Little and Stow, 1979).

The materials were irradiated as 3 mm diameter discs at temperatures of ∿ 380$^o$, 420$^o$, 460$^o$ and 615$^o$C to a damage level of ∿ 23 dpa (NRT) in the Dounreay Fast Reactor. Thermal controls received 8250 hour ageing treatments at 460$^o$ and 600$^o$C.

Precipitates were unambiguously identified on carbon extraction replicas by combined CBED and EDX procedures from a minimum of ten particles, using a Phillips EM400T electron microscope operating at 100 kV.

## 3. Results

Initial structures of the steels consisted of tempered martensite, alone (steel B), or associated with δ-ferrite (steel A). Precipitated phases included $M_{23}C_6$, $M_2X$ (typically as $Cr_2N$), and (in steel B only) NbC, as previously described (Little and Stoter, 1982).

Irradiation resulted in coarsening and/or dissolution of existing precipitate types, together with the formation of several new phases. Considering specifically phosphorus-rich variants, two intragranular phosphide precipitate types, identified as $M_3P$ and MP, were detected. Crystallographic and compositional details are as follows:

(i) The body-centred-tetragonal $M_3P$ phase (a = 0.984 nm, c = 0.490 nm) was present only in steel A, forming at irradiation temperatures of 420° and 460°C as thin ∿ 1 µm long sheets, with preferential nucleation on pre-existing oval platelets of $Cr_2N$, as shown in Fig. 1(a). The [100] zone axis of the CBED pattern is illustrated in Fig. 1(b) and exhibits an unusual but very characteristic 4-fold rotational symmetry without any mirror symmetries, and corresponding to the $I\bar{4}$ space group of $Cr_3P$. An EDX spectra, shown in Fig. 2(a) gives a derived composition of M = 60Cr; 8-12Fe; 0.3-0.6Ni and P = 29 (at.%), confirming identification as $M_3P$, and close to $Cr_3P$.

(ii) The orthorhombic MP phase (a = 0.368 nm, b = 0.683 nm, c = 0.602 nm) occurred only in steel B, forming at irradiation temperatures of 460° and 615°C as globules ∿ 0.2 µm in diameter (460°C), or as larger irregular particles ∿ 0.7 µm wide (615°C). CBED patterns for a range of zone axes arranged in approximate sterographic projection are presented in Fig. 3, and confirm the Pnma space group for MP. The corresponding EDX spectra is given in Fig. 2(b), and the derived composition (M = 20Cr; 29Fe; 2.5-4Ni; 2.5V; 4-5Mo; 1.5Nb; 0.3As and P = 35) reveals significant incorporation of additional elements into this phase.

Phosphide compositions were not significantly dependent on irradiation temperature; however, measured lattice parameters fell outside ranges cited in the literature for $Cr_3P$ (a = 0.911 to 0.918 nm, c = 0.446 to 0.456 nm) or the monophosphides CrP and FeP (a = 0.310 to 0.328 nm, b = 0.559 to 0.622 nm, c = 0.501 to 0.573 nm) (Rundqvist, 1962a), probably as a consequence of the additional element substitutions. There is also the possibility in the MP phase of substitution of P by C, N or B, not detectable by EDX (Rundqvist, 1962b).

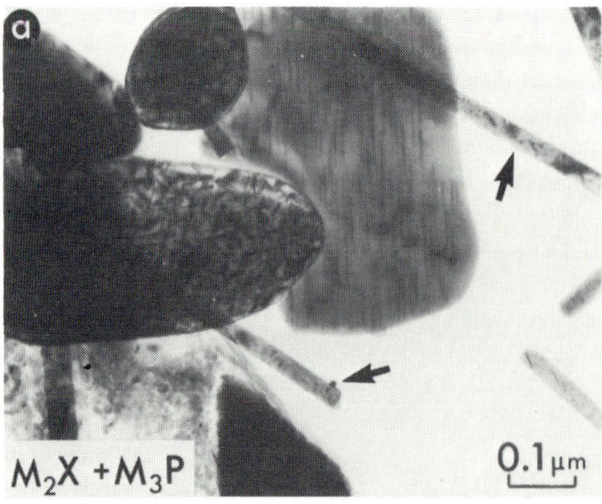


Fig. 1(a) $M_3P$ phase (arrowed) nucleating on $M_2X$ plates.

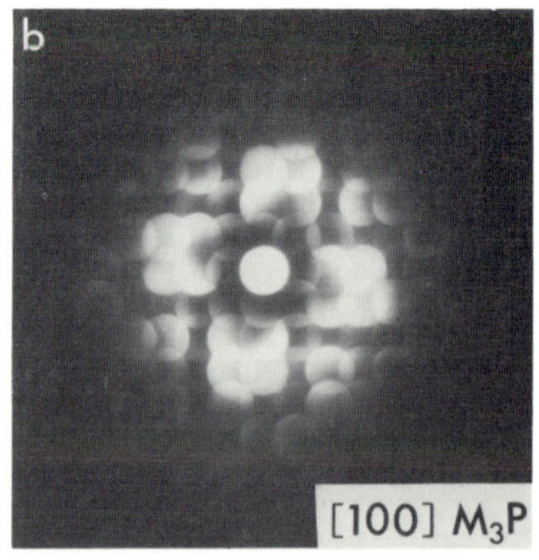


Fig. 1(b) [100] zone axis CBED pattern for $M_3P$ phase.

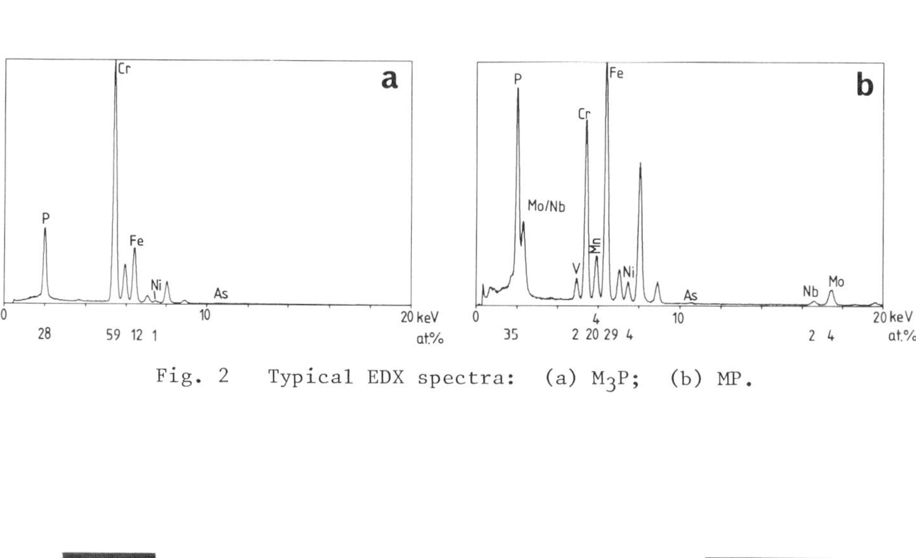


Fig. 2 Typical EDX spectra: (a) $M_3P$; (b) MP.

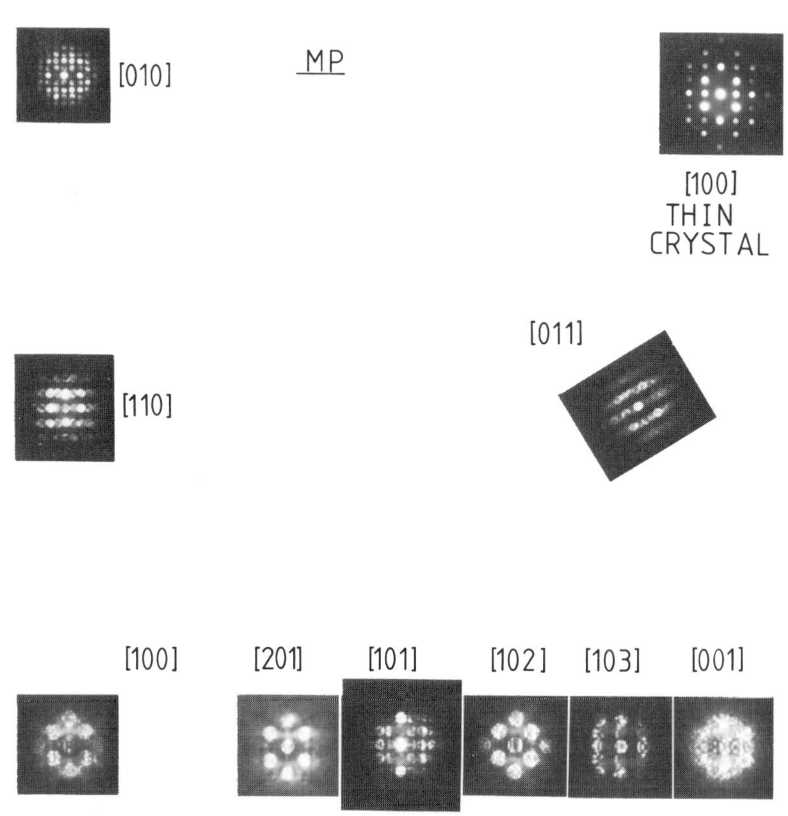


Fig. 3 Zone axis map of CBED patterns for MP.

## 4. Discussion

It is well established that point defect fluxes generated during irradiation can result in non-equilibrium solute segregation to, or away from, point defect sinks (e.g. grain boundaries, precipitate interfaces, etc.) - the direction of solute flow depending on the solute-point defect binding energy (e.g. Russell, 1984). Furthermore, solute enrichment at sinks can result in irradiation-induced precipitation if local solubility limits are exceeded, leading to the formation of new phases not expected from bulk equilibrium considerations.

The observation of phosphide precipitates in irradiated 12%Cr steels containing small bulk levels of phosphorus is consistent with strong irradiation-induced segregation, indicating high phosphorus-point defect binding energies in ferritic steels. Furthermore, since P is an undersized solute in α-iron (King, 1966), it is likely that P is strongly bound as a mixed-dumbell self-interstitial, by analogy with undersize solute behaviour in f.c.c. alloys (Lam et al., 1978).

The substitution of additional alloying elements into the MP phase may imply interaction and cosegregation of P with elements such as Mo, V and Nb (in addition to Cr); analogous interactions have been suggested to explain temper embrittlement effects (e.g. Eyre et al., 1982). The data also suggest that such interactions lead to the formation of phosphides with enhanced thermal stability; thus, the MP variant containing Mo, V and Nb is the preferred phase in the alloyed 12%Cr steel, and possesses metastability (under irradiation) to 615$^{o}$C.

Whilst phosphides were detected only after neutron irradiation, their formation would appear likely whenever strong segregational processes are operative; hence, temper embrittled steels may be worthy of close examination in this respect. The formation of intragranular phosphides during irradiation of ferritic steels may have beneficial effects in suppressing temper embrittlement by matrix depletion of phosphorus.

## References

Eyre B.L., Edwards B.C. and Titchmarsh J.M., 1982, "Advances in Physical Metallurgy and Applications of Steels", (Metals Society).
King H.W., 1966, J. Nucl. Mater., 1, 79.
Lam N.Q., Okamoto P.R. and Wiedersich, H., 1978, J. Nucl. Mater. 74, 101.
Little E.A. and Stoter L.P., 1982, ASTM Spec. Tech. Publ. No. 782, 207.
Little E.A. and Stow D.A., 1979, J. Nucl. Mater., 87, 25.
Rundqvist S., 1962a, Arkiv För Kemi, 20 (7), 67.
Rundqvist S., 1962b, Acta Chem. Scand., 16, 287.
Russell K.C., 1984, Progr. Mater. Sci., 28, 229.

*Inst. Phys. Conf. Ser. No 78: Chapter 8*
*Paper presented at EMAG '85, Newcastle upon Tyne, 2–5 September 1985*


# An analytical electron microscope study of grain boundary segregation in irradiated ferritic steels

G J Mahon, A W Nicholls, I P Jones, C A English* and T M Williams*
Department of Metallurgy and Materials, University of Birmingham, P.O. Box 363, Birmingham, B15 2TT, U.K.
*Materials Development Division, AERE Harwell, Oxfordshire, OX11 ORA, U.K.

## 1. Introduction

Ferritic stainless steels are currently being considered for use as fast reactor wrapper materials as a consequence of their resistance to void swelling (Little and Stow 1979). In contrast to their austenitic counterparts, investigations into the effect of irradiation on phase stability and solute segregation are limited, and a clear picture of the precise behaviour of the various solutes commonly found has yet to be established.

Radiation-induced solute segregation occurs as a consequence of the vacancies and interstitials generated during irradiation of the metal. These are eliminated from the crystal lattice by mutual recombination or by absorption into either dislocations, causing climb, or into other sinks such as grain boundaries or precipitate-matrix interfaces. Associated with these defect fluxes are solute atom fluxes resulting from either the migration of mobile defect-solute complexes, or differences in the various tracer diffusion coefficients (inverse Kirkendall effect).

In this work two ferritic steels based on an iron-12% chromium binary alloy were irradiated using electrons and heavy ions, both of which cause damage at much greater rates than those obtained in fast reactors. Solute redistribution to planar grain boundary sinks was measured using STEM based microanalysis.

## 2. Experimental

The first of the two steels, which are both low carbon, has the composition (wt.%) Fe-12Cr-0.5Ni-0.5Mn. This steel is designated alloy A. The other (alloy B) is identical except for the addition of 1 wt.%Si and 2 wt.%Mo.

3mm discs were punched from rolled strip and heat treated for 3 hours at 980$^{o}$C followed by furnace cooling to 750$^{o}$C where they were held for 2 days to produce a ferritic microstructure. The discs were then profiled in a jet-polisher using a solution of 65% ethanol, 32% butoxyethanol, and 3% perchloric acid at 60 volts and room temperature, followed by polishing to perforation in 70% ethanol, 20% glycol and 10% perchloric acid at 10 volts, also at room temperature.

Electron irradiations were carried out at Birmingham using the AEI EM7 High Voltage Electron Microscope (HVEM) operating at 1 MeV to a dose of 10 dpa at 2-3 x $10^{-3}$ dpa.s$^{-1}$ on thin foils at 450, 550 and 650$^{o}$C. Heavy ion irradiations were performed using 52 MeV $Cr^{2+}$ ions in a Variable Energy

Cyclotron (VEC) at AERE Harwell at the same three temperatures. In this case unthinned discs were irradiated to 10 dpa at a dose rate of $2\text{-}3 \times 10^{-4}$ dpa.$s^{-1}$, producing a peak radiation damage region 5μ from the irradiated surface. This layer is removed by vibropolishing in a 0.05μ alumina suspension and back polishing from the other side of the disc. As a control, specimens were thermally aged at the three temperatures for 10 hours, which corresponded approximately to the time required for the heavy ion irradiation.

Microanalysis of irradiated and thermally aged specimens was performed in a Philips EM400T with a STEM/EDX facility. The specimen was tilted until the foil normal was directed towards the EDX detector and, more importantly, the grain boundary of interest was oriented parallel to the incident electron beam. This was not always possible due to the magnetic nature of the specimens since as the tilt was increased correction for astigmatism became more difficult, thus reducing the spatial resolution of the electron beam.

Additional problems also exist with the detection of certain elements within these specific alloys. The complex nature of the background at the low energy end of the spectrum makes background fitting difficult, hence affecting the counts in the silicon window. For manganese, detection of less than 1 wt.% is not possible since the $MnK_\alpha$ peak lies almost directly under the $CrK_\beta$ peak. In the case of Nickel some contribution to the intensity in the window may be coming from the tail of the adjacent $FeK_\beta$ peak.

Characteristic intensities were extracted from the spectra using the EDAX 9100/60 analysis system and Z corrected compositions obtained using thin film software available on the University main computer. From several matrix values obtained under identical collection conditions 92% confidence limits have been calculated for each element and these figures have then been used as the errors.

## 3. Results

The results of this investigation are given in Table 1 and summarised below. The sense only of segregation is given due to the difficulties, mentioned previously, in accurately measuring the composition. The amounts of segregation varied from several times to just outside the 92% confidence limits.

### 3.1 Alloy A

Under electron irradiation chromium depletion was observed at the grain boundary at 450, 550 and 650°C. The nickel concentrations, however, all fell within the experimental errors indicating no radiation-induced segregation of this element. This was not the case with heavy ion irradiation. At 450°C there was no observable segregation, whereas at 550 and 650°C there was chromium depletion and nickel enrichment. Thermal ageing at 650°C for 10 hours produced no detectable solute redistribution at grain boundaries.

### 3.2 Alloy B

Under electron irradiation at 650°C chromium is depleted from the grain boundary whilst at 450 and 550°C chromium enrichment occurs. Silicon and molybdenum enrichment were detected at all three temperatures and nickel

<u>Alloy A</u> Fe-12Cr-0.5Ni-0.5Mn

| | Electron Irradiation | Heavy Ion Irradiation | Thermal Ageing for 10 hours |
|---|---|---|---|
| 450°C | Cr$\downarrow$. Ni- | Cr, Ni- | |
| 550°C | Cr$\downarrow$. Ni- | Cr$\downarrow$. Ni$\uparrow$ | |
| 650°C | Cr$\downarrow$. Ni- | Cr$\downarrow$. Ni$\uparrow$ | Cr, Ni- |

<u>Alloy B</u> Fe-12Cr-0.5Ni-0.5Mn-1Si-2Mo

| | Electron Irradiation | Heavy Ion Irradiation | Thermal Ageing for 10 hours |
|---|---|---|---|
| 450°C | Cr, Mo, Si$\uparrow$. Ni- | Cr, Mo-. Si, Ni$\uparrow$ | |
| 550°C | Cr, Mo, Si, Ni$\uparrow$ | Cr, Mo, Si, Ni$\uparrow$ + precipitation. | Cr, Mo, Si$\uparrow$. Ni- + precipitation. |
| 650°C | Cr$\downarrow$. Mo, Si, Ni$\uparrow$ | Cr, Mo, Si, Ni$\uparrow$ + precipitation | Cr, Mo, Si$\uparrow$. Ni- + precipitation. (After 1 hour Mo, Si$\uparrow$. Cr, Ni-). |

Table 1 Solute redistribution at grain boundaries after irradiation to 10 dpa at three temperatures or thermal ageing only. $\uparrow\equiv$solute enrichment, $\downarrow\equiv$solute depletion, - no change.

enrichment at 550 and 650°C. After heavy-ion irradiation precipitation was observed along most grain boundaries at 550 and 650°C. Significantly higher solute levels were detected at the grain boundaries after VEC irradiation as compared with HVEM irradiation. The general trends, however, are similar to the electron irradiation case with a few exceptions. Both nickel and silicon are enriched at all three temperatures, whereas chromium and molybdenum are enriched at 550 and 650°C only, with no significant enrichment at 450°C. Thermal ageing at 650°C for 1 hour led to slight enrichment of molybdenum and silicon, after 10 hours gross enrichment of chromium, molybdenum and silicon had occured at the grain boundaries leading to the precipitation of both Mo/Si and Mo/Cr rich phases (Fig. 1). Similar precipitation was also observed after 10 hours at 550°C although the amount of precipitation was not as great. The effects of thermal ageing are very similar to, and almost indistinguishable from, those of heavy ion irradiation.

Fig. 1 Cr/Mo/Si rich precipitate in alloy B after ageing at 650°C for 10 hours.

## 4. Discussion

The results from the Mo/Si containing steel (alloy B) appear to be dominated by grain boundary precipitation. The speed of this is increased by both ion and electron irradiation. The compositions of the precipitates observed after 10 hours' annealing at 650°C are similar to those found by Little and Stoter (1982) in a 12%Cr-Mo-V steel after annealing at 600°C. It is to be expected, but has not yet been shown, that the compositions of the precipitates will be modified by irradiation as was found by Little and Stoter. It is evidently not possible to compare meaningfully the results with those of, for example, Sethi and Okamoto (1981) who investigated the effects of Mo and Si on solute segregation in a 20/18 austentic steel.

The simpler ferritic steel (alloy A) showed no grain boundary segregation effects after a purely thermal heat treatment at 650°C, and the segregation observed under irradiation may therefore be ascribed with some confidence to non-equilibrium radiation driven segregation. The chromium depletion is as expected for a fast diffusing oversize element and may, as in the austenitic case, be assigned to the inverse Kirkendall mechanism, or alternatively to some distaste the oversize Cr atoms may have for interstitials when compared with their average solvent neighbours. The segregation of Ni atoms towards the boundary observed for the two higher temperature ion irradiations is more puzzling, however. In ferritic steels, in contrast to austenitic steels, nickel is a faster diffuser than the solvent iron (Kucera and Stransky 1982). It is also likely to be slightly oversized (Pearson 1958) rather than undersize as it is in an austenitic steel. These are two good reasons why nickel might be expected to segregate away from the grain boundaries. Although there is some evidence of appreciable Ni-vacancy binding energies in ferritic steels (Möslang et al. 1983), in view of the inexplicable discrepancy between the electron and ion-irradiations (see Table 1), the existence of nickel segregation under ion-irradiation needs to be further investigated. It is possible that the lack of nickel segregation in the HVEM is allied to radiation induced diffusion out of the irradiation area, as predicted and detected by Lam et al. (1983), but control irradiations of this alloy in grain centres failed to show any such effect, possibly because of segregation to radiation created defects.

The results of this study indicate that in Alloy A genuine radiation induced solute segregation can be measured, but by adding small amounts of molybdenum and silicon the alloy is pushed out of the single α-phase field thereby causing thermally induced precipitation to dominate the solute redistribution.

## References

Kucera I and Stransky K 1982 Mater. Sci. and Eng. 52 1.
Lam N G, Leaf G K and Minkoff M 1983 J. Nucl. Mater. 118 248.
Little E A and Stoter L P 1982 ASTM STP 782 p.207.
Little E A and Stow D A 1979 J. Nucl. Mater. 87 25.
Möslang A, Albert E, Recknagel E, Weidinger A and Moser P 1983 Hyper. Inter. 15/16 409.
Pearson W B 1958 A Handbook of Lattice Spacings and Structures of Metals and Alloys. Pergamon Press, London.
Sethi V K and Okamoto P R 1981 Phase Stability during Irradiation. Metallurgical Society of AIME, Warrendale p.109.

*Inst. Phys. Conf. Ser. No 78: Chapter 8*
*Paper presented at EMAG '85, Newcastle upon Tyne, 2–5 September 1985* 

# Boron redistribution during solidification of a two phase brass

N R Gregg, P J Goodhew, and P M Budd

Department of Materials Science and Engineering, University of Surrey, Guildford GU2 5XH.

## 1. Introduction

Most metals are fabricated from the cast columnar and equiaxed state. Grain refinement by innoculation involves making additions to the melt promoting a more equiaxed grain structure which has a lower susceptibility to cracking and a more even distribution of any second phase present. Thus grain refined metals can withstand greater deformation rates during fabrication and possess improved mechanical properties.

## 2. Grain Refinement of αβ Brass with Al-4B Masteralloy

60-40 brass specimens with 0.8 wt% Al-4B masteralloy additions were cast with a 50-100K superheat into a graphite mould 38mm diameter by 100mm length. The non-refined cast structure is shown in the upper half of Figure 1 with columnar crystals radiating from the centre to the initial outer layer of fine chill crystals. The refined brass has a similar layer of chill crystals but then a coarse equiaxed structure is observed leading to a very fine equiaxed structure in the centre of the specimen. Metallographic and microprobe studies proved the refined brass to solidify wholly as β subsequently precipitating α at the original grain boundaries. The original β grain structure was shown to have a typical grain size of about 60μm in the very fine centre area. This represented a reduction of grain size of the order of X1000.

## 3. Microscopical Investigation

Standard metallographic techniques revealed not only the grain structure described but also a fine precipitate passing through the α grains and presumably at the original β grain boundaries, which was further revealed by electropolishing in 40% orthophosphoric acid at 2.2V (Figure 2). This precipitate had a feathery form and in addition certain globular artifacts were seen. Deeper electropolishing and subsequent extraction replication revealed further information (Figures 3 and 4). The precipitate was seen to grow from a globular centre with a coral or highly branched form. This structure repeatedly produced diffuse ring diffraction patterns and was therefore amorphous in nature.

## 4. TEM Microanalysis

Standard EDS analysis of extraction replicas of the glass on copper grids was undertaken. Many grain refined castings were investigated which significantly reduced the number of elements consistently found in the glass, the purity of the initial melt being somewhat variable. The use of aluminium grids ruled out copper as a necessary constituent of the glass and windowless EDS analysis proved that oxygen is always present. A small boron peak was recorded (Figure 5) only from the regions which overhang the edge of the carbon support film by more than 2μm (Figure 4), otherwise the boron peak was not observed presumably because of absorption in the support film and the overlapping of the adjacent boron and carbon peaks.

EELS analysis of the glass was used to obtain a boron to oxygen ratio and found to be on average 20:1 atom ratio (Figure 6). This was then used along with the average windowless EDS analysis to give the following analysis.

| Element | B | O | Al | P | S | Ca | Fe | Zn |
|---|---|---|---|---|---|---|---|---|
| wt% | 89.7 | 6.4 | 2.3 | 0.1 | 0.4 | 0.2 | 0.5 | 0.4 |

The accuracy of the EELS analysis with respect to the boron to oxygen ratio was investigated by analysing $B_2O_3$ on a carbon support film. The ratio was seen to vary between 0.67 and 0.9, the stoichiometric being 0.67. The errors associated with this analysis thus may vary by as much as ±30%, but even if the above analysis was at the boundary of these errors, the glass must be extremely boron rich. The strength of the boron and oxygen peak observed in the EELS and windowless EDS analysis illustrated the low sensitivity of the latter technique to boron. The presence of boron was also confirmed by SIMS, LIMA and Auger analysis. However these techniques do not have the resolution necessary when investigating such a fine artifact and are not so quantifiable.

## 5. Discussion of Results

Analysis of the glass has shown that boron, aluminium, oxygen and other minor constituents segregate to the growing β crystal:liquid interface, where it is the last phase to solidify. The strong grain refining effect is produced by either constitutional undercooling or growth restriction mechanism. Constitutional undercooling has been discussed by Flemings and also Elliot but since pure aluminium or pure boron addition alone has no great effect, it seems that the glass plays a critical role. A proposed mechanism is illustrated in Figure 6, which relies on a solute enriched layer produced as a result of rejection ahead of the growing β crystal then restricting the diffusion of copper and zinc atoms to the crystal and thus lowering the growth velocity. This would increase the melt undercooling and allow other nucleants in the melt to become operative. The presence of the coarser equiaxed outer grain structure can be explained by the greater cooling rate experienced in this region resulting in a greater growth velocity not permitting the build up of a stable growth restricting layer.

Work is continuing in the light of the analysis obtained.

## Acknowledgements

The work was sponsored by the London and Scandinavian Metallurgical Co. Ltd., Rotherham and the SERC. The help of Miss D. Chescoe and staff of the Microstructural Studies Unit at the University of Surrey is gratefully acknowledged. Thanks are also due to Miss V. Kohler for the availability of LIMA facilities at Cambridge.

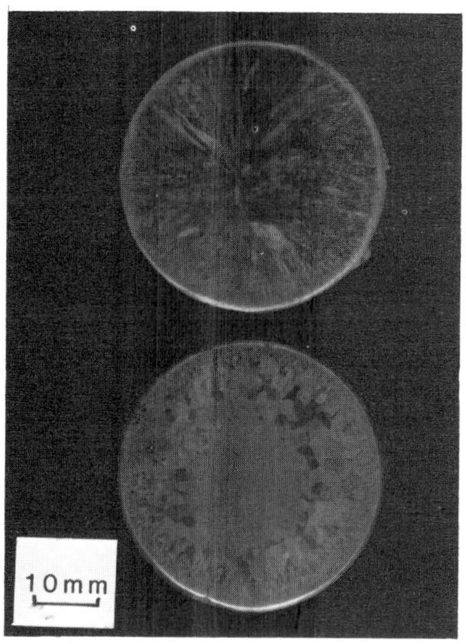


Fig.1 Standard 60-40 Brass and grain refined 60-40 Brass cast with a 0.8% wt% Al-4B addition (below).

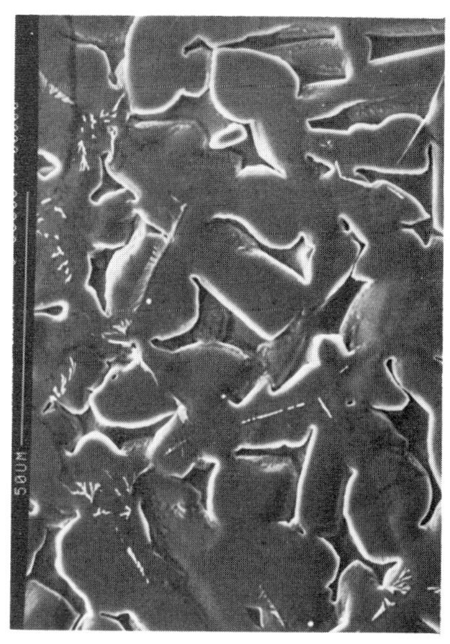


Fig.2 Electropolished grain refined Brass.

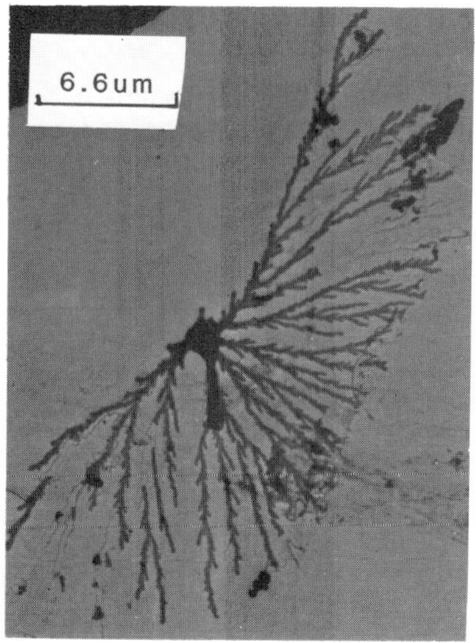


Fig.3 Extraction replica of the fine glass structure seen in Fig.2.

Fig.4 Extraction replica of the glass showing conditions under which an EDS Boron peak was recorded.

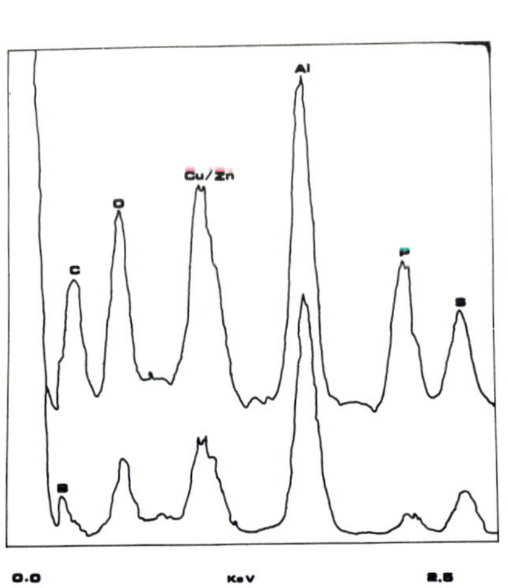


Fig.5 Standard and windowless TEM EDS analysis of the glass illustrating how the Carbon peak absorbs the Bordon peak.

Fig.6 EELS spectra of the glass.

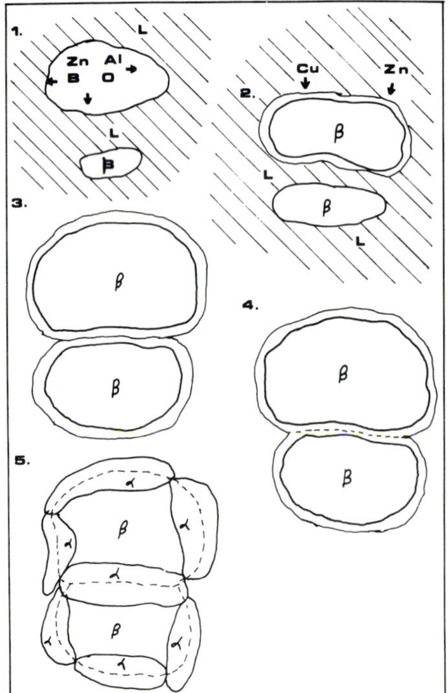


Left - Fig.7

1. Zinc, Aluminium, Boron, Oxygen and other minor constituents segregate ahead of the solidification front.
2. Here they form a solute enriched layer which impedes the diffusion of Copper and Zinc to the solid.
3. Grains meet.
4. Glass forms which branched structure.
5. Precipitation of α at former β grain boundaries. Glass runs through centre of α grains.

## Literature

Flemings, M.C., Solidification processing, New York 1974.

Elliot, R, Eutectic Solidification Processing, London 1983.

Gould et al, Trans Amer. Found. Soc. 1969, 68 258.

Backerud et al. Solidification Characteristics of Some Copper Alloys, INCRA, 1982.

Cibula A. JIM 1953, 82 513.

# STEM–EDS x-ray microanalysis of grain boundary segregation in structural steels

P Doig and P E J Flewitt

Central Electricity Generating Board, South Eastern Region, Scientific Services Department, Gravesend, Kent, DA12 2RS, U.K.

## 1. Introduction

Chemical composition differences at the grain boundary region in ferritic and austenitic steels influence creep deformation, weldment reheat cracking and stress corrosion cracking susceptibility of electrical power plant components. Non-equilibrium segregations to grain boundaries have been observed in a range of alloy systems following quenching or cooling below a critical rate (Westbrook 1963, Williams et al 1976, Doig and Flewitt 1981, 1985). In general, these segregations increase in extent with increasing initial solution treatment temperature and it is considered that segregation of solute arises from the diffusion and decomposition of vacancy-solute complexes (Doig and Flewitt 1981, 1985). STEM-EDS X-ray microanalysis enables measurements to be undertaken on thin foil specimens to a spatial resolution that is sufficient to establish such solute segregations to grain boundaries (Doig et al 1981, Doig and Flewitt 1983). Segregations of substitutional Cr and Sn which occur during fast cooling 2¼%Cr1%Mo, 2¼%Cr1%Mo0.08%Sn and type 316 stainless steels have been measured in thin foils using the STEM-EDS X-ray microanalysis technique together with solute segregations that cause reheat cracking in low alloy steel boiler steam chest weldments. These are compared with the predictions of a theoretical analysis based on the migration of vacancy-solute atom pairs to grain boundary sinks.

## 2. Experimental Procedure

A nominal 2¼%Cr1%Mo steel and a similar composition steel with an addition of 0.08%Sn were prepared by induction melting under an argon atmosphere. Specimens, 4mm diameter and 25.5mm long, were austenitised at 1423K, 1323K and 1223K for 7.2 x $10^3$s before quenching into water. Commercial AISI type 316 austenitic stainless steel samples were similarly heat treated before argon gas quenching. In addition a sample was cut from a power station boiler steam chest structural weldment heat affected zone (HAZ) with a nominal composition ½%Cr½%Mo¼%V and containing a trace impurity level of Sn. Chemical analyses for these steels are shown in Table 1. Thin foils were prepared and examined as described elsewhere (Doig and Flewitt 1981). X-ray microanalysis was carried out at an electron accelerating voltage of 100kV with an incident electron probe ≃8nm diameter using a conventional tungsten filament. The analysis was performed on foils of ≃100nm thickness to give a recorded X-ray count rate of ≃$10^3$ counts $s^{-1}$. Foil thickness measurements were estimated from X-ray count rates using a precalibration established for

TABLE 1 Chemical Analysis of the Steels Used (wt.%) (bal. iron)

| | C | S | P | Si | Cr | Mn | Ni | Mo | V | Sn |
|---|---|---|---|---|---|---|---|---|---|---|
| 2¼%Cr1%Mo | 0.07 | 0.02 | 0.02 | 0.38 | 2.1 | 0.53 | 0.1 | 1.00 | - | - |
| 2¼%Cr1%Mo0.08%Sn | 0.09 | 0.01 | 0.005 | 0.46 | 2.2 | 0.46 | 0.15 | 1.00 | - | 0.08 |
| Type 316 | 0.04 | 0.02 | 0.019 | 0.48 | 18.0 | 1.64 | 10.8 | 2.65 | - | - |
| Weldment | 0.13 | 0.013 | 0.019 | 0.27 | 0.35 | 0.49 | 0.17 | 0.58 | 0.037 | 0.01 |

fixed microscope operating conditions and direct measurements of foil thickness. Spectra were recorded for 100s and analysed using the LINK RTS2 FLS program for thin foil microanalysis.

## 3. Results and Discussion

Transmission micrographs showed that the 2¼%Cr1%Mo and the 2¼Cr1%Mo0.08%Sn steel specimens in the water quenched condition had a martensite microstructure with prior austenite grain sizes ≃100, ≃80 and ≃60μm (mean linear intercept) corresponding to the austenitising temperatures of 1423K, 1323K and 1223K respectively. There was no evidence of carbide precipitation either at the prior austenite or lath boundaries. In the case of the type 316 stainless steel specimens, the initial heat treatment followed by an argon gas quench produced austenitic microstructures with grain sizes similar to those for the low alloy ferritic steel specimens and transmission imaging of the boundaries showed no evidence of $M_{23}C_6$ carbide precipitation. Within the HAZ of the ½%Cr½%Mo¼%V steel weldment the microstructure comprised ferrite laths with a distribution of fine carbide precipitates.

Point X-ray microanalyses were undertaken on the foil specimens at various positions both on the grain boundaries and along a line normal to the boundary into the adjacent grain. Only segregations of Cr were detected in the 2¼%Cr1%Mo and type 316 steels whereas for the 2¼%Cr1%Mo0.08%Sn steel Sn was the only segregated species evident. In the case of the HAZ weldment specimens, enrichment in Sn was observed in the region of the prior austenite grain boundaries. The data are summarised in Table 2 and provide clear experimental evidence for the segregation of Cr and Sn.

A theoretical analysis based upon conjugate vacancy-solute atom migration to a grain boundary sink has been described previously (Doig and Flewitt, 1981, 1985). Here the thermodynamic driving force for the solute segregation is the excess concentration of vacancies retained during the period of cooling from the higher temperature. This analysis has been employed to predict the magnitude and spatial extent of quench induced solute segregations to the grain boundaries in these steels. The composition profiles, measured using STEM-EDS X-ray microanalysis, are degraded by the finite size of the analysing electron probe and the scattering of this beam as it penetrates the thin foil. The extent of this may be quantitatively evaluated from a consideration of the microanalysis parameters (Doig and Flewitt 1985) and as such it is possible to predict the profiles for the specific experimental conditions. Here vacancy-solute binding energies of 0.5eV and 0.75eV were established for Cr and Sn respectively. The resulting calculated

TABLE 2 X-ray Microanalysis Results Across Grain Boundaries (wt.%)

| (Material) Element | Quench Temp. (deg K) | Position (distance : nm) g.b | 20 | 50 | 100 | 200 | Area* | Accuracy ** (std.dev.) |
|---|---|---|---|---|---|---|---|---|
| (2¼%Cr1%Mo) | 1423 | 3.23 | 2.10 | 2.01 | 2.00 | 2.11 | 2.10 | 0.15 |
| chromium | 1323 | 2.54 | 1.09 | 2.11 | 2.23 | 2.06 | 2.10 | 0.15 |
| | 1223 | 2.17 | 1.95 | 2.02 | 2.08 | 2.13 | 2.10 | 0.15 |
| (2¼%Cr1%Mo0.8%Sn) | 1423 | 3.81 | 0 | 0 | 0 | 0 | 0 | 0.25 |
| tin | 1323 | 1.30 | 0 | 0 | 0 | 0 | 0 | 0.25 |
| | 1223 | 0 | 0 | 0 | 0 | 0 | 0 | 0.25 |
| (Type 316) | 1423 | 21.2 | 19.2 | 18.9 | 17.6 | 17.6 | 18.0 | 0.3 |
| chromium | 1323 | 19.3 | 18.5 | 18.3 | 17.8 | 17.7 | 18.0 | 0.3 |
| | 1223 | 19.2 | 18.3 | 18.0 | 18.0 | 17.8 | 18.0 | 0.3 |
| (Weldment) tin | | 0.95 | 0.9 | 0.8 | 0.5 | 0.2 | 0 | 0.2 |

** Derived from the LINK RTS2 FLS statistic. * k-factor adjusted in RTS2 FLS program to give area analysis equal to the bulk chemical analysis; Sn k-factor derived from an $Fe_3Sn$ standard.

profiles for these binding energies and higher quench temperatures are given in Figure 1(a) to (c) for the first three steels considered in Table 2. In the case of the weldment the measurements reveal a pronounced segregation of tin to the prior austenite grain boundaries of the HAZ and this is one of the impurity elements known to promote the formation of grain boundary cavities which lead to reheat cracking. The profile shown in Figure 1(d) was evaluated using a simple thermal transient computed from the assumed welding parameters and shows a good corrrelation with the measured profile, Table 2. Unfortunately there remain limitations in predicting the extent and magnitude of these segregations in weldments due to uncertainties associated with both the welding thermal cycle and the vacancy-tin binding energy.

## 4. Conclusions

Non-equilibrium segregations to grain boundaries have been measured using STEM-EDS X-ray microanalysis of foils from a range of steels including a service component weldment. These profiles agree with predictions of a theoretical analysis based upon a vacancy-solute atom flux.

## 6. References


Doig P and Flewitt P E J 1981 Acta Metall. 29 1831.

Doig P and Flewitt P E J 1983 Microsc. Spectrosc. Electron. 8 193.

Doig P and Flewitt P E J 1985 Solute-Defect Interaction: Theory and Experiment, ed. S.Saimoto, Queens Univ. Canada (in press).

Doig P, Lonsdale D and Flewitt P E J 1981 Met. Trans. 12A, 1277.

Westbrook J H 1964 Metall. Rev. 9 415.

Williams T M, Stoneham A M and Harries D R 1976 Metal Sci. 10 14.


Acknowledgement: This paper is published with the permission of the Executive Director, CEGB, South Eastern Region.

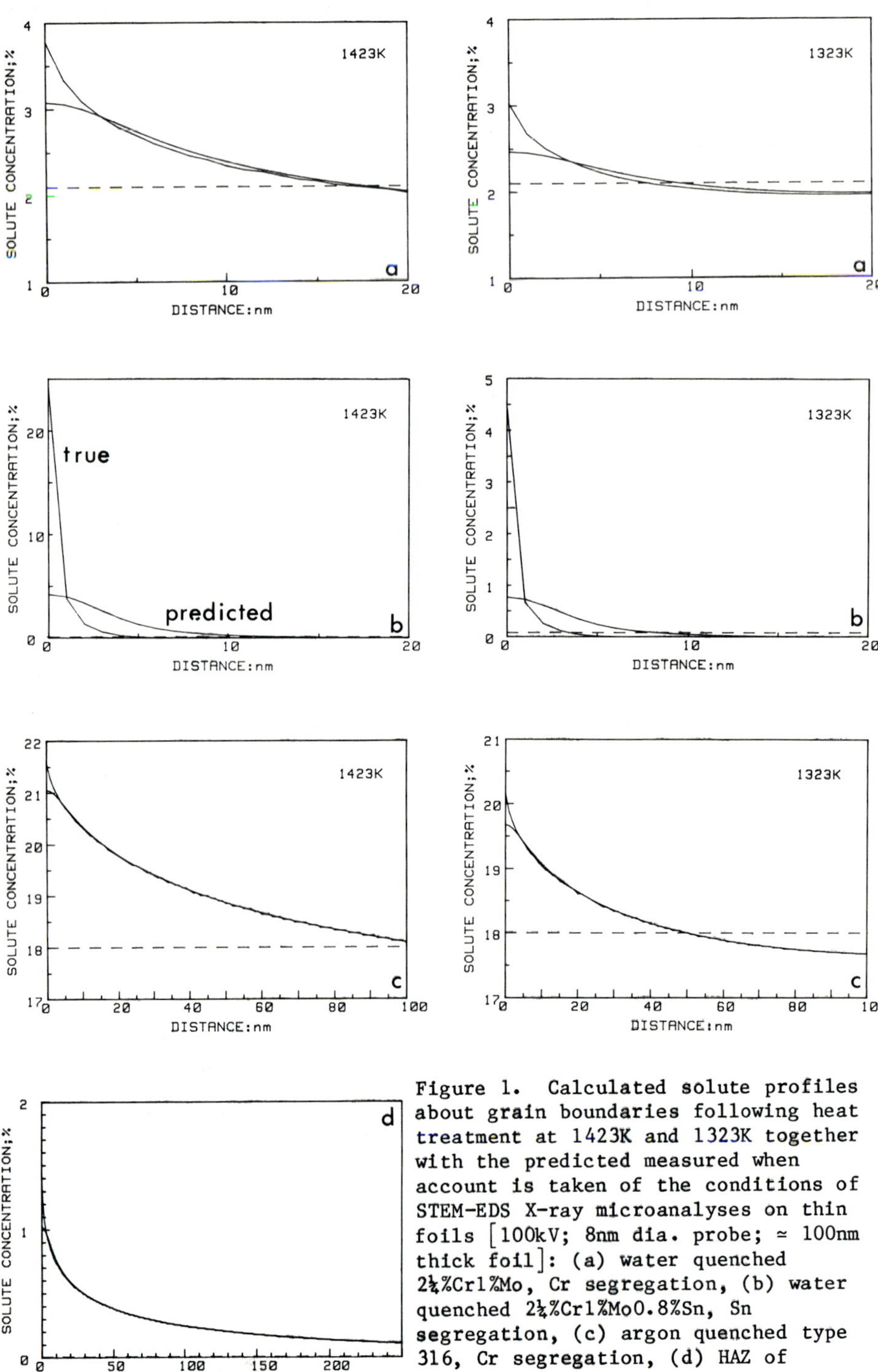


Figure 1. Calculated solute profiles about grain boundaries following heat treatment at 1423K and 1323K together with the predicted measured when account is taken of the conditions of STEM-EDS X-ray microanalyses on thin foils [100kV; 8nm dia. probe; ≃ 100nm thick foil]: (a) water quenched 2¼%Cr1%Mo, Cr segregation, (b) water quenched 2¼%Cr1%Mo0.8%Sn, Sn segregation, (c) argon quenched type 316, Cr segregation, (d) HAZ of weldment, Sn segregation.

Paper presented at EMAG '85, Newcastle upon Tyne, 2–5 September 1985

# The influence of specimen preparation technique on an analytical electron microscopy study of partitioning in a 1.3 wt% chromium eutectoid steel

G Cliff, A Z Mohamed, N Ridley and G W Lorimer
Department of Metallurgy and Materials Science, University of Manchester/UMIST, Grosvenor Street, Manchester M1 7HS, U.K.

## 1. Introduction

Analytical electron microscopy studies of thin foils and extraction replicas have shown that during the austenite/pearlite transformation in eutectoid steels containing chromium there is preferential partitioning of the alloying element to the cementite ($Fe_3C$) phase at the reaction front. Practical problems which must be overcome to obtain reliable partitioning data include limitations in spatial resolution associated with the fineness of the lamellar pearlite, absorption and fluorescence effects within the specimen, and the influence of solute-rich surface films. (Ridley and Lorimer 1983; Lorimer et al 1977; Nockholds et al 1979). In the present work the influence of specimen preparation technique on partitioning measurements obtained for pearlite formed in a 1.3wt%Cr eutectoid steel has been investigated using single and two stage extraction replicas taken from specimens etched under various conditions, and also electropolished thin foils subjected to ion-beam cleaning for various times (Figures 1 and 2).

## 2. Experimental

The studies were made on a high purity alloy of composition: Fe-0.8wt%C-1.38wt%Cr. Specimens of thickness 1mm were austenitised for 30 minutes at 1050°C in a pure argon atmosphere and then partially transformed to pearlite (about 30% vol.) by isothermal transformation in a lead bath for two minutes at 720°C, prior to water quenching.

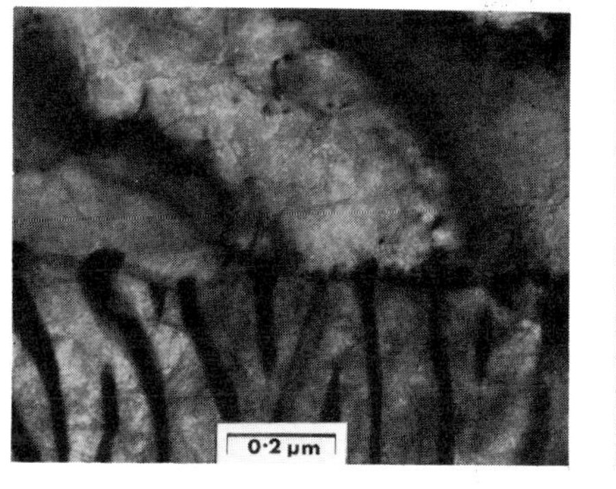


(a)

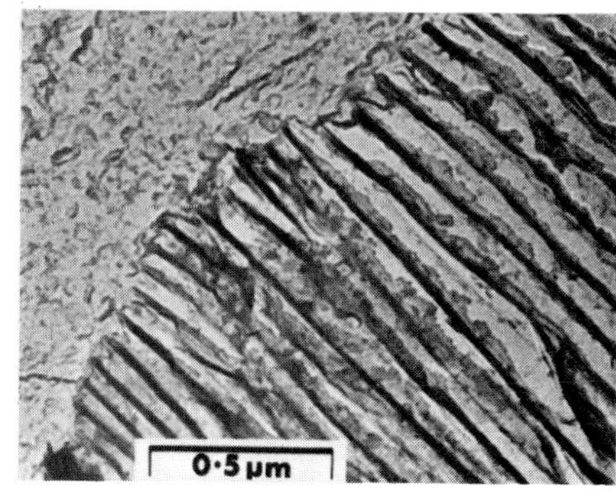


(b)

Figure 1. Lamellar pearlite and the reaction front as seen in (a) a thin foil (b) a single-stage replica.

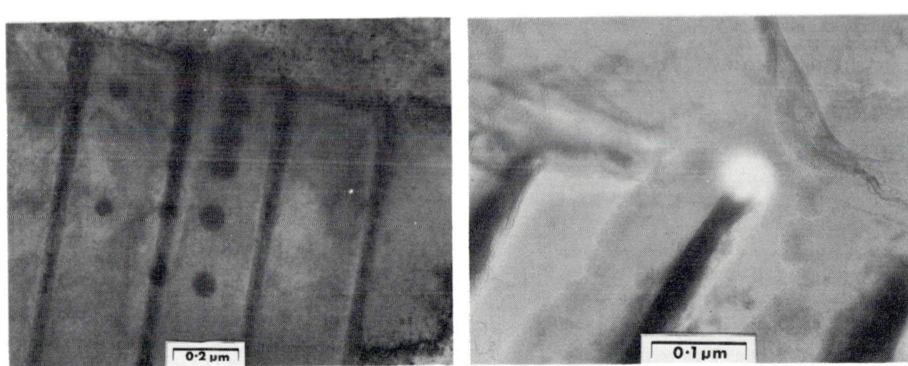


(a) (b)

Figure 2 (a) Electron beam contamination spots located in the cementite and ferrite constituents of a thin foil. (b) Position of the analysis probe in cementite in a single-stage replica (double exposure to show the probe as a white spot).

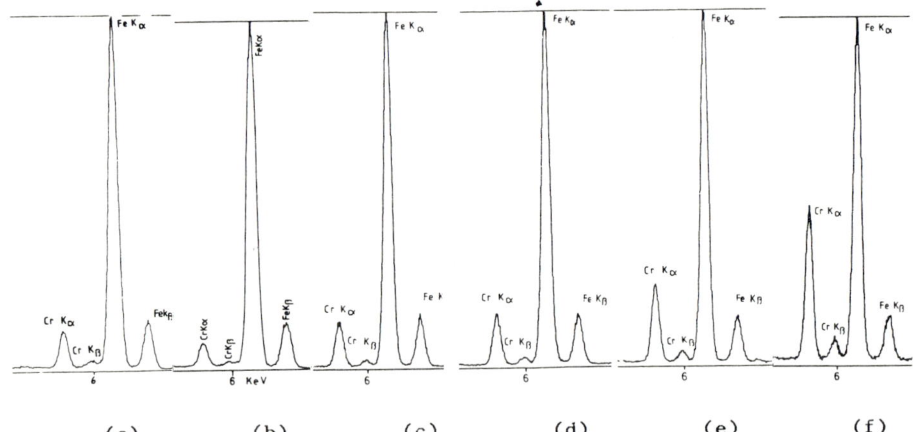


(a) (b) (c) (d) (e) (f)

Figure 3 X-ray spectra from cementite in

(a) thin foil before cleaning (7.68 wt%Cr) showing effect of surface enrichment of Cr and beam spreading into ferrite.

(b) thin foil after ion-beam cleaning for 20 mins. (5.45wt%Cr) showing a low Cr concentration due to beam spreading into ferrite.

(c) single-stage replica; etched 4 mins. in 2% nital (8.48 wt%Cr).

(d) two-stage replica (8.33 wt%Cr)

(e) single-stage replica etched in 10% nital; left for 3 days (17.3 wt%Cr).

(f) single-stage replica etched in electropolishing solution; left for 3 days (33.4 wt%Cr).

Discs for TEM were prepared and polished in a solution of 8% perchloric acid, 38% 2-butoxyethanol, 54% ethanol at 60V and -20°C using a twin jet polisher, washed in ethanol and air dried. The thin foils obtained were examined in the electropolished condition, and after ion-beam cleaning in a vacuum of $10^{-5}$ torr, or in an ultra high vacuum of $10^{-9}$ torr. Single-stage carbon replicas were prepared from metallographically polished specimens which were etched for various times in 2% nital, 10% nital or in the above electropolishing solution, while two-stage carbon replicas were also obtained from specimens etched in 2% nital.

All the specimens were examined at 120kV on a Philips EM400T analytical electron microscope equipped with an $LaB_6$ filament.

## 3. Discussion

The single-stage and two-stage replica results, Figure 3c and d, respectively, were in very good agreement and identical with analyses obtained from lamellae of cementite protruding from the edge of ion-beam cleaned thin foil specimens (8.2wt%Cr). Not only were these results mutually consistent but the experimental data showed a low degree of scatter. However, if the replicas were washed in their etching solutions for too long a period of time then leaching led to the removal of iron and to a considerable over-estimation of the Cr concentration, Figure 3d and e. In the case of thin foils two problems were evident: surface enrichment of chromium and the spread of the analysis probe from the cementite into the adjacent ferrite Figure 3a. The magnitude of the first problem was reduced by ion-beam cleaning Figure 3b. The second problem can be further complicated by astigmatism introduced into the electron probe by the magnetic specimen.

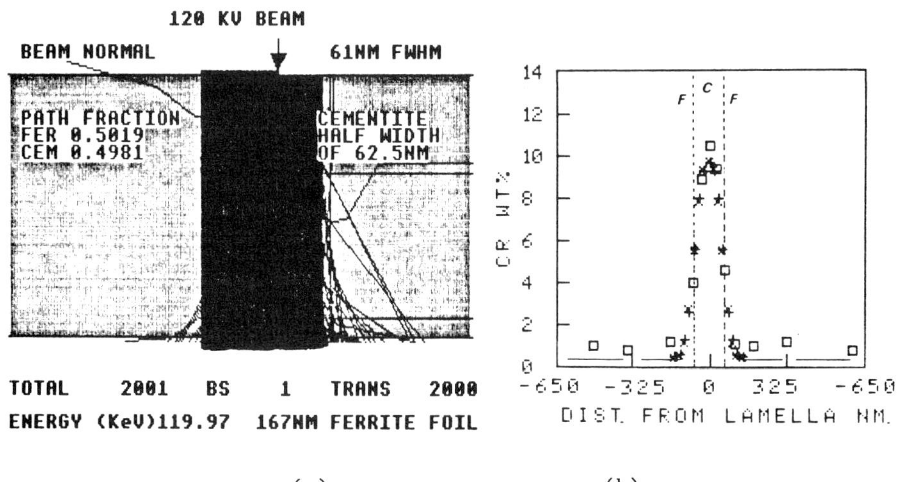


(a) (b)

Figure 4 (a) Computer model of the interaction of the electron beam with the specimen. (b) Computed results (+ and x) of the analyses shown in figure 4a compared with the measured data (□).

A possible way to evaluate the analysis of a thin foil specimen is to model the beam/specimen interactions using Monte Carlo procedures and to extract the true composition, or solute profile, from the measured data iteratively, by an appropriate deconvolution procedure. Figure 4a is a model of a sample 167nm thick containing a cementite lamella 61nm wide. The results obtained using the Monte Carlo modelling (represented by the symbols + and x) are compared with the measured data (open boxes) in Figure 4b; the two results agree well.

The geometry of the specimen shown in Figure 4 is fairly simple. In many two-phase microstructures complex geometries may be produced during the preparation of the thin foil. It is necessary to know this geometry in every detail, complemented by accurate modelling, if meaningful deconvolution is to be carried out.

## 4. References

1 Lorimer G.W. Al-Salman S.A. and Cliff G. 1977, Developments in Electron Microscopy and Analysis ed. D.C. Misell (London: Institute of Physics), pp. 369-372.

2 Nockholds C. Nasir M.J. Cliff G. and Lorimer G.W. 1979, Developments in Electron Microscopy and Analysis ed. T. Mulvey (London: Institute of Physics) pp.417-420.

3 Ridley N. and Lorimer G.W. 1981, Quantitative Microanalysis with High Spatial Resolution ed. G.W. Lorimer, M. Jacobs and P. Doig (London: Metals Society) pp.80-84.

*Inst. Phys. Conf. Ser. No 78: Chapter 8*
*Paper presented at EMAG '85, Newcastle upon Tyne, 2–5 September 1985* 

# AEM quantification of phase compositions in SiAlON-YAG ceramics

H. Hohnke[1], L. F. Allard[2], M. E. Mochel[3], and T. Y. Tien[1]

1. Dept. of Materials Engr, Univ. of Michigan, Ann Arbor, MI 48109
2. American Cyanamid Co., Stamford, CT 06904
3. Materials Research Lab, Univ. of Illinois, Urbana, IL 61801 USA

Silicon nitride ceramics based on SIALON-YAG compositions have been shown to crystallize garnet (nominally $Y_3Al_5O_{12}$) from a glassy matrix phase upon heat-treatment of specimens hot-pressed with $Y_2O_3$ as a sintering aid (Lewis et al., 1980; Hohnke et al., 1981). It is of interest to accurately determine phase compositions in such materials in order to fully understand crystallization behavior and the development of microstructure. Upon initial inspection, these ceramics pose particular difficulties for quantitative AEM microanalysis. These difficulties are due in part to the extremely fine grain and intergrain dimensions, and to the potential for spectral artefacts to be generated by secondary fluorescence effects. The latter problem has been demonstrated in Ni-Ti analyses, for example (Angelini and Bentley, 1984), and was expected to be critical in the characterization of the garnet phase since both BSE's and yttrium x-rays can excite Si which is present in large concentration in bulk regions of ion-milled samples. The present report describes details of the quantification scheme developed for analysis of these specimens, and the experimental tests used to accurately characterize secondary fluorescence effects. The analysis techniques are applied to quantification of phases in hot-pressed and annealed specimens containing nominally 5 eq.% Al in the $\beta$-SIALON phase and 20 w/o garnet in annealed specimens.

## Experimental

Specimens were analyzed in detail on three different instruments in different laboratories. The results obtained on a Philips 420T AEM with a Tracor-Northern 5500 EDS system are presented here. They are illustrative of the general trend of the data observed on all three instruments.

Spectra were acquired on STEM POINT mode, using a $LaB_6$ emitter, 20 microamps beam current, less than 10 nm diameter beam, with the specimen tilted 20 degrees towards the detector to give a 40 degree take-off angle. The specimen holder was designed to minimize systems peaks from material surrounding the specimen (Ryan et al., 1984). A thick Pt 70 micron CL aperture was used for all analyses.

## Results and Discussion

The hot-pressed specimens showed $\beta$ -SIALON crystals with hexagonal cross-sections 0.05 to 0.5 micron in diameter retained within an amorphous matrix phase. After 20 hours at 1350$^\circ$C, the outstanding characteristic of the microstructure was the development of a large-scale network of garnet which formed as a continuous single crystal within the matrix (a phenomenon first described by Lewis, 1980). Figure 1 shows bright and dark-field TEM images of this structure, with the dark-field image recorded using the garnet reflection shown in Figure 1c. An EDS spectrum from this phase, recorded to ensure that no direct overlap with an adjacent $\beta$ grain was likely, is shown in Figure 2a. A minor Si peak is seen in this spectrum, compared to a spectrum of stoichiometric YAG shown in Figure 2b. A model specimen was devised to determine whether this Si might result from secondary fluorescence effects. This specimen consisted of an ion-milled slice of $SiO_2$, having a holey carbon film supported over the hole in the foil, with crushed grains of pure YAG dispersed over the holey film. Hole count spectra were inconsequential. Hole counts contributed no detectable Si peak when added to a pure YAG spectrum having relatively few counts when compared to typical sample spectra. Spectra were acquired with the beam on YAG grains very near the foil edge, and in every instance exhibited no detectable Si peak (using TN's MTF program for data reduction). Thus it was concluded that in the Philips instrument under the analysis conditions used, the minor Si peak observed in spectra from the garnet phase represented Si actually present in the garnet phase in annealed specimens.

Quantitative results were obtained using the Cliff-Lorimer ratio technique (Cliff and Lorimer, 1975), with $k_{YAl}$ and $k_{SiAl}$ factors determined by obtaining 10 400-second analyses on crushed grains both from a pure YAG specimen and from a stoichiometric feldspar (Adularia, $KAlSi_3O_8$), well-characterized by electron microprobe techniques. Both k factors were applied to quantification of their standard materials, and the SiAl k factor was tested against stoichiometric Spodumene ($LiAlSi_2O_6$). Results of these analyses are shown in Table I to illustrate the relative precision of the determination of these factors. The yttrium L lines were chosen for quantification first because mass absorption coefficients for both Y L and Al K in garnet are nearly equal, resulting in no substantial variation in Y/Al intensity ratios until crystal thicknesses resulting in detector dead-times of 50% are reached. Second, the L lines are more intense than K lines in thin foils, and counting statistics are improved. The coefficient of variation ($\sigma/\bar{x}$) in $Y_K$/Al intensity ratios was more than three times as great as the coefficient of variation in $Y_L$/Al intensity ratios (4.3% to 1.3%). Since the thin film criterion is satisfied for Al and Si in the $\beta$-phase at thicknesses of 200 nm or so, it is clear that absorption was not a factor in the analysis of these materials.

Quantitative analyses for the $\beta$ and matrix phases present in each of the three specimens are also shown in Table I. Figure 3 shows EDS spectra typical of the matrix phases in the hot-pressed and 10 min. annealed specimen (compare to Figure 2a). The matrix in the 10 min. annealed sample was largely crystallized into a single crystal network similar to that seen after a long anneal. However two matrix compositions were observed: some garnet was observed, and a single-crystal phase determined quantitatively to be consistent with a $Y_2AlSiO_5N$ composition (perhaps a metastable precursor to garnet) was seen. The Al concentration in the $\beta$-phase was also shown to decrease upon annealing. These quantitative results differed somewhat from those obtained in the study by Lewis (1980), but the trends were similar. It is probable that Si diffuses from a metastable crystalline matrix into the $\beta$ grains with a concurrent diffusion of Al from $\beta$ into the matrix, until a nearly pure YAG phase results. This accounts for the observed decrease in Al in $\beta$ grains after a long-time anneal. Since no divalent cations were observed in the garnet phase to balance charge, it is likely that Y and/or Al vacancies or N substitution for O account for charge balance. UTW-EDS and EELS analyses have thus far been inconclusive in answering this question.

## References

Angelini P and Bentley J 1984 Proceedings EMSA p582

Cliff G and Lorimer G W 1975 J. Micros. 103 p203

Hohnke H and Tien T Y 1981 Progress in Nitrogen Ceramics ed F L Riley (Martinus Nijhoff) pp 111-120

Lewis M, Bhatti A R, Lumby R J and North B 1980 J. Mat. Sci. 15 pp 103-113

Ryan K G, Flower N E and Presland M R 1984 J. Micros. 134(3) pp 281-289

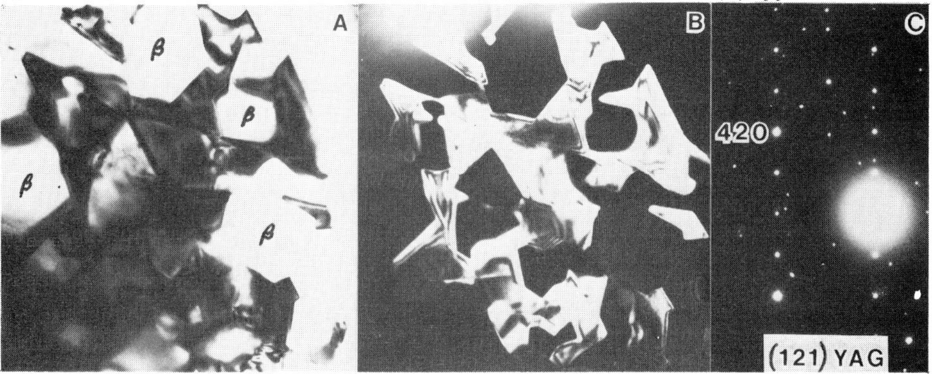


Figure 1. Bright-field (A) and (420)YAG dark-field (B) images showing extensive single-crystal garnet network in 20 hour annealed specimens.

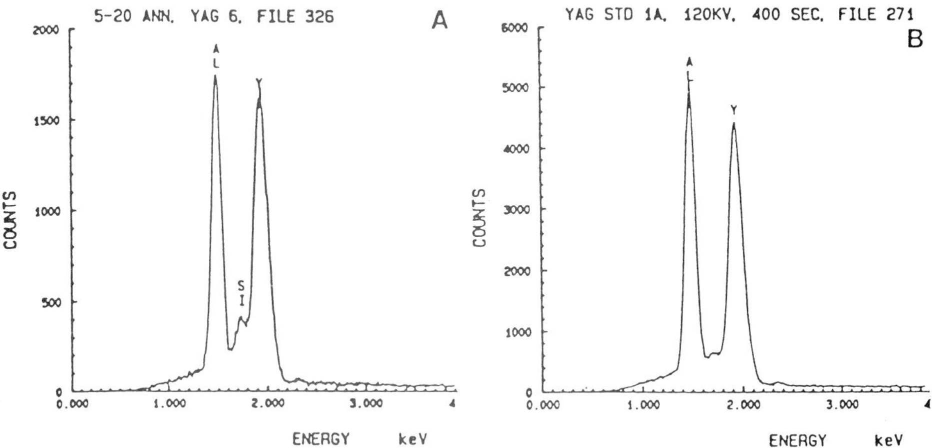


Figure 2. Typical EDS spectra from (A) matrix phase in 20 hour annealed specimens and (B) pure YAG standard material.

TABLE I COMPOSITION SUMMARY

| SAMPLE | PHASE | $C_{Al}$ X | $C_{Al}$ .95CL* | $C_{Si}$ X | $C_{Si}$ .95CL | $C_Y$ X | $C_Y$ .95CL | n |
|---|---|---|---|---|---|---|---|---|
| 5-20 HP | β | 9.71 | 2.14 | 90.28 | 2.14 | *** | | 14 |
| | GLASS | 16.22 | 0.26 | 16.92 | 0.58 | 66.85 | 0.66 | 13 |
| 5-20 ANN 10 MIN. | β | 11.07 | 2.63 | 88.92 | 2.63 | *** | | 16 |
| | MATRIX | 12.03 | 0.47 | 11.67 | 0.41 | 76.30 | 0.59 | 17 |
| 5-20ANN 20 HR | β | 7.35 | 0.97 | 92.64 | 0.97 | *** | | 25 |
| | GARNET | 32.50 | 0.45 | 1.15 | 0.41 | 66.38 | 0.29 | 19 |
| YAG STD (YAG CALC) | | 33.59 (33.60) | 0.25 | *** | | 66.40 (66.40) | 0.25 | 10 |
| ADULARIA STD (ADULARIA CALC) | | 24.26 (24.25) | 0.18 | 75.73 (75.75) | 0.18 | *** | | 8 |
| SPODUMENE STD (SPODUMENE CALC) | | 32.65 (32.45) | 0.96 | 67.35 (67.55) | 0.95 | *** | | 10 |

* 95% CONFIDENCE LIMIT

(CONCENTRATIONS EXCLUDE OXYGEN AND NITROGEN)

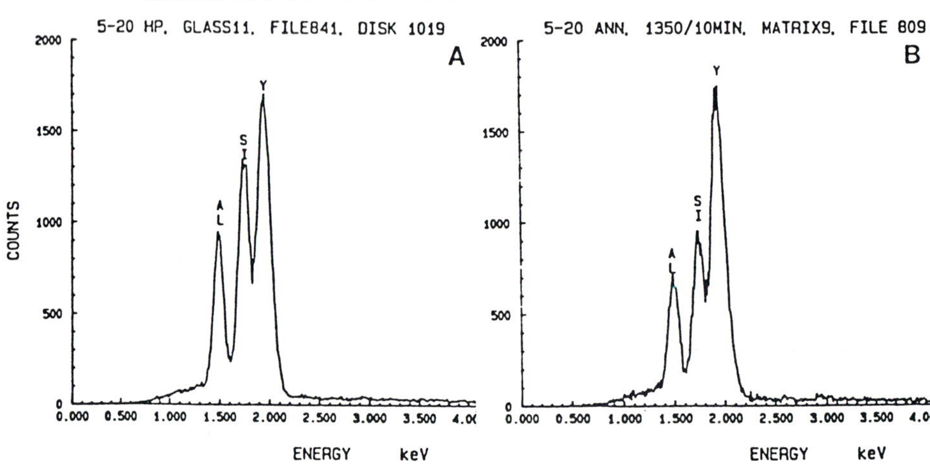


Figure 3. Typical EDS spectra from (A) the glass phase in hot-pressed specimens, and (B) the matrix phase in the 10 minute annealed specimens.

*Inst. Phys. Conf. Ser. No 78: Chapter 8*
*Paper presented at EMAG '85, Newcastle upon Tyne, 2–5 September 1985*


# The role of sodium desorption from silicate glasses during x-ray microanalysis

D G Howitt, W A P Nicholson, W P Jones and B E Launder

Department of Mechanical Engineering, University of California, Davis California 95616
Department of Natural Philosophy, The University, Glasgow
Department of Chemical Engineering, Imperial College of Science and Technology, London
Department of Mechanical Engineering, University of Manchester Institute of Science and Technology, Manchester

The depletion of sodium and other elements from a characterizing electron probe is often observed in analytical electron microscopy and is particularly prevalent in ceramics and glasses. The process is undoubtedly due to some form of radiation enhanced diffusion or desorption although the particulars of the driving forces which give rise to it and the mechanisms by which it occurs are not understood.

A determination of how the diffusion and desorption processes are rate controlling can be achieved by quantifying the loss of the appropriate atomic species as a function of time. The accumulation of the x-ray counts for sodium in a soda borosilicate glass is shown in figure 1. The sample was evaluated as a thin section in the Vacuum Generators HB5 STEM at a beam current of about 5nA and a probe size of about 200Å diameter.

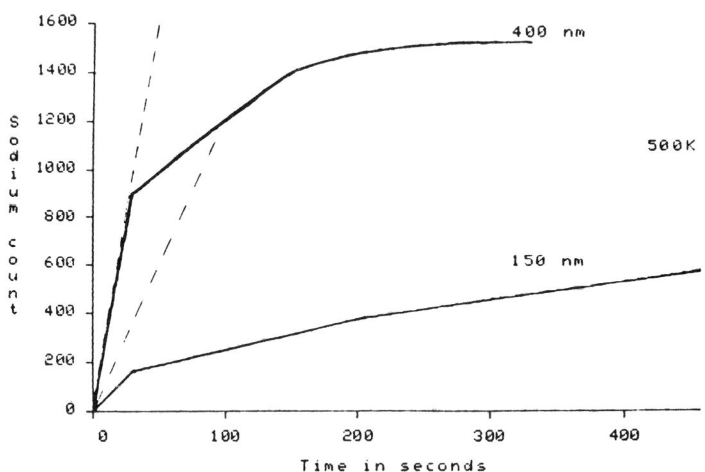


*Figure 1. The accumulation of sodium x-ray counts from a borosilicate glass containing 2% sodium.*

In the case where there is no depletion of sodium from within the probe, the total counts would obviously continue to increase (indicated by the dotted lines in the figure). However, for the case shown there is ultimately a characteristic value of the total signal which is independent of any continued electron irradiation.

The number of x-ray counts that are collected in the microscope is directly proportional to the number of atoms within the volume of the probe (N), which in turn is modified by the efficiency of the x-ray detection system. In the case of the k shell ionization of sodium where absorption and multiple scattering are ignored, the effective cross-section behind the x-ray detector will, at best, be given by

$$\sigma_x = 0.02 \cdot \sigma_k \cdot \omega \cdot \frac{I_{k_\alpha}}{I_{k_\alpha} + I_{k_\beta}}$$

where $\sigma_k$ is the cross-section for x-ray production, $\omega$ is the fluorescent yield, and the bracketed term is the fraction of emission devoted to $k^\alpha$. For these thin foil experiments this translates to an approximate electron dose to obtain one x-ray count of

$$D = \frac{1.6 \times 10^6}{N} \text{ coulombs cm}^{-2}$$

The calculation of the diffusion and desorption behavior of sodium in the glass in response to radiation damage from the probe can be conveniently determined using finite element methods, and this can then be used to predict the way in which the x-ray counts accumulate.

The geometry of a circular probe incident on a thin specimen can be solved in two dimensions, and changes to the local sodium concentration can be realistically induced from the various types of diffusion in the interior elements and the desorption from those at the surfaces. The introduction of atomic sources and sinks is used to simulate the production of defects in the irradiated regions, and the diffusion enhancement is also correlated with the defect concentrations. The variation in the sodium x-ray counts, calculated where the activation energy for defect production is 0.2 eV, is shown in figure 2 by way of example, as are some of the concentration profiles responsible for it in figure 3.

The sodium diffusion coefficient in borosilicate glasses has been determined to be in the range from $10^{-18}$ to $10^{-23}$ $m^2 sec^{-1}$, Frischat (19), and the lower value was used for the calculation in figure 2. A cross-section for the damage process of $10^{-25}$ $m^2$ was also used, which corresponds to a radiolytic yield of 0.005. This is larger than the radiolytic yield for the production of oxygen bubbles in glass under electron irradiation and the damage cross-section for sodium release should not exceed this value. Under these restrictions, regardless of the extent of desorption, it is not possible to account

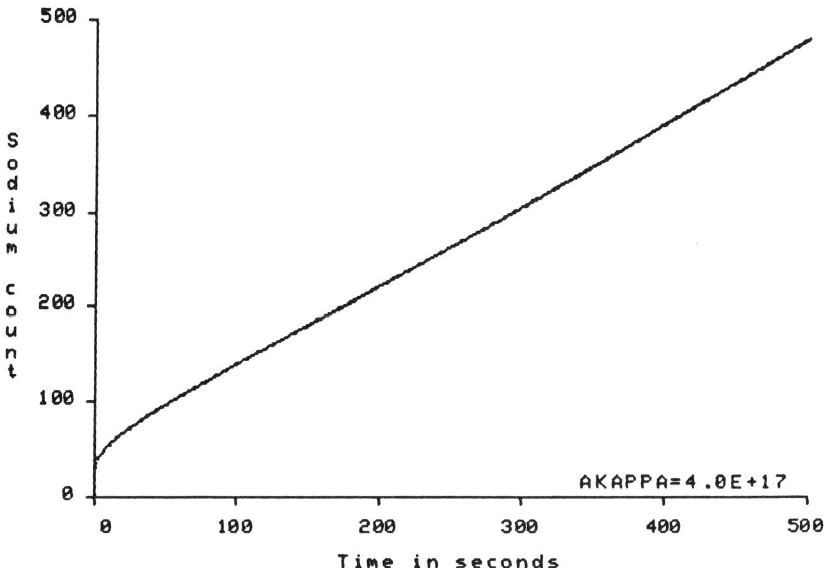


*Figure 2. The accumulation of sodium counts from a 2000Å thin foil calculated for the glass used in the experiments. The calculation assumes that there is enhanced diffusion in the irradiated region and the formation energy for the diffusing species is 0.2 eV.*

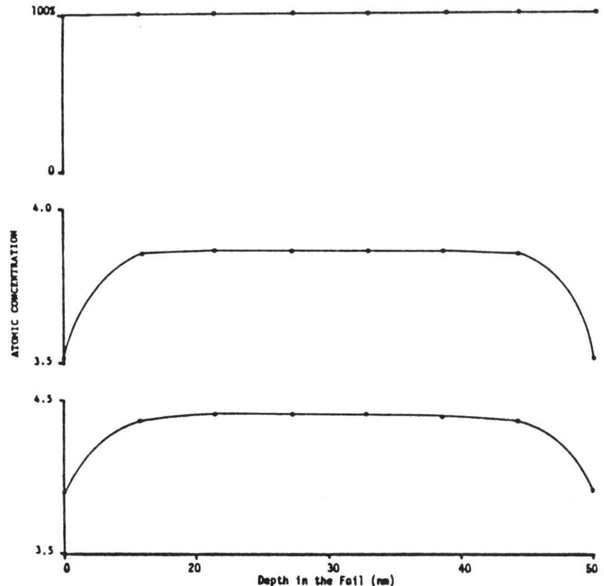


*Figure 3. The change in sodium concentration in the thin foil calculated at intervals of 300 seconds during the irradiation.*

for the rapid loss of sodium because of the slow transport of sodium to the surfaces. In addition, unless the damage cross-section is substantially larger than $10^{-25}$ $m^2$, enhanced diffusion of all the sodium in the material, not just that associated with radiation damage, must occur.

In this case, since there will be no enhanced diffusion outside the electron irradiated volume, it can be concluded that most of the sodium is lost to desorption and that the diffusion coefficient increases substantially within the confines of the electron probe by the time all the sodium is lost.

## References

Howitt, D. G., Analytical Electron Microscopy, Eds. Joy, D.C., Romig, A.R., and Goldstein, J.I., Plenum NY, (1985).

Frischat, G. H., Ionic Diffusion in Oxide Glasses, Trans. Tech. Publications, Ohio, (1975).

## Acknowledgements

The authors would like to thank Barry Dinsley for technical assistance and Professor R. P. Ferrier for supplying laboratory facilities. Support for the work was provided by the Department of Energy Contract No. DE-AT03-79ER10437 and the Science and Engineering Research Council.

# EELS analysis of a metallic glass

R S Payne and P M Budd

Department of Materials Science and Engineering, University of Surrey, Guildford GU2 5XH.

## 1. Introduction

Metallic glasses have an amorphous structure. It has been suggested that this randomly packed atomic structure will make them more resistant to irradiation damage than crystalline metals and alloys. If this proves to be the case, metallic glasses would be useful in fusion reactor environments, where they would be subject to a flux of alpha particles and fast neutrons.

During helium implantation experiments with $Cu_{50}Zr_{50}$, blisters of 1µm to 30µm diameter have been found to occur at high doses. In this paper we report some TEM measurements of the blister lid thickness using EELS.

## 2. Results and Discussion

From the calculated range of 100keV, $He^+$ ions in $Cu_{50}Zr_{50}$, blister lids were expected to be approximately 350nm thick. However high angle SEM surveys of the edges of exfoliated blisters consistently suggested a thickness of 95±10nm. An example can be seen in Fig.1. To examine lids away from their edges, some were removed using softened acetate sheet. The lids were coated with a carbon film and left on a TEM grid by dissolving the acetate sheet in acetone vapour. Fig.2 shows a TEM micrograph of one such lid, still containing helium bubbles.

The thickness of a specimen can be estimated from the low loss region of an EEL spectrum using the equation (Joy, 1979):

$$t = \lambda_p p(1)/p(0)$$

where p(1)/p(0) is the ratio of the intensity of the first plasmon peak to the zero-loss peak. $\lambda_p$ here, is the mean free path for plasmon excitations in our metallic glass system. A value of $\lambda_p$ was obtained from reference specimens of $Cu_{50}Zr_{50}$. These were prepared from the same metallic glass used for the implantation experiments, by electrochemical thinning with a 33% solution of $HNO_3$ in Methanol. These specimens of course contained no gas bubbles, so their thickness could be measured using contamination spots. Fig.3 shows a contamination spot at 45° tilt, with specimen thickness given simply by $2^{-1/2}d$. By following the collection of EEL spectra with contamination spot measurements, we obtained a value of 136±14nm for $\lambda_p$ in our material with a 120keV electron beam. This seems to be a reasonable value, if we consider that aluminium under the same conditions has a plasmon mean free path of 130nm. The data leading to this result is contained in Table 1.

TABLE 1

| Spec. thickness(nm) | 643 | 505 | 605 | 748 | 707 | 578 | 580 | 500 | 590 |
|---|---|---|---|---|---|---|---|---|---|
| $\lambda p$ (nm) | 1436 | 1128 | 1189 | 1470 | 1388 | 1381 | 1491 | 1290 | 1509 |

Returning to the blister lids we found that the value for $\lambda p$ gave thicknesses which were on average 110±30nm as Table 2 shows.

TABLE 2

| Total thickness(nm) | 200 | 100 | 143 | 95 | 108 | 112 | 119 | 160 | 200 | 140 | 165 |
|---|---|---|---|---|---|---|---|---|---|---|---|
| Carbon film (nm) | 50 | 20 | 20 | 20 | 20 | 20 | 20 | 25 | 25 | 55 | 55 |
| Blister lid (nm) | 150 | 80 | 123 | 75 | 88 | 92 | 99 | 135 | 175 | 85 | 110 |

This result is in good agreement with the SEM results. It seems probable that this anomalous lid thickness results from the presence of an oxide layer on specimens prior to irradiation. Implanted helium would tend to migrate to the bottom of such a layer. Blisters would then be produced with a lid thickness corresponding to that of the oxide.

During the accumulation of EELS and windowless x-ray data on $Cu_{50}Zr_{50}$ several points of interest were noted, mainly linked with oxidation. Firstly, an oxide film of considerable (≃10nm) thickness was found to be present on the surface of any specimen prepared for TEM. This was indicated by a decreasing ratio of oxygen to copper atoms, with increasing specimen thickness during quantitative EELS analysis of $O_K$ and $Cu_L$ edges (Maher, 1979 and Williams, 1984), see Fig.4. Confirmation of the existance of an oxide film came from windowless x-ray analysis (Goodhew, 1985) and electron diffraction. Fig.5 shows line scans of diffraction patterns taken from a) a thin area and b) a thicker area. Near the specimen edge the oxide contribution to the diffraction pattern is significant in intensity. Secondly the low loss EELS for $Cu_{50}Zr_{50}$ metallic glass looks consistently different to those of specimens crystallised in an EM hot stage. Fig.6 shows spectra from a) an amorphous sample, b) a crystallised and c) a thin specimen of zirconium. It seems probable that the similarity in shape of 6b and 6c is due to considerable amounts of $ZrO_2$ being present. Lastly we have found that the collection of an EEL spectrum can affect the state of a metallic glass specimen. Fig.7 shows the effect of a condensed beam of 120KeV electrons with 'spot size 2' from a Philips EM400T microscope. The crystallites took about one minute to grow. During the collection of EELS then, atomic mobility is significantly increased by beam heating and displacement damage. We have also found an increase of about 3% per minute in oxide thickness, to occur at EEL analysed areas.

## 3. Conclusions

1) EELS can be used to estimate the thickness of $Cu_{50}Zr_{50}$ metallic glass using the plasmon mean free path of 136±14nm.
2) The blister lid thickness was uniformly less than expected. This was probably caused by the presence of an oxide film on the specimen surface.
3) This metallic glass can suffer a change of state in a condensed electron beam ie. during the collection of EELS or windowless X-ray spectra. It also suffers an increase in thickness of surface oxide.

References

Goodhew P.J. 1985, these proceedings.
Joy D.C., 1979, "Introduction to Analytical Electron Microscopy" Ed. Hren J.J., Goldstein J.I. and Joy D.C. (Plenum) pp 223-44.
Maher D.M., 1979, "Introduction to Analytical Electron Microscopy" Ed. Hren J.J., Goldstein J.I. and Joy D.C. (Plenum) pp 259-94.
Williams D.B., 1984 "Practical Analytical Electron Microscopy in Materials Science". (New Jersey: Philips Electron Optics).

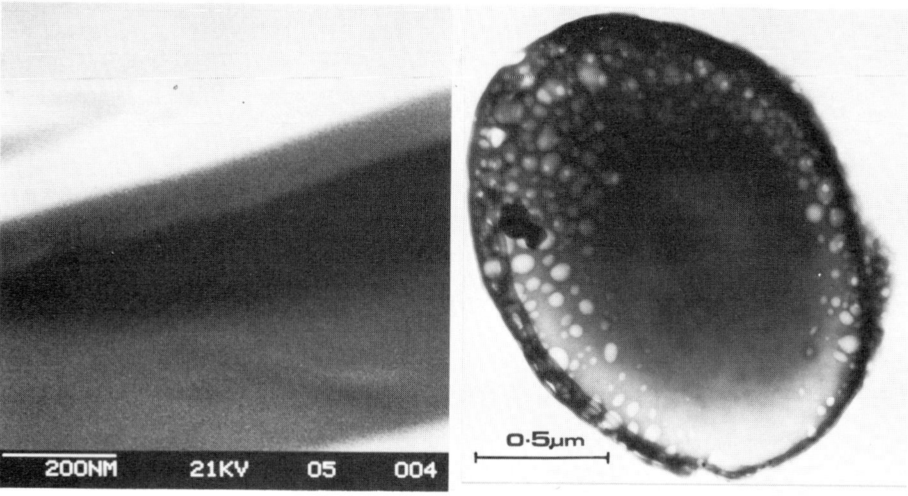


Fig.1.Exfoliated blister lid edge in an SEM at 80° tilt.

Fig.2.Exfoliated blister lid in the TEM; note the lid still holds gas bubbles.

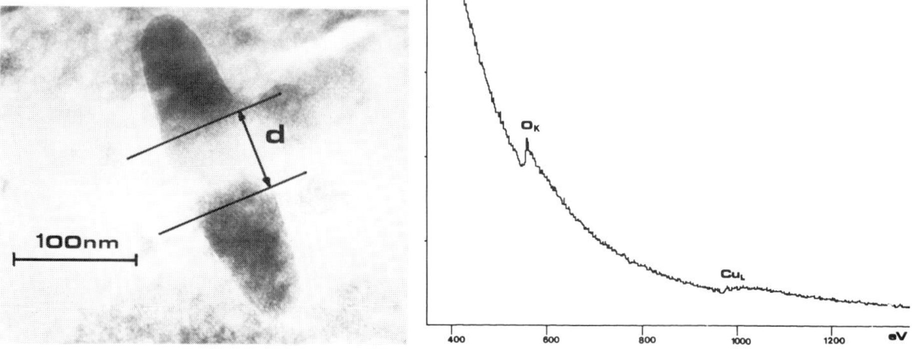


Fig.3.Contamination spot, as used to find the thickness of $Cu_{50}Zr_{50}$ reference specimens.

Fig.4.$O_K$ and $Cu_L$ edges as used to show the fall in oxygen with respect to copper atoms as thickness increased.

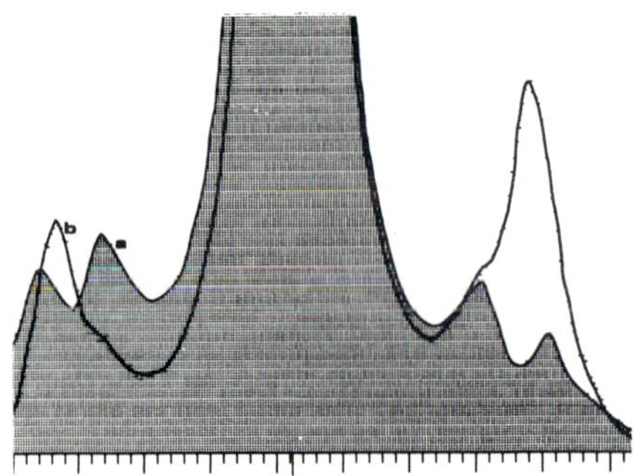


Fig.5.Line scan of amorphous rings from electron diffraction pattern of $Cu_{50}Zr_{50}$:a)thin and b)a thicker area where the $ZrO_2$ contribution becomes less significant.

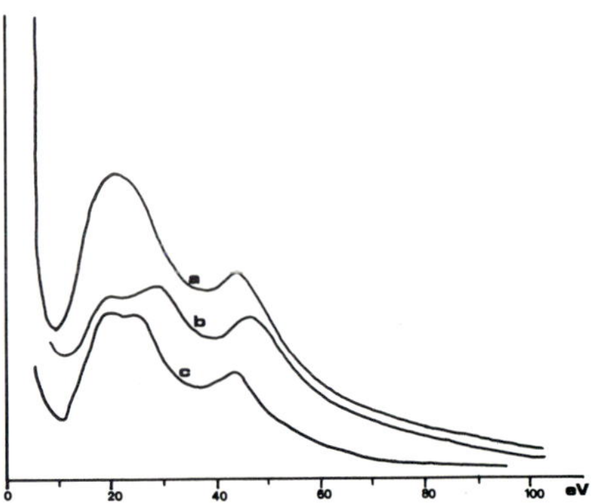


Fig.6.Low loss EELS from a)amorphous $Cu_{50}Zr_{50}$, b)a crystallised specimen and c)a thin specimen of zirconium.

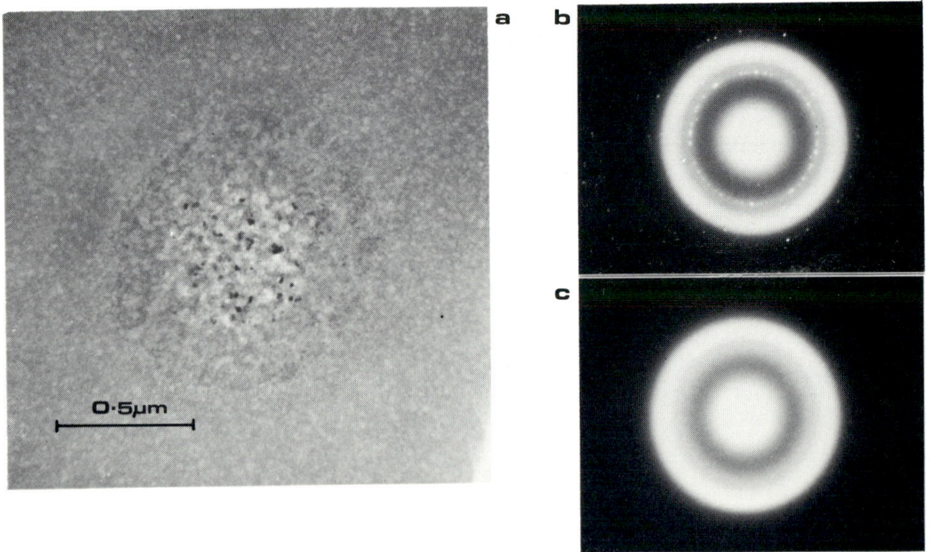


Fig.7.Crystallisation induced by EELS collection:a)a TEM micrograph showing crystallites in the region where the condensed beam was directed, b)electron diffraction pattern from the EELS area, c)electron diffraction pattern taken nearby, showing no crystallisation

*Inst. Phys. Conf. Ser. No 78: Chapter 8*
*Paper presented at EMAG '85, Newcastle upon Tyne, 2–5 September 1985*


# Collection of airborne pollution particles for analytical electron microscopy

L M Brown, P M Dickson, P A Hay-Jahans, J M Macaulay, M J Northcott, H Spooner, H B Timimi and P R Ward
Cavendish Laboratory, Madingley Road, Cambridge. CB3 0HE

## 1. INTRODUCTION

The study of airborne particles and aerosols by electron microscopy has a long history, for which reference should be made to the book 'Particulate Clouds' (1964) and to various papers in the journal 'Atmospheric Environment'. However it seems that since the advent of analytical electron microscopy little work has been carried out. Analytical electron microscopy should permit the convenient identification of a wide range of solid particles down to nanometre sizes, provided the particles are sufficiently damage- resistant.

The problem with electron microscopy is acquisition of the sample. Most current methods are based on drawing a known volume of air through a 'millipore' filter, which is a fibrous membrane composed of cellulose esters, soluble in acetone. It is thus possible to collect the particles and to dissolve away the filter to leave the particles on a Formvar support suitable for electron microscopy.

Our aim in this project has been to develop a less cumbersome method, suited to monitoring the particles by analytical electron microscopy but at the same time compatible with the established methods based on the millipore filter. We therefore designed a simple electrostatic collector which can be used at the same time as the filter and which hardly affects its operation. This paper describes preliminary results based on samples of MgO smokes and heavily polluted air in the Park Street multi-storey car park in Cambridge.

## 2. DESCRIPTION OF COLLECTOR

Fig. 1 shows a version of the collector. An air sampler type L2SF supplied by Rotheroe and Mitchell Ltd. draws air at a controlled rate through a millipore filter. Mounted in front of the filter is a simple pair of wire electrodes held on a close-fitting plastic hood and energised by a Brandenburg model 428B battery-powered HT supply. Fig. 2 shows the wires mounted in position, about 1 mm apart. The corona discharge between the wires and their potential is monitored by a digital voltmeter, and the whole assembly with its rechargable battery is easily portable. For some experiments, air flow less than the lowest of which the pump is capable is achieved by doubling the filters and electrodes to collect two samples in parallel, as shown in fig. 1.

After the sample has been taken, a jig enables the wires to be clamped between two rings, the outer of which is 3 mm diameter to fit into standard cartridges in the electron microscope. Fig. 3 shows the wires as they appear in the SEM at low magnification, where the larger particles can be

* An account of final-year projects undertaken by physics students

analysed. The wires in their mount can then be transferred to the STEM (in this case, an HB5 from Vacuum Generators) for analysis at high spatial resolution.

In addition to providing samples of particles clinging to the wires, the collector provides a sample on the millipore filter, collected at the same time. Although we have not attempted any chemical analysis of the filter paper, the used filters show a dark circle which gives a qualitative indication of the level of pollutant collected during exposure.

It is important to be able to clean the wires effectively. Lens tissues following degreasing has been found to work, although an ultrasonic bath is superior. It is necessary to prepare control wires which are stored and handled in parallel with the sampling wires, but which are not mounted in the pump inlet. The control wires give an indication of the 'zero' of particle density.

## 3. EXPERIMENTS ON MgO SMOKES

In order to assess the operation of the collector, a series of experiments has been carried out on MgO smoke obtained by burning MgO ribbon in a cardboard box into which the collector can be inserted. Often more particles are found on the negative wire than on the positive one, implying that the freshly-formed MgO particles are on average positively charged: see Fig. 4. This result agrees qualitatively with previous observations (1964). It is possible to quantify the observation by measuring the relative amounts of magnesium on the positive and negative wires. For collection conditions with a voltage of between ± 100 V and ± 200 V on the wires and air flows of around 1 litre per minute, the ratio of magnesium on the positive wire to that on the negative wire is between 0.6 and 0.8.

It is usually found from short exposures that the coverage of the wires is uniform, as much MgO adhering to the downstream surface of the wire as to the upstream surface. When the coverage is very high, after long exposures, strings of particles can be found following the field lines between the wires, suggesting that the characteristic strings of particles are formed as a result of the electric fields experienced by them as they are collected, and not by coagulation in the air stream far from the collector.

It is possible to estimate the overall efficiency of the collector by comparing the weight gain of the filter paper with the estimated weight of MgO collected on the wires. In fact, the weight is mostly carried by the larger particles visible in the SEM, so the weight can be estimated without a detailed statistical analysis of the very small particles observable in the CTEM. The result of such experiments for collection conditions as described above is

$$E = \frac{\text{weight of MgO on the wires}}{\text{weight of MgO on filter}} \sim 0.1\% \qquad (1)$$

This collection efficiency is achieved with 50 μm wires across an aperture 1 cm diameter. If the wire presented a cross-section for capture equal to its length multiplied by its diameter, the expected value of the collection efficience E would be 1.3%; from which it is clear that many particles follow streamlines around the wire and fail to make contact with it, or perhaps do not stick to it. However, particles, once they are on the wire, adhere to it fairly strongly. There is no evidence that handling the wire

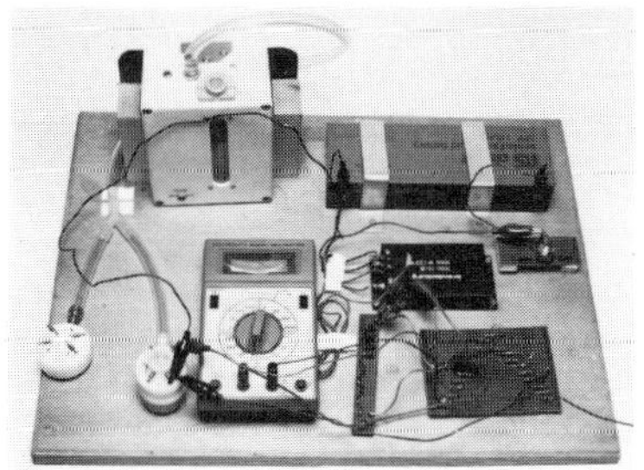

Fig. 1 Portable sampler and power supplies.

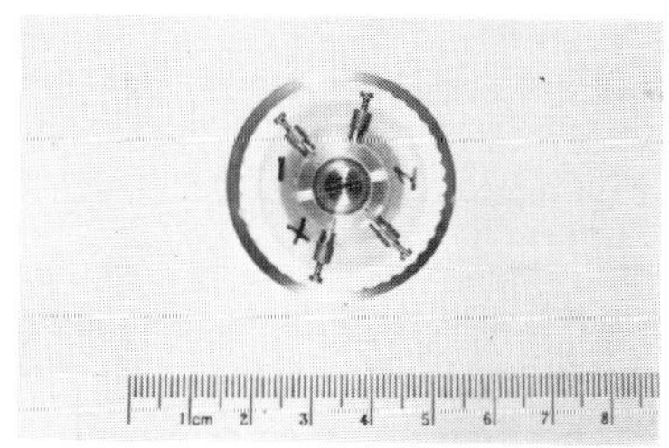


Fig. 2 Mount for wires and filter paper.

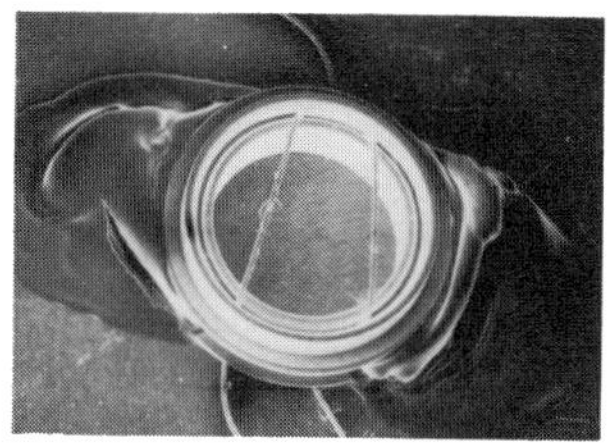

Fig. 3 Wires mounted for microscopic examination (SEM, wires about 1 mm apart).

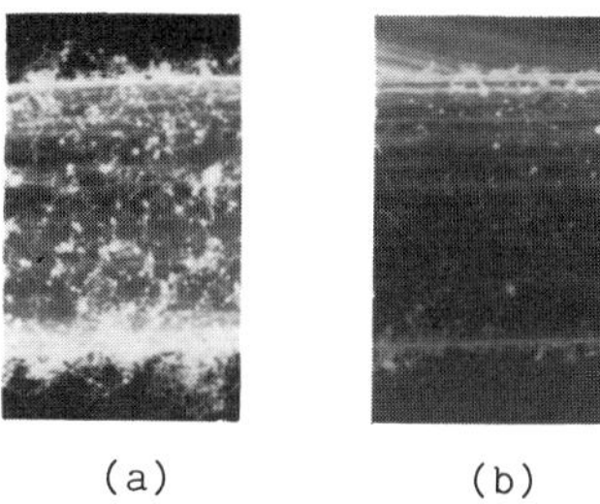


Fig. 4 MgO smoke on negative wire (a) and on positive wire (b). (SEM, wire diameter 50 μm).

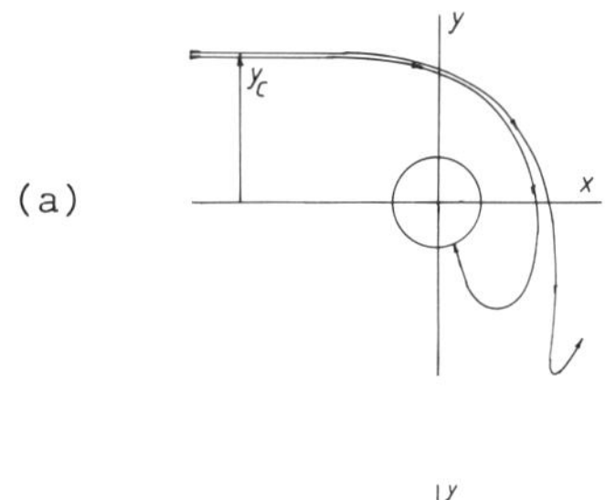


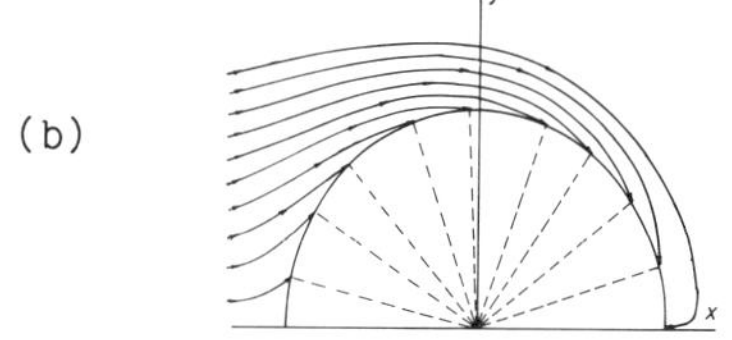


Fig. 5 Theoretical particle trajectories (a) just bracketing a critical trajectory (b) showing uniform deposition for extensive capture.

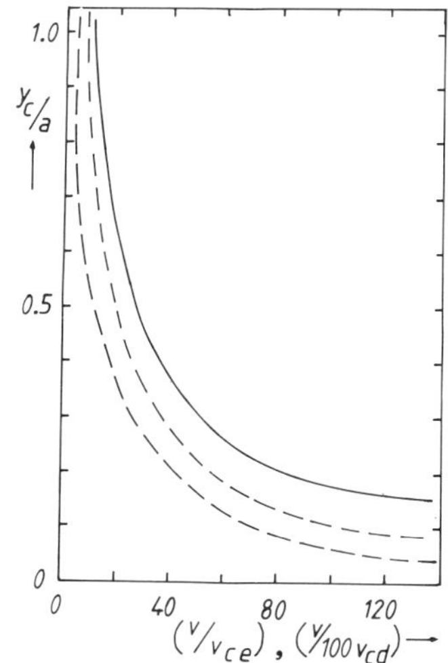


Fig. 6 Theoretical capture cross-section for electrophoretic forces (broken curves indicate scatter band) and dielectrophoretic forces (solid curve).

looses particles. Therefore most of the particles follow streamlines around the wire and are not collected. This view is supported by the fact that practically nothing is collected if no voltage is applied to the wires.

## 4. THEORY OF COLLECTION EFFICIENCY

It is essential to get a simplified understanding of the collection mechanism to relate the distribution of collected particles to the distribution in the air sample. It seems probable that van der Waals' forces act at very short range to produce adhesion between a particle and the wire on contact, but that they do not otherwise affect the particle trajectory. Other assumptions are as follows: the electric field due to the collector wires is essentially that for an infinitely long single wire; the viscosity and relative permittivity of the air are unaffected by the particles; the air flow is laminar; and - most important - only Stokes forces and electrical forces act on a particle because its inertia can be neglected.

It is easy to justify these assumptions, except perhaps the last one. If one imagines a particle of radius $r_o$ following a streamline with velocity v passing a wire of diameter a, it is subject to time-dependent electrical forces with a large Fourier component at a maximum frequency v/a. The inertial forces will be of order (mass of particle) multiplied by the square of the frequency, whereas the Stokes forces will be of order (mobility of particle) multiplied by the frequency. We can thus estimate the ratio of Stokes forces to inertial forces.

$$R = \frac{\text{Stokes force}}{\text{Inertial force}} = \frac{6\pi\eta r_o}{m.(2\pi v/a)} \qquad (2)$$

where $\eta$ is the viscosity of air and m is the mass of the particle. Typical air velocity in these experiments $v \approx 0.2\ \text{ms}^{-1}$. We find that the value of R exceeds unity for masses less than about $10^{-13}$ kg, corresponding to particle sizes of about 2 μm. It thus seems that the largest particles travelling closest to the axis of the wire might be affected by inertia, but that for the greater part of the particle population inertia can be neglected.

With these assumptions, one can write a computer program for the BBC B which plots particle trajectories. We are concerned with two types of electrostatic forces, the electrophoretic force $F_e$, which is radial and is given in terms of the potential of the wire V, the charge on the particle, q, the half-distance between the wires d, and the distance from the centre of a wire r by

$$F_e\ (\text{radial}) = \frac{qV}{2\ \ln(2d/a)\ r} \qquad (3)$$

and the dielectrophoretic force which acts on a neutral particle of permittivity $\varepsilon_r$

$$F_{di}\ (\text{radial}) = \frac{\varepsilon_o(\varepsilon_r - 1)(4\pi r_o^3/3)V^2}{4\{\ln(2d/a)\}^2\ r^3} \qquad (4)$$

In the computer, a particle is released on a streamline whose distance from the streamline passing through the centre of the wire is y. If y is large, capture does not occur, but if y is small enough, capture occurs. The critical value of y for capture $y_c$ is the capture cross-section per unit length for a particular value of air velocity, voltage on wires, particle charge, etc. Fig. 5a shows an example of a critical trajectory, and fig. 5b shows the pattern of trajectories for extensive capture i.e. for conditions in which all values of y lead to capture. From fig. 5b it can be seen that uniformly spaced values of y produce uniform coverage of particles on the wire, as many particles being captured on the downstream side of the wire as on the upstream side, in agreement with experimental observation. The problem now is to find a dimensionless presentation of the results. Because inertia is unlikely to be important, it is easy to calculate the time for capture required for a particle acted upon by Stokes forces and be electrostatic forces alone. The time available for capture is essentially d/v, and equating the two one can calibrate a critical velocity $v_c$ which should normalize the results. We find

for electrophoretic forces, $$v_{ce} = \frac{V\,q}{6\pi\eta r_o d\,\ln(\frac{2d}{a})} \qquad (5)$$

for dielectrophoretic forces, $$v_{cd} = \frac{2\varepsilon_o(\varepsilon_r - 1)r_o^2V^2}{9\eta d^3[\ln(\frac{2d}{a})]^2} \qquad (6)$$

It is now possible to plot $y_c$ normalized to the wire radius as a function of air velocity normalized to $v_c$. Fig. 6 shows the results for electrophoretic and dielectrophoretic forces, and fig. 7 for dielectrophoretic forces. A wide range of voltages, charges, and particle sizes have been studied, and the normalization seems to work adequately although not perfectly. Using figs. 6 & 7 it is possible to estimate the collection efficiency for different types and sizes of particles. As is clear from the figures, low air velocities greatly increase the capture cross-section, which rises to infinity as the velocity approaches zero.

If, as is suggested by Green and Lane (1964), the MgO particles are charged on average by about 10 electronic charges, we find that $v_c$ for 1 μm particles is about 4 cm $s^{-1}$ for electrophoretic fores, and about 20 μm $s^{-1}$ for dielectrophoretic forces. Using figs. 6 & 7, we find that $y_o/a$ is about 0.1 for both forces, from which we can deduce a value of the collection efficiency as defined in equation (1) of about 0.1% - in very close agreement with observation. Because the dielectric forces will attract particles to both wires, whereas the electrical forces will attract particles to the negative wire only, it follows that if all the particles are positively charged we expect to find a ratio of about ½ of them on the positive wire to those on the negative wire. This is again roughly consistent with experimental observations.

Note that for electrophoretic forces, the collected size distribution is biased towards small particles, whereas for dielectrophoretic forces it is biased towards large particles.

## 5. COLLECTION OF CAR PARK AIR

Over a period of two years during January and February, particles were collected and their X-ray spectra taken in the SEM. Additionally, some particles were analysed by EELS, which is effective for small particles in the proximity of the collecting wires. It is difficult to present the data in a useful way; the following general results are based on a total of about 150 particles:

1. No evidence can be found for preferential collection of any component of particle population on either the positive or negative wire: the pollution of the car park appears to be on average electrostatically neutral.
2. On wet days, substantial numbers of NaCl cuboidal particles are collected - up to 15% of the particles.
3. About one-third of the particles are mainly organic in nature and cannot yet be identified.
4. About one-sixth of the particles are lead-containing. These show in addition a chlorine signal, or both a bromine and chlorine signal. This is consistent with the lead salts found in vehicle exhaust by macroscopic methods (Harrison et al 1980). The lead particles are often agglomerated in an organic-looking matrix.
5. The remaining particles are minerals containing Si, Al, Ca, S and Fe in varying amounts. Prominent among these are small rust particles ($Fe_2O_3$) identified by EELS and shown in fig. 8.
6. Using the collection efficiency for dielectrophoretic forces estimated from figs 6 and 7, it is possible to make a very rough estimate of the amount of lead salts in the air of the car park. It is about 0.3 mg $m^{-3}$. This is ten times larger than the average concentration of smoke in urban air, as measured by the Department of the Environment in the mid-seventies (D.O.E. Pollution Report 1978). It is somewhat larger than particulate concentrations quoted for a 'moderately polluted environment' (Harrison et al 1980). However the figure derived here must be confirmed by more detailed observation; it undoubtedly varies with time and position of the collector and we have not yet studied this.

## References

Digest of Environmental Pollution Statistics, D.O.E. Pollution Report No. 4, H.M.S.O. 1978, p. 9.

Green, H.L. and Lane, W.R. 'Particulate Clouds' 2nd Ed., E. & F.N. Spon Ltd., London 1964.

Harrison, R.M. and Laxen, D.P. 'Metals in the Environment', Chemistry in Britain 16 (1980) pp. 316-320.

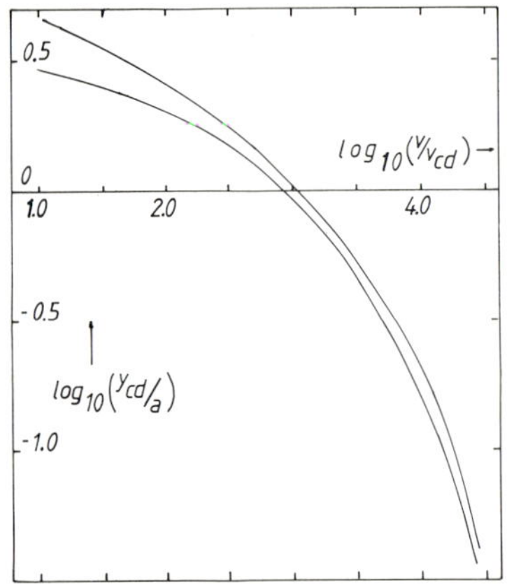


Fig. 7 Theoretical cross-section for dielectrophoretic forces (curves indicate scatter band).

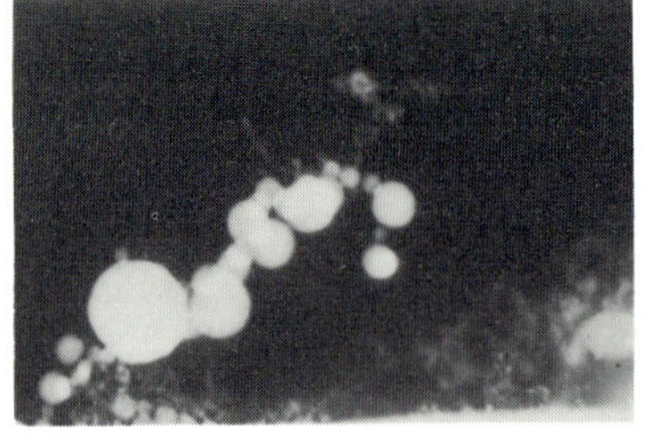

Fig. 8 $Fe_2O_3$ particles (STEM ADF, largest particle about 0.3 μm diameter).

# Correlation between cathodoluminescence and the composition of a $Tb^{3+}$ doped phosphor

J Yuan*, P Caro†, L M Brown*

* Cavendish Laboratory, Madingley Road, Cambridge, CB3 OHE
† Laboratoire des Elements de Transition dans les Solides C.R. No. 210
1 Place A Briand, 92190 Meudon Bellevue, France

## 1. Introduction

Cathodoluminescence (CL) as a means of material characterization has great appeal as it is sensitive to the electronic structure of the solid. However in many cases it is hampered by low quantum efficiency and difficult signal interpretation. Rare earth doped compounds are more promising because of their brightness and their sharply defined optical emission spectrum. Richards and Trigg (1981) and Pennycook et al (1981) have studied rare earth phosphors in a SEM and a STEM respectively. Here we combine the bulk methods of X-ray diffraction and photoluminescence (PL) and the microanalysis techniques of CL and X-ray fluorescence, to study a complicated system. Correlation is obtained between the CL signal, crystal structures, their chemical composition and activator distribution in sub-micron scale. These results demonstrate the potential of the techniques for phosphors and other opto-electronic materials.

## 2. Experimental Techniques

The compound was obtained from the Laboratoire des Elements de Transition dans les Solides. It is in the form of powder with grains about a micron in size. X-ray powder diffraction (fig. 1a) showed two major sets of Bragg peaks. The broad peaks were indexed as having a structure of hexagonal rare earth sesquioxide ( a = 3.84 Å and c = 6.66 Å). The strong narrow ones were identified as that of cubic $Y_2O_3$. There were also some small peaks which are attributed to $La_2O_2S$. The mercury-lamp-excited PL spectrum (fig 1b), obtained at room temperature, shows a line structure that is typical of rare earth ion luminescence.

The sample was examined in a dedicated VG-HB5 STEM equipped with a CL detection system (see Berger et al 1985 for its detailed description). Spectral resolution of 0.5 nm in wavelength can be obtained, with the size of the exit slit of the spectrometer set at 0.25 mm. Quasi-monochromatic digital mapping of CL distribution in the sample with submicron spatial resolution can also be obtained. Slit sizes as large as 1.25 mm were used, provided no neighbouring lines overlapped, to obtain maximum light output. The STEM also provides an energy dispersive X-ray spectrometry (EDS) with facility for digital mapping of individual elemental peaks. Two approaches were adopted. Firstly a finely focused electron beam was stationed on a grain, CL and EDS spectra were collected simultaneously (fig. 2). Secondly monochromatic CL maps of two interesting grains were obtained, together with their elemental EDS mappings (figs. 4 & 5).

## 3. Results and Discussion:

a) The powder contains two major phases in about equal amounts and a third minor phase. The EDS measurement showed that the system contains elements Y, La and S. The major phase with narrow Bragg peaks is identified as cubic $Y_2O_3$. For the other major phase with broad Bragg peaks, the possible stable phases are $Y_2O_2S$, $La_2O_2S$ and $La_2O_3$, all with the same structure. Since La is usually detected together with S, the second phase cannot be an oxide. The unit cell parameters of the phase lie between the values for $Y_2O_2S$ and $La_2O_2S$. Leskela and Niinisto (1978) have shown that La and Y can easily form a solid solution $(La_xY_{1-x})_2O_2S$ with unit cell parameter a continuous function of x. From EDS analysis it is also found that La and Y coexist in most of the grains. Thus we may conclude that the second major phase is $(La_xY_{1-x})_2O_2S$. The mean value of x deduced from the phase diagram (Leskela and Niinisto 1978) is about 0.28. The number of lines due to the third phase is not sufficient for structure identification. These lines are attributed to the $La_2O_2S$ phase because they all match the strong Bragg peaks of that phase. The fact, that in most grains examined all three elements are found to be co-existent, shows that there is more than one phase in a individual grain.

b) The PL spectrum from the powder contains a line structure which can largely be attributed to $Tb^{3+}$ ions in the oxysulfide phase. There are also some unaccountable lines, notably that at 515 nm. The CL spectra from individual grains shows a similar line structure with a complex variation of the different line intensities from one grain to another. In addition there is a broad blue and UV band emission from some of the grains. Assuming that more than one luminescent center is responsible for such a spectrum and the variation in the spectrum is mainly due to the variation in these centers from grain to grain, two lines entirely due to the same center should have the same intensity ratio for all the grains examined. This is indeed so (see fig. 3) for the line transitions at 543, 548, 554, and 615 nm which correspond to the $^5D_4 \rightarrow ^7F_j$ transitions of the $Tb^{3+}$ ions in $Y_2O_2S$ phases. (The first two lines might also be partially due to emission from the $La_2O_2S$ phase. The present instrument is not capable of resolving their differences. Neverthless the distinction can be made at the $^5D_4 \rightarrow ^7F_3$ transition where the emission wavelength for $Y_2O_2S$ and $La_2O_2S$ respectively are 615 and 623 nm. Both lines are observed. Only the line at 615 nm is correlated with other $Tb^{3+}$ lines. This indicates that the center is mostly $Tb^{3+}$ in $Y_2O_2S$). Similarly the line emissions at 515 and 538nm belong to another center. The third one is that of broad band emission.

c) Despite the fact that the grains examined are of different sizes and shapes, some correlation of CL spectra and chemical composition can be found for most grains. Firstly it is clear that the broad band only exists in yttrium rich grains (fig. 2). The evidence from the CL and EDS digimaps (fig. 4) also shows that the region associated with the band emission is deficient in S and La, so the broad band is probably the matrix luminescence of the $Y_2O_3$ phase. Such emission has been observed in slightly La-contaminated $Y_2O_3$ cubic phases. Also there is no trace of the line emission at 537 nm which is characteristic of $Tb^{3+}$ ions in $Y_2O_3$. Since the CL quantum effeciency of $Tb^{3+}$ ion in the $Y_2O_3$ is about half of that in the $Y_2O_2S$ at 1% doping level (Buchanan et al 1968), all these suggest that the $Tb^{3+}$ ions do not enter the oxide phase. Thus the odd CL spectrum (e) in fig. 2 is due to the beam hitting solely the oxide phase.

d) It is also clear that the intensities of the sharp line emissions due to the $Tb^{3+}$ ions in the oxysulfide phase decrease as the La content of the particle increases. The broad nature of the Bragg peaks of that phase suggest that it contains a large number of faults or strains. Electron microscope study of the mixed oxysulfide phase (Leskela and Niinisto 1979) revealed that large strains exist as the material is made of $[YO]_n^{n+}$ and $[LaO]_n^{n+}$ layers of different c-parameters. Thus the quenching of the $Tb^{3+}$ emission in that phase may be associated with the increasing strains in the phase of high La content. Digimaps of grains (fig 4 & 5) show that the $Tb^{3+}$ lines are from the S-rich region, thus comfirming its oxysulfide structure. Looking closely at fig 4, there are two brightest regions for these line emissions: one in the middle and the other in the bottom. Obviously the middle spot is thicker than the bottom one, but it is less luminescent. It is noted that the region in the center has a higher La to Y ratio than the bottom one. Thus a mixture of Y and La in a oxysulfide matrix seems to quench the $Tb^{3+}$ luminescence.

e) The line emissions at 515 and 538 nm seem also to come from the mixed oxysulfide phase. In fig. 5, we look at a grain which shows a relatively large CL contribution from these emissions. Because the interference of the broad band emission with the digimap of CL line emission at 515 nm, we study the digimap obtained from the line emission at 538 nm. Comparision of different CL maps shows that they are distributed in a similar way to that of other $Tb^{3+}$ line emissions. The centre causing them could be due to another rare earth impurity in that phase.

## 4. Conclusions

We have determined the structure of the sample as mainly a mixture of $Y_2O_3$ and $(La_{0.28}Y_{0.72})_2O_2S$. The oxide and the oxysulfide are found to be segregated inside micron-size grains in regions which are a few thousand Angstroms across. The oxide is undoped and shows a broad band emission. The oxysulfide is doped with the $Tb^{3+}$ ions and shows a characteristic line spectrum. High mixtures of La into $Y_2O_2S$ quenches the luminescence. The mechanism for the quenching is believed to be the strain introduced in such crystals. In the oxysulfide phase some anomalous line transitions (at 515 and 538 nm) are also observed. The origin of these lines is still not very clear. Because we do not know the conditions for preparing this powder, it is unclear how such a complex system arises. However, it is obvious that high resolution STEM with a CL output has a role to play in characterising phosphors. Maximum information is obtained from a combination of macro- and micro-scopic investigations.

Acknowledgement We like to thank Mme. O.K. Moune-Minn for performing the PL measuremnt. J. Yuan thanks the SERC for a CASE studentship.

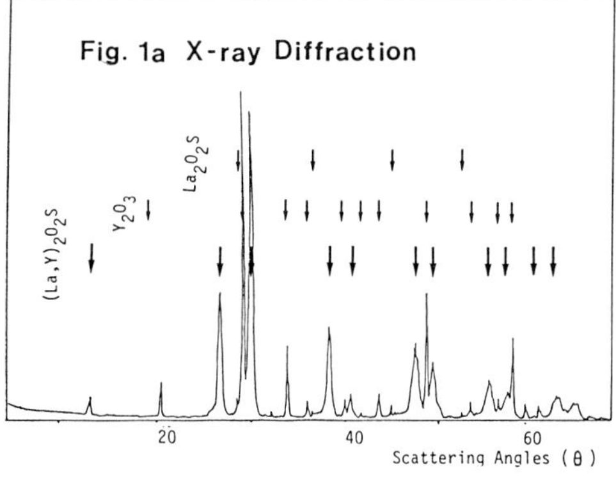


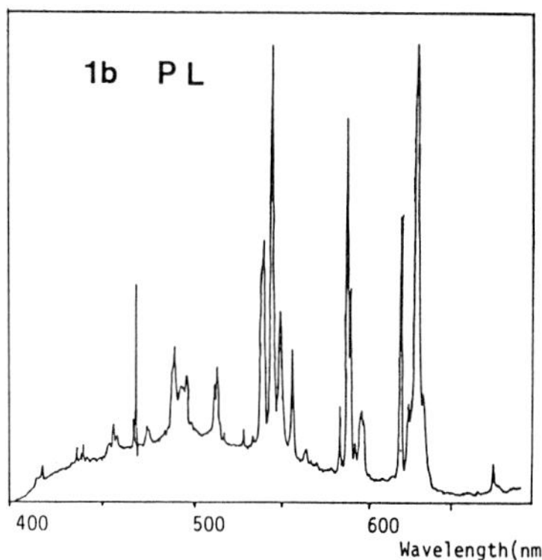


Figure 1 Powder Chracterization By Bulk Technique: X-ray Diffraction (a) and PL (b)

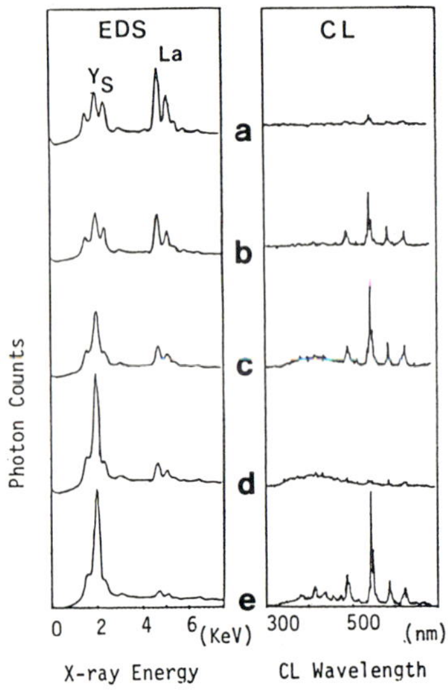


Figure 2 X-ray fluorescence (EDS) and CL spectra from individual grains

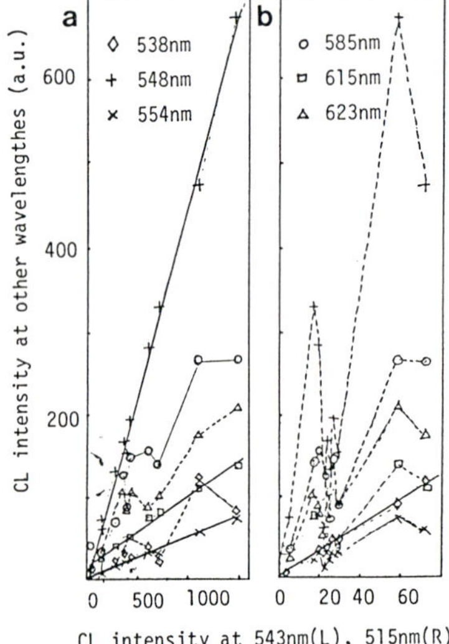


Figure 3 Correlation, in CL spectra of different grains, between lines at (a) 543nm, (b) 515nm and other lines.

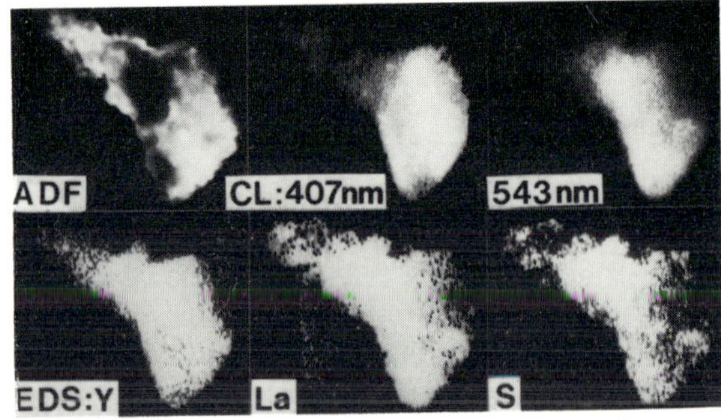


Figure 4 A typical grain containing phase segeration imaged in STEM by annular dark field (ADF), CL at 407nm (broad band), 543nm ($Tb^{3+}$ emission) and EDS at Y, La and S.

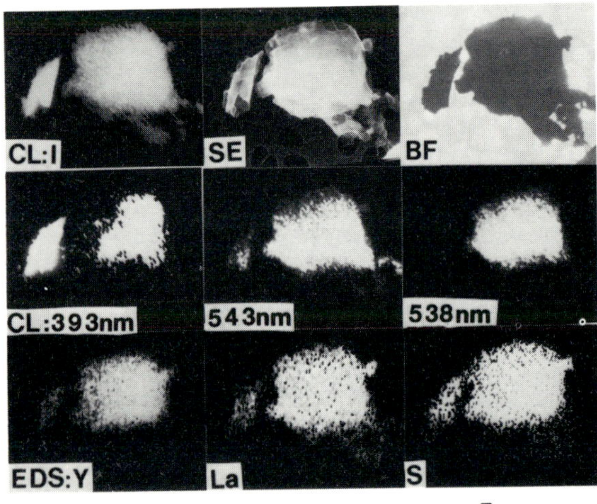


Figure 5 A grain with relatively high emission at 515nm, 538nm imaged in STEM by bright field (BF), secondary electron (SE), CL at integrated output (I), at 393nm (broad bånd), 543nm ($Tb^{3+}$ emission), 538nm (origin unclear), and EDS at Y, La and S.

## References

Buchanan R.A. K.A.Wickersheim J.L.Weaver E.E.Anderson 1968 Appl.Phys. 39,9,4342

Berger S. D.McMullan J.Yuan L.M.Brown 1985 (this conference).

Leskela M., L.Niinisto 1978 J. Solid State Chem. 19,245

Leskela M. L.Niinisto H.Dexpert Y,Charreire 1979 Mat.Res.Bull.14,455

Pennycook S.J. H.Dexpert Y.Charreire 1981 J.Microsc.Spectro.Electro.6,637

Richards B.P., A.D.Trigg 1981 Inst. Phys. Conf. Ser.,61 Ch.5,227

# Atomic resolution electron microscopy in perspective

David J Smith

Centre for Solid State Science and Department of Physics, Arizona State University, Tempe, Arizona 85287, USA.

## 1. Introduction

The possible attainment of atomic resolution in the electron microscope essentially provided the motivation for the construction of various higher voltage electron microscopes during the last decade, mostly in Japan but also in the UK (Cambridge) and the US(Berkeley). Although mechanical and electrical instabilities often limited the ultimate performance of these instruments (Smith et al, 1982), several proved capable of sub-2Å resolution and the microstructures of a wide range of metallurgical, ceramic and inorganic materials were revealed for the first time on the atomic scale. The successes of these studies have attracted world-wide attention, and several commercial manufacturers have recently started production of medium voltage (300-400kV) microscopes in response to the escalating demand for these machines. The results from these atomic resolution electron microscopes have certainly had a dramatic impact and are clearly starting to fulfil the expectations of microscopists. The objective of this review is to examine whether the scientific value of the results has so far justified the considerable expense involved in their purchase. We first consider the latest advances in instrumentation and practice before summarising some of the recent applications. Likely future developments are also briefly discussed.

## 2. Recent advances in instrumentation and practice

Steady advances in pole-piece design, coupled with reduced dimensions for specimen stages and holders, have recently led to small, but significant, improvements to the resolution figures of conventional 100 and 200kV microscopes (eg. at 200kV, $C_s$ = 0.8mm, giving an interpretable resolution figure of 2.2Å). A new 120kV microscope operating in the "second zone" with $C_s$ = 0.3mm is also now available (Yanaka et al, 1983). Manufacturers have further increased the maximum operating voltages of their production-line instruments and, at the present time, microscopes are available commercially which operate at 300kV (Philips, Hitachi) and at 400kV (JEOL). Furthermore, these machines all have aberration coefficients of around 1mm so that resolution figures of better than 2Å are predicted. Latest results indicate that these figures are, in fact, starting to be achieved on test samples although it is not yet clear whether such performance levels will be obtained on a routine basis with normal specimens. It is comparatively simple to record micrographs from thin amorphous films when no tilting is required and slight incident beam misalignment does not adversely affect the optical diffractogram (since the primary effect is only to alter its phase which is not visible). It is far more complicated to orientate a crystalline material and the incident beam direction so that these are parallel with the optic axis of the objective lens. Failure to achieve a satisfactory alignment of these parameters results in a variety of anomalous effects which can easily cause erroneous interpretation of image detail (Smith et al,1983 ). Indeed, some materials have an extraordinary sensitivity towards tilt effects because of their crystal symmetry in the beam direction - see Fig.1 (Smith et al,1985a). The retrieval of reliable specimen information from experimental micrographs thus requires a skilled, patient and knowledgeable operator with a considerable appreciation of electron diffraction and imaging theory. Image simulations are highly recommended whenever image details are related to sample characteristics, particularly those close to the resolution limit.

Whilst conventional wisdom accepts that any image detail beyond the interpretable resolution limit of a particular microscope can not usually be related in a meaningful way to local specimen features, recent work has delineated some circumstances wherein such interpretation could be justified. It is clear, for example, that the so-called "dumbbell" images of Si and other semiconductors (eg. Izui et al,1978-79) are of incorrect spacing and result from {002}- and {111}-type reflections in thicker crystal regions (Hutchison et al,1981). Nevertheless, the positions of individual atomic columns at surfaces (Marks,1984) and at interfaces (Saxton and Smith,1985) can sometimes be located to better than one-tenth of the microscope resolution limit. In the special case of rigid-body displacements at $\Sigma$=3 twin boundaries in gold and copper, the relative bulk lattice positions on either side of the boundary could be located to within an error of 0.04Å (Stobbs et al,1984; Wood et al,1984). Finally, it is appropriate here to note the suggestion of Humphreys and Spence(1981) that a coherent electron source could be used to provide genuine specimen information beyond the resolution limit, albeit in reverse phase due to the effect of the contrast transfer function: this could subsequently be rendered interpretable by image processing. Except for a few isolated cases (Saxton et al,1977;Saxton and Smith,1985), this possibility does not appear to have been explored.

## 3. Applications of atomic resolution imaging

The real objective of atomic resolution imaging is to study local irregularities and structural inhomogeneities which cannot be characterised by other techniques. Thus attention is focussed on the atomic configurations at defects such as dislocations and stacking faults, or at interfaces and surfaces. Extension of the interpretable resolution limit to below 2Å means that it becomes possible not only to resolve the tunnels in block oxide materials, which is possible at 100kV, but also, for example, to visualise i individual columns of atoms in low index zones of most metals (<110> , <100>); ii the cation column locations in many oxides; and iii the projections of atom pairs in many tetrahedrally-based materials such as semiconductors, carbides and sulphides. In the examples described below, it is assumed that there is a direct one-to-one correspondence between the features of the experimental micrographs and the projected crystal structure and that, unless noted, the black spots in the images roughly correspond, at least to within about 0.2Å, with the positions of cation columns. This assumption has usually been confirmed by image simulations but, for the sake of brevity, these details are omitted here.

A major thrust of high resolution imaging has been towards elucidating the atomic structure of dislocations, interfaces and grain boundaries in semiconductors. This work was effectively initiated by the 500kV studies of grain boundaries in germanium by Krivanek et al (1977) who showed that the boundary consisted of five- and seven-membered rings of germanium atoms in alternating columns. Subsequent work, coupled with careful comparisons of simulated images, has enabled the structures of dislocation cores in Si (Bourret et al,1983; Hutchison,1983) and in Ge (eg. Bourret et al,1983) to be determined unambiguously. In the case of compound semiconductors, such as CdTe, the situation is complicated by the presence of two atomic species. Partial dislocations can be readily identified as being either Shockley or Frank partials, corresponding to intrinsic or extrinsic stacking faults respectively, but the exact nature and configuration of the atoms around the fault terminations still remain to be determined.

More recent efforts have been concentrated on grain boundaries and interfaces, particularly those between epitaxial layers, such as Si, silicides or fluorides, and either insulators such as sapphire or semiconductors such as Si. It is salutory to note that as recently as 10 years ago, the boundary between silicon-on-sapphire(SOS), for example, was believed to consist of either a thick (200Å) amorphous layer or else some sort of intermediate layer phase (Hutchison,1984a).The development of edge-on techniques (Krivanek et al,1978) enabled transverse cross-sections of the various boundaries to be imaged and, in the case of the SOS boundary, it was later found that these were atomically smooth (Hutchison,1984;Smith et al,1984a) though sometimes with steps or dislocations (Ponce,1982). Reference should be made to several recent review papers for further details of these and other observations(Cherns,1984;Gibson,1984;Hutchison 1984b).

Atomic resolution imaging has been just as effective when applied to the characterisation of defects in oxides and ceramic materials but it it far beyond our scope here to survey these achievements: reference should be made to Hutchison(1984b) for further details. A typical crucial example, taken from the work of Bursill and Smith(1984) is shown in Fig.2. Fig.2a is an experimental 500kV micrograph showing two pairs of pentagonal bipyramidal columnar defects in a niobium-doped tungsten trioxide,$W(Nb)O_{2.933}$, and Fig.2b is a structural model drawn directly from the image. Location of all the cation columns to a very high accuracy is possible with the assistance of image simulations(Evans et al,1985). Careful analysis of experimental images of other crystallographic shear defects in the same material have enabled them to be identified as being of intrinsic nature, essentially implying a vacancy nature for the small defects responsible for the $WO_{3-x}$ phase (Bursill and Smith,1984). It was therefore interesting that a similar analysis of terminations in slightly-reduced rutile samples indicated an extrinsic nature, in turn suggesting an interstitial cation defect (Smith et al,1984b). A striking example of an incoherent phase boundary between the 3C and 6H polytypes of SiC is shown in Fig.3. This was recorded at the optimum defocus so that it it possible (Smith and O'Keefe,1983), by working in from the thinnest edge regions of each phase, to deduce the tetrahedral configurations in the region of the interface and hence gain some insight into the transformation mechanism.

The surface is a special interface of a solid and a variety of EM methods are now available for imaging surfaces. One of the most effective is the profile imaging technique (Marks and Smith,1983). Under optimum imaging conditions, with a thin sample, the profile image reveals the projected atomic structure along the surface.The materials first studied were small metal particles (Marks and Smith,1983) and extended Au foils (Smith and

Marks 1985) as shown in Fig.4. The technique has since been extended to spinel catalyst particles and the effects of catalytic use have been investigated (Briscoe and Hutchison,1983; Hutchison and Briscoe,1985). The technique is highly effective for observing surface diffusion and some striking events involving structural rearrangements and atomic column hopping have been reported on small gold particles (see Bovin et al, these proceedings). Another phenomenon recently revealed by this technique is the "metallisation" of certain oxide surfaces under intense electron irradiation (Smith et al,1985b) presumably due to electron-stimulated desorption of oxygen - an example is shown in Fig.5.This result clearly implies that care must be used when imaging oxides by this technique.

## Future developments.

Since atomic resolution electron microscopy is really still in its infancy it is actually rather difficult to pass a quantitative judgment on it at this time. Qualitatively, the results already being obtained, such as those outlined above, provide good grounds for future optimism. Meanwhile, a number of things can be done to ensure that the highest possible level of performance is maintained over extended periods. For example, a low-light-level TV camera for image viewing substantially increases the productivity and quality of actual picture taking. Attachment of a TV also allows dynamic events to be followed, and recorded when required,in real time.Moreover, it is then possible to connect the output into a suitable image processing system and thereby provide on-line adjustments to the focussing, stigmating and alignment controls (Saxton et al,1983). This would certainly facilitate direct structure determinations since, at present, even the most skilled operators can not routinely adjust these variable parameters to the accuracy required. There has been a trend towards better vacuums ($10^{-7}$ torr) which almost makes history of system contamination. However, for credibility in surface science terms, ultra-high-vacuums are mandatory; a number of dedicated machines can be anticipated in the very near future which will provide atomic resolution, with UHV and *in-situ* heating/cleaning facilities. An exciting array of surface science experiments could then be feasible. Even without these facilities, numerous physical and chemical processes could be followed in real time at the atomic level. These could include, for example, the development of corrosion or oxidation, crystal growth , epitaxy, effects of nonstoichiometry, precipitation phenomena, ion implantation and so on. In this author's opinion, we are entering a golden age in the history of electron microscopy!

## References

Bourret A, Desseaux-Thibault J and Lancon F 1983 J.Physique **44** C4-15
Bourret A and Bacmann J J 1985 Inst.Phys.Conf.Ser. **78** in press
Briscoe N and Hutchison J L 1983 Inst.Phys.Conf.Ser. **68** 249
Bursill L A and Smith D J 1984 Nature **309** 319
Cherns D 1984 Proc. 42nd.Ann.Meet.EMSA (Detroit) p.376
Evans R, Bursill L A and Smith D J 1985 Optik, in press
Gibson J M 1984 Ultramicroscopy **14** 1
Humphreys C J and Spence J C H 1981 Optik **58** 125
Hutchison J L 1983 J. Physique **44** C4-3
Hutchison J L 1984a Ultramicroscopy **15** 51
Hutchison J L 1984b Electron Microscopy 1984 (Budapest) p.181
Hutchison J L, Anstis G R, Humphreys C J and Ourmazd A 1981 Inst.Phys. Conf. Ser. **61** 357

Hutchison J L and Briscoe N 1985 Ultramicroscopy, in press
Izui K, Furuno S, Nishida T and Otsu H 1978-79 Chemica Scripta 14 99
Krivanek O L, Sheng T T and Tsui D C 1978 Appl.Phys.Letts. 32 437
Krivanek O L, Isoda S and Kobayashi K 1977 Phil.Mag. 36 931
Marks L D 1984 Surface Sci. 143 495
Marks L D and Smith D J 1983 Nature 303 316
Ponce F A 1982 Appl. Phys. Letts. 41 371
Saxton W O, Pitt A J, Mistry A and Howie A 1977 Inst.Phys.Conf.Ser.36 119
Saxton W O, Smith D J and Erasmus S J 1983 J.Microscopy 130 187
Saxton W O and Smith D J 1985 Ultramicroscopy, in press
Smith D J, Camps R A and Freeman L A 1982 Inst. Phys.Conf.Ser. 61 381
Smith D J and O'Keefe M A 1983 Acta Cryst. A39 838
Smith D J, Saxton W O, O'Keefe M A, Wood G J and Stobbs W M 1983 Ultramicroscopy 11 263
Smith D J, Freeman L A, McMahon R A, Ahmed H, Pitt M G and Peters T B 1984a J.Appl.Phys.56 2207
Smith D J, Bursill L A and Blanchin M G 1984b Phil.Mag. A 50 473
Smith D J, Bursill L A and Wood G J 1985a Ultramicroscopy 16 19
Smith D J, Bursill L A, Marks L D and Peng Ju Lin 1985b Proc. 43rd.Ann. Meet. EMSA (Louisville) p.256
Smith D J and Marks L D 1985 Ultramicroscopy 16 101
Stobbs W M, Wood G J and Smith D J 1984 Ultramicroscopy 14 145
Wood G J, Stobbs W M and Smith D J 1984 Phil.Mag. A50 375
Yanaka T , Yonezawa A, Oosawa K, Suzuki S, Nakamura O and Watanabe M 1983 Proc.41st.Ann.Meet.EMSA (Phoenix) p.312

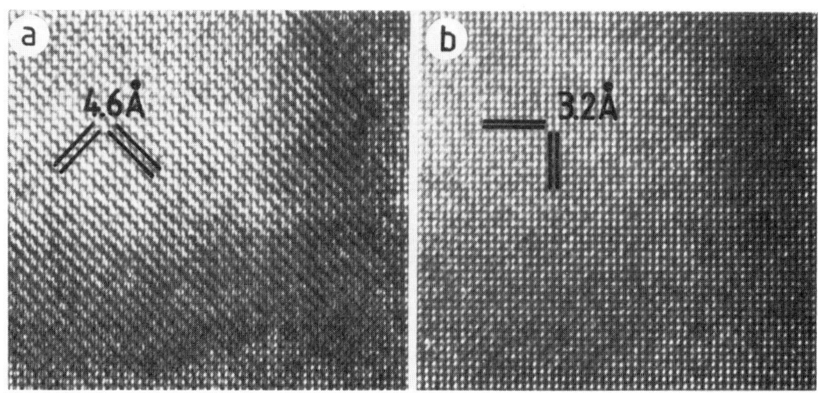


Fig.1 HREM images of $TiO_2$ showing the extreme sensitivity of image detail to slight changes in beam tilt ( < 0.2 mrad )

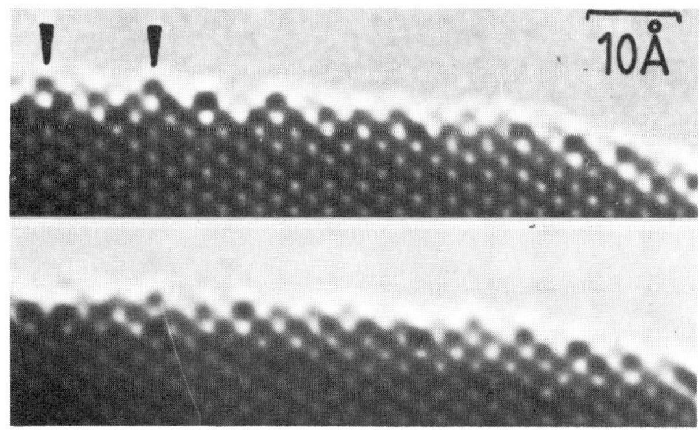


Fig.4 Successive 500 kV images (time separation ~15s) of an extended Au (110) surface showing the movement of entire atomic columns(arrowed)

Fig.2 500kV image of $(W,Nb)O_{2.933}$ showing two pairs of pentagonal bipyramidal columnar defects, together with structural models.

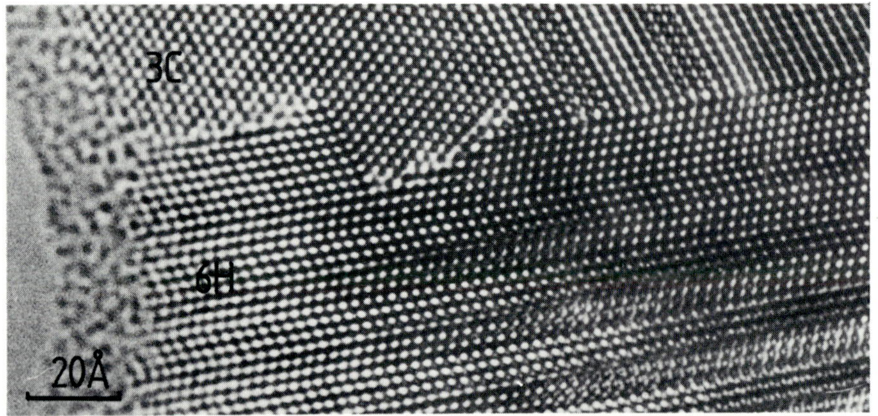


Fig.3 Boundary between 3C and 6H polytypes of SiC imaged at optimum defocus (black contrast at tetrahedral positions)

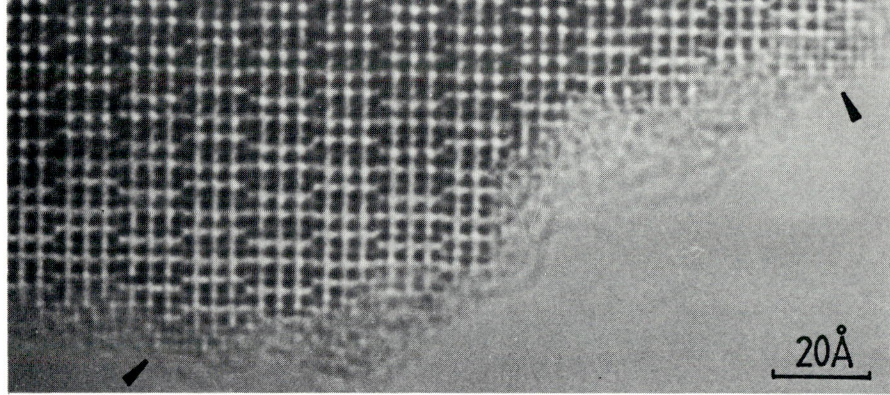


Fig.5 Edge of $Ti_2Nb_{10}O_{29}$ crystal showing development of metallic patches (arrowed) as oxygen is desorbed from the surface.

*Inst. Phys. Conf. Ser. No 78: Chapter 9*
*Paper presented at EMAG '85, Newcastle upon Tyne, 2–5 September 1985*


# High resolution electron microscopy of aluminium-based icosahedral quasicrystals

R Portier(1), D Shechtman(2), D Gratias(1), J Bigot(1) and J W Cahn(3).
(1) C.E.C.M./C.N.R.S. 15,rue G. Urbain 94400-Vitry/France.
(2) Dept of Mat. Eng., Israel Institut of Technology, Technion,32000 HAIFA, Israel.
(3) Center for Mat. Sci., N.B.S.,Gaithersburg, MD-20899, U.S.A.

## Introduction

Some rapidly solidified binary and ternary alloys exhibit a long range ordered structure with no translational periodicity (Shechtman et al 1984, 1985,Zhang et al to appear) but with a discrete Fourier spectrum characteristic of Almost-periodicity (Besicovith 1932). These have been recently called "quasi-crystals" ( Levine et al 1985) and can be described by the Cut and Projection Method (C.P.M.) (Duneau et al 1985, Elser to appear, Kalugin et al 1985) which is a generalized discrete version of the earlier hyperspace description used for continuous density functions of incommensurate structures (Janssen et al 1984). The particular topological properties of such aperiodic networks are best observed by direct imaging of the quasi-lattice in the electron microscope. It is shown here that the relevant characteristics of quasi-periodicity are present in the images which can be interpreted independently from the actual organization of the atomic species within the quasi-periodic framework.

## Experimental

The observations reported here were made on rapidly solidified ribbons of the Al6Mn alloy. Electron microscopy was performed on a JEOL 200CX provided with the Cs=1mm pole piece and the ± 10° goniometer stage. The HR images along a five-fold axis are shown on Fig.1 for different defocus values and objective apertures .Although the images show no rigourous five-fold symmetry (only the Patterson of the structure does show an exact icosahedral symmetry) numerous homogeneously distributed pentagons and decagons are observed at atomic scale. White dots in Fig.1-a fit remarkably well with the simple projection of the quasi-lattice generated by C.P.M.; this is confirmed by Fig.2 which corresponds to the very important two-fold orientation.

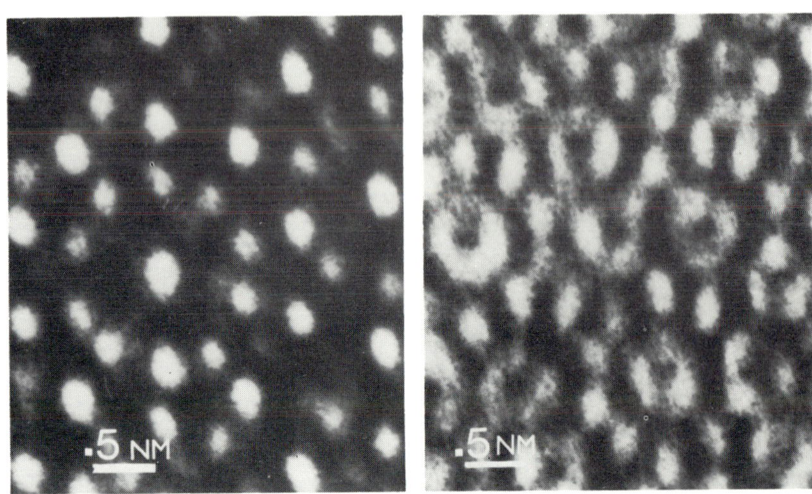


Figure 1: (a) 5-fold orientation at -90 Nm defocus with 0.08 $Nm^{-1}$ objective aperture. (b) -170Nm defocus, same aperture.

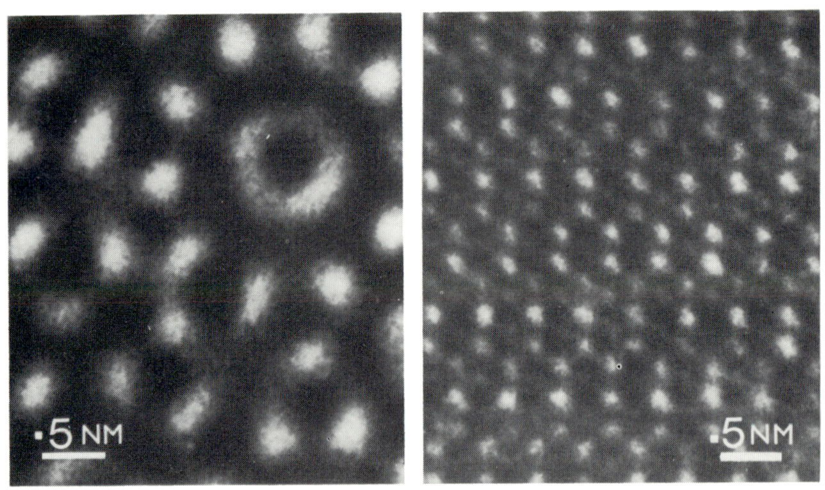


Figure 1: (c) 5-fold orientation at -150 Nm defocus with 0.03 $Nm^{-1}$ objective aperture.

Figure 2: 2-fold orientation at -50 Nm defocus with 0.05 $Nm^{-1}$ aperture.

Here too,the projected quasi-lattice nodes and intensity modulations of the images correspond: neither dynamical effects nor phase-changes due to aberrations can modify the intrinsic topology of the images. The unit length of the edge of the elementary rhombohedra ( Mackay 1982, Kramer et al 1984) constituting the basic tiles for aperiodic space filling is .46 nm; this parameter gives a unit length of .65 nm for the 6-dimensional primitive hypercubic generating lattice. These values strongly suggest that the actual structure contains certainly more than one atom per unit cell; not only some of the Al-Mn distances in the usual crystalline phases are known to be abnormally short (Cooper et al 1966) but the intrinsic volumes of the rhombohedra are far too large to be consistent with the experimental density (Kelton et al 1985) if occupied by a single atom (an average of 4 atoms is expected). From the strict mathematical point of view, the C.P.M. is not able to make any distinction between any two homothetic quasi-lattices which are in the ratio $\tau^3$ (where $\tau$ is the golden mean). This property is directly observed in both diffraction and images of the two-fold orientation /see Portier et al 1985 for a detailed discussion/ whereas an apparent $\tau$ scaling is observed on the five-fold orientation.

Convergent Beam Patterns of the different principal orientations all show a uniform contrast within the discs. This effect might be explained by the fact that the reciprocal quasi-lattice being a dense Z-modulus (see below), each intensity within the discs results from an infinite number of contributions of diffracted beams. An additional plausible explanation is the fact that actual quasi-crystals have a short correlation length (Bancel et al 1985) due to an imperfect quasi-periodicity in the material which would weaken the effect of the excitation parameters in the dynamical diffraction.

## Discussion

The very challenging problem which has to be solved is the description of the atomic structure. Electron microscopy is considered as being a suitable tool for collecting informations about the positions of the atoms and some models (Hiraga et al 1985, Guyot et al 1985, Knowles et al 1985) have already been proposed based mostly on only the five-fold orientation. There are, however, serious difficulties in interpreting the HR images: the dynamical calculations depend crucially on the number of allowed diffracted beams (Cornier et al this issue) which, contrary to the case of crystals, form a dense set. The so-called "small divisor" problem (Belissard 1982) in the almost periodic diffraction Hamiltonian makes the perturbation expansion diverge, leading to approximate solutions which are critically dependent on the chosen cut-off (Cornier et al this issue) . For an optimal cut-off, the simulated dynamical images of a simple quasi-lattice with an average atom at each quasi-lattice node fit remarkably well the experimental ones (Cornier et al this issue). Hence, HR images of relatively close packed structures are essentially governed by the topology of the lattice and are quite insensitive to the atomic motif. Indeed, a specific result of the C.P.M. is that the Fourier components of the potential are multiplied by the Fourier transform of a generic cut function which does not depend on the atomic species. This term is so predominant that chemically different icosahedral phases -like those in Shechtman et al 1985 , Zhang et al 1985- do show very similar intensity distributions although they are constituted of different atoms. These reasons show that HR microscopy will be of limited use for the structure determination of metallic quasi-crystals.

## References

Bancel P A , Heiney P A , Stephens P W , Goldman A I and Horn P M Phys Rev. Lett. 54, 2422 (1985)

Bellissard J Lectures Notes in Phys. 153, 356 (1982) Springer Verlag

Besicovith A "Almost periodic Functions", Cambridge ,1932

Cooper M and Robinson K Acta Cryst. 20, 614 (1966)

Cornier M ,Portier R and Gratias D , this issue.

Duneau M and Katz A Phys. Rev. Lett. 54, 2688 (1985)

Elser V , Preprint to appear in Acta CrystA

Guyot P and Audier M Phil. Mag B, L15 (1985)

Hiraga K , Hirabayashi M , Inone A and Masumoto A Sc. Reports of the Res. Inst. Tohoku Univ. A32,2, 309 (1985)

Janssen T and Janner A Physica 126A, 163 (1984)

Kalugin P A , Kitaev A Y and Levitov L S JETP Lett. 41, 145 (1985)

Kelton K F and Wu J W Appl. Phys. Lett. 46, 1059 (1985)

Knowles K M , Greer A L , Saxton W O and Stobbs W M Phil. Mag. B , L31 (1985)

Kramer P and Neri R Acta Cryst. A40, 580 (1984)

Levine D and Steinhardt P J Phys. Rev. Lett. 53, 2477 (1985)

Mackay A L Physica 114A, 609 (1982)

Portier R , Shechtman D , Gratias D and Cahn J W J. Mic. Spect. Elec. 10, 107 (1985)

Shechtman D , Gratias D and Cahn J W C.R.A.S. 300 Serie II, 909 (1985)

Shechtman D , Blech I , Gratias D and Cahn J W Phys. Rev. Lett. 53, 1951 (1984)

Zhang Z ,Ye H Q and Kuo K H Preprint to appear in Phil. Mag. B

# Atomic level interpretation of quasicrystal microstructure

K M Knowles, W O Saxton, W M Stobbs and A L Greer

Department of Metallurgy and Materials Science, Pembroke St., Cambridge, CB2 3QZ.

## 1. Introduction

High resolution EM at a number of laboratories has confirmed that the recent discovery by Shechtman et al. (1984) of precipitates in rapidly cooled aluminium alloys whose electron diffraction patterns exhibit icosahedral symmetry must be interpreted in terms of a fascinating 'quasicrystalline' organisation that involves extended ordering without translational periodicity. The diffraction patterns are also characterised by an infinite hierarchy of reflections, differing in scale by the golden ratio $(1+\sqrt{5})/2$, and there has been concern that this hierarchical arrangement might also dominate the images, making them insufficiently sensitive to test different atomistic interpretations of the structures permitted mathematically. We demonstrate here that a structure derived from a mathematically perfect quasilattice, but with clearly inappropriate atomic occupancy, does indeed match observed diffraction patterns reasonably well, but leads to strongly model-sensitive images. We also present some of our own observations of the icosahedral phase in an $Al_{86}Mn_{14}$ alloy using the Cambridge University 600kV HREM.

## 2. Results and discussion

Figs. 1 to 4 show high resolution images of the icosahedral phase down the 5-fold, 3-fold and 2-fold axes, and a mirror axis normal to the 5-fold axis; in terms of the six vectors $\underline{e}_i$ for i=1 to 6, which lie along the six five fold axes of this phase, the zones for these images can be written as $\langle\underline{e}_1\rangle$, $\langle\underline{e}_1+\underline{e}_2\rangle$, $\langle\underline{e}_1+\underline{e}_2+\underline{e}_3\rangle$ and $\langle 2\underline{e}_1+\underline{e}_2+\underline{e}_3\rangle$ respectively. These images were obtained from the thinnest parts of the regions studied, so as to minimise dynamical effects, which are expected to be particularly confusing given the hierarchical nature of the quasicrystal ordering. Like other images reported previously (e.g. Portier et al, 1985; Bursill and Peng, 1985), they confirm the quasiperiodicity qualitatively, but the projection problem still hampers direct interpretation.

We have previously considered possible appropriate modifications of normal crystal structures such as that of $Al_6Mn$ (Knowles et al, 1985). The model explored further here is based on Conway's description of 2-D and 3-D tilings (unpublished work): two orthogonal three dimensional spaces V and W are defined with respect to a six dimensional space divided into (6-D) cubes; each cube is said to 'hit' the V space, and generates an allowed quasilattice point in the 3-D tiling, if and only if its projection onto W (which is in fact a rhombic triacontahedron) contains the origin. Possible mathematical descriptions of quasiperiodic tiling have also been given by Kramer and Neri (1984) and Duneau and Katz (1985).

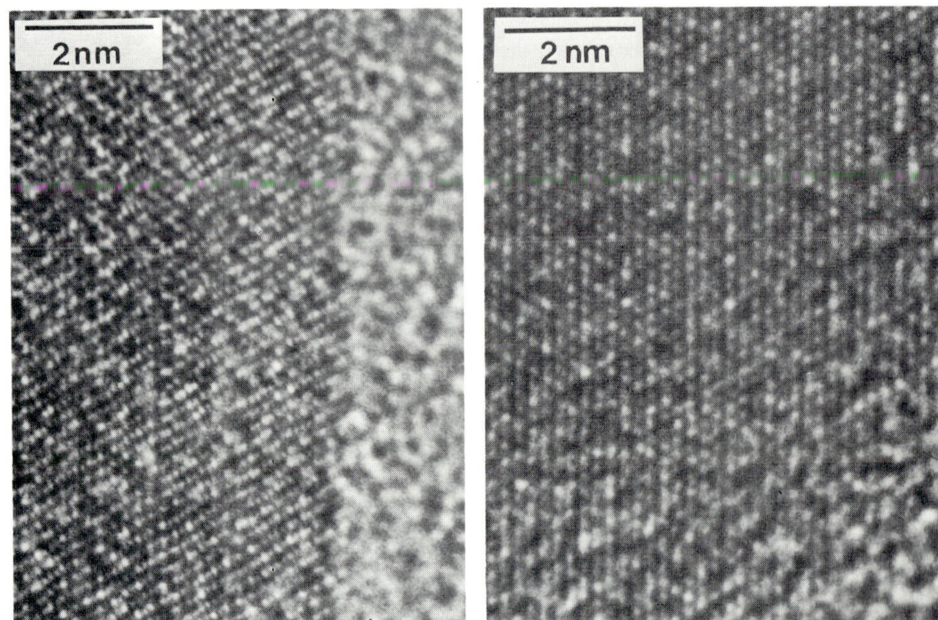


**Fig. 1: Thin Al-Mn quasicrystal: HREM image down 2-fold axis**

**Fig. 2: Image down 3-fold axis**

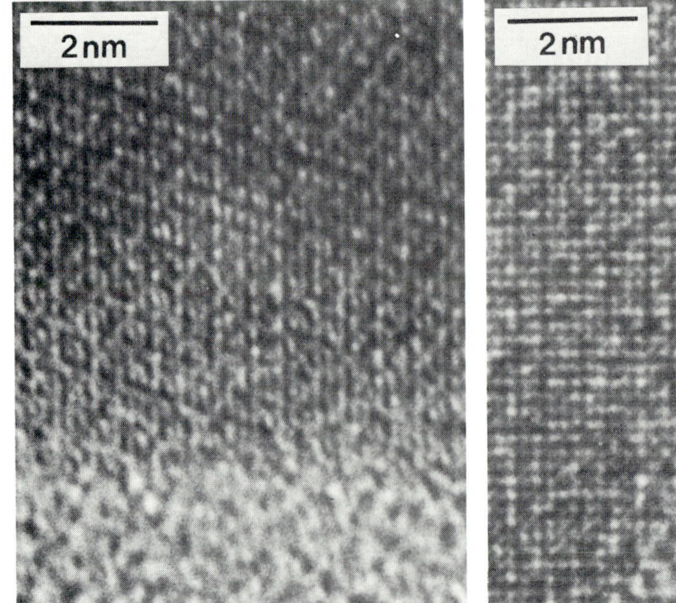


**Fig. 3: Image down 5-fold axis**

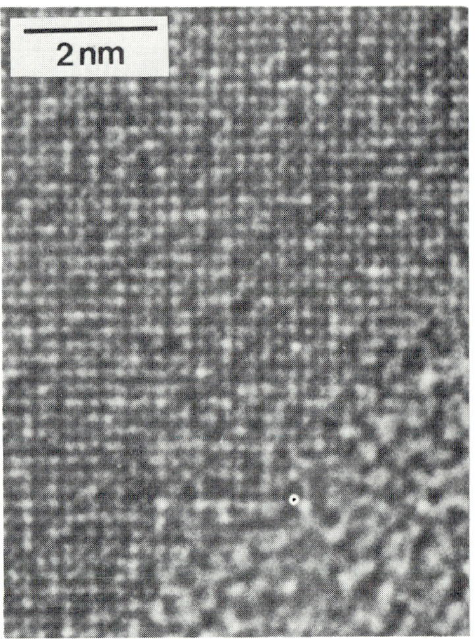


**Fig. 4: Image down mirror axis normal to 5-fold axis**

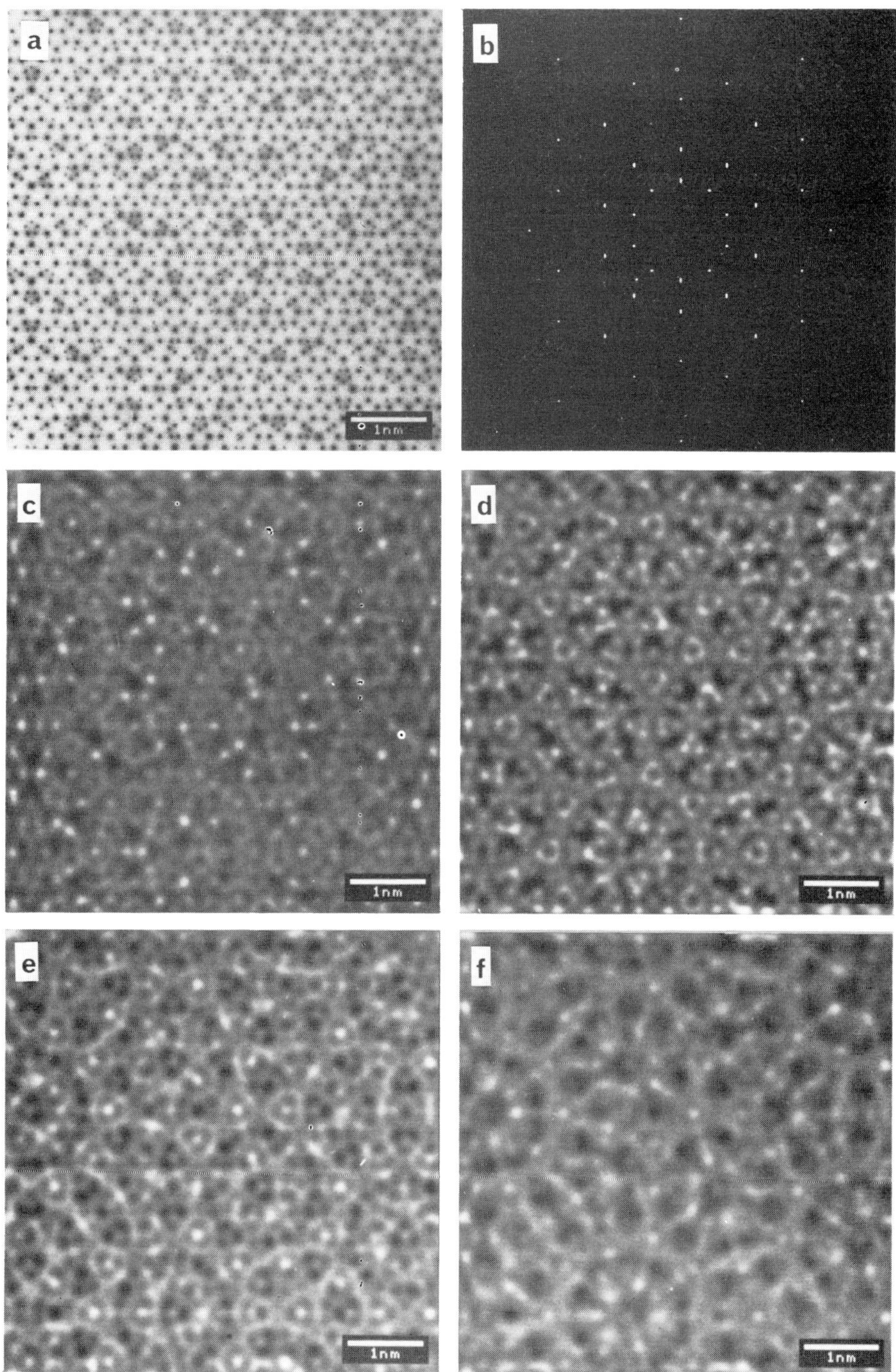


Fig. 5: Quasicrystal modelling: (a) projected potential; (b) diffraction pattern; (c–f) simulated images at 0.72, 1.22, 1.58 and 1.87Sch underfocus

The image simulations presented in fig. 5 are for a structure in which an aluminium atom lies at all the 6000 vertices lying within a 5.6nm square region of a 1.6nm slice of the tiling; the imaging conditions are those of the 600kV HREM.* The atom positions form a network filled by two distinct shapes of rhombohedron (Mackay, 1982), with a side 0.279nm long; they are unrealistic in that anomalously short atom-atom distances are encountered along the body diagonals of the flatter rhombohedra. The diffraction pattern from this array matches the observed diffraction pattern (fig. 6) quite well, showing 10-fold symmetry with the innermost set of strong reflections falling at a spacing of 0.206nm. The images show the expected sensitivity to defocus, and exhibit quasiperiodicity of the correct type. They do not reveal individual projected atomic positions, but they are model-sensitive: further simulations (not shown) of a structure where only a systematic subset of the vertices has been retained, have a very different appearance from those in fig. 5.

The situation is therefore much more promising than that encountered with the modelling of 'amorphous' materials, where the image simulations are discouragingly insensitive to the precise model used. The next step must evidently be the determination of the appropriate occupancy rules for the quasilattice, using chemical as well as mathematical considerations; it is now clear that high resolution imaging will provide a viable test of possible rules.

Acknowledgements. We would particularly like to thank Prof. J.H. Conway for his help on the mathematical properties of 2D and 3D Penrose tiling. We are also grateful to the SERC for financial support and Prof. D. Hull for the provision of laboratory facilities.

References

Bursill L A and Peng Ju Lin 1985 Nature 316 50
Duneau M and Katz A 1985 Phys. Rev. Lett. 54 2688
Knowles K M, Greer A L, Saxton W O and Stobbs 1985 Phil. Mag. B52 L61
Kramer P and Neri R 1984 Acta Crystall. A 40 580
Mackay A L 1982 Physica Acta 114 609
Portier R, Shechtman D, Gratias D and Cahn J W 1985 J. Microsc. Spectrosc. Electron. 10 107
Shechtman D, Blech I, Gratias D and Cahn J W 1984 Phys. Rev. Lett. 53 1951

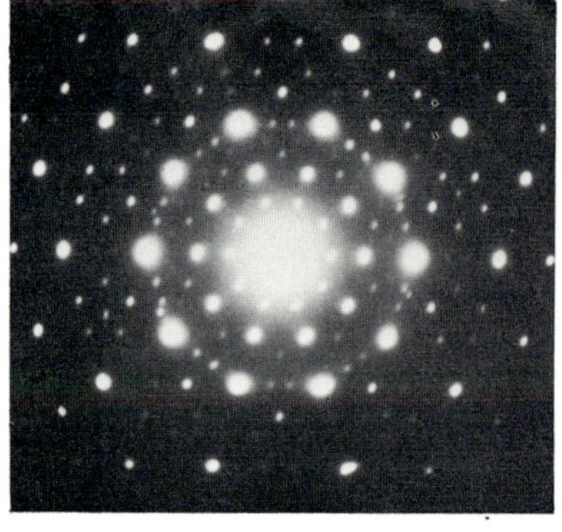

* 500 kV, $C_s$=2.7 mm
$\alpha$=0.15 m rad
$\Delta$=16 nm

Fig. 6: Electron diffraction pattern of Al-Mn precipitate down the 5-fold axis; the brightest reflections correspond to a period of 0.206nm.

*Inst. Phys. Conf. Ser. No 78: Chapter 9*
*Paper presented at EMAG '85, Newcastle upon Tyne, 2–5 September 1985*


# TEM studies of lithium intercalation in the solid-state battery cathode, titanium disulphide

D R Johnson, J L Hutchison

Dept. of Metallurgy and Science of Materials, Oxford University, Parks Road, Oxford OX1 3PH

## 1. Introduction

Much attention has been given to rechargeable batteries, incorporating an alkali metal anode, solid electrolyte, and a cathode of an intercalation compound, such as a transition metal dichalcogenide or oxide (Whittingham 1978). Cell operation requires the insertion, conduction and discharge of alkali ions within interconnecting vacant lattice sites of the host cathode, preferably without structural rearrangement. In the system $M_x^+TiS_2$ $(0 \leq x \leq 1)$, where $TiS_2$ has a layered structure, reversible reaction with $M^+$ = Na or K has demonstrated microstructural complexity over cell cycling, due to the formation of new polytype phases between the layers of $TiS_2$, observed as discontinuities in their Voltage-Composition curves. However, in $Li_xTiS_2$ $(0 \leq x \leq 1)$ an ideal single phase reaction with smooth V/x has been reported (Dahn et al 1980). In our study we use TEM to investigate the microstructure of this system in order to elucidate problems experienced in practical cells, and to assign a mechanism for intercalation.

## 2. Experimental

Nominally stoichiometric $TiS_2$ having grain sizes of <10μm was intercalated with Lithium by two routes: (a) cells of Li/electrolyte/$TiS_2$ were cycled using either potential stepping, or constant current modes, under micro-processor control. Electrolyte used was either $LiBF_4$ in propylene carbonate (p.c) or tetrahydrofuran (T.H.F.) (liquid state), or $LiI/Al_2O_3$ (2:1) pressed as thin pellets (<0.1mm) (solid state). (b) Lithium was chemically inserted using n-butyllithium (Dines 1975). Initial and prepared materials were stored in an argon atmosphere glove box, whereby they were dispersed using p.c. or T.H.F. onto T.E.M. grids containing holey carbon support films. Lattice imaging on $Li_xTiS_2$ $(0 \leq x \leq 1)$ was performed at 200kV using a JEOL 200cx with a $C_s$ = 1.2mm objective lens and diffraction contrast TEM was carried out using a JEOL 100B, 100kV instrument.

## 3. Results and Interpretation

The structure of $TiS_2$, shown in $[11\bar{2}0]$ in Fig. 1a consists of planes of close packed Ti (abc) and S (ABC) atoms separated by a "conduction plane" ([ ]) for $M^+$ insertion. The stacking sequence can therefore be regarded as AbC[ ]AbC, giving an overall symmetry of 1T ($P\bar{3}m\ell$)having lattice parameters a = b = 3.406Å and c = 5.695Å (Dahn et al. 1980). Previously reported insertion of lithium into octahedral sites of the conduction planes leading to the stacking sequence AbC[b]AbC (Fig. 1b) is consistent with a single phase reaction, in which the layer separation is increased by up to 10%, to give a = b = 3.46Å and c = 6.187Å (Dahn et al 1980), However, sodium insertion into $TiS_2$ generates several polytypes which are outlined in Fig. 1c–e. In Fig. 1c every $N^{th}$ layer is preferentially intercalated to produce $N^{th}$ order staging. In Fig. 1d and e, separation of the intercalated layers is sufficient to promote sliding of the basal planes by + or − $a^*/_3$, to give the 3R (I) and (II) types of AbC[b]CaB[a]BcA[b] and AbC[a]BcA[b]CaB[c] stacking respectively.

The lattice image in Fig. 2a of $TiS_2$ taken down the $[11\bar{2}0]$ direction, shows a relatively perfect crystal in which conduction planes lie across the page. Since individual atoms are not resolved, the unit cell in Fig. 1a and for changes in symmetry therein, may be represented by the rectangular spot pattern on the image. Lattice parameter measurements from the image and electron diffraction patterns gives a = b = 3.4Å and c = 5.7Å.

Fig. 2b represents a $[11\bar{2}0]$ lattice image of $Li_{0.3}TiS_2$ prepared from nbu Li and cells having $LiI/Al_2O_3$ or $LiBF_4$/THF electrolyte. Changes in contrast over N layers are explained by $N^{th}$ order staging where N = 1,2,3 in an irregular array and c = 5.7Å and 6.2Å in "normal" and "intercalated" layers respectively. At compositions below $Li_{0.3}TiS_2$ values for N ranging from 4 to 10 have been observed in lattice images and confirmed by superlattice reflections, in electron diffraction patterns.

$Li_{0.3}TiS_2$ prepared in cells having $LiBF_4$ in commonly used p.c, as an electrolyte, is represented in Fig. 2d. Again, contrast variations across the layers are due to staging, but intercalated layers have become expanded to values of c ≈ 9 and 11.5A. Confirmation is given in Fig. 2d, where in a stage 2 phase, layers can clearly be seen opening inwards from the surface. Additional contrast in the centre of the layers is believed to be an electron optical, rather than structural feature.

Further insertion of Li up to $Li_{0.7}TiS_2$ using this electrolyte results in extrusion of precipitates from the particle (which may also remain as blisters within the layers, similar to those seen in $Na_xTiS_2$ (Cherns and NgO 1983)). In Fig. 2e showing $Li_{0.7}TiS_2$, the stage 1 crystal has developed a complex microstructure, in which the 1T phase is offset by intergrowths of a 'skew' 3R phase type, having a revised layer separation of C ≈ 6.3Å. Phase boundaries, as expected are represented by faults in the layer stacking and associated partial dislocations (arrowed). Dark field images of $Li_{0.4}TiS_2$ along the c- axis in Fig. 2f use $(11\bar{2}0)$ and $(10\bar{1}0)$ type reflections to confirm the presence of an increasing density of these faults having b = $^a/_3\langle 10\bar{1}0\rangle$ type partials, which is consistent with phase transformation dislocations. The apparent increase in polycrystallinity, demonstrated in mottled dark field images and ringed diffraction patterns, would be expected where a ~50% lattice parameter mismatch exists.

The mechanism for microstructural change in this type of cell would appear to arise from the cointercalation of propylene carbonate with lithium in order to expand the layers sufficiently to facilitate sliding of the basal planes. Yamamoto et al (1984) have recently reported an 8.8Å spacing for propylene carbonate with $Li_xTiS_2$, but they assign it to a $Li_x(H_2O)TiS_2$ phase through contamination. In our study, propylene carbonate (p.c) which was vacuum distilled, dried over molecular sieves (3Å) and combined in argon with anhydrous $LiBF_4$, gave no trace of water in infra-red spectrophotometric analysis.

Recent experiments with $Li_xTiS_2$ ($x \sim 1$) prepared using the solid electrolyte, indicates the retention of the 1T phase, in which none of the microstructural complexity experienced with p.c. cells is observed. Lattice parameters from lattice images and electron diffraction give a = b = 3.4Å and c = 6.2Å.

Faulting which is observed in the layers appears as terminating conduction planes, forming edge or vacancy/interstitial loop dislocations of a density comparable with nbuLi prepared material (Cherns and NgO 1983). Diffraction contrast experiments using $(11\bar{2}0)$ type reflections and the '$g \cdot b=0$' criterion gives a Burgers vector $b = \frac{1}{2}\langle 0001\rangle$. However, apart from these features, crystallinity is much improved over p.c/$LiBF_4$ cell types.

## 4. Conclusions

Microstructural complexity in $Li_xTiS_2$ is found to depend on the method of Li insertion. Electrolytes using solvents of typically high dielectric constant (for solvation) lead to cointercalation which greatly affects structural integrity, electrical and ionic conductivities, in $TiS_2$. Solid-state, nBuLi or 'THF' routes for insertion lead to dislocations having $b = \frac{1}{2}\langle 0001\rangle$ which probably form by the 10% lattice parameter mismatch that exists with staging. Small crystals (<100Å$^2$) appear to have few such dislocations, which would imply that cells containing solid state electrolyte and cathodes of stoichiometric $TiS_2$ having a small particle size would operate ideally.

## Acknowledgements

This work was supported under a C.E.G.B. studentship (D.R.J.).

## References

Whittingham M S 1978 Prog Solid st. chem 12 1-72

Dines M B 1975 Mat Res Bull 10 287

Dahn J R, Mckinnon W R, Haering R R, Buyers W J L, Powell B M 1980 Can J. Phys 58 207-213

Cherns D, NgO G P 1983 J. Solid st. chem 50 7

Yamamoto T, Kikkawa S, Kaizumi M 1984 J. electrochem soc Vol 131 No 6 1343-5

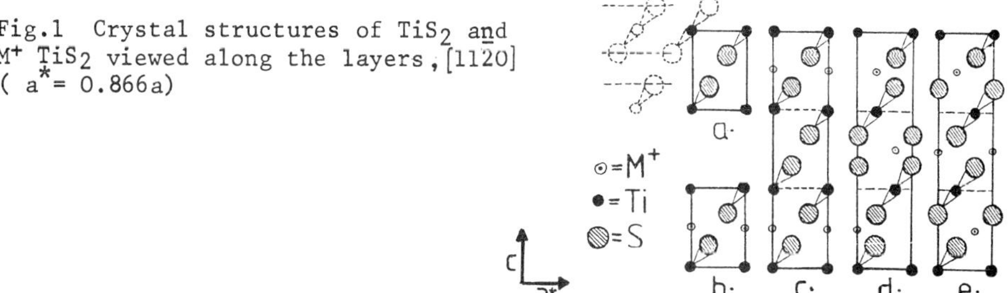


Fig.1 Crystal structures of $TiS_2$ and $M^+TiS_2$ viewed along the layers, $[11\bar{2}0]$ ( $a^* = 0.866a$)

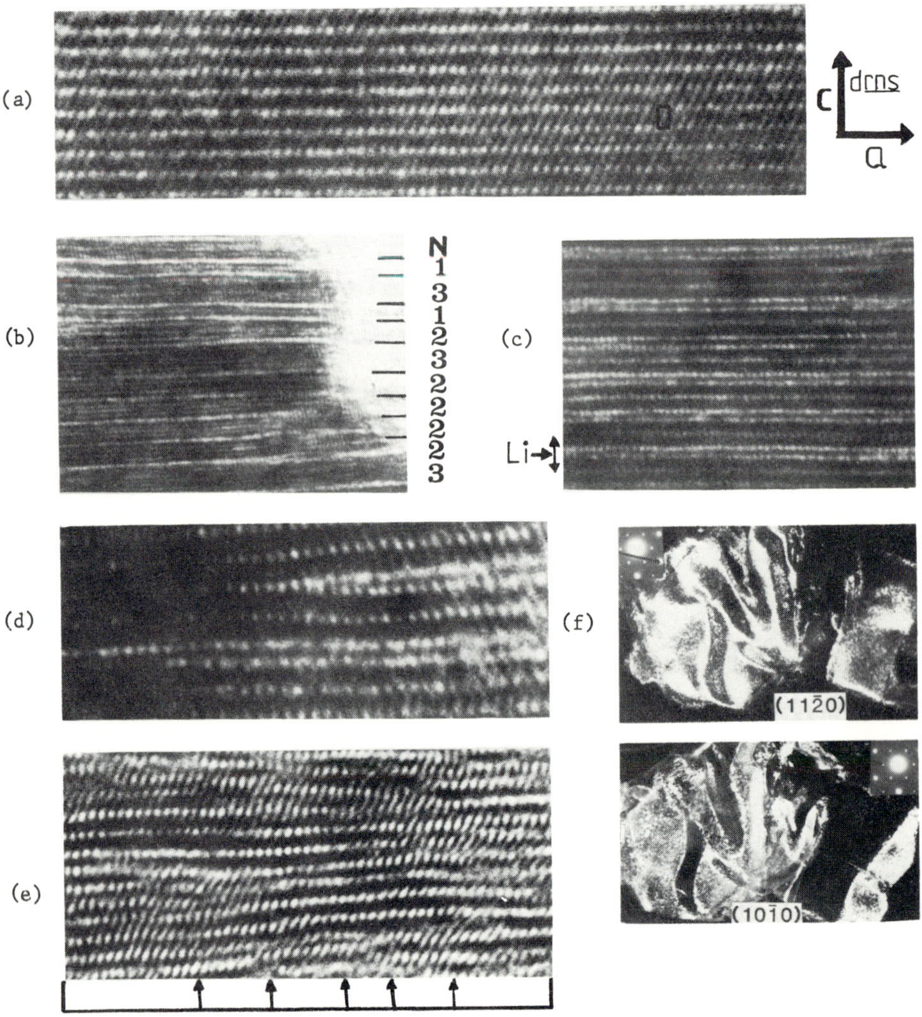

Fig.2 Lattice images(a-e) and diffraction contrast(f) of $TiS_2$ and $Li_xTiS_2$ (a) $TiS_2$ (b) $Li_{0.3}$ $TiS_2$ - solid electrolyte prep. (c)" " - liquid electrolyte(p.c) and $<Li_{0.3}$ $TiS_2$ in (d). (e)$Li_{0.7}$ $TiS_2$ as in (c).(f)"Stacking fault"contrast along the c-axis on $Li_{0.4}$ $TiS_2$, as in (c). Lattice images were recorded at or near optimum defocus and information cut-off 0.39 $Å^{-1}$ along[$11\bar{2}0$].

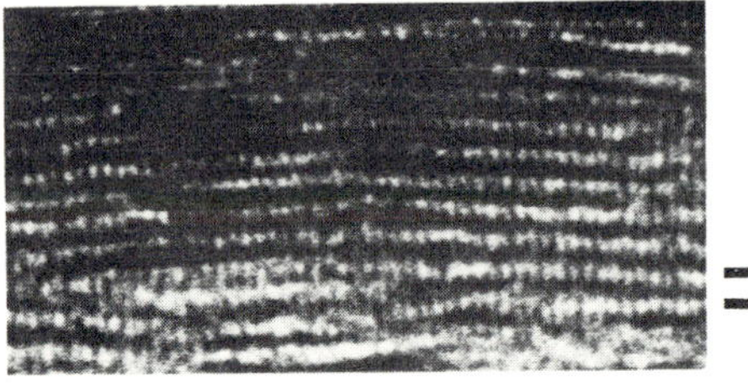


Fig. 3 Lattice image of approx $Li_1$ $TiS_2$ using solid electrolyte representing terminating conduction planes due to edge and vacancy/ interstitial(X) loop dislocations.

*Inst. Phys. Conf. Ser. No 78: Chapter 9*
*Paper presented at EMAG '85, Newcastle upon Tyne, 2–5 September 1985* 

# HREM study of lead substitution sites in the apatite structure

E F Brès, J L Hutchison* and W G Waddington*

Unité de Recherches INSERM U157, UFR d'Odontologie, 1 Place de l'Hôpital, 67000 STRASBOURG. FRANCE.
* Department of Metallurgy and Science of Materials, University of Oxford, ParksRoad, Oxford OX1 3PH, UK.

Lead as well as other metals can be readily incorporated into the hydroxyapatite unit cell (chemical formula : $Ca_6(2)Ca_4(1)(PO_4)_6(OH)_2$) where it replaces the calcium ions (Lacout et al 1984, Andres-Verges et al 1983). The study of the mechanisms of substitution of these metals into the hydroxyapatite structure is important for understanding their incorporation in the calcified tissues of vertebrates and for the industrial purification of precious metals such as Ti, V and U. A step towards the understanding of the substitution of Pb for Ca in mineral and biological apatites is the determination of the detailed structural location of the Pb substituted atoms in the simpler systems provided by the synthetic apatites. Hydroxyapatite crystallises in the hexagonal system and has the space group $P6_3/m$ (Kay et al 1964); the Ca atoms can occupy two non equivalent positions in the hydroxyapatite unit cell. The Ca (1) atoms are on the four fold positions and the Ca (2) atoms are located on the six fold positions. The most interesting case of study is then the 60% substitution where the Pb atoms either fill all the Ca(2) positions or are distributed statistically in both Ca sites depending on the nature of the chemical bonds between the cations and the other ions in each site.

In this paper we have considered : (i) the theoretical feasibility of a subsequent study concerning the detection of the Pb substitution sites inside the hydroxyapatite unit cell using HREM, and (ii) the study of the mineral Nasonite of chemical formula : $Pb_6Ca_4(Si_2O_7)Cl_2$.
(i) The theoretical feasibility of the differentiation by HREM of crystals of same chemical composition where in one case (case P) all the lead atoms are distributed preferentially in the Ca (2) sites ($Pb_6(2)Ca_4(1)(PO_4)_6(OH)_2$) and in the other case (Case E) where all the lead atoms distributed evenly in both sites ($Pb_{3.6}(2)Ca(2)_{2.4}Pb_{2.4}(1)Ca_{1.6}(PO_4)_6(OH)_2$) has been investigated for the case of two commercially available electron microscopes, <u>i.e.</u> : the Jeol 200CX microscope (V=200kV, $C_s$=1.2mm, $\Delta$ =50Å) and the Jeol 4000EX microscope (V=400kV, $C_s$=0.5mm, $\Delta$ =30Å) at 2.5 and 1.4Å resolution respectively. The images shown in Fig.1 were calculated at Scherzer focus for a variety of crystal thickness values using the SHRLI suite of computer programmes described by O'Keefe and Buseck (1979). It is clear from Fig.1 that a progressive variation of image contrast can be observed for both structures up to a maximum thickness of 38.7Å in the 200kV case and 77.3Å in the 400kV case, in the latter case images of the P and E structures can be differentiated and the difference of cation distribution can be shown.

(ii) The mineral Nasonite possesses 60% Pb and 40% Ca and was shown by Giuseppetti et al (1971) to possess the same space group as hydroxyapatite ($P6_3/m$). However, in our study we have observed strong deviations (i) from the proposed space group, (ii) from the hexagonal symmetry. The presence of 00l (l being odd) reflections in the diffractogram in Fig 2 and the strong (001) lattice plane shown in Fig 3 forbid the existence of the $6_3$ screw axis characteristic of the apatite structure in the sample studied Vainshtein (1981). Furthermore, a deviation from the hexagonal symmetry is seen in the diffraction pattern in Fig 4 where the angles between the (100), ($0\bar{1}0$) and ($1\bar{1}0$) planes are all different from the 60° angle characteristic of the hexagonal structure, the maximum distortion encountered in the study of diffraction patterns with the 200CX is 2°, less than the deviations observed. Nasonite clearly does not possess the hexagonal structure and it is impossible for us to carry out any match with the images obtained on an electron microscope and images calculated with the atomic positions proposed by Giuseppetti et al (1971). This prevents us carrying out any proper structural evaluation as first intended. It is clear that the change in space group of the Nasonite is not due to the incorporation of the lead atoms in the structure but to the inherent structure of the mineral, since lead substituted hydroxyapatite crystals were shown to conserve their original space group (Engel and Klee 1972 and Verbeck et al 1982). Further work will be developed concerning the two aspects of this study: (i) structure of substituted apatite crystals (especially the biological ones) and (ii) structure and space group of nasonite.

Acknowledgements :
The authors are thankful to Mr. P.Dunn from the Smithsonian Institution (Washington D.C. USA) for providing the Nasonite specimen (ref : C2661-1, New Jersey) and to Mrs. Kempler (Strasbourg) for typing the manuscript.

References :

Lacout J L , Assarane J , Trombe J C 1984 C.R.Acad. Sci. Paris. 298 173
Andres-Verges M , Hijes-Rolando F J , Valenzuela-Calahorro C and Gonzales-Diaz P F 1983 Spectrochemica Acta Vol.39A, N°12, 1977
Kay M I , Young R A and Posner A S 1964 Nature 4963, 1050
O'Keefe M A and Buseck P R 1979 Trans. Am. Crystallogr. Assoc. 15, 27
Giuseppetti G , Rossi G and Tadini C 1971 The American Mineralogist Vol.56, 1175.
Vainshtein B K 1981 Modern Crystallography I (Springer : Berlin) pp 248-9
Engel G and Klee W E 1972 J. Solid-St. Chem. 5, 28
Verbeck R M H , Lassuyt C J , Heijligers J J M , Driessens F C M and Vrolitk W G A 1981 Calcif. Tiss. int. 33, 243

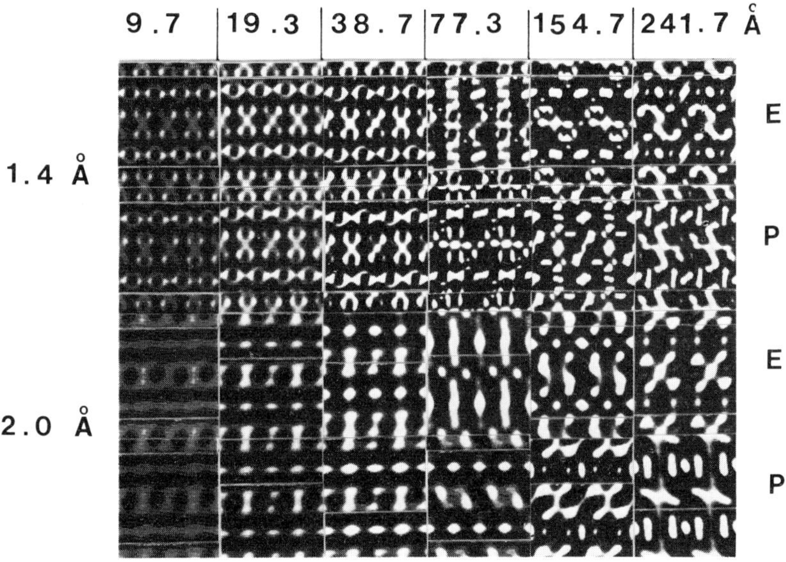


Fig.1 : HREM calculated images corresponding to one unit cell of lead substituted apatite view along the [$2\bar{1}\bar{1}0$] zone. The thickness increase is shown in abcissa and the resolution of the microscope used is shown in ordinate.

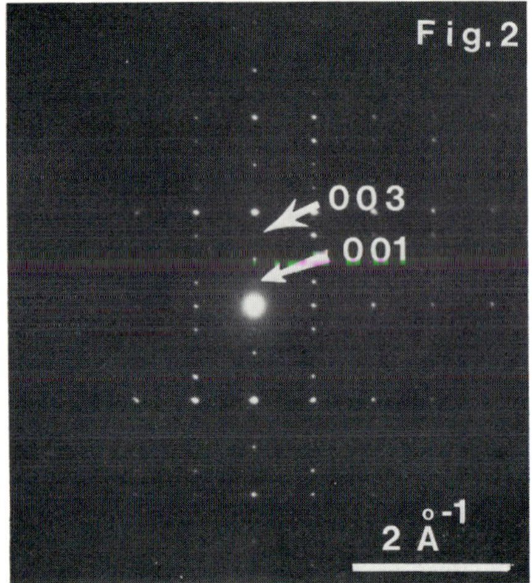


Fig 2.

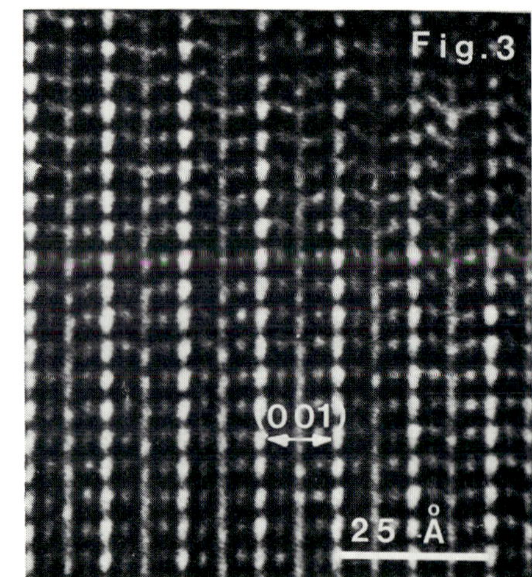


Fig 3.

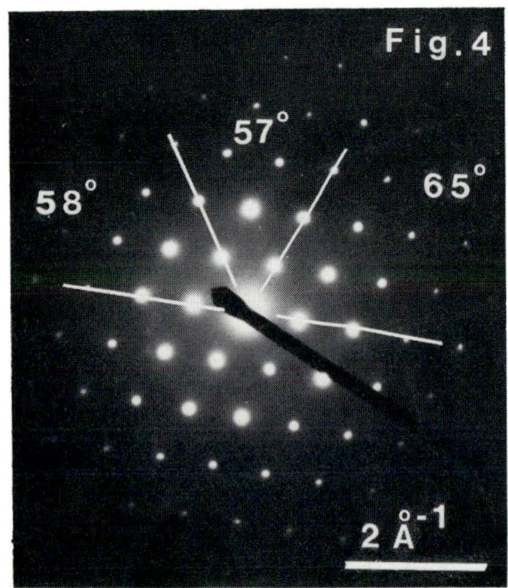


Fig 4.

Fig.2 : Electron diffraction pattern of a Nasonite crystal viewed down the [01$\bar{1}$0] zone.

Fig.3 : Micrograph corresponding to the electron diffraction pattern in Fig.2.

Fig.4 : Electron diffraction pattern of a Nasonite crystal viewed down its c axis.

*Inst. Phys. Conf. Ser. No 78: Chapter 9*
*Paper presented at EMAG '85, Newcastle upon Tyne, 2–5 September 1985*



# An HREM investigation of praseodymium oxides related to $Pr_7O_{12}$

L Eyring*, David J Smith+ and G J Wood+

*Department of Chemistry and +Centre for Solid State Science, Arizona State University, Tempe, Arizona 85287, U.S.A.

## 1. Introduction

In the composition region $Pr_2O_3$-$PrO_2$ there is evidence for a homologous series of oxides with the generic formula $Pr_nO_{2n-2}$ ($n \leq 4 \leq \infty$, n=integer) (Eyring, 1980). The structure of $Pr_7O_{12}$ has been determined by means of powder neutron diffraction (Von Dreele et al, 1975). (The same basic crystal structure also occurs in the cerium and terbium oxides as well as in several of the fluorite-stabilised ternary oxides involving $ZrO_2$ and $HfO_2$ doped with alkaline earth or rare earth oxides of the same overall metal-to-oxygen ratio). The fundamental structural element is an oxygen vacancy pair across the body diagonal of the coordination cube of the Pr atoms, and some associated atomic relaxations of the cations. The vacancy pairs are canted at about 70° across the $\{135\}_F$ a-c planes of the primitive unit cell that contain all the six-coordinated metal atoms. This planar element (or folded planar element in the case of the even members) is believed to be the entity that is common to all members of the two series as illustrated in Fig. 1 (Eyring, 1980) but this hypothesis can not be tested by conventional structural analysis because extended single crystals of these materials generally cannot be prepared. The motivation for the present study is the possibility that atomic-level HREM might enable the structures to be determined.

## 2. Methods

Praseodymium oxide powder (99.999%) was treated hydrothermally with nitric acid to produce small oxide single crystals (McKelvy and Eyring, 1983). The desired $Pr_7O_{12}$ composition was then obtained by control of the temperature and oxygen pressure of the environment both i by reduction to $Pr_7O_{12}$ from the highly oxidised hydrothermal crystals ($PrO_2$); and ii by reduction to $PrO_{1.61}$ followed by reoxidation to the final product. The influence of the preparative history on the nature of the final equilibrated product could then be followed. Micrographs were recorded on a JEM 4000EX, operated at 400kV ($C_s$=1mm) and equipped with a double-tilt top-entry specimen holder. Images were calculated on a local VAX 11/750 computer with standard multislice and imaging programs to simulate respectively the dynamical electron scattering within the crystal and the electron-optical operating conditions of the microscope.

## 3. Results and Discussion

Examination of the nominal composition $Pr_7O_{12}$ obtained in reduction and in oxidation indicated some differences despite the high oxygen and electron mobility in $Pr_7O_{12}$. Crystals prepared in reduction have a greater number of patches of oxidised intergrowths and less material clearly reduced below $Pr_7O_{12}$, whereas that produced in oxidation had some patches of higher oxides but with a greater remnant of the c-type sesquioxide ($P_4O_6$). These differences were presumably due to hysteresis in the preparations similar to that which has been observed in the two-phase regions of the $PrO_x$-$O_2$ system (Knittel et al, 1975). Fine intergrowths of the structures $Pr_7O_{12}$, $Pr_8O_{14}$ and $Pr_9O_{16}$ at the unit cell level were commonly observed. These sometimes gave rise to disordered stacking faults although for many examples, such as that shown in Fig. 2 which contains an intergrowth of 7 unit cells wide of $Pr_9O_{16}$, the boundary was coherent.

It was interesting that in these carefully prepared $Pr_7O_{12}$ specimens, a number of regions of other compositions such as $Pr_2O_3$, and $Pr_{12}O_{22}$ were identified. In addition, many other patches were observed with structures which did not belong to the homologous series; we hope to deduce these in a later part of this investigation. The differences between the preparations were blurred by the effect of the electron beam irradiation since this could reduce or oxidise the specimen depending on the beam intensity and the period of its application. Fig. 3 shows the development of a novel structure, temporarily assigned as 4b since it appears to be a sesquioxide.

A major part of this study has been to ascertain the sensitivity of electron image contrast to structural variations associated with the proposed homologous series. It was thus important to determine whether the superstructure contrast seen in electron micrographs (eg. Fig. 4) results primarily from the missing oxygen atoms or from the relaxation of cations. Images were simulated for a) the basic fluorite structure, $PrO_2$; b) $Pr_7O_{12}$ as determined by neutron diffraction; c) $Pr_7O_{12}$ with oxygen atoms added at vacancy sites; and d) $Pr_7O_{12}$ with all oxygen atoms removed. These simulations are shown in Fig. 5 for the approximate electron-optical conditions used in recording Fig. 4. It is clear that all the simulated images are similar in appearance to the basic fluorite structure for thin crystals and reveal little of actual structural variation. In thicker crystals, the superstructure becomes obvious and can be generated without regard for oxygen atoms. The superstructure contrast appears to be as much dependent upon cation relaxations as upon oxygen vacancies. Supercells can be measured directly from the micrographs due to the periodic variation in structure. However, because crystal potential variations are very small, it appears that detailed interpretation of image contrast requires a significantly greater knowledge and control of the prevailing local imaging and specimen conditions than is normally available.

This work has been supported by NSF Grant DMR-8108501 and used the facilities of the Center for High Resolution Electron Microscopy funded in part through DMR-8306501. The specimens were expertly prepared by Michael McKelvy.

References

Eyring L 1980 In Science and Technology of Rare Earth Materials (Academic Press) pp. 99-118.

Knittel D R, Pack S P, Lin S H and Eyring L 1977 J. Chem. Phys. 67 134.

McKelvy M and Eyring L 1983 J. Cryst. Growth 62 625.

Von Dreele R B, Eyring L, Bowman A L and Yarnell J L 1975 Acta Cryst. B31 971.

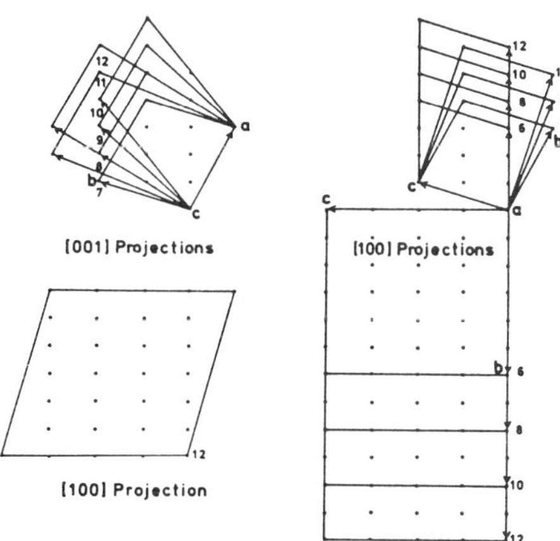


Fig. 1 Schematic representation of unit cell projections for the series $Pr_nO_{2n-2}$ in the <21$\bar{1}$> fluorite zone (see Eyring, 1980).

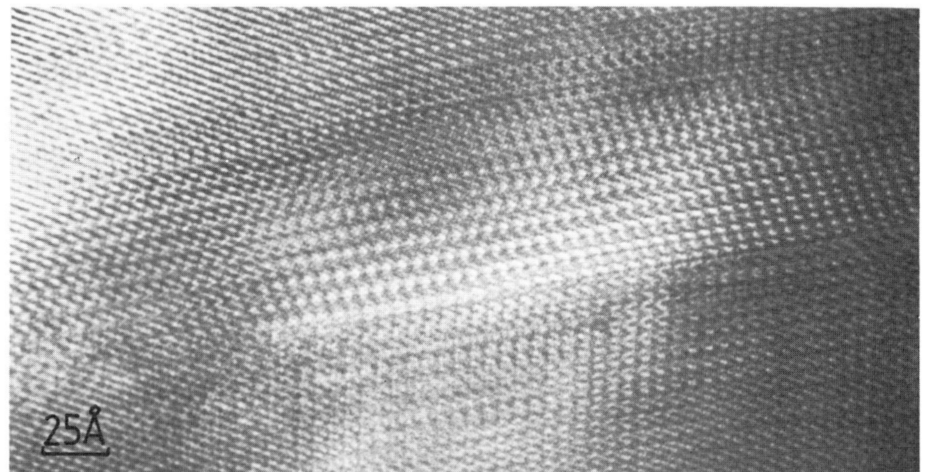


Fig. 2 HREM image showing a coherent intergrowth of $Pr_9O_{16}$, seven unit cells wide, within a crystal of $Pr_7O_{12}$.

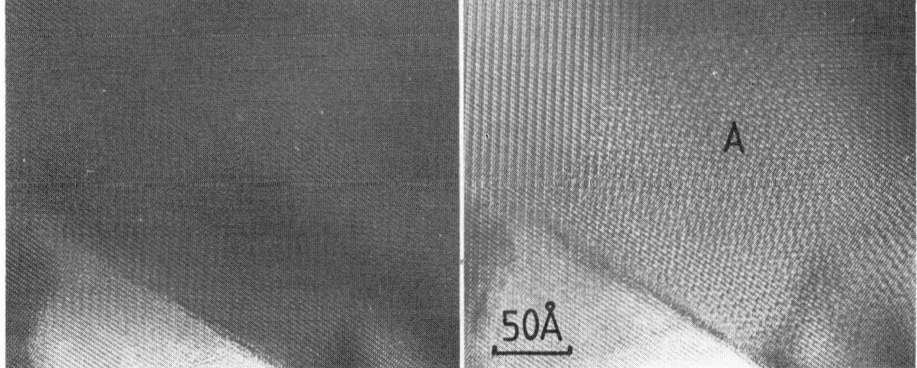


Fig. 3 HREM images showing fine compositional intergrowths and change of structure induced by electron irradiation. Area labelled A in (b) appears to be a novel type of sesquioxide, assigned as 4b.

Fig. 4 Region of $Pr_7O_{12}$ in the $<21\bar{1}>_F$ zone recorded close to optimum focus, showing variations in local image contrast.

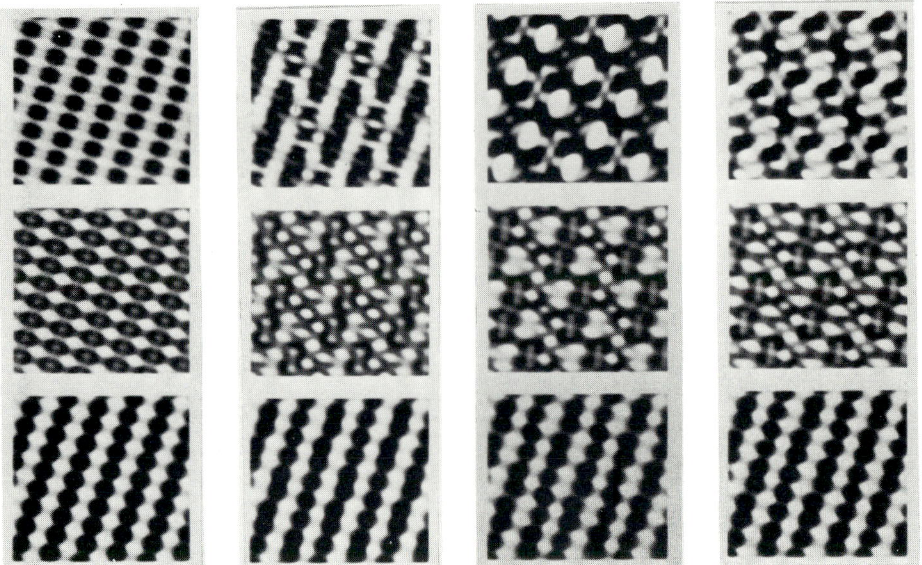

Fig. 5. Simulated images for the four structures (a) to (d), described in text, at thicknesses H=34, 67 and 200Å. Common parameters: $<21\bar{1}>_F$ zone, $\Delta Z$=2.25Å, HV=400kV, $C_s$=1.0mm, $\Delta f$=-450Å, $\alpha$=0.8 mrad, $\sigma$=100 Å

*Inst. Phys. Conf. Ser. No 78: Chapter 9*
*Paper presented at EMAG '85, Newcastle upon Tyne, 2–5 September 1985*



# Structure of grain boundary dislocations and steps in {211} $\Sigma = 3$ germanium bicrystals as studied by HREM

A BOURRET[1] and J J BACMANN[2]
[1] DRF/Service de Physique/ S and [2] IRDI/Département de Métallurgie de Grenoble/LECM, CENG - 85 X - 38041 GRENOBLE Cédex France

Abstract. A high energy grain boundary (GB) such as {211} $\Sigma$=3 in germanium contains many linear defects. High resolution electron microscopy reveals that there are i) coherent steps either pure ledge or associated with an elastic field ii) grain boundary dislocations (GBD), b = 1/6<211>, associated with steps iii) GBD, b = 1/2<110> associated with a change in the rigid-body-displacement of {211} $\Sigma$=3 GB. The step height and extension of these defects allow us to conclude that the GB surface energy is independant of the habit plane close to {211}. It is also shown that GBDs energy is mainly due to the elastic part and not to the step part in such a high energy GB.

## 1. Introduction

Steps and grain boundary dislocations (GBDs) are thought to be responsible for grain boundary (GB) migration (Smith et al. 1980). This is evident for pure steps but is also true for GBDs as they are generally associated with steps due to crystallographic constraints (the GB structure should be similar on both sides of the defect). The importance of a good knowledge of the defect geometry has been emphasized by King and Smith (1980). They have given a theoretical analysis of the relationship between GBDs and steps, showing the importance of being able to measure independently the GBD Burgers vector $\vec{b}$, and the step height h. Experimentally this can only be performed using high resolution electron microscopy (HREM) : step heights are only few atomic interplanar distances and cannot be detected by conventional strain contrast electron microscopy.
A recent experimental work using HREM on perfect {211} $\Sigma$=3 GB in germanium gives its atomic structure (Bourret et al. 1985). This structure has a large rigid body translation $\vec{t} = 1/11[1\bar{1}1] + 1/20[21\bar{1}]$ and is unique. This particular GB having a relatively high energy contains a high defect density and is particularly suitable for this first experimental work on GBDs and steps.

## 2. Geometrical character of GBD in {211} $\Sigma$=3 GB.

Following the geometrical analysis proposed by King and Smith (1980) the defect characteristics are first determined by measuring the step vector, $\vec{s}$. This vector, measuring the displacement of equivalent sites of the GB plane introduced by the defect, can be obtained directly on HREM pictures. The step vector can be determined in reference to crystal I giving $\vec{s}_I$ or to crystal II giving $\vec{s}_{II}$ (fig. 1). The Burgers vector is then given by the relation :

$$\vec{b} = \vec{s}_I - \vec{s}_{II}$$

All Burgers vectors have been defined in respect to crystal I. The possible [011]b– component along the observation axis is undetermined, therefore for non edge dislocation a positive [011] component has been arbitrarily selected. The other equivalent b–vector can be obtained by adding –1/2[011]. As shown by

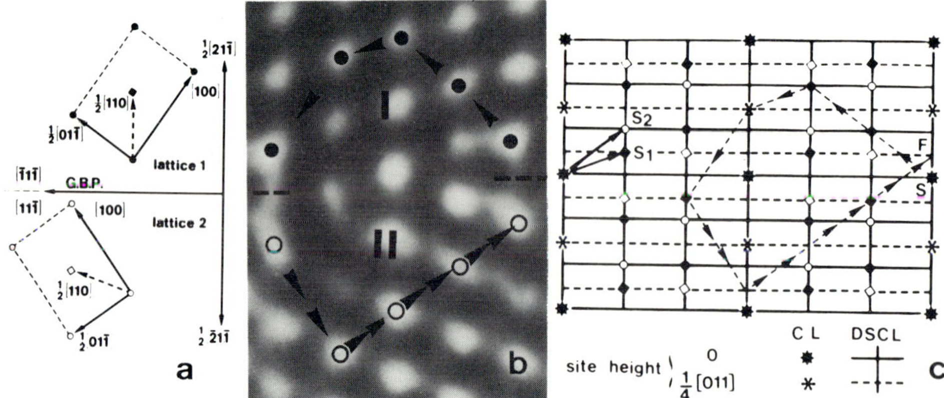


Fig. 1. a) Relationship between primitive cells of two fcc lattices in $\Sigma=3$ coincidence orientation ; $\theta$ = 70°53 around [011]. b) Closed circuit around an experimental GBD on a {211} $\Sigma=3$. The GB is crossed at equivalent points on each side of the defect. c) Step vectors $s_I$ and $s_{II}$ and closure failure in the related DSC lattice. The Burgers vector $\overrightarrow{FS}$ of the experimentally observed GBD is $\vec{b}$ = 1/6[$\bar{1}$12] and is a 30° dislocation.

Bollmann (1970) GBD Burgers vector should belong to the $\Sigma=3$ DSC lattice. The smallest four possible $\vec{b}$s are either the three 1/6<211> normal to the [1$\bar{1}$1] axis, or the 1/3[1$\bar{1}$1]. The rigid body translation $\vec{t}$, if similar on both sides of the defect does not change the Burgers vector. However if a GBD is associated with a translation defect, separating two variants of the structure with +or-$t_{1\bar{1}1}$, the total Burgers vector is then given by $\vec{b} + 2\,\vec{t}_{1\bar{1}1}$.

The step height associated with the defect is measured in respect to crystal I and is given by :

$$h = \vec{s}_I \cdot \vec{n}$$

where $\vec{n}$ is a unit vector normal to the GB plane. This step height can be decomposed into two parts :

$$h = h_o + Nc$$

where $h_o$ is the smallest step height which is crystallographically inevitable, c is the minimum coherent step height, and N an integer. The coherent step height is obtained by drawing the coincident lattice (CL) of $\Sigma=3$. This lattice is hexagonal with [1$\bar{1}$1], 1/2[01$\bar{1}$] and 1/2[011] vectors as lattice parameter. When projected on the (011) plane the minimum coherent step height c is 1/4[21$\bar{1}$] or 0.35 nm.

A GBD is geometrically characterized by the two independant parameters $\vec{b}$ and N. In addition the defect core structure is characterized by measuring its extension parallel to [1$\bar{1}$1], a, ie. the length of modified structure in comparison to the perfect {211} $\Sigma=3$ GB.

## 3. Experimental results

Germanium Czochralski grown bicrystals are cut in (011) slices and observed along the [011] common axis. HREM is performed at 200 kV with the following experimental parameters : $C_s$ = 1.05 mm, beam divergence = 0.015 rd, and objective aperture = 5 nm$^{-1}$. The {211} $\Sigma=3$ GB contains a large density of linear [011] defects varying from 0.5 to 2.10$^7$ m$^{-1}$. Apart from microcrystal formation which is ignored in this paper, four different defect types are found : coherent steps not associated with GBD, 1/6[21$\bar{1}$] pure edge GBD, 1/6<112> 30° GBDs and GBDs including $2t_{1\bar{1}1}$ translation (table I).

Table I. Characteristics of 21 defects analyzed in a {211} Σ=3 GB in germanium. The given parameters are : the Burgers vector for GBDs ; $h_o$, the smallest step height due to the GBD ; h, the experimental step height ; N, the number of minimum coherent step height ; a, the extension of the defect.

| Defect type | $h_0$(nm) | h(nm) | N | a(nm) | Remarks |
|---|---|---|---|---|---|
| Coherent step (A) | | 2.1 to 9.0 | 6 to 26 | 0 | {111} facet |
| Coherent step (B) | | 0.35 | 1 | 1.3 | smooth step |
| ± 1/6[21$\bar{1}$] edge | 0.12 | 0.12 | 0 | 2.5 to 6.5 | Type B step |
| | 0.12 | 0.47 | 1 | 2.3 to 4.2 | " |
| | 0.12 | 0.58 | 2 | 4.6 | " |
| 1/6<112> 30° | 0.12 | 0.12 | 0 | 0.3 to 1.3 | " |
| | 0.12 | 0.58 | 2 | 2.3 to 2.6 | " |
| | 0.12 | 4.6 | 13 | 0 | Type A step |
| 1/6[4$\bar{1}$1]-2/11[1$\bar{1}$1] | 0* | 0.69 | 2 | 4.3 | ± $t_{1\bar{1}1}$ |
| 1/2[101]-2/11[1$\bar{1}$1] | 0 | 0.92 | 3 | 1.3 | domains |

* when referred to crystal II

3.1 Coherent steps not associated with GBD (pure ledge)

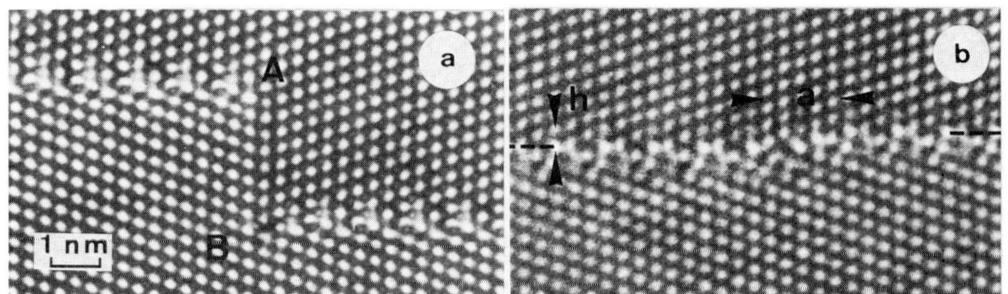


Fig. 2. a) coherent step including a {111} Σ=3 facet. A dislocation dipole is created at the intersections A and B in order to accomodate the rigid-body-translation difference on {111} and {211} plane. b) Coherent step (h = 0.345 nm and a = 1.3 nm) not associated with any dislocation or elastic relaxations.

Two types are present. Type A is formed by a coherent {111} Σ=3 step perpendicular to the original GB plane (fig. 2a). The smallest observed step height is 6c. Type A steps are always associated with an elastic field, the rigid body translation being equal to $\vec{t}$ at the GB interface and zero along the {111} facet. A dislocation dipole $+\vec{t}$, $-\vec{t}$ is therefore formed at the two intersections. The GB energy is given by :

$$W_1 = \frac{\mu}{2\pi(1-\nu)}\, 0.04\, a_o^2 \log \frac{h}{r_o} + \frac{\mu}{2\pi(1-\nu)}\, 0.015\, a_o^2 + K_1 h + 2W_c \qquad (1)$$

where $a_o$ is the Ge lattice parameter, $r_o$ the inner cut-off radius, $\mu$ the shear modulus, $K_1$ the surface energy of the coherent $\Sigma$=3 twin and $W_c$ the core energy of the dislocation dipole.

Type B is formed by a smooth step (fig. 2b) with a large extension but no associated elastic relaxation. Its energy is simply given by the length and surface energy change introduced during the rotation of the GB plane :

$$W_2 = K'\sqrt{a^2 + h^2} - K_2 a \quad (2)$$

where K' is the surface energy of the new inclined interface and $K_2$ the surface energy of {211} $\Sigma$=3.

3.2 1/6 [21$\bar{1}$] pure edge GBDs

These GBDs are the most frequently observed (fig. 3). As shown in table I, for the same $\vec{b}$, several N values are found. The defect extension, a, is neither correlated to h or N and may, for the same geometrical defect (b and N fixed), vary by a factor more than two. The $\vec{b}$ sign is random showing that these GBDs are not due to a small misorientation from the $\Sigma$=3 twin position. The residual misorientation as measured by X-rays is ~ $3.10^{-4}$ rd and may introduce a maximum GBD density of $10^6$ m$^{-1}$, one order of magnitude smaller than the observed defect density.Similarly the step height sign is also random with no tendency to an average rotation of the GB plane. The GBD energy is given by :

$$W_3 = \frac{\mu}{4\pi(1-\nu)} \frac{1}{6} a_o^2 \mathrm{Log} \frac{R}{r_o} + K' \sqrt{a^2+h^2} - K_2 a + W_{3c} \quad (3)$$

with R the outer cut-off radius.
Occasionaly this GBD is dissociated into two 30° adjacent GBDs following the reaction :

1/6[21$\bar{1}$] → 1/6[1$\bar{1}\bar{2}$] + 1/6[121]

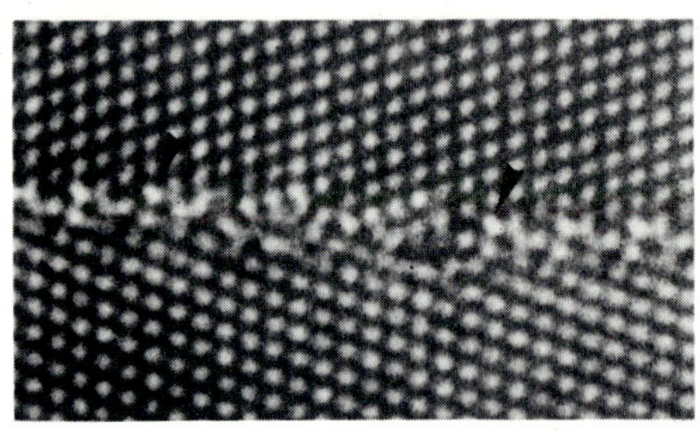

Fig. 3 A $\vec{b}$ = 1/6[$\bar{2}\bar{1}$1] edge GBD associated with a step vector $s_I$ = 1/2[$\bar{1}\bar{2}$1] ; a = 4.6 nm and h = 0.35 nm.

Fig. 4 A b = 1/6[$\bar{1}$12] 30° GBD with a step vector $s_1$=1/2[$\bar{1}\bar{1}$2]. a=2.3 nm and h = 0.58 nm (type A step)

3.3 1/6<112> 30° GBDs

These GBDs have a screw component and in the average are less extended than the pure edge GBDs (fig. 4). They present however the same characteristics and in particular the non reproduceable extension, a, for similar defects. The GBD energy is given by :

$$W_4 = \frac{\mu}{4\pi}\left(\frac{3}{4} + \frac{1}{4(1-\nu)}\right)\frac{1}{6}\, a_o^2 \,\mathrm{Log}\,\frac{R}{r_o} + K'\sqrt{a^2+h^2} - K_2 a + W_{4C} \qquad (4)$$

This GBD can be associated with a type-A step. In this case the 1/6<112> 30° GBD reacts with one of the dislocation $\pm\vec{t}$ present at the facet intersections. The GBD is located at the dislocation wich partially relaxes the $[21\bar{1}]$ component of the Burgers vector. For instance a $\vec{b} = 1/6[\bar{1}12]$ GBD is observed at the intersection where $t_{21\bar{1}} = +1/20[21\bar{1}]$. As a consequence the GBD core structure is not extended and cannot be distinguished from one dislocation of the dipole.

3.4 GBDs including $2t_{1\bar{1}1}$

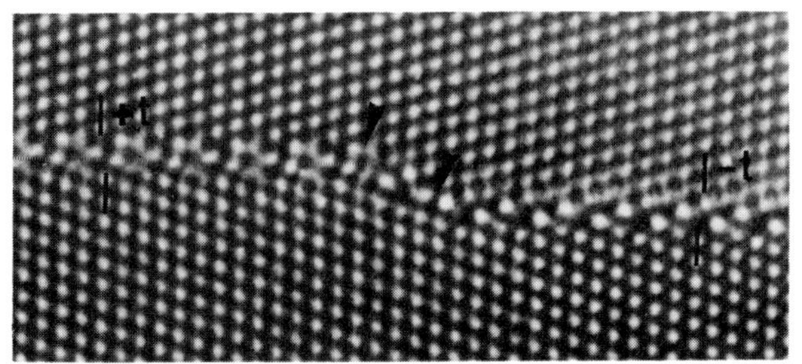


Fig. 5 A $\vec{b} = 1/2[101]-2t_{1\bar{1}1}$ GBD associated with a step $\vec{s}_1 = [\bar{1}\bar{1}1]$. Note the change in the sign of translation $t_{111}$ when crossing the GBD ; a = 1.3 nm and h = 0.92 nm.

Interfacial dislocations limiting domains with the same structure but symmetry related have been suggested by Pond et al; (1983). They should have a $\ddot{b}$ = 2/11 $[1\bar{1}1]$. Such dislocations are not present. Instead it is observed that GBDs with b = 1/2<110> of either crystal I or II are associated with a sign change of $\vec{t}_{1\bar{1}1}$ (fig. 5). Table I gives the characteristics of two types : either 90° or 60° dislocation with a $1/3[1\bar{1}1]$ component. (Note that $b = 1/6[4\bar{1}1]_I$ is simply $1/2[0\bar{1}1]_{II}$). It should be noted that the perfect dislocation $1/2<101>_{I \text{ or } II}$ is not split into two DSC dislocations by formation of $1/3[1\bar{1}1]$ and $1/6[121]$ GBDs. In fact the $1/3[1\bar{1}1]$ is partially relaxed by adding $-2t_{1\bar{1}1}$ or $-2/11[1\bar{1}1]$ creating a change in the rigid body translation sign along $[1\bar{1}1]$. The energy of the edge type GBD is :

$$W_5 = \frac{\mu}{4\pi(1-\nu)}\, 0.23\, a_o^2 \,\mathrm{Log}\,\frac{R}{r_o} + K'\sqrt{a^2+h^2} - K_2 a + W_{5C} \qquad (5)$$

4. Discussion

A coherent step may have two structures depending of its height. Considering type B it is easily deduced by energy minimization of $W_2$ that a and h should obey the relation :

$$\frac{h}{a} = \left[\frac{K'}{K_2}\right]^2 - 1$$

The experimental value of a/h is 3.7 showing that $K' = 1.04\, K_2$. The surface energy of the GB is only weakly affected by small change in the GB plane close to {211}. As a consequence $W_2$ can be approximated by

$$W_2 \sim K_2 \frac{h^2}{2a} \text{ or } K_2 \frac{h}{7.5} \qquad (6)$$

Therefore $W_1$ increases linearly with h while $W_2$ varies as Log h (the energy introduced by the (111) Σ=3 twin can be neglected). Therefore type B should be stable for small h and type A for high h values. The stability limit is experimentally of the order of h = 2nm. Applying formula (1) and (6) with $a_0$ = 0.565 nm, $\mu = 5.6\ 10^{10}$ $Jm^{-3}$, $r_0 = 0.5$ nm and assuming $K_1 \ll K_2$, one can deduce an order of magnitude for $K_2$ ie : $K_2 \sim 0.9\ J\,m^{-2}$. This value is comparable with calculated energies of GBs in Germanium which are in the range 0.3 to 1.4 $J.m^{-2}$ (Moller et al. 1985, Pond et al. 1983, Bourret et al. 1985).
These values for K' and $K_2$ when introduced in GBDs energies (3-5) show that the elastic term is always dominating by at least an order of magnitude the term introduced by the step formation. This explains why the step height can be larger than the minimum step $h_0$, and why the extension of the defect can vary without affecting the total energy. These conclusions are coherent with an easy step formation in a high energy GB.
All GBDs Burgers vectors have a component normal to the GB plane so they cannot glide in this plane. However, in addition to GBDs, coherent steps are also present and they may move by a conservative mechanism. As a consequence in addition to the climb mechanism of GBDs, GB migration could also involve displacement of coherent steps without any climb and long range point defects migration. To be effective however this mechanism assumes that GBDs would not attract all the coherent steps : this seems reasonable in view of the present observation of few isolated pure ledge steps. As a conclusion movement of pure ledge steps not involving climb may well play an important role in the migration of high energy GBs.

References :

Bollmann W. 1970 Crystal defects and crystalline interfaces - (Berlin : Springer - Verlag)
Bourret A., Billard L. and Petit M. 1985 4th Oxford Conf. on Microscopy of semi-conducting materials (to be published)
King A.H. and Smith D.A. 1980 Acta Cryst. A36 335
Moller H.J. and Singer H.H. 1985 Polycrystalline semiconductor Ed. G. Harbeke (Berlin : Springer-Verlag) pp. 18-26
Pond R.C., Bacon D.J. and Bastaweesy A.M. 1983 Inst. Phys. Conf. Ser. No 67 (London, The Institute of Physics) pp. 253-8
Smith D.A., Rae C.M.F. and Grovenor C.R.M 1980 Grain boundary structure and kinetics (American Society for Metals) pp. 337-7.

*Inst. Phys. Conf. Ser. No 78: Chapter 9*
*Paper presented at EMAG '85, Newcastle upon Tyne, 2–5 September 1985*


# Interface structures in long-period superlattices based upon gamma-brass type alloys

A J Morton

CSIRO Division of Chemical Physics, Clayton, Victoria, 3168, Australia

## 1. Introduction

A common feature of alloys whose basic structure is of the γ-brass (space group $I\bar{4}3m$) or a closely related type is the formation of long-period superstructures composed of regular arrays of inversion boundaries. Some understanding of the formation and stability of these superstructures has been gained by studies of the variation of the superstructure long-period with composition and of the crystallography of the inversion domains and their associated domain boundaries. (Morton 1975, 1978; Van Sande et al. 1980). Two distinct types of domain interface are present in these long-period superstructures. These are (i) the inversion domain boundaries that are an integral part of the superstructure, and (ii) the superstructure domain boundaries that separate regions in which the long-period structure has developed along different crystallographic directions. In this study conventional TEM and HREM have been used in conjunction with image computation to characterise both interface types in the alloy Cu-36 at.%Al.

## 2. Superstructure Orientations in Cu-36 at.%Al

Before proceeding to the study of the superstructure domain boundaries a survey of the superstructure orientations in the Cu-36 at.%Al alloy was carried out. As shown in fig.1 two quite distinct orientations were found; one with long-period parallel to a <110> direction of the basic structure (fig.1a); and, less frequently, the second orientation with long-period parallel to <225> as shown in fig.1(b). Only the <110> form of the superstructure had been observed in the Cu-Zn, Ni-Zn and Pd-Zn γ-phases (Morton 1975, 1978). The occurrence of the <225> superstructure in

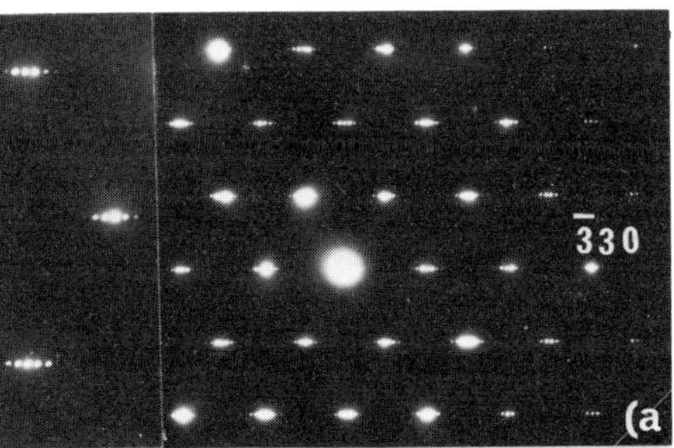


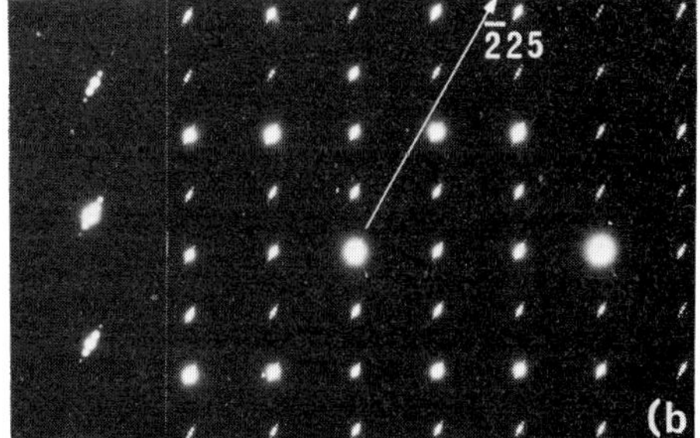


Fig. 1. Electron diffraction patterns from Cu-36 at.%Al showing (a) the <110> and (b) the <225> forms of the superstructure.

the present γ-phase alloy has considerable significance for our understanding of these long-period structures. Previous explanations of the stability of the long-period structures have been based upon the matching of split Brillouin zones of the superstructure with the Fermi surface (Morton, 1977). The prominent faces of the Brillouin zone for the γ-brass type structures are of the {110} and {411} types and hence matching of the Fermi surface to split <110> zones could be envisaged. The <225> form of the superstructure is not so readily explained by such a mechanism.

## 3. The Domain Boundary Structures

The <110> form of the long-period structure can give rise to both $60^{\circ}$ and $90^{\circ}$ domain boundaries whereas boundaries between <225> and <110> domains can be of four types $30.5^{\circ}$, $60.5^{\circ}$, $68.3^{\circ}$ and $90^{\circ}$. <225> domains were relatively rare and the current study is confined mainly to <110> domains. The dislocation structure of the domain boundaries has been determined using a combination of $\underline{g}.\underline{b}$ rules and image computation. The orientation of the basic γ-phase structure in adjoining domains is essentially the same and hence most image computations have been performed using the program PCTWO designed for the computation of theoretical electron microscope images of dislocations within a single crystal. Some image computations were also made with a version of this program modified to allow the computation of theoretical images of dislocation lying within a grain boundary and imaged under simultaneous two-beam diffraction conditions (Humble and Forwood 1975)

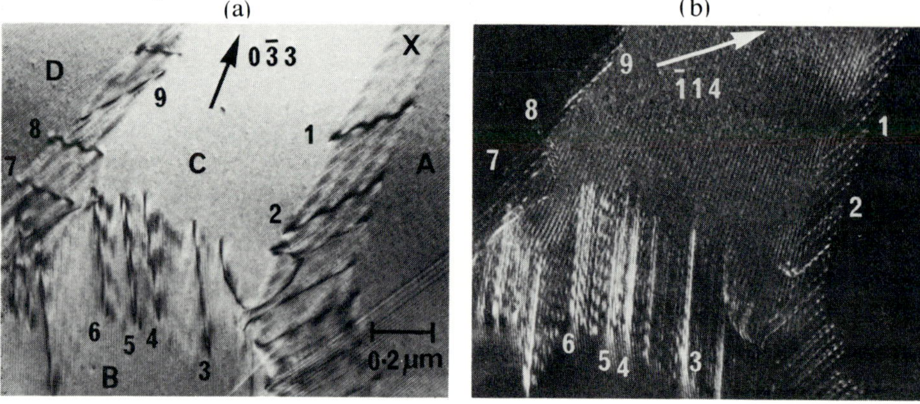


Fig. 2. Bright field and weak-beam dark field images of $60^{\circ}$ domain boundaries in Cu-36 at.%Al. Beam directions are: (a)[166], (b)[20 16 1].

Figure 2. shows a region in the Cu-36 at.%Al alloy in which four superstructure domains are present. The domains have long-period parallel to: $[\bar{1}10]$, domains A and D; $[\bar{1}01]$, domain B; and [011], domain C, and hence the boundaries are all of the $60^{\circ}$ type. The domain boundaries in fig. 2. are typical of the $60^{\circ}$ boundaries observed in that they are not confined to any one crystallographic plane but in relatively flat areas (e.g. near facet X) appear to have a regular dislocation structure. Dislocations are also associated with steps in the boundary or changes in boundary plane.

The facet X of the boundary between domains A and C is parallel to (011), i.e. parallel to the inversion boundaries in domain C. The defects forming the regular boundary structure have line direction $\underline{u}=[1\bar{1}1]$ and show strong contrast for $g=\bar{1}14$ (fig.2 b) but only very weak contrast for all diffracting vectors in the [111] zone. Hence these defects have $\underline{b}$ // [111] and image computations suggested that the Burgers vector of these dislocations was $\underline{b}=[111]/6$. The spacing of the dislocation structure in facet X corresponded to one dislocation for every three periods of the long-period structure which terminate at the boundary. Planar sections of the $90^{o}$ boundaries observed were found to have the same relationship between the long-period structure and the domain boundary periodicity. Similar observations have been made previously for the superstructure domain boundaries in γ-brass (Morton 1974).

Those dislocations in fig. 2. which are additional to the regular boundary structure were found to have a variety of Burgers vectors. For example defects 1,2,7 and 8 were found to have $b=[\bar{1}10]/3$ while defects 3,4 and 9 have Burgers vectors $\underline{b}=[001]/3$, $\underline{b}=[010]/3$ and $\underline{b}=[0\bar{1}1]/3$ respectively. Defects 5 and 6 were composed of pairs of $\underline{b}=[001]/3$ dislocations separated by 26nm and 11nm respectively. The Burgers vectors of defects 1-9 determined above represent the total Burgers vector of each of these defects. However the $\bar{1}14$ weak-beam image (fig. 2 b) shows that most of these dislocations are dissociated. For example in this image defects 5 and 6 show four distinct lines of contrast whereas defects 1-4 and several other dislocations in the boundaries show two closely spaced images. The $\underline{b}=[111]/6$ dislocations which form the regular boundary structure are not dissociated.

The variety of dislocation Burgers vectors exhibited by the additional dislocations in the $60^{o}$ domain boundaries was matched by those present in the $90^{o}$ boundaries. A typical area of such a boundary between domains with long-periods parallel to [110] and $[1\bar{1}0]$ is shown in fig. 3. The

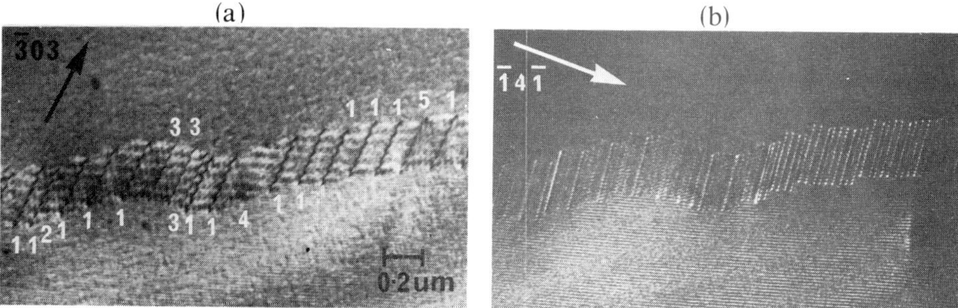


Fig. 3. Bright field and weak-beam dark field images of a $90^{o}$ domain boundary. Beam directions are: (a)[929], (b)[749].

line direction of the dislocations is $\underline{u}=[001]$ ie. the line of intersection of the (110) and $(1\bar{1}0)$ planes and, as indicated in fig. 4(a), five different Burgers vectors are represented in the area shown. This is in addition to the $\underline{b}=[111]/6$ dislocation forming the regular boundary structure which are clearly visible in the $\bar{1}4\bar{1}$ weak-beam image in fig. 3 (b). Matching experimental and computed images of dislocation type 1 with $\underline{b}=[1\bar{1}0]/3$ are shown in fig. 4. The remaining dislocation types were found to have Burgers vectors: $\underline{b}=[101]/3$; $\underline{b}=[100]/3$; and $\underline{b}=[1\bar{1}1]/3$; and $\underline{b}=[1\bar{1}\bar{1}]/3$ for types 2-5 respectively.

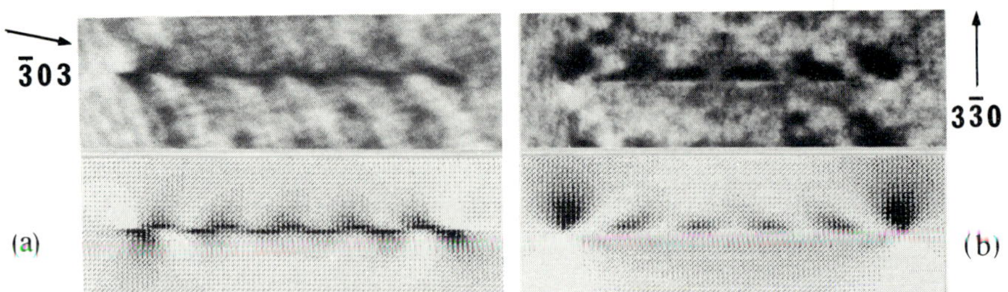


Fig. 4. Matching experimental and computed images of a dislocation with b=[1$\bar{1}$0]/3. Beam directions are: (a)[929], (b)[554].

## 4. The Inversion Boundary Structure

Fig. 5 shows a high resolution electron micrograph of the long-period structure in Cu-36 at.%Al taken along the [111] zone axis. In the area chosen for this micrograph the inversion boundaries are not viewed edge-on but run obliquely through the foil ( 20nm foil thickness). That is regions in the micrograph correspond to areas of the specimen where plus and minus domains overlap. The very good resolution exhibited in all areas of the micrograph including the regions of domain overlap is a strong indication that the inversion boundaries are essentially strain free. The relative displacement of the unit cells from one domain to another is also clearly shown in fig. 5.

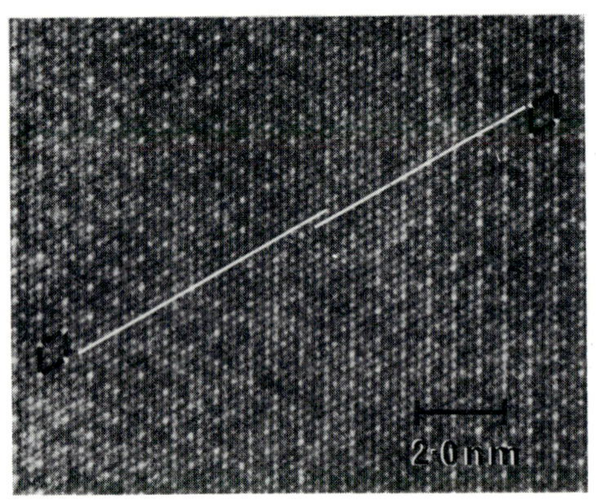


Fig. 5. High resolution electron micrograph of superstructure in Cu-36 at.%Al showing the displacement which occurs at the inversion boundaries. Beam direction [111].

## References


Humble P. and Forwood C.T. 1975 Phil.Mag. 31 1011
Morton A.J. 1974 Phys.Stat.Sol.(a) 23 275
Morton A.J. 1975 Phys.Stat.Sol.(a) 31 661
Morton A.J. 1977 Phys.Stat.Sol.(a) 44 205
Morton A.J. 1978 Acta Metall. 27 863
Van Sande M., Van Landuyt J., Avalos-Borja M., Torres Villasenor C., and Amelinckx S. 1980 Mater.Sci.Eng. 46 167.


## Acknowledgements


The author thanks Dr. J.L. Hutchinson for his advice and assistance in obtaining fig. 5.

*Inst. Phys. Conf. Ser. No 78: Chapter 9*
*Paper presented at EMAG '85, Newcastle upon Tyne, 2–5 September 1985*



# The measurement of rigid body expansions at twist boundaries

K B Alexander, E M Grant, A S Bor* and W M Stobbs

University of Cambridge, Department of Metallurgy and Materials Science Pembroke Street, Cambridge CB2 3QZ, *On leave from Metallurgical Engineering Department, Middle East Technical University, Ankara, Turkey.

Recent measurements (Pond et al. 1974, Forward and Clareborough 1985, Papon et al. 1985) and calculations (Pond et al. 1979, Bristowe and Crocker 1978) of the rigid body displacements away from CSL orientations indicate that a considerable expansion can exist normal to the grain boundary plane. Such an expansion, leading to an increase in free volume, is very important with respect to segregation. (001) twist boundaries with the boundary edge-on are ideal for measuring the rigid body displacement normal to the boundary and the effect of a systematic variation in twist angle on the expansion can also be studied.

Twist boundaries were made using procedures similar to those of Lamarre and Sass (1983). Cleaved single crystals of salt were welded, under an applied load, at ∿ 898K. A surface perpendicular to the grain boundary plane was polished and approximately .5 μm of silver was deposited onto the bicrystal surface at 513K, followed by 30 nm of gold at 473K. Prior to and during the first ∿ 10 nm of deposition, the sample was bombarded with electrons accelerated through 350 - 500 V; both this and a careful silver annealing treatment improved the consistency with which the original boundary was reproduced in the gold film. The epitaxial layers were floated off the salt substrate in water and the silver was dissolved in dilute nitric acid. The geometry of the twist boundaries is shown in Figure 1.

Figure 2 shows (001) twist boundaries, portions of which are curved and faceted. In some cases, low angle boundaries disappeared for distances considerably larger than the average defect spacing (Figure 2b). Measurements of the local twist angles showed variations along the boundary. The portion between areas of differing twist angle must elastically accommodate the differences in local twist angle. This could be simply the result of the lack of constraint inherent in the thin film geometry used, but suggests that, even for these low angles, the boundary energies may not be a smooth function of the defect spacing. This possibility requires further investigation. Relatively flat portions of the (001) twist boundary can be found as shown in Figures 3a and 3b. The primary screw dislocation networks observed have a spacing of 2.95 nm, which is consistent with the measured local twist angle of $\Theta = 5.2°$.

For the HREM study of the rigid body expansion, rigorous specimen orientation and illumination conditions must be met. The planes common to both grains must be in a systematic orientation and the deviation parameter and contrast on both sides of the boundary should be identical. This helps to ensure that there are no differences in fringe position relative to the projected structure on either side of the boundary. Several factors make the HREM observations very difficult. As shown in Figure 4, the boundary often disappears near the edges of the samples where HREM observations tend to be made. In the required (002) systematic orientation, the low angle boundaries are nearly invisible, and this makes it very difficult to determine exactly where the boundary is. Specimen contamination resulting from the manufacturing procedure was also found to be severely limiting.

Figure 5 shows two micrographs take using the Cambridge HREM. In Figure 5a, the boundary plane is not exactly parallel to the (002) planes, and the boundary is readily identified. This image would be inherently unsuitable for analysis because of the contrast changes across the boundary. Furthermore, if the contrast had been suitable, the observed displacement (parallel to (002)) would have been only a component of the true expansion. In Figure 5b, the exact boundary position is much less obvious. The contrast across the boundary, though weak, is much more appropriate for the determination of the rigid body expansion. The image in Figure 5b, which was from a $\Theta = 14.5°$ twist boundary, was digitized and the lattice fringe positions were analysed (Wood et al. 1984, Stobbs et al. 1984). A varying window of fringe positions was used to calculate the expansion at the boundary. This facilitated the identification of those fringe groups which were not affected by prefential etching near to the boundary, spatially variable misorientations across it, or local contrast changes due both to Fresnel fringe effects and the strain fields associated with boundary defects. The starting fringe positions used were further away from the boundary than those used in the twin boundary case since the Fresnel Fringe effects would be larger and there exist local strain fields from the defects in the boundary. The data set with the lowest overall standard deviation corresponded to an expansion of .027 nm ± .012 nm. An analysis of all the data sets showed that the expansion values obtained tended to cluster at a plateau level (Figure 6) which in each case corresponded to the smallest standard deviation for a given starting fringe Taking into account the added confidence lent by the above result, the best estimate for the expansion is between .026 and .056 nm, which represents .066 - .14 of the (001) gold spacing. Calculations by Bristowe and Crocker (1978) for copper and nickel (001) twist boundaries give an expansion of .069 - .082 of the (001) spacing for copper and .127 - .146 of the (001) spacing for nickel in low $\Sigma$ twist boundaries.

Despite the experimental difficulties involved, the consistency of the experimental and theoretical results is sufficiently encouraging that we will now investigate the variation of the rigid body expansion and the associated dislocation structure as a function of twist angle and segregation state.

Acknowledgements

We are grateful to Risø National Laboratory (EMG), the British Council (ASB), SERC (WMS) and NATO (KBA) for funding, and we acknowledge Prof. D. Hull for provision of laboratory facilities.

References

Bristowe P D and Crocker A G, 1978 Phil. Mag. A 38 487
Forwood C T and Clareborough L M, 1985 Phil. Mag. A 51 589
Lamarre P and Sass S L, 1983 Scripta Metall. 17 1141
Papon A-M, Petit M and Bacman J J, 1984 Phil. Mag. A 49 573
Pond R C, Smith D A and Clark W A T, 1974 J. Micros. 102 309
Pond R C, Smith D A and Vitek V, 1979 Acta Metall. 27 235
Stobbs W M, Wood G J and Smith D J, 1984 Ultramic. 14 145
Wood G J, Stobbs W M and Smith D J, 1984 Phil. Mag. A 50 375

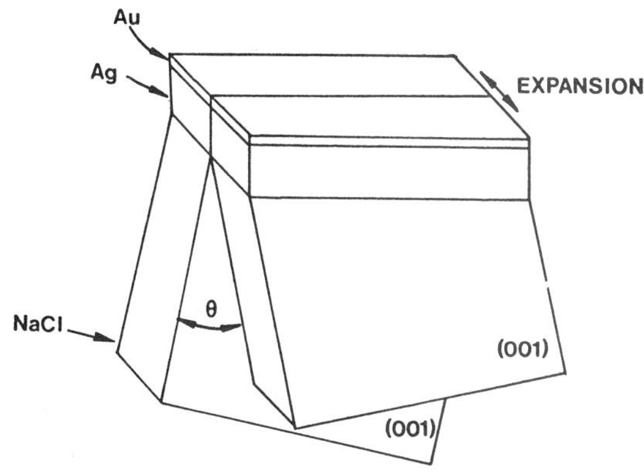


Figure 1 Schematic of the twist boundary geometry (not to scale) (001) planes in both crystals are parallel to each other and to the boundary plane.

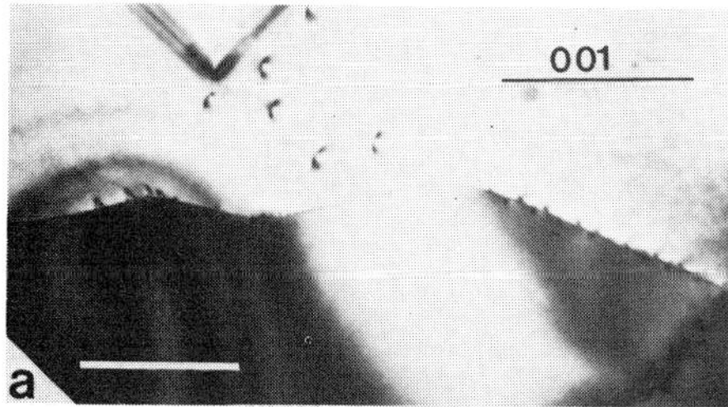

Figure 2 TEM micrographs of (001) twist boundaries. The (001) trace is marked in both cases. a) $\theta=18^{o}$, bright field micrograph; the marker represents 100 nm. b) $\theta=5.2^{o}$, weak beam micrograph; the marker represents 50 nm.

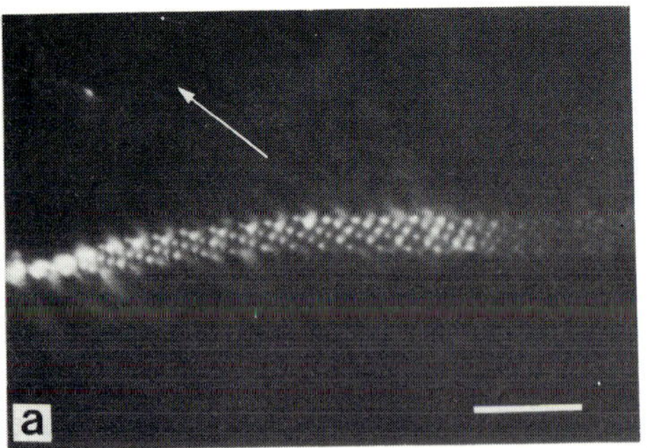


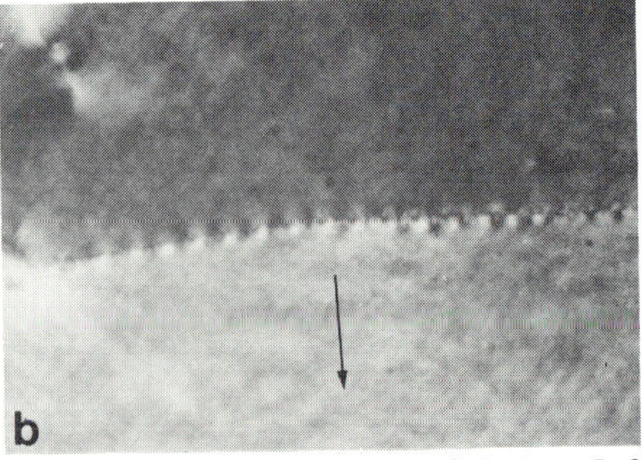


Figure 3 TEM weak beam micrographs of a (001) twist boundary with $\Theta = 5.2°$ a) the boundary is tilted and $\underline{g}$ = (022) b) the boundary is edge-on and $\underline{g}$ = (020). The marker represents 20 nm.

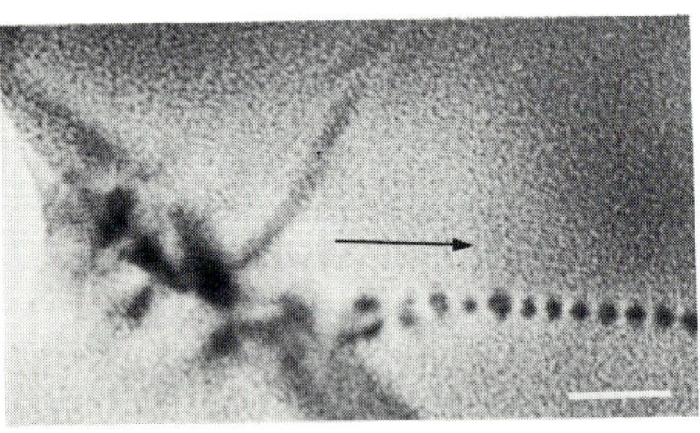

Figure 4 TEM BF of an edge-on twist boundary ($\Theta = 5.2°$) near to the foil edge, with $\underline{g}$ = (020). The marker represents 20 nm.

Figure 5 HREM micrographs showing (002) fringes in both grains. a) $\Theta = 14.5°$ b) $\Theta = 14.5°$. The marker represent 2 nm.

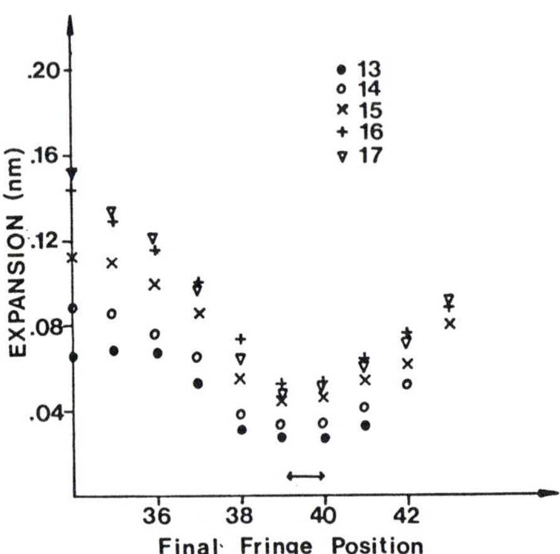


Figure 6 The rigid body expansion calculated for various starting fringe positions. The marked region is the location of the lowest standard deviation for any given starting fringe. Starting fringe position = 13 is the case with the most accuracy as it includes the most fringes.

*Inst. Phys. Conf. Ser. No 78: Chapter 9*
*Paper presented at EMAG '85, Newcastle upon Tyne, 2–5 September 1985* 

# High resolution study of long period structures in $Ti_{1+x}Al_{3-x}$ alloys: a devil's staircase?

. Loiseau[1], R Portier[2], G Van Tendeloo[3], F Ducastelle[1]
ONERA, BP 72, 92322, Châtillon Cedex, France
CECM/CNRS, 15 rue G. Urbain, 94400 Vitry, France
University of Antwerp (RUCA), groenenborgerlaan 171, B2020 Antwerp, Belgium

## . Introduction - Main properties of the APB structures

ong period antiphase boundaries structures (APB structures) have been extensively studied by X-Ray, electron diffraction, and recently by high resolution microscopy. These structures are based on the $L1_2$ structure (type $Cu_3Au$) or on the $L1_0$ structure (type CuAu) with regular antiphase domains long the long period axis c which is a cube direction of the $L1_2$ structure. PB structures can be described by the set of the successive lengths xpressed in units $a_0$, parameter of the $L1_2$ lattice) of domains limited by he APB. The average domain size M is then a very useful parameter for this escription : M can be an integer, a rational as well an irrational number. n this latter case, the period is incommensurate. In the other cases, M is ritten M = p/q where p and q have no common divisor ; p is the period of the equence of domains and q is the number of APB per period. M can be easily etermined on the electron diffraction patterns, from the positions of the uperstructure spots. This gives the mean structure. The exact APB distribution in real space is determined from high resolution images (see fig.3) Portier et al 1980, Loiseau et al 1985).

PB structures are mostly observed in noble metal alloys such as $Ag_3Mg$, CuAu, $u_3Mn$, $Au_3Cd$, $Au_3Zn$... (Ogawa 1974, Van Tendeloo et al 1977, Portier et al 980), that are FCC in the disordered state at high temperature and transform at lower temperature into ordered states. In general, the mean domain size M depends on the composition of the alloy : for example, in the Cu-Au alloys, M is found to vary continuously in the range [35,65] %at Au (Jéhanno & Pério 1964, Guymont et al 1979), so that M is in general an irrational number while in the Ag-Mg alloys, M varies discontinuously in the range [20,26] %at Mg, taking only rational values between 2 and 5/3 (Portier et al 1980).

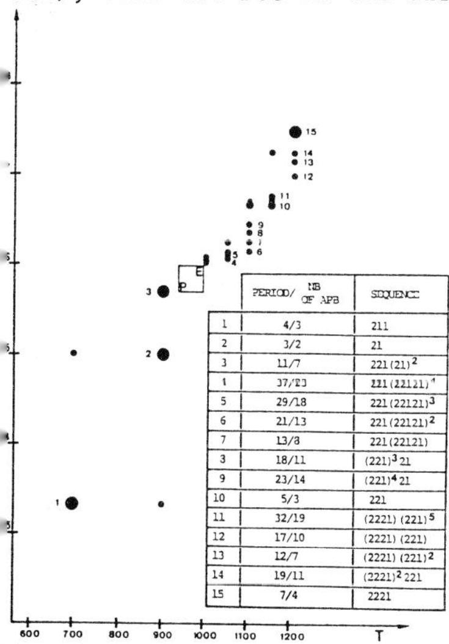


| | PERIOD/ NB OF APB | SEQUENCE |
|---|---|---|
| 1 | 4/3 | 211 |
| 2 | 3/2 | 21 |
| 3 | 11/7 | 221(21)$^2$ |
| 4 | 37/23 | 221(22121)$^4$ |
| 5 | 29/18 | 221(22121)$^3$ |
| 6 | 21/13 | 221(22121)$^2$ |
| 7 | 13/8 | 221(22121) |
| 8 | 18/11 | (221)$^3$21 |
| 9 | 23/14 | (221)$^4$21 |
| 10 | 5/3 | 221 |
| 11 | 32/19 | (2221)(221)$^5$ |
| 12 | 17/10 | (2221)(221) |
| 13 | 12/7 | (2221)(221)$^2$ |
| 14 | 19/11 | (2221)$^2$221 |
| 15 | 7/4 | 2221 |

*ig. 1 Variation of the mean omain size M with temperature n the Ti-72%Al alloy.*

The existence of APB structures is not restricted, however, to noble metal alloys and has been recently reported in the $Ti_{1+x}Al_{3-x}$ alloys (Miida et al 1984 , Loiseau et al 1985). In these alloys, which remain ordered up to the melting point, the APB structures exhibit quite original and stimulating properties. The stoichiometric $TiAl_3$ alloy has the well known $DO_{22}$ structure which is the simplest APB structure (M =1) and all other APB structures, with 4/3 < M < 2, are observed in off-stoichiometric alloys, in the range [69,73] %at Al. The domain of existence of the APB structures is given in Loiseau et al (1985). An important point is that M varies not only with the composition of the alloy but also with the temperature T. The variation of M with T in the alloy containing about 72%at Al is shown in fig. 1. One can distinguish between two domains of variation : the first one at low temperature (T<900°C), where the values of M are simple rational numbers, 4/3, 3/2, 11/7, and the second one at high temperature (T>1000°C), where M takes simple rational (5/3,7/4) as well complex rational values (like 37/23). By means of high resolution microscopy, we have studied the nature of the variation of M with T and the nature of the mixing of the APB domains. We have shown, in particular, that, in the high temperature range, the mixing contains a kind of disorder which may be due to a thermal effect.

## 2. Nature of the mixing of the domains

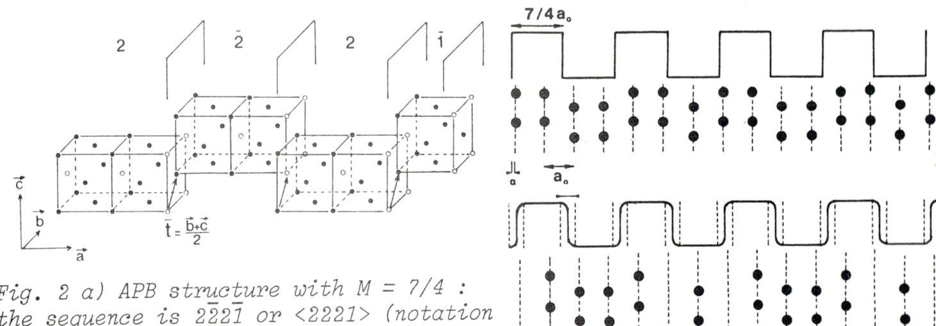


*Fig. 2 a) APB structure with M = 7/4 : the sequence is $2\bar{2}2\bar{1}$ or <2221> (notation of Fisher and Selke 1979) (●:Al; o:Ti). b) The perfect step function and the projected M = 7/4 structure along [010]only the Ti atoms (●) are shown. c) The weakly smoothed step function and the corresponding projected structure : in the planes indicated by the arrows the atomic postions become undetermined. These planes correspond to the diffuse planes on the HR images (see fig. 3).*

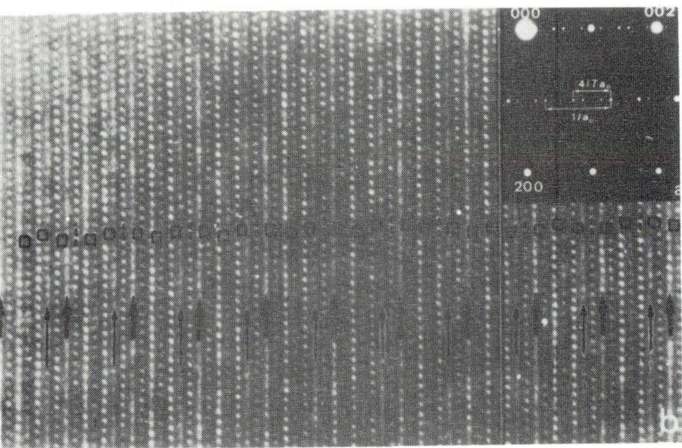


*Fig. 3 a) Diffraction along a (010) zone axis. b) High resolution image of the M=7/4 structure projected along [010].(Δf=-60nm;Δz=12nm). The Ti columns appear as white dots. Note the presence of diffuse planes indicated by arrows, which are due to jogging antiphase boundaries.*

For a given value of M, many mixings of domains are possible. For instance, the value M = 7/4 could correspond to the arrangements <2221>, <3211>, <4111>... As a general result, which is not restricted to the Ti-Al alloys the arrangements are experimentally always uniform in the sense defined by Fujiwara (1957) : in this kind of arrangements, the APB tend to be as equidistant as possible, the ideal spacing being M. Such an ideal distribution can be represented by a regular step function of period 2M, where the discontinuities of the function define the ideal positions for the APB. The real distances between APB however can only take integer values 1 or 2, in units of $L1_2$ cell, 1 and 2 being the integer the closest to M (1<M<2). The structure given in fig. 2 for M = 7/4 is the straightforward result of this simple rule and is independant of the origin : the uniform mixing is thus <2221>.

It is easy to verify the uniformity of the sequence of domains on the high resolution image of the structure M = 7/4 (fig. 3). This may be more delicate in the case of very long periods which correspond to intricate arrangements of domains 1 and 2. To resolve this difficulty, we decompose a sequence into uniform mixings of elementary sequences which are themselves uniform mixings. For this decomposition, we use the branching process for constructing the rational numbers between 1 and 2, based on the principle involved in the construction of the so-called Farey series (Hardy and Wright 1979). This algorithm detailed in Loiseau et al (1985) and in Loiseau (1985) is illustrated in fig. 4. Let us consider at a given step n two neighbouring rational numbers R = a/b and R' = c/d, associated with the sequences A and B : at step n+1, they generate the rational number (a+b)/(c+d) and the corresponding uniform mixing is the sequence AB. Let us now consider any rational number Γ : in the algorithm, it originates from two simpler parents associated with the sequences A and B, which are easily identifiable on the HR images. The sequence corresponding to Γ is an uniform mixing of A and B. To verify the uniformity criterion, it is thus sufficient to verify that the mixing of A and B is uniform. This method has been used in the cases of very long periods and is illustrated on some examples in Loiseau et al (1985).

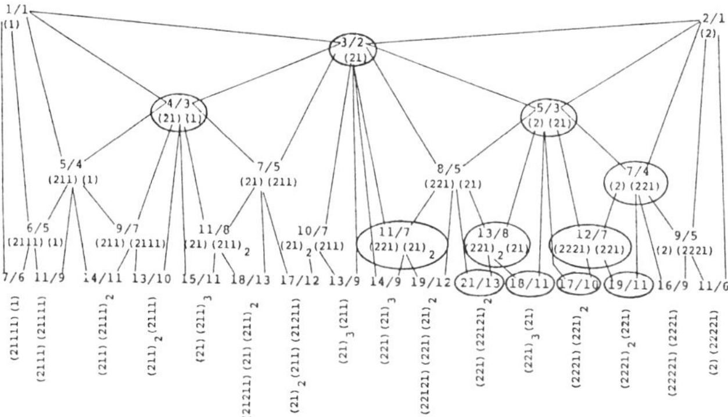

*Fig. 4 Branching process for constructing the rational numbers between 1 and 2 with the corresponding uniform mixings noted between brackets. The observed structures in the Ti-Al alloys are encircled.*

In such situations, two kinds of cases should be considered : this distinction involves the notion of complexity of a rational number, which is based on its continued fraction representation. The number of levels of the fraction is finite for rational numbers and infinite for irrational numbers. In this sense,a rational number is complex if the number of levels is large. A complex number between two simple rational numbers associated with the sequences A and B corresponds to a sequence that involves an intricate mixing of A and B and can then be delicate to identify. The determination of such sequences is then limited to an approximant of the rational number, obtained by truncating the continued fraction. On the contrary, even if the period is large, the sequence remains simple if the associated rational is very close to a simple rational of sequence A: it can be written $A_nB$ (n, large integer) and it can be easily read off on the images. One can thus appreciate the different limitations attached to the identification of a structure: the first one is of course the length of the period and the second one is the complexity of the arrangement.

## 3. Defects observed in the sequences : a thermal effect

The structures observed in the high temperature range contain some defects that correspond to jogs of some APB. These jogs are described in detail in Loiseau et al (1985). They affect a single atomic plane: in this plane, close to the jog, the projected atomic columns are no more monoatomic so that the contrast becomes diffuse. Such diffuse planes are identified for M = 7/4 (fig. 3). Computer simulations have been performed to verify that this diffuse contrast is not an artifact but really a consequence of a different atomic occupation in this plane. Furthermore these jogs only occur in well-defined APB i.e. they transform 1 into 2 or 2 into 1. The diffuse planes are periodic with the period of the perfect structure. We have shown that these defects can be accounted for using a slightly smoothed step function(fig.2). The discontinuities extend now over a small interval $\varepsilon$. The positions of the atoms in the planes indicated by arrows, which are the closest to the discontinuities, are no more unambiguously defined, since the APB may be anywhere in the interval $\varepsilon$.These planes then contain a kind of disorder, which perturbs the contrast and corresponds to the diffuse planes observed on the images. The physical origin for this kind of defect is expected to be a thermal effect. Indeed, the structures observed at low temperature do not contain diffuse planes (Loiseau, to be published), whereas in the high temperature range, diffuse planes are always observed. The softening of the step function describing the structures increases with the temperature.

## 4. Conclusion

To understand the evolution of the APB structures with the temperature in the Ti-Al alloys, it is necessary to develop a thermodynamic theory. Recent approaches in this field are discussed by de Fontaine and Kulik (1985) and by Loiseau et al (1985) : one of these is based on the study of the ANNNI model which is an Ising model with short range competing interactions in one direction (Selke 1984). At finite temperature, many(an infinite number in fact) structures can be stabilized which may have very long periods and which are uniform.This model also accounts for the softening of the step function when the temperature increases.

de Fontaine D and Kulik J 1985 Acta Metall. 33 n°2 145
Fujiwara K 1957 J. Phys. Soc. Japan 12 7
Guymont M and Gratias D 1979 Acta Cryst. A35 181
Hardy G and Wright E M 1979 in An Introduction to the Theory of Numbers Fifth Edition (Clarendon Press, Oxford) p 23
Jéhanno G and Pério P 1964 J. Phys. 25 966
Loiseau A, Van Tendeloo G, Portier R, Ducastelle F 1985 J. Phys. 46 595
Loiseau A 1985 Thèse Université Paris VI
Miida R and Watanabe D 1984 in Phase Transformations in Solids Th. Tsakalakos Ed. (North Holland, New-York) p 247
Ogawa s 1974 in Order-Disorder Transformationns in Alloys H. Warlimont Ed. (Springer Verlag, Berlin) p 240
Portier R, Gratias D, Guymont M and Stobbs W M 1980 Acta Cryst. A36 190
Selke W and Fisher M E 1979 Phys. Rev. B20 257
Selke W 1984 in Modulated Structures Materials Th. Tsakalakos Ed. (Martinus Nijoff Publishers, Dordrecht) p 23
Van Tendeloo G and Amelinckx s 1977 Phys. Status Solidi (a) 43 533

*Inst. Phys. Conf. Ser. No 78: Chapter 9*
*Paper presented at EMAG '85, Newcastle upon Tyne, 2–5 September 1985*


# High resolution electron microscopy of grain boundaries in a mullite/$ZrO_2$ composite

J M[a] RINCON*, G THOMAS.
Department of Materials Science and Min.Eng.and Lawrence Berk. Lab., University of California, Berkeley, CA,94720, USA.

J S MOYA, P PENA and S DE AZA.
Instituto de Cerámica y Vidrio, C.S.I.C. Arganda del Rey,Madrid.

Abstract.- It has been proved that zirconia additions to ceramic matrices can increase the toughness of the ceramic composites either by transformation toughness or by grain boundary solid solution mechanisms. Preliminary results by HREM in a mullite/zirconia ceramic composite with CaO additions show that a coherent relation exists between lattice fringes corresponding to (110) planes of mullite and monoclinic zirconia, having a corresponding periodicity of 6/2 spacings of zirconia with respect to mullite.

## 1. Introduction

High toughness ceramic composites in the system mullite/zirconia/anorthite have recently been obtained with $K_{IC}$ values of $\sim$4.5 MPa $m^{1/2}$ (Pena et al 1985). Observations of these high toughness zirconia materials by High Resolution Electron Microscopy (HREM) can provide valuable information about the presence of phases at grain boundaries, defects and ordering of the lattice and ordering effects at the interphases.

The aim of this work is to study the mullite/$ZrO_2$ grain boundaries in order to confirm the presence of solid solution between mullite and $ZrO_2$ as has been previously reported by Dinger et al (1984). To achieve this objective a mullite/zirconia/anorthite composite has been investigated by using HREM.

## 2. Materials and Methods

A mullite/$ZrO_2$/anorthite composite was prepared by reaction sintering of a zircon, alumina and calcium carbonate mixture.The sample was obtained by the method described elsewhere (Pena et al 1985) with 1 mol% CaO and heat treated at 1450°C for 2 hours.

HREM observations were performed using a Jeol-200 CX transmission electron microscope equipped with the high resolution pole piece and a ± 10° double tilt stage. Thin foils were prepared in a regular way; i. e. by wafering, grinding, ion milling and carbon coating, as is usual in ceramics preparation for transmission electron microscopy.

* J.M[a]. RINCON on leave from the Instituto de Cerámica y Vidrio, C.S.I.C. Arganda del Rey, Madrid, Spain.

## 3. Results and Discussion

Low magnification observations made by the authors in a previous paper (Rincon et al 1985) show : a) round zirconia grains inside the mullite crystals (intragranular $ZrO_2$), with shaped,twinned grains between the mullite grains ( intergranular $ZrO_2$); b) anorthite crystals which are also twinned, as it is normally found in feldspar structures (Wenk 1976) and c) some areas of glassy phase generally close to anorthite crystals.

Figure 1a is a TEM image showing two mullite crystals and a particle of zirconia with round and edged boundaries.

Figure 1b shows with more detail at higher magnification the mullite lattice fringes ($d_{(110)}$= 0.55nm) corresponding tothe periodicity of (110) planes and Figure 1c shows the cross fringes of 0.508 nm corresponding to the periodicity of (100) planes and 0.514 nm corresponding to (010) in the monoclinic zirconia crystal.

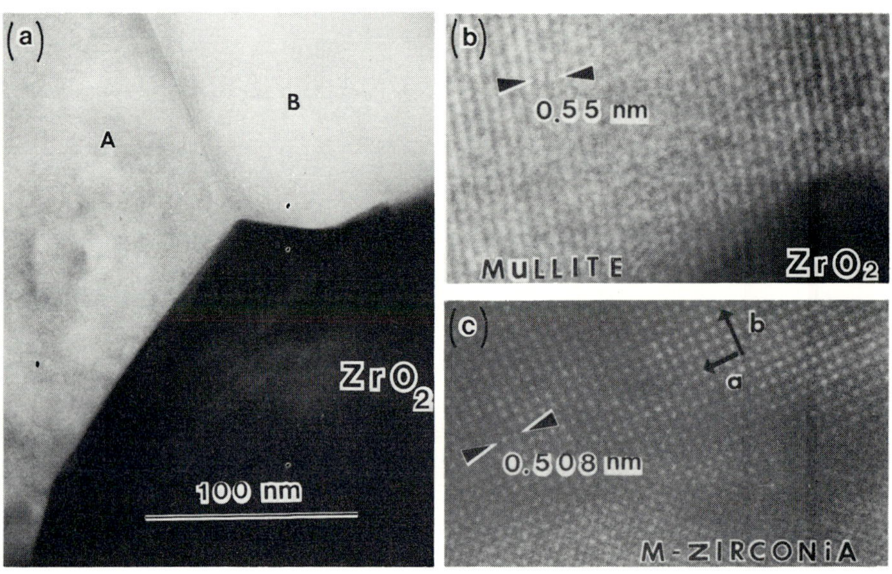


Fig. 1. a) High resolution electron micrograph at lower magnification showing two mullite and a $ZrO_2$ crystals interphases.
b) lattice fringes of mullite corresponding to (110) planes
c) [001] zone axis lattice-image of monoclinic $ZrO_2$.

At higher magnification lattice planes of the mullite continue to the boundary and some extend inside the zirconia, indicating a coherent relation between both crystals (Fig. 2). As Fig. 2a clearly shows the coherent relation between mullite and zirconia grains is only observed in the grain (A). In the mullite grain (B) oriented 1° with respect to the former one (Fig. 1) this phenomenon is not observed. It seems that the mutual lattice orientation mullite-zirconia is a critical parameter in order to obtain coherency between both crystals.

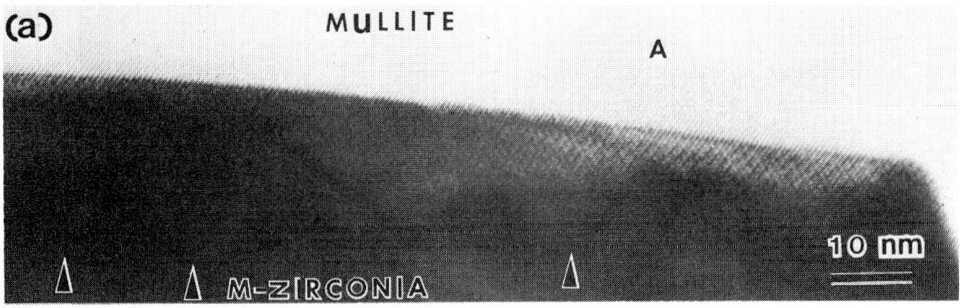


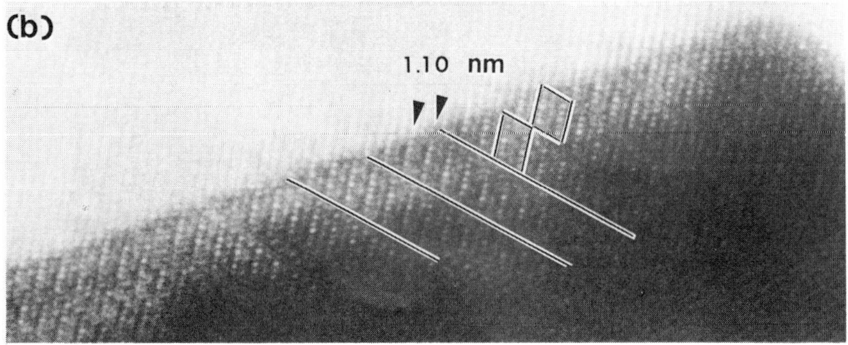


Fig. 2a. High resolution image showing the mullite-zirconia interface with irregular zig-zag domains (triangles) inside the zirconia.

Fig. 2b. The same area at higher magnification showing regular blocks of mullite lattice fringes inside the zirconia grain.

Domains of closure are visible at the end of the triangles (arrowed), indicating that domains nucleate at boundaries growing inside the crystal due to heterogeneous nucleation produced at the interface or by the previously detected solid solution effect (Dinger 1984). With more careful observation of the same boundary at higher magnification (Figure 2b) regular intervals of mullite lattice fringes are seen in zirconia, with three fringes for each block. This observation can be in agreement with the previously observed solid solution effect of mullite in zirconia by analytical electron microscopy (Rincon et al 1985); conversely, inside the mullite and close to the boundary no change in the lattice fringes exists.

In other areas shown in Figure 3a and b a granular structure characteristic of the amorphous or glassy phase close to twinned monoclinic $ZrO_2$ grains can be seen. Spacing of lattice fringes in zirconia is approximately 0.36 nm which corresponds to the periodicity of (110) planes of the zirconia. Twinned areas with domains similar to the ones observed by Van Tendeloo et al (1982) in ceramics of the $ZrO_2$-ZrN system are easily observed.

In summary results here reported by HREM can support the solid solubility of mullite inside the zirconia boundary in mullite/$ZrO_2$ high toughness ceramic composites. This effect can possibly be due to the matched orientation between both crystals in some grain boundaries.

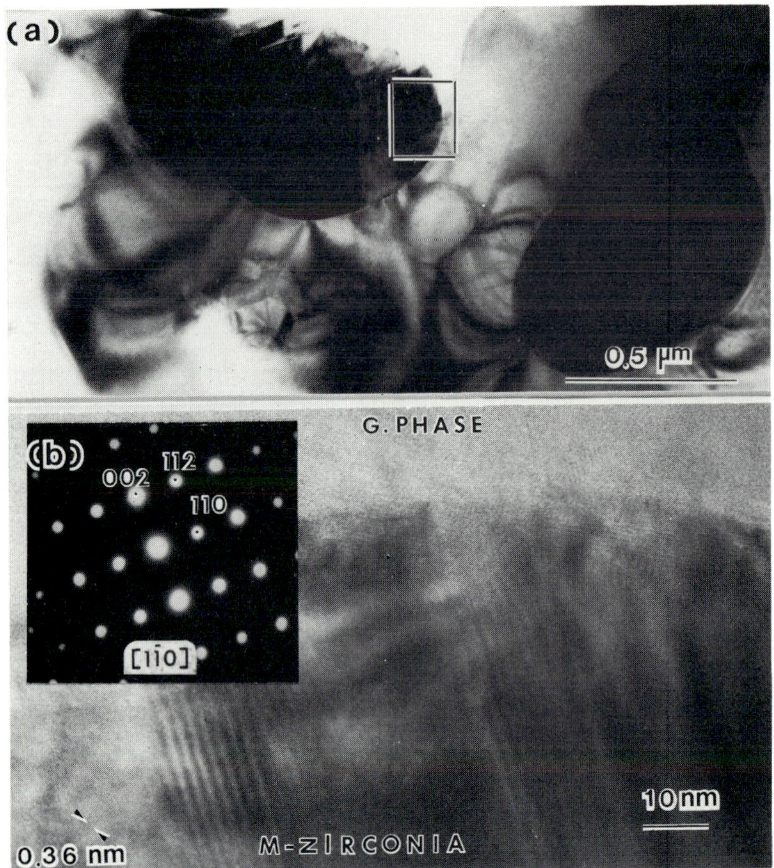


Fig.3.a) Low resolution electron micrograph of a different area and b) high resolution electron micrograph with insert of electron diffraction with [110] zone axis of monoclinic $ZrO_2$.

## Acknowledgment

Thanks are due to NSF support (Grant DMR-83-1317239) and M. Sarikaya and National Center for EM. In addition thanks are due to R. S. Rai and G.Van Tendeloo for valuable discussions.

## References

Dinger T, Krishnan K M, Thomas G, Osendi M I, and Moya J S 1984 Acta Metall.32 1601

Pena P, Miranzo P, Moya J S, and Aza S 1985 J.Mat.Sci.20 2011

Rincon J M[a], Thomas G, Pena P, Moya J S, Aza S 1985 Science of Ceramics 13 in press (Orleans: Journal de Physique-Colloques).

Van Tendeloo G, Anders L, and Thomas G 1982 Journal de Physique 43 c4-411

Wenk H R, 1976 Transmission Electron Microscopy in Mineralogy (Berlin, New York, London: Springer-Verlag) pp 248-65

*Inst. Phys. Conf. Ser. No 78: Chapter 9*
*Paper presented at EMAG '85, Newcastle upon Tyne, 2–5 September 1985*


# High resolution electron microscopy of MBE grown GaAs/AlAs superlattices

LAVAL J Y , DELAMARRE C , DUBON A ,* SCHIFFMACHER G , de SAGEY G **, GUENAIS B ,***
* UA 450 CNRS-ESPCI - 10, rue Vauquelin, PARIS
** ER 210, CNRS MEUDON BELLEVUE, FRANCE
*** Département MPA CNET LANNION B FRANCE

## 1. Introduction

The electronic and optical properties of GaAs/$Ga_{1-x}Al_xAs$ superlattices are strongly related to the structural defects present in the heterostructures and to the structure and composition of the interfaces. We have described previously the nature and distribution of structural defects in the heterolayers (Delamarre et al 1985). In the present paper high resolution TEM is used to study at the atomic level the interface structure of GaAs/AlAs superlattices which were chosen because they give a much better contrast between layers.

## 2. Experimental

The heterostructures were prepared by molecular beam epitaxy (MBE) by CNET Lannion (A. Regreny et al 1985). Alternate layers of AlAs and GaAs were grown on a {100} plane of monocrystalline GaAs substrates heated at 690°C. {110} cross-sections perpendicular to the growth plane were prepared by imbedding them in epoxy resin followed by ion milling and then observed by conventional and high resolution electron microscopy on a Jeol 100 C equipped with a high resolution top entry goniometer stage. Double crystal X-Ray diffraction data led to layer thicknesses :
$e_1$ (AlAs) = 4.5 $a_0$ $e_2$(GaAs) = 4 $a_0$ $a_0$ (AlAs) $\simeq$ $a_0$(GaAs) $\simeq$ 0.566 nm
Image computation were done with the multislice method (O'Keefe et al 1975, Skarnulis et al 1976) on an Amdahl V7 at C.I.R.C.E. (Orsay). Simulation of multislices was performed along <110> axis with no deviation to the exact projection. The thickness of each slice was 0.4 nm and a total of 10 to 40 slices were considered. The electron optical parameters corresponded to
$c_s$ = 0.7 mm $r_c$ = 10 nm $\alpha = 2.10^{-3}$ rd.
In order to limit the number of beams contributing to the image the $d^*$cut off value was taken $\leq 13\ nm^{-1}$ and very weak beams were neglected (those for which $I/I_t < 2 \times 10^{-6}$).

## 3. Results and discussion

From dark field images formed with 200 spots it was inferred that there is no noticeable roughness or asymmetry at either interface. However, the quality of the interfaces at the atomic level can only be assessed using high resolution TEM. Accordingly, high resolution images were obtained with the electron beam exactly parallel to the<110> zone axis, so that the transmitted beam, four 111 and two 200 diffracted beams formed the image.

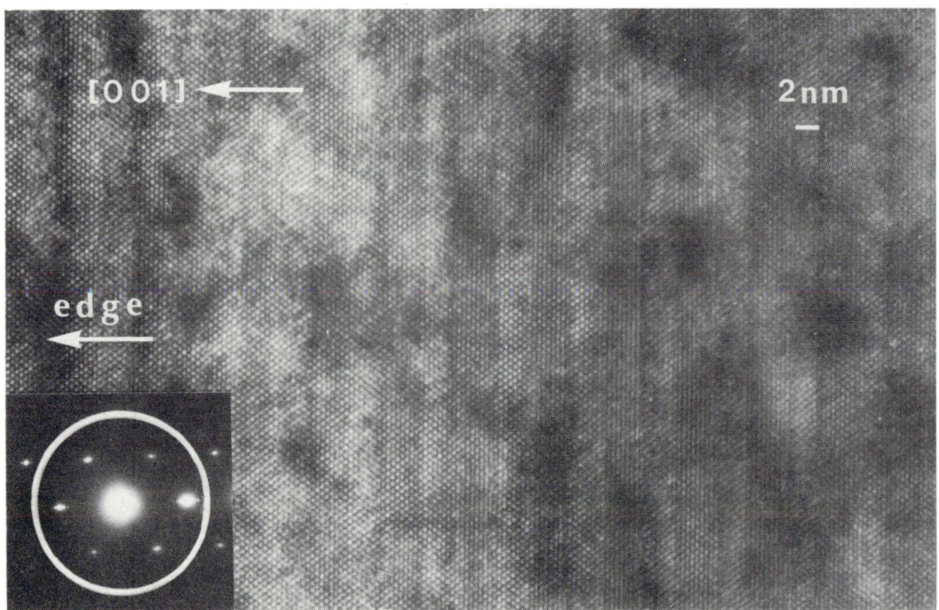


Fig.1a : 110 high resolution image of AlAs/GaAs superlattice (100 KV, axial illumination, the growth direction is arrowed)

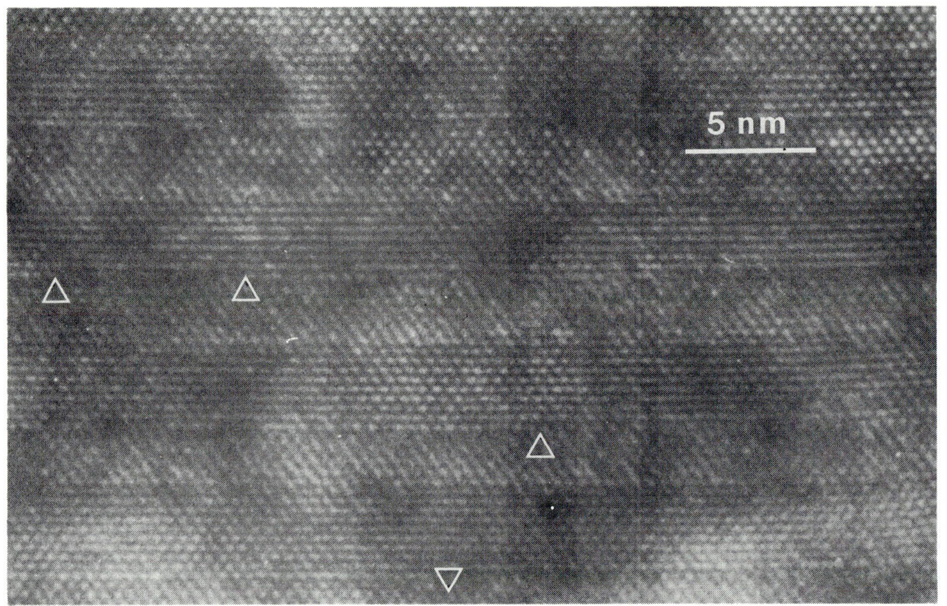


Fig.1b : HREM image (same conditions) showing reticular steps (arrowed) at interfaces.

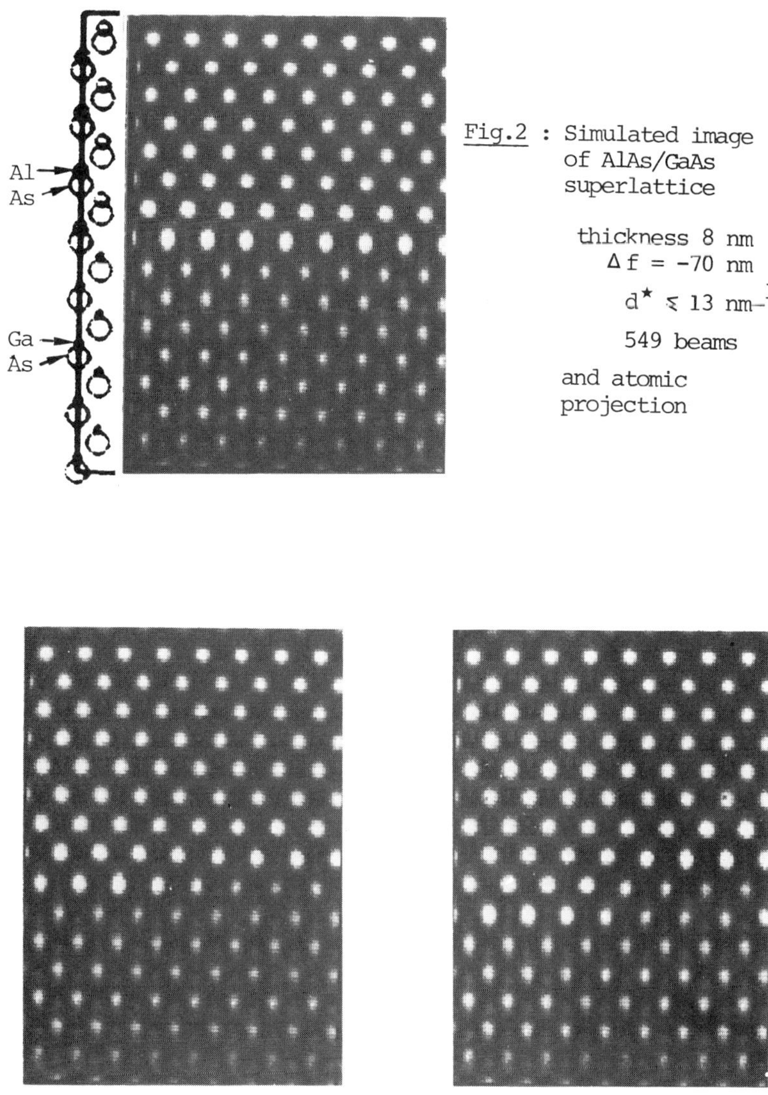


Fig.2 : Simulated image of AlAs/GaAs superlattice

thickness 8 nm
$\Delta f = -70$ nm
$d^{\star} \leqslant 13$ $nm^{-1}$
549 beams

and atomic projection

a b

Fig.3 : Simulated image of AlAs/GaAs superlattice containing a step :
a) one reticular plane (0.283 nm) (2565 beams)
b) two reticular planes (0.566 nm) (2859 beams)

From Fig. 1a, it is seen that there is a very good lattice continuity both within the heterolayers and at the interfaces with no evidence for localized structural defects. Furthermore it was previously found that the density of the structural defects is very low and that they essentially consist of stacking faults and 60° dislocations.

The only noticeable changes specific to interfaces lie in "white spot" intensities on the observed images as shown with arrows Fig. 1b. In order to interpret such discontinuities in terms of Ga and Al atom distribution and atomic steps it is essential to compare observed and simulated images. Thus we calculated the image contrast introduced at the interface by considering first a perfect supercell. For practical reasons two kinds of supercells with orthogonal axes were compared :

a complete one matching X-ray data i.e. $a = b = a_0 \quad c = 17\ a_0$ (136 atoms)

and a simpler one : $e_1 = e_2 = 4\ a_0 \quad a = b = a_0 \quad c = 8\ a_0$ (64 atoms)

Then we considered a new supercell containing one $2\ a_0 \sqrt{2} = 1.6$ nm long step. This step which was periodically repeated every other interface was assumed to pass vertically through the foil and its height was taken either as one or two reticular planes. This led to supercells with dimensions :

$a = a_0 / \sqrt{2} \quad b = 4\ a_0 \sqrt{2} \quad c = 17\ a_0$ (544 atoms) and

$a = a_0 / \sqrt{2} \quad b = 4\ a_0 \sqrt{2} \quad c = 8\ a_0$ (256 atoms)

The best agreement with the electron images is obtained for a foil thickness of 8 nm. In this condition the optimal defocus is -70 nm. Fig. 2 corresponds to the simulated image obtained for these optimized conditions on the perfect superlattice. There is a very good fit between the computed and observed images cf. Fig. 1. In GaAs/AlAs superlattices the contrast changes abruptly at each interface which allows the discontinuity at steps to be imaged. This does not seem to be the case for ternary GaAlAs/GaAs heterolayers (Wright and Williams 1985). Fig. 3 a and b compares the simulated contrast corresponding to both step heights and it is seen that the images match the atomic structure of the steps, thus it should be possible to distinguish steps of one or two reticular planes. The present observations do not rule out the possibility that both single and double steps are present, but up to now single steps only have been unambiguously identified.

## 4. Conclusion

The comparison of observed and simulated images allows one to ascertain the existence of atomic growth steps at interfaces of GaAs/AlAs MBE grown superlattices. Since no structural defects have been found at interfaces it seems likely that the X (e, hh) excitonic peak linewidth of the photoluminescence spectra is directly related to these atomic steps (Weisbuch et al 1981) and that they play a major role (Deveaud and al 1985), depending on their average length (Bastard et al 1984), on the luminescence quantum yield of these heterostructures.

## References

Bastard G , Delalande C , Meynadier M H , Frijlink P M , Voos M 1984 Phys. Rev. B 29 12

Delamarre C, Dubon A , Laval J Y , Guenais B , Emery J Y 1985 MRS proceedings Strasbourg, J. Physique, to be published

Deveaud B., Regreny, Emery J. Y., Chomette A., J. Appl. Phys. to be published

O'Keefe M A , Sander J V 1975 Acta Cryst. A31 303

Regreny A , Emery J Y , Baudet M , Poudoulec A 1985 idem Delamarre C al

Skarnulis A J , Ijima S , Cowley J M , 1976, Acta Cryst. A32 799

Weisbuch C , Dingle R , Gossard A C , Wiegmann W 1981 Sol. State Com. 38, 709

Wright A C and Williams J O 1985 Material letters 3 80

*Inst. Phys. Conf. Ser. No 78: Chapter 10*
*Paper presented at EMAG '85, Newcastle upon Tyne, 2–5 September 1985*


# Crystal structure determination of two ternary Al–Fe–Si compounds using geometric projection of electron diffraction patterns

P Liu and G L Dunlop
Department of Physics
Chalmers University of Technology
S-412 96 Göteborg, Sweden

## 1. INTRODUCTION

Fe and Si are present in all aluminium alloys either as impurities or as intentional additions. Depending upon alloy composition and solidification rate a number of intermetallic phases can form (Miki et al 1975; Simensen and Vellasamy 1977; and Westengen 1982), which influence a number of properties including recrystallisation, texture, formability, strength and ductility.

A commercially pure direct-chill cast Al-0.25Fe-0.13Si alloy has been investigated by transmission electron microscopy (TEM) and two previously unreported phases have been identified. The crystal parameters of these compounds were determined by a method of geometric projection from a series of selected area diffraction (SAD) patterns in the TEM. The chemical composition of the phases was also determined by energy dispersive X-ray analysis (EDX) in the TEM.

## 2. EXPERIMENTAL

The composition of the investigated alloy is shown in Table 1. The solidification rate of the investigated samples was 10°C/s. Both as-cast material and samples heat treated 24 hours at 600°C were investigated. Specimens for electron microscopy were prepared by jet polishing in an electrolyte consisting of 25% $HNO_3$ in methanol at 30 V and -30°C. The metallographic examination was performed in a JEOL 200 CX TEM/STEM electron microscope operated at 200 kV and equipped with a Link 860 System energy dispersive X-ray spectrometer. The specimens for X-ray analysis were mounted in a graphite holder in order to minimize the effect of spurious X-rays. The results were made quantitative using the Link RTS2/FLS computer program which applies the Cliff-Lorimer thin film approximation (Cliff and Lorimer 1975) with experimentally determined $k_{xSi}$ values (Wirmark 1984), utilizes stored standard spectra of pure elements for peak stripping and corrects for absorption. The SAD experiments were performed using a double tilt holder capable of ± 45 degrees.

The following method for constructing geometrical projections from SAD patterns was used (Kuo et al 1983). When the crystal is tilted around the closest packed reciprocal lattice direction from one pole to another, the corresponding specimen tilt angle, i.e. the angle between two successive SAD patterns, is calculated from the angles shown on the goniometer for X- and Y-tilt respectively. This new SAD pattern is then projected back to a projection plane perpendicular to the axis of rotation after a rotation of the pattern equal to the experimental tilt angle. In this manner, a series of SAD patterns with the same tilt axis is successively projected on to the same projection plane perpendicular to the tilt axis. Thus a reciprocal lattice plane is obtained. If the same procedure is repeated with tilting around a new axis not parallel with the previous tilt axis, two non-coplanar reciprocal lattice planes are obtained. From these planes the reciprocal lattice can then be constructed.

## 3. RESULTS AND DISCUSSION

The general microstructure in the as-cast condition is shown in Fig. 1 (a).$\alpha$-AlFeSi (bcc, a = 12.56 Å (Cooper, 1967)) and the finer, $q_1$-AlFeSi were precipitated along the cell boundaries (Fig. 1 (a)). The crystal structure of $q_1$-AlFeSi was determined by obtaining two series of 12 SAD patterns by tilting the specimen around each of the two perpendicular axis $\bar{R}_1$ and $\bar{R}_2$ shown in Fig. 1 (b). The corresponding reciprocal lattice planes related to the two tilt axes are shown in Fig. 2. Since the tilt axes are perpendicular to each other, so are the two reciprocal lattice planes. The reciprocal unit cell constructed from these two planes and the corresponding crystal unit cell are shown in Fig. 3. $q_1$-AlFeSi has a c-centered orthorombic structure with the following lattice parameters: a=12.7Å b=36.2Å c=12.7Å.

After heat treatment for 24 hours at 600°C the $q_1$-AlFeSi phase disappeared in favour of $q_2$-AlFeSi. The same geometric projection method was used to show that this phase has a simple monoclinic structure with following lattice parameters: a=12.5Å b=12.3Å c=19.3Å $\beta$=109°

Agreement between experimental and calculated d-plane spacings and angles was very good for both phases. The difference between experimental and calculated angles was less than 1.2 degrees and the relative error of the d-plane spacings was always smaller than 1 per cent.

The chemical compositions of the two phases as determined by STEM/EDX are shown in Table 2. It can be noted that the compositions of both phases and also two of their lattice parameters are very similar. Accordingly, it seems likely that $q_1$-AlFeSi transforms "in-situ" to $q_2$-AlFeSi during heat treatment.

X-ray diffraction patterns for some crystal systems, such as the monoclinic, are generally complex and can seldom be solved without the aid of a computer. It can be seen from the present work that with use of the geometrical projection method SAD patterns can be readily indexed and the crystal parameters of an unknown phases easily determined.

## 4. ACKNOWLEDGEMENTS

Discussions with Dr.T.Thorvaldsson and H. Westengen are gratefully acknowledged as is the financial support of the Swedish Board for Technical Development. Special thanks are given to Dr. H.Q. Ye, Dr. Q.B. Yang and Prof. K.H. Kuo who also participated in valuable discussions.

## 5. REFERENCES

Cliff G and Lorimer G W 1975 J. Microscopy 103 203

Cooper M 1967 Acta Cryst. 23 1106

Kuo K H et al. 1983 Electron Diffraction Pattern with its Applications of Crystallography (in Chinese). (Beijing: Science Press) 241

Miki I Kosuga H and Magahana K 1975 J. Jap. Inst. L. Met. 32 1

Simensen C J and Vellasamy R 1977 Z. Metallkde 68 428

Westengen H 1982 Z. Metallkde 73 360

Wirmark G Thorvaldsson T and Nordén H 1984 Proc. Inst. Phys. Conf. Ser. 68 71

Table 1. Composition of experimental alloy (wt.%)

| Fe | Si | Mn | V | Zn | Cr | Cu | Ga |
|---|---|---|---|---|---|---|---|
| .270 | .125 | .005 | .008 | .016 | .001 | .004 | .007 |

| Mg | Ni | Pb | Sn | Ti | Zr | Al |
|---|---|---|---|---|---|---|
| .001 | .002 | .002 | .001 | .003 | .001 | balance |

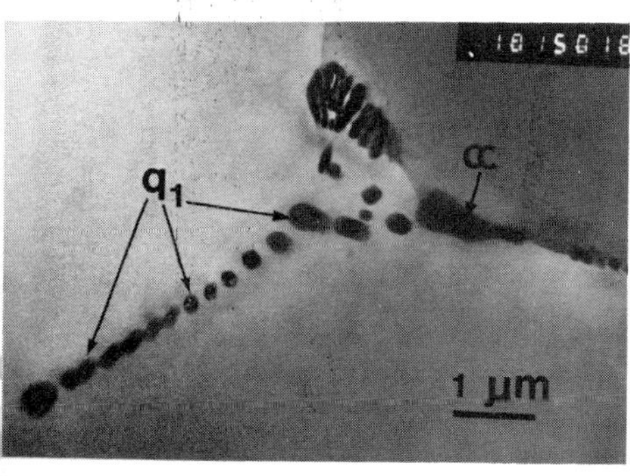


(a)

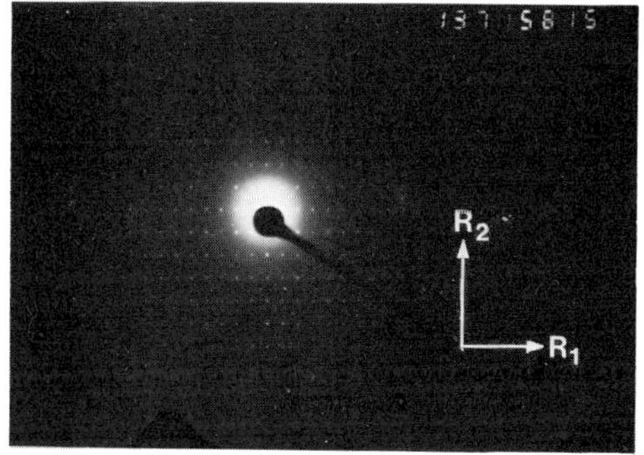


(b)

Fig1(a) α-AlFeSi and $q_1$-AlFeSi precipitates in the cell boundaries of as-cast material.

1(b) SAD pattern of $q_1$-AlFeSi showing the two closest packed reciprocal directions $\bar{R}_1$ and $\bar{R}_2$.

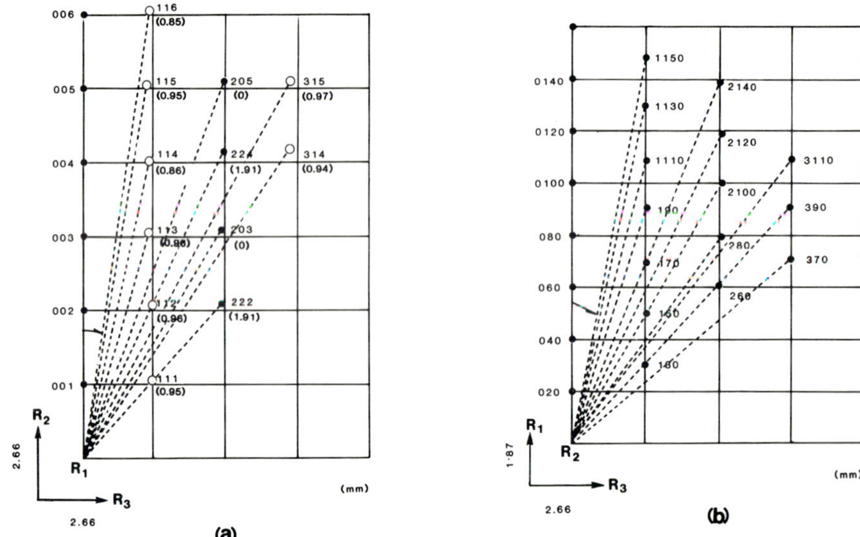


Fig 2 Reciprocal lattice plane obtained by projection of SAD patterns obtained by tilting around $\bar{R}_1$(a) and $\bar{R}_2$(b).

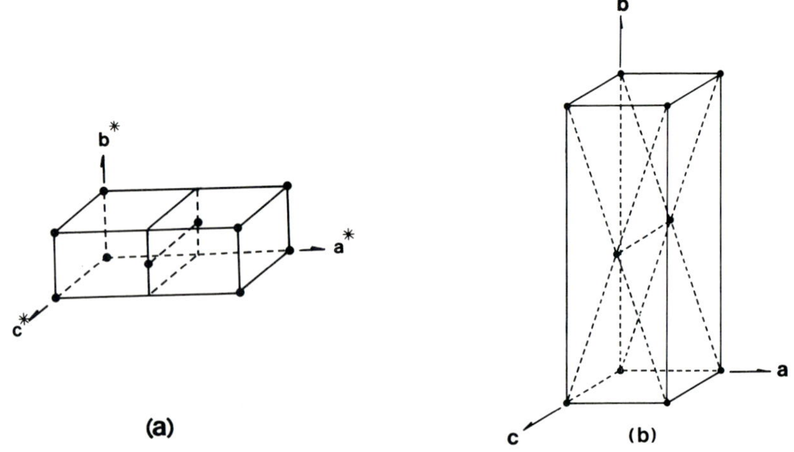


Fig 3 (a) Reciprocal lattice unit cell of $q_1$-AlFeSi constructed from two planes of the type shown in Fig 2.

(b) Corresponding crystal lattice unit cell for $q_1$-AlFeSi.

Table 2. Composition of both ternary AlFeSi compounds as determined by STEM/EDX (wt.%).

| | Al | Fe | Si |
|---|---|---|---|
| $q_1$-AlFeSi | 70.5±2.0 | 25.8±2.0 | 3.7±0.6 |
| $q_2$-AlFeSi | 68.7±0.6 | 27.8±0.6 | 3.4±0.3 |

*Inst. Phys. Conf. Ser. No 78: Chapter 10*
*Paper presented at EMAG '85, Newcastle upon Tyne, 2–5 September 1985* 

# The use of selected area channelling in the STEM for the study of local texture development during grain growth

E M Grant, A J Bourdillon and W M Stobbs
University of Cambridge, Department of Metallurgy and Materials Science
Pembroke Street, Cambridge CB2 3QZ

The advantages of neutron diffraction methods over X-ray or EM approaches for the measurement of textures are well known (see e.g. Szpunar 1984). An essential disadvantage of the technique becomes apparent when attempts are made to relate the texture development during annealing to changes in the grain size distribution. This is often necessary, however, to clarify the relevant mechanism of normal or anomalous grain growth (see e.g. Distl et al. 1983) and information is then required about the texture changes of specific components in the grain size distribution.

The kinetics of grain growth and texture development for copper have been measured by neutron diffraction using the DR3 reactor at Risø (Grant et al. 1984). The results obtained indicated relatively stronger retention of the <111> fibre component of the texture than the <100>, and a parallel microstructural investigation revealed complex changes in the grain size distribution. The results were thus of a type for which a knowledge of any grain size distribution/texture relationship would be of value. Here we describe the application of backscattered channelling techniques to the problem.

Johnson Matthey "Specpure" copper of T.M.I. $< 10^{-5}$, as previously examined (Grant et al. 1984), was cold drawn to 20% reduction in area and then recrystallised for one hour at 400°C before being subjected to a series of grain growth anneals. During these anneals the characteristic tendency was for a few grains to grow relatively rapidly to a size of ~200μm but then to stop growing despite being surrounded by much smaller (~ 10μm) grains. Further annealing in general resulted in a greater number but not a significantly increased size of these larger grains. The size distribution and spacings involved ruled out respectively the convenient application of standard SEM and TEM methods. Accordingly, a relatively old STEM system on a Philips 400T was modified to provide the data required from bulk specimens, though it should be noted that modern Philips STEM systems allow for the formation of backscattered selected area channelling patterns (BSCPs). The simple modifications required for our system involved being able to apply alternatively a scan to the beam tilt for the formation of a BSCP, or a scan to the deflection for the formation of a backscattered image. So that BSCPs could be reliably obtained from small regions (< 1μm in diameter) the main requirement was to ensure purity of beam tilt and shift, this being limited by the aberrations of the condensing lens system as a function of the tilt angles required. It was found to be convenient to set the eucentric height, and the objective lens current, using a thin copper foil and then to set the backscattered detector amplifiers for the formation of a BSCP from a thick region of this foil. With the thin foil replaced by the bulk specimen to be examined, it was then necessary only to focus the BSCP using the specimen height adjustment. It was found that BSCPs of 20$^{o}$ full angular width could be obtained from regions of ~5µm in diameter, but reducing this rocking angle to 5$^{o}$ allowed patterns to be obtained from sub-micron sized grains at reasonable scanning rates, and with the best contrast being obtained at 40kV.

For the determination of the orientation of a large number of grains, as required here, it proved most convenient to stay in the channelling pattern mode of operation. Starting from a fiducial mark, the specimen could be traversed, each orientation encountered then being recorded on a channelling map. The traverse position was noted to the nearest 1µm each time the pattern changed, so that the grains whose orientation had been measured could be identified by correlation with a backscattered SEM image of the same area.

The orientations obtained for a succession of numbered grains along a 500µm line on the starting material, as-recrystallised at 400$^{o}$C, is shown in figure 1, and, while no non-random correlations of adjacent grain orientations are apparent, the overall distribution correlates reasonably with the inverse pole figure obtained by neutron diffraction methods (figure 2). The overall grain orientation distribution found after a 1 hour treatment at 460$^{o}$C showed that it is specifically the larger developing grains (★) which tend to have near (111) orientations (figure 3) (confirming results obtained by standard SEM methods). More interestingly, the analysis of the orientations of the small grains surrounding the larger grains suggested a non-random orientation distribution of these retained neighbours. The grain orientations shown in figure 4 were obtained for the grains marked on the SEM image in figure 5.

The statistics involved in this type of analysis require large amounts of information to be obtained fairly painlessly, and the method described is capable of this. Our preliminary results appear to indicate that the apparently anomalous grain growth tends to stop when the small grains which surround a large grain are at relatively low angles of misorientation either to it or to its primary twin orientations. The reasons for this will require further TEM analysis of possible reasons for the different relative developments of primary recrystallisation twins from grain to grain in the initial texture.

Acknowledgements

We are grateful to Drs N. Hansen and D. Juul Jensen, and to Prof. B. Ralph for useful discussions; to the Risø National Laboratory for funding; and to Prof. D. Hull for the provision of laboratory facilities.

References

Distl J S, Welch P I and Bunge H J, 1983 Scripta Metall. 17 975

Grant E M, Juul Jensen D, Ralph B and Hansen N, 1984 Proc. 7th Int. Conf. on Textures of Materials ed. C M Brakman et al. (Zwijndrecht: Netherlands Society for Materials Science) pp239-244

Szpunar J A, 1984 J. Mater. Sci. 19 3467.

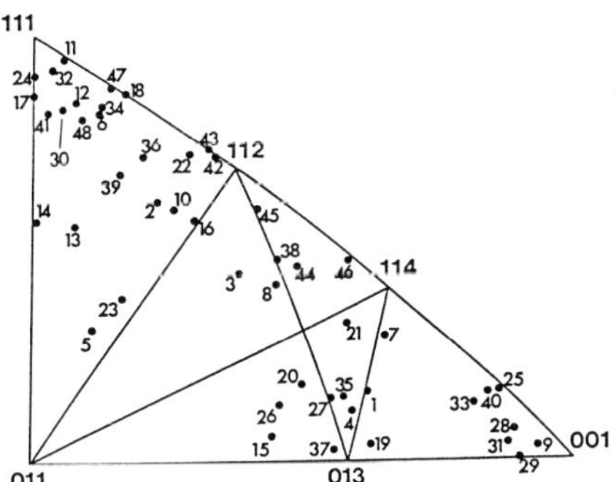


Figure 1 Unit triangle showing orientations of grains encountered on a 500μm traverse of the starting material (as-recrystallised at 400°C). The grains were numbered in sequence as they were found.

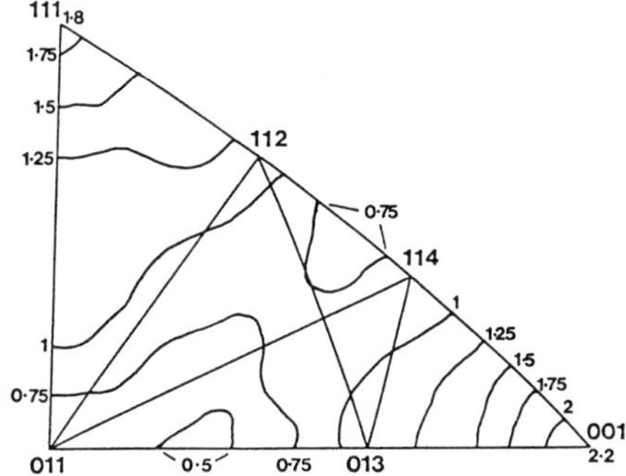


Figure 2 Inverse pole figure of the starting material. Contour levels are in multiples of random intensity.

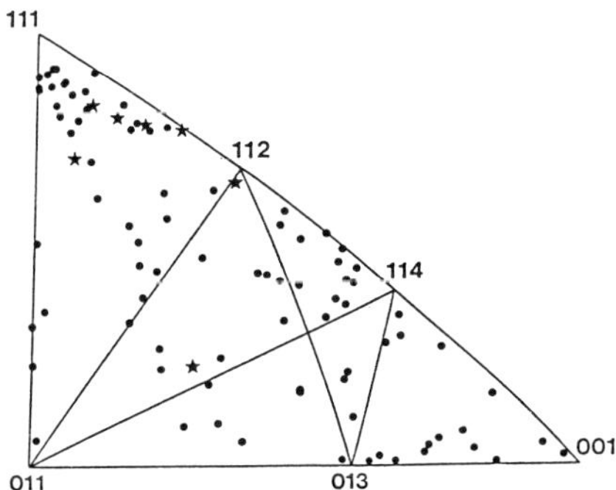


Figure 3 Unit triangle showing orientations of all grains encountered during several traverses of a specimen annealed for 1 hour at 460°C. Grains over 100μm in size are indicated ★ .

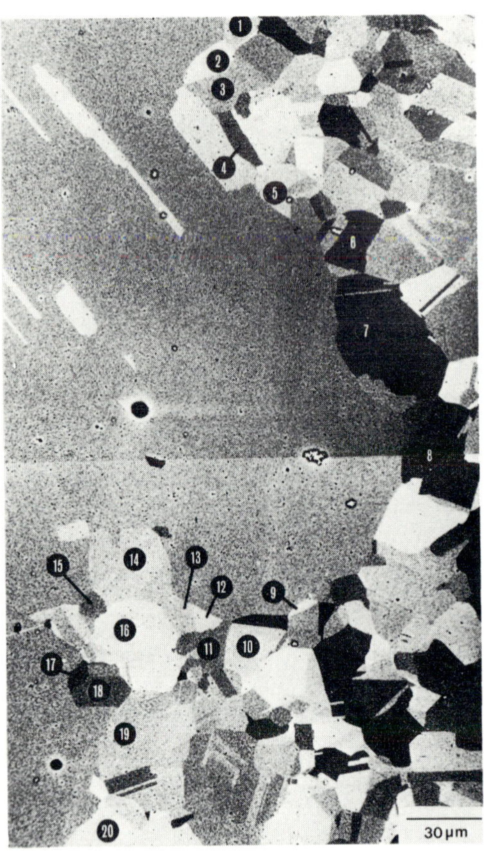


Figure 4 Backscattered SEM micrograph showing a large grain, the primary twins within it and its immediate smaller neighbours. The orientations of the numbered grains are shown in Figure 5.

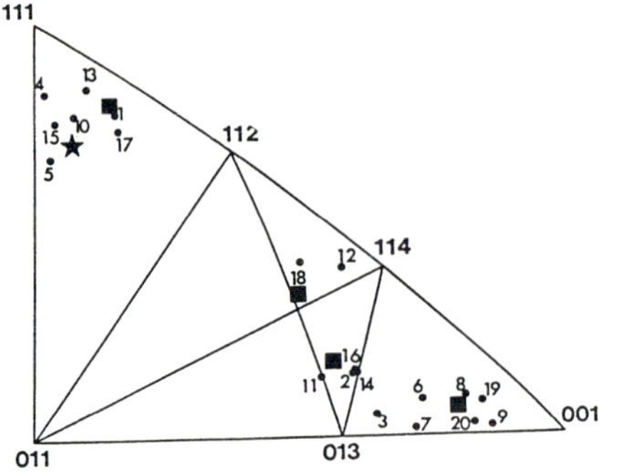


Figure 5 Unit triangle showing orientations of the large grain (★), the four primary twins it could form (■), and the 20 neighbours shown in Figure 4.

*Inst. Phys. Conf. Ser. No 78: Chapter 10*
*Paper presented at EMAG '85, Newcastle upon Tyne, 2–5 September 1985*



# Early stages of the omega transition in a Ti–20 wt% Mo alloy

A S Bor* and W M Stobbs

Department of Metallurgy and Materials Science
Pembroke Street Cambridge CB2 3QZ

*On leave from Metallurgical Engineering Dept.
Middle East Technical University Ankara Turkey

Ti-20 wt% Mo alloy is typical of many omega (ω) phase forming alloys in that electron diffraction patterns of the material in the quenched state exhibit complex and controversial diffuse scattering (de Fontaine et al 1971). The scattering has for example been variously argued to be due either to static or dynamic lattice displacement waves, whether transverse or longitudinal, associated with the inherent instability of the bcc lattice and possibly linked to charge density wave - lattice displacement coupling (Cook 1975) or to precursor nucleation effects for the omega phase particles growing with <111> prolate morphology (Kuan and Sass 1976) or simply to stacking effects for various subvariants of the omega structure (Borie et al 1973).

We describe here some of the results of a more extended investigation the aims of which were to characterize the nucleation and growth of the omega phase in this alloy and to clarify whether or not the diffuse scattering could be unambiguously associated with lattice distortions or precursory effects.

The alloy examined was prepared by multipass argon arc melting followed by warm rolling and a homogenization at 1100°C. Electron microscopy foils were prepared by standard electropolishing techniques using a 6 % perchloric acid solution in butanol and methanol at ~25 V and ~-50°C from ~1 mm thick discs which had been solution treated at 850°C in the β phase field prior to quenching. Despite the quench being rapid, it is probable that some ageing cannot be avoided during the process and this is emphasised by the finding that specimen left for only five hours at room temperature could be readily distinguished from a freshly quenched specimen in that the former exhibited more well developed omega phase particles than the latter. It was also found that the electron beam accelerated the ageing process so, when examining freshly quenched specimens, particular care was taken to spend as little time as possible in the examination of any one area.

(110) diffraction patterns of the as quenched, lightly aged and aged material are shown in figures 1a, 1b and 1c, respectively, while diffraction patterns of a quenched alloy at (113) and at (112) are shown in figures 1d and 1e, respectively. The β→ω transition can most easily be described as the <111> collapse of two $(111)_\beta$ planes to form a hexagonal unit cell with atoms ideally at 0,0,0; 1/3,2/3,1/2; 2/3,1/3,1/2. Three

subvariants on each of four systems are possible depending on which two of the three (111) planes in the triple stacking sequence collapse. The omega diffraction patterns exhibit reflections at ±1/3<112>, the development of which upon ageing is clear by comparison of figures 1a and 1b, and it can be seen for example that their more localized nature in figure 1b prevents off plane reflections (typified by their varied projection on tilting) such as are seen at -1/2,-1/2,5/6 in figure 1a from appearing in figure 1b. Equally while the typical <112> diffuse scattering may be seen to become relatively weaker on ageing its extent does not decrease making it immediately difficult to associate either with particle shape or with internal faulting effects for growing omega phase nuclei. It is this diffuse streaking which has been argued to be associated with lattice distortions and while this was found to be supported by a tendency for the strength of streaks to become greater at increased g the generally greater strength of tangential to radial streaking (see figure 1) is qualitatively suggestive in that this is indicative of scattering by transverse distortions. By the examination of a number of different normals the diffuse scattering was shown to be in the form of (111) sheets and generally thus remarkably similar to that seen in the $\beta$ phase of marmem alloys such as Cu-Zn-Al, Ni-Ti or Ni-Al (Stobbs 1979, Georgopoulos 1981). Although it differs in detail in, for example, the rumpled geometry of the strong diffuse streaking seen at (113), figure 1d.

The essential problem in delineating the differing or related origins of the diffuse streaks and omega reflections lies in differentiating the real space imaging behaviour of a phase with generally locally variable lattice spacing from that of a discretely nucleated particle of a second phase with a necessarily uniform (even if strained) lattice embedded in a parent matrix. Recipes for this are well known: appropriately tilted dark field images of a second phase should exhibit stereo effects and would not be expected to be aperture size dependent beyond some lower limit markedly larger than the unit cell size (Stobbs 1984). As in many investigations of this type it is essential to avoid the potentially confusing effects of amorphous contamination. Here this can be accomplished by imaging using aperture positions other than 1 on the streaks and omega reflections as shown in figure 1f beyond the first major diffuse ring for carbon. This is emphasized by comparison of figures 2a and 2b with aperture positions 1 and 2 in figure 1f, respectively. While for both images a phase and a distortion should be imaged the speckle necessarily uniquely associated with carbon contamination may be seen clearly at the edge of the specimen in figure 2a but is virtually absent from figure 2b.

The images of room temperature aged material shown in figures 3a, 3b, 3c and 3d were formed using aperture positions 2, 3, 4 and 5 respectively in figure 1f. Comparison of figures 3a and 3b demonstrates that speckle is indeed associated with both the streaks and the omega reflections. It may also be seen both that not all the scattering associated with the omega phase spot position is aperture size limited (compare figures 3a and 3d) and that,as is further confirmed by examination of figures 3b and 3c, much of the speckle is inherently associated with the streaks alone. It proved difficult to gain further evidence for the continued existence of lattice distortions after the nucleation of the omega phase, as indicated above, either by $2\frac{1}{2}$ D or 3 D techniques. This proved to be because not only do the distortions remain after omega phase nucleation (so that reduced streak intensity on ageing is associated with a decreased volume fraction of parent phase rather than reduced distortion amplitudes) but also because, as examined at room temperature under electron irradiation, particles of omega phase were continuously forming, dissolving and re-forming up to sizes 2 nm. This is demonstrated by the temporal sequence of micrographs of the same area shown in figure 4.

The evidence presented, which has been further confirmed by a variety of other techniques, indicates that the streaking is associated with distortions and that these distortions, inherent for this bcc lattice, remain in existence after nucleation and omega phase growth has commenced. This suggests that the distortions, despite being dynamically unstable, are not directly associated in a precursory way with omega phase nucleation.

One of us, ASB, is indebted to the British Council for support while on leave and we are grateful to Prof.D.Hull for the provision of laboratory facilities.

references

Borie B, Sass S L and Andreassen A 1973 Acta Crystall. A29 585,594
Cook H E 1975 Acta Metall. 24 1027,1041
de Fontaine D, Paton N E and Williams J G W 1971 Acta Metall. 19 1153
Georgopoulos P and Cohen J B 1981 Acta Metall. 29 1535
Kuan T S and Sass S L 1973 Acta Metall. 24 1053
Sikka S K, Vohra Y K and Chidambaram R 1982 Prog. in Mat. Sci. 27 245
Stobbs W M 1984 J.Micr. 136 137
Stobbs W M 1979 ICOMAT'79 526

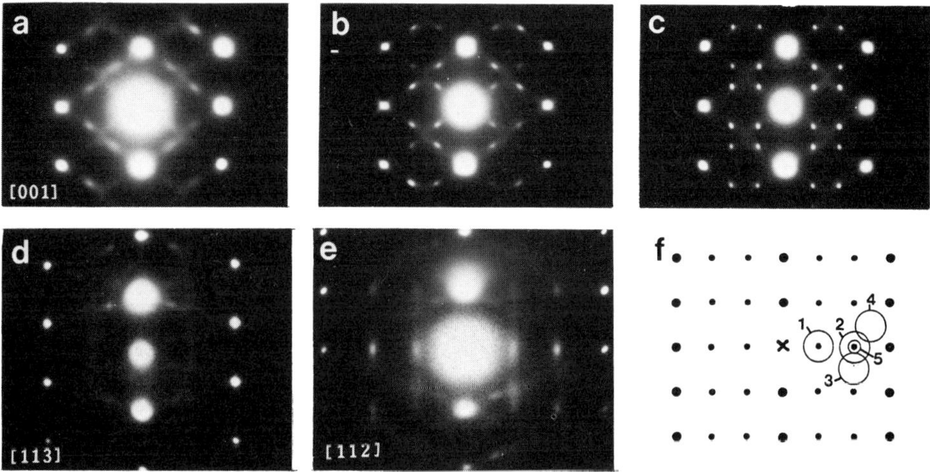


Figure 1. Diffraction patterns of: quenched (a), (d) and (e) ; aged at 25°C for 50 hrs (b) ; and aged at 370°C for 2 hrs (c) material.

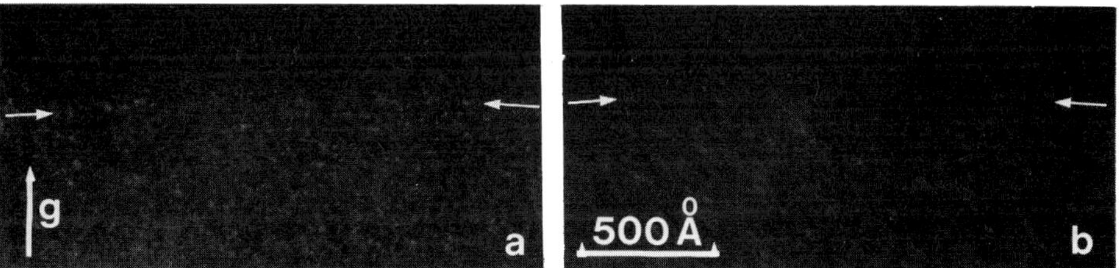


Figure 2. Dark field images of quenched material using aperture positions 1 (a) and 2 (b) as given in figure 1f. Foil edge, above which speckle due to contamination is apparent in (a), is marked.

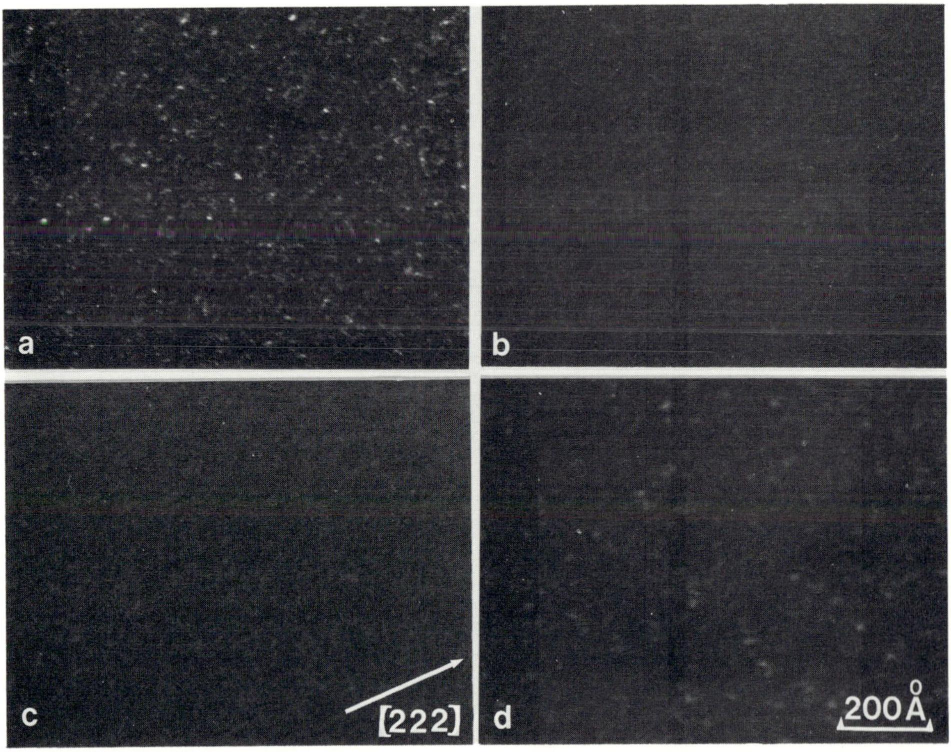

Figure 3. Dark field images of the same area of the quenched material using aperture positions 2 (a) ; 3 (b) ; 4 (c) ; and 5 (d) as given in figure 1f. The common [222] direction is marked.

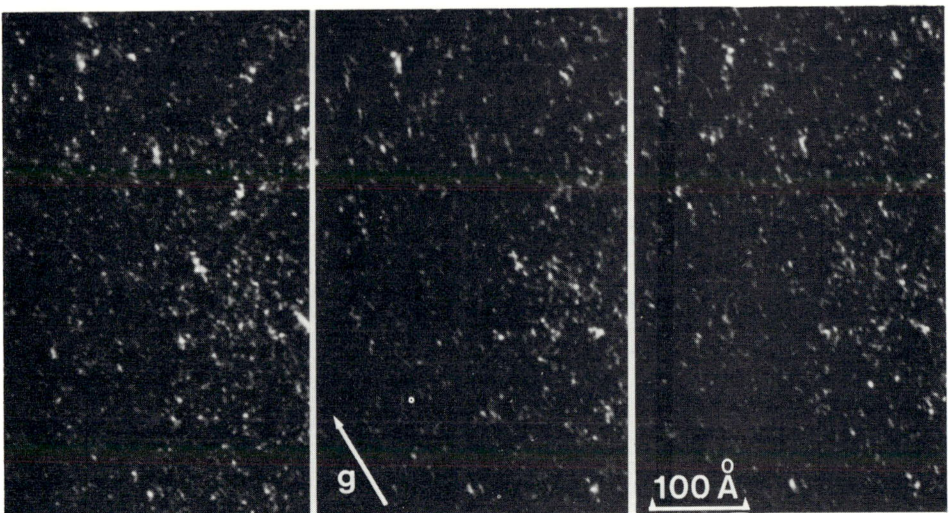


Figure 4. The 60 sec interval time sequence of (0002) omega dark field images of 70 hrs room temperature aged material.

*Paper presented at EMAG '85, Newcastle upon Tyne, 2–5 September 1985*

# Microtextures in two-phase aluminium alloys deformed at elevated temperatures

P N Kalu and F J Humphreys

Department of Metallurgy and Materials Science, Imperial College,
Prince Consort Road, London SW7 2BP.

## 1. Introduction

The effect of second-phase particles on the high temperature deformation of metals is complex. Gliding dislocations may interact with particles, generating extra dislocations, leading, at large particles to the formation of local lattice rotations. (e.g. Humphreys 1985), and the particles may also hinder the movement of both high and low angle boundaries at high temperatures. Thus the microstructure, mechanical properties and texture of a two-phase alloy may be quite different to that of a similar, but single-phase alloy.

This paper reports part of a wide ranging study of the effect of non-deforming particles on the deformation at elevated temperatures of aluminium, and discusses the use of an electron beam microtexture technique to investigate the conditions under which lattice rotations are formed at particles.

## 2. Experimental

Aluminium-silicon alloys were heat treated to produce dispersions of silicon particles, as shown in the table below.

| Designation | Wt% Silicon | mean particle diameter μm | volume fraction % |
|---|---|---|---|
| S | 0.01 | - | 0 |
| E | 0.21 | 0.52 + 0.31 | 0.14 |
| B | 0.6 | 2.84 + 1.10 | 0.56 |
| D | 1.6 | 9.68 + 3.25 | 1.64 |

Compression specimens were deformed at temperatures between 293 and 723K at strain rates of between $\sim 10^{-6}$ and $\sim 10^{-1}$ $s^{-1}$. The tests were terminated by direct water quenching of the specimens in order to preserve the deformed microstructure. Microstructures were examined by optical and electron microscopy. The spread of lattice orientations both in the matrix, and adjacent to the particles was measured in a JEOL 100CX TEMSCAN usingthe microtexture technique (Humphreys 1983a,b).

Using this technique, a diffraction pattern is obtained from an area of a thin foil, typically of diameter 5 to 50μm, using the microscope in a standard diffraction mode. A selected Debye ring of the diffraction pattern, typically 111, is then scanned at 1.5 degree intervals over a transmission detector, the process being repeated automatically as the eucentrically mounted specimen is tilted over a range of ± 50 degrees. The data is collected and processed by a microcomputer to produce a semiquantitative pole figure of the area selected. The procedure may be reversed, so that a texture component, selected on the VDU may subsequently be imaged in dark field. The technique is particularly suitable for analysing specimens in which large orientation gradients are present.

## 3. Results

Results are presented for alloys containing particles of diameter 0.5μm to 10μm, deformed at 523K at various strain rates.

The stress-strain curves of the alloys are parabolic up to a strain of ~0.05, and the work hardening rates, expressed as the parabola coefficient are shown in figure 1 as a function of strain rate. The lattice misorientations occurring at the particles were found to be a function of the strain, strain rate and particle size. A typical 111 pole figure from an area containing two 10um particles is shown in figure 2a. The spread of lattice misorientations is ~30 degrees, which may be compared with the lattice misorientations of ~8 degrees in an adjacent matrix area of the same size, as shown in figure 2b. A bright field micrograph of the particles is shown in figure 2c, and figure 2d is a dark field micrograph obtained using the texture component X of figure 2a, which is misoriented 14 degrees from the matrix.

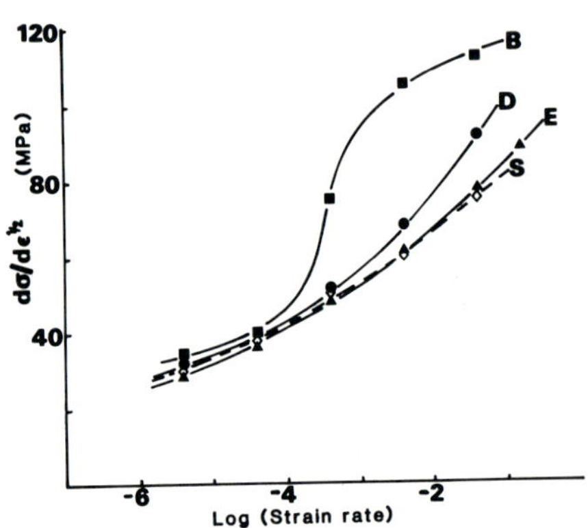


Fig 1. Work hardening rates at 523K.

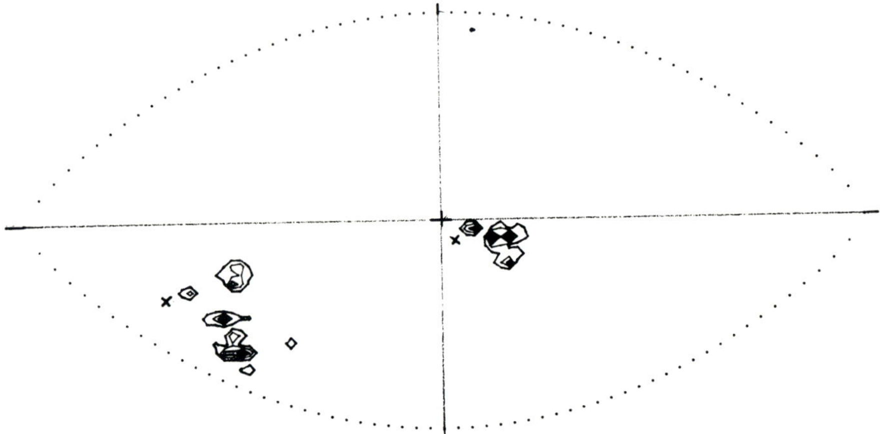

Fig. 2a. 111 pole figure from an area containing the particles shown in fig 2c.

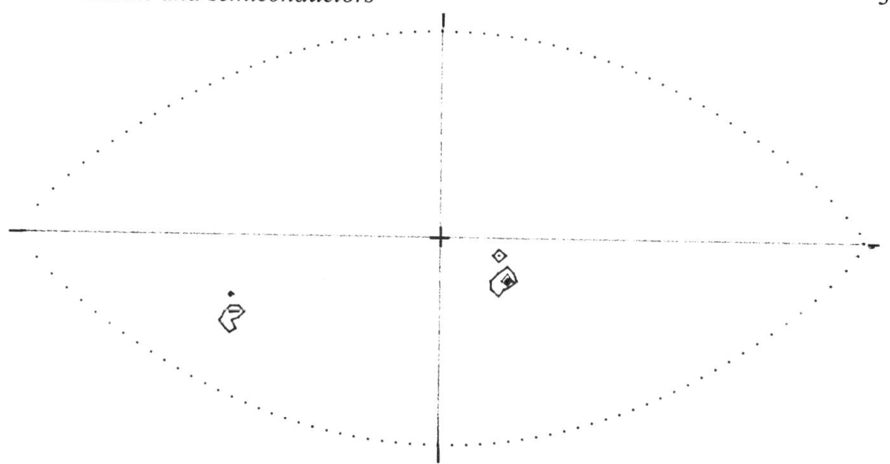

Fig. 2b. 111 pole figure from the matrix adjacent to the particles of fig 2c.

5 um

Fig. 2c. The area from which fig 2a was obtained.

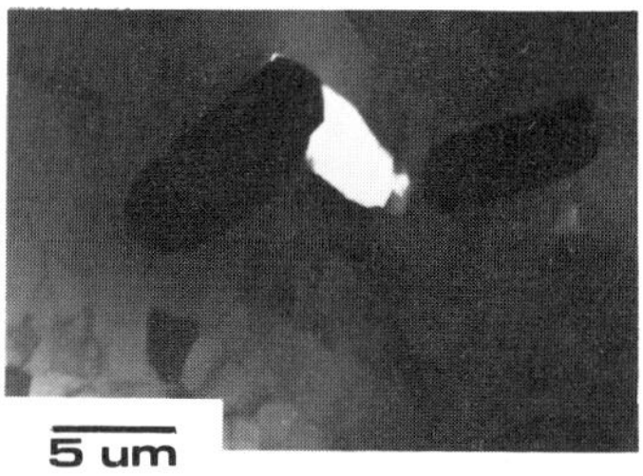

Fig. 2d Dark field micrograph using texture component X of fig. 2a.

The cumulative misorientations measured from such pole figures, for alloys B and D deformed 50%, are shown in figure 3.

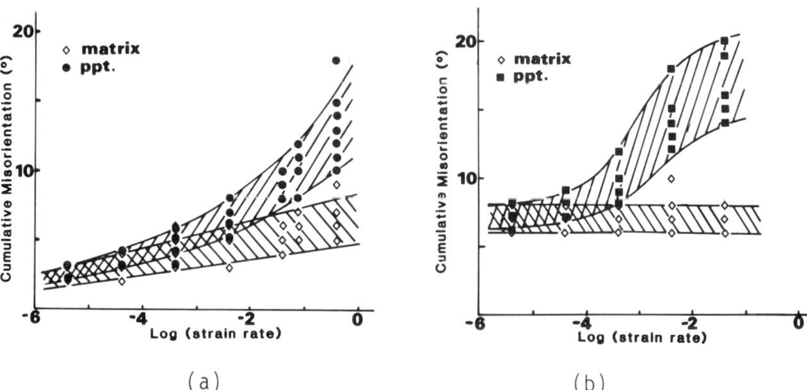


(a) (b)

Fig. 3 Strain rate dependence of the orientation spread at the particles and matrix after deformation at 523K in alloy B (a) and alloy D (b)

It can be seen that there is some scatter in the results. This is partly due to the sectioning effect whereby some misoriented regions are removed by electropolishing, and partly due to some variation in particle size in the alloys. Mean values of misorientation for particles of different size in these alloys are shown in figure 4.

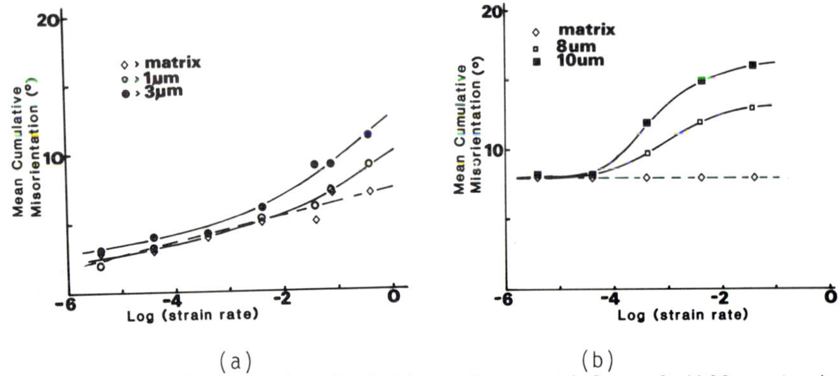


Fig. 4 Mean cumulative misorientations for particles of different sizes in a) Alloy B, and b) Alloy D.

## 4. Discussion

It can be seen from figure 4 that, at low strain rates, the lattice misorientations at the particles are similar to those of the matrix, and figure 1 shows that under similar conditions, the work hardening rates of the single phase alloys are similar to the two-phase alloys. It is reasonable to assume that at low strain rates, the dislocations are able to avoid the particles by climb, without leaving any dislocation debris. It is however, noticeable that even at the lowest strain rates, the lattice misorientations in the matrix of alloy B are greater than those in alloy D. The critical strain rate above which dislocation accumulate at the particles, the work hardening rate increases, and lattice misorientations occur, is seen to decrease with increasing particle size. Various theories of dislocation climb around particles ( see e.g Brown 1979), suggest that at a constant temperature, the critical strain rate should be inversely proportional to the cube of the particle size. A preliminary analysis of these results is in accordance with these predictions.

## 5. References

Brown L M 1979 Proc 5th Int Conf on Strength of Metals and Alloys, 1551-1572

Humphreys F J 1983a Textures and Microstructures, 6, 45-62.

Humphreys F J 1983b EMAG83 Inst Phys Conf series no.68, 283-288.

Humphreys F J 1985 In Dislocations and Properties of Real Materials, Institute of Metals, London, 175-204.

*Inst. Phys. Conf. Ser. No 78: Chapter 10*
*Paper presented at EMAG '85, Newcastle upon Tyne, 2–5 September 1985*



# The electron charge distribution of $\beta'$ NiAl

Alan G Fox

School of Engineering, The Polytechnic, Wulfruna Street, Wolverhampton, WV11LY, UK, and
Center for Advanced Materials, Lawrence Berkeley Laboratory, University of California, Berkeley, CA 94720, USA

## 1. Introduction

The use of the Critical Voltage Effect, $V_c$, in HEED to assist in producing accurate sets of structure factors for elements has been outlined by Smart and Humphreys (1980) who produced finely detailed electron density maps for several cubic elements. Fox (1984), and in another paper in these proceedings, Fox and Fisher (1985) have shown how $V_c$ measurements can detect electron charge distribution changes arising from the alloying of cubic elements in binary primary solid solutions. For intermediate phases, especially the electron compounds whose stability is based on specific electron/atom ratios, a set of accurate structure factors is very important for producing electron density maps which will give hitherto unknown solid-state bonding information about these alloys. In addition, with the increasing importance of HREM in materials characterization, O'Keefe and Fox (1985) have found that it is often necessary to have good sets of structure factors for both elements and compounds in order to make accurate computer simulations of HREM images. The object of the present work is to show how $V_c$ measurements on the 3/2 electron compound $\beta'$NiAl (50 at % Ni) can produce accurate low angle structure factors for this alloy, which combined with higher angle values deduced from best pure element atomic scattering factors (form factors) give a complete structure factor determination. $\beta'$NiAl was chosen because it is one of the few intermediate phases for which a complete x-ray structure factor analysis has been made, and so a direct comparison with previous work is possible.

## 2. Theory

The structure factors, F, for an ordered stoichiometric B2($\beta'$) phase, comprising A and B atoms, such as $\beta'$NiAl, are given by:

$$F = f_A \exp(-B_A \sin^2\theta / \lambda^2) \pm f_B \exp(-B_B \sin^2\theta / \lambda^2) \qquad (1)$$

where the + sign refers to a fundamental reflection and the − sign to a superlattice reflection, $f_{A(B)}$ are the form factors for atoms A(B) and $B_A$ and $B_B$ their Debye-Waller factors. $\theta$ is the Bragg angle and $\lambda$ the electron (or x-ray) wavelength. Fox (1983) produced a model for $B_A$ and $B_B$ to analyse $V_c$ measurements in B2 alloys, but in this case an accurate set are available from the work of Georgopolous and Cohen

(1977). Equations (1) and a suitable computer programme (see e.g., Fisher, 1968) were used in conjunction with Doyle and Turner (1968) free atom values of $f_A$ and $f_B$ to produce a complete set of theoretical $V_c$'s for all low angle systematic rows up to (211). $V_c$'s in higher order rows are all > 3MeV and out of the range of any HVEM. These results are shown in Table 1.

Table 1. Theoretical and Experimental room temperature $V_c$'s below 3MEV for $\beta$'NiAl

| Systematic Row (hkl) | $V_c$'s (Theoretical)(kV) | | $V_c$'s (Experimental)(kV) |
|---|---|---|---|
| 100 | $V_c^{300}$ | 510 | $690 \pm 50^{(1)}$ |
| | $V_c^{400}$ | 1819 | $1600 \pm 20^{(2)}$ |
| 110 | $V_c^{220}$ | 479 | $457.5 \pm 2.5^{(1)}$ |
| | $V_c^{330}$ | 1164 | not detected |
| | $V_c^{440}$ | 2696 | not detected |
| 111 | $V_c^{333}$ | 1530 | $1550 \pm 100^{(2)}$ |
| 210 | $V_c^{630}$ | 2419 | Difficult to measure (see text) |
| 211 | $V_c^{422}$ | 3183 | >3MeV |

Sources of data (1) from Fox (1983) (2) from present work.

## 3. Experimental

Thin foils of $\beta$'NiAl for TEM were prepared from 3 mm discs (0.25 mm thick) using a TENUPOL proprietary twin-jet electropolisher operating at 70V (room temperature) with a solution of 5% perchloric acid/95% acetic acid. The $V_c$'s below 1 MeV were measured using the convergent beam technique on an A.E.I. EM7 HVEM, and those >1 MeV using Kikuchi line technique on the 3 MeV HVEM at the Laboratoire D'Optique Electronique du CNRS, Toulouse, France. $V_c$'s 330 and 440 were not detected and it is sometimes the case that higher order $V_c$'s in rows not containing superlattice reflections are unobservable (see e.g. Lally et al., 1972). The 630 reflection showed a minimisation between 2 and 3 MeV but because this reflection belongs to the superlattice and is faint, it is difficult to see. $V_c^{422}$ is > 3 MeV. The experimental results are shown together with those predicted by theory in Table 1.

## 4. Results and Discussion

A set of experimental low-angle structure factors for $\beta$'NiAl were then computed and these are shown in Table 2 (a) along with those calculated from the free atom form factors of Doyle and Turner (1968) and those interpolated by curve fitting the form factors of Smart and Humphreys

(1980) (Ni) and Inkinen et al (1970)(Al), which give good agreement with the pure element Ni and Al $V_c$ measurements of Thomas et al (1974). The theoretical and experimental structure factors from the x-ray work of Cooper (1963) for Ni-50.76 at % Al are shown in Table 2(b).

Table 2. Theoretical and Experimental Structure Factors for $\beta$'NiAl at room temperature.

| | (a) from present work | | | (b) from Cooper (1963) for Ni 50.75 at % Al | |
|---|---|---|---|---|---|
| (hkl) | Free atom | pure element | Expt. | Free atom | Expt. |
| 100 | 13.30 | 13.58 | 13.53 ± 0.05 | 13.27 | 13.08 ± 0.08 |
| 110 | 28.21 | 28.13 | 28.08 ± 0.10 | 28.81 | 28.83 |
| 111 | 10.29 | 10.33 | 10.30 ± 0.22 | 10.79 | 10.80 ± 0.14 |
| 200 | 23.13 | 23.06 | 22.60 ± 0.12 | 23.86 | 22.80 ± 0.23 |
| 210 | 8.31 | 8.39 | | 8.92 | 8.79 ± 0.11 |
| 211 | 19.67 | 19.59 | | 20.34 | 20.46 ± 0.16 |
| 220 | 16.93 | 16.90 | | 17.64 | 17.52 ± 0.25 |
| 300 | 6.15 | 6.14 | | 6.74 | 6.85 ± 0.11 |
| 310 | 14.71 | 14.75 | | 15.64 | 15.61 ± 0.20 |
| 311 | 5.45 | 5.37 | | -- | -- |
| 222 | 12.92 | 13.03 | | 13.95 | 14.15 ± 0.28 |

It should be noted that the low-angle structure factors measured by the $V_c$ technique are much more accurate than the x-ray results, because, in addition to the statistical errors on these shown, there is an estimated additional 1.3% error on each result arising from establishing an absolute scale. Nonetheless, the trends for both sets of results agree remarkably well, in particular both predict a significant reduction in the 200 structure factor as a result of alloying. The reason for this will be discussed shortly. Because of the good agreement between the structure factors obtained by interpolation between the best pure element form factors and the experimental results (except for the 200 reflection) it is reasonable to suppose that all values of F with (hkl) $\geq$ (111) are accurately given by these. The 100 and 111 superlattice structure factors agree very well with those obtained from the best pure element form factors, and this immediately suggests that there is no transfer of charge between the atoms in this alloy as discussed by Cooper (1963), i.e., the bonding is not ionic. Cooper (1963) also Fourier analysed the results of Table 2(b) to produce theoretical free atom and experimental (110) section electron density maps for $\beta$'NiAl, and these results are shown in Figure 1(a) and 1(b).

It is clear from these maps that the theoretical outer electron distributions overlap in the close packed, (111), direction which is a direct result of the low alloy 200 structure factor, i.e., there is a build up of electron charge at the midway position between Ni and Al atoms in the nearest neighbour direction. This suggests that the bonding in this alloy is essentially covalent.

Figure 1. The theoretical and experimental outer electron distributions in β'NiAl, (110) section, calculated from the structure factors of Table 2(b) (From Cooper, 1963-courtesy Philosophical Magazine).

(a) Theoretical (free atom)

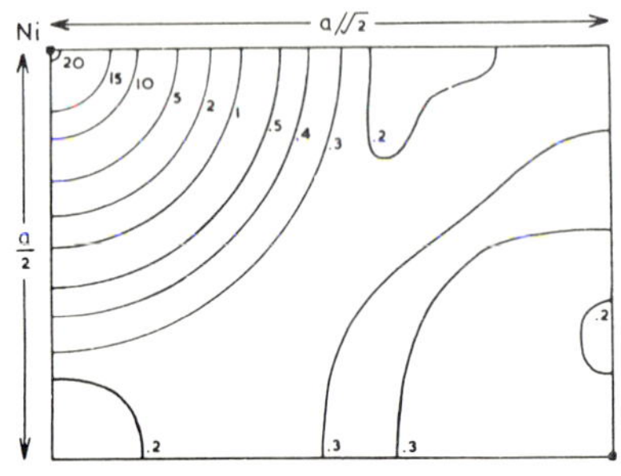


(b) Experimental

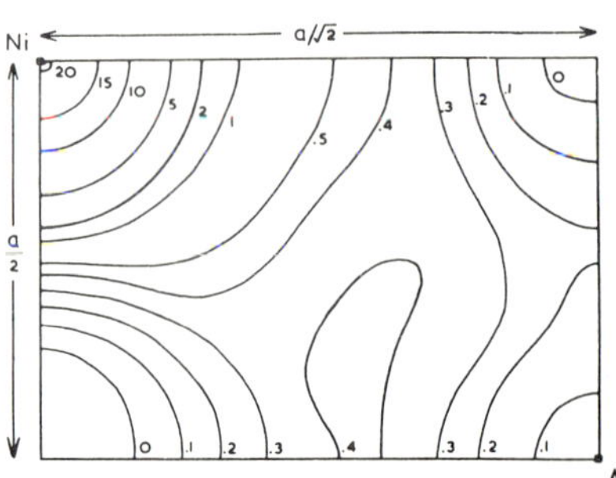


## 5. Acknowledgements

The author would like to thank Professor R.E. Smallman, Head of the Department of Metallurgy and Materials, University of Birmingham U.K. for the use of the A.E.I. EM7 HVEM, Dr. B. Jouffrey, Director of the Laboratoire D'Optique Electronique du CNRS, Toulouse, France for use of the 3 MeV HVEM and Dr. K.H. Westmacott of the National Center for Electron Microscopy, Lawrence Berkeley Laboratory, Berkeley, CA 94720 U.S.A. for the use of the Kratos 1.5 MeV HVEM. Thanks are also due to the Director of the Polytechnic Wolverhampton for the provision of laboratory and computing facilities and to Mr. J.B. Page for assistance with the computing. This work was supported in part by the Director, Office of Basic Energy Research, Office of Basic Energy Sciences, Materials Sciences Division of the U.S. Department of Energy under Contract No. DE-AC03-76SF00098.

## References

Cooper M J 1963 Phil. Mag. 8 811
Doyle P A and Turner P S 1968 Acta. Cryst. A24 390
Fisher P M J 1968 Jap. J. Appl. Phys. 7 191
Fox A G 1983 J. Phys. F. 13 1593
Fox A G 1984 Phil. Mag. B. Vol50 No4 477
Fox A G and Fisher R M 1985 these proceedings
Georgopolous P and Cohen J B 1977 Scripta. Met. Vol11 No2 147
Inkinen O Pesonen A and Paakkari T 1970 Ann. Acad. Sci. Fenn. AVI 344
Lally J S Humphreys C J Metherell A J F and Fisher R M 1972 Phil. Mag. 25 321
O'Keefe M A and Fox A G 1985 unpublished work
Smart D J and Humphreys C J 1980 in Electron Microscopy and Analysis 1979 (Int. Phys. Conf. ser. No31) Chap3 211
Thomas L E Shirley C G Lally J S and Fisher R M 1974 in High Voltage Electron Microscopy (London and New York:Academic Press) 52

*Inst. Phys. Conf. Ser. No 78: Chapter 10*
*Paper presented at EMAG '85, Newcastle upon Tyne, 2–5 September 1985*


# TEM of dislocation structure near regions of surface damage in Al

P Charsley and L J Harris

Department of Physics, University of Surrey, Guildford, GU2 5XH, U.K.

## 1. Introduction

Previous work on copper fatigued in reverse bending has shown that persistent slip bands (PSBs) form much earlier near regions of surface damage (indentations) than for regions well away from indentations (Charsley, Puttick and White 1981, White 1984). The observed PSB distribution cannot be explained by surface topography changes or by residual stresses due to the indent, however it may be attributed to the dislocation arrangements produced by indenting (Charsley and White 1985). The work on copper was confined to surface studies of PSB formation. In order to examine the associated dislocation structures work has begun on aluminium, and preliminary observations of the microstructure near indents which have not been fatigued are presented here. The main experimental difficulties are the production of TEM specimens which enable the dislocation structure to be studied at known and various positions near indents.

## 2. Experimental

The surfaces of polycrystalline aluminium of 99.99% purity (90 μm thick) and a single crystal of high purity aluminium (2 mm thick) grown by the Bridgeman technique, were prepared by mechanical polishing and electro-polishing in a nitric acid/methanol solution. The orientation of the single crystal was $\sim 5^\circ$ from [013] and had a shape shown in Fig. 1a which provided a region of uniform surface cyclic strain during reverse bending, suitable for the study of PSB formation near indents. After annealing for 1 hr at 425°C, a diamond pyramid indenter was used to produce indents of diagonal lengths (d) 56 μm (30 gm load) and 36 μm (10 gm load), in the single crystal and polycrystalline specimens respectively. Before cutting discs around these indents by spark erosion, a section was spark eroded from the indented region in the single crystal and its thickness was reduced to $\sim$ 0.2 mm by mechanically polishing the non-indented side. Discs were prepared for TEM by jet thinning from the non-indented side only. This generally gave rise to a perforation which covered the major part of the indent, but enabled the dislocation structure in the resulting thin areas to be related to the indent edges by using appropriate low magnifications. Specimens were examined using a 200 CX microscope operated at 200 kV using both bright field and weak beam techniques.

## 3. Results

An example of the PSB distribution near an indent in the fatigued single crystal is shown in Fig. 2. PSBs have not nucleated within the pit or directly at its edges. The distance from the indent centre to the closest PSBs is d (i.e. ∿ 60 μm): a second group of PSBs has formed at 2.1d (i.e. ∿ 120 μm). The dislocation structure around an indent within a single grain in polycrystalline material has been studied at the following distances from the indent centre, 0.5d (i.e. at the edge of the indent), 1.2d, 2.3d and 3.3d (see Fig. 1b). In the single crystal material the dislocation structure has been studied at 0.5d only.

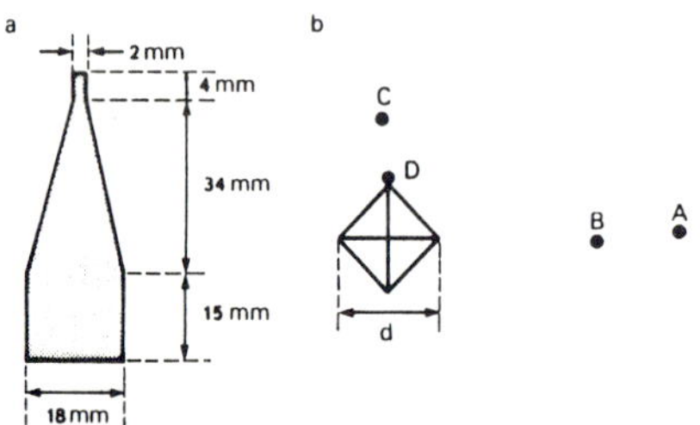


Fig. 1(a) Shape and dimensions of the single crystal. (b) Positions of areas at 3.3d(A), 2.3d(B), 1.2d (C) and 0.5d(D) relative to the indent in the polycrystalline material.

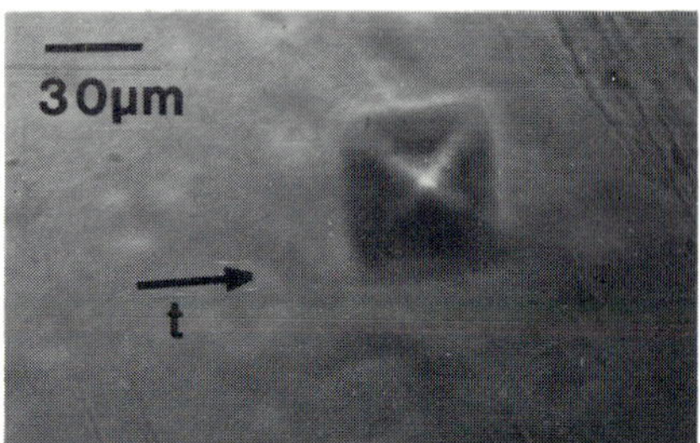


Fig. 2 Optical micrograph of PSBs developed near an indent in a fatigued Al single crystal. The direction t is the specimen axis.

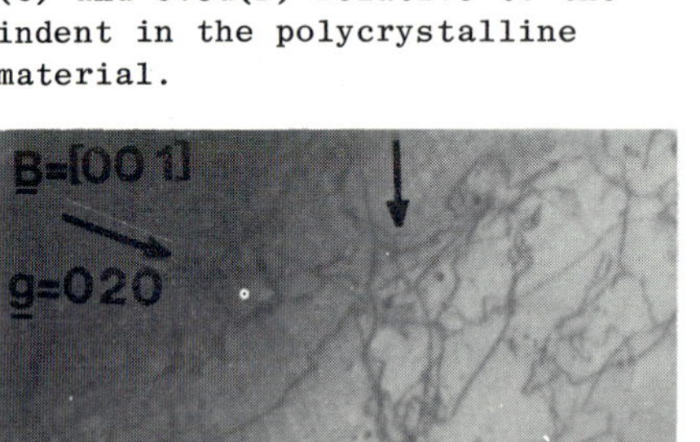


Fig. 3 Bright field image of dislocation structure at A (polycrystal).

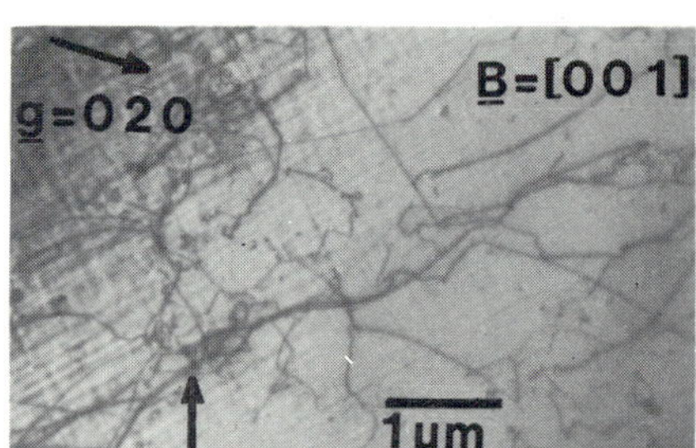


Fig. 4 Bright field image of dislocation structure at B (polycrystal).

At both 3.3d (Fig. 3) and 2.3d(Fig. 4) the dislocation arrangement consists of loose tangles of dislocations with areas of low dislocation density between the tangles, however the number of dislocations present in the tangles appears higher at 2.3d. At 1.2d (Fig. 5) dislocation tangles are again evident but the density of dislocations is much higher than at 3.3d and 2.3d mainly because the dislocation free areas are very much smaller. Examination of micrographs taken using different operating reflections (g) show that in all three areas dislocations with at least 4 Burgers vectors ($\underline{b}$ = ± [110], $\underline{b}$ = ± [$1\bar{1}0$], $\underline{b}$ = ± [011] or ± [$01\bar{1}$] and $\underline{b}$ = ± [101] or ± [$10\bar{1}$]) are present. The majority (≳ 75%) of dislocations at 3.3d have Burgers vectors $\underline{b}$ = ± [110] with the vertical grouping of dislocations (indicated by the arrow) mainly of this type, and those approximately parallel to [$\bar{1}10$] mainly with $\underline{b}$ = ± [011] or ± [$01\bar{1}$]. At 2.3d the majority

($\sim$ 70%) of dislocations which are approximately parallel to $[\bar{1}10]$ have Burgers vectors $\underline{b}$ = ± [110]; in the arrowed vertical grouping approximately 70% have Burgers vectors $\underline{b}$ = ± $[1\bar{1}0]$. Slip traces parallel to $[\bar{1}10]$ and [110] indicating that movement of dislocations has occurred are apparent both at 3.3d and 2.3d. At 1.2d the majority ($\sim$ 80%) of dislocations have Burgers vectors $\underline{b}$ = [110] or ± $[1\bar{1}0]$ and there is some grouping of dislocations along the directions indicated by the arrows. At the corner of the indent (0.5d) an elongated cell structure lying approximately parallel to $[\bar{1}10]$ is apparent (Fig. 6). Dislocations present in this area have the following Burgers vectors: $\underline{b}$ = ± $[1\bar{1}0]$; $\underline{b}$ = ± $[10\bar{1}]$; $\underline{b}$ = ± [110] or ± [101] and $\underline{b}$ = + [011] or ± $[01\bar{1}]$.

Cell structures are also formed at 0.5d in the single crystal (Figs. 7 & 8). The cells are elongated and are of variable size. Typically the cell width is $\sim$ 0.5 µm. Comparison of images taken using different values of g showed that the cell walls were made up of a number of different Burgers vectors. At E in Fig. 7 dislocations present have Burgers vectors $\underline{b}$ = ± [011].

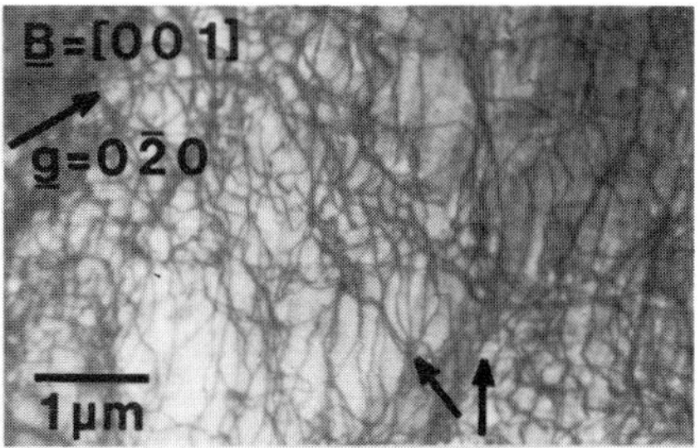


Fig. 5 Bright field image of dislocation structure at C (polycrystal).

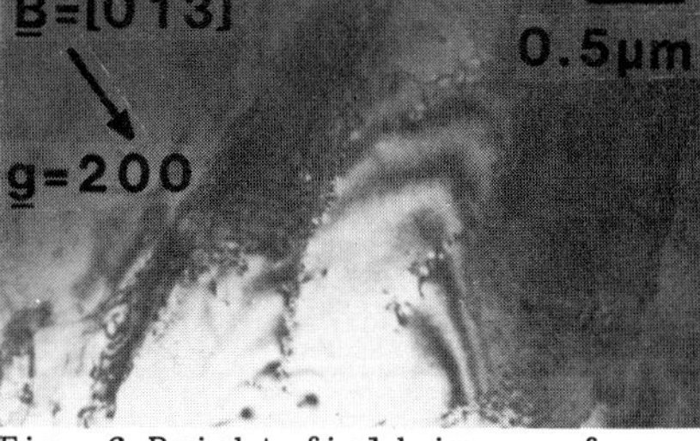


Fig. 6 Bright field image of dislocation structure at the corner of the indent (D) in polycrystalline material.

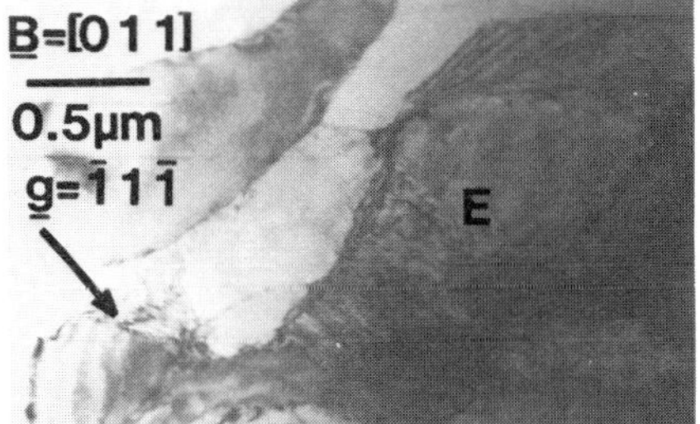


Fig. 7 Bright field image of dislocation structure at the corner of the indent in single crystal material.

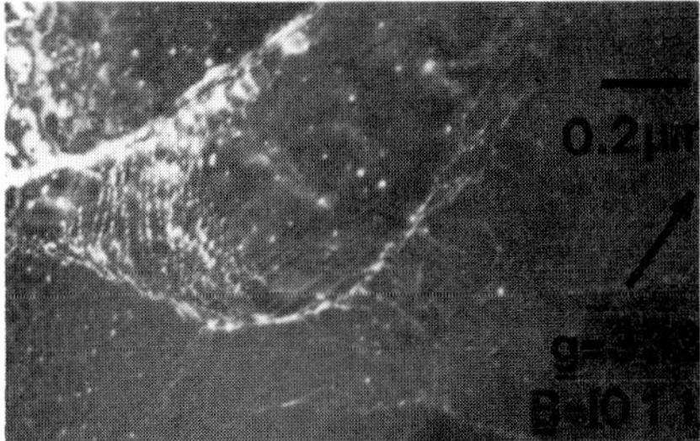

Fig. 8 Weak beam dark field image of part of the area in Fig. 7.

Generally the areas between the cell walls are relatively free of dislocations. Misorientation between the cells is indicated by the change in contrast across the cell walls. The presence of internal stress is indicated by these misoriented cells and by the point that dislocation walls appear curved. The absence of slip traces at 0.5d is an indication that dislocations cannot readily move in these areas.

## 4. Conclusion

The results presented show the following:

(i) The dislocation structure in areas near an unannealed indent where PSBs are observed to form during fatigue in reverse bending, appears to consist of dislocation tangles, the density of dislocations decreases with distance from the indent centre.

(ii) Near the edge of an unannealed indent where no PSBs are observed to form during fatigue testing, the dislocation arrangement consists of cell structures, and evidence of internal stress is apparent.

## 5. Acknowledgements

The authors would like to acknowledge financial support by the Ministry of Defence (R.A.E., Farnborough).

## 6. References

Charsley P. and White J.D.M 1985, Proc. 7th Int.Conf. on Strength of Metals and Alloys, Ed. H.J McQueen, J.P Bailon, J.I Dickson, J.J Jonas, M.G Akben, 1483.
Charsley P, Puttick K.E and White J.D.M 1981, Scripta Met. 15 1201-1204.
White J.D.M 1984, Ph.D Thesis, University of Surrey.

*Inst. Phys. Conf. Ser. No 78: Chapter 10*
*Paper presented at EMAG '85, Newcastle upon Tyne, 2–5 September 1985*



# The structural characterisation of multilayered Cu/NiPd films at the atomic level

C S Baxter and W M Stobbs
University of Cambridge, Department of Metallurgy and Materials Science
Pembroke Street, Cambridge CB2 3QZ

The interest generated in multilayered metal films with small (1-5nm) modulation wavelengths ($\lambda$) is largely associated with the anomalous variations of a variety of their physical properties, for example their elastic moduli (Baral et al. 1985), which arise at wavelengths of 1-3nm. Explanations for any common origin for all of these anomalies range from arguments associated with Fermi surface modifications due to the insertion of extra Brillouin zone boundaries to the effects of changes in coherency strain associated with the presence or otherwise of misfit dislocations. It would thus appear that unequivocal modelling of the behaviour of multilayers requires their structural characterization at the atomic level. We have previously provided recipes for how this difficult problem might be tackled using a variety of TEM techniques and have given examples of the type of data which can be obtained for Au/Ag multilayers with small relative misfit (Baxter and Stobbs 1985). Here we present some of our initial results on the $Cu/Ni_xPd_{1-x}$ system. This system was specifically chosen for examination since the layer misfit can be altered by varying the Ni:Pd ratio so that the effects of coherency can be examined for a given chemistry. The data presented is for a $Cu/Ni_{0.5}Pd_{0.5}$ system deposited by Dr R.E. Somekh using a magnetron sputter deposition technique (Somekh et al. 1984). The films were grown epitaxially on freshly cleaved (001)NaCl. Cross-sectional specimens were prepared for examination in the TEM by embedding the films in copper plate after removal from their substrates, and ion beam thinning polished sections at 3kV and liquid $N_2$ temperature.

Multilayers with $\lambda$ = 1.5-5.0nm (i.e. spanning the range in which the magnitude of any anomaly would be expected to show major changes) were first examined by weak beam techniques to determine whether or not the dislocation content showed any systematic variation. While dislocation loops associated with ion beam damage in the surface layers of the edge-on specimens were evident in all films and tend to be seen in specific layers (figure 1), no interfacial arrays of dislocations were observed in any of the specimens although the misfit for the equiatomic Ni:Pd ratio, as nominally deposited, would be expected to be 3.5%.

As has been discussed elsewhere (Baxter and Stobbs 1985) the determination of the remanant elastic strains in the multilayer cannot be carried out by a qualitative lattice fringe spacing approach, principally because of the fineness of the layers, and requires the comparison of through-focal series with full image simulations. The first step in such a determination is the measurement of any interdiffusion between the layers which might have occurred during growth in order to provide the requisite data to formulate an appropriate atomic model for high resolution image simulations. EDS techniques are of course inapplicable and the interdiffusion was assessed by the comparison of the low resolution (0.4nm) "Fresnel like" contrast profiles produced as a function of defocus with image simulations for varying degrees of interdiffusion. Four of a through focal series of 32 images for a 4.4nm wavelength film are shown in figure 2a and the related simulations of the series showing the best common agreement are shown in figure 2b. Given the large number of independent parameters involved both in the layer model and in the characterisation of the imaging conditions the uniqueness of any fit obtained is bound to be questionable. Frequently, however, certain features of a given series could be individually sensitive to a particular parameter and this gave confidence to the overall approach. For example we are confident that the specimen shown contained ten atomic layers of Cu alternating with 14 of NiPd, and surprisingly it seems that only the two atomic layers on either side of the interface are significantly intermixed. The sensitivity of the fitting procedure in the latter context may be judged by comparison of figures 2a and b with figure 2c for which intermixing is allowed over a further two planes at the interface.

Clearly we can be reasonably confident that the chemical composition of the deposited layers can be determined to sufficient accuracy for the production of an elastically strained structural model for high resolution image simulation. High resolution through focal image series were obtained for a variety of layer spacings using the Cambridge HREM. An example of an image from such a series is shown in figure 3. Images of this type were enlarged and then digitised at a resolution of approximately 24 pixels per fringe for subsequent characterisation of lattice fringe spacing variation. Fringe profiles were obtained by summing sections of the images parallel to the interfaces and SEMPER routines (Saxton et al. 1979) were used for their analysis. Figure 4 shows the deviations of a series of 21 fringe positions from the line of least squares fit for a region with layer spacings as shown. Any periodic variation associated with the retained elastic strain would be expected to be revealed (at least at some defocus) as a periodic variation in the deviation, but none is apparent. That no periodic deviations were observed at any defocus is surprising given that such behaviour is apparent in some graphs, such as figure 5, obtained similarly for simulated images of a coherent multilayer of the expected misfit. While relaxations of the coherency strains at the foil surfaces as discussed by Treacy et al. (1985) occur, these effects would not be expected to remove the periodic spacing variation altogether for the specimen thicknesses and layer spacings examined. As is demonstrated by figure 6 the Cambridge HREM is able to transfer information to 0.18nm though this spacing is best imaged under second broad band transfer conditions. In general however transfer at this spacing will tend to occur under conditions at which the transfer function is varying rapidly. We are accordingly examining whether more reliably analysable fringe spacing data can be obtained using non-axial conditions as might be expected for the reasons discussed by Hall et al. (1983).

Acknowledgements

We are grateful to Dr R.E. Somekh for the preparation of the specimens examined and to the SERC and Johnson Matthey for financial support. We also thank Prof. D. Hull for the provision of laboratory facilities.

References

Baral D B, Ketterson J B and Hilliard J E 1985 J. Appl. Phys. 57 1076

Baxter C S and Stobbs W M 1985 Ultramicroscopy 16 213

Hall D J, Self P G and Stobbs W M 1983 J. Microscopy 130 215

Saxton W O, Pitt T J and Horner M 1979 Ultramicroscopy 4 343

Somekh R E, Barber Z H, Baxter C S, Donovan P E, Evetts J E and Stobbs W M 1984 J. Mat. Sci. Lett. 3 217

Treacy M M J, Gibson J M and Howie A 1985 Phil. Mag. A 51 389

Fig.1. Weak beam image of a 4.4nm wavelength Cu/NiPd film. Note that point defects (arising from ion beam damage) are visible but arrays of interfacial dislocations are not apparent.

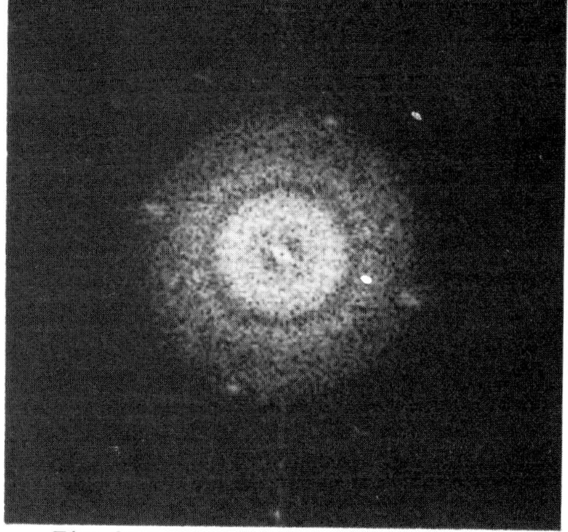

Fig.6. Computed diffractogram from the edge of the specimen shown in Fig.3. - the prominent spots are equivalent to spacings of about .18 nm. Imaging conditions are $C_s$ = 2.7 mm, defocus = -110nm.

Fig.3. Lattice image (obtained using the Cambridge HREM at 500kV) of a 1.6nm wavelength Cu/NiPd film, as used for fringe position analysis. The layers are not immediately apparent under these imaging conditions but the layer normal is parallel to [001] and there are nine fringes per wavelength. The fringe spacing is about 0.18nm.

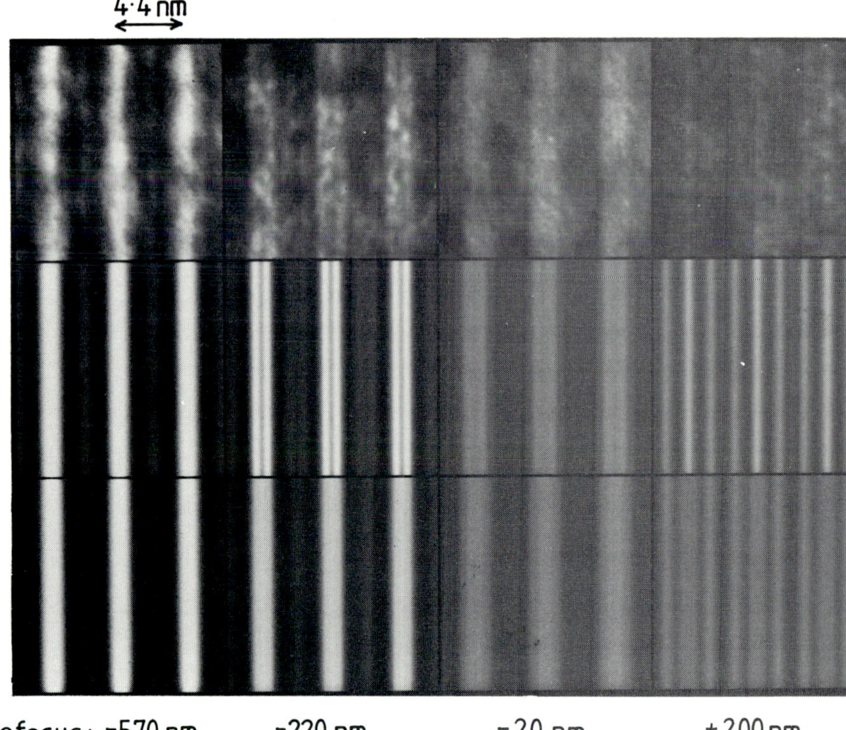


Fig.2. a) Through focal series of bright field images of a 4.4nm wavelength Cu/NiPd film. b) Equivalent multislice simulations for a model containing two intermixed layers at each interface (at 22nm thickness). c) Simulations for a model containing four intermixed layers at each interface. Note the sensitivity of features such as the split fringes visible at -220nm defocus to the change in interdiffusion.

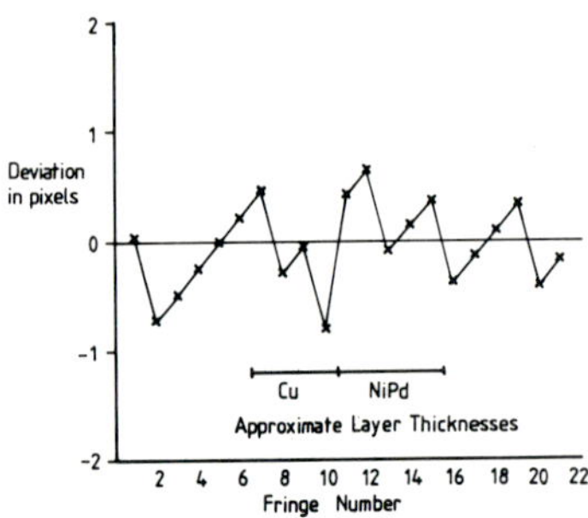


Fig.4. Deviations of measured lattice fringe positions from the line of least squares fit using positions obtained from the specimen shown in Fig.3. Note that no layer periodicity is detectable.

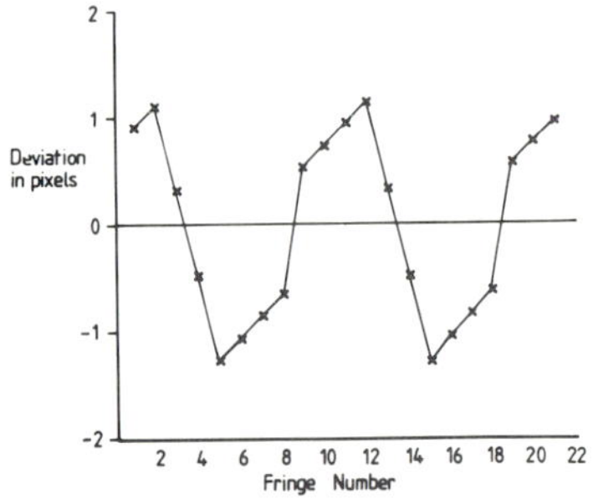


Fig.5. Deviations of fringe positions in a simulated model from the line of least squares fit using the same analysis procedure as for Fig.4. In this case the periodicity can be clearly seen.

# Effect of microalloying additions on the coarse grained heat-affected zone in HSLA steels

R TAILLARD*, T MAURICKX**, J FOCT*
* Laboratoire de Métallurgie Physique- U.S.T.L. - 59655 Villeneuve d'Ascq Cedex - France
** Service Métallurgie - Usinor Dunkerque - 59381 Dunkerque Cedex, France

## 1. Introduction

The knowledge of metallurigcal welding mechanisms is of primary importance for the elaboration of weldable high strength low-alloyed (H.S.L.A.) steels. During welding, along the fusion line, the base metal is subjected to a series of transformations which are successively:

- the ferrite-austenite transformation during heating,
- precipitate dissolution with the subsequent growth of austenite grains,
- the austenite transformation which depends on the cooling rate and on the steel's hardenability.

In order to reduce austenite grain growth and thus, the steel's hardenability, S. Kanazawa et al (1976) have recommended the addition of $15.10^{-3}$ wt % titanium to the steel and the adjustment of the nitrogen content to $5.10^{-3}$ wt %. These two conditions are necessary to obtain a homogeneous and fine titanium nitride precipitate distribution which is stable at high temperature (~1650 K).

The aim of the present work is to characterize the effect of microadditions of titanium and the influence of an additional microaddition of niobium on the microstructure of the coarse-grained heat-affected zone (H.A.Z.).

## 2. Materials and experimental procedures

The composition of the two Ti-microalloyed steels is 0.1 wt % C, 1.5 wt % Mn, 0.5 wt % Si, 0.5 wt % Ni, 0.05 wt % Al and 0.004 wt % N but alloy A contains 0.010 wt % Ti only while alloy B contains both 0.013 wt % Ti and 0.026 wt % Nb. The steels are investigated, both, in the 1183 K (910°C) normalised condition, and, after a simulated H.A.Z. thermal cycle. The applied thermal cycle is defined by a 1623 K (1350°C) peak temperature and by a 300 s cooling time from 973 K (700°C) to 573 K (300°C). The investigations are performed by light and thin foil transmission electron microscopy.

## 3. Results

In the normalised state, the two steels exhibit the same grain size of equiaxed ferrite-perlitic structure. Figure 1 displays the low intragranular precipitate density in alloy A. These particles are aligned in groups of 5 or 6. They adopt a 50 nm mean edge-length cubic morphology and have a face centred cubic crystallographic structure with an "a" lattice parameter close to 0.43 nm. Figure 2 shows the two distributions of intragranular precipitates in alloy B. The densely precipitated distribution consists of 7 nm mean-size spheroidal particles with a f.c.c. (a∿0.43nm) or a tetragonal (a∿0.4 nm; c∿0.88nm) crystal structure. The sparsely precipitated distribution is composed of 80 nm mean edge-length cubic precipitates with a f.c.c. crystal structure (a∿0.43 nm).

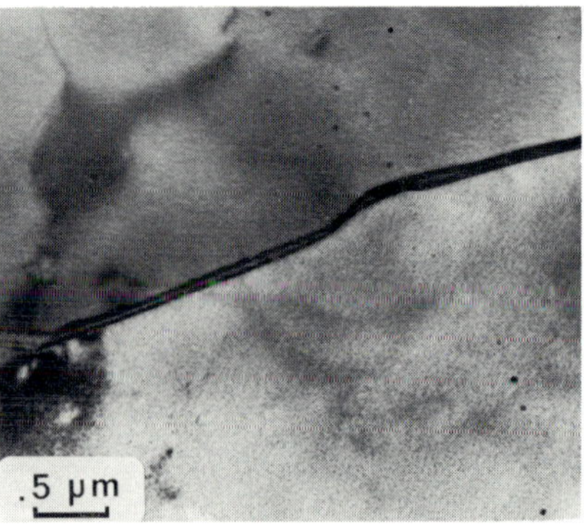


Figure 1 : precipitation state in normalised alloy A

It is worth noting that the microstructures of the alloys are drastically modified by the simulated H.A.Z. thermal cycle. Figure 3 exhibits light microscope observations. Alloy A (Figure 3(a)) presents an acicular structure in which the austenite grain boundaries are delineated by proeutectoid ferrite. By contrast, alloy B has a structure (Figure 3(b)) which seems exclusively acicular. In the two steels, the estimated austenite mean grain size is close to 100µm. Previous observations are confirmed by transmission electron microscopy. Alloy A microstructure (Figure 4) consists of equiaxed ferrite, and acicular ferrite with zones of martensite and retained austenite which were termed "M.A." constituent by Biss and Cryderman (1971).

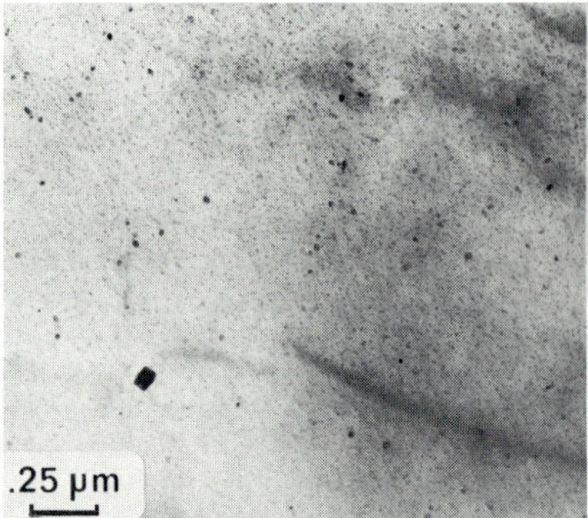


Figure 2 : Intragranular precipitation in normalised alloy B

Alloy B microstructure is composed of acicular ferrite with blockly "M.A." constituent (Figure 5(a,b)) and of upper bainite with interlath "M.A." constituent (Figure 5(c)). EDX microanalysis shows that the metallic contents of the ferrite matrix and of the "M.A." constituent are not significantly different. In the simulated H.A.Z., the intragranular precipitate state consists of a low density of f.c.c. precipitates (a∿0.43 nm). These particles present a cubic morphology with a 60 nm and 80 nm mean edge length in alloy A and B respectively. The calculations of Easterling (1983) indicate that these precipitates are probably titanium nitride particles because of their high thermodynamic stability. Whatever the applied heat treatment, the two alloys exhibit an equivalent cementite intergranular precipitation state.

## 4. Discussion

The different microstructures simulated H.A.Z. of alloys A and B may be related to hardenability differences and are consistent with "C.C.T." diagrams obtained in welding conditions. Those distinct quenching behaviours may stem from various metallurgical parameters which are a priori:

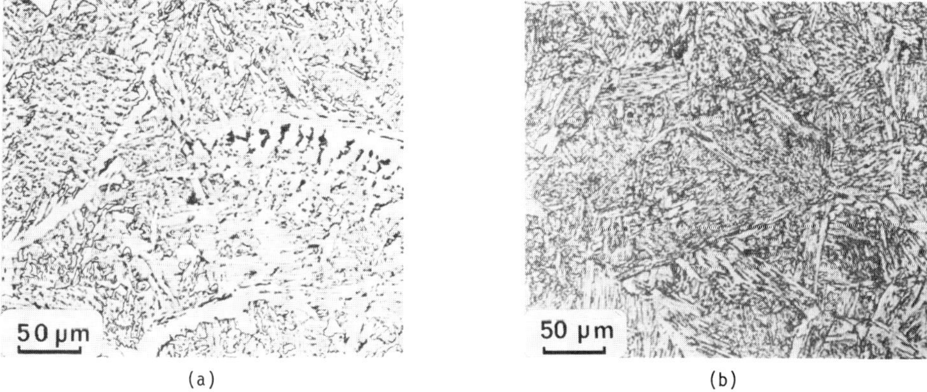


(a) (b)

Figure 3 : Heat-affected zone microstructure a) alloy A b) alloy B

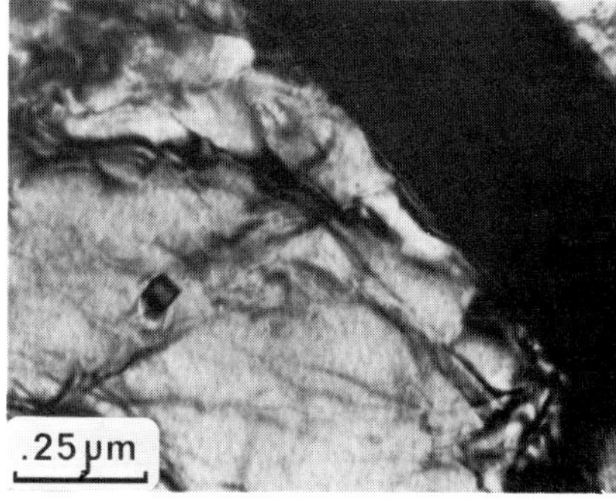


Figure 4 : Heat-affected zone microstructure of alloy A (T.E.M. bright field image)

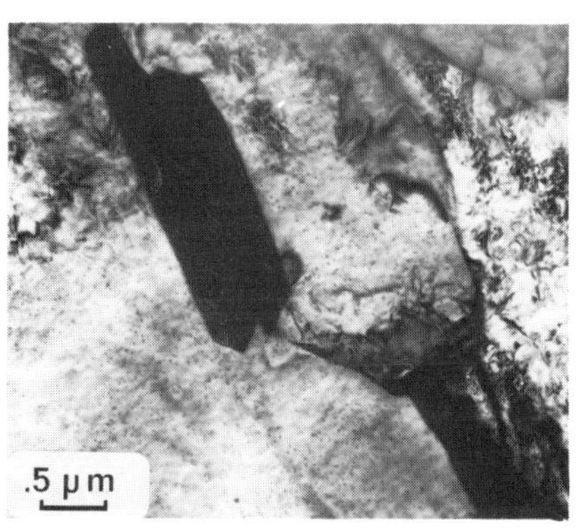


a) blocky morphology

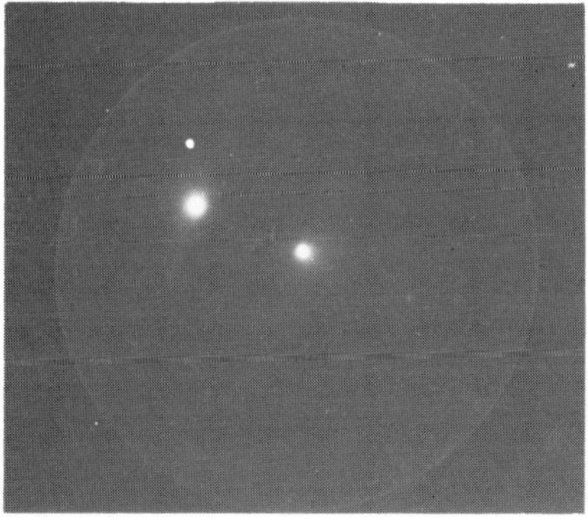

b) μμ diffraction pattern |$\bar{1}12$| austenite reciprocal lattice plane

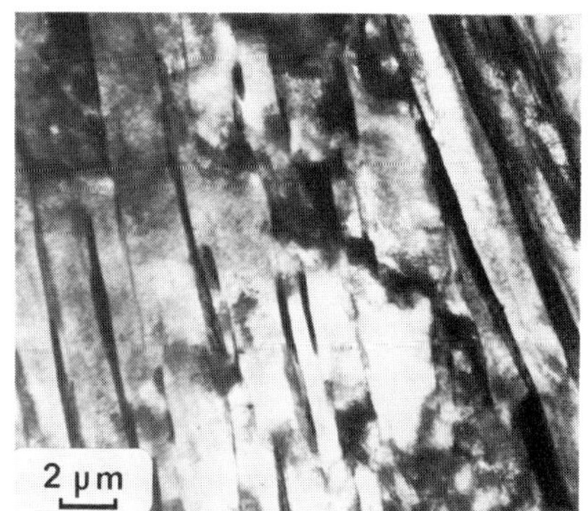


c) interlath morphology

Figure 5 : "MA" constituents aspect in alloy B

- an austenite grain size difference,
- a substitutional difference in solid solution composition,
- the nucleation of intragranular ferrite.

As the austenite grain size of the two steels in the coarse-grained H.A.Z. is equivalent, this parameter is not the cause of the hardenability difference. By contrast, the second parameter seems likely. The free (Nb, Ti) solute content in alloy B is probably increased by the welding operation because of dissolution of the small precipitates without any subsequent precipitation. The non-observation of proeutectoid ferrite in alloy B agrees well with the slowing down action of the Nb free solute content on the proeutectoid ferrite formation already noticed by Thomas et al (1980). In other respects, the heteregeneous intragranular ferrite nucleation on precipitates may take a significant part in the cooling transformation. According to this hypothesis, a more equiaxed structure would exist in alloy A than in alloy B, where the higher (Ti, Nb) free solute content would tend to increase the activation energy for the ferrite nucleation on precipitates with the subsequent growth of ferrite laths from the austenite grain boundaries.

## 5. Conclusion

The T.E.M. observations are essential in order to explain the coarse-grained H.A.Z. hardenability difference between a Ti and a Ti-Nb microalloyed steel.

The lower hardenability of the microalloyed steel containing only Ti seems to result, both from the lower free (Ti, Nb) solute content of the alloy an from the existence of precipitation stable to high temperature. These two parameters promote the ferrite formation.

## 6. References

Biaa V, Cryderman R L 1971 Met. Trans. 2 2267.

Easterling K 1983 Introduction to the Physical Metallurgy of Welding (Butterworths) p 121-3.

Kanazawa S, Nakashima A, Okamoto K and Kanaya K 1976 Trans. I.S.I.J. 16 486.

Thomas M H, Rigsbee J M, Heheman R F 1980 Proc. 18th Int. Conf. on Heat Treatment of Materials ASM.

*Inst. Phys. Conf. Ser. No 78: Chapter 10*
*Paper presented at EMAG '85, Newcastle upon Tyne, 2–5 September 1985* 

# The formation of a/2 ⟨111⟩ dislocation loops in ferritic steels

R N Alexander+, M H Loretto+, C A English* and E A Little*

+ Department of Metallurgy and Materials, University of Birmingham, P.O. 363, Birmingham B15 2TT.

* Materials Development Division, AERE, Harwell, Oxfordshire, OX11 0RA

## 1. Introduction

Ferritic steels exhibit a high resistance to radiation induced void swelling (e.g. Little and Stow 1979), and are thus of interest as candidate materials for wrapper and first wall applications in fast breeder and fusion reactors respectively. The low swelling however remains an enigma, but has been attributed to the microstructure (Bullough et al 1981), which is unique, a low capture bias at dislocations (Sniegowski and Wolfer 1984) and various chemical factors (e.g. Little 1979).

The microstructure is unique because large interstitial loops of both b = a/2<111> and b = a<100> nucleate and grow during irradiation. Previously, it has been assumed that both species of loop result from the shear of an a/2<110> loop nucleus (Eyre and Bullough 1965). The relevant reactions are given as:

$$a/2\langle 110\rangle + a/2\langle 110\rangle \rightarrow a\langle 100\rangle$$

$$a/2\langle 110\rangle + a/2\langle 001\rangle \rightarrow a/2\langle 111\rangle$$

The driving force in both cases is the removal of the high energy stacking fault. The shear has been predicted to occur when the nucleus contains about 16 atoms only (Bullough and Perrin 1968). This mechanism for loop production is thus unlikely to be confirmed by direct observation.

The relative proportion of a/2<111> loops is reported to increase with increasing chromium content (Gelles 1982) and with decreasing temperature (Eyre and Bullough 1965).

An 'in situ' study of radiation damage created by electrons in a High Voltage Electron Microscope, has revealed that a/2<111> loops can definitely be produced by an alternative mechanism to that mentioned above. Observations of this mechanism in operation are reported here.

## 2. Experimental

A series of high purity binary Iron Chromium alloys of composition 1, 5, 10 and 15% Chromium were commissioned by the U.K.A.E.A., Harwell from Materials Research SA. These were rolled to ∿ 125 μm thick and annealed in argon for 2 hours at 950°C followed by a further anneal of 200 hrs. at 750°C. This resulted in a large grained dislocation free microstructure.

T.E.M. specimens were produced by a two stage process of jet profiling followed by bench polishing. This technique resulted in very thin specimens thus reducing the ferromagnetic astigmatism.

Irradiations were carried out at 1 MeV in the High Voltage Electron Microscopes of the University of Birmingham and the U.K.A.E.A. Harwell. The temperatures of the irradiations were in the range 350-650°C.

## 3. Results

At dose rates (calculated using beam current data, and assuming a cross section of 60 barns) that are low ($\sim 10^{-4}$ dpa/second) when compared with those available in electron irradiations, a dislocation microstructure containing almost entirely a<100> loops was induced in all the alloys. Such a microstructure is shown in figure 1a. These loops do not interact attractively, though a network will form if a/2<111> loops are also present.

When the high chromium alloys (10-15%) chromium are irradiated at high dose rates ($> 10^{-3}$ dpa/second) the loop populations are reversed and a/2<111> loops predominate. This is shown in figure 1b. The dose rate does not have such a striking effect on the low chromium alloys.

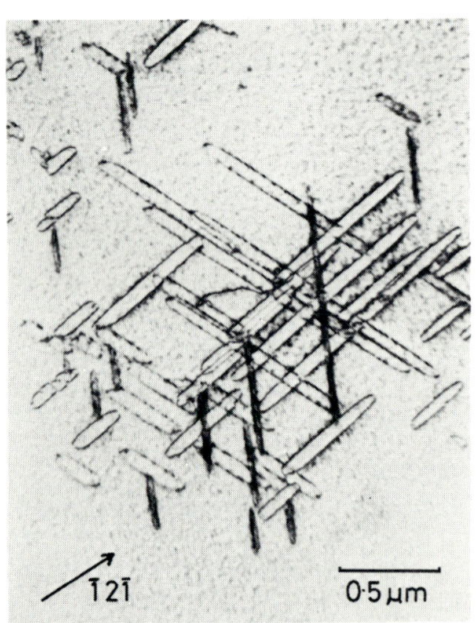


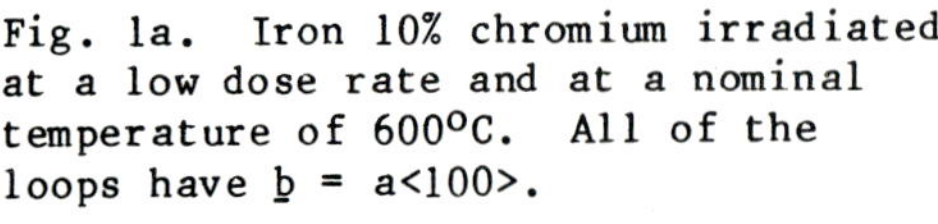

Fig. 1a. Iron 10% chromium irradiated at a low dose rate and at a nominal temperature of 600°C. All of the loops have b = a<100>.

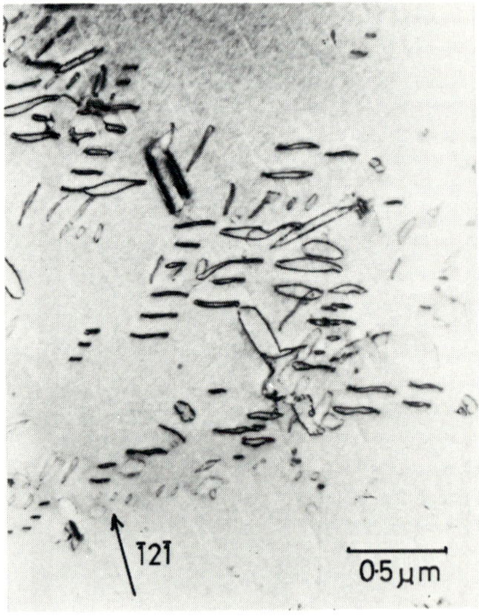


Fig. 1b. Iron 15% chromium irradiated at a high dose rate and at a nominal temperature of 650°C. The groups of three loops with b =a/2<111> are easily discernible.

These a/2<111> loops were observed to 'punch out' in **groups of** three from some central nucleus. These groups are easily discernible in figure 1b, while the sequence of micrographs in figure 2 capture the process in operation. The three a/2<111> loops have the same Burgers vector, and once they have been formed no further loops are added to the group. Indeed the central nucleus is destroyed during the process and in its place may be found the central loop of the three. The other two loops have moved away from the point of nucleation on their glide cylinders, and as the loops grow this initial spacing is increased. High loop nucleation densities are created by this punching even at high temperatures where low nucleation densities would be expected.

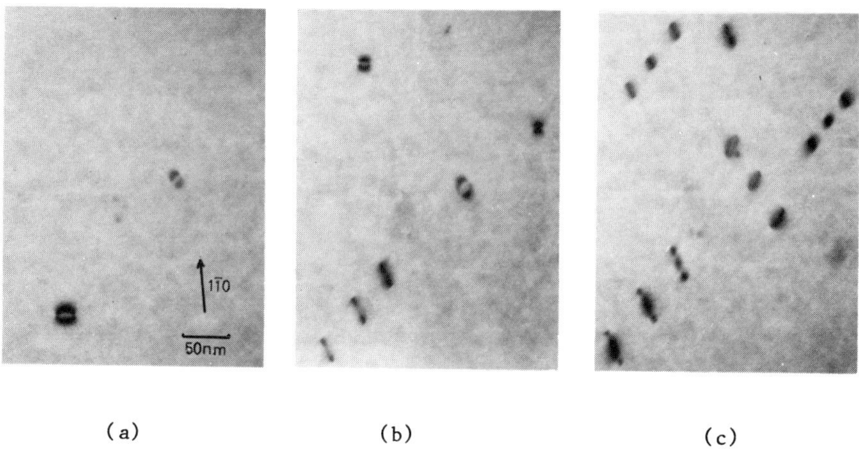


(a) (b) (c)

Fig. 2. A series of micrographs taken during irradiation of iron 15% chromium. The nominal temperature was $650^{\circ}$C. All of the loops shown have $\underline{b}$ = a/2<111>. There is a delay of approximately 20 seconds between the micrographs and irradiation began $\sim$40 seconds before the first micrograph.

The nucleus, which grows to produce a diffraction contrast image of about 200 Å before punching has a directional strain field. Evidence for this may be obtained from figure 2 where the lines of no contrast on different nuclei vary for a given $\underline{g}$ vector. No change to the nucleus has been observed during cooling, and analysis of the strain field is in general consistent with it being <111>, though many irrational results have also been recorded.

Punching of this kind occurs occasionally even in the 1% chromium alloy, but attempts to create a/2<111> loops in pure iron have so far been unsuccessful.

Alterations in temperature within the range defined earlier do not alter the propensity for this mechanism to operate. The nucleation density is however affected. The irradiation environment, which has been found to be important when irradiating ferritic steels, does not appear to affect the operation of this punching mechanism.

## 4. Discussion

The creation of dislocation loops by prismatic punching is by no means a novel observation, but it is not a mechanism that has previously been considered as relevant to the clustering of point defects during irradiation. The observations reported here are however unambiguous evidence for punching.

In order to explain this punching it is necessary to believe that the nucleus shown in figure 2a is a three dimensional cluster of interstitials. Such a cluster would until recently have been considered unlikely, but since the observation of interstitial tetrahedra, which it is thought grow from a three dimensional cluster rather than by way of dissociation from a planar loop, (Hardy and Jenkins 1985) a three dimensional interstitial cluster is now more acceptable. The stress created by the cluster is relieved by prismatic punching in the same way that stress is relieved around a growing precipitate. However in the present case the series of self interstitial loops produced contain within them the point defects which were previously in the cluster. In this way the point defects in the cluster have condensed out into loops which can then glide on their glide cylinders and absorb further defects by climb in the normal manner.

If three-dimensional clusters are to form it is necessary to propose that collapse of a point defect cluster to a platelet is in some way hindered. The fact that such a collapse is difficult may also be inferred from the high nucleation doses required to create visible defects in irradiated ferritic alloys (Robertson et al 1982).

The observations reported in the present paper show that the reactions given in the introduction do not fully describe the mechanisms of loop formation in ferritic materials and that a/2<111> loops are produced frequently if not exclusively by a process of prismatic punching.

## References

Bullough R and Perrin R C 1968 Proc. Roy. Soc. A305 541.
Bullough R, Wood M H and Little E A 1981 ASTM Spec. Tech. Pub. 725, 593.
Eyre B L and Bullough R 1965 Phil. Mag. 11 31.
Gelles D S 1982 J. Nucl. Mater. 108-109 515.
Hardy G and Jenkins M L 1985 submitted for publication.
Little E A 1979 J. Nucl. Mater. 87 11.
Little E A and Stow D A 1979 J. Nucl. Mater. 87 25.
Robertson I M, Jenkins M L and English C A 1982 J. Nucl. Mater. 108-109 209.
Sniegowski J J and Wolfer W G 1984 Proc. "Topical Conference on Ferritic Alloys for use in Nuclear Energy Technologies", Eds. J W Davis and D J Michel, (Met. Soc. AIME) pp 579.

*Inst. Phys. Conf. Ser. No 78: Chapter 10*
*Paper presented at EMAG '85, Newcastle upon Tyne, 2–5 September 1985*



# Identification of a new intermetallic phase in a P/M superalloy

Y P Lin and J W Steeds

H H Wills Physics Laboratory, University of Bristol, Bristol BS8 1TL

The microstructure of a Ni-based powder metallurgical (P/M) superalloy, AF115, after creep and stress rupture testing, has been investigated. Carbon extraction replicas were examined mainly in a Philips EM430 electron microscope. Various intermetallic precipitates were observed, including a previously unreported phase. This phase, with a composition of (at %) 41% Cr, 20% Co, 12% Ni, 16% Mo and 11% W, occurs at boundaries between $\gamma'$ and $\gamma + \gamma'$ grains. Due to the presence of extensive faults, selected area electron diffraction pattern for the precipitate, Fig. 1, shows only approximate 6-fold symmetry. However, by using a 20 nm probe, the convergent beam electron diffraction (CBED) patterns obtained could be indexed in terms of a hexagonal P-lattice with a=12.6 and c=4.6Å. Patterns for the [0001] and [10$\bar{1}$0] axes, Fig. 2 and 3, show whole pattern symmetries of 6mm and 2mm respectively. By referring to Buxton et al (1976), it follows that the crystal has a point group of 6/mmm, Table 1. The possible space groups are P6/mmm, P6/mcc, P63/mcm or P63/mmc. Since Figs. 2 and 3 show no evidence of absent reflections, it is concluded that the space group of the crystal is P6/mmm. Perusal of the literature failed to identify any known phase with the above space group and cell dimensions. However, similarities in terms of rings of strong reflections were apparent between the [0001] pattern, Fig. 2, and the [001] pattern of the σ phase (The Bristol Group, 1984). In describing the σ phase as kagomé tiled with tiling units containing four hexagons, Frank and Kasper (1959) pointed out that other tiling units exist, including a larger hexagonal unit containing seven hexagons. This hypothetical structure, consisting of a $3636+6^23^2+6^3$ (6:3:1) net at z=0, a $3636+6^23^2+6^3$ (1:1:1) net at z=½ and two $3^6+3^2434$ (1:6) nets at z=$^1/_4$ and $^3/_4$, is consistent with our cell dimensions and space group and Multislice calculations based on this assumption generated diffraction patterns closely resembling our experimental results. In a parallel work, the structure of an apparently identical phase, but with different composition, has been deduced by high resolution electron microscopy as the hexagonal kagomé tiled phase of Frank and Kasper (Li and Kuo 1985). Our high resolution image, Fig. 4, is in close accord with theirs and shows clustering of seven bright spots corresponding to the tiling units with each spot representing the hexagonal region of comparatively low electron density, although irregularities exist due to faulting (such as indicated). It is concluded that the atomic arrangement of this newly observed phase is that of a hypothetical structure proposed by Frank and Kasper (1959) and is named the F phase in their honour (Li and Kuo 1985 and Lin and Steeds 1985).

The authors wish to thank the SERC for financial support and Rolls Royce Ltd, for provision of sample material.

References

Bristol Group 1984 Convergent Beam Electron Diffraction of Alloy Phases compiled by Mansfield J F (Bristol and Boston: Adam Hilger)

Buxton B F, Eades J A, Steeds J W and Rackham G M 1976 Phil.Trans.R.Soc. 281 171

Frank F C and Kasper J S 1959 Acta Cryst. 12 483

Li D X and Kuo K H 1985 Acta Cryst. B41 in press

Lin Y P and Steeds J W 1985 ibid

Table 1. Deduction of the 6/mmm point group symmetry for the F phase (after Buxton et al 1976).

| Axis | pattern symmetry | possible diffraction groups | possible point groups |
|---|---|---|---|
| [0001] | 6mm | 6mm | 6mm |
| | | $6mm1_R$ | 6/mmm |
| [10$\bar{1}$0] | 2mm | 2mm | $\bar{6}m2$ |
| | | $2mm1_R$ | 6/mmm |

Fig. 1. [0001] selected area diffraction pattern showing approximate 6-fold symmetry.

Fig. 2. [0001] CBED pattern showing 6mm whole pattern symmetry.

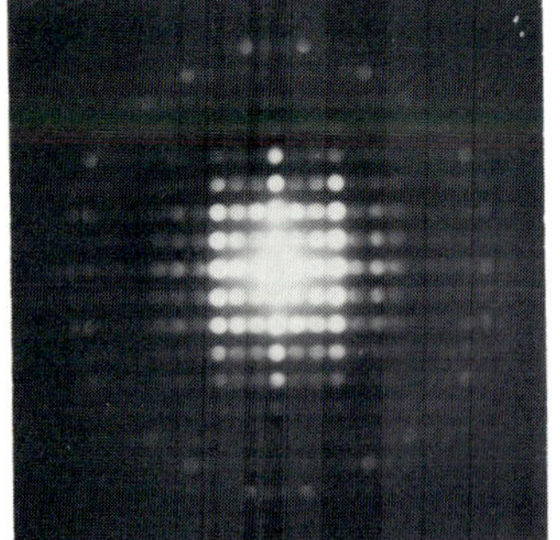

Fig. 3. [10$\bar{1}$0] CBED pattern showing 2mm whole pattern symmetry

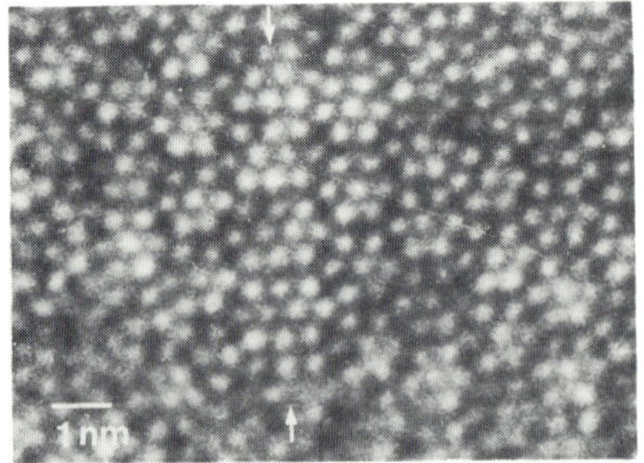


Fig. 4. High resolution image of F phase. Arrows indicate fault line.

# A quantitative investigation of weak-beam contrast from intrinsic faults in silicon

I G Salisbury and R S Timsit*

Department of Metallurgy and Science of Materials, University of Oxford, Parks Road, Oxford OX1 3PH, UK.

* Alcan Laboratories, Kingston, Ontario, Canada.

Large-scale changes in mean fault intensity on reversing g in weak-beam were reported for intrinsic defects by Chen and Thomas (1974). Recently Wilson and Cockayne (1985) found that a multislice calculation could predict an asymmetry between ± g pairs, an improvement over kinematical and two-beam column approximation computations which the workers attributed to better modelling of the atomic potential. In the present work, the intensity of an intrinsic edge fault in [001] epitaxial silicon, having a displacement, R, of a/3 $[\bar{1}\bar{1}1]$, has been measured for a variety of angles between R and the beam direction, B, using imaging conditions close to g(2g), for the two accessible 113's for which the fault shows strong contrast in bright field; $31\bar{1}$ and $3\bar{1}1$.

A full explanation of this will be given elsewhere, but two factors were largely responsible for the choice; 1) the g's give different magnitudes for g.R, 5/3 and 1/3 respectively, while maintaining $\alpha$ and $s_g$ constant and 2) for both reflections the fault may be tilted from vertical (i.e.R orthogonal to B) at [011] to around 63.1° at B close to $[2\bar{1}5]$ or 32.5° at $[\bar{2}\bar{1}5]$. (This may be seen clearly from a Kikuchi map, e.g. Smallman and Loretto (1975). The normalized intensity $\langle C_{in} \rangle$, defined:

$$\langle C_{in} \rangle = (\langle I_{in} \rangle - \langle I_b \rangle)/\langle I_b \rangle$$

where $\langle I_{in} \rangle$ is the fault intensity and $\langle I_b \rangle$ the background intensity, is shown as a function of fault angle for the different reflections in Fig.1. Missing from the figure is the intensity when imaged using $31\bar{1}$, since the fault never exceeded background, and when it could be discerned at all, $\langle I_{in} \rangle$ was actually less than $\langle I_b \rangle$. A similar phenomenon has been reported by Self et al. (1982).

Caution is required when interpreting the results in Fig.1, since the experimental background includes an inelastic contribution due to the finite size of the objective aperture (Cockayne et al., 1984).

Two features are clear, however; 1) there is a change in intrinsic fault intensity with angle to the beam, and 2) different values of g.R give, at the same fault angle and phase factor, different asymmetries on reversing g. It may be that the angle between g and the fault normal has some significance. Further calculations from the new multislice model would here be of interest.

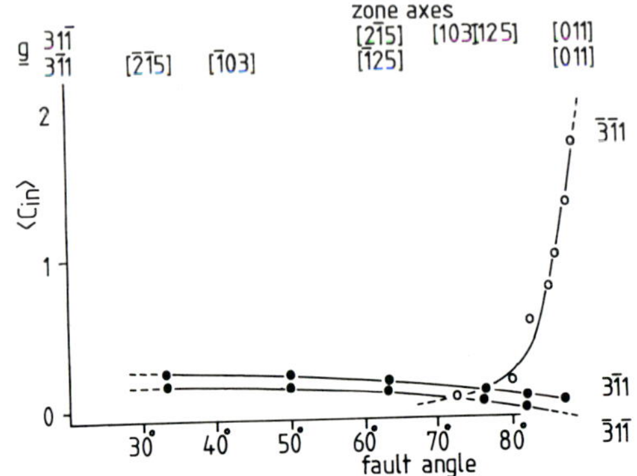


Fig.1. The dependence of the mean fringe intensity of an intrinsic a/3 $[\bar{1}\bar{1}1]$ fault, imaged in $\underline{g}$(2g), upon angle.

Chen L J and Thomas G 1974 phys. stat. sol.(a) 25 193

Cockayne D J H, Pirouz P, Lui Z, Anstis G and Karnthaler 1984 phys. stat. sol.(a) 82 425

Smallman R E and Loretto M H 1975 Defect Analysis in Electron Microscopy (London: Chapman and Hall) pp118

Self P G, Shaw M P and Stobbs W M 1982 phys. stat. sol.(a) 73

Wilson A R and Cockayne D J H 1985 Phil. Mag.A 51 341

*Inst. Phys. Conf. Ser. No 78: Chapter 10*
*Paper presented at EMAG '85, Newcastle upon Tyne, 2–5 September 1985*



# An analytical electron microscopic study of the dry formation of $Cu_2Se$–CdSe heterojunctions

C P McHardy and A G Fitzgerald

Carnegie Laboratory of Physics, The University, Dundee DD1 4HN

## 1. Introduction

The $Cu_2S$-CdS heterojunction solar cell is one of the most extensively studied. The most common method of producing this heterojunction is the "wet method" described by Chopra and Das (1983). This involves dipping a CdS film into a bath of CuCl, the $Cu_2S$ being produced by a reaction of the form

$$2CuCl + CdS \longrightarrow Cu_2S + CdCl_2.$$

However, te Velde (1974) has shown that a solid state reaction of a similar form between CdS and evaporated CuCl will take place at an elevated temperature. The structural aspects of this "dry method" of producing the $Cu_2S$-CdS heterojunction have been studied by Fitzgerald and McHardy (1985) and by De Vos et al (1983).

This paper reports on the results of an investigation of the reaction between evaporated CuCl and thin films of CdSe in an attempt to produce $Cu_2Se$-CdSe heterojunctions which also have potential applications in solar energy conversion.

## 2. Experimental

Thin films of CdSe were prepared by vacuum evaporation onto glass substrates. CuCl was then deposited on these films without breaking vacuum. The composite films were than examined as prepared and after annealing at various temperatures. The analytical techniques used were reflection electron diffraction (RED), in a transmission electron microscope, and Auger electron spectroscopy (AES) in conjunction with sputter depth profiling in an ultra-high vacuum scanning electron microscope (UHV-SEM).

## 3. Auger Electron Spectroscopy

Figures 1 and 2 show Auger electron spectra obtained after etching CdSe/CuCl for various periods of time. For the freshly prepared junction (fig. 1), only copper and chlorine are detected on the surface, with some carbon and oxygen contamination, implying that some of the deposited CuCl has not reacted with the CdSe. However, at a particular depth, a region containing mainly copper and selenium is found, indicating that a reaction has taken place, even at room temperature. Profiling further into the film, the underlying CdSe layer is revealed.

For the annealed junction (fig. 2), the surface consists only of $CdCl_2$ with traces of carbon and oxygen contamination, indicating that all of the deposited CuCl has reacted. A more extensive region containing copper and

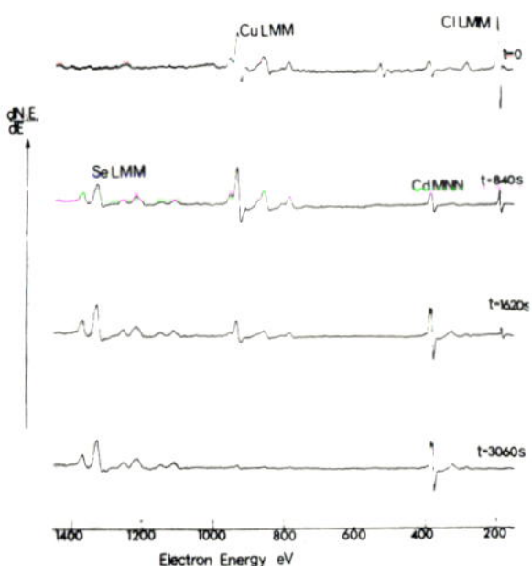


Fig. 1. Auger electron spectra obtained from an as-prepared CdSe/CuCl film after ion etching for the times indicated.

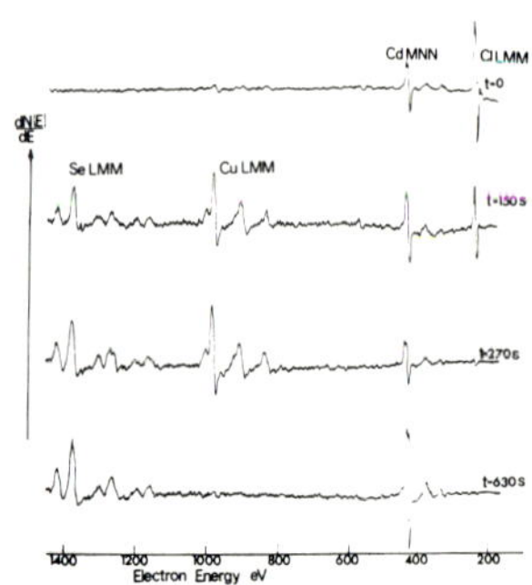


Fig. 2. Auger electron spectra obtained from a CdSe/CuCl film annealed at 185°C, after ion etching for the times indicated.

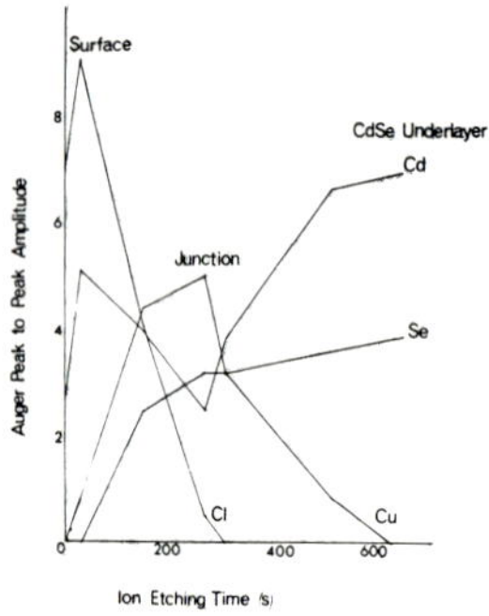


Fig. 3. Plot of Auger peak to peak amplitude versus ion etching time for a CdSe/CuCl film annealed at 185°C.

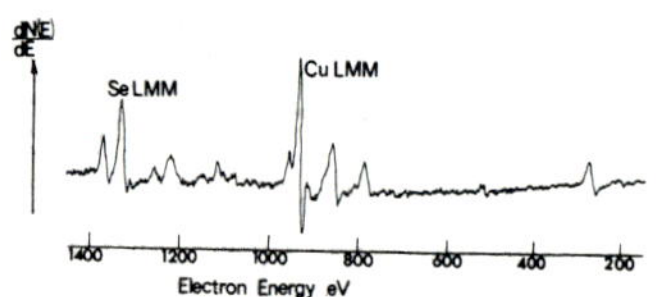


Fig. 4. Auger electron spectrum from the surface of a $Cu_2Se$-CdSe heterojunction prepared at 185°C and washed in methanol to remove the $CdCl_2$ surface layer.

selenium is formed, and the boundary between this layer and the underlying CdSe is sharper than the junction obtained in films which have not been annealed. The form of the junction obtained after annealing is illustrated most clearly by a plot of Auger peak amplitude against ion etching time, as shown in fig. 3. An Auger electron spectrum taken from a CdSe/CuCl film that has been annealed and subsequently methanol washed to dissolve the surface $CdCl_2$ layer is shown in fig. 4. This spectrum consists of copper and selenium peaks, with a very small cadmium peak and some carbon contamination. This suggests that the reaction product is a selenide of copper.

While Auger depth profiling shows that a reaction takes place between evaporated CuCl and CdSe, producing a chloride of cadmium and a selenide of copper, the products of the reaction cannot be specifically identified and information about the structural aspects of the reaction cannot be obtained.

## 4. Reflection Electron Diffraction

A reflection electron diffraction pattern from CdSe (fig. 5), shows a hexagonal structure, with preferred orientation. The (0001) axis is perpendicular to the glass substrate. A similar oriented growth of CdSe on glass has been observed by Dhere et al (1977). Electron diffraction patterns from freshly prepared and annealed CdSe/CuCl films are shown in fig. 6. The pattern from the freshly prepared film has been identified as CuCl, whereas the pattern from the annealed film has been identified as $CdCl_2$. The diffraction pattern from a CdSe/CuCl film which has been annealed and washed in methanol to remove the surface $CdCl_2$ layer is shown in fig. 7. All of the spots can be indexed as the cubic form of $Cu_2Se$. This is supported by an Auger spectrum from this film (fig. 4). The electron diffraction pattern from this film shows evidence of preferred orientation. Some of the spots in the pattern can be indexed as belonging to either $Cu_2Se$ or the CdSe underlayer.

Similar diffraction patterns to those obtained before and after a methanol wash have been obtained by selected area reflection electron diffraction at points across the crater produced by depth profiling an annealed CdSe/CuCl film. This result suggests that any ion beam modification of CdSe/CuCl films during ion etching for Auger depth profiling is minimal.

## 5. Conclusions

Using a combination of RED and Auger depth profiling, it has been shown that evaporated CuCl reacts with CdSe in a solid state reaction of the form;

$$2CuCl + CdSe \rightarrow CdCl_2 + Cu_2Se.$$

The results of this investigation show that it is possible to produce $Cu_2Se$-CdSe heterojunctions by the "dry method".

## References

Chopra, K.L. and Das, S.R. Thin Film Solar Cells 1983 (New York: PLenum) pp 349-390.

Dhere, N.K., Parikh, N.R. and Ferreira, A. Thin Solid Films, 1977, 44 83.

De Vos, A., Stevens, K., Vandendriessche, L. and Burgelman, M. Solar Cells 1983, 8, 33.

Fitzgerald, A.G. and McHardy, C.P. Surface Science 1985, 152/153, 1255.

te Velde, T.S., Energy Conversion, 1975, 15, 111.

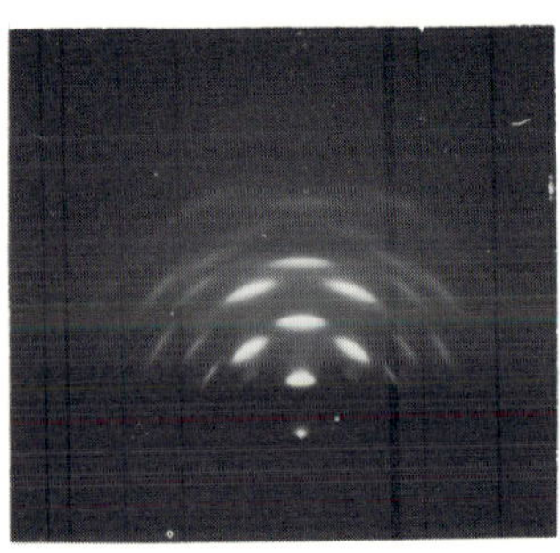

Fig. 5. Reflection electron diffraction pattern from a CdSe film deposited on glass.

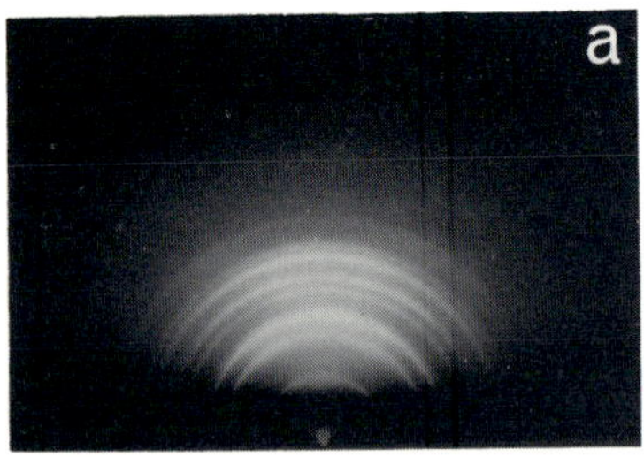


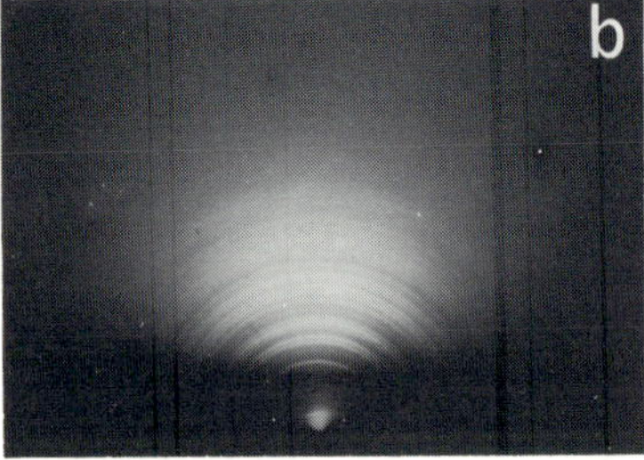


Fig. 6. Reflection electron diffraction patterns from a CdSe/CuCl film (a) freshly prepared, and (b) annealed at 185°C.

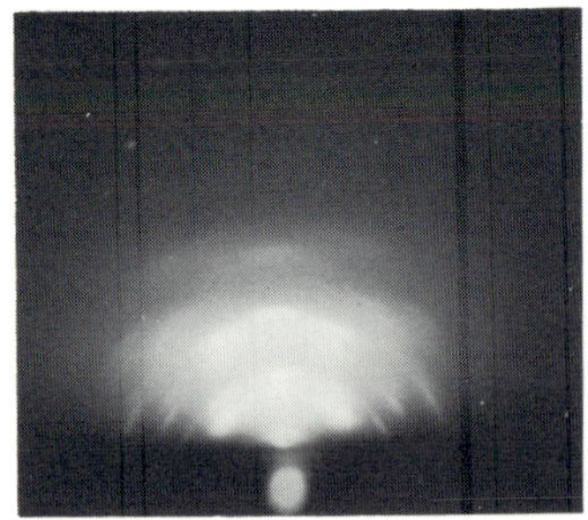

Fig. 7. Reflection electron diffraction pattern from a $Cu_2Se$-CdSe heterojunction after washing in methanol to remove the $CdCl_2$ surface layer.

# Tantalum-iridium Schottky barrier contacts on gallium arsenide

C B Boothroyd, R E Somekh and W M Stobbs
University of Cambridge, Department of Metallurgy and Materials Science
Pembroke Street, Cambridge CB2 3QZ

There are many possible applications for Schottky barrier contacts to GaAs if they could withstand temperatures of the order of 1000K for short periods during processing. For this reason much work has been done on the use of metallic glasses as diffusion barriers between gold contacts and GaAs (Boothroyd et al. 1983; Todd et al. 1984). In order to optimise the barrier design, increasing the temperature that it can withstand, it is necessary to establish which of the possible breakdown mechanisms is dominant as well as how each might affect the electrical properties. Mechanisms for breakdown can include glass crystallisation, bulk diffusion of Au through the barrier and reactions at either interface. Here we present the results of an examination of the structural changes occuring during breakdown of amorphous tantalum-iridium barriers for gold/gallium arsenide contacts.

The substrates used were 5µm MOCVD n on $n^+$ GaAs wafers, cleaned and etched prior to deposition. 110nm of Ta-64 at%Ir was sputtered onto this substrate and then a contact layer of Au was similarly deposited in the same vacuum system. Samples were then subjected to 1 hour anneals under vacuum at 750, 850 and 950K. Cross-sectional and plan-view transmission electron microscopy specimens were prepared by argon ion milling techniques, as described in Newcomb et al. (1985). While the cross-sectional specimens allowed the gross morphology of any reactions to be determined as a function of depth, the plan-view specimens were more useful for the characterisation of the phases formed.

Considering firstly the as-deposited form of the barrier a dark field micrograph (figure 1a) of the Ta-64%Ir in this state shows it to consist of an amorphous matrix containing a few crystals of up to 50nm grain size. Their crystal structure approximates most closely to that of $TaIr_3$, though both the 111 and 200 d-spacings appear to be larger than might be expected (figure 1b). Small pits of diameter 3nm are present throughout the Ta-Ir, as may be seen in figure 1c. At the Ta-Ir/GaAs interface the Ta-Ir, viewed in plan in figure 1d, has a streaked appearance showing coarse lines parallel to [100] GaAs and finer lines at 45$^o$ to these in one direction only, suggesting that the originally etched GaAs surface was slightly off cube normal. The tendency for the observed pitting (figure 1c) to be crystallographically related to the substrate suggests that there should be greater difficulty in forming efficient barriers on vicinal rather than truly (001) GaAs.

Annealing at 750K resulted in only very few changes in the form of the behaviour. No reaction was observed between Ta-Ir and either the Au or the GaAs but the crystals in the amorphous Ta-Ir could now be seen to be differently developed as a function of depth, being larger and more widely separated nearer to the GaAs (figures 2a and b). The Au, however, recrystallizes at this annealing temperature as is seen in the cross-sectional micrograph shown in figure 2c. In this region the Au grain size is 100-200nm, and there are clear indications of the Au starting to facet and "ball-up".

The first signs of reaction appear after 1 hour at 850K, when the gold has "balled-up" on the surface and new phases are formed at the top surface of the Ta-Ir (figure 3a) though these cannot at present be identified as known intermetallics or oxides. The reactivity was not uniform over the whole specimen, some areas showing coarser microstructures (figure 3b) and is undoubtedly to some degree oxidation dependent. Diffraction patterns from such regions (figure 3c) demonstrated that the structure of the $TaIr_3$ now formed was, unusually, undistorted. This might be explained if tantalum oxide had formed in this area, decreasing the Ta:Ir ratio in the matrix towards the stoichiometric value of 1:3. Though some increased crystallinity of the Ta-Ir was observed after annealing at this temperature near the GaAs interface, no new phases were generally observed to be formed here. However when examined optically (figure 3d), isolated, large (~30μm) rectangular crystals could be seen to have grown deeply into the GaAs, presumably having formed where pinhole defects in the Ta-Ir film allowed direct ingress of Au.

After 1 hour at 950K the Au and Ta-Ir were observed to have reacted completely with the GaAs, to produce a variety of phases with grain sizes of ~1μm (figure 4), one of which has a cubic F lattice with lattice parameter 5.59A.

The structural results obtained can be preliminarily correlated with Schottky barrier measurements made by Kelly et al. (1983). They found no change in barrier height after a 620K anneal for 25 hours had occured, whereas a further 2.5 hours anneal at 770K caused an increase in the ideality factor. This second anneal is intermediate between our 1 hour 750K and 1 hour 850K anneals. For the lower of these treatments few structural changes were observed whereas for the latter, gross Au migration and surface reaction were commonplace.

In conclusion, while the data we have obtained indicates that a number of rather interesting solid state reactions can occur for a Au/Ta-Ir/GaAs system, it is clear that the technological use of supposedly amorphous Ta-Ir as a barrier does not depend on the total absence of a degree of crystallinity. At the same time it would appear to be important to cap the Ta-Ir to prevent oxidation reactions, although even when this is done interactions of the capping layer and GaAs at pinholes can rapidly negate the barriers effectiveness. In this context our results also suggest that pinholing of the Ta-Ir as deposited is likely to be more of a problem on vicinal than on truly (001) GaAs surfaces.

Acknowedgements

We are grateful to Professor D. Hull for the provision of laboratory facilities and acknowledge financial support from the SERC and GEC Laboratories, who also provided the GaAs substrates.

## References

Boothroyd C B, Newcomb S B and Stobbs W M 1983 IOP Conf. Ser. No. 68 ed. P Doig (Bristol:IOP ) pp441-444

Kelly M J, Todd A G, Sisson M J and Wickenden D K 1983 Electron. Lett. 19 474

Newcomb S B, Boothroyd C B and Stobbs W M 1985 J. Microsc. in press

Todd A G, Harris P G, Scobey I H and Kelly M J 1984 Solid State Electron. 27 507

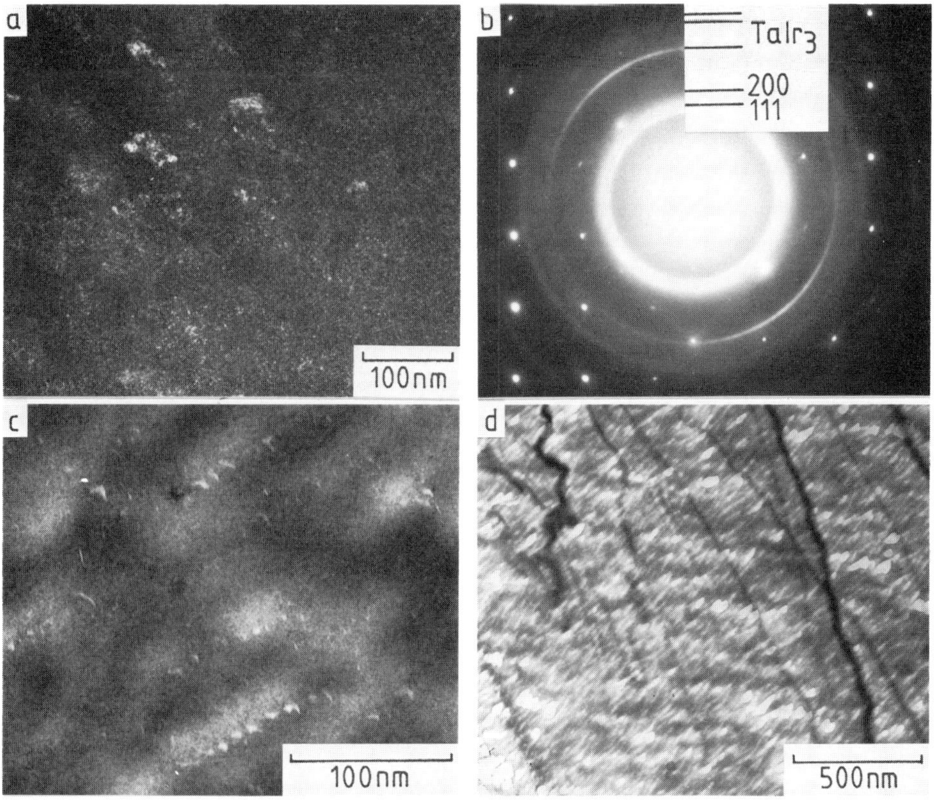

Fig.1. As-deposited; a) dark field and b) diffraction pattern of Ta-Ir, note larger 111 and 200 d spacings, c) small pits and d) lines visible at the Ta-Ir/GaAs interface.

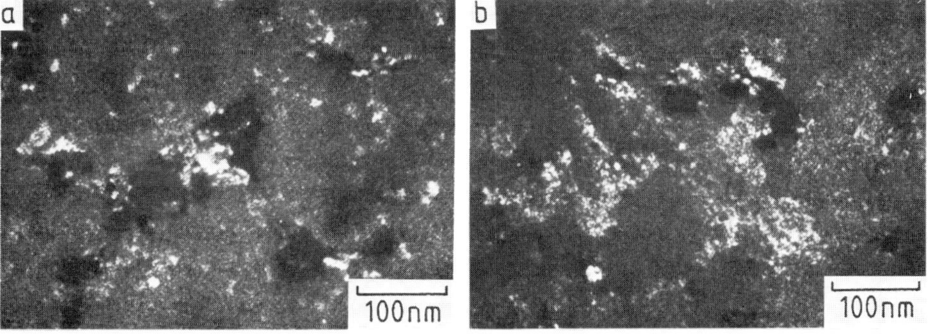

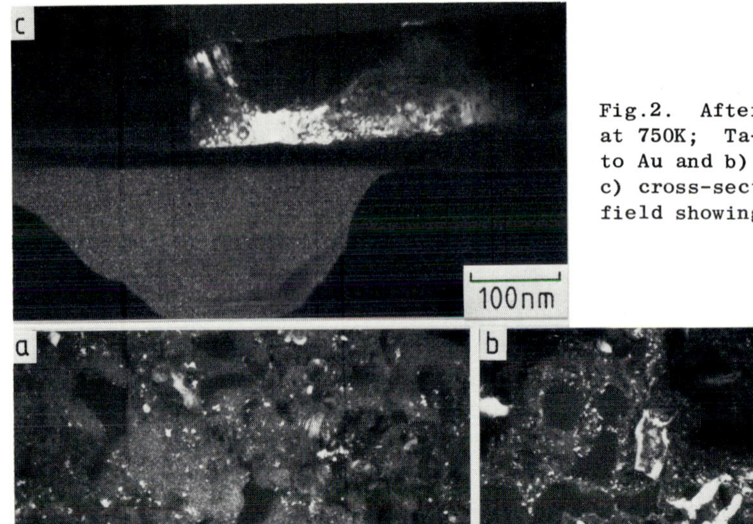


Fig.2. After annealing at 750K; Ta-Ir a) near to Au and b) near to GaAs, c) cross-sectional dark field showing Au and Ta-Ir.

Fig.3. a) fine and b),c) coarse microstructures and diffraction pattern from 850K annealed specimen. d) Optical micrograph showing 'balled-up' Au and rectangular crystals.

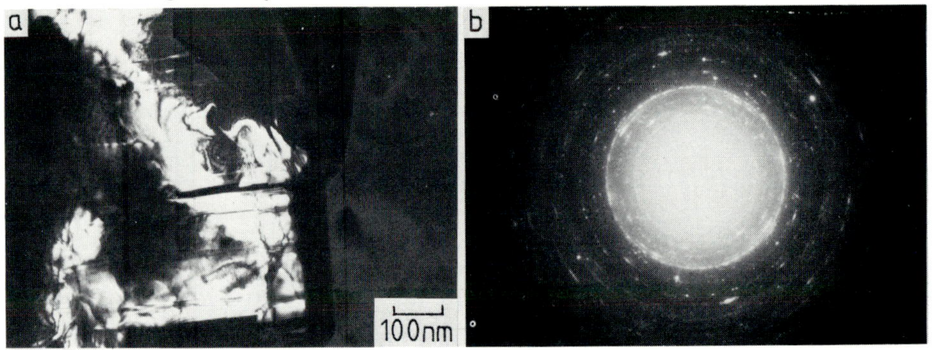


Fig.4. a) large faulted grains and b) diffraction pattern from 950K annealed specimen.

*Inst. Phys. Conf. Ser. No 78: Chapter 11*
*Paper presented at EMAG '85, Newcastle upon Tyne, 2–5 September 1985*


# Electron crystallography of n-paraffins

Douglas L Dorset
Electron Diffraction Dept., Medical Foundation of Buffalo, Inc., 73 High St., Buffalo, NY 14203 USA

## 1. Introduction

Linear paraffins often have been studied as models for both polyethylene and membrane phospholipids and, though such compounds were among the first examined by quantitative transmission electron diffraction techniques (Rigamonti 1936), the projected view of many overlapping atoms down the long chain axes afforded by solution growth is probably the least favorable for distinguishing subtle structural changes. In recent years, however, this crystal habit constraint has been overcome with the development of elegant epitaxial crystallization techniques for polymethylene-containing compounds by J C Wittmann and coworkers (1978ab, 1981, 1984). Lath-like crystals (grown e.g. on naphthalene or benzoic acid) thus permit a structurally more informative view onto the longest molecular axis and have already facilitated quantitative electron diffraction structure analysis of polymers and lipids (Dorset and Pangborn 1982, Moss et al. 1985). Results from the most "simple" crystal structures of paraffins in particular have given some of the most useful data about various solid state phenomena.

Electron diffraction data from epitaxially crystallized n-paraffins have been shown to be useful for quantitative crystal structure analysis. After establishing (Moss et al. 1984) that the epitaxially grown crystal form of $nC_{36}H_{74}$ is the orthorhombic polymorph determined by Teare (1959) and not the lower energy monoclinic form (Shearer and Vand 1956), we have analyzed even chain homologues from $nC_{32}H_{66}$ to $nC_{60}H_{122}$ to confirm that these all pack in the same unit cell (Fig. 1b) (Dorset 1985a) The one odd-chain paraffin studied by us, i.e. $nC_{33}H_{68}$, was found (Dorset, 1985a) to crystallize in the twinned monoclinic unit cell described by Piesczek et al. (1974) (Fig. 1a) and not the lower energy orthorhombic form (Nyburg and Potworowski 1973).

## 2. Structure analysis of imperfect crystals

With the quantitative basis discussed above we have used such crystals to study various phenomena, e.g.:

### 2.1 chain melting

Below a $C_{40}$ chain length, even chain paraffins are found to transform to a hexagonal phase before melting. Although diffraction patterns from solution grown lozenge crystals can be used to follow this phase change, the electron diffraction intensity data can be explained by at least

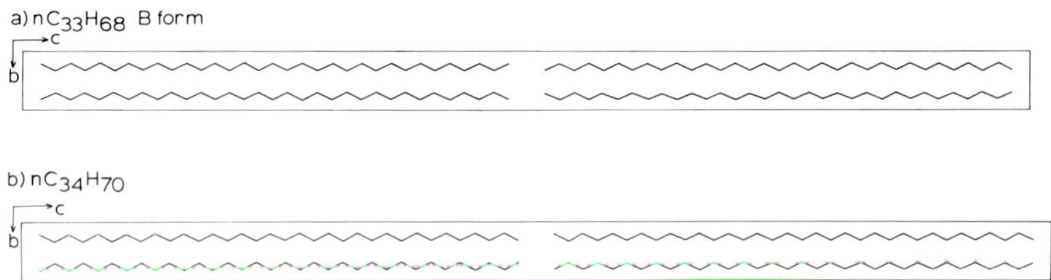


Fig. 1 Crystal structure of n-paraffins in projection down a~7.5Å (a) The B form of odd chain paraffins, space group Aa. This higher energy form is found in the electron diffraction structure analysis of epitaxially crystallized $nC_{33}H_{68}$. (b) The orthorhombic form of even chain paraffins experimentally found for chain lengths from $nC_{32}$ to $nC_{60}$. Note the similarity of end plane packing to the structure in (a).

three models, all of which degenerate into a methylene rotor in this projection (Dorset 1974). On the other hand, analysis of electron diffraction data from $nC_{36}H_{74}$ heated to near this transition demonstrates (Dorset et al. 1984a) that a chain defect model (Blasenbrey and Pechhold, 1967) accounts for the structural change better than alternatives based on a rigid chain rotor (Hoffmann 1952) or a well defined chain helix (D'Ilario and Giglio 1974). The continuous diffuse scattering data from these crystals also demonstrates the stability of these kink defects, especially in diffraction patterns taken below 16°K (Dorset et al. 1985).

## 2.2 radiation damage

Assessment of electron beam damage to paraffin crystals has often been made with electron diffraction patterns from solution-crystallized samples. Indeed, the phased intensity data can be used to calculate the structural change which accompanies this damage (Dorset and Zemlin 1985). Again, in a view down the long chain axes, it is difficult to arrive at a mechanism for this phenomenon. Damage studies on epitaxially crystallized samples, however, demonstrate that the mechanism is "quasi-thermal" (Dorset et al. 1984b), i.e. due to the production of trans vinylene groups (Patel 1975) and the attendant increase of chain flexibility (Craven and Sawzik 1983), the damage mechanism resembles the pre-melt transition, although the chain defects are different.

## 2.3 solid solution formation

Although binary paraffin solid solutions have been often studied by powder X-ray diffraction techniques (Mnyukh 1960), single crystals of these have never been studied before recent work with epitaxially crystallized samples (Dorset 1985b). As shown in Fig. 2, a plot of lamellar spacing vs concentration closely resembles the X-ray data. Electron diffraction intensity data (Fig. 3) e.g. from a 1:1 $nC_{32}H_{66}$ and $nC_{36}H_{74}$ solid solution, however, have been used for a quantitative crystal structure analysis, giving a model very similar to Fig. 1b, but with an exponential falloff of chain end atom occupancy factors. The

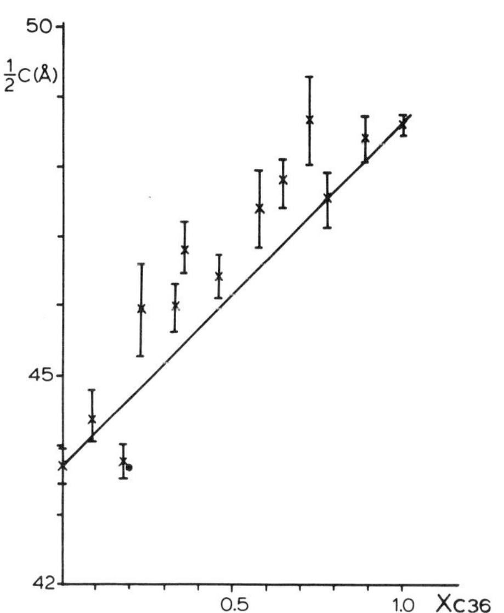


Fig. 2 Electron diffraction lamellar spacing vs mole fraction $X_{c36}$ of the largest chain length component for the solid solutions of $nC_{32}H_{66}$ in $nC_{36}H_{74}$. The lamellar spacings were determined from epitaxially crystallized samples giving patterns as shown in Fig. 3.

Fig. 3 Electron diffraction (0kℓ) pattern from a epitaxially crystallized solid solution (1:1) of $nC_{32}H_{66}$ in $nC_{36}H_{74}$. The inindices of major reflections and lamellar spacing match the values for $nC_{34}H_{70}$. However the inner 00ℓ reflections are truncated much as those for heated paraffin crystals, i.e. the creation of chain end voids by heating (via kinks) or in solid solutions causes similar changes in the diffraction pattern.

disputed (Mnyukh 1960) existence of odd-even solid solutions, moreover, has been demonstrated with similar data from $nC_{33}H_{68}/nC_{36}H_{74}$ binary crystals (Dorset 1985b). Such solid solutions are justified by the close similarity in symmetry of the crystal packings in Figs. 1a and 1b and undoubtedly would be disallowed if the respective lowest energy crystal polymorphs were combined. Recent "lattice images" of these solid solutions support many of the conclusions of the electron diffraction studies, as will be discussed by McConnell et al. (1985). Although our observations support previous data (Asbach and Kilian 1970) that continuous solid solutions occur when the chain length difference is no longer than four carbon atoms, recent work in our laboratory on the binary system $nC_{30}H_{62}/nC_{36}/H_{74}$ indicate the phase diagram is complex and may not be a simple eutectic. Electron diffraction data (Fig. 4) and direct lattice images (Fig. 5) at intermediate mole ratios suggest that two crystalline phases coexist, which are aligned the same way and in intimate contact. Preliminary analyses indicate the layer packing is alternately composed of the shorter and longer components.

Fig. 4 Electron (0kℓ) pattern from a 1:1 binary $nC_{30}H_{74}$:$nC_{36}H_{74}$ system. The inner 00ℓ reflections indicate breakdown of the normal packing symmetry found in pure orthorhombic paraffins or their solid solutions (see arrow). The lamellar spacing $d_{001}$ = 90.35Å may indicate an alternating long chain-short chain sequence.

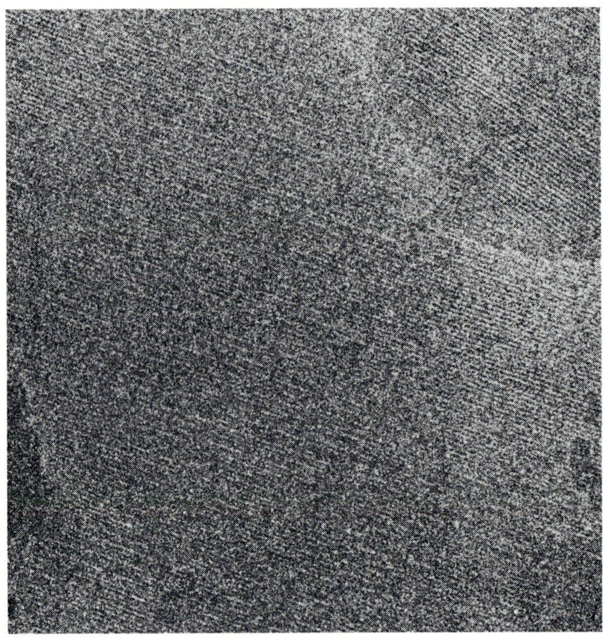

Fig. 5 Lattice image from epitaxially crystallized fused 1:1 mixture of $nC_{30}H_{62}$ with $nC_{36}H_{74}$. The spacing of single lines is approximately 45Å. Note the presence of two lattice types. (Photographed at initial magnification of 20,000x on Kodak DEF-5 X-ray film).

2.4 crystal defects

The detection of point defects in the chains ('kinks') has been discussed above. The identification of edge dislocations has been possible without recourse to moiré magnification in recent direct lattice image work on both solution and epitaxially crystallized paraffin. From collaborative work with F Zemlin and coworkers (1985), quasi-optically filtered e.m. images of monolamellar $nC_{44}H_{90}$ lozenges clearly show edge dislocations which correspond to model image calculations. Image data from epitaxially crystallized samples (collaboration with C McConnell and J R Fryer) similarly reveal such dislocations in the orthogonal direction (McConnell etal. 1985).

## 2.5 linear polymer crystal texture

The major differences in physical behavior between polyethylene and linear paraffins are thought to be due to the chain end fold regions in the infinite polymer (e.g. Wunderlich 1973). Recently, however, it was discovered that the observed correspondence of measured hk0 diffraction intensity from monolamellar paraffin lozenges to their theoretical values predicted by n-beam dynamical theory (Dorset 1980) is no longer found for chain lengths beyond $C_{44}$ (Dorset 1985c). Visualization of crystal texture by bright field low dose imaging at low

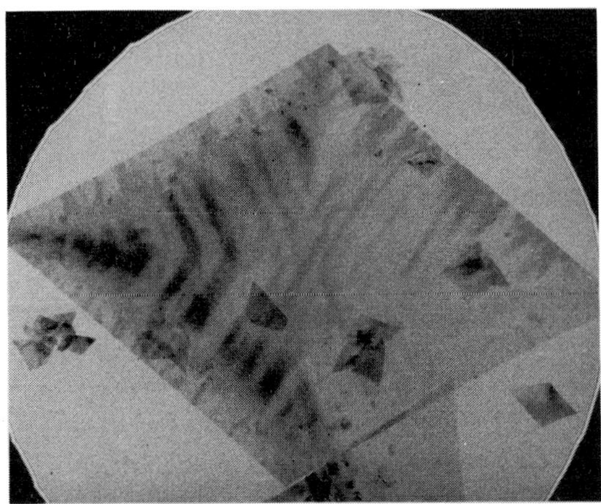

Fig. 6 Bright field electron image of a lozenge crystal of $nC_{82}H_{166}$ grown from hot hexane solution. Note the Bragg contrast striations along <130> and the sectorization of the crystal.

magnification (Fig. 6) reveal sharp Bragg contrast detail unlike the gradually undulating bend contours found in shorter chain crystals. These lines are directed along <130> compared to the similar <1$\bar{3}$0> lines found for collapsed polyethylene lozenges (Bassett et al. 1963). Although the paraffin chains are not folded (e.g. data from epitaxial crystals) there is a sectorization of the crystals. Since no surface detail can be resolved by oblique heavy metal shadowing, low dose lattice images from $nC_{50}H_{102}$, $nC_{60}H_{122}$ and $nC_{82}H_{166}$ have been obtained by C. McConnell and will be analyzed for twin planes and defect content to explain this similarity to polyethylene which cannot be accounted for by fold plane polarization.

## 3. Conclusions

Despite the fact that the X-ray crystal structures of n-paraffins have been known for over 30 years, quantitative electron diffraction and microscopy recently have made considerable progress in the description of crystal imperfections with single crystal data largely inaccessible to X-ray techniques. The analogy of paraffin packing to that of the infinite polyethylene polymer appears to be closer than previously suspected. Thus, with the advantages presented by two microcrystallization techniques, which often give orthogonal views of the same crystal structure, the further study of model paraffin systems will continue to provide further structural insights into the physical characteristics of heterogeneous and/or defect containing linear chain aggregates.

## References

Ashbach G I and Kilian H G 1970 Ber Bunsenges. Phys. Chem. **74** 814
Bassett D C, Frank F C and Keller A 1963 Phil Mag. **8** 1739
Blasenbrey S and Pechhold W 1967 Rheol. Acta. **6** 174
Craven B M and Sawzik P 1983 J. Lipid Res. **24** 784
D'Ilario L and Giglio E 1974 Acta Cryst. **B30** 372
Dorset D L 1974 Chem. Phys. Lipids **13** 133
Dorset D L 1980 Acta Cryst. **A36** 592
Dorset D L and Pangborn W A 1982 Chem. Phys Lipids **30** 1
Dorset D L, Moss B, Wittmann J C and Lotz B 1984a Proc. Natl. Acad. Sci. USA **81** 1913
Dorset D L, Holland F M and Fryer J R 1984b Ultramicroscopy **13** 305
Dorset D L, Moss B and Zemlin F 1985 J. Macromol. Sci. Phys, in press
Dorset D L and Zemlin F 1985 Ultramicroscopy, in press
Dorset D L 1985a J. Polym. Sci. Polym. Phys. Ed. in press
Dorset D L 1985b Macromolecules in press
Dorset D L 1985c J. Macromol. Sci. Phys. in press
Hoffmann J D 1952 J. Chem. Phys. **20** 541
McConnell C M, Dorset D L and Fryer J R These proceedings
Mnyukh Yu V 1960 Zh. Strukt. Khim. **1** 370
Moss B, Dorset D L, Wittmann J C and Lotz B 1984 J. Polym. Sci. Polym. Phys. Ed. **22** 1919
Moss B, Dorset D, Wittmann J C and Lotz B 1985 J. Macromol. Sci. Phys. in press
Nyburg S C and Potworowski J A 1973 Acta Cryst. **B29** 347
Patel GN 1975 J. Polym. Sci. Polym. Phys. Ed. **13** 351
Piesczek W, Strobl G R and Malzahn K 1974 Acta Cryst. **B30** 1278
Rigamonti R 1936 Gazz Chim Ital **66** 174
Shearer H M M and Vand V 1956 Acta Cryst. **9** 379
Teare P W 1959 Acta Cryst. **12** 294
Wittmann J C and Manley R St J 1978 J. Polym. Sci. Polym. Phys. Ed. **15** 1089
Wittmann J C and Manley R St J 1978 J. Polym. Sci. Polym. Phys. Ed. **16** 1891
Wittmann J C and Lotz B 1981 J. Polym. Sci. Polym. Phys. Ed. **19** 1837, 1853
Wittmann J C, Hodge A M and Lotz B 1984 J. Polym. Sci. Polym Phys. Ed. **21** 2495
Wunderlich B 1973 Macromolecular Physics Vol 1 (New York: Academic) p 253
Zemlin F, Beckmann E, Reuber E, Zeitler E and Dorset D L 1985 Science **229** 461

*Inst. Phys. Conf. Ser. No 78: Chapter 11*
*Paper presented at EMAG '85, Newcastle upon Tyne, 2–5 September 1985*



# Epitaxial crystallisation of long chain molecules: Morphological and structural investigations and their applications

J C Wittmann and B Lotz
Institut C. SADRON (CRM-EAHP), CNRS-ULP Strasbourg, 6 rue Boussingault, 67083 Strasbourg Cedex FRANCE

## 1. Introduction

Considerable effort has been expended recently in structural electron microscopic studies of organic materials to overcome (a) their high radiation sensitivity and (b) the often unfavourable crystal orientations obtained by most of the conventional crystal growth methods. The use of low dose techniques to limit radiation damage has allowed, for example, detailed structural studies to be made on condensed aromatic hydrocarbons using high resolution electron microcospy (Fryer 1978, Fryer and Smith 1982). Epitaxy, on the other hand, has been used advantageously as an alternative growth technique to obtain crystals of low and high molecular weight compounds in an orientation more suitable for structural analysis. In the case of long chain molecules in particular, this technique gives access to crystal projections normal to those obtained by solution or melt growth.

During the past few years, we have been using electron microscopy and electron diffraction to study the crystallographic interactions between various synthetic polymers and long chain molecules on organic substrates. Investigations on polymer-organic solvent mixtures have rapidly revealed the formation of polymer eutectics and many examples of epitaxial crystallization onto the freshly grown solvent crystals (Wittmann and St John Manley 1977). Building on these earlier results, the epitaxy of paraffins and various polymers, including paraffinic segments in their repeating units onto chemically as well as crystallographically related organic substrates, was investigated in a more systematic manner (Rickert et al. 1978, Wittmann and Lotz 1981, Wittmann et al. 1983) and the general principles of polymer epitaxial growth on organic substrates established. These principles were shown recently to apply to polyolefins with a helical conformation (Lotz and Wittmann 1984) and, interestingly, to numerous alkane chain lipids (Dorset et al. 1983).

The present article will consist of a short survey of past and present efforts to understand polymer-organic substrate interactions and, with the help of this powerful growth technique to produce well developed and well oriented polymer crystals adequate for structural and morphological EM investigations.

## 2. Representative organic substrates

To be efficient, an organic substrate has to fulfil different requirements namely : adequate chemical structure, melting temperature at least equal to that of the deposited material, good wetability, adequate morphology...

Furthermore, a good knowledge of the structure of the exposed substrate surface is particularly desired if a detailed analysis of the underlying epitaxial relationships is to be carried out. Although a wide variety of organic compounds partly or totally comply with such requirements, only substrates belonging to the two classes of aromatic hydrocarbons and aromatic carboxylic acids will be considered here.

## 2.1 aromatic hydrocarbons

Aromatic hydrocarbons are expected to be good solvents for chemically different polymers and low molecular weight substances. In addition, their crystalline structure is well known, and by varying the number of phenyl rings they cover a wide range of melting temperatures.

Table I displays the melting temperatures together with the crystallographic parameters of the monoclinic unit cells of the lower melting members of : (a) condensed aromatic hydrocarbons (from naphthalene to 1,2 benzanthracene) and (b) linear polyphenyls (from biphenyl to quaterphenyl) which form two parallel series of chemically and structurally related compounds.

TABLE I : Unit-cell parameters and repeat distances of interest for selected condensed aromatic hydrocarbons and linear polyphenyls

| Hydrocarbons | $T_m$(°C) | a(Å) | b(Å) | c(Å) | β(deg) | d(Å) inter-molecular | 110 inter-row spacing (Å) |
|---|---|---|---|---|---|---|---|
| Naphthalene | 80 | 8.235 | 6.003 | 8.658 | 122.9 | 5.09 | 4.85 |
| Anthracene | 216.2 | 8.562 | 6.038 | 11.184 | 124.7 | 5.24 | 4.93 |
| Phenanthrene | 101 | 8.46 | 6.16 | 9.47 | 97.7 | 5.23 | 4.98 |
| Chrysene | 255.6 | 8.386 | 6.196 | 25.20 | 116.2 | 5.21 | 4.98 |
| 1-2 benzan-thracene | 162 | 7.95 | 6.50 | 12.12 | 100.5 | 5.13 | 5.03 |
| Biphenyl | 71 | 8.12 | 5.63 | 9.51 | 95.1 | 4.94 | 4.63 |
| p-terphenyl | 210 | 8.08 | 5.60 | 13.59 | 91.9 | 4.91 | 4.60 |
| Quaterphenyl | 320 | 8.05 | 5.55 | 17.81 | 95.8 | 4.89 | 4.57 |

Perusal of Table I shows that these hydrocarbons are nearly isostructural in the ab section or (001) plane. Indeed, since the long molecular axis is oriented approximately parallel to the c axis, the corresponding c parameter increases proportionally to the number of phenyl rings present, whereas the a and b parameters corresponding to the lateral packing of these almost planar molecules remain fairly constant. The structure is a layered one with the (001) plane being an easy cleavage plane and the predominant growth face of the platelet-shaped hydrocarbon crystals. The phenyl rings protruding from these large (001) surfaces are aligned along rows parallel to the two equivalent [110] and $[1\bar{1}0]$ directions at an angle of nearly 70°. Also given in Table I are the structurally important intermolecular distances along these rows and the inter-row distances on the substrate surfaces.

2.2 aromatic carboxylic acids

Alteration of the solvent properties or wetability of the aromatic hydrocarbons without large change in the surface structure can be easily brought about by substitution on the phenyl rings with different substituent groups (X = $CH_3$, OH, Cl ...) in para or ortho position. More drastic changes in properties can be further introduced by covalently bonding various functional groups e.g. carboxylic acids. The properties of these organic acids can in turn be altered in a controlled fashion through salt formation. The corresponding two organic series, the simplest members of which are benzoic acid and sodium benzoate, respectively, offer an even larger choice of efficient and versatile substrates for the long chain molecules than the hydrocarbon series and detailed crystal structure data are readily available. In both series, the carboxylic groups or the metal ions form an inner polar layer shielded by two apolar layers of aromatic or substituted aromatic rings. These apolar surface layers make only weak Van der Waals contacts. They correspond again to well developed faces of the crystals and consequently are the active surfaces for the deposited molecules. It can easily be conceived that unlike inorganic substrates, such organic layered crystals can be "structurally" adjusted to a given deposit.

## 3. Growth procedures

In practice, various growth procedures have been devised based on the type of E.M investigations performed and, whether the organic substrate is used as a solvent for the long chain compound or whether this substrate remains in the solid state during the entire growth process.

A most powerful and rapid method which has been successfully applied to linear polymers (Wittmann and Lotz 1981) as well as paraffins and phospholipids (Dorset et al. 1983) makes use of very dilute solutions or mixtures of the deposit in the molten substrate taken as a solvent. This method has a distinct advantage over more conventional ones in that it provides uncontaminated substrates grown in-situ. Typically, a thin layer of solution prepared directly between two glass slides by co-melting crystals of the two species, is either quenched to room temperature or moved slowly along a temperature gradient until the two-stage solidification process has reached completion. The vacuum sublimation or selective dissolution of the substrate leaves very thin layers of oriented material on both slide faces which are then treated as usual for E.M observations. In an alternative method a C-coated grid is dipped directly into the molten substrate/deposit mixture (Fryer 1981).

In addition, whenever the substrate crystals were electron beam resistant, the orientation of the deposit and the epitaxial relationships were established by selected area electron diffraction on very thin substrate/deposit bi-crystals grown and deposited from solution (Wittmann and Lotz 1981).

## 4. Epitaxial growth of polyethylene on aromatic hydrocarbons : The substrate surface structure and the lattice matching criterion

Crystallization of polyethylene (PE), the paradigm of linear polymers with a planar zigzag conformation, on anthracene and p-terphenyl (selected among the numerous members of the condensed aromatic hydrocarbon and linear polyphenyl series respectively for their relatively high melting temperature and comparatively good radiation resistance) yield oriented thin films consisting primarily of a network of doubly oriented rod-like PE lamellae The epitaxial relationships derived from the electron diffraction patterns of polymer/hydrocarbon bicrystals can be summarized as follows :

(a) $[001]_{PE}$ // $\langle 110\rangle_{Anthracene}$ (b) $[001]_{PE}$ // $\langle 110\rangle_{p\text{-}Terphenyl}$

$(100)_{PE}$ // $(001)_{Anthracene}$ $(110)_{PE}$ // $(001)_{p\text{-}Terphenyl}$

In both cases the PE chains are oriented parallel to the substrate (001) planes in the crystallographically equivalent [110] and [1$\bar{1}$0] directions, accordingly forming two sets of chain folded lamellae seen edge-on and oriented ca. 70° to each other. A noteworthy difference exists in the PE contact planes with the (001) surfaces of the two organic substrates despite the near isomorphism of the latter.

The proposed mechanism of epitaxy rests on a two-dimensional lattice matching and involves the deposition and orientation of the polymer chains parallel to the 110 "furrows" formed by the rows of aromatic rings on the (001) substrate contact surface. The spacing between the furrows dictates the type of PE chain packing at the interface and hence the nature of the contact plane (Wittmann and Lotz 1981). The best possible match is thus realized between the 4.94 Å interchain distance in the (100) PE plane and the anthracene 4.93 Å inter-row distance on one hand, and between the 4.45 Å interchain distance in the other densely packed (110) PE plane and the slightly shorter 4.60 Å p-terphenyl inter-row distance (cf. Table I). Probably less important, a second close lattice match, on comparing the intermolecular distance in the <110> directions and nearly twice the PE chain axis repeat of 2.54 Å, is also seen to exist for both systems. Identical two-dimensional matching of lattice parameters occurs with other aromatic hydrocarbons since all members of a given series are isomorphous in the basal (001) surface.

5. Generalization to other substrates and long chain molecules

As indicated, tailored changes can be made in the structure and/or properties of organic aromatic substrates either by substitution on the phenyl rings or by covalent linkage of carboxyl groups.
Such changes have several definite advantages :

- most important, the substrate surface symmetry may be lowered so that monodirectional orientations of large deposited crystals can be induced. The epitaxial growth of different polymers in a single orientation on benzoic acid crystals for example is consistent with the lack of symmetry elements in the exposed (001) plane (Wittmann et al. 1983).

- the wetability or compatibility can be adjusted and the polymer/substrate interactions increased as is the case with polyamide-benzoic acid systems or fluorinated polyolefins epitaxially crystallized on surfaces rendered "fluorophilic" by use of F-substituted benzene rings.

- a wider range of substrate lattice periodicities is available allowing very different chain packings at the interface and therefore different crystal orientations or even different crystal structures to be obtained for a given sample (Rickert et al. 1978).

On the other hand, despite chemical differences, polyethylene and linear aliphatic polyesters, polyamides, paraffins (Wittmann et al. 1983) as well as a large number of shorter chains with polymethylene segments including biological alkane chain lipids (Dorset et al. 1983) were shown to behave quite similarly when solidified in the presence of organic crystals. As underlined by Dorset et al. (1983), the oriented growth seems indeed to be insensitive to the polymethylene chain length (from octyl segments to PE chains), the presence, disposition and size of substituents, the chemical nature of the linkage etc ... The crystal structure of all these compounds being largely determined by the presence of paraffinic moieties in a planar zigzag conformation, their structural resemblance to polyethylene is probably sufficient to result in similar epitaxies on a given substrate.

A more striking feature is the recent demonstration of the efficiency of the same organic substrates towards structurally completely different polymers such as polyolefins with a helical conformation like isotactic polypropylene which has methylene side groups (Lotz and Wittmann 1984) or selenium, an inorganic chain molecule (Wittmann and Lotz 1984). These unexpected epitaxies are explained by unit cell parameters close to those of PE for the latter inorganic polymer (trigonal Se unit cell with a = 4.355 Å and and c = 4.949 Å) and for polypropylene by the existence in densely packed planes of rows of methylene groups with a characteristic 5 Å periodicity comparable to the interchain distances of PE and linear polyesters.

## 6. Applications and future trends

Although still in its infancy, current attempts reveal the great potential of the epitaxial growth technique of long chain molecules and its general utility in E.M investigations dealing with :

(a) the structure of chain folded lamellae ; for instance, low and wide angle electron diffraction may be combined or gold decoration of the stacks of edge-on lamellae used to determine the crystallographic nature of the fold plane or to analyze the variation of lamellar thickness and crystal perfection with thermal treatments (Wittmann and Lotz 1981).

(b) the determination of unit cell dimensions and symmetry in particular of polymorphs ; the existence of very large domains with a unique cell orientation makes it possible to obtain single-crystal diffraction patterns in various projections normal to the chain axis.

(c) Quantitative electron diffraction structure analysis or "electron crystallography" ; a great deal of work has been performed by Dorset and his associates on epitaxially crystallized alkanes or alkane derivatives including lipids and polymers which amply demonstrates the importance and potential interest of epitaxy for preparing samples suited for total crystal structure analysis by electron diffraction (Dorset et al. 1983, Moss et al. 1984, Dorset 1985) as well as for deformation or phase transformation studies (Dorset 1985).

(d) High resolution electron microscopy of molecular crystals ; the lattice images obtained recently by Fryer (1981) on paraffins epitaxially crystallized on naphthalene and by Isoda (1984) on poly(p-xylylene) oriented via epitaxial polymerization on NaCl deserve great interest in that they provide unique information at a molecular level on long chain compounds.

The epitaxial synthesis or "epipolymerization" used by Isoda was first applied to the growth of poly(sulfur nitride) or $(SN)_x$ single crystals on alkali halides (Rickert et al. 1980). This innovative procedure which combines epitaxial crystallization and solid state polymerization offers, in conjunction with the use of organic substrates, real new prospects for the controlled preparation of highly oriented, near-perfect polymer single crystals. Interestingly enough, the various monomers to which this growth technique may be applied (di-p-xylylene, $S_2N_2$, diacetylenes etc...) very often give rise to polymerized crystals which, besides possessing interesting optical or electronic properties, are of fairly high radiation resistance and therefore well suited for electron crystallography and high resolution electron microscopy.

## References

Dorset D L, Pangborn W A and Hancock A J 1983 J. Biochem. Biophys.Methods 8 29

Dorset D L 1985 J. Electron Microsc. Tech. 2 89

Fryer J R 1978 Acta Crystallogr. A 34 603

Fryer J R 1981 Inst. Phys. Conf. Ser. n°61 pp 15-22. Paper presented at EMAG Cambridge 7-10 Sept.

Fryer J R and Smith D J 1982 Proc. Roy. Soc. Lond. A 381 225

Isoda S 1984 Polymer 25 615

Lotz B and Wittmann J C 1984 Makromol. Chem. 185 2043

Moss B, Dorset D L, Wittmann J C and Lotz B 1984 J. Polym. Sci., Polym. Phys. Ed. 22 1919

Rickert S E, Baer E, Wittmann J C and Kovacs A J 1978 J. Polym. Sci., Polym. Phys. Ed. 16 895

Rickert S E, Lando J B, Hopfinger A J and Baer E 1979 Macromolecules 12 1053

Wittmann J C and Manley R St John 1977 J. Polym. Sci., Polym. Phys. Ed. 15 1089

Wittmann J C and Lotz B 1981 J. Polym. Sci., Polym. Phys. Ed. 19 1837,1853

Wittmann J C, Hodge A M and Lotz B 1983 J. Polym. Sci., Polym. Phys. Ed. 21 2495

Wittmann J C and Lotz B 1984 J. Mat. Sci. 19 1439

*Inst. Phys. Conf. Ser. No 78: Chapter 11*
*Paper presented at EMAG '85, Newcastle upon Tyne, 2–5 September 1985* 

# Electron diffraction studies of polymer–substrate interfaces

D Vesely and G Ronca

Dept. of Materials Technology, Brunel University, Uxbridge, Middlesex UB8 3PH

## 1. Introduction

The nucleation of crystalline lamellae in polymer melts is in nearly all cases heterogenous. Impurities and additives are always present in commercial grades of polymers, some of them are added intentionally to increase the rate of crystallisation. However the nucleating mechanism is not well understood and crystallographic information on the molecular arrangement at the interface between the nucleating particle and the crystalline lamella of the polymer is needed. Unfortunately, polymers and often also the nucleating agent, are very sensitive to the electron beam and it is experimentally difficult to obtain a diffraction pattern before the crystallinity is destroyed. Special techniques and precautions are thus required and in this work STEM microdiffraction has been used. This technique is described together with some preliminary results on the crystallographic correlation between the polymer lamellae and the substrate on which they have nucleated.

## 2. STEM microdiffraction technique

In order to obtain crystallographic information from an interface the diffraction area must be small and simultaneously the spots on the diffraction pattern sharp. The conventional selective area diffraction technique is not suitable for beam sensitive materials because an area much larger than the area used for the formation of the diffraction pattern is exposed to the beam and it is thus very difficult to control the total exposure of the specimen. On the other hand STEM microdiffraction is far more convenient, accurate and reliable, and its use for polymers is paramount.

The crystallinity of the specimen decays exponentially with exposure and it is thus essential to utilise the very early stages of degradation. It is a matter of convenience to make a definition of the critical exposure of the specimen $D_{cs}$, beyond which we regard the specimen as damaged. For good quality diffraction patterns the critical exposure is, for most polymers, smaller than $10C/m^2$. The volume of the specimen must therefore be large enough so that the diffracted beam has sufficient current density $I_{gf}$ to expose the film to the required contrast during the exposure time $\tau_{cf}$. The relationship between the critical exposure of the film $D_{cf}$ and the critical exposure of the specimen $D_{cs} = I_o \tau_{cs}$ for thin specimens can be written as:

$$D_{cf} = I_{gf} \tau_{cf} < \frac{C}{L^2 \beta^2} \frac{I_o \tau_{cs} A t^2}{E_o} \left| \frac{F_g}{V_c} \right|^2 f(s_g, t)$$

where C is a constant which also includes the losses of electrons in the imaging system, L - camera length, $\beta$ - incident angle (or beam convergence), $I_o$ - current density of the incident beam, $\tau_{cs}$ - critical exposure time for the specimen, A - irradiated area, t - specimen thickness, $E_o$ - accelerating voltage, $F_g$ - structural factor, $V_c$ - volume of the unit cell, $S_g$ - deviation from Ewald sphere. The function $f(s_g,t)$ describes the contrast conditions and cannot be selected or calculated as most polymer specimens have low crystallinity and are polycrystalline in nature. Similarly the specimen thickness cannot be predicted or measured as the specimens are very uneven. For this reason a sufficiently large diffraction area 'A' must be selected so that the above equation is satisfied. In most cases an area of 2μm in diameter proved to be the most convenient.

In order to obtain diffraction patterns from areas which were not accidentally pre-irradiated and are of the required size the procedure described below was used. The size of the microdiffraction area is controlled by the CL2 apperture and by the CL2 and OL currents. For each CL2 current there is one OL current for which the diffraction is in focus (minimum convergence) or the image is in focus (convergent beam). This is shown in Fig. 1, where the size of the irradiated area is also indicated, for 120μm CL2 apperture. It is apparent that from the best image condition (at point A) it is possible to obtain a 2μm probe diameter just by increasing the CL2 current. The probe diameter can be changed in steps by using a different apperture size. In our case spot sizes 8, 2, 1 and 0.5μm could be obtained with appertures 500, 120, 60 and 30μm respectively. Alternatively a different setting of the OL and CL2 currents can be used, but it is not recommended to change the OL current, since the off centre position of the probe will cause the diffraction to no longer correspond to the area selected on the CRT. For this reason the first method has been used and the whole procedure can be described as follows: The specimen was scanned at low current density in STEM mode. The magnification, the beam current and the scanning time were selected in such a way, that no area received an exposure larger than $1C/m^2$. When a suitable area was identified, the electron beam was cut off above the specimen. The microscope was then re-aligned for microdiffraction using a 5-digit lens current monitor and pre-calibrated charts. The electron beam was thus used only to expose the diffraction pattern on a fast X-ray film. The major difficulty in recording the two diffraction patterns simultaneously was in our inability to evaluate quickly their relative intensities and for this reason not all attempts have been successful. Nevertheless the results described below represent about 20 diffraction patterns for each polymer/nucleant system.

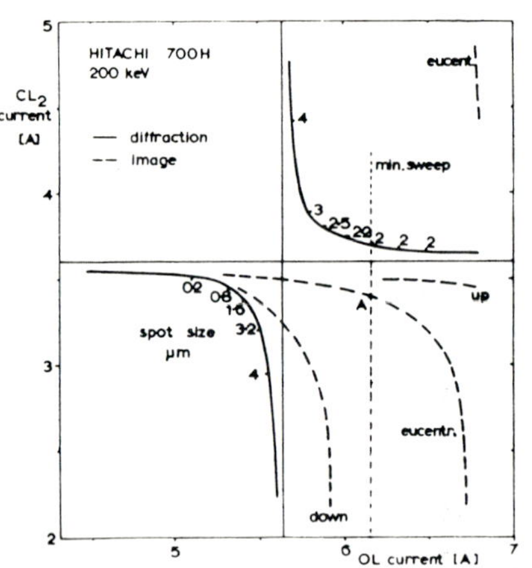


Fig. 1 The second condensor lens current andthe objective lens current settings for focussed diffraction and focussed image (for specimen at eucentric, down and up positions) are shown.

## 3. Polymer-substrate interfaces

The specimens were prepared by solvent casting, followed by melt crystallisation of the polymer. Small amounts of the nucleating agent were added into the solution, so that most spherulites contained a nucleating particle in their centre. A typical case is shown in Fig. 2a where polypropylene (PP) has been nucleated on a talc particle. A small area of 2μm in diameter, from which the microdiffraction has been taken is indicated.

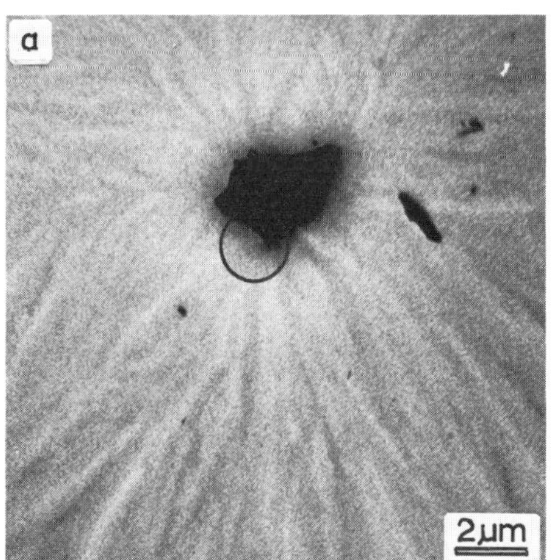


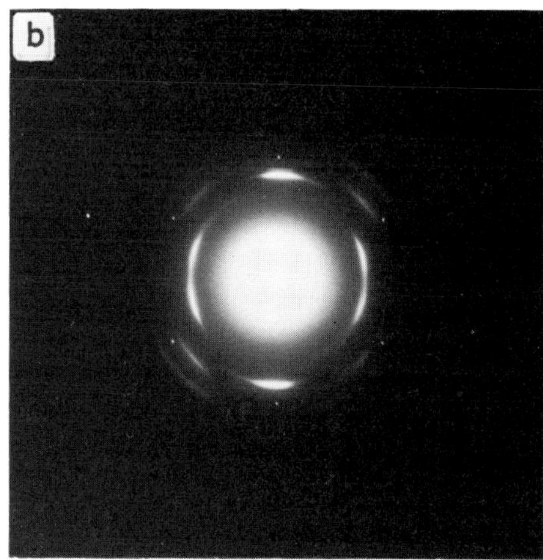


Fig. 2 A STEM micrograph of a talc particle, in the centre of a PP spherulite is in (a). The corresponding microdiffraction in (b) is oriented to show its true relationship to the spherulite lamellae. The six small spots correspond to the diffraction pattern from the talc particle.

The corresponding diffraction pattern is in Fig. 2b. A simple correlation between the diffraction spots from the polymer and from the nucleant is apparent. The diffraction patterns are re-drawn and indexed in Fig. 3 for talc and boron nitride (BN) with polyethylene (PE), polypropylene (PP) and polyhydroxybutyrate (PHB). It is obvious that there is a very good crystallographic correlation for all six systems in the (001) plane, however we do not now the (hk0) plane on which the epitaxy occurs. From thermodynamical considerations of crystal growth it might be logical to expect, that a crystal will nucleate on the growth plane. For polymers this plane would be the plane in which the molecular chains are folded. All other planes would have much higher energy and it is thus unlikely, that they can be nucleated. In most polymers the plane of growth is perpendicular to the axis of the growing lamella and we can expect that this plane will be parallel to a low index cleavage plane on the nucleating particle. The likely combinations of planes are listed in Table A, together with the number of repeat polymer chain spacings which correlate with the substrate spacings. So far we have not considered the correlation along the c axis, which is likely to be less important. Nevertheless the correlation of molecular spacings with the substrate spacings along the c axis is excellent for PP(2 and 1) and PHB (1 and 3) and only slightly worse for PE (4 and 6) on talc and BN respectively. The angle of the c axis is however not always 90°, as talc has a monoclinic structure and BN is hexagonal. PP has always nucleated in the monoclinic form with the angle of the c axis identical to talc. PE is tetragonal but has nucleated on the (200) plane of the talc particle. It is thus necessary to assume 10° tilt along the b axis for PE/talc and along the a axis for PP/BN and PHB/talc systems.

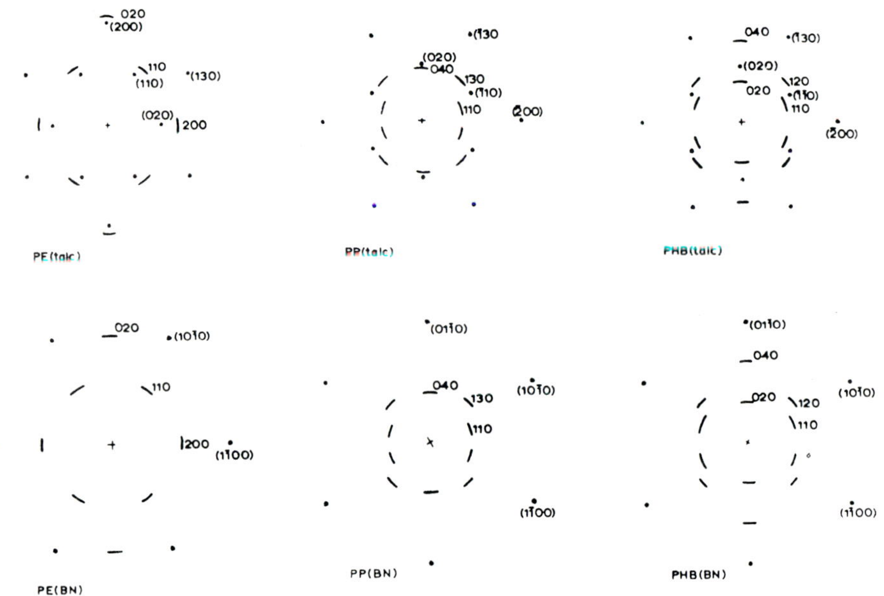


Fig. 3 The calculated and indexed diffraction patterns of PE, PP and PHB (arcs) are shown together with the diffraction patterns from talc and BN (spots) in their observed mutual orientation.

Table A

| Nucleant | Polymer | Epitaxial planes | Correlation |
|---|---|---|---|
| Talc | PE | 200/020 | 5 |
| | PP | 020/040 | 4 |
| | PHB | 020/020 | 11 |
| BN | PE | 11$\bar{2}$0/020 | 3 |
| | PP | 10$\bar{1}$0/040 | 3 |
| | PHB | 10$\bar{1}$0/020 | 3 |

## 4. Conclusion

The microdiffraction technique described here proved to be a very valuable tool in obtaining new information on the polymer-substrate interfaces. It was possible to use specimens with the geometry and thermal history which closely resembled the industrial material. The analysis of diffraction patterns from PE, PP and PHB nucleated on talc and BN has shown that a close epitaxial match between the nucleating particle and the polymer is a necessary requirement for nucleation. Any mismatch in the inteface will result in the formation of defects and thus in an increase in the interfacial energy. From this point of view talc is likely to be a poorer nucleant than BN especially for PHB, as has been confirmed recently by DSC measurements (unpublished results).

*Inst. Phys. Conf. Ser. No 78: Chapter 11*
*Paper presented at EMAG '85, Newcastle upon Tyne, 2–5 September 1985* 

# Convergent beam electron diffraction from organic crystals

R Vincent

Physics Department, University of Bristol, Bristol BS8 1TL

Structure analysis by convergent-beam electron diffraction (CBED) using a probe focussed on to the specimen is inevitably limited to crystals resistant to radiation damage. However, the current density is drastically reduced by illumination with a defocussed probe, where the CBED discs become equivalent to multiple bright- and dark-field images, with a spatial resolution equal to the probe diameter. Unless the specimen is a perfect, unbent, parallel-sided slab, all precise symmetry information is lost. We can recover most of this information, both by limiting the area illuminated, and also by a large increase in the convergence angle, where the overlapping discs of adjacent reflections are excluded by insertion of a small SA aperture in the probe image plane (Tanaka 1980). These large-angle CBED patterns are formally equivalent to bend contour images from a crystal uniformly bent in two dimensions with a small radius of curvature. Any physical bending of the crystal distorts or stretches the patterns, but is unlikely to obscure symmetry-related features around a zone axis, provided that the thickness is constant. This technique has been tested on sublimed platelets of the aromatic crystal, p-terphenyl (monoclinic, space group $P2_1/a$). In Fig. 1, the absence of a 100 mirror (vertical) in the [001] pattern is demonstrated by comparison of the [102] and [$\bar{1}$02] patterns, where the diffraction geometry is almost identical ($\beta = 92°$), but structure factors are radically different. The space-group is confirmed in Fig. 2, where CBED patterns from kinematically forbidden 100 and 010 reflections on the [001] axis contain examples of the Gjønnes-Moodie dark cross (see also Jones and Williams 1975). The beam illuminated an area ~5μm diameter with a probe current of 0.02 to 0.05 nA, equivalent to current densities between $1 \times 10^{-4}$ and $2.5 \times 10^{-4}$ A $cm^{-2}$, or observation times of 500 to 200s for 10% of the critical dose (0.5 A s $cm^{-2}$ at 100 kV).

The maximum angular field ($2\theta$) in large-angle patterns is obtained by a small reduction in the C2 lens current away from crossover, so that the SA aperture occupies the disc of least confusion (diameter 2r), slightly above the Gaussian image plane (see Fig. 3 and Klemperer and Barnett 1971). For the EM430 in nanoprobe mode, the maximum convergence angle is 6.1°, reduced to $2\theta = 4.3°$ by insertion of a 5μm SA aperture. Substituting these values in $r = \frac{1}{4} C_s \theta^3 M$, where the objective magnification, M is 40x, we find an effective $C_s$ of 4.8mm, more than double the objective $C_s$ of 2.1mm, due to contributions from the probe-forming lenses. Overall, the 5μm aperture represents a useful balance between the area illuminated ($\propto r^2$), and the angular field ($\propto r^{1/3}$ , as in Fig. 4).

References

Jones W and Williams J O 1975 J.Mat.Sci. 10 379
Klemperer O and Barnett M E 1971 Electron Optics 3rd ed. (London:Cambridge University Press) pp 169-81
Tanaka M 1980 J.Electron Microsc. 29 408

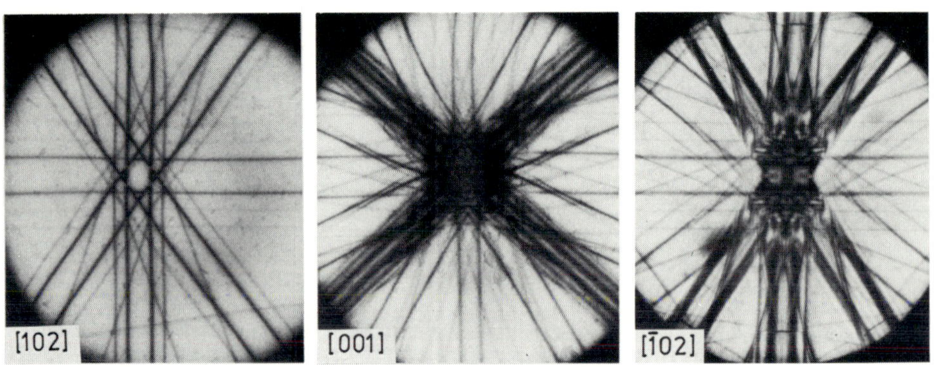


Fig. 1 Large-angle bright field CBED patterns taken from three zone axes parallel to the (010) mirror plane (horizontal) in p-terphenyl at 300 kV.

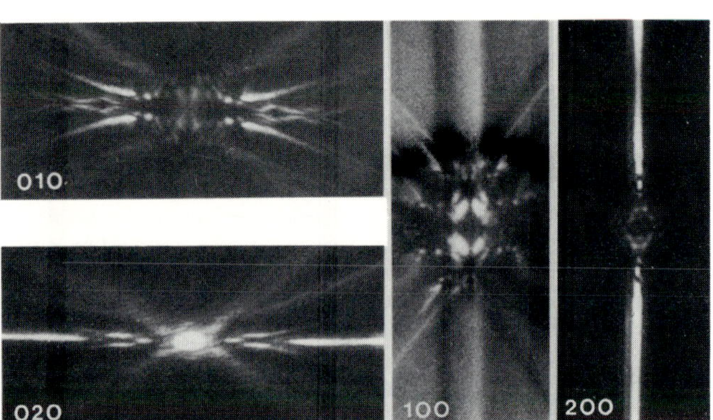


Fig. 2 Dark field CBED patterns taken from the [001] axis in p-terphenyl at 300 kV. The Gjønnes-Moodie dark crosses in the 010 and 100 reflections are consistent with space group $P2_1/a$.

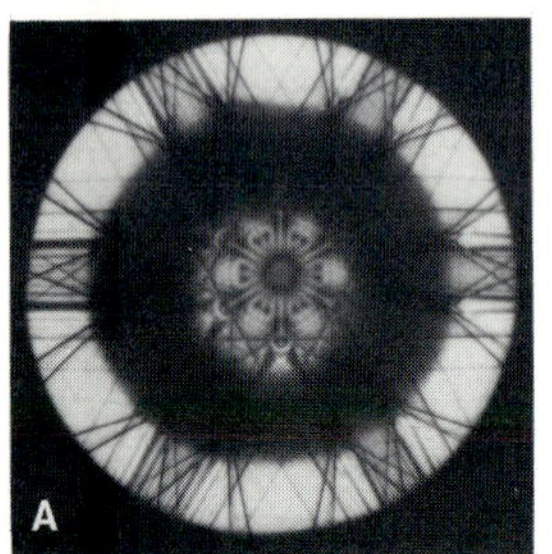


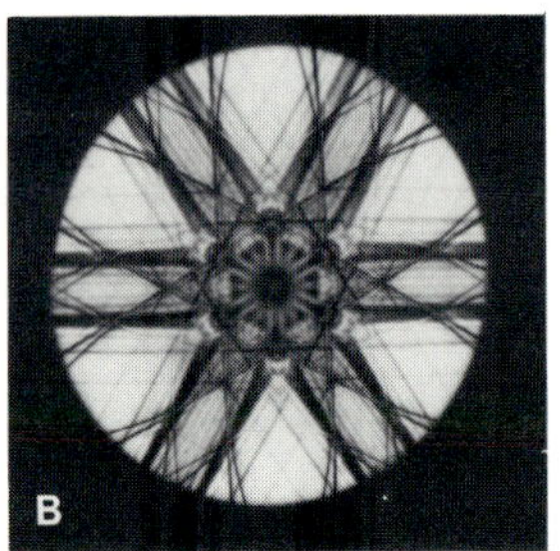


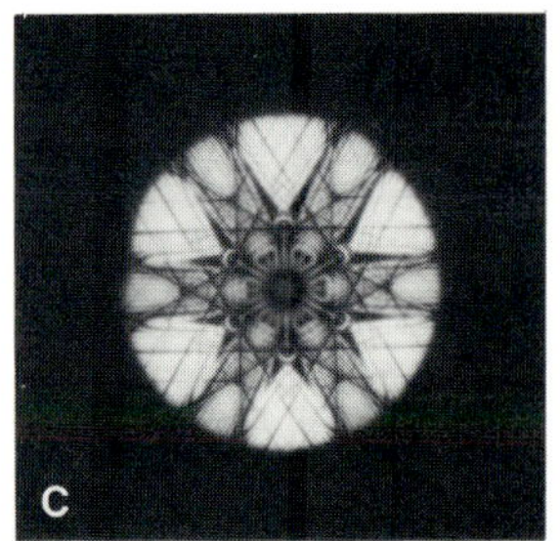


Fig. 3 (above) Variation of angular field in large-angle CBED patterns with small increments in C2 lens current. Relative to the optimum setting (b),the SA aperture lies above and below the disc of least confusion in (a) and (c), respectively.

Fig. 4 (opposite) Variation of the maximum transmission angle (2θ) with SA aperture diameter (2r) in the EM430.

# Tilted illumination studies of crystalline polymers

J R White

Department of Metallurgy and Engineering Materials, University of Newcastle upon Tyne, Newcastle upon Tyne, NE1 7RU

## 1. Introduction

The beam sensitivity of polymer crystals prevents the use of many standard diffraction contrast TEM procedures. Diffraction contrast is very sensitive to the direction of the electron beam in relation to the crystal axes, and a change in orientation of a fraction of a degree can produce changes in contrast which contain structural information. With specimens that are not damaged in the electron beam, the electron microscope can be set at a magnification and illumination level suitable for visual inspection and the crystals examined while using a goniometer stage to alter the relative direction between the electron beam and the crystal. Such a procedure cannot be attempted with most polymer crystals because they degrade rapidly under these conditions.

It is common to conduct CTEM studies of polymer crystals at sub-visual levels. Diffraction contrast information can be obtained using dark field (DF) conditions and taking several pictures each with a different diffracted beam. For very sensitive polymers, such as polyethylene, this requires the use of low magnification (to minimise the electron intensity at the specimen) and as a consequence the resolution is low, limited by the detection system (e.g. the film grain size) (Thomas and Ast 1974, White 1975). Under such conditions the deterioration in electron optical resolution caused by the use of shifted aperture ('simple') DF conditions is not important as long as the chosen diffracted beam is not at too high an angle from the objective lens axis. It is often advisable to use simple DF conditions when recording a series of images in different diffracted beams for the shifted aperture technique is generally slightly quicker than the alternative tilted illumination DF method, aiding specimen preservation. In the ('high resolution') tilted DF method the illumination is tilted until the chosen diffracted beam passes along the axis of the objective lens, minimising aberrations, and the incident electron beam impinges on the specimen in a different direction for each DF image. With simple DF the electron beam impinges on the specimen in the same direction for each image and this should be taken into account when interpreting the images and when comparing the two DF techniques.

Recently a hybrid method has been introduced in which a small illumination tilt, different to that required for high resolution DF, is applied using the tilt controls, and DF (or bright field, BF) obtained by shifting the objective aperture (White and Thomas 1985). In this way a series of images using slightly different diffraction conditions can be recorded for subsequent analysis. This is more convenient than tilting with a goniometer and the tilt angle can be determined with high accuracy.

## 2. Experimental

Results will be presented for three types of specimen, for which detailed preparation procedures are described in the literature.
Polyethylene (PE) films, drawn from the melt, were provided by D C Yang and were made from solution cast film by a development of the method introduced by Petermann (Petermann and Gohil 1979, Yang and Thomas 1984). The results described here were obtained using drawn film that was subsequently annealed for 2h at 127°C then carbon coated before inspection in the TEM.
Poly(p-phenylene benzobisthiazole) (PBT) fibres heat-treated by S R Allen as part of the programme described by Allen et al (1981) were placed between adhesive tape and thinned by repeated separation. A thin carbon film was evaporated onto the fibre fragments, before releasing from the adhesive tape by submersing in chloroform, and pieces picked up on electron microscope grids.
Poly(butylene terephthalate) (4GT) crystals (α form) were grown within films of polyether-polyester random block copolymers containing 73% by weight of 4GT and 27% of poly(tetramethylene ether glycol). Single crystals of 4GT cannot be easily grown from the homopolymer but isolated lamellae formed from the copolymer melt when cooled from 250°C to the crystallization temperature of 203°C. The specimens were prepared by R M Briber and further details of the preparation technique and the morphology of the films are given in Briber and Thomas (1985).

All observations were made on a JEOL 100CX microscope equipped with a side entry goniometer and operated at 100keV using low beam intensities. PE and PBT were imaged using simple DF. Some areas were imaged successively selecting a different diffracted beam or group of beams (as happens when two or more beams occur too close together to separate with the objective aperture). Other areas were imaged in one or two diffracted beams only but using different illuminating beam tilts to record a series of images. With the polyether-co-polyester films the 4GT crystals were large and the areas selected for photography gave single crystal patterns, permitting selection of single diffracted beams for (simple) DF. With PE some series of BF images were recorded using a different beam tilt for each exposure. Tilt angles were measured by the method described by White and Thomas (1985) from doubly exposed electron diffraction patterns showing the shift obtained on applying the tilt excitation (figure 1).

## 3. Results and Discussion

### 3.1. PE

PE films produced by drawing from the melt have a fibre orientation with the [002] axis parallel to the draw direction. Although the crystalline lamellae are quite extensive (often more than a micron) perpendicular to the molecular axis, the diffracting regions are very small, of the order of 30nm across, making identification of individual crystals imaged under different diffraction conditions quite difficult when restricted to low magnification-low resolution operation. Cross comparison of images recorded under different diffraction conditions is normally most easily accomplished by superimposing the negatives on a light box.

That different diffracting regions join up to form such well-defined lamellae suggested that the diffraction contrast might be accounted for if the crystals were curved or twisted, and the purpose of the observations made here was to test this hypothesis. If the crystals are bent about the [002] axis or twisted about the [hk0] axis then on tilting the illumination in the appropriate direction the diffracting region should move. Many areas were examined using different tilts and no evidence was found for extensive bending or twisting. In DF the crystals changed intensity when viewed using different diffraction conditions (obtained either by tilting the illumination or by changing from the reflection with diffraction vector g to that with diffraction vector -g) but the diffracting region did not shift in the manner expected of a larger bent or twisted crystal. When comparing images with a large relative tilt, many microcrystals were found to be visible in both, but many were visible only in one. Those visible in only one reflection were randomly distributed and there was no tendency for progressive imaging at neighbouring sites as required if bending or twisting were present (e.g. figure 2). The same conclusion was drawn from BF observations. With BF the angular range within which a particular crystal was visible was smaller than in DF as a consequence of the inferior contrast in BF. Thus, BF imaging of the microcrystals is much more sensitive to tilt and the ability to provide small controlled tilts is even more valuable than with DF imaging.

### 3.2. PBT

The PBT fibres contain small coherently diffracting regions that appear to line up into microfibrils which form parallel bundles. It was attempted to investigate the continuity of the diffracting regions by examining the effect on DF of tilting the illumination. As with PE it was found that the diffracting regions were either active or not and did not show evidence for twisting about the fibre axis.

### 3.3. 4GT

The images of 4GT crystals grown in polyether-copolyester films contain a series of straight lines parallel to the (010) plane. These have been interpreted by Briber and Thomas (1985) as cracks, possibly caused by differential shrinkage between the crystal and the surrounding matrix. In DF the crystal image consists of a complicated yet fairly regular pattern of alternating dark and bright regions, in parts showing discontinuity across (010) boundaries and in other parts being continuous across them. By imaging the same area in several diffracted beams it was shown that the different regions were crystalline and that the DF contrast was a consequence of small local differences in orientation rather than segregation of crystalline and non-crystalline material, figure 3.

## References

Allen S R, Filippov A G, Farris R J and Thomas E L 1981 J. Appl. Polym. Sci. 26 291.
Briber R M and Thomas E L 1985 Polymer 26 8.
Petermann J and Gohil R M 1979 J. Mater. Sci. 14 2260.
Thomas E L and Ast D G 1974 Polymer 15 37.
White J R 1975 Polymer 16 157.
White J R and Thomas E L 1985 J. Mater. Sci. 20 2169.
Yang D C and Thomas E L 1984 J. Mater. Sci. 19 1098.

Fig 1. Double exposure diffraction pattern showing the effect of a tilt of $0.29^{\circ}$ in the [001] direction (x-direction) in PE.

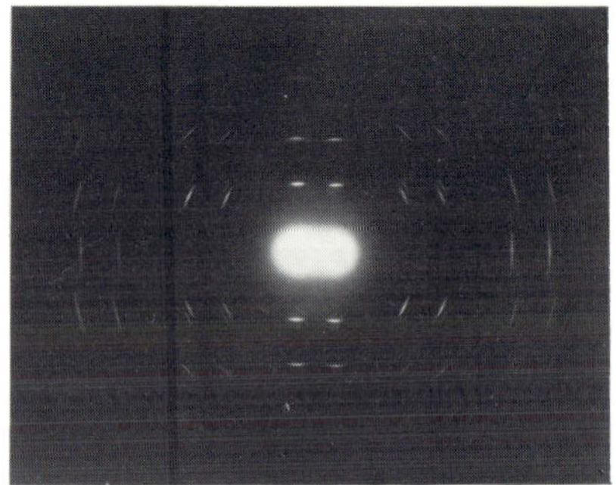

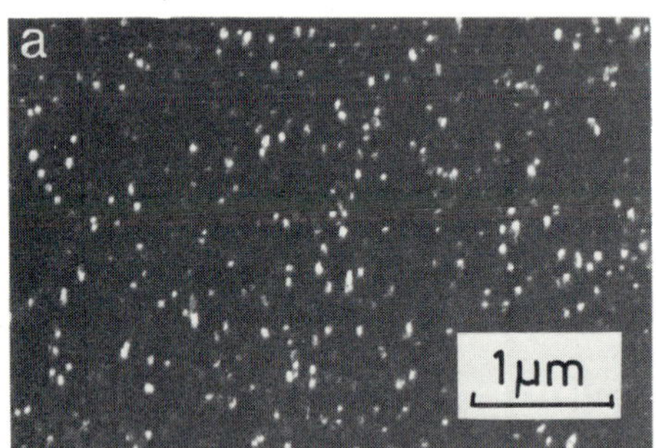

Fig 2. Drawn, annealed PE imaged in DF using (a) the 110 and 200 reflections and (b) the $\bar{1}\bar{1}0$ and $\bar{2}00$ reflections. The [001] direction is horizontal.

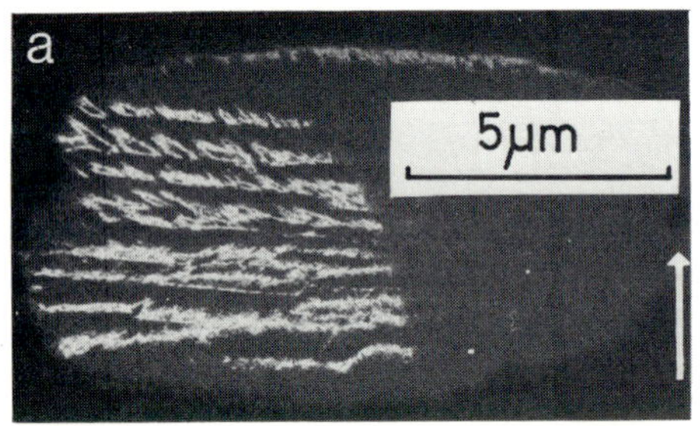


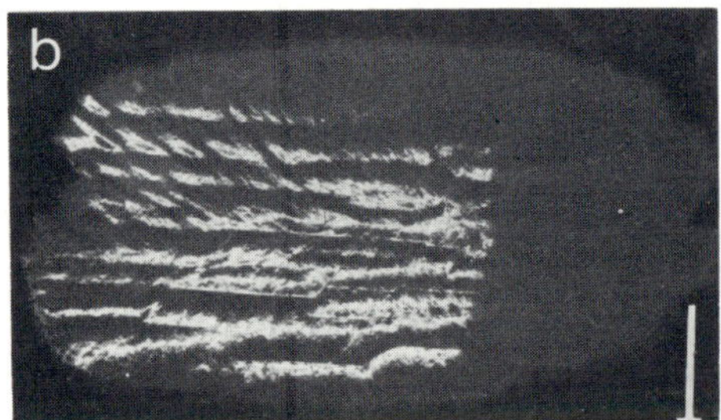

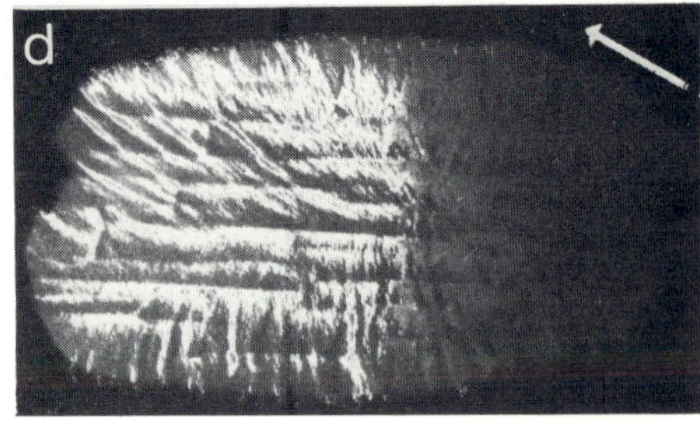


Fig 3. 4GT crystal within polyether copolyester film imaged in simple DF using reflections in directions indicated by the arrows: (a) 010; (b), (c) $0\bar{1}0$; (d) 100. (a) (b) and (d) untilted illumination; (c) illumination tilted $\sim 0.35^{\circ}$ in the [010] direction.

Inst. Phys. Conf. Ser. No 78: Chapter 11
Paper presented at EMAG '85, Newcastle upon Tyne, 2–5 September 1985 

# Direct lattice imaging of crystalline paraffins

*C H McConnell,* J R Fryer,**D L Dorset, ***F Zemlin

*University of Glasgow, Glasgow G12 8QQ. **Medical Foundation of Buffalo, Inc, 73 High Street,Buffalo, N.Y. ***Fritz-Haber Institut, Faradayweg 4-6. 1000 Berlin 33.

## Introduction

Paraffins provide an interesting series of specimens for electron microscope studies, being in effect short-chain polymers and in some ways analogous to lipids and other biological substances. The structures of the different polymorphs of various paraffins have been determined by X-ray diffraction methods (Shearer and Vand 1955, Teare 1958) and the spiral growth of the $nC_{36}H_{74}$ crystal from solution was studied by electron microscopy by Dawson and Vand (1950).

Aliphatic hydrocarbons are extremely beam-sensitive, but it has been proved possible to obtain structural images of monolamellar crystals of $nC_{44}H_{90}$ at a resolution of 0.25nm using a low-dose technique (Zemlin et al 1985). The 001 lattice of $C_{36}H_{74}$ was imaged by Fryer, 1981.

In the present work, paraffins of various chain lengths were crystallised in two orientations, allowing lattice images to be made of the lamellar spacings as well as of the orthogonal (001) projection of the chain packing.

## 1. The hk0 Orientation.

Single, monolayer paraffin crystals grown from hexane solution were recorded in the hk0 orientation using the helium-cooled superconducting microscope at the Fritz-Haber-Institut,Berlin. Chain-lengths studied were $C_{82}H_{166}$, $C_{60}H_{122}$, $C_{44}H_{90}$ and $C_{36}H_{74}$, plus an example each of an odd-even and an even-even solid solution. Low magnification images of the lozenge-shaped crystals, grown from solution, of long-chain paraffins show striations in the ⟨130⟩ direction, as well as sectorisation similar to that observed with polyethylene (Dorset, 1985(b), Bassett and Keller,1962).

To investigate this effect further, images were recorded with a magnification of 60,000X, using a minimum dose technique and with the specimen cooled to 15K. It was found that a satisfactory image could be obtained applying an electron dose of $12e^-/Å^2$ to the specimen. It has been found that under these conditions, at 2.5Å resolution, the diffraction intensities which contribute to the image do not change their relative intensity until beyond $14e^-/Å^2$ (Dorset and Zemlin, 1985).

The electron diffraction pattern of $C_{36}$ in fig.1 is characteristic of all the paraffins. A low-dose image is also shown (fig.2). The periodic information cannot be seen by eye but the light optical diffraction pattern (inset) shows a resolution of 2.5 Å .

Selected areas of the micrographs were enlarged by a factor of 5, then digitised using a flat-bed Datacopy microdensitometer with a 25 μ aperture. The images were then filtered using the "IMAGIC" programme of van Heel, (1981). A filtered image is shown in fig.3. Variations in contrast seem to indicate small changes in the surface structure of the crystal about 4nm apart, although these are too small to account for the ⟨130⟩ striations, the cause of which is still under investigation. As is expected, the filtered images in this projection of the solid solutions are similar to those of the pure compounds.

2. Epitaxially Crystallised Solid Solutions.

The formation of solid solutions by crystallising two different chain lengths together in certain proportions has been reviewed by Mnyukh (1960) who stated the following conditions for their formation:

(a) The form and dimensions of the two components must be similar. In the case of paraffins the chains must be nearly the same length.

(b) The symmetry of the binary crystal structure must be the same as or lower than that of the pure component crystals.

Series of solid solutions of paraffins have been studied by Dorset (1985 a and b) who has constructed melting and freezing point curves to show that while the restriction on chain length difference appears to hold for the formation of true solid solutions, it is possible to combine a paraffin with an even number of carbon atoms in the chain with one with an odd number to make a solid solution. This is because the polymorphs involved are similar high energy crystal forms; fractionation is expected in the lowest energy forms, however.

In order to see the chain length directly, the Okl plane of the paraffin crystal must be projected on to the image screen. This is achieved by epitaxial crystallisation on benzoic acid, as described by Wittmann et al, (1983). Images were recorded using a JEOL 100B microscope at room temperature keeping the electron dose very low and using the dark field deflector switch to focus on an area of crystal adjacent to the one to be photographed. The paraffins chosen were $C_{32}/C_{36}$ and $C_{33}/C_{36}$ solid solutions in various proportions, as examples of even-even and odd-even solutions. Images were recorded with a magnification of 17,000X and their 001 spacings measured by optical diffraction, which in some cases extended to 3rd order. An example is shown in fig.4 of a 9:1 solution of $C_{33}H_{68}$ and $C_{36}H_{74}$, along with its optical diffraction pattern. Fig.5 (a and b) shows graphs of the observed spacings against the proportion of the larger component. The results appear to support Dorset's observation of continuous solution formation. For $C_{32}/C_{36}$ the chain lengths lie above the straight line joining the values for the two pure components. It appears that the longer chain dominates and incorporates the shorter at all ratios. With the $C_{33}/C_{36}$ system the results lie around the line, but because of the rather coarse sample of this curve, further work with intermediate compositions is needed to give a better indication of the behaviour of these solutions.

The images, once filtered, show defects in the lattices (fig.6). Pairs of defects are common, and the lattices are seen to bend to accommodate the strain. Some areas of crystal appear quite free of defects, but the concentration of defects does not appear to change much with the composition of the solid solution.

## Conclusion

The above results show that important information about lattice defects and crystal texture can be directly obtained by electron microscopy from such beam-sensitive materials as the paraffins. So far we have examined both pure and polydisperse systems as a model of polyethylene crystal packing and anticipate future work of this kind on microcrystals of the polymer itself.

Thanks to the National Science Foundation, U.S.A. who partially supported this work (DMR81-16318) and to SERC for grant support (CHM).

## References

Bassett, D.C. and Keller A, Phil. Mag. 7th Series (1962) 1553.
Dawson, I.M. and Vand V, Proc.Roy.Soc. 206A (1951) 555.
Dorset, D.L. 1985 (a) Macromolecules (in press)
Dorset, D.L. 1985 (b) These proceedings.
Dorset, D.L and Zemlin, F Ultramicroscopy 1985 (in press).
Fryer, J.R. Inst. Phys.Conf. Ser 61 EMAG 1981.
Mnyukh, Yu V. J.Struct. Chem. (USSR) 1 (1960) 346.
Shearer, H.M.M. and Vand, V. Acta.Cryst. (1956) 9 379.
Teare, P.W. Acta.Cryst. (1959) 12 294.
Van Heel, M and Keegstra, W. Ultramicroscopy 7 (1981) 113.
Wittmann, J.C. ; Hodge, A. and Lotz, B. J.Polym.Sci(Polym.physics) 21 (1983) 2495.
Zemlin, F.; Reuber, E.; Beckmann, E.; Zeitler, E. and Dorset, D.L. Science 229 (1985) 461.

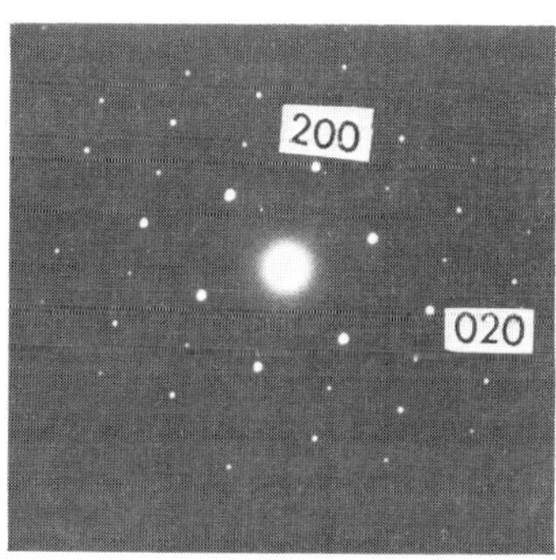


Fig. 1
Electron diffraction pattern of hk0 projection of $nC_{36}H_{74}$

Fig. 2
Low-dose image of hk0 projection of $nC_{36}H_{74}$ with optical transform, showing 020 reflection.

Fig.3 Filtered image of hk0 projection

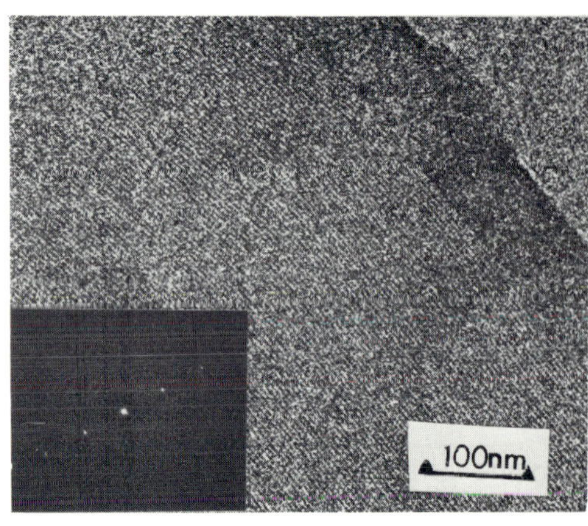


Fig.4 9:1 solution of $C_{33}/C_{36}$

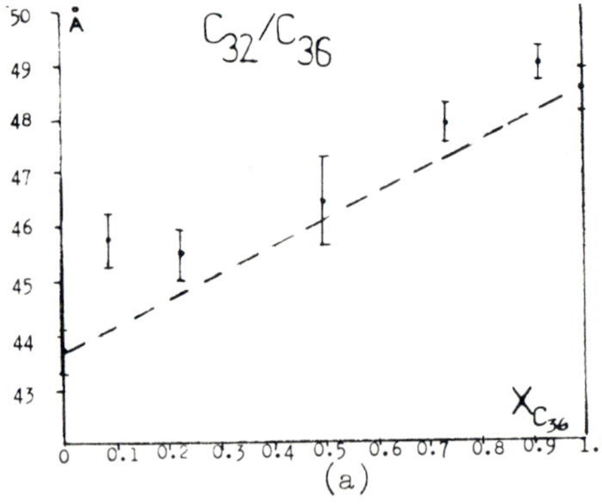


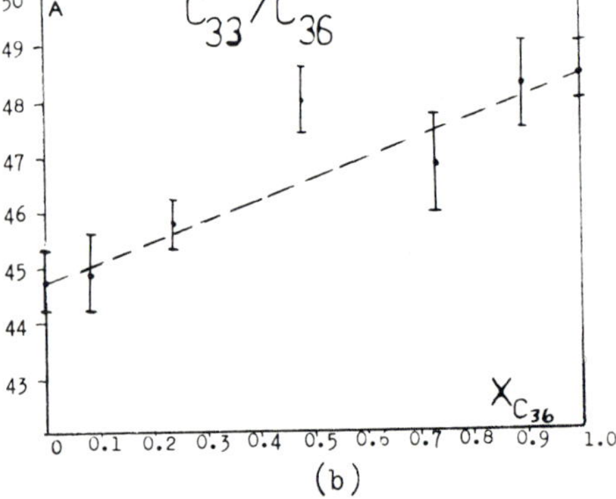


Fig.5 (a) and (b) 001 lattice spacing of solid solutions, plotted against mole fraction of $C_{36}H_{74}$

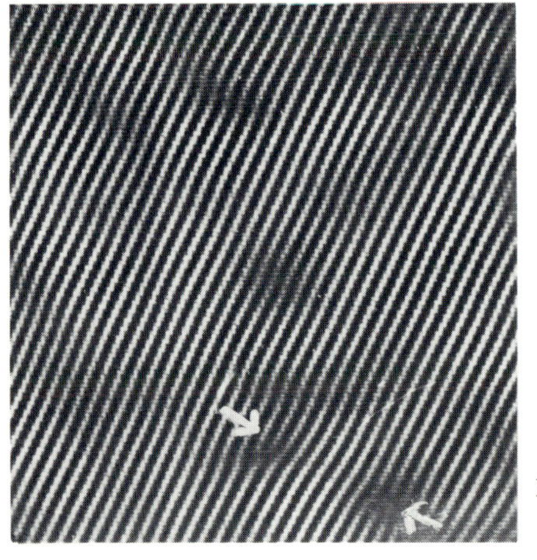

Fig.6 Filtered image of a solid solution, with defects

*Inst. Phys. Conf. Ser. No 78: Chapter 11*
*Paper presented at EMAG '85, Newcastle upon Tyne, 2–5 September 1985*



# Electron microscope studies of a mono-AZO pigment

G R Duckett, J R Fryer and T Baird

Chemistry Dept., University of Glasgow, Glasgow G12 8QQ

One of the difficulties in obtaining structural information by x-ray diffraction techniques is that of the crystal size required to give a measurable intensity of scattering of the x-ray beam. The failure to grow suitable crystals, even in gels, of such systems as air-sensitive metal hydroxides and halides and organic pigments, means little information about their solid state structures is available. The use of computer image-processing procedures for electron microscope images offers the potential for the acquisition of structural data from such crystals.

There is little difficulty, unless the crystals are very beam sensitive, in obtaining high resolution images of crystals of 50 nm in size. An industrially important material which only occurs in very small crystals is the red monoazo pigment, calcium 4B toner (pigment red 57, CI. No. 15850).

This pigment is widely used in the colouring of printing inks, plastics and household paints. The colouristic properties however, are largely determined by its mode of preparation. Two distinct red shades are exhibited by the Ca4B toner dependent upon the crystal water content. At a high water content the pigment has a distinctly yellow shading, whereas at a lower level the shade becomes bluer. Analysis has shown that the yellow shading is associated with a crystal water content of two moles, whilst the bluer shading has only one mole of associated water and that this water is in some way discreetly bound to the pigment molecules.

The purpose of the present work is to obtain structural information about the crystals in order to determine the possible sites where the water molecules may be bound.

## Experimental and Results

The preparation of the pigment involves the coupling of amine sulphonic and β-oxynaphthoic acids and the subsequent laking of the pigment as the metal salt of calcium, manganese, strontium or barium. Calcium was used in this case (fig. 1.)

Several methods of preparing samples, suitable for electron microscopical examination were tried, the most successful being simply to suspend some Ca4B crystals in dried ethanol and then spray the suspension on to carbon-coated copper grids by microspray. This produced a reasonably dispersed sample with limited crystal agglomeration. For the purpose of a later three-dimensional reconstruction of the crystals from electron microscope images, a tilting stage was constructed, which allows a greater range of angles than that afforded by the microscope goniometer. All micrographs were obtained on a JEOL 1200EX electron microscope, fitted with a $LaB_6$ filament, using minimum dose techniques.

As can be seen from plate 1, the external morphology shown is that of thin plate-like crystals revealing, when viewed 'side-on', a lattice periodicity of 1.8 nm. At this resolution, the electron beam sensitivity of the crystals presents little problem, however, at a higher resolution information is more rapidly lost. Optical diffraction patterns of low dose images have shown structural preservation of 0.3 nm and it was these micrographs that were used in the digital processing.

Electron diffraction patterns, both polycrystalline and single crystal have been obtained, and from their symmetry it would appear that the crystals are monoclinic. Lattice spacings obtained by x-ray powder diffraction correlate well with those obtained in the electron diffraction patterns. An estimate of the unit cell parameters has been made;

Monoclinic : a = 0.84 nm, b = 0.58 nm, c = 1.8 nm, $\beta$ = 92°.
Density = 1.66 g $cm^{-3}$ : 2 molecules per unit cell.

The low dose images were scanned on an optical bench to select areas for processing. Digitisation was carried out on a Joyce Loebl, mark 3CS, flat-bed microdensitometer using a 20 μm aperture on a 150 point square raster.

Processing of the images has been confined to Fourier and real space averaging. 'Windowing' of calculated image transforms has led to images, as shown in plate 2, being obtained. The power spectrum calculated from such images compares favourably with optical transforms of the original low dose image (: inserts on plate 2.).

From the few x-ray determined crystal structures of similar compounds (Whitaker), it would seem reasonable to assume that the bulk of the molecule will be planar or nearly so. In all cases, the hydrazo link in the molecule is present in the keto tautomer and the resulting intramolecular hydrogen bonding constrains both ring systems to the same plane. Solid state n.m.r. spectra of the caesium analogue of the 4B toner has shown only one metal ion environment, suggesting a 'head to tail' arrangement of the molecules. From the structural information obtained on the Ca4B toner to date, the pattern visible in plate 2 is consistent with a molecular arrangement as shown in figure 2. Further processing of many images is required in order to validate such a suggestion however, including accurate transfer function estimation for the images concerned; at present difficult due to the small picture size being considered.

From this preliminary work however, it seems clear that, once the problems of the Glasgow Semper installation have been solved, the use of image processing techniques show the promise of structural information being obtained for this class of materials.

Acknowledgement

The authors would like to thank Dr. W.O. Saxton for his help with the installation of SEMPER. The project is an SERC-CASE award, in conjunction with CIBA-GEIGY (UK) LTD.

References

Saxton W.O., Pitt. T.J. and Horner M. 1979 Ultramicroscopy 4 343-354.
Whitaker A., 1978 JSDC 431-435.

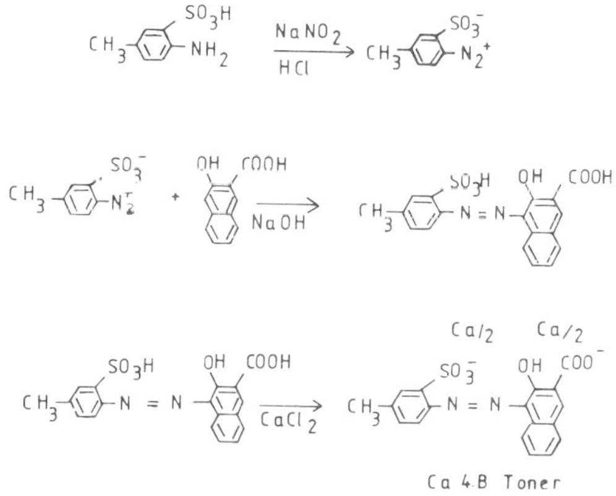

$CH_3C_6H_3(SO_3H)NH_2 \xrightarrow[HCl]{NaNO_2} CH_3C_6H_3(SO_3^-)N_2^+$

$CH_3C_6H_3(SO_3^-)N_2^+ + C_{10}H_6(OH)COOH \xrightarrow{NaOH} CH_3C_6H_3(SO_3H)-N=N-C_{10}H_5(OH)COOH$

**Figure 1**

$CH_3C_6H_3(SO_3H)-N=N-C_{10}H_5(OH)COOH \xrightarrow{CaCl_2} CH_3C_6H_3(SO_3^-)-N=N-C_{10}H_5(OH)COO^-$ (Ca/2, Ca/2)

Ca 4.B Toner

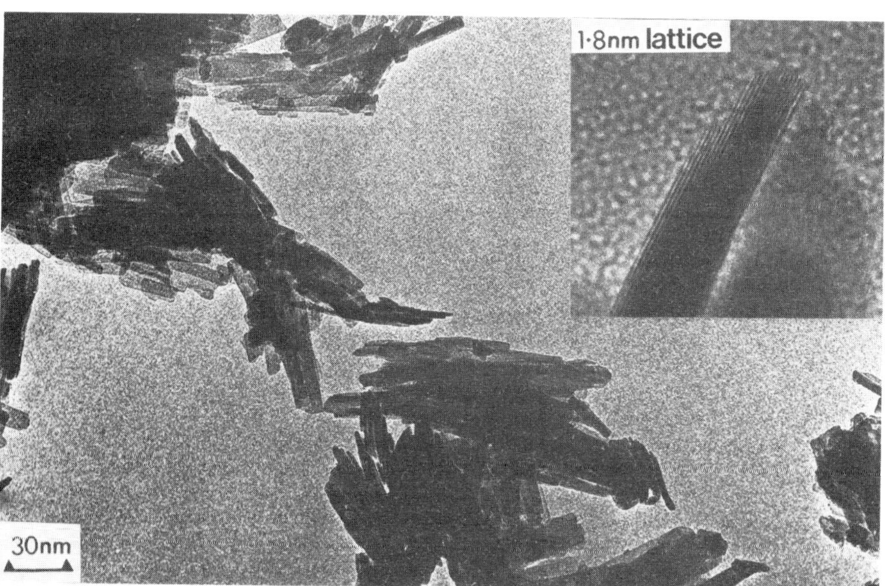


**Plate 1**

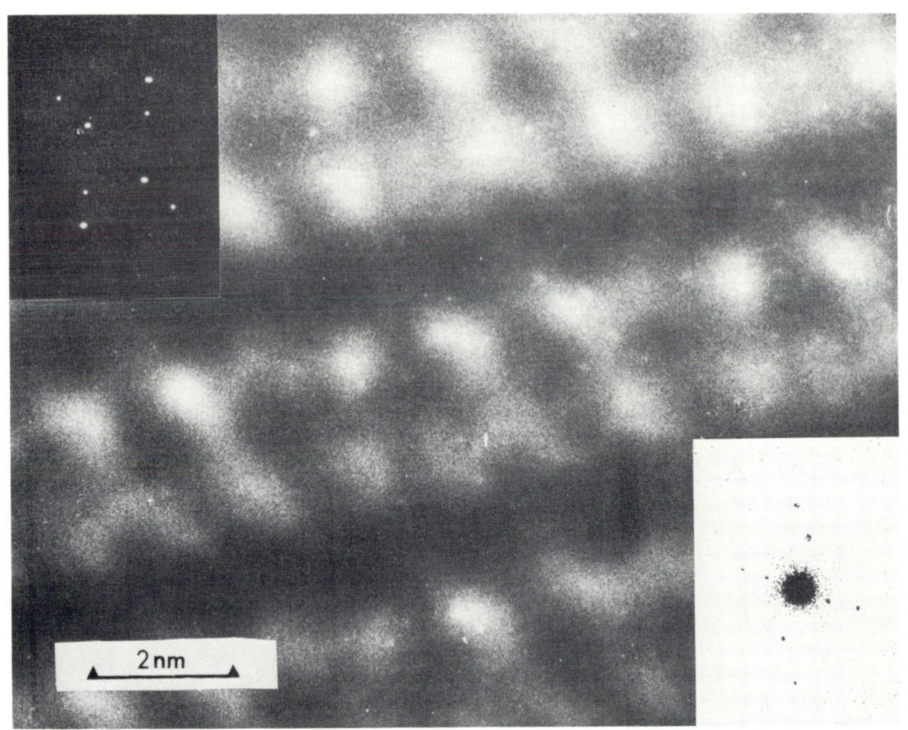


**Plate 2**

**Figure 2**

*Inst. Phys. Conf. Ser. No 78: Chapter 11*
*Paper presented at EMAG '85, Newcastle upon Tyne, 2–5 September 1985*



# Structural arrangements in polymeric phthalocyanines

J R Fryer, Chemistry Department, University of Glasgow, Glasgow, G12 8QQ, Scotland.
M E Kenney, Chemistry Department, Case Western Reserve University, Cleveland, Ohio 44106, U.S.A.

Phthalocyanines are of considerable interest as electronic and photoelectric materials because of their chemical and thermal stabilities and ready availability. Doped phthalocyanines exhibit high electrical conductivities and in some cases metallic behaviour has been observed. Similar conductivities are shown by polymeric phthalocyanines where the central metal atoms are linked by oxygen or fluorine atoms forming a continuous backbone with the phthalocyanine rings at right angles to the backbone. Nohr et al (1981) has shown that iodine doped aluminium/fluorine phthalocyanine polymer has a conductivity up to $5\Omega^{-1}cm^{-1}$.

As synthesised polymers of silicon/oxygen phthalocyanine -(SiPcO)n- and aluminium/fluorine phthalocyanine-(AlPcF)n- Kenney et al(1962,1980), gave a microcrystalline product that was not suitable for X-ray diffraction analysis. Other techniques showed the schematic structure to be as shown in Fig 1 but no accurate structural information was obtained. These compounds proved sufficiently electron beam stable for structural imaging at 3A.

Specimens were prepared by evaporation on to a cleaved KCl(100) substrate maintained at 230C. The epitaxial film was backed with carbon, floated off on water and picked upon a grid. Total evaporation of the (AlPcF)n was obtained but some residual carbonised polymer of (SiPcO)n remained in the evaporation boat. Non epitaxial specimens were made by spraying a slurry of the microcrystals in propan-2-ol on to a carbon covered grid. For high resolution examination specimens had a further layer of carbon evaporated to encapsulate the specimen and thereby reduce radiation damage-Fryer and Holland(1984).

Electron microscopy employed a JEOL1200EX-120KeV- instrument fitted with an image intensifier so that high resolution images were obtained at magnifications of 120,000X under minimal exposure conditions on Kodak X-Ray film. Some samples were also examined at 500KeV using the Cambridge High Resolution microscope.

The epitaxially prepared specimens crystallised with the chain axis perpendicular to the substrate. An example of (AlPcF)n is shown in Fig 2. There is no resolution of the quatrefoil shape of the phthalocyanine molecule but the individual molecular columns are clearly defined. The specimen was not tilted so this micrograph shows that the Al-F-Al bond between the phthalocyanine moieties is at 90° to the plane of the phthalocyanine as represented in the diagram in Fig 1.We have shown-O'Keefe et al(1983)-that even a resolution of 7A is sufficient to resolve the shape of the phthalocyanine molecule.However,both the 120KeV and 500KeV microscopes were unable to resolve the shape,although the latter was achieving a resolution of 2A.We therefore conclude that the phthalocyanine moieties are staggered with respect to each other,unlike Fig 1 in which they are eclipsed.

Disorder is evident within the crystalline islands but the similarity of contrast over the whole area indicates that the vertical chain length of the polymer is fairly constant.The image of the non-epitaxial material-Fig 3-shows breaks in the lattices indicating that the polymer chain length is 120-200A.

Unit cell constants of (AlPcF)n based on a tetragonal unit cell were obtained by electron diffraction and had values a=b=13.37A and c=3.6xA.Thus for a chain length of 150A,x-the number of molecules per unit cell-=42.

The (SiPcO)n polymer did not sublime easily and therefore the crystalline film obtained may consist of the shorter chain length molecules in the product.The non-epitaxial material-Fig 4- consisted of agglomerates of small crystals in all orientations as shown by the 13A lattices that represent the intermolecular separation.There is a large range of different chain lengths but no other structural differences between small and large crystals indicating that the material is chemically homogeneous.The epitaxial film was thin but had a structure very similar to the (AlPcF)n as shown in Fig 5.

This tetragonal based structure-within the limits of resolution-has also been observed in the silicon/oxygen naphthocyanine polymer and the copper cyanophthalocyanine polymer reported previously-Fryer and Kinnaird(1983).The b axis projection is similar to that of a-copper phthalocyanine which has a lower lattice energy than the more tightly packed β-phase.The polymeric chain would prevent the attainment of the stacking sequence and hence these compounds would have sufficient space within their lattices for high concentrations of halogen dopant.

Acknowledgments

We thank the Office of Naval Research and the Science and Engineering Research Council for support of this work and Mr R.Camps of the High Resolution High Voltage microscope at Cambridge for his cooperation and assistance.

References

Fryer JR and Holland F.1984.Proc.Roy.Soc.Lond.A393,353.
Fryer JR and Kinnaird TA.1983.Inst.Phys.Conf.Ser.68,27.
Kenney ME and Joyner RD.1962 Inorg.Chem.1,717.
Kenney ME,Linsky JP,Paul TR and Nohr RS.1980.Inorg.Chem.19,31
Nohr RS,Kuznesof PM,Wynne KJ,Kenney ME and Siebenman PG. 1981. J.Am.Chem.Soc.103,4371.
O'Keefe M,Fryer JR and Smith DJ.1983.Acta Cryst.A39,838.

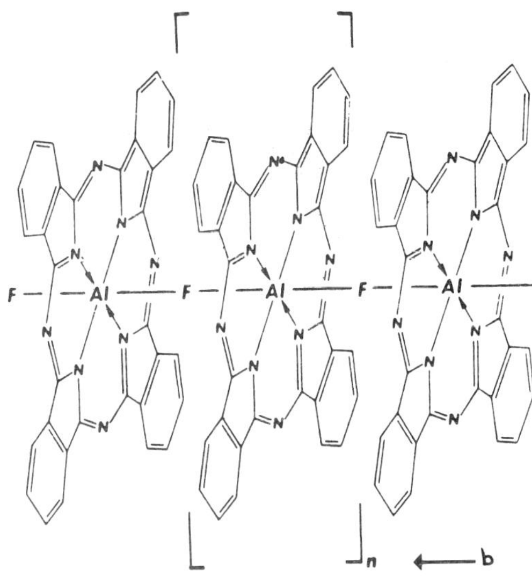


Fig.1.Schematic diagram of (Al Pc F)n

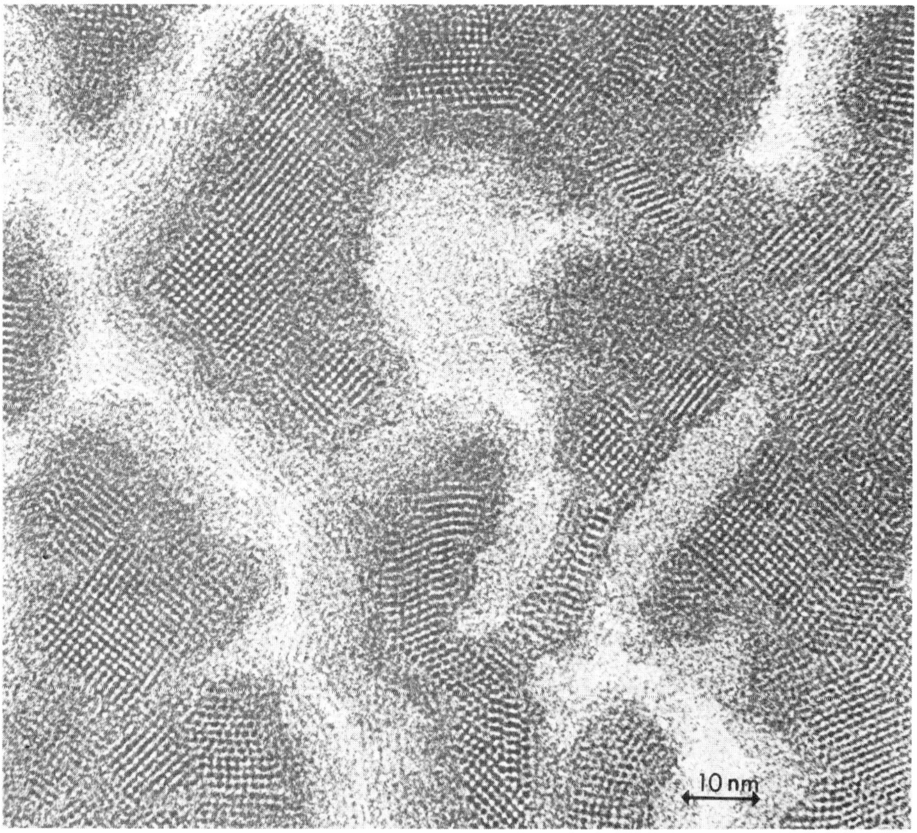

Fig.2.Epitaxial film of (AL Pc F)n.b axis normal to substrate.

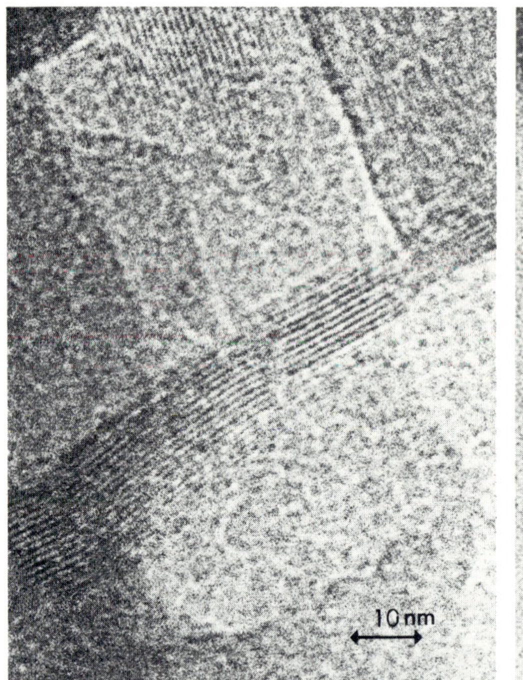


Fig.3.Acicular (Al Pc F)n lattice breaks indicating chain length.

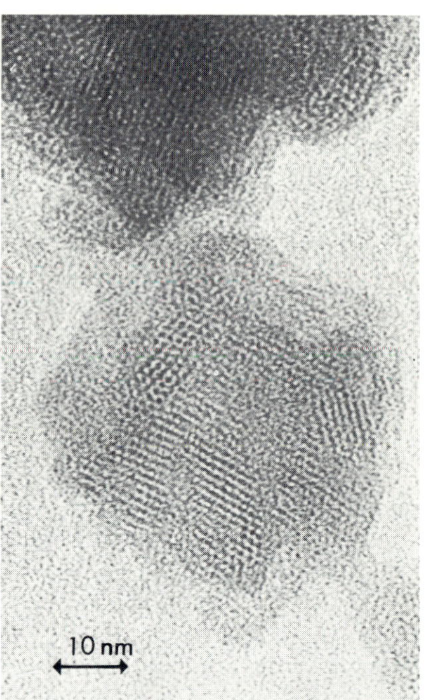


Fig.4.As polymerised (Si Pc O)n.

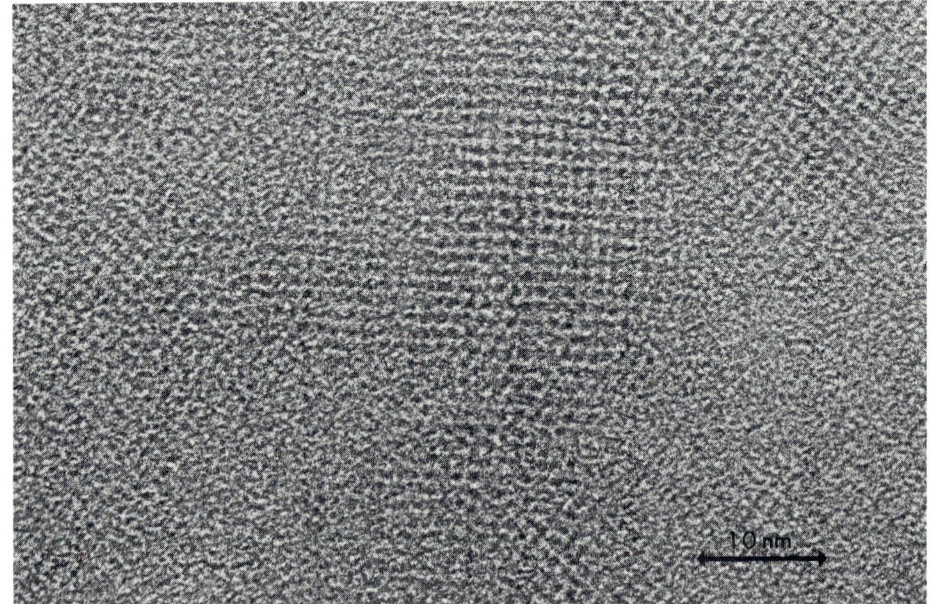


Fig.5.Epitaxial film of (Si Pc O)n.

*Inst. Phys. Conf. Ser. No 78: Chapter 12*
*Paper presented at EMAG '85, Newcastle upon Tyne, 2–5 September 1985*


# The role of the TEM in the study of oxidation and reduction reactions

H M Flower

Department of Metallurgy and Materials Science, Imperial College, London, SW7 2BP

## 1. Introduction

In its simplest form an oxidation or reduction of a solid by a gas is controlled by the thermodynamic driving force provided by the chemical free energy difference between reactants and product(s) and proceeds at a rate determined by the kinetics of the transport of reactants across the product layer which forms at the gas/solid interface. If the product is a uniform and continuous solid with homogeneous transport properties the reaction will be controlled by solid state diffusion of the reactant species through it. However, such is rarely the case since the product generally contains short circuit diffusion paths - pores, cracks, grain boundaries and dislocations and bulk diffusion may not be rate controlling. Additionally the concentration (and hence activity) gradients produced across the product layer may alter the thermodynamics such that reaction products unpredicted by consideration of the overall reactant activities may be produced. Variations in the chemistry and microstructure of the parent solid at the reaction front (e.g. grain boundaries, second phase particles, solute depleted precipitate free zones) may also modify the thermodynamics and kinetics of the reaction locally. This may have important consequences for the mechanical integrity of a structural material which averaged kinetic data could not predict. Some of the possible features associated with a reaction product layer on a polycrystalline parent material are shown schematically in Figure 1.

In order to understand such oxidation and reduction reactions it is, therefore, necessary to characterise the reaction product microchemically and microstructurally and to determine the influence of parent microstructure/chemistry upon that of the product. The transmission electron microscope (TEM) is the most powerful single instrument available in which this information can be directly determined. In this brief paper it is not possible to review the TEM imaging and microanalytical methods of relevance to oxidation and reduction studies. Since the use of TEM in this field has been constrained by the practical problem of producing suitable electron transparent sections the advantages and limitations of the available preparative methods will be discussed and knowledge of analytical electron microscopy will be assumed.

## 2. Thin Foil Methods

Figure 2 shows, schematically, a number of thin foil preparative methods designed to provide the information noted above. A is the simplest to carry out and a very large number of studies of oxidation have used this technique either oxidising the foil ex situ or in the limited vacuum of the microscope. It is well suited to the study of the earliest stages of product growth and the influence of parent microstructure. Experimentally great care is needed in preparation to ensure that the foil surface is clean and free from films left by the electrolyte used in thinning. In a modern analytical TEM the presence of surface films can be determined and characterised using X-ray energy dispersive and electron energy loss analysis methods. Recently Kuroda et al (1983) have used the technique to study the early stages of oxidation of Fe-Cr alloys. An interesting feature of that work is the use of secondary electron imaging to determine whether the oxide nucleii grow out from or into the metal. They were able to show that whereas wustite and magnetite grow out, via cation diffusion, the spinel grows in under oxygen short circuit diffusional control.

In situ studies carry a further advantage that the progress of the reaction can be directly observed and kinetic data (e.g. rate of growth of product nuclei) can be obtained. A gas reaction cell (GRC) is required to accommodate the widest range of gas pressures and compositions and a wide gap objective lens pole piece is required to provide the space needed for the cell and its controls. This is most readily achieved in a high voltage electron microscope (HVEM) where the additional penetrating power extends the thickness of material to be examined and the maximum gas pressure which can be used without an unacceptable loss in image contrast due to inelastic scattering. (Flower et al 1974). Currently the GRC method is being increasingly adopted to study both catalysts and catalysis, notably by Gai and co-workers in Oxford and Baker et al in the USA. Gai (1983) has shown that at the reaction temperature partial reduction of $V_2O_5$ catalyst surface leads to a number of microstructural changes of importance in catalysis which are only stable under limited ranges of temperature and gas environment and cannot, therefore, be studied by post reaction microscopy. The action of a catalyst may also be directly observed. For example Baker et al have observed the gasification of graphite, in situ, in a variety of gaseous environments catalysed by potassium (Mims et al 1984)) and by barium (Baker et al 1984). In the latter study at 1000$^{o}$C the BaO particles catalyse oxidation only locally moving with the oxidation front and producing channels running into the graphite along <11$\bar{2}$0> directions. Measurement of channel growth rates over a range of temperatures permitted the activation energy for the process to be determined. Such information cannot be obtained by any other means and GRC studies are making a substantial contribution to the understanding of catalysts and catalysis. The use of a HVEM does, however, create certain problems at low homologous temperatures where irradiation damage (ionisation and displacement) under the electron beam can modify the chemical reaction in nature or rate. This was exploited by Christodoulou and Flower (1980) where increased reactivity under the electron beam was used to study the formation of hydrogen gas bubbles at grain boundaries in aluminium alloys during reaction with water vapour: the results were used in the development of a model for hydrogen embrittlement in these materials. The presence of a GRC and heating stage in the microscope inevitably compromises some of the normal TEM facilities of specimen translation and tilting and diffraction patterns angular range and renders X-ray E.D. analysis impossible. Additionally confusion of image/ diffraction information can result from the presence of product layers on both surfaces of the sample and solute depletion effects can modify the longer term reaction behaviour of thin foils.

There is a need, therefore, for additional, ex situ foil preparation methods. For the study of the early stages of reaction product nucleation and growth method B in Figure 2 offers considerable advantages. It is relatively simple, produces a foil with the product on one face only and can provide films thin enough to permit high resolution imaging to be carried out (Flower and Swann 1974). Thinning may be carried out chemically or electrochemically or by ion beam erosion depending upon the chemical nature of the parent material.

In the case of thicker product films confusion of overlapping image/diffraction and microchemical information renders the plan section foil inappropriate. A transverse section (C in Figure 2) across the reaction interface provides microstructural/chemical information from the parent through to the product and permits determination of how these parameters vary with distance from the reaction front. Since parent and product are necessarily chemically dissimilar chemical or electrochemical dissolution is not practical for producing transverse section foils. Ultramicrotomy can be used but is very limited in the range of materials which can be cut: anodically or thermally oxidised aluminium alloys provides an ideal system (e.g. Furneaux et al 1978). It carries the considerable disadvantage that the original metal microstructure is destroyed by the deformation involved in the cutting operation. Ion beam thinning avoids this (Tuck 1977) and can be employed on a wide range of materials. The introduction of practical ion beam thinning devices for TEM specimen preparation has had as great an impact in the study of minerals, ceramics and semiconductors as electropolishing had in metallurgy nearly three decades ago.

The specific application of ion beam thinning to transverse section preparation was adopted first in semiconductor research by Abrahams and Buiocchi (1974) and is ideally suited for the study of real device structures which are multilayer in form. It is now very widely used as examination of any recent volume of Applied Physics Letters will demonstrate. An interesting recent example concerns the formation of buried dielectric layers in silicon by ion implantation of an oxidising gas (O or N) (Hemment et al 1985). Secondary ion mass spectroscopy of annealed samples implied the existence of a stoichiometric $Si_3N_4$ layer with a well defined interface with the Si: however, it is difficult to quantify depth by this surface technique. A transverse TEM section (Figure 3) immediately provides the necessary depth scale and confirms the sharpness of the $Si/Si_3N_4$ interfaces. It additionally reveals that the nitride is predominantly polycrystalline although there are some amorphous regions. It was also observed that in contrast to oxygen implanted samples the top Si layer remains monocrystalline and relatively defect free.

The application of transverse sectioning by ion beam milling to more conventional oxidation involving thermally oxidised metal was pioneered by Manning and Rowlands (1980). The technique involves resin mounting of the oxidised samples, mechanical grinding and polishing down to a section 30-50μm in thickness followed by ion beam thinning. The sections were held in cartridges designed to fit into both the ion beam thinner and the electron microscope sample holders; thus obviating the need to handle directly the fragile thin foils. Subsequently a number of variations on the original Manning and Rowlands method have been developed and applied principally to the study of iron alloy oxidation. However, the most major advance has been the use of "dimpling" the sample prior to ion beam thinning. The section is ground and polished as before to ~100μm. The central region of the 3mm disc sample is then mechanically polished down to 5-20μm in a precision "dimpler". The advantages are twofold: firstly the rim of the disc remains thick providing mechanical strength and facilitating handling and secondly the time required for ion beam thinning, typically many hours, is substantially reduced. The technique is described in detail by King et al (1984) and Bravman and Sinclair (1984) and "dimplers" are now commercially available. Transverse sectioning has been developed from the specialist technique of one or two research groups to a readily available method which can be carried out in any electron microscopy laboratory.

In all the studies undertaken the principal advantages have been the ability to identify local structure and chemistry. For example in a recent study of oxidation of a 20%Cr25%Ni0.7%Nb0.9%Mn0.56%Si stainless steel in $CO_2$ (Bennett et al 1984) the oxide was shown to concist of three layers: an outer layer of spinel $Mn(FeCr)_2O_4$, a central layer of $Cr_2O_3$ which is enriched in Fe and Mn close to a continuous inner layer of amorphous $SiO_2$ only 12.5nm in thickness (Figure 4). The importance of the minor elements Mn and Si is immediately evident: it is proposed that the $SiO_2$ forms beneath the previously produced $Cr_2O_3$ layer primarily by grain boundary diffusion from the steel and once formed it limits outward cation transport and thus controls the thickening of the outer oxide layers.

Most recently the transverse sectioning technique has been applied to the oxidation of dilute Mg-Al alloys in $CO_2$ (Morris 1985). At high homologous temperatures the oxide growth becomes unstable and large spike shaped intrusions of oxide grow deeply into the metal (Figure 5). X-ray ED analysis indicates that Al partitions to the oxide and becomes depleted in the parent metal although thermodynamically MgO should form in preference to any aluminium containing oxide. Transverse sections taken near the base of the oxide spikes provide the basis for explaining this anomaly.Figure 6 shows the flocculent and highly porous MgO oxide film with a highly irregular interface with the MgO.8%Al matrix oxidised at 550$^o$C. X-ray ED analysis confirms the total depletion of the matrix in Al at this point and confirms the principal product of oxidation to be MgO. However, local regior of Al concentration exist within the oxide and Al rich particles are observed in the parent metal within ~1µm of the oxide. X-ray ED analysis indicates a composition 70%Al30%Mg. Electron energy loss spectra from the parent Mg exhibit a small oxygen edge due to the thin MgO layer on the foil surfaces: however, this totally disappears in spectra from the particles and a carbon loss edge is prominent (Figure 7). Detailed chemical and microdiffraction analysis shows the particles to be a hexagonal phase (C = 0.581mm a = 3.37mm) of approximate composition $Al_4Mg_2C_3$. These observations permit a model for breakaway oxidation to be developed. Initially only MgO is produced on oxidation at 550$^o$C. Mg vapour phase oxidation is involved and results in a highly porous flocculent film. As the film thickens gas transport is limited and the $CO/CO_2$ ratio at the reaction front in the oxide near the metal increases. Eventually carbon deposition by the Boudouard reaction occurs. At this high carbon level internal carburisation of aluminium becomes thermodynamically possible and this provides the driving force for the growth of the spikes into the metal. The carbides are then consumed by the advancing oxide producing regions of local Al concentration observed within the oxide.

## 3. Conclusions

The ability to determine detailed microstructural, crystallographic and microchemical data with high spatial resolution makes TEM an extremely powerful tool for the study of oxidation and reduction reactions. Care is needed in the choice of thin foil preparation method to obtain the maximum benefit from the TEM. Backpolishing to a reacted surface is preferred for the study of nucleation and early stages of growth of the product and transverse sectioning is ideal for the study of thicker films. In situ reaction using a GRC/HVEM combination is most valuable for determining the microkinetics of the process and for investigating catalyst microstructure under true reaction conditions and for studying the catalysis itself.

References

Abrahams M S and Buiocchi C J 1974 J.Appl.Phys.45 3315
Baker R T K, Lund C R F and Chudzinski J J 1984 J.Catalysis 87 255
Bennett M J, Desport J A and Labun P A 1984 UKAEA Harwell Report AERE-R 11034
Bravman J C and Sinclair R 1984 J.Elec.Mic.Tech.1 53
Christodoulou L and Flower H M 1980 EMAG'79 (Inst.Phys.) 313
Flower H M and Swann P R 1974 Acta Met.22 1339
Flower H M, Tighe N J and Swann P R 1974 High Voltage Electron Microscopy (Academic Press) 383
Furneaux R C, Thompson G E and Wood G C 1978 Corr.Sci.18 853
Gai P L 1983 Phil.Mag A 48 359
Hemment P L F, Peart R F Yao M F, Stevens K G, Chater R J, Kilner J A, Meckinson D, Booker G R and Arrowsmith R B 1985 Appl.Phys.Lett.46 952
King W E, Peterson N L and Reddy J F 1984, Proc.9th Int.Conf.Metal Corr. Toronto 4 28
Kuroda K, Labun P A, Welsch G and Mitchell T E 1983 Oxid Met. 19 117
Manning M I and Rowlands P C 1980 Brit.Corr.J. 15 184
Mims C A, Chludzinski J J, Pabst J K and Baker R T K 1984 J.Catal.88 97
Morris A J 1985 PhD Thesis, U.London
Tuck C D S 1977 Corr.Sci.17 777

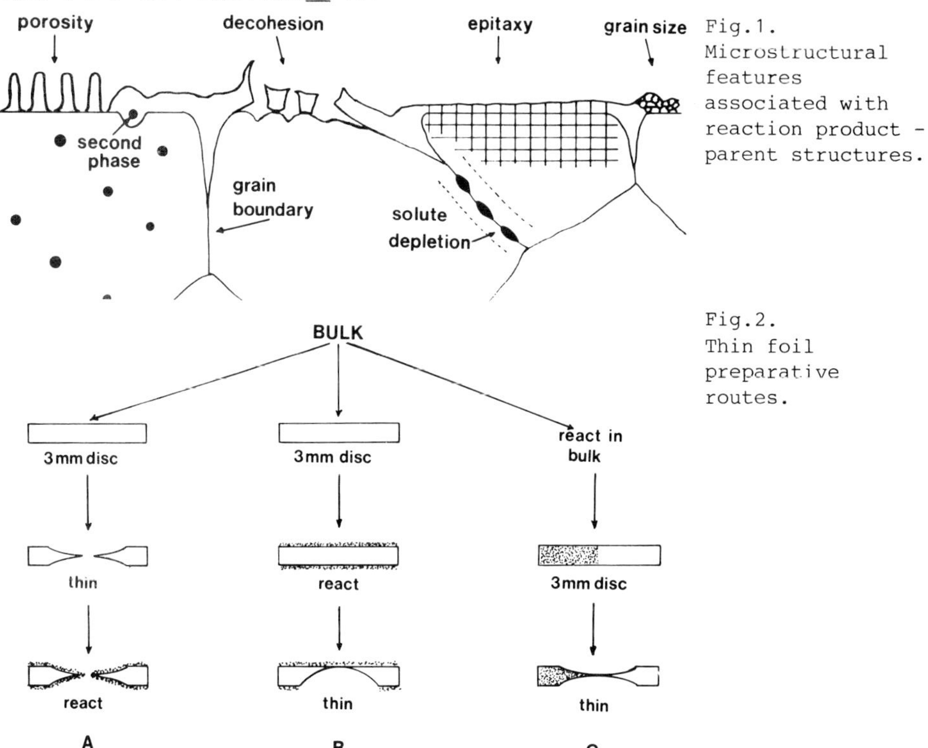


Fig.1. Microstructural features associated with reaction product - parent structures.

Fig.2. Thin foil preparative routes.

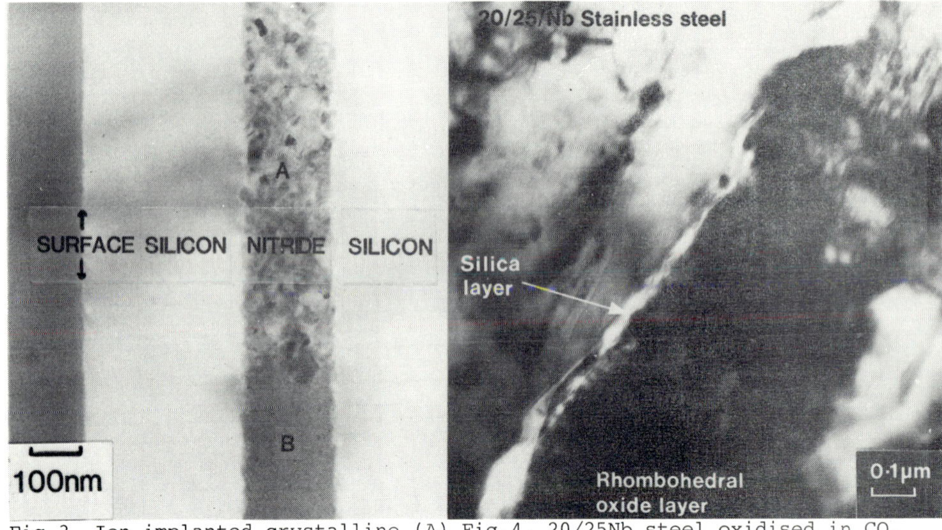


Fig.3. Ion implanted crystalline (A) and amorphous (B) nitride layer in Si annealed 2 hr/1200C.

Fig.4. 20/25Nb steel oxidised in $CO_2$ 825C/4900 hr.

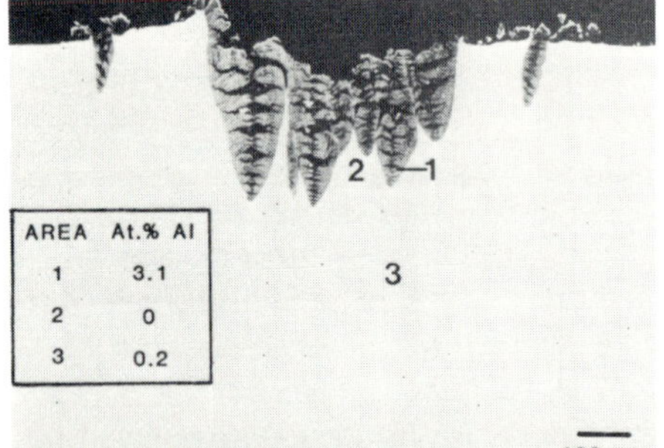


Fig.5. Oxide spikes produced by reaction of Mg0.8Al in $CO_2$ at 550C.

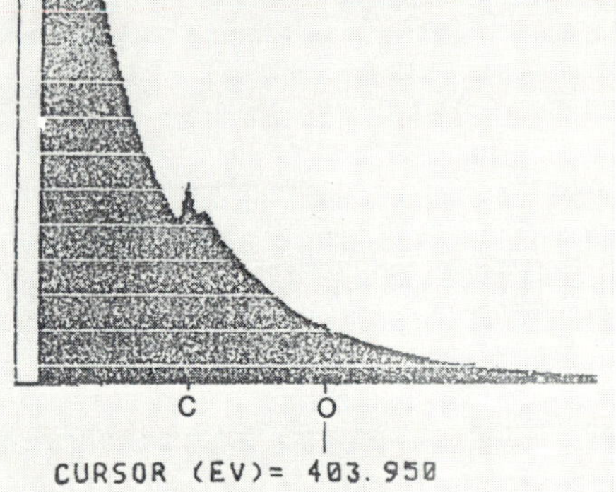


Fig.7. EELS spectrum of carbide particle in Mg0.8%Al oxidised in $CO_2$.

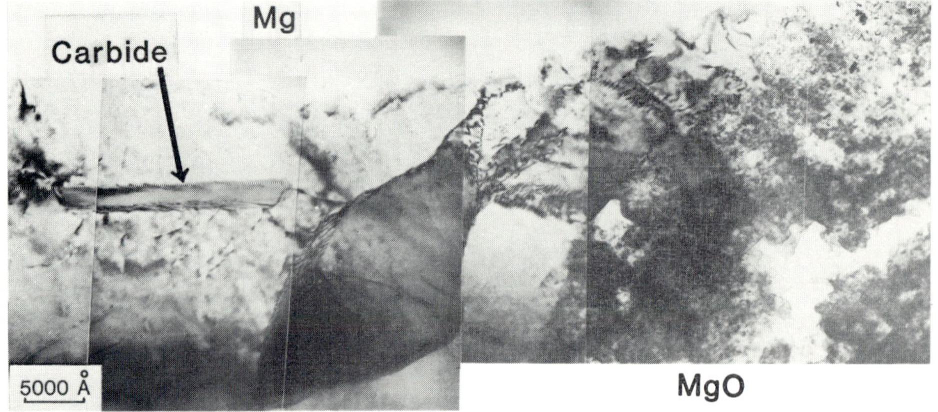


Fig.6. Mg0.8%Al oxidised in $CO_2$ at 550C. Transverse section foil.

*Paper presented at EMAG '85, Newcastle upon Tyne, 2–5 September 1985*

# Twinning and precipitation reactions in the oxidation of iron

S B Newcomb and W M Stobbs

Department of Metallurgy and Materials Science
University of Cambridge, Cambridge CB2 3QZ. U K

## 1. Introduction

The high temperature oxidation of iron results in the formation of a layered scale consisting of wüstite (FeO), magnetite ($Fe_3O_4$) and haematite ($Fe_2O_3$). While the growth of the majority of the scale is controlled by the outward diffusion of cations, oxidation at the $Fe_3O_4/Fe_2O_3$ interface is assumed to be dominated by an inward ingress of anions (e.g. Hauffe, 1965). In general, however, the mechanisms by which the inward movement of a variety of oxidants may occur through developing scales remain undetermined (e.g. Newcomb et al. 1985a). It is not clear, for example, how anions and even $CO_2$ can move through magnetite readily and yet cannot through fine grained and inward grown chromium rich spinel (Stobbs et al. 1985). In the oxidation of iron, the scaling process may be modified when blistering or spalling occurs, under which conditions the oxidation reactions become dominated by an inward flux of oxygen. In these circumstances both the $Fe_3O_4$ and $Fe_2O_3$ layers thicken at the expense of underlying FeO (Newcomb and Stobbs, 1985) and $Fe_2O_3$ precipitates may be observed in the $Fe_3O_4$. Here we report some preliminary results of a TEM study on the form of these precipitates as examined in cross-section and prepared using techniques which have been described previously (Newcomb et al. 1985b). The structure of magnetite/haematite interfaces is of particular importance in indicating how different oxidation and reduction processes may proceed.

## 2. Results and Discussion

A low magnification bright field micrograph of a typical $Fe_3O_4 - Fe_2O_3$ intergrowth region as formed at 850°C in a spalled part of the scale is shown in figure 1a. The region consists of fine laths of $Fe_2O_3$ which are of variable width (50-250nm) and length (1-5μm). The diffraction pattern taken from this area (figure 1b) shows a $(110)_{Fe_3O_4}$ and two $(21.0)_{Fe_2O_3}$ zones and demonstrates the orientation relationship:

$$(111)_{Fe_3O_4} \,||\, (0001)_{Fe_2O_3} \text{ and } [0\bar{1}1]_{Fe_3O_4} \,||\, [10\bar{1}0]_{Fe_2O_3}$$

The above orientation relationship is consistent with one of the four deduced by Becker et al. (1977) for these materials and indicates that growth occurs across a common close packed oxygen plane. The elongated precipitates clearly have preferential growth directions and in figure 1a intersect one another by an angle of approximately 70° In this case haematite precipitation has occurred on $(\bar{1}1\bar{1})$ and $(11\bar{1})$ magnetite planes, as indicated in figure 1.

While the formation of haematite occurs in a preferential direction, the subsequent growth of the higher oxide is itself of interest. Figure 2, for example, shows a fine haematite lenticular precipitate which is approximately 1 $\mu$m length. In this region dislocations in the surrounding magnetite are particularly evident, and these may be important in relieving any stress associated with the oxidation process. More developed regions of precipitation were found to consist of multiple laths of haematite so that alternating bands of both oxides were commonly observed. A typical area is shown in figure 3a where the parallel growth of haematite laths has occurred and where both the surrounding matrix and intervening magnetite are in the same orientation. A region showing a similar intergrown morphology is shown in figure 3b, which is a dark field micrograph imaged with a $(02\bar{2})_{Fe_3O_4}$ reflection. Interfacial dislocations are now particularly evident (as arrowed) and may be regarded as steps with associated strain fields which are confined to the interface (Christian, 1975). While the elongated morphology of the precipitates is itself indicative of a preferential growth direction for the haematite, the interfacial dislocations may provide the required diffusion path for the inward flux of oxygen. Both the chemistry and structure of this interface are currently being investigated using high resolution techniques.

Further investigation of the haematite laths indicated that they were twinned. Figure 4a is a dark field micrograph demonstrating the morphology of the twins which in this case is particularly evident in the thicker of the two lamellae. The residual contrast from dislocations should be noted. Twinning was confirmed by selected area diffraction and found to be occuring on $(\bar{1}014)$ planes: figure 4b, for example, shows a twinned $(31.1)_{Fe_2O_3}$ normal where the surrounding magnetite is at $(\bar{1}12)$, as indicated. Here the growth of the twins is likely to be controlled by matrix diffusion and interface mobility. In other regions more extended lateral growth of the haematite was found. The oxide shown in figure 5a now has a comparatively large grain size (~2$\mu$m) and a dislocation content of approximately $5 \cdot 10^8 cm^{-2}$ was observed. The interface between the haematite and an adjacent grain of magnetite of low dislocation content is arrowed. Sub-grain boundaries in the haematite were also apparent and these may have been formed as a result of the rearrangement of the dislocations so that stresses within the oxide are relieved by creep. Stresses may have also caused cracks to be initiated within the haematite, an example of which is shown enlarged in figure 5b (as arrowed). The presence of such microcracks, which frequently emanated from the $Fe_2O_3 | Fe_3O_4$interface itself, would obviously facilitate the ingress of oxygen into this part of the scale while their continued propagation could also lead to grosser mechanical disruption.

While the results presented here are of a preliminary nature, we have demonstrated that the diffusion of oxygen to the $Fe_2O_3 | Fe_3O_4$ interface is facilitated by the form of the local microstructure at different stages of the precipitation reaction. This in itself highlights the need for a characterization of this interphase boundary structure.

Acknowledgements

We are grateful to Professor D. Hull for the provision of laboratory facilities and to CERL (Leatherhead) for financial support.

## References

Becker P, Heizmann J J and Baro R 1977 J. Appl. Cryst. 10 77.
Christian J H 1975 Theory of Phase Transformations in Metals and Alloys (Oxford: Pergamon) p.284.
Hauffe 1965 Oxidation of Metals (New York: Plenum).
Newcomb S B and Stobbs W M 1985 J. Microsc. In press.
Newcomb S B, Stobbs W M and Metcalfe E 1985a Phil. Trans. Roy. Soc. In press.
Newcomb S B, Boothroyd C B and Stobbs W M 1985 b. J .Microsc. In press.
Stobbs W M, Newcomb S B and Metcalfe E 1985 Phil. Trans. In press.

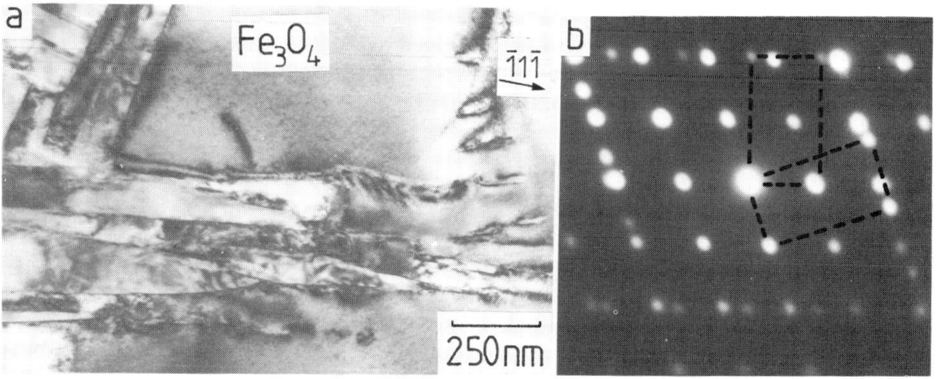


Figure 1. a) b.f. micrograph of $Fe_2O_3$ precipitates formed in $Fe_3O_4$ at 850°C. b) d.p. showing the orientation relationship of the two habits of $Fe_2O_3$ with $Fe_3O_4$

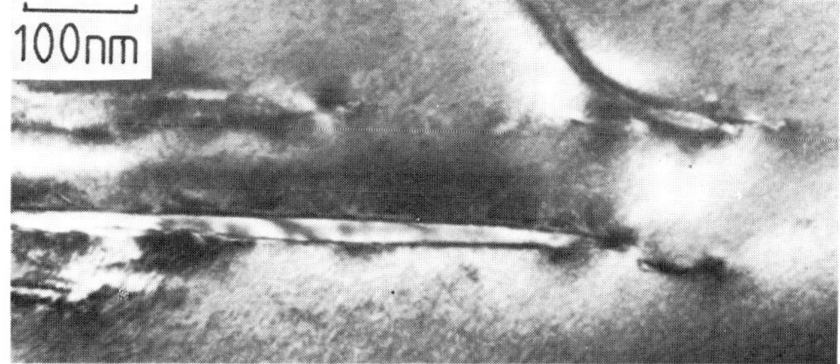


Figure 2.
b.f. micrograph of lenticular $Fe_2O_3$ oxide growing into $Fe_3O_4$

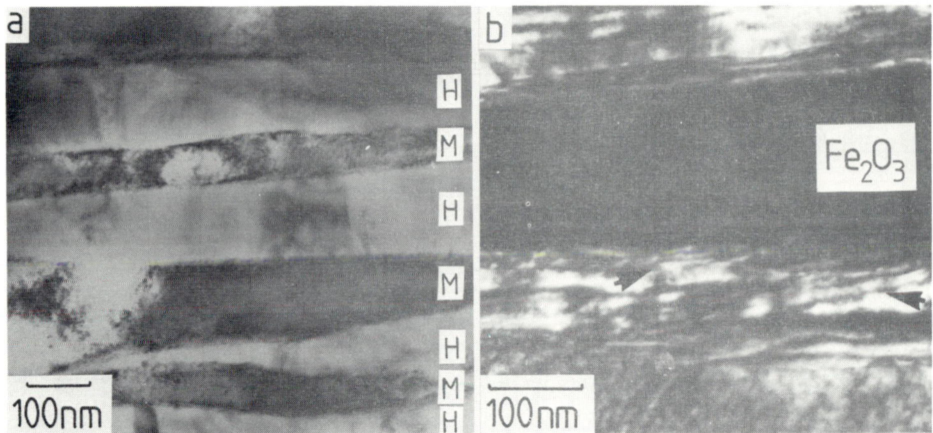


Figure 3. a) b.f. micrograph of zone of layered magnetite (M) and haematite (H). b) magnetite d.f. image showing evidence of interface dislocations.

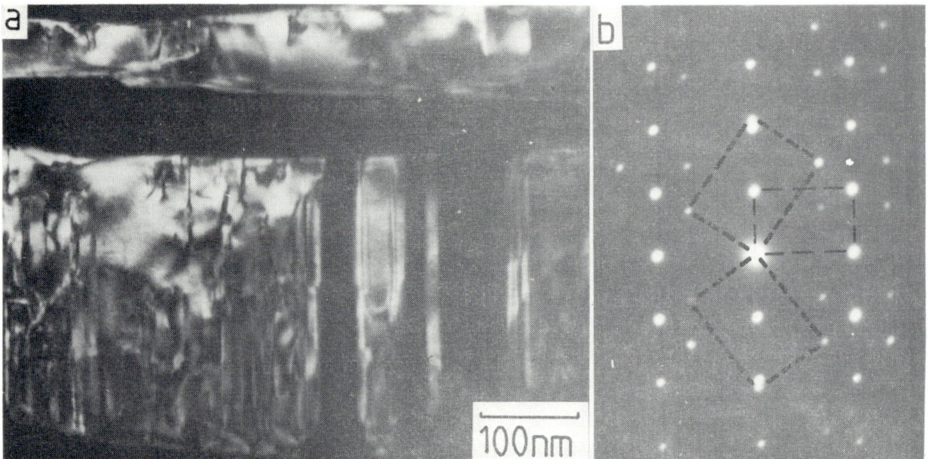


Figure 4. a) d.f. micrograph showing twinned nature of $Fe_2O_3$. b) d.p. from region a).

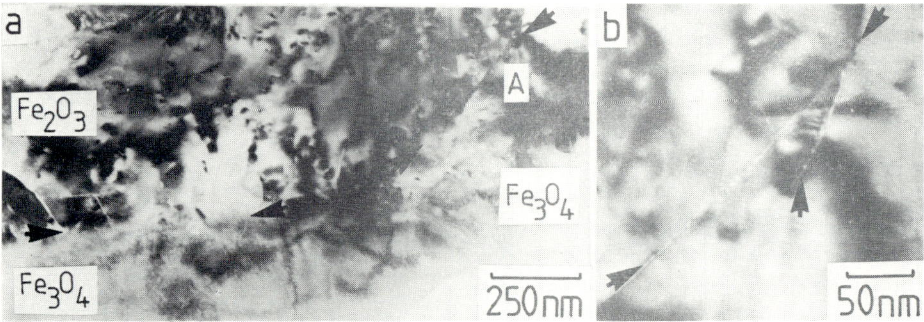


Figure 5. a) b.f. micrograph of bulk $Fe_3O_4/Fe_2O_3$ interface. The region exhibiting cracks at A is enlarged in b).

*Inst. Phys. Conf. Ser. No 78: Chapter 12*
*Paper presented at EMAG '85, Newcastle upon Tyne, 2–5 September 1985*


# Oxidation studies using EELS and EDX in the TEM

G J Tatlock, A G Baxter, R W Devenish and J S Punni

Department of Metallurgy and Materials Science, University of Liverpool, P.O. Box 147, Liverpool L69 3BX

## 1. Introduction

Much valuable information has been obtained about oxidation phenomena by the examination of cross-sections through metal oxide interfaces in the scanning electron microscope or electron probe microanalyser. Their spatial resolution is sufficient to determine diffusion profiles, oxide compositions for thick scales etc. However the analytical transmission electron microscope is often needed in order to investigate detailed oxidation mechanisms, especially in samples with thin scales or where interfacial reactions or internal precipitation is occurring. Results are presented in this paper which highlight the advantage of the use of electron energy loss spectroscopy and energy dispersive X-ray analysis in the TEM during oxidation studies of Fe 9Cr 1Mo steels and Ni-Al alloys.

## 2. Internal Oxidation in Ni-Al Alloys

Although the oxidation of nickel containing small amounts of aluminium has been studied extensively (see e.g. Hindam and Smeltzer (1980)), questions still remain about the precise oxidation mechanisms. Alloys containing 2w/o Al exhibit a characteristic distribution of internal oxide particles after oxidation in air or oxygen atmospheres, and a typical example is shown in the optical micrograph in Fig. 1. Sectioning of the sample and subsequent ion beam thinning gives TEM samples which allow higher resolution images to be taken such as those shown in Figs. 2 - 4. EDX analysis and electron diffraction suggest that most of the rods or platelets are composed of $NiAl_2O_4$ which contains a large number of planar defects as shown in Fig. 3. Some precipitates were identified as $Al_2O_3$ particles (Fig. 4) but these were much less common than previous lower resolution studies had suggested, even after allowing for differences in oxidation conditions, and only a few of the alumina particles were connected to the spinel rods. Hence the mechanism whereby rods or particles of $Al_2O_3$ grow into the matrix and subsequently transform into $NiAl_2O_4$ as the surface NiO scale thickens needs to be re-examined. Small amounts of Ni were sometimes detected in the $Al_2O_3$ and overlap effects between the particles and matrix could have led to misleading results.

However under these conditions EELS proved particularly useful. Firstly the higher spatial resolution of the technique often removed any overlap ambiguities and secondly even particles which were too small for conventional analysis could be readily characterized by examining the fine structure on the oxygen edge in the energy loss spectrum. Figs. 5 and 6 show examples of EELS spectra taken from $NiAl_2O_4$ and $Al_2O_3$ precipitates.

The features in the fine structure were unique and sufficiently pronounced to allow particle identification over a range of specimen thicknesses. The improved resolution during EELS analysis is easily seen in the aluminium EDX and oxygen EELS line profiles shown in Fig. 7, which were both recorded while the electron beam was being scanned across an oxide precipitate. (Both oxygen and aluminium levels were below the levels of detection in the surrounding nickel matrix). Only rarely was any evidence seen of the possible transformation of $Al_2O_3$ to spinel, but one such example is shown in Fig. 8. The centre of the particle is very rich in aluminium while the outer edges have a nickel/aluminium ratio characteristic of $NiAl_2O_4$.

## 3. Fe 9Cr 1Mo Steels Oxidised in Air and CO2

Another oxide system which is currently under much investigation is that formed on the boiler tube steel composed primarily of iron with 9% chromium and 1% molybdenum additions. The use of EELS to investigate the early stages of oxide formation on this steel have been reported previously (Tatlock et al 1984), hence only work on cross sections through the metal oxide interface produced after oxidation for several thousand hours will be discussed here. The main problem for analysis under these conditions is sample preparation, and the fact that the resulting thin samples are both delicate and magnetic! However useful TEM samples can be obtained (see e.g. Fig. 9), but once again detailed analysis casts doubt on some of the models of breakaway oxidation after prolonged exposure to $CO_2$/CO atmospheres. In particular, very little free carbon was observed within the oxide scale near the oxide/metal interface, although the distinct particles which were formed exhibited a graphitic nature as shown in the fine structure of the EELS spectrum in Fig. 10. Small regions were also found which contained both carbon and mixed spinel oxides. However what did appear to be more significant was the growth and change of morphology of the carbides in the subscale region after prolonged exposure to a $CO_2$/CO environment.

Once again the higher spatial resolution of the TEM is essential in examining the oxidation mechanisms since thermodynamics would suggest that in the Fe Cr system different types of oxide would be formed in a systematic way varying from almost pure $Cr_2O_3$ through iron - chromium spinels to iron oxides. However detailed analysis shows that oxides of very different compositions are stable even in adjacent grains, and one example of this is shown in Fig. 11, where the three grains have widely varying chromium content. Further examples of this type are under active investigation; but the main aim of this paper is to emphasise that the recent advances in high spatial resolution analysis are beginning to shed more light on some complex oxidation mechanisms while casting doubt on other seemingly well understood phenomena.

## 4. Acknowledgements

The provision of oxidized samples from CERL Leatherhead and the financial support of SERC are gratefully acknowledged.

## 5. References

Hindam H M and Smeltzer W W 1980 J. Electrochem. Soc. 127 1622

Tatlock G J, Baxter A G, Devenish R W and Hurd T J 1984 Analytical Electron Microscopy - 1984 (San Francisco:San Francisco) pp 227-230

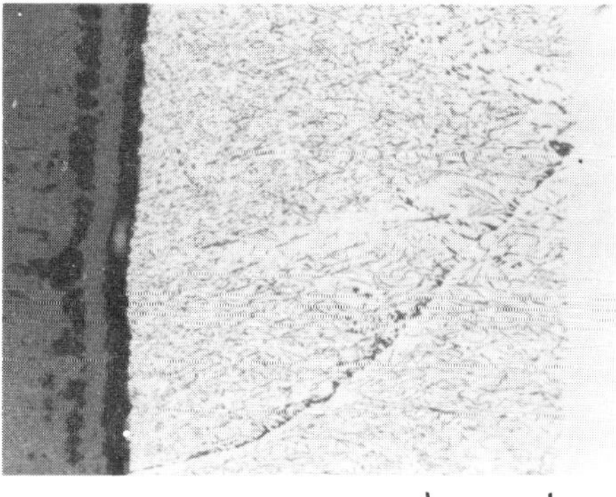


Fig. 1. Optical micrograph of a cross section through a Ni 2% Al alloy after oxidation in air for 100 hrs at 1000°C.

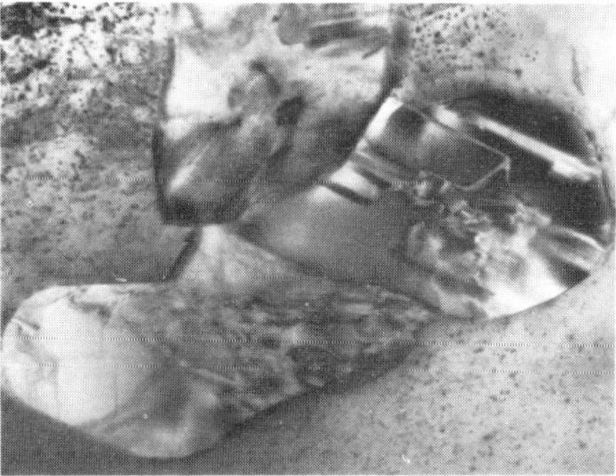


Fig. 3. TEM micrograph of a section through Ni-Al spinel rods showing the large number of planar defects.

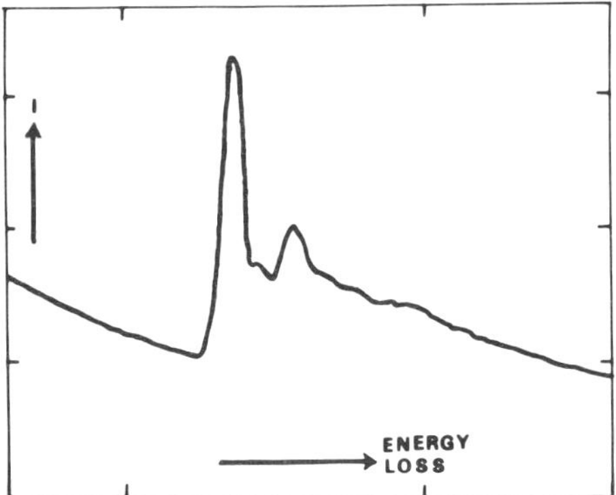


Fig. 5. The oxygen edge in an EELS spectra taken from a $NiAl_2O_4$ particle.

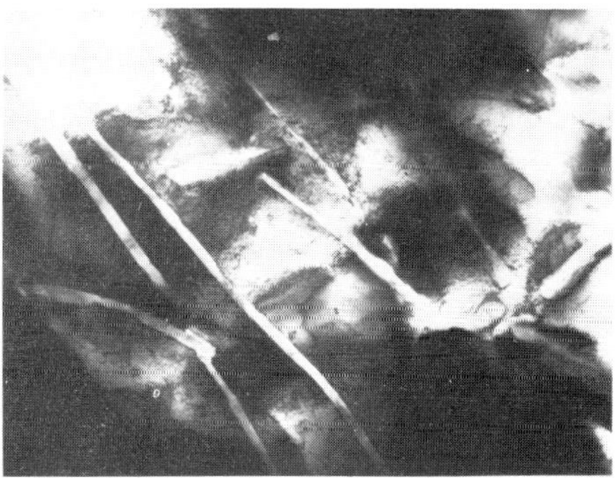


Fig. 2. TEM micrograph of Ni-Al spinel rods.

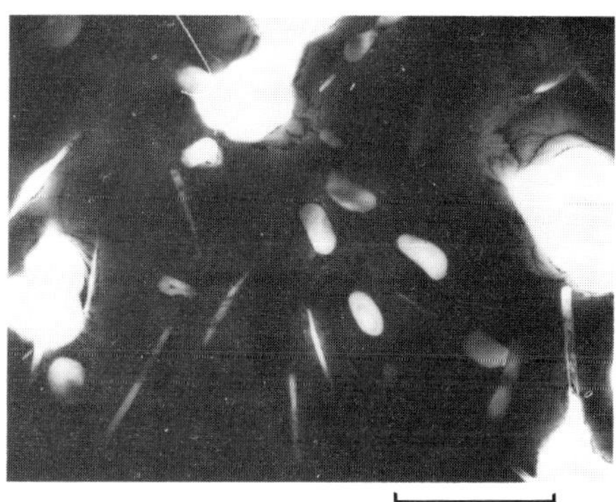


Fig. 4. $Al_2O_3$ particles in a nickel matrix.

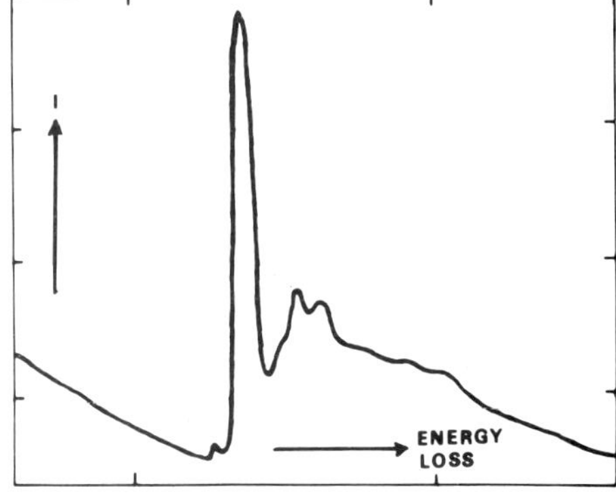


Fig. 6. The oxygen edge in an EELS spectra taken from an $Al_2O_3$ particle.

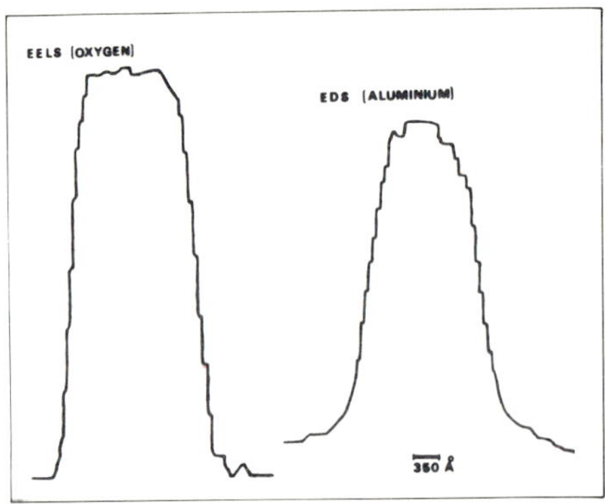


Fig. 7. EDX Aluminium and EELS oxygen line profiles across a spinel particle.

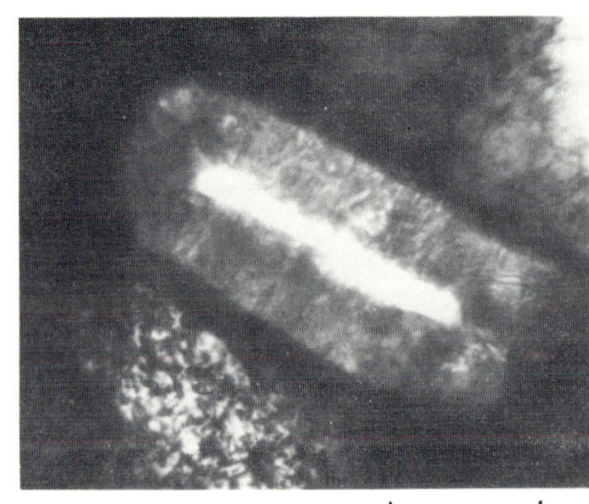


Fig. 8. Compound oxide particle in Ni 2%Al.

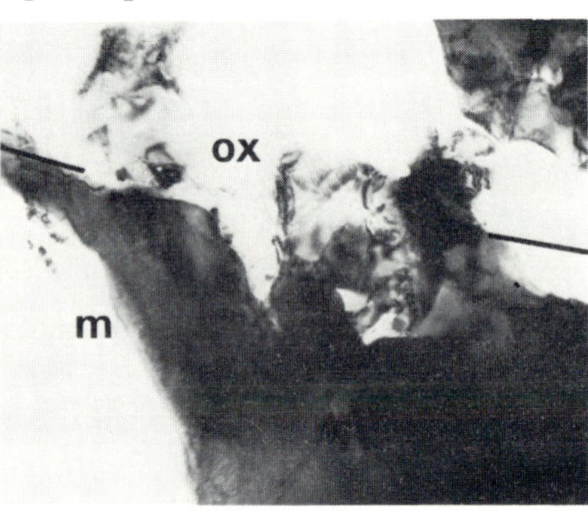


Fig. 9. Thinned section through a metal oxide interface in Fe 9Cr 1Mo oxidized for 12,000 hrs. at 550°C in a $CO_2$ 1%CO mixture.

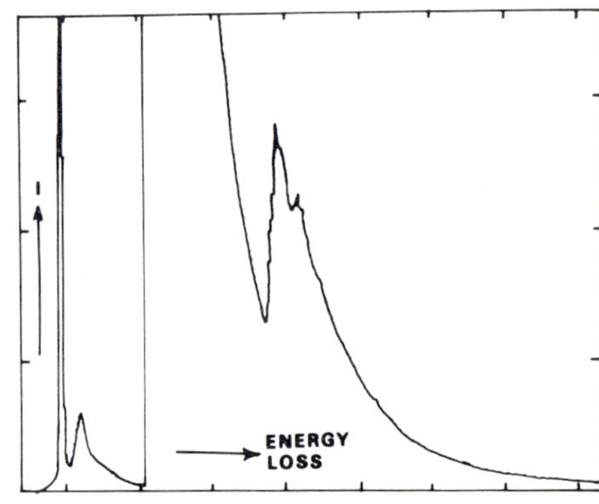


Fig. 10. EELS spectra taken from a carbon particle embedded in the oxide close to the metal oxide interface in Fe 9Cr 1Mo.

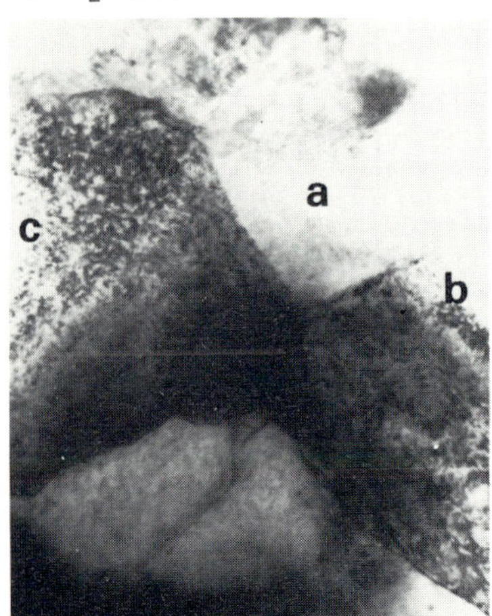


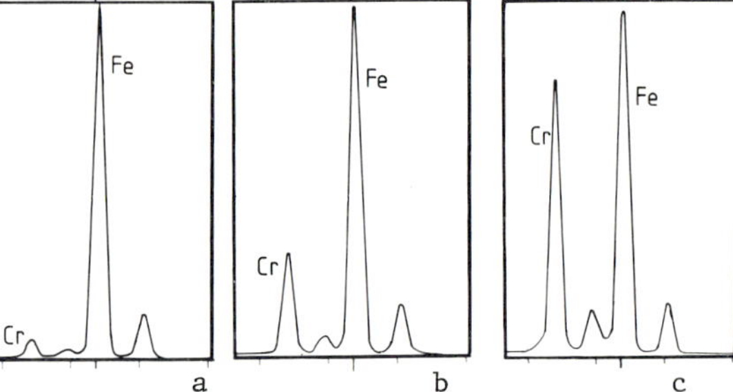


Fig. 11. Electron micrograph and EDX spectra of oxides on Fe 9Cr 1Mo.

*Inst. Phys. Conf. Ser. No 78: Chapter 12*
*Paper presented at EMAG '85, Newcastle upon Tyne, 2–5 September 1985*


# Growth and microstructure of oxides in a dilute Ni–Al alloy at high temperatures

N J Simms and J A Little
University of Cambridge, Department of Metallurgy and Materials Science
Pembroke Street, Cambridge CB2 3QZ

## 1 Introduction

The high temperature oxidation behaviour of dilute Ni-Al alloys with >0.5wt%Al has been the subject of several investigations at various partial pressures of oxygen (e.g. Smeltzer et al. 1981, Stott et al. 1977) but lower concentrations of Al in the alloy have received little attention. The studies at higher concentrations have characterised the morphology of the oxides formed and have indicated a trend in the morphology of the internal oxides which depends upon Al content, as well as the temperature of oxidation and the partial pressure of oxygen. (See also Shida et al. (1982)). This study was undertaken to confirm these trends at a lower Al content (0.25wt%Al) and also to study the effect of oxidising both cold worked and annealed samples on the oxide morphologies.

## 2 Experimental

The alloy was made by induction heating Ni and Al (supplied by William Rowland Ltd.) under vacuum in an $Al_2O_3$ crucible. Specimens for oxidation were prepared from hot-rolled strip 2mm thick, and then cut to the same size (3cm x 2cm) before being metallographically polished to 1um on both sides. Three types of specimen preparations were carried out:-

1) Anealing "in situ" in flowing argon after polishing.
2) Annealing in vacuum after "120 grit" polishing, before completing polishing process.
3) Polishing with no annealing.

All oxidations were carried out using a Stanton Redcroft Thermal Balance with specimens suspended by Pt-13%Rd wire in dry flowing oxygen at 1 atmosphere pressure. All runs were started by heating to the required temperature in flowing argon before introducing the dry oxygen. Cross-sections of oxidised specimens were mounted in conductive bakelite and polished to 1um finish (no etch). TEM specimens were prepared by jet-polishing, using 10% perchloric acid, 60% ethanol, 30% glycerol solution at 50V, 100mA.

## 3 Results

The weight gain versus time results are illustrated in figure 1. These show that cold work increases the oxidation rate at a given temperature (c.f. Pirin et al. 1984; work on pure Ni), although the log-log plot indicates that the behaviour is not straightforwardly parabolic in both cases. It can be seen that the curves converge at long time and it is suggested that a grain size/boundary effect may be responsible for this relative difference, which disappears with time as grain growth and recrystallisation in the oxide occurs (Atkinson 1981).

The external oxide consists of a duplex scale: a dense dark green outer layer, and a porous light green inner layer, figure 2. The outer layer consists of columnar grains of NiO, which EDX indicates contains very little Al. The inner layer contains many small oxide grains and pores. EDX indicates a higher Al content here, and $NiAl_2O_4$ particles would be expected (Smeltzer et al. 1981) which have been incorporated from the internal oxidation zone as the external oxide grows. X-ray diffraction of the spalled external oxide reveals the rhombohedral form of NiO which has transformed from the high temperature f.c.c. B2 structure. The rhombohedral phase has a distorted B1 structure with $\alpha = 60.06^{\circ}$ at $18^{\circ}C$ (Rooksby, 1943) and thus transformation strains will be small compared to those produced due to vacancy clustering and void formation. This led to frequent spalling of the external oxide, being most pronounced in the thicker oxides where internal stresses in the oxide film were more difficult to accomodate upon cooling, spalling occuring at about 500-$600^{\circ}C$.

There was preferential internal oxidation along the alloy grain boundaries at all temperatures tested being most pronounced at the lower temperatures and thus confirming the trends of other work (Shida et al. 1982). The precipitation was almost continuous along the alloy grain boundaries with an adjacent precipitate free zone (figure 3). The backscattered SEM image shows apparently "spherical" internal oxide precipitates close to the alloy - external oxide interface becoming larger, and more acicular further into the alloy. These "spherical" precipitates have been previously identified as $NiAl_2O_4$ (Stott et al. 1982) whilst the more acicular precipitates are believed to be $Al_2O_3$. The small size of the precipitates makes unique identification using EDX or WDX on a polished SEM specimen very difficult and further investigations have been made using TEM. Two types of precipitate have been found (figure 4) although EDX indicates that both contain both Ni and Al.

It thus appears that the preparation produced a thin region close to the external oxide/internal oxide interface where only $NiAl_2O_4$ precipitates exist. These two types of precipitate are believed to be particles which have grown either within the grains (X) or along the grain boundaries (Y). Growth of internal oxide particles within a grain involves the generation of a phase of specific volume $65Å^3$/Ni atom from the nickel alloy of volume $11Å^3$/Ni atom. Precipitates grow as tabular (111) platelets twinned on (111). There is a strong tendency for the boundaries of such twins to be parallel to the 211 planes normal to the platelets (see e.g. Pashley and Stowell, 1963) and this can be clearly seen in the micrograph. The (111) diffraction pattern confirms the existence of the twin (figure 5). The configuration of the platelet w.r.t. the electron beam means that the twin is not readily detected because the different orientations give rise to identical diffraction patterns. However double diffraction effects between allowed twin reflections and matrix reflections in the higher order Laue zones [e.g. 1/3(151) + (111) = 1/3(224)] account for the extensive and symmetric diffraction pattern recorded. The second type of precipitate is a rod-like precipitate which does not show the extensive twinning and has therefore grown under less constraint. This can occur more readily at the grain boundaries and triple points of the alloy.

## Conclusion

A cold worked Ni-Al alloy oxidises at a faster initial rate than an annealed alloy due to rapid grain boundary diffusion of nickel ions. As oxide grain growth and recrystallisation occurs the two rates become comparable. The external oxide consists of an outer columnar and an inner porous layer which spalls upon cooling and transforms to the rhombohedral form of NiO. The internal $NiAl_2O_4$ precipitates are of two forms. One forms within grains and growth strains are minimised by extensive twinning whilst the grain boundary precipitates are less constrained and grow in a rod-like manner.

## Acknowledgements

One of us (NJS) is grateful to the SERC for financial support and we thank Professor D. Hull for the provision of laboratory facilities.

## References

Atkinson A, Taylor R I, Hughes A E 1981 High Temperature Corrosion Ed. R A Rapp (NACE, Houston) pp110-114
Pashley D W, Stowell M J 1963 Phil. Mag. 8.2 1605
Pirin J C, Morvan J, Mairey D 1984 Acta Met. 32 2203
Rooksby H P (1943) Nature 152 304
Shida Y, Stott F H, Bastow B D, Whittle D P, Wood G C 1982 Oxid. Metals 18 93
Smeltzer W W, Hindam H M, Elrefaie F A 1981 High Temperature Corrosion Ed. R A Rapp (NACE, Houston) pp251-257
Stott F H, Shida Y, Whittle D P, Woods G C, Bastow B D 1982 Oxid. Metals 18 127
Stott F H, Wood G C 1977 Corr. Sci. 17 647

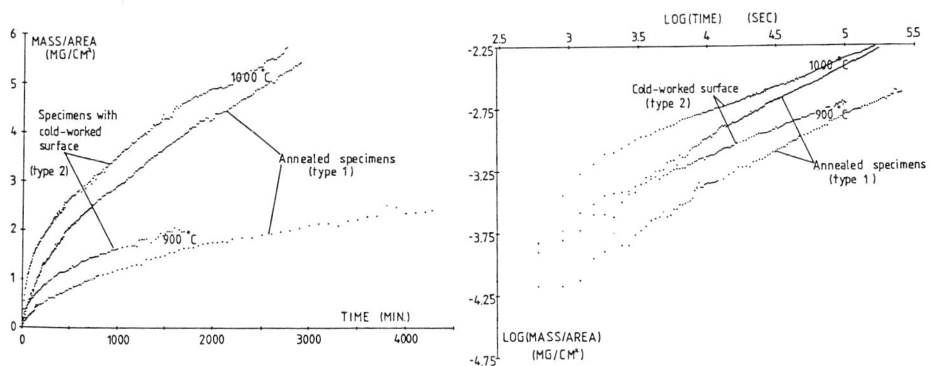


Fig.1. Kinetic curves for oxidation of Ni-0.25wt%Al in oxygen.

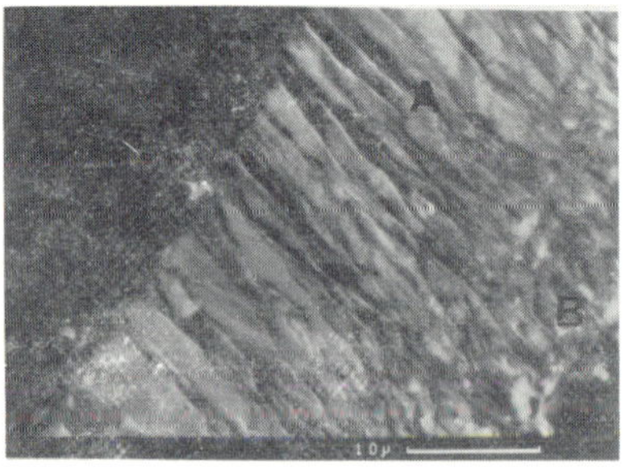


Fig.2. Duplex external oxide: A is dense outer layer, B is porous inner layer.

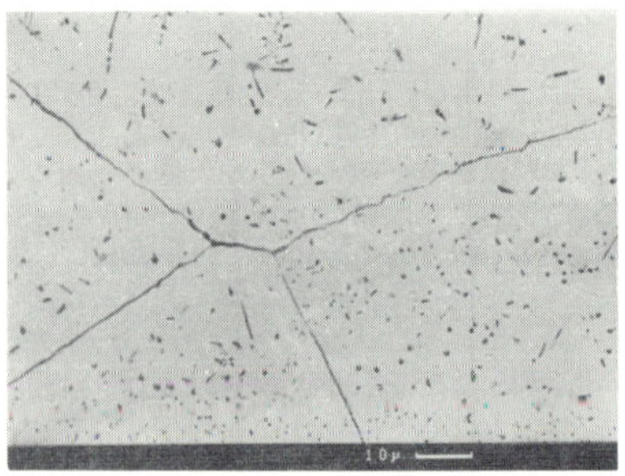


Fig.3. Backscattered SEM image of internal oxidation zone.

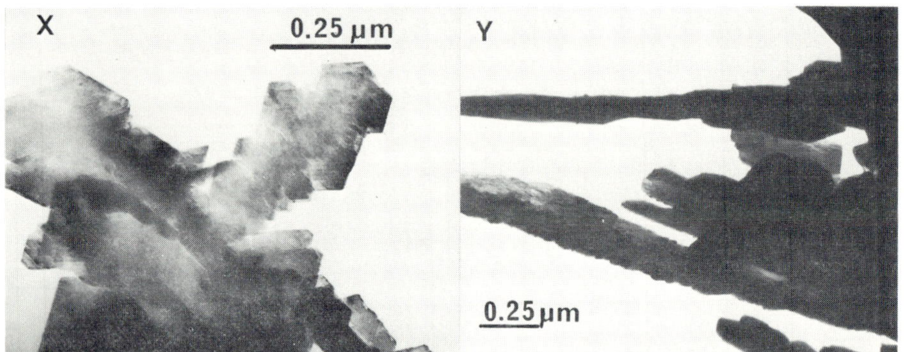


Fig.4. TEM images of two types of internal oxide particles.

Fig.5. Typical (111) diffraction pattern from tabular particle.

*Inst. Phys. Conf. Ser. No 78: Chapter 12*
*Paper presented at EMAG '85, Newcastle upon Tyne, 2–5 September 1985* 

# Superstructure and surface images of uranium oxide ($U_4O_9$)

L A Bursill[*+], Peng Ju Lin[*+], David J Smith[*] and I E Grey[x]

*Centre for Solid State Science, Arizona State Univ., Tempe, AZ, 85287 USA
+School of Physics, Univ. of Melbourne, Parkville, Vic. 3052, Australia
xCSIRO Div. of Mineral Chemistry, Port Melbourne, Vic. 3207 Australia

## 1. Introduction

The crystal structure of $U_4O_9$ is known to be cubic above $80^oC$ (α-phase) but may be rhombohedral below this temperature (β-phase) (for phase diagram, see Fig. 1 of Bevan et al, 1985): the initial purpose of this study was to find evidence for a rhombohedral phase. There is also some possibility that stoichiometric $U_4O_9$ does not exist, since the composition range at room temperature is $U_4O_{9-y}$, 0.020<y<0.060, which corresponds to O:U = 2.235 to 2.245 (Van Lierde et al, 1970). Hence the presence of any defect structures would be of interest. During this study, some novel and unexpected surface phenomena were found. This report describes observations both of the bulk superstructure and the surface topography.

## 2. High-Resolution Images of $U_4O_9$

High-resolution images of $U_4O_9$ for the [100], [110] and [111] zone axes are shown in Figs. 1a-c. These were recorded on the JEM-4000EX instrument, recently installed at Arizona State University, which has an effective resolution near optimum focus (400kV, Δf = -400Å, $C_s$ = 1.0mm) of ~1.7Å. The thickness variation of the images was observed to be quite remarkable and they also proved to be sensitive to crystal and beam tilts of <0.5 mRad (cf. Smith et al, 1983; Smith et al, 1985).

## 3. Computer Simulation of Electron Diffraction Intensities and Images

The β-phase, s.g. $I\bar{4}3d$, gives fundamental and superlattice intensities based on a 21.9Å unit cell composed of 4x4x4 fluorite-type subcells (Blank and Ronchi, 1968). The zero order (001) reciprocal lattice sections of $UO_2$ and $U_4O_9$ are compared in Figs. 2a,b. Discrete clusters of six U-centred square antiprisms ($U_6O_{37}$), surrounded by eight 10-coordinated U sites, form the fundamental building units of $U_4O_{9-y}$ (Bevan et al, 1985). There are 12 such clusters per unit cell, yielding $U_{256}O_{572}$, or $UO_{2.2344}$, as illustrated schematically in Fig. 3 for the [001] projection. This latter gives minimum overlap of individual clusters so image analyses for this case were given first priority. (The numbers in Fig. 3 refer to the z-coordinates of the cluster centers, in sixteenths of 21.9Å)

In order to optimise HREM imaging of the superstructure, the variations of the electron diffraction intensities of $UO_2$ and $U_4O_9$ with crystal thickness for the most significant subcell and superstructure reflections were calculated and these are compared in Figs. 4a-d. The transmitted (000) beams diverge for thickness H>40-50Å (Fig. 4a). It is significant that for $UO_2$ (Fig. 4b), the subcell beams {020} and {022}, are comparable with (000) for 200Å<u><</u>H<u><</u>400Å, and 400Å<u><</u>H<u><</u> 500Å, respectively, whereas the subcell beams for $U_4O_9$ never approach (000) in intensity (Fig. 4c). Superstructure intensities for $U_4O_9$ are a further order of magnitude weaker, the strongest being (015) which is significant only for H<u>></u>300Å (Fig. 4d). The observed electron diffraction intensities were generally consistent with these calculations for a range of crystal thicknesses. Similarly, the through-thickness series of computed images of $U_4O_{9-y}$ showed the topological features of the experimental images and, as expected from Fig. 4a, images of $UO_2$ and $U_4O_9$ differed only for H $\geq$ 50Å. Our tentative conclusion is that the structural model for $U_4O_9$ of Bevan et al provides a reasonable basis for the interpretation of the HREM images of thicker crystals (H > 50Å).

For thicknesses in the approximate range 0 < H < 50Å, the experimental image contrast often showed topological deviations compared with calculations, implying some disordering of the $U_6O_{37}$ (or similar) clusters, or perhaps some loss of oxygen, where the structures might have reverted locally to $UO_2$. The latter might be due to ejection of oxygen atoms by electron irradiation or due to chemical reductions. For H $\leq$ 50Å, the images are equally consistent with calculations for $UO_2$ or $U_4O_9$. For H > 50Å, even semi-quantitative image-matching is made difficult by (i) the sensitivity of the images to beam and crystal tilt; (ii) uncertainty about the electron form factors, including charge state considerations, which should be used for U (cf. Zakharov et al, 1983); and (iii) neglect of inelastic scattering effects on the image contrast (cf. Ahn et al, 1985).

## 4. Surface studies

Surface profile images of $U_4O_9$ specimens indicated that the surfaces were remarkably clean and that surface facetting and the rearrangements of steps and ledges were readily visible (Fig. 5). It was significant that (110) and (100) vicinal faces gave markedly different contrast, indicative of quite different surface activity and relaxations. Atomic diffusion occurred along the surface, as shown in Figs 6a,b, and it was again noteworthy that the extent and rapidity of the step rearrangements and surface reconstructions depended on the particular surface. These observations are explicable in terms of different binding energies for different faces. The surface mobility presumably represents a response to the thermal energy deposited in the sample by inelastic scattering of the incident electrons, with the non-equilibrium fracture surface tending to equilibrate at elevated temperature. Moreover, direct transfer of momentum to individual ions could lead to cooperative rearrangements. Definitive interpretation of these profile images, in particular quantifying the extent of any surface contractions, must necessarily await multi-slice image calculations but it is already clear that the technique provides unique information about local surface structure.

## 5. Catalytic Gasification of Carbon

In some cases it appeared that loss of oxygen from $U_4O_9$ during electron irradiation provided a ready source of oxygen for the catalytic gasification of carbon (or hydrocarbons) inside the electron microscope. Thin layers of carbon contamination initially present in some areas were burnt away during observation. Real-time video recordings at atomic resolution are being studied to provide models for the mechanisms and kinetics of such catalytic reactions.

This work was supported by the Australian Research Grants Scheme, the University of Melbourne and the Facility for High Resolution Electron Microscopy, within the Center for Solid State Science at Arizona State University (NSF Grant DMR-8306501).

References

Ahn C C, Krivanck O L and Wood G J 1985 Ultramicroscopy, in press

Blank H and Ronchi C 1968 Acta Cryst. A 24 657

Bevan D J M, Grey I E and Willis B T M 1985 J. Sol. State Chem. in press

Smith D J, Saxton W O, O'Keefe M A, Wood G J and Stobbs W M 1983 Ultramicroscopy 11 263

Smith D J, Bursill L A and Wood G J 1985 Ultramicroscopy 16 19.

Van Lierde W, Pelsmaekers J and Lecoq-Roberts A 1970 J. Nucl. Maters. 37 276

Zakharov N D, Gribeluk M A, Vainshtein B K, Rozanova O N, Uchida K and Horiuchi S 1983 Acta Cryst. B39 575

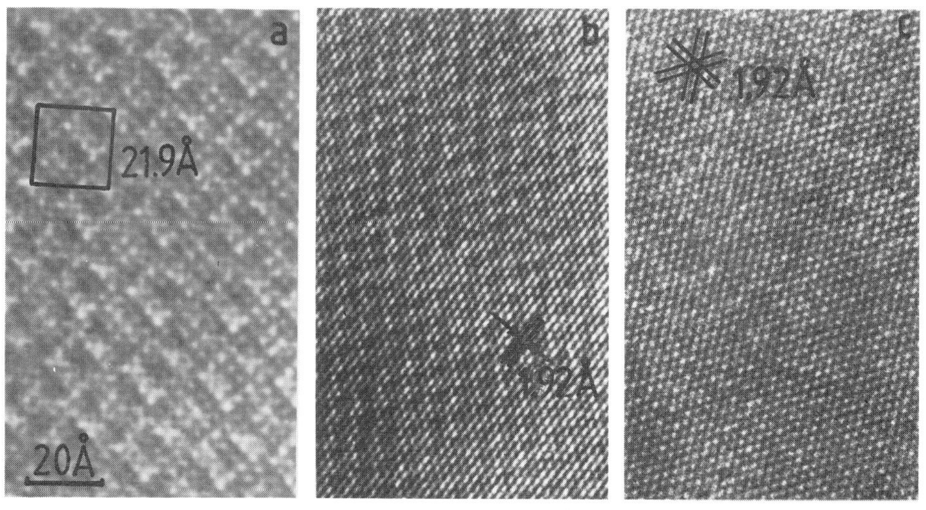


Fig. 1 HREM images of $U_4O_9$. (a) [100]; (b) [110]; (c) [111].

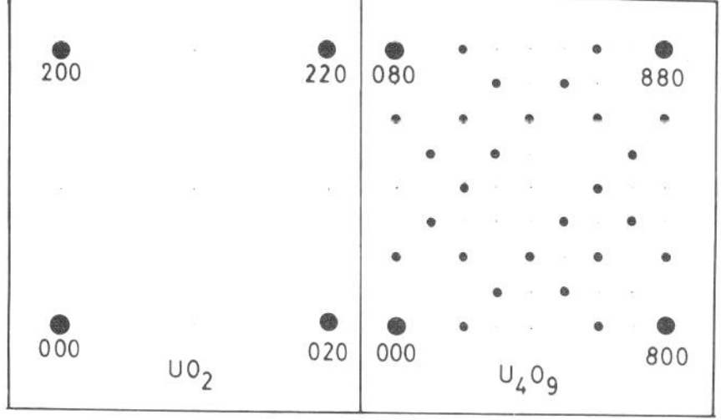


Fig.2 Comparison of (001) reciprocal lattice sections of (a) $UO_2$ ($\underline{a}$ =5.45Å) and (b) $U_4O_9$ ($\underline{a}$ = 21.9Å).

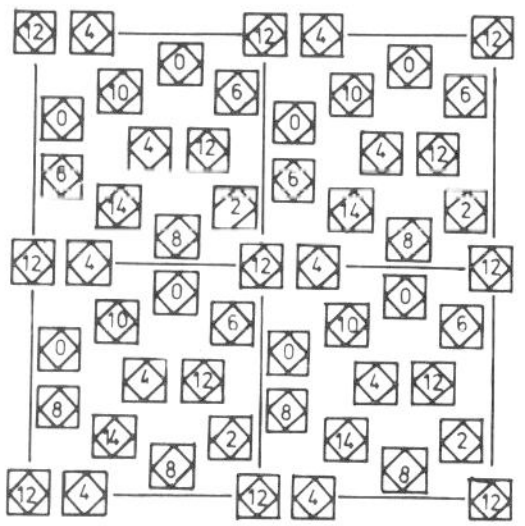

Fig. 3 Distribution of $U_6O_{37}$ clusters within fluorite-type matrix proposed for $\beta$-$U_4O_{9-y}$.

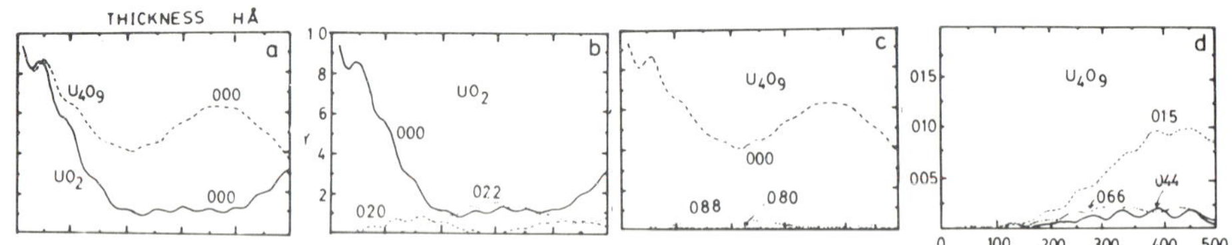


Fig. 4 Intensity of (a) transmitted, (b,c) subcell and (d) superstructure beams of $UO_2$ and $U_4O_9$ as a function of crystal thickness.

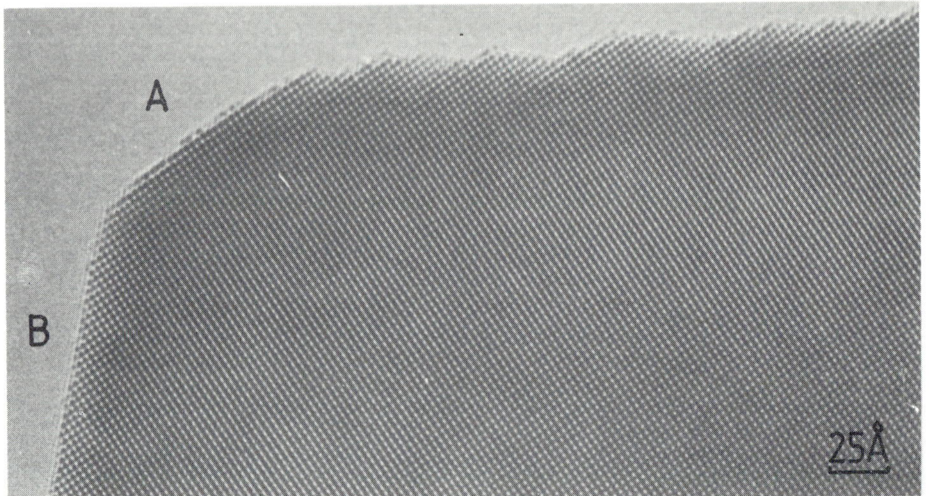


Fig. 5 Surface profile imaging of $U_4O_9$ crystal in [100] projection. Note variations in edge contrast and apparent relaxations for (110)surface (B).

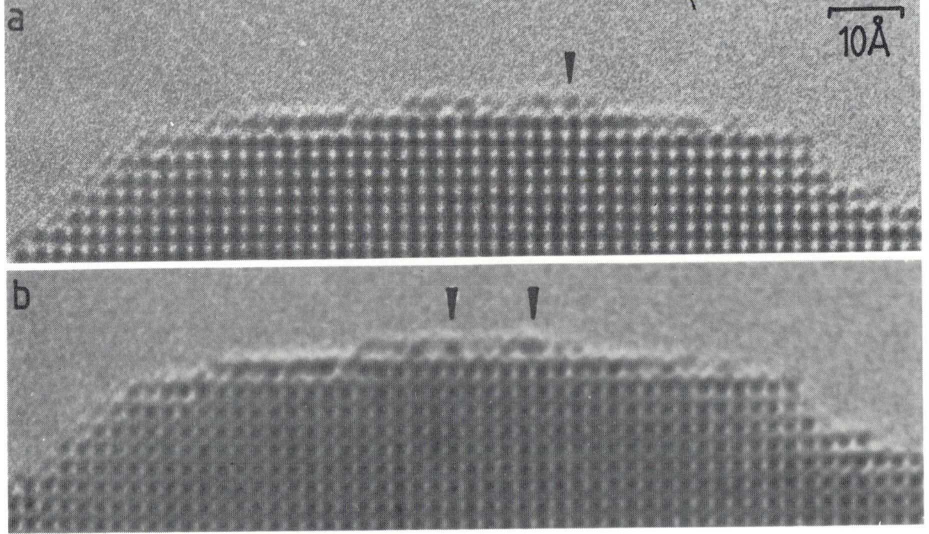


Fig. 6 Pair of surface profile images showing rearrangements which occurred during observation (time difference: 30s).

*Inst. Phys. Conf. Ser. No 78: Chapter 12*
*Paper presented at EMAG '85, Newcastle upon Tyne, 2–5 September 1985* 

# Electron microscopy of microdomains in perovskites

J M Gonzalez-Calbet, M Vallet-Regi and M A Alario-Franco

Departamento de Quimica Inorganica, Facultad de Quimicas,
Universidad Complutense, 28040 Madrid, Spain

## 1. Introduction

A few solids showing $A_3B_3O_8$ composition (A = Ca, La; B = Fe, Mn, Ti) can be described as an ordered intergrowth between the perovskite ($ABO_3$) and the brownmillerite ($A_2B_2O_5$) structures (Megaw 1946, Bertaut et al 1959). The existence of microdomains in some of these 'anion deficient'-perovskite-derived solids has been recently demonstrated (Alario-Franco et al 1983a,b). In this way $Ca_2LaFe_3O_8$, i.e. $A_2A'B_3O_8$, when oxidised in air at 1673 K, shows, as observed by electron diffraction and microscopy, three types of microdomains. In each of them, the multiple perovskite axis oriented at random in one of the three space directions.

We report in this paper the study, by ED and TEM, of two more examples of microdomain textured solids which, having big amounts of oxygen vacancies, show simple x-ray diffraction patterns.

## 2. Experimental

Solids with nominal composition $A_3B_3O_8$ ($Ca_3Fe_2TiO_8$ and $Ca_3Mn_{1.8}Fe_{1.2}O_8$), having an orthorhombic unit cell ($a \approx a_c\sqrt{2}$, $b \approx 3a_c$, $c \approx a_c\sqrt{2}$, where $a_c$ is the cubic perovskite unit cell parameter), were heated in air at 1673 K for 24 hours. The obtained black materials were quenched to room temperature.

According to the x-ray diffraction data, all the products appear to have a cubic cell related to the perovskite structure with the following parameters:

| | |
|---|---|
| $Ca_3Fe_2TiO_{8+x}$ | a = 3.827(1) |
| $Ca_3Mn_{1.8}Fe_{1.2}O_{8+y}$ | a = 3.776(1) |

Electron diffraction and microscopy were carried out with a SIEMENS ELMISKOP 102 electron microscope, fitted with a double tilting (± 45°) goniometer stage. Specimens for observation were prepared by grinding under acetone; a drop of suspension was then transferred to a holey carbon support film. Individual crystals were aligned so that the electron beam was parallel to one of the main orientations of the perovskite substructure, namely $[100]_c$.

## 3. Results and discussion

### 3.1 $Ca_3Fe_2TiO_{8+x}$

The formal substitution of some titanium by iron in $CaTiO_3$ gives a phase with the $Ca_3Fe_2TiO_8$ composition (Grenier et al 1976). As it has two different cations at the B-sites, the behaviour of this solid under oxidation is similar to that of the calcium-lanthanum ferrite (Alario-Franco 1983b). Although x-ray diffraction investigations suggest the presence of a cubic unit cell, electron diffraction and microscopy show that a three-dimensional microdomain texture is developed (Fig 1). Each of these microdomains appears to have a structure related to the $Ca_3Fe_2TiO_8$ sample with the b axis alternating at random in one of the three space directions.

### 3.2 $Ca_3Mn_{1.8}Fe_{1.2}O_{3+y}$

When this sample is heated in air at 1673 K, two sets of microdomains, having three types of domains each are formed. Six types of microdomains are then present (Fig 2): three of them have the $Ca_2Fe_2O_5$-type structure (Bertaut et al 1959), with their b axis, equal to $4a_c$, distributed at random in each of the three space directions. The other set contains also three types of domains, but their structure is of the $CaMnO_3$-type (Pepopelmeier et al 1982) with $b = 2a_c$ ($a_c$ being the usual cubic perovskite parameter ~ 3.9 Å).

If this sample is heated under stronger oxidising conditions, pure oxygen at 1273 K, another new set of microdomains is produced. They have the structure of the $A_3B_3O_8$ phase and the multiple perovskite axis, which is now $3a_c$, is again located at random. However, not at all the above solid can be oxidised and, besides the newly formed $A_3B_3O_8$-type set of domains, there remains the previous six sets of domains of $CaMnO_3$ and $Ca_2Fe_2O_5$ in some crystal regions. In all, then, nine sets of microdomains appear in the crystals of this solid as observed by electron microscopy. Some of these sets are, however, indistinguishable, so that lattice fringes of only three kinds appear on Figure 3.

According to the above results, microdomain formations seems to be the essential characteristic of the oxidation process in both $A_2A'B_3O_8$ and $A_3B_2B'O_8$ systems. However, the difference observed in the material having some manganese at the B-sites seems to be due to the tendency of $Mn^{3+}$ to have a pyramidal environment (Pepopelmeier et al 1982, Reller et al 1984, Nguyen et al 1984) while $Fe^{3+}$ prefers to stay in tetrahedral and octahedral coordinations. So, in the microdomain textured ferrites all the oxygen excess must be situated in the domain walls (Alario-Franco 1983a,b), but some oxygen could be employed to convert square pyramids into octahedra in those perovskite-type solids containing some manganese at the B-sites.

## Acknowledgments

We thank L Puebla and A Garcia for technical assistance.

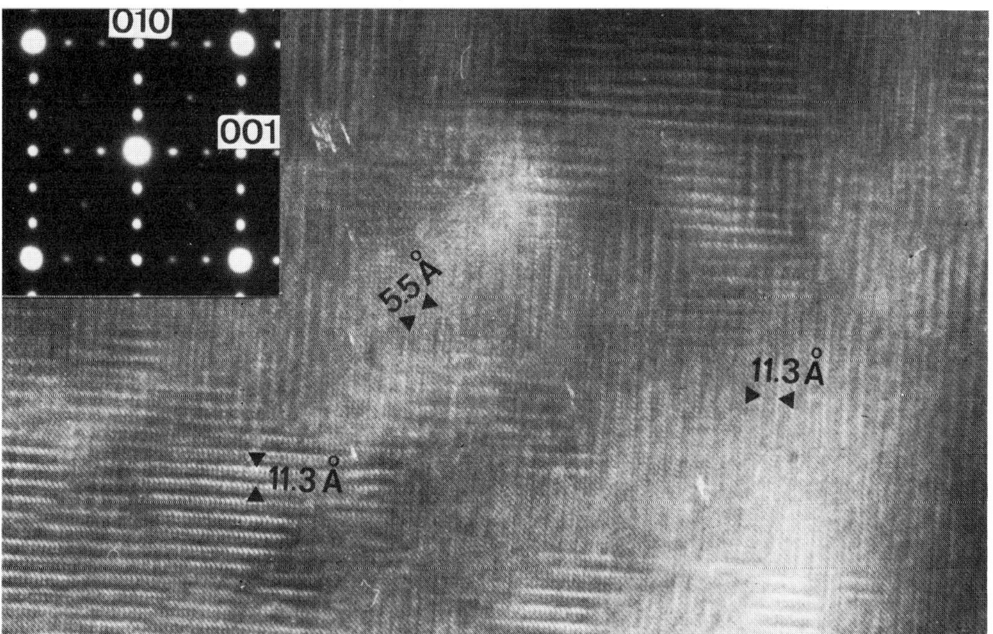


Figure 1. Electron micrograph of the $Ca_3Fe_2TiO_{8+x}$ sample along the [001] zone axis. It can be seen that the perovskite long axis is oriented at random in one of the three space directions.

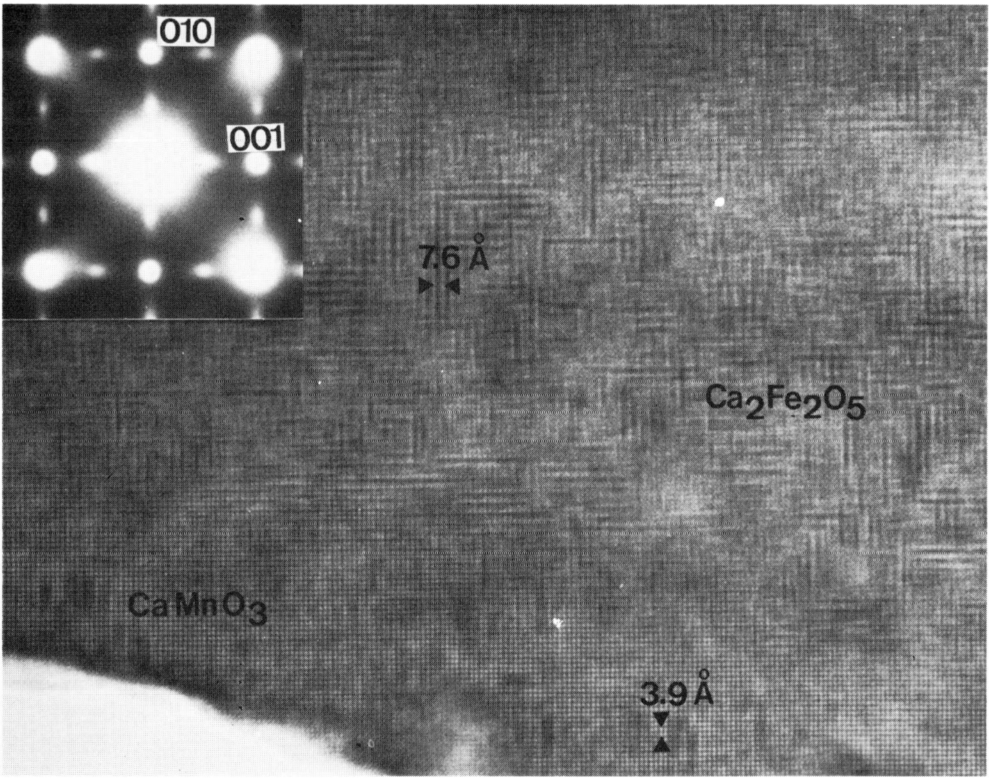

Figure 2. Electron micrograph of the $Ca_3Mn_{1.8}Fe_{1.2}O_{3+y}$ sample corresponding to the $[001]_c$ zone axis. $Ca_2Fe_2O_5$-like domains and $CaMnO_3$-like domains are intergrown.

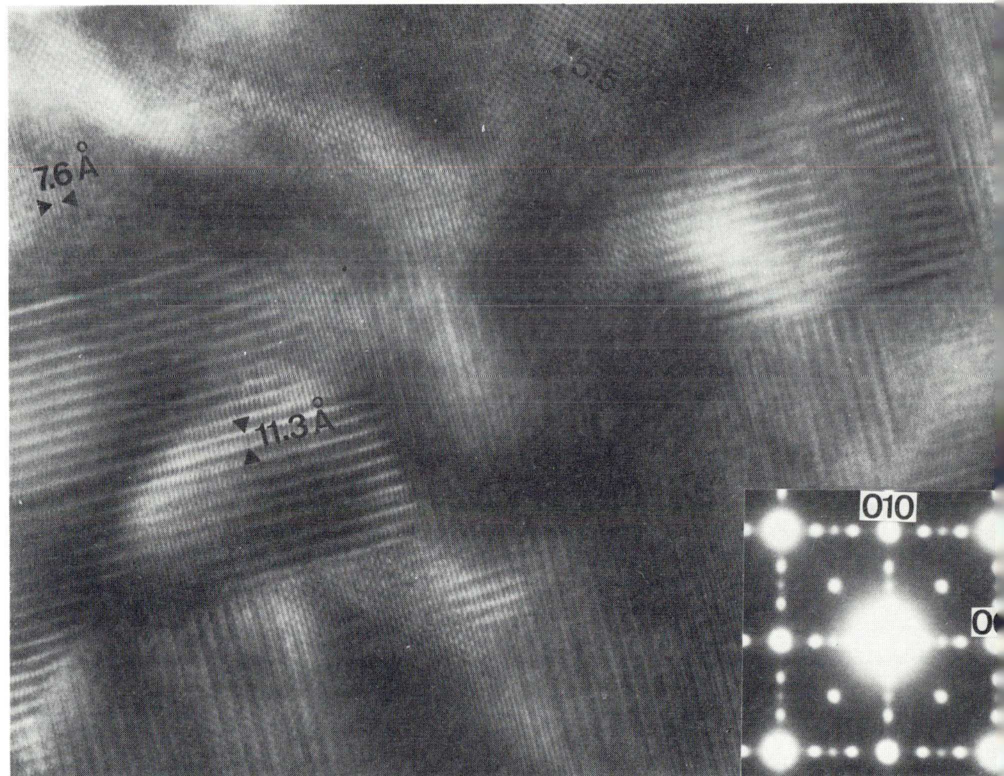


Figure 3. Electron micrograph of the oxidised calcium manganese-ferrite along the $[001]_c$ zone axis. Nine microdomains are present. (Three different spacings are shown in the figure.)

## References

Alario-Franco M A, Gonzalez-Calbet J M, Vallet-Regi M and Grenier J C, 1983a, J. Solid State Chem. 49 219

Alario-Franco M A, Henche M R J, Vallet M, Calbet J M G, Grenier J C, Wattiaux A and Hagenmuller P, 1983b, J. Solid State Chem. 46 23

Bertaut E F, Blum P and Sagnieres A, 1959, Acta Crystallogr. 12 149

Grenier J C, Darriet J, Pouchard M and Hagenmuller P, 1976, Mater. Res. Bull. 11 1219

Megaw H D, 1946, Proc. R. Soc. 58(2) 133

Nguyen N, Calage Y, Varret F, Ferey G, Caignaert V, Hervieu M and Raveau B, 1984, J. Solid State Chem. 53 398

Pepopelmeier K R, Leonowicz M E and Longo J M, 1982, J. Solid State Chem. 44 89

Reller A, Thomas J M, Jefferson D A and Uppal M K, 1984, Proc. R. Soc. A 394

# Decomposition of iron nitrides on a surface treated steel

F Mohammed, B Billon* and A Hendry

Wolfson Laboratory, Department of Metallurgy and Engineering Materials, University of Newcastle upon Tyne, Newcastle upon Tyne, NE1 7RU

* Now at Ecole Nationale Superieure d'Arts et Mètiers Lille, France

## 1. Introduction

Surface hardening of steels by nitriding is a well established process for improving the wear resistance of engineering components. Ferritic treatments are normally employed but in recent years an increasing demand has arisen for surface hardening by nitriding of austenitic stainless steel. Previous work in the Wolfson Research Group has established the nature of the surface layers on nitrided stainless steel (Billon and Hendry (1984)) and the inter-relation of microstructure with mechanical properties and the rate of nitriding (Billon and Hendry (1985)). However, industrial practice frequently places the nitrided component in a high-temperature environment where the nitrogen potential of the gas phase is much lower than that under which the layers were formed on the steel surface. The way in which these layers change will then dictate the subsequent mechanical behaviour of the surface.

## 2. Results and Discussion

The microstructure of AISI 316 stainless steel (17.8w/o Cr, 12w/o Ni, 2.4w/o Mo, 0.02w/o C) nitrided in ammonia at temperatures above 700°C has been described by Billon and Hendry (1985) and consists of a surface layer of $\epsilon$-$Fe_2N_{1-x}$ and an inner layer of $\gamma$ + CrN. Regions of ferrite ($\alpha$) with precipitation of $Cr_2N$ are found in the inner layer at 800°C and at 700°C some $\gamma'$ - $Fe_4N$ may be found near the surface. $\gamma'$ is not stable above 690°C in pure iron but the presence of nickel in the stainless steel stabilises $\gamma'$ by forming a solid solution $(Fe, Ni)_4N$.

Samples were nitrided at temperatures above 700°C and aged for times up to 24h at either 500°C or 700°C in inert atmospheres. The surface layers formed during nitriding in ammonia are in equilibrium with an effective nitrogen potential of several thousand atmospheres and therefore when aged in inert gas at atmospheric pressure there is a strong driving force for both decomposition of the iron nitrides at the surface and for de-nitrogenation of the austenite matrix. Chromium nitrides are much more stable than those of iron and are therefore unlikely to redissolve in the early stage of ageing.

Ageing at 500°C. Figure 1a shows the structure of a sample nitrided, cooled to 500°C in argon and then immediately quenched. The thin surface layer of ε is no longer present and the formation of γ' is confirmed by X-ray diffraction. The inner layer is unchanged γ + CrN. After 24h at 500°C however (Figure 1b) a pronounced layer of γ' - $Fe_4N$ has formed as a surface white layer which shows that although decomposition of the highest nitrogen content nitride, ε, is rapid the de-nitrogenisation process is sufficiently slow to allow γ' to form. Once again the inner layer structure is unchanged and is shown in the TEM micrographs of Figure 2. If the sample after nitriding is allowed to cool to 500°C in the nitriding gas before ageing for 24h in inert gas, then the white layer of γ' is again stable and appreciably thicker (Figure 1c) than in samples cooled in inert gas (Figure 1b). Cooling in high nitrogen potential in this way also results in retention of some ε on the surface.

From these experiments it can be concluded that decomposition of ε occurs rapidly on removal of the nitriding gas potential, but only at high temperature. If ε persists on cooling to 500°C (in the nitriding gas) then subsequent decomposition is slow; indeed all changes proceed slowly at 500°C as the γ' formed by cooling appears to be very slow to decompose, denitrogenation is slow and no appreciable ageing of the inner layer occurs.

Ageing at 700°C. Samples were nitrided in ammonia at 700°C and then quenched and aged in inert gas at 700°C. The as-nitrided structure of an ε outer layer with an inner layer of γ + CrN ages rapidly at this temperature. After times less than 1 hour the iron nitride layer has completely decomposed leaving a structure of γ + CrN throughout the nitrided surface. Denitrogenation also proceeds to some degree as changes in the X-ray unit-cell dimensions of austenite are detected on ageing. However, as at 500°C this process appears to be slow and the nitrogen concentration profile across the hardened layer remains constant. The predominant microstructural change on ageing at 700°C is the appearance of grain boundary precipitation along the prior austenite grain boundaries of the un-nitrided structure. This is shown clearly in the BEI images of Figure 3. This precipitation is believed to be γ' - $Fe_4N$ although positive confirmation by TEM has not been obtained due to difficulties of specimen preparation.

The appearance of grain boundary γ' may be ascribed to the slow rate of denitrogenation described above which results in a levelling off of the nitrogen concentration gradient within the nitrided layer. At any stage of the ageing process, nitrogen can escape from the surface by combining to form gas molecules

$$\underline{N} + \underline{N} \rightleftharpoons N_2$$

and can also diffuse inward, down the concentration gradient, toward the centre of the sample. Nitrogen recombination to form molecules is known to be a slow step in the process and therefore nitrogen diffusion is not rate controlling. On ageing at 700°C therefore the nitrogen concentration in the surface layer becomes more nearly uniform and γ' can precipitate at greater depth in the layer. The extent of nitrogen penetration is clearly seen in the BEI image of Figure 3.

Ageing in inert gas at 500°C was found to have no effect on the mechanical properties of the nitrided layer. However, the grain boundary precipitation which occurs at 700°C results in severe embrittlement of the layer. As can be seen from the topographical BEI image in Figure 3, the grain boundaries present an extensive crack network which will operate whenever a bending stress is applied to the surface.

## 3. References

Billon B and Hendry A 1984, EMAG 83, Inst. Phys. Conf. Series No 68, p 293.

Billon B and Hendry A 1985, Surface Engineering, 1, 114 and 125.

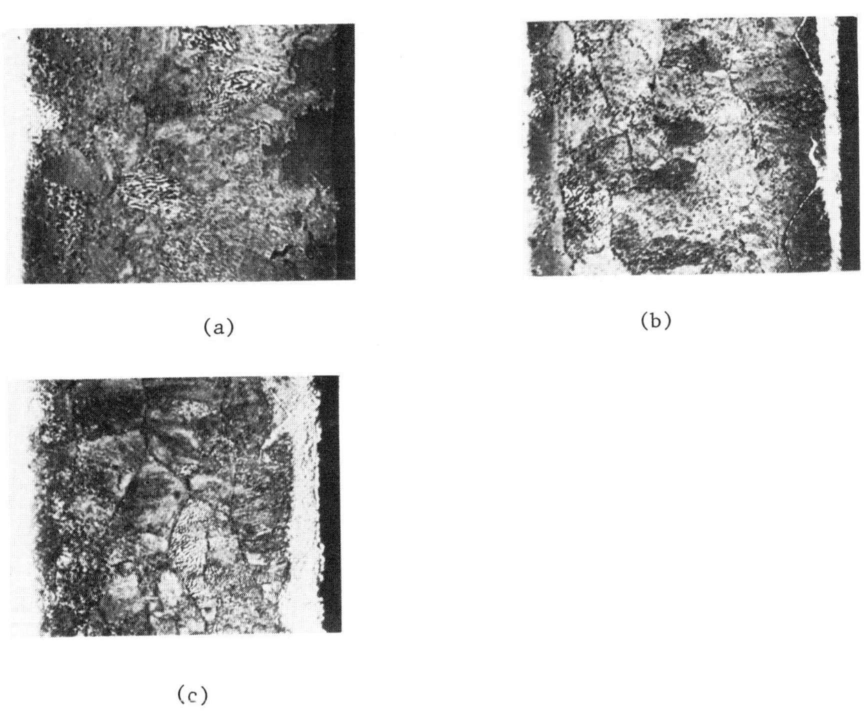

(a) (b) (c)

Figure 1. Optical micrographs of aged AISI 316.
(a) cooled to 500°C in Ar and quenched
(b) cooled to 500°C in Ar and aged 24h
(c) cooled to 500°C in $NH_3$ and aged 24h in Ar.

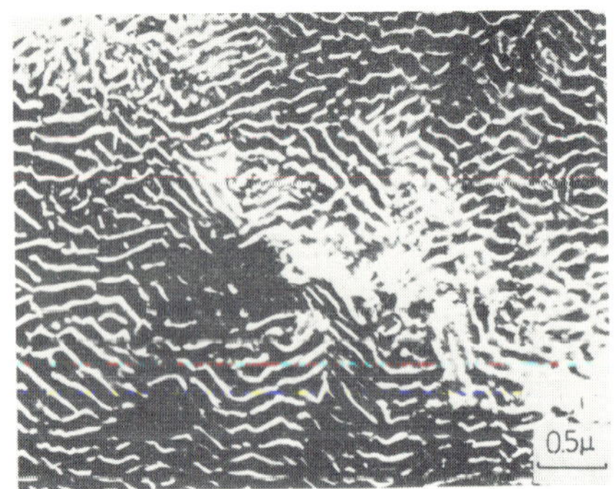


Figure 2. TEM bright field of the inner layer of an as-nitrided sample. $\gamma$ + CrN.

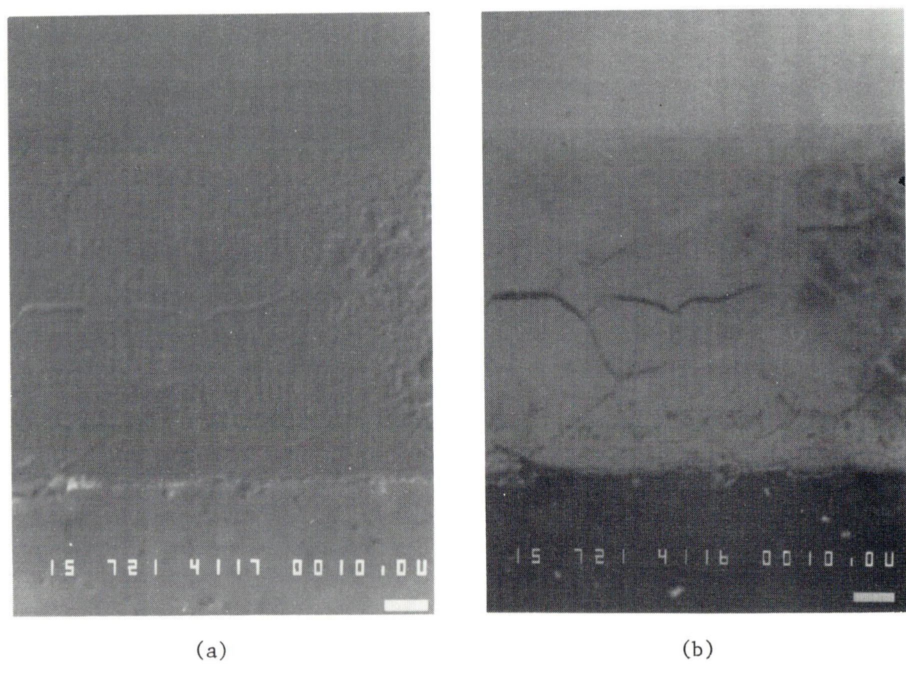


Figure 3. SEM back-scattered electron images of a sample aged at 700°C.
(a) topographical image and
(b) compositional image.

*Inst. Phys. Conf. Ser. No 78: Chapter 12*
*Paper presented at EMAG '85, Newcastle upon Tyne, 2–5 September 1985*



# Metal and metal carbide microcrystals obtained by catalytic co-disproportionation: A CTEM study

M AUDIER Laboratoire de Thermodynamique et Physico-Chimie Métallurgiques
ENSEEG - Domaine Universitaire, BP.75 - 38402 SAINT MARTIN D'HERES - FRANCE

## I- INTRODUCTION

A few years ago, Baker et al (1978) demonstrated that catalytic carbon growth occurs by bulk diffusion of carbon atoms through catalytic particles, in exothermic reactions such as the catalytic decomposition of hydrocarbons and the disproportionation of carbon monoxide. On the basis of this mechanism, further studies could be attempted in order to clarify, for example, the generalization of such a mechanism for the different morphological and crystallographic characteristics of the catalytic particles.

For several years, my colleagues and I have tried to carry out a detailed study of catalytic carbon deposition upon different aspects (i.e. thermodynamic, kinetic and microscopic) and especially from the CO-disproportionation reaction ($2CO \rightarrow C+CO_2$) which presents the advantage of being solely a catalytic reaction over a large range of temperature, eg. the thermal decomposition of CO never occurs, which is not the case with hydrocarbons. The aim of this paper is to show from some examples, how in connection with thermodynamic and kinetic results, CTEM studies of the CO-disproportionation solid products-catalyst-carbon— can be very helpful in furthering the understanding of the characteristics and mechanisms of catalytic carbon deposition (i.e. mechanism of carbon growth, of fragmentation of the metal catalyst, of metal$\rightarrow$carbide and carbide$\rightarrow$metal transformations...)

## II- CATALYTIC CARBON GROWTH

a) <u>morphological and crystallographic characteristics of some catalytic particles</u>

Thermodynamic and kinetic studies have shown that the CO-disproportionation occurs differently depending on whether the experimental conditions of temperature and partial pressures Pco, $Pco_2$ are carbiding or not carbiding for the metal catalyst (Audier, Coulon, Bonnetain (1983)) :

- Under non-carbiding conditions, filamentous catalytic carbon in the form of simple tubes with conical particles at one end or in the middle of carbon bitubes have been selectively prepared by CO-disproportionation catalysed by Fe, Co, Ni, FeCo alloys and FeNi alloys of various compositions and structures (bcc,fcc and hcp). Such particles present several very well defined morphological and crystallographic characteristics i.e. their metal-gas interfaces are (100) for the bcc structure, (111) for fcc and $(10\bar{1}2)$ for hcp (fig.1). According to the structure of the metal catalyst, biconical particles in the middle of carbon bitubes keep the same crystallographic orientations with respect to the filament axis as simple cones : [110] for the fcc structure and [100] for bcc. However the most interesting characteristic of these bicones is the existence of a subgrain boundary at the level of their basal plane. These subgrain boundaries are (110) in the case of the fcc structure (fig.2) and probably (100) for the bcc structure (Audier, Oberlin, Coulon (1982)).

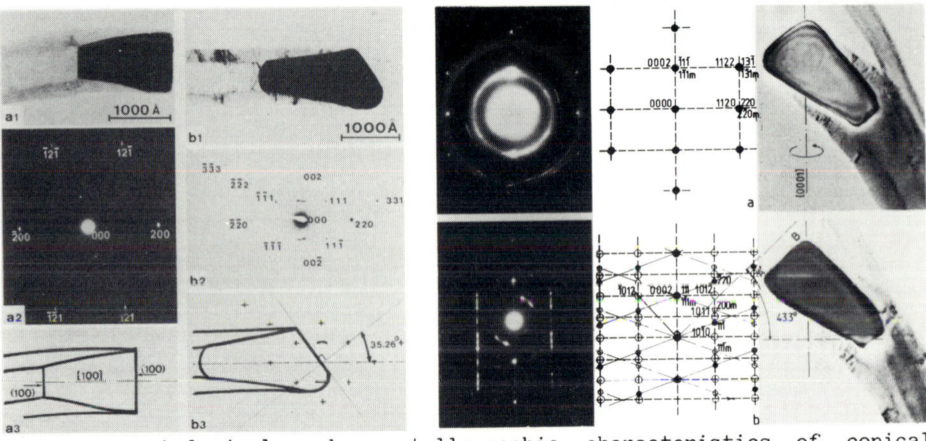

Fig.1: morphological and crystallographic characteristics of conical particles of bcc, fcc and mixed (hcp-fcc) structure.

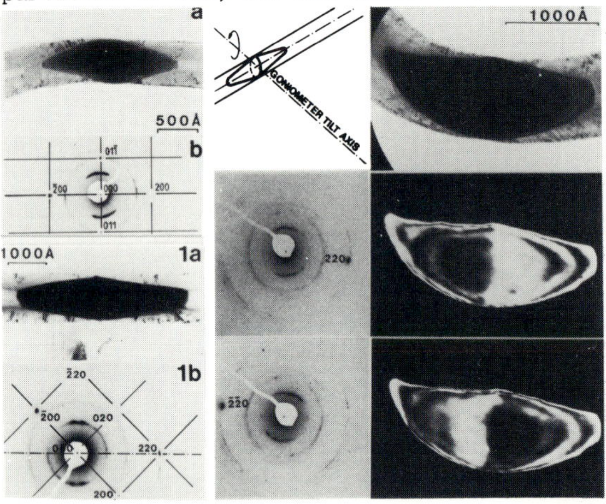

Fig.2: Characteristics of biconical particles of bcc structure (a,b-the [100] axis coincides with the carbon tube axis) and fcc structure (1a,1b -the [110] axis coincides with the carbon tube axis). From bright and dark field images with g=220,$\bar{2}\bar{2}0$ of a fcc bicone,it has been established that bicones are formed of a combination of two single cones with a (110) subgrain boundary as interface. (see Audier,Oberlin and Coulon (1982)).

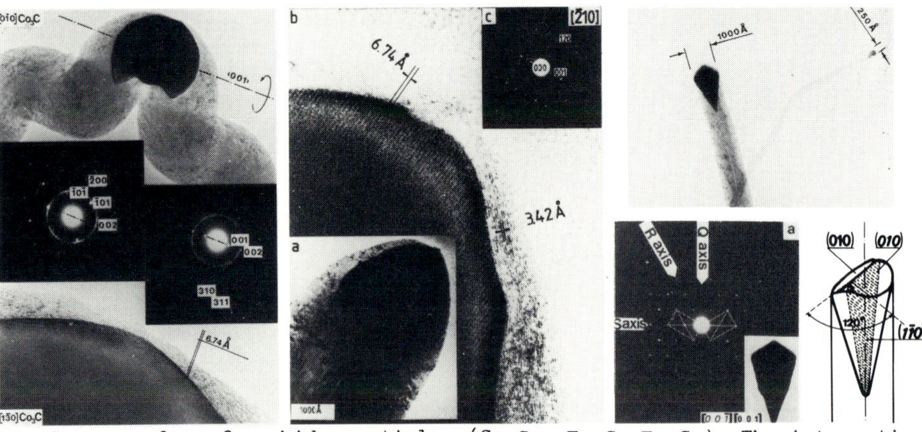

Fig.3 : examples of carbide particles ($Co_3C$, $Fe_3C$, $Fe_7C_3$). The interesting characteristic of these particles is that the accessibility of gas to the metal carbide is possible either on a line (e.g. the edge of a (001) $Co_3C$ or $Fe_3C$ plane) or on a surface (i.e. twinned $Fe_7C_3$ particle).

- Under carbiding conditions, CTEM studies have revealed that microcrystals of several different carbides are formed during the disproportionation of CO at 300°C ⩽ T ⩽ 500°C on iron and cobalt (alloy carbides have not yet been studied). These metal carbides are : $\theta Fe_3C$ and $Co_3C$, $\chi$ $Fe_5C_2$ and $Co_5C_2$, $Fe_7C_3$, some particles present a twinned $M_5C_2$ structure and others, an intergrowth of both structures $M_3C$-$M_5C_2$ (Audier, Bowen, Jones (1983)). There are a large variety of morphologies : some examples are shown on fig.3 ; one can see that these particles present either a line or a surface as carbide-gas interface.

b) mechanism of the catalytic carbon growth

In a previous study of the disproportionation of CO on iron-cobalt alloys, we have established the kinetic law of the growth of carbon with the idea that the bulk diffusion of carbon atoms through the catalytic metal is controlled by the chemical potentials of the gas carbon phase $CO$-$CO_2$ and of the carbon (1983). It was one of the two possible interpretations of the mechanism of Baker et al (1972,1975...) where one can consider that the diffusion is driven either by a temperature gradient, resulting from the exothermicity of the reaction, or just by an isothermal concentration gradient (Rostrup-Nielsen, Trimm(1977)). Recently, Coulon and Audier (1985) have measured carbon deposition rates, using $CO$-$CO_2$ and $CH_4$-$H_2$ gas mixtures and a FeNi alloy catalyst which was fragmented (see later about the catalyst fragmentation) in a previous experiment: between 500 and 650°C and with gas mixture compositions not too far from thermodynamic equilibrium the rate of carbon deposition depends on the chemical potential of the carbon in the gas mixtures, irrespective of their nature ($CO$-$CO_2$ or $CH_4$-$H_2$). As CO-disproportionation and $CH_4$ decomposition are respectively highly exothermic and highly endothermic, the possibility, previously claimed (Baker et al), that diffusion might be driven by a temperature gradient is ruled out under these conditions. In fact, this supports strongly the idea that the rate limiting step is the bulk diffusion of carbon driven by an isothermal gradient with local equilibria at both metal-carbon and metal-gas interfaces.

Consequently, in the case of carbon tubes with a conical particle at one end (fig.1 and 4) one can propose that the dissociative chemisorption of the carbon monoxide (or methane) occurs on the free metal surface leading to bulk diffusion of carbon atoms through the metal. Thus precipitation of the carbon atoms at the metal-carbon interface causes the formation of aromatic layers parallel to the side surface of the conical particle as can

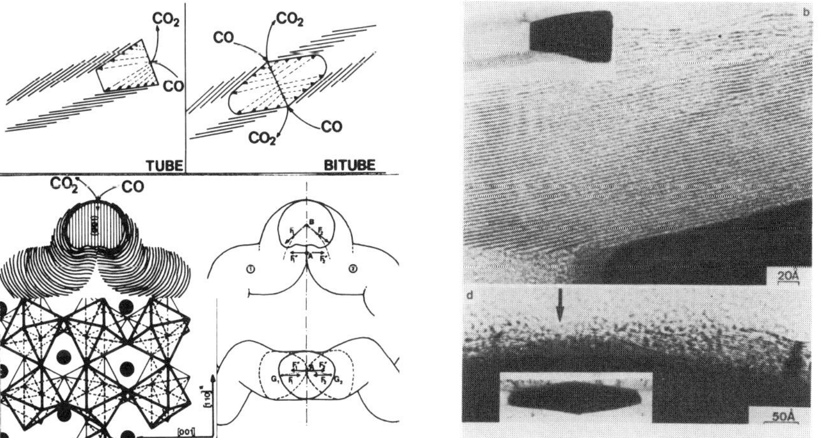


Fig.4 : Mechanisms of carbon growth for tube, bitube and carbide particle with two spiral carbon filaments (inset shows structural representation of the cementite as proposed by Anderson and Hyde (1974).

be seen by high resolution imaging (fig.4). In the case of a biconical particle which yields a bitube, the dissociative chemisorption of the gas can occur at the line of emergence of a subgrain boundary of the metallic bicone. As the diffusion coefficient along a subgrain boundary is about $10^4$ times higher than the bulk one, then carbon atoms can diffuse at first very quickly into the subgrain boundary and afterwards symmetrically through both parts of the bicone. As before, carbon aromatic layers are formed and stacked parallel to the side surface of the bicone (fig.4). The bitube formation is an additional argument supporting the theory of carbon diffusion driven by a concentration gradient rather than by a temperature gradient since the heat is evolved not at the subgrain boundary but at the emerging line. In the case of carbide particles and, for example, a carbide particle with two spiral carbon filaments, the dissociative chemisorption of the gas can occur at the edge of a (001) plane of the carbide ($\theta Fe_3C$ or $Co_3C$). Carbon atoms diffuse along the (001) plane and through both parts of the particle. The structural representation of the cementite proposed by Anderson and Hyde (1974) allows one to understand that the interstitial diffusion of carbon atoms is practically equivalent both parallel to and perpendicular to the (001) planes (carbon atoms jump through the windows formed by four octahedra)(fig.4). Again the precipitation of the carbon atoms at the carbide-carbon interface causes the formation of aromatic layers parallel to the surface of the particle. Because of the contact between both spiral carbon filaments (point A in fig.4) spiral formation could be explained by the existence of a force couple.

## III- PHASE TRANSFORMATION OF CATALYTIC PARTICLES

a) metal $\rightarrow$ carbide transformation

When the CO-disproportionation is started under non-carbiding conditions and finished under carbiding conditions, the carbon deposition rate, initially high, is observed to decrease with the change of experimental conditions until slowly reaching a constant value. This kind of experiment has been carried out with Fe and Co and CTEM studies of the resulting solid product show that metal particles are transformed into metal carbide particles. We have explained elsewhere how hcp cobalt particles are transformed into a tetragonal cobalt carbide phase via an intermediate orthorhombic $Co_2C$ phase (Audier and Guyot (1985)). In this paper, one can examine another case of phase transformation ; i.e. how a bcc Fe particle (as shown in fig.1) is transformed into a $\theta Fe_3C$ particle (fig.5) which presents a (110) $Fe_3C$-gas interface and some morphological characteristics different from those of a bcc particle : - two assumptions concerning the transformation will be made and shown to be in agreement with the observations. The first assumption, that the carbide particle is obtained by transformation of a bcc particle, requires that the (100) bcc-gas interface transforms into a (110) $Fe_3C$-gas interface. The second assumption is that the main step in the mechanism of the transformation $Fe \rightarrow Fe_3C$ can be considered as a periodical twinning $(12\bar{1})$ $[1\bar{1}\bar{1}]$ of the bcc phase as proposed by Andrews (1963) combining these assumptions, we see that the (100) metal-gas interface is transformed into a plane with $[5\bar{2}1]$ as a normal (fig.5). Then, the crystallographic orientation relationships between $\alpha Fe$ and $\theta Fe_3C$ ($(001)\theta Fe_3C // (12\bar{1})\alpha$ ; $[100]\theta Fe_3C // [101]\alpha$ ; $[010]\theta Fe_3C // [1\bar{1}\bar{1}]\alpha$) allows one to determine the correspondance between this $[5\bar{2}1]$ bcc direction and the $(110)Fe_3C$ plane. The drawing of fig.6 shows that the morphological transformation of a bcc particle into a $Fe_3C$ particle is in good agreement with the crystallographic characteristics of both bcc and carbide particles justifying the two assumptions made earlier.

b) Carbide $\rightarrow$ metal transformation

The thermal decomposition of carbide particles has been carried out directly within the microscope using a heating goniometer stage (but such a decomposition can occur during catalytic CO-disproportionation under

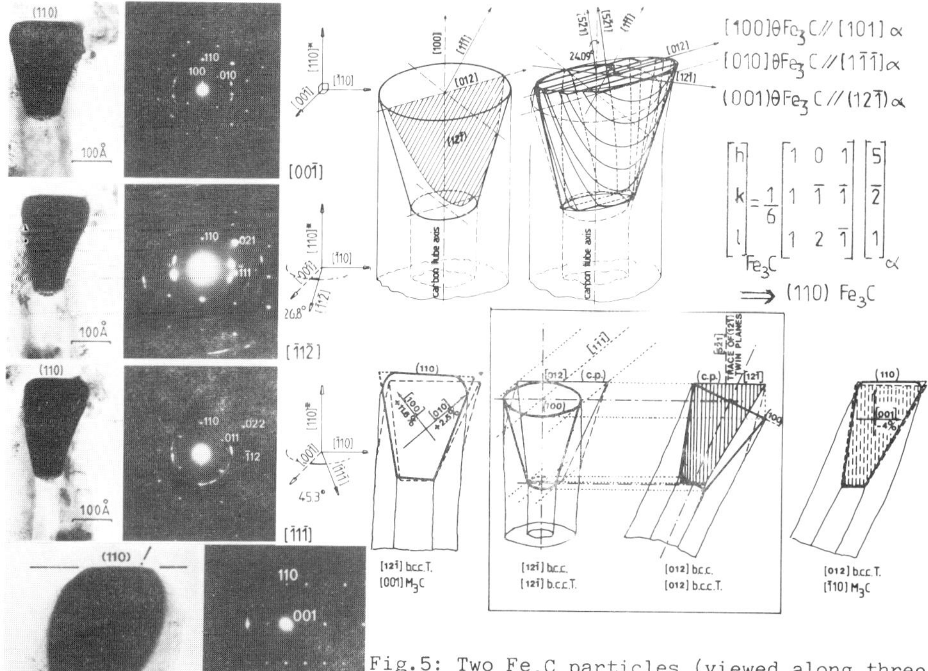


Fig.5: Two $Fe_3C$ particles (viewed along three zone axes for the first one and along the [$\bar{1}$10] zone axis for the second one) resulting of the carburization of a bcc Fe particle as shown in Fig. 1. (Above) schematic drawing of the periodical multiple twinning of a bcc particle - the resulting [5$\bar{2}\bar{1}$] direction corresponds to a (110) $\theta Fe_3C$ plane from the epitaxial relationships $\alpha$Fe-$\theta Fe_3C$ (below) morphological and crystallographic changes of the Fe→$\theta Fe_3C$ transformation of a conical particle.

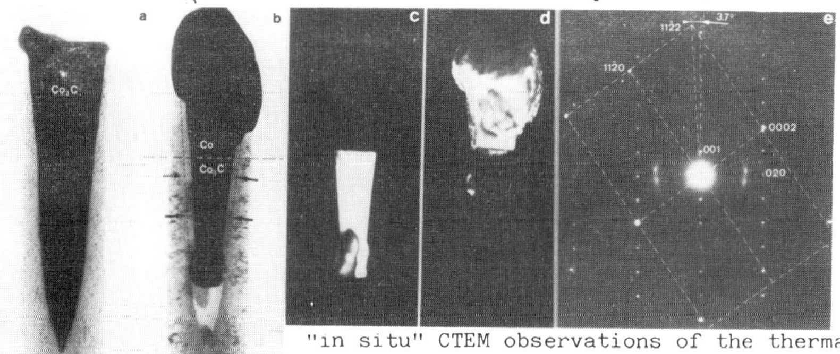


Fig.6: "in situ" CTEM observations of the thermal decomposition of $Co_3C$ particle (a) before heating (b) after heating (c,d) dark field images of both phases $Co_3C$, Cohcp (e) corresponding diffraction pattern.

appropriate experimental conditions). The example presented here is the thermal decomposition of the carbide $Co_3C$ which, from another point of view, was interesting to study because the epitaxial relationship between Co hcp and $Co_3C$ have not been determined before : during the heating of a $Co_3C$ particle oriented along a [100] zone axis, we have observed the nucleation of metallic hcp cobalt on the carbide-gas interface and we have noted a reduction in the diameter of the carbide particle (fig.6). This means that an opposite diffusion of metal and carbon has occured between both carbide-gas and carbide-carbon interfaces. In this case the epitaxial relationship is $[100]\,Co_3C // [1\bar{1}00]$ Cohcp ; $(001)Co_3C$ at 3.7° from $(11\bar{2}2)$Cohcp.

## IV- POSSIBLE MECHANISM OF CATALYST FRAGMENTATION

In our experiments, the catalytic disproportionation of CO leads, at first, to a fragmentation of the metal catalyst because the catalyst is used in the form of filings or shavings. This phenomenon of fragmentation probably involves a bulk diffusion of carbon through the metal (Audier, Coulon, Bonnetain (1983)). Consequently parallel phenomena studied in metallurgy could be helpful in understanding this fragmentation. For instance, we have remarked that some shapes of fragmented particle (conical) look similar to the dendritic precipitates which occur in some iron-carbon alloys. It is therefore tempting to think that the fragmentation of a conical particle from the bulk metal is due, for example, to the dendritic recrystallisation of a supersaturated metal-carbon solid solution. The fragmentation would occur when a first aromatic layer of carbon precipitates at the level of the dendrite-matrix interface (fig. 7).

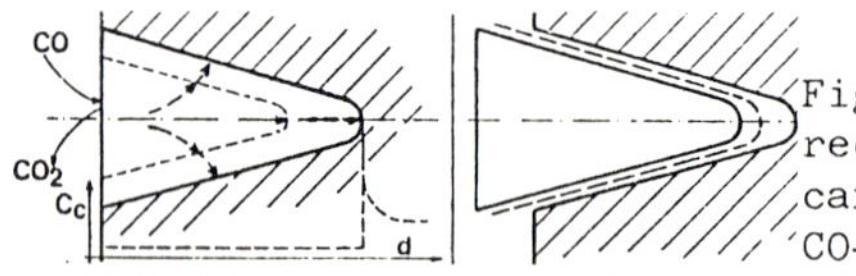


Fig. 7 : Schematic drawing of the dendritic recrystallisation of a supersaturated metal carbon solide solution in contact with a $CO-CO_2$ gas phase. The carbon concentration profile ($C_c$ vs d) along the dendrite axis presents a local supersaturation at the dendrite-matrix interface, which can lead to the precipitation of a first carbon layer and therefore to the fragmentation of the dendrite .

## V- CONCLUDING REMARK

It has to be noted that these CTEM characterisations and resulting interpretations for the mechanisms of carbon growth and catalyst fragmentation are in good agreement with thermodynamic predictions and kinetic results. Such studies have allowed us to propose kinetic laws of the carbon growth and of the catalyst fragmentation in the case of non-carbiding conditions ($T, Pco, Pco_2$). However, further studies are needed in order to explain such kinetic laws on the basis of physical theories (for instance, we believe that the kinetic law of carbon growth must follow the first Fick's law of diffusion but it is difficult to demonstrate). Concerning the phenomenon of catalyst fragmentation, we must still find good arguments in support of the possible mechanism we have proposed.

## REFERENCES

M.AUDIER,A.OBERLIN and M. COULON J. of Cryst.Growth 55(1981)549;57(1982)524.
M.AUDIER, P. BOWEN and W. JONES J. of Cryst.Growth 63(1983)125;64(1983)291.
R.T.K. BAKER and P.S. HARRIS Chem. and Phys. of Carbon, vol.14, p.83 Ed. M.Dekker, New-york (1978).
M.AUDIER, M.COULON and L.BONNETAIN Carbon 21,2,93; 21,2,99; 21,2,105 (1983).
R.T.K.BAKER et al, J.Catal 26,51 (1972); 30,86 (1975); 37,101 (1975) and Carbon 13,17 (1975); 13,245 (1975); 13,211 (1975).
J.ROSTRUP-NIELSEN and D.L.TRIMM J.Catal, 48,155 (1977)
M.AUDIER and M.COULON Carbon 23,3,317 (1985).
S.ANDERSON and B.G.HYDE J.Solid State Chem 9,92 (1974).
M.AUDIER and P.GUYOT J.Microsc.Spectrosc.Electron. 8 (1983) 261; 10 (1985) 8; 10 (1985) 17.
K.W.ANDREWS Acta Met. 11 (1963) 939.

*Inst. Phys. Conf. Ser. No 78: Chapter 12*
*Paper presented at EMAG '85, Newcastle upon Tyne, 2–5 September 1985* 

# Structural rearrangements in small metal particles at atomic resolution

J-O Bovin*, L R Wallenberg* and David J Smith+

*Inorganic Chemistry 2, Chemical Center, University of Lund, P.O. Box 124, S-221 00, Lund, Sweden
+Centre for Solid State Science, Arizona State Univ., Tempe, AZ 85287 USA

## 1. Introduction

Real-time video-tape recording, combined with the improved resolution of modern high-resolution electron microscopes (HREMs), has proven very helpful in studies of dynamic events on the atomic scale in crystalline materials, such as defect motion in gold foils (Hashimoto et al, 1980) and defect annealing in cadmium telluride (Sinclair et al, 1981). The profile imaging method (Marks and Smith, 1983) can provide atomic-level information about surfaces, and crystal growth, row by row, of atoms, on a {111} gold surface has been recorded in real time (Wallenberg et al, 1985). The present paper describes the rapid structural rearrangements, the surface twinning, and the presence of atomic clouds outside specific surfaces, which have been observed in small (1-30nm) gold particles at spatial resolutions of better than 2Å in real-time.

## 2. Experimental details

The gold crystals used in these observations originated from 55-gold-atom clusters (Wallenberg et al, 1985). The accompanying ligands were evaporated by the influence of the electron beam leaving the clusters initially distributed randomly on the holey carbon support film. A JEM-4000EX HREM in the top-entry configuration ($C_s$=1 mm), and equipped with a Gatan 622 TV system fibre-optically coupled to a YAG screen, was used for imaging. The accelerating voltage was either 350 or 400kV, with an electron-optical magnification typically of 600,000x and current densities at the sample of 20-25 A/cm$^2$. Recursive filtering using a Quantex digital image processing system-typically over 4 frames-was sometimes used to reduce noise and improve image contrast.

## 3. Results and Discussion

The 55-atom clusters, originally arranged in a cubeoctahedron, were unstable once the ligands were removed, and microcrystals immediately started growing and coalescing with other particles. (Images were usually obtained from crystals protruding over holes in the carbon film to avoid interference from the substrate.) Structural rearrangements of these small particles then occurred very rapidly. Many were in almost constant motion, pulsating and changing their shapes and orientations relative to the incident beam direction. In extreme cases, crystals transformed from single crystal to bi-crystal to pentagonal multiply-twinned to icosahedral multiply-twinned all within the period of a few seconds.

Interesting examples can be seen in Figs. 1 and 2. Fig. 1a shows a multiply-twinned decahedral particle viewed along the <110> zone axis, with the twin planes accentuated by the notches at the edges of the tetrahedral segments. Within 0.3s, the twin planes disappeared and the crystal continued the process of changing into other shapes. The single crystal shown in Fig. 1b appeared about 10s later. A larger particle which contains a twin lamellae, 5 lattice spacings wide, is shown in Fig. 2. This crystal is again viewed along the <110> direction and the twin planes are parallel to {111}. This lamellae was stable for several seconds before continuing to other configurations, an example of which is shown in Fig 2b, recorded some 5s later.

The rapidity of the changes in shape and orientation to the particles was size-dependent, generally only being observed in particles ~50Å or less in diameter. The changes were affected by the electron beam current density since spreading of the beam seemed to slow down the motion considerably. Another contributing factor was the opportunity for heat dissipation: better contact with the support clearly reduced the dynamic behaviour. A detailed investigation of the experimental parameters affecting the almost liquid-like behaviour of these particles is in progress.

Twinning on {111} surfaces was commonly observed and very often appeared to be part of the particle growth mechanism. The particle visible in Fig. 3 has an internal twin plane (marked A in 3a) with two vicinal {111} surfaces on each side of the twin plane. In 3b the horizontal surface layer has moved collectively to the right into twin positions, i.e. hcp positions. Using the coordinates for the fcc unit cell, this corresponds to a shift from (1, 1/2,1/2) to (1/3, 1/3, 1/3), a distance of 1.67Å, assuming perfect close-packings. Figs. 3b-d show how a new column, marked B, is formed adjacent to the point where the internal twin plane reaches the surface. The contrast of the column gradually gets darker as more atoms are added. When the column is apparently completed, the surface layer moves back into ccp-positions as shown in 3e. Fig. 3f shows an atomic column, formed at the extension of the internal twin plane, which was only partly occupied during the exposure. This particular sequence of events was recorded within a period of three seconds.

Clouds of rapidly moving gold atoms were often observed outside specific surfaces. These clouds ranged out to ~9Å off the surface and apparent exchange of atoms between the clouds and the surfaces have been seen. The crystal in Fig. 4 is approximately 75Å in diameter and has a localised cloud outside the lower {100} surface. Clouds have been observed outside {100}, {110} and {331} surfaces but never outside the close-packed {111} surfaces. Details of the cloud-surface interactions have been described elsewhere (Bovin et al, 1985).

This work was supported by the Swedish National Science Research Council (E-EG 3914-110), the National Swedish Board for Technical Development (DNR 84-3515), the National Science Foundation (DMR-830871) and used the facilities of the Arizona State University Regional Facility for High Resolution Electron Microscopy, supported by NSF Grant DMR-8306501.

References


Bovin J-O, Wallenberg L R and Smith D J 1985 Nature in press

Hashimoto H, Yokota Y, Takai Y, Endoh H and Kumao A 1978-1979 Chemica Scripta 14 125

Marks L D and Smith D J 1983 Nature 303 316

Sinclair R, Ponce F A, Yamashita T, Smith D J, Camps R A, Freeman L A, Erasmus S F, Smith K C A, Nixon W C and Catto C J D 1982 Nature 298 127

Wallenberg, L R, Bovin J-O and Schmid G 1985 Surf. Sci. 156 256


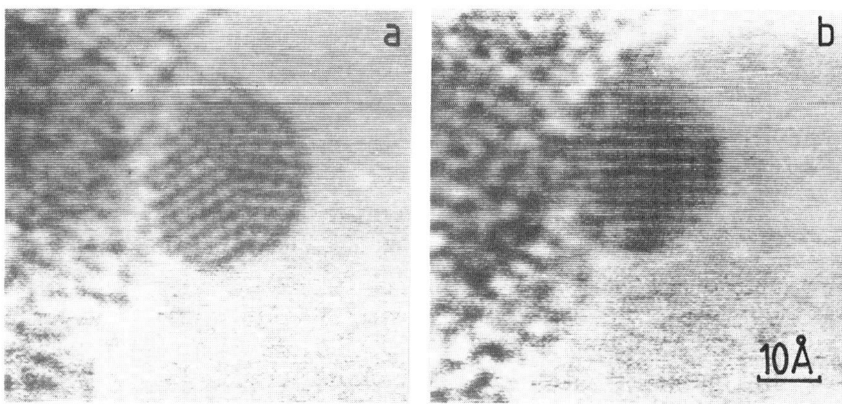


Fig.1 a. Small decahedral particle of gold viewed along <110> ; b. Same particle 10s later, in single crystal form, now viewed along <100>.

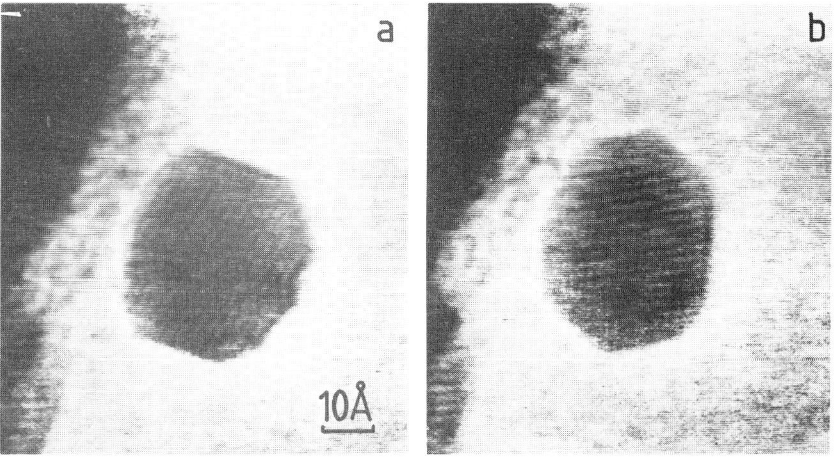


Fig.2 a. Gold particle containing a twin lamellae.
b. Same particle in different shape, 5s later.

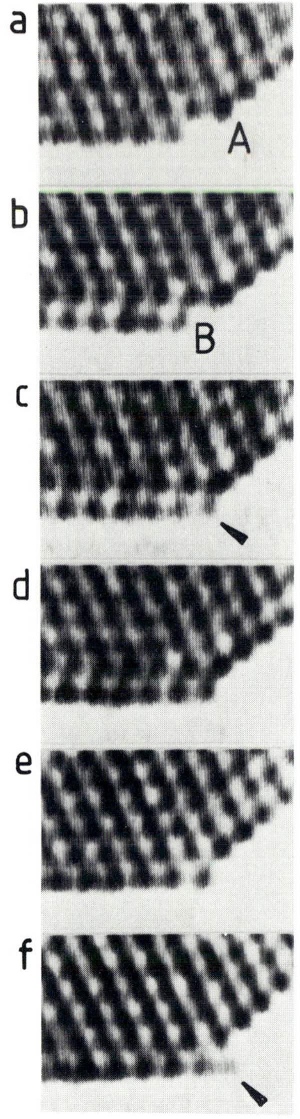


Fig.3 Time series showing surface twinning(see b,c and d) which occurs during particle growth. Note extra partial columns arrowed in c and f. Total elapsed time of 3s.

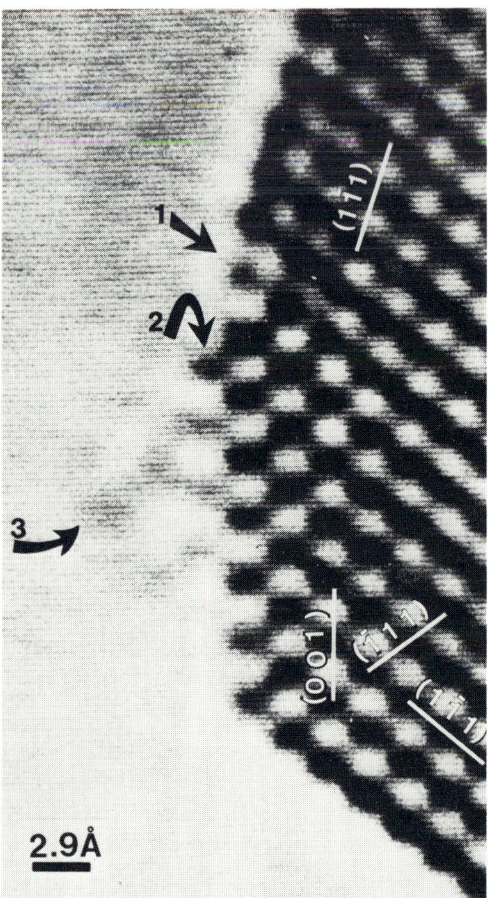


Fig.4 Large Au crystal (75Å) showing cloud of gold atoms outside a (001) surface. Outer limit of cloud is marked 3. Image averaged over 0.13s.

# HREM image contrast from supported metal particles

P L Gai, M J Goringe, J C Barry, W G Waddington and E D Boyes
Department of Metallurgy and Science of Materials, University of Oxford, Parks Road, Oxford OX1 3PH.

High resolution electron microscopy (HREM), image calculations and small probe diffraction/microanalysis have been used to characterise small supported metal particles used in a variety of industrial processes including catalysis, and metal/substrate interfaces. With HREM however, surface and bulk structural detail in supported partcles is not always directly interpretable; being obscured in some cases by the support contrast. Identification of very small particles ( <5nm) thought to be the active species in catalysis is also often difficult. We have therefore carried out image calculations for model silver, platinum and copper particle catalysts (<5nm diam.) supported on amorphous alumina and carbon substrates, to examine the effect of support thickness on the visibility and replication of structure in the images of small particles.

Model catalysts were prepared by evaporating the metals on to the amorphous substrates. The diffraction contrast and HREM images (with CBED/microdiffraction) were recorded using a JEOL 200CX (200keV, Cs=1.2mm, chromatic focus spread FS=5nm and resolution R=0.24nm), JEM 2000FX (with parallel AEM/HREM, 200keV, Cs=2.3mm, R=0.28nm) and JEM 4000EX (400keV, Cs=1mm, FS=3nm). Figs. 1(a) and (b) show experimental diffraction contrast (DC) and HREM images of $Ag/Al_2O_3$ at 200keV. Figs. 2(a) and (b) show experimental HREM images of $Pt/Al_2O_3$ at 400keV. It can be seen from Figs 1b and 2 that edge definition is blurred and that the effect is more pronounced at 200keV. Image calculations were carried out using an Intellect 200 system for model 55 atom cuboctahedral and icosahedral shapes since energy calculations (Mackay 1962) show that clusters of 55 atoms are relatively stable; and for 309 atom particles.

The calculations were done at both 200keV and 400keV; at various defoci, with and without supports. The particles were oriented along their two-fold axis. Slice thicknesses were 0.24nm for Ag/Alumina and 0.23nm for Pt/Alumina and 128x128 arrays were used in the calculations to give a real space sampling of 50 points per nm. The 55-atom particles (1.2nm diam.) were sectioned into 5 slices and 309 atom particles (2.4nm diam.) into 11 slices. The amorphous layers were simulated by placing 68 Al and 102 O atoms in the 2.45nmx2.45nmx0.24nm cell to give a density equivalent to crystalline alumina; and in 2.35nmx2.35nmx0.23nm for Pt/alumina using the appropriate number of Al and O atoms, so that in the calculations the arbitrary substrate cells were a multiple of the lattice parameters of the respective metals . Similarly carbon scattering matrices were generated for Ag/C using 198 carbon atoms. 20 different amorphous slices were used in each case to closely relate to real supports, with substrate thicknesses from 1 to 20nm. Fig.3 shows Ag/Alumina (55 atom cuboctahedral particle) at 200 keV:(a) without support (at optimum defocus df=-66nm giving black atoms), and (b) with the support (of 2 nm), at df=-66nm. Ag/carbon images with a 4.8nm support at 200 keV are shown in Fig.4(a) and (b) at defoci of -66nm and -76 nm respectively. Fig.5 shows

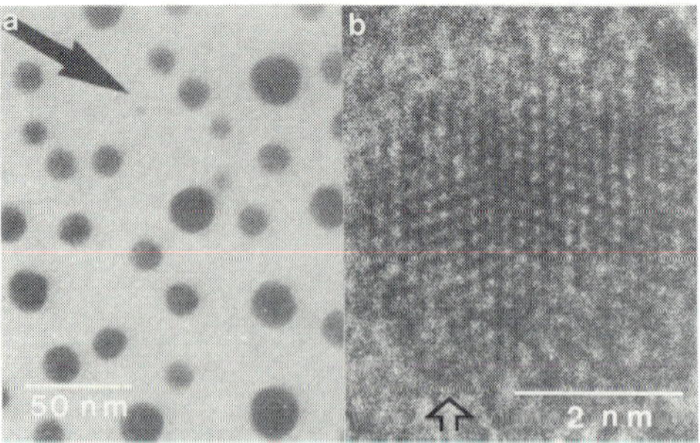


Fig.1 Ag/amorphous alumina 200 kV expt.(a) DC (b)HREM.

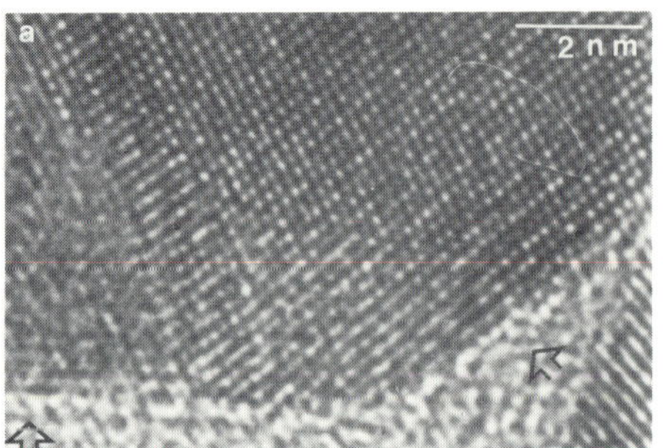


Fig.2 (a,b) Pt/alumina, 400 kV experimental HREM images.

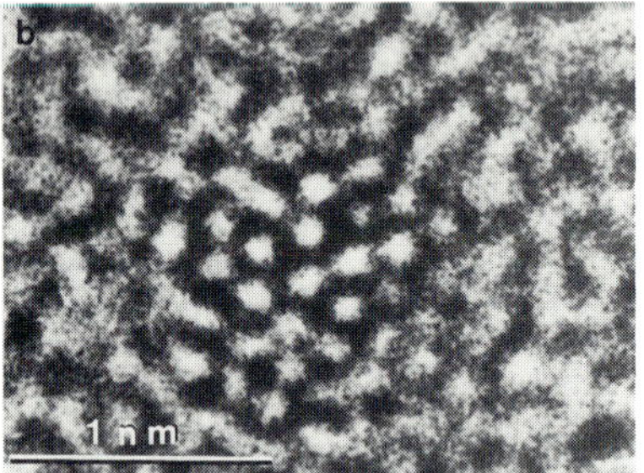


2b Experimental HREM, 400kV

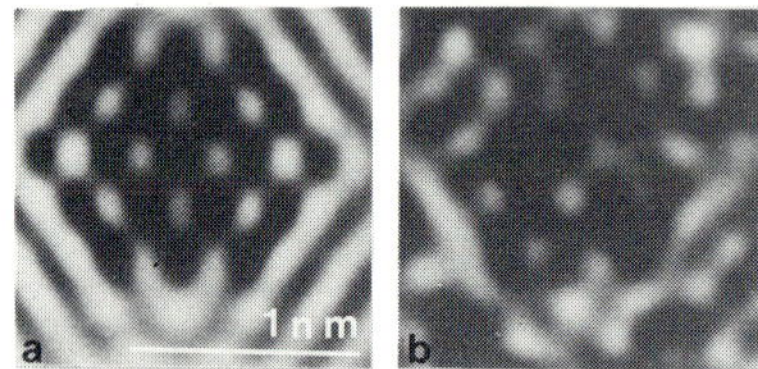


Fig.3 Calculated Ag/Al2O3, 200 kV:a)without(b)with support

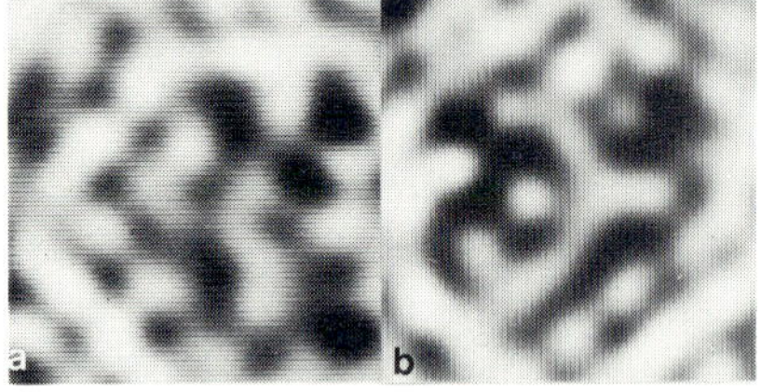


Fig.4 Ag/carbon,200kV,(support 4.8nm):(a) -66nm (b)-76nm

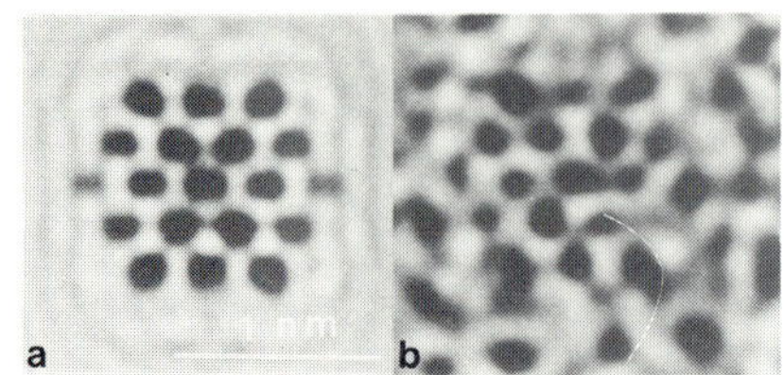


Fig 5 Calculated Ag/alumina 400 kV

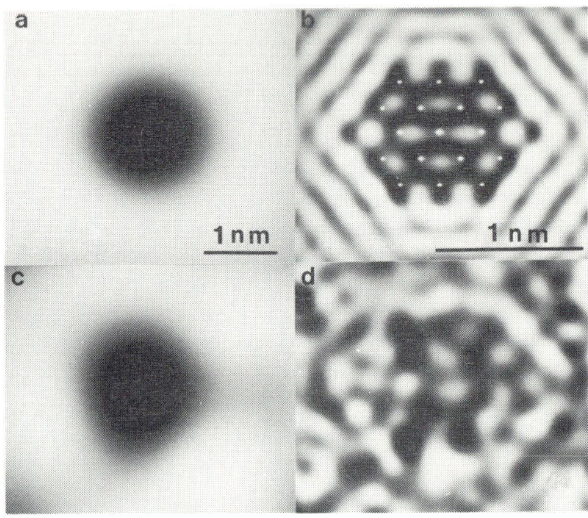


Fig.6 Calculated Pt/alumina 200 kV unsupported:a)DC (b)HREM with atoms; supported (4.6nm thick):c)DC (d,e) HREM at df -66nm, -66nm and -76nm.

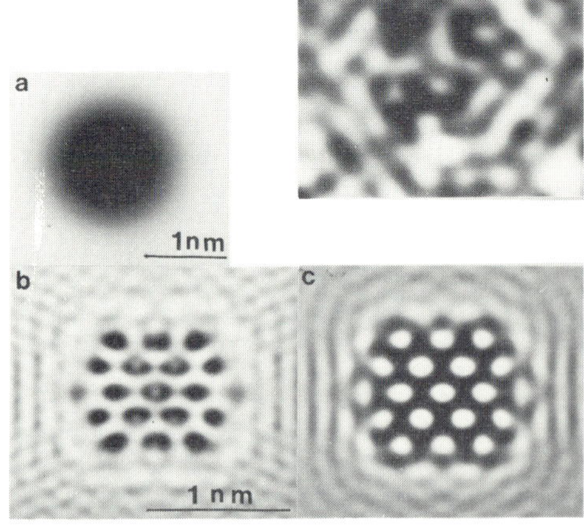


Fig.7 Pt/alumina 400 kV calculations unsupported, (a)dif.cont. (b)HREM (df -48.7nm) (c)white atoms (-63.6nm).

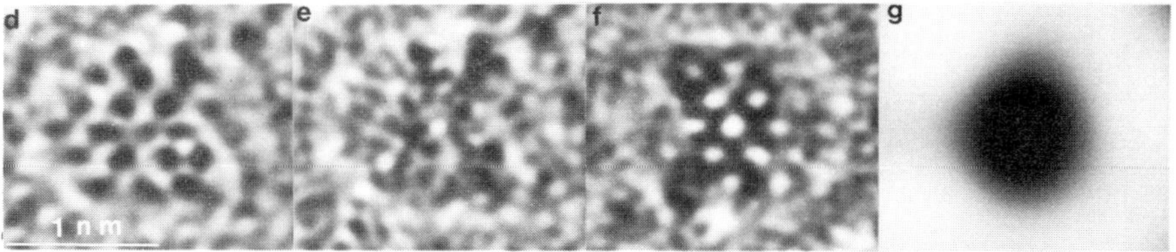


Fig.7 Images with support (4.6nm thick):(d) df -48.7nm (e) -56nm (f) -63.6nm (g) diffraction contrast image, df -48.7nm, 400 kV.

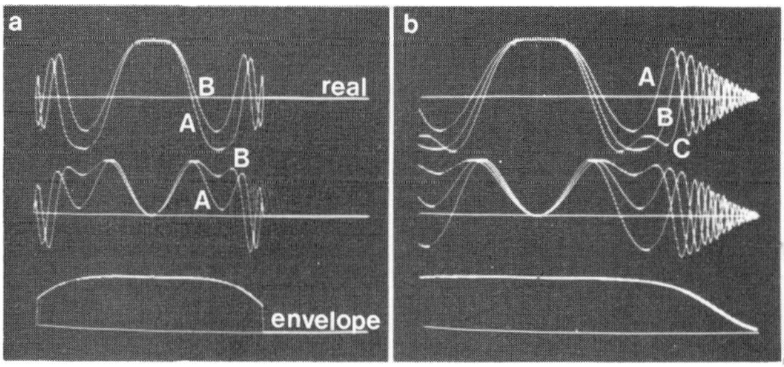

Fig.8 CTF profiles with real &-imag. parts
(a) 200 kV,ap .5A-1;df-76nm(A); -66nm(B).
(b) 400 kV,ap 1A-1;df(nm)-48.7(A);-56(B);-63.6(C).

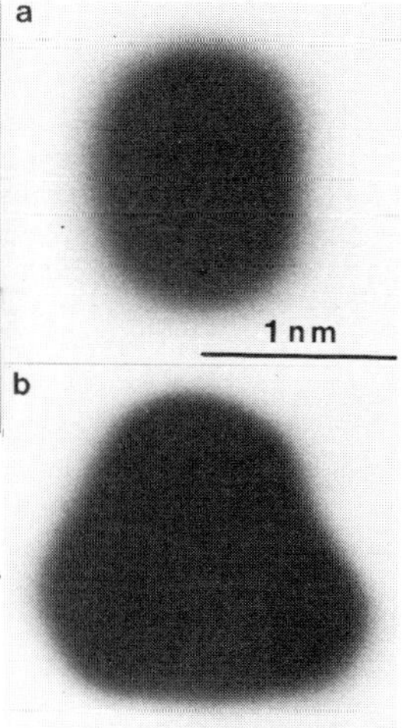

Fig.9 DC images of 309-atom Ag/alumina:(a)without(df -40nm) (b)with 5 nm support,df -40nm.

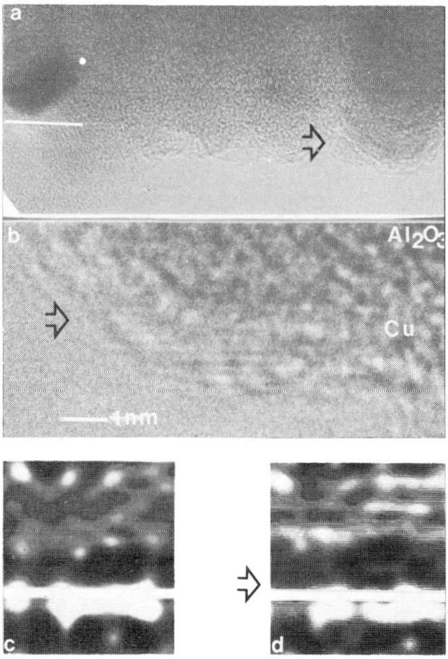

Fig.10 HREM of sintered real Cu/Al2O3 (a) Cu-alumina edge boundary (arrowed) (b) boundary at higher magnification. Calc. images (c) df -60nm (d) -90nm (200kV, substrate thickness 10nm).

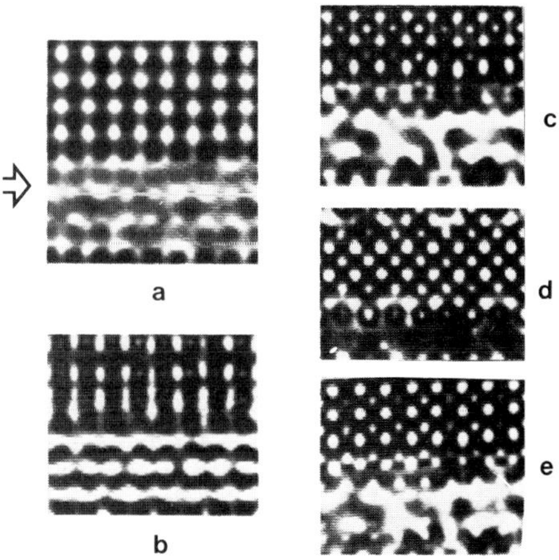

Fig.11 Cu-Alumina interfaces; 400kV (a)df-48nm;support thickness (t)=2.5nm (b)df-20, t=2.5 (c)df-45,t=7.5 (d) df-80, t=10 and (e)df-40.5,t=7.5nm.

Ag/alumina calculated images at 400 keV:(a) without the support (df -48nm) and (b) with 2.4 nm support (df -37nm). Fig. 6 shows a Pt/Alumina particle at 200keV :with (a) and (b) DC and HREM images without the support. 6 (c,d and e) are DC and HREM results with the support (4.6nm thick) at defoci of -66nm,-66nm and -76 nm. Fig.7 shows calculations for Pt/alumina at 400 keV : an unsupported particle is shown in (a) DC (b) HREM (df -48.7nm) and (c)HREM showing white atoms at df -63.6 nm. Supported particle images (4.6nm support) at defoci of -48.7 nm, -56nm and -63.6nm are shown in (d),(e) and (f) respectively. 7(g) shows a DC image at df=-48.7 nm.

The calculations show progressive variations in contrast with defocus and support thickness; and improved contrast on the carbon support. The better visibility of the supported particles at a greater underfocus than the optimum (e.g. at df=-76nm in Fig.4b,6e and in Fig.7f) is explained in terms of the contrast transfer function profiles at 200keV and 400keV (Fig.8). It is also seen that the particle images are obscured by the support contrast, but 400 keV is more favourable for their visibility. In general the results illustrate the difficulties in identifying very small particles and in interpreting their structure only from HREM images. In diffraction contrast although identification is easier, particles appear to change shape with the addition of the support as shown in Figs.6&7, and for the 309-atom particle of Ag/alumina at 400 keV (Fig.9). This clearly has implications for determining accurately the size and shape of these small particles.

Metal/substrate interfaces: metal particle/substrate interfaces play a key role in sintering during operation. We have considered the feasibility of simulating the structures of some simple interfaces observed by HREM/AEM during sintering, with particles forming a boundary at the edge of the support. Initially, a Cu/Alumina boundary formed during H2 sintering was calculated (Fig.10 a & b), using the SHRLI suite of multislice programmes . Calculations of structure factors for (001)Cu/alumina show that significant beams occur out to $d^* \geqslant .2nm^{-1}$ with a multislice 'window' $\geqslant .2nm^{-1}$. Hence ideally we should form phase gratings using $d^*max \geqslant .4nm^{-1}$. Since the phase grating is formed using a 128x128 FFT this imposes a limit on the size of artificial superlattice chosen to represent Cu/alumina boundary structure. In the present study, we have chosen a value of d*max=.4nm-1, thus limiting the maximum superlattice projection distances (in real space) to be < 1.6nm. For Cu/amorphous alumina each calculation used 7 phase gratings stacked in a random manner with the proviso that no phase grating should appear next to itself in the stacking sequence. Each of the 7 superlattice cells used measured 1.446nm x 1.479nm x .3615nm. 49% of each cell by volume was filled with a regular array of Cu atoms, the placing of which was identified from cell to cell. The remaining volume was filled with 14 Al and 21 O atoms distributed randomly but with no atom within 0.185nm of any other. Slice thickness of 0.3615nm, number of beams in the ms aperture = 3025 (d*max=4) and thicknesses (t) upto 20nm were used. Calculation of a single image of the interface required 8.4 hours on the ICL 2988 mainframe computer. Fig.10 (c) and (d) show images at 200keV at df -60nm and -90nm with an objective aperture out to 0.1nm-1. Fig.11 shows the interface images at 400 keV. It is seen that the results at 400 keV are more informative, but care is still necessary in interpreting the images.

We thank the SERC for supporting this work.

Reference : Mackay A L 1962 Acta Cryst 15 962.

Paper presented at EMAG '85, Newcastle upon Tyne, 2–5 September 1985

# The structure of supported platinum catalyst particles

P J F Harris

Department of Physical Chemistrty, University of Cambridge, Lensfield Road, Cambridge, CB2 1EP England

## Introduction

Supported metal catalysts are of major industrial importance and have been the subject of intensive research over a very long period, yet many fundamental questions concerning their activity, selectivity and stability remain unanswered. Transmission electron microscopy can undoubtedly make an important contribution in these areas by providing direct images of the supported particles, although the technical difficulties associated with studying real catalysts are considerable. In this study conventional bright field TEM has been used to investigate two aspects of the structure of alumina-supported platinum particles. These relate to the mechanism of particle growth, which is a serious cause of catalyst deactivation, and to the effect of adsorbates on particle structure, which could be important in poisoning and promotion. The specimens used here were prepared by a novel technique based on the sol-gel process (Harris et al 1983) which enables thin self-supporting films of catalysts to be prepared directly on microscope grids.

## Fast Growing Particles

In earlier studies (Harris et al 1983; Harris 1984) it was shown that heat-treating platinum/alumina specimens in air at 600°C and above resulted in the appearance of abnormally fast-growing particles. These crystallites, which probably grow by interparticle transport of $PtO_2$ vapour, exhibited a wide variety of structures the most common of which appeared to be the triangular plate. Fig. 1 shows a number of these particles in a specimen heated in air at 800°C for 6 hours. Similar anisotropically growing particles had previously been observed by Wynblatt (1976) in "model" platinum/alumina catalysts and have also been studied by Cosandey et al (1983). Microdiffraction analysis has suggested that the plates contain twin boundaries parallel to the large faces, but the detailed mechanism of growth has not yet been established. In the present work a number of micrographs of very large triangular plates tilted to an edge-on position were recorded, in an attempt to achieve an insight into the growth mechanism. One of these micrographs is reproduced in Fig. 2, and this clearly shows the presence of two parallel twin boundaries. Fig. 3 shows a similar particle in which the twin boundaries are less obvious, but where a re-entrant surface feature can clearly be seen. These observations suggest that the rapid growth involves the "twin-plane re-entrant edge mechanism". This mechanism was first described by Wagner (1960), who showed that the presence of two parallel twins in a cubic crystal produces a self-perpetuating system of grooves which provide preferential nucleation sites. Fig 4 shows a cross-section of a fast-growing particle according to Wagner's model. We have assumed here for simplicity that the particle is completely bounded by {111} faces; in practice the edges and corners were always rounded. Although re-entrant edges were not observed in all plate-like particles, this is to be expected since during growth the particles would not always maintain their equilibrium morphologies. In addition to the plate-like particles discussed here various other types of fast-growing structures could be identified, which will be descibed in detail elsewhere.

## Strong Particle Faceting

Particles heat treated in air often exhibited some degree of faceting, particularly the rapidly-growing particles, but the edges and corners were invariably rounded. It has recently been demonstrated that strong faceting of the particles can be induced by heat treating the catalyst in a vacuum furnace (Harris 1985). This is illustrated in Figs. 5 and 6 which show respectively a catalyst heated in air at 700°C for 1 hour and a catalyst heated at 900°C for 30 minutes in a vacuum furnace following a similar air heat treatment. During the vacuum heat treatment the pressure was $10^{-5}$ Torr. The formation of sharp facets was probably a consequence of the deposition of carbon onto the particles due to cracking of residual hydrocarbons in the vacuum: in fact a contaminating layer could often be seen on the surfaces of the particles, as well as on parts of the support. It is well established that adsorption can induce structural rearrangements of metal surfaces, and the phenomenon has been studied widely using various surface techiques. The question arises as to why strong faceting was not observed during heat treatments in air, since LEED studies have shown that oxygen adsorption can cause surface faceting of single crystal platinum (Blakely and Somorjai 1977). The explanation is probably that particle growth is very rapid during air-heating, so particle shape is dictated largely by kinetics rather than thermodynamics. Particle growth during vacumm heating, on the other hand was very slow so thermodynamics prodominate in determining the surface structure. An examination of vacuum-heated particles indicated that the facets were generally either {111} or {100} planes. Fig. 7 shows a faceted particle containing a single twin boundary. The shape of this particle suggests it is completely bounded by {111} facets, with the structure shown in Fig. 8. In Fig. 9 a similar particle is shown in which the twin boundary can be more clearly seen, while Fig. 10 shows a strongly faceted single crystal. In this case there appear to be some {100} as well as {111} facets.

Adsorbate-induced faceting could have important implications for catalysis by metal particles, since the transformation of surface morphology could produce major changes in selectivity. Of course the heavy carbon contamination which occurred during the experiments reported here would result in deactivation of the catalyst, but surface studies have shown that faceting can be induced by very much smaller amounts of adsorbate. It is quite possible that faceting, or the related phenomenon of adsorbate-induced reconstruction, could explain certain types of catalyst promotion (McCarroll 1975) as well as catalyst poisoning (Somorjai 1972). Clearly this is an area where further electron microscopy could be of great value.

## Acknowledgements

The author thanks the SERC for a Postdoctoral Fellowship, AERE Harwell for provision of materials, and Prof. J M Thomas for many discussions.

**References**

Blakely D W and Somorjai G A 1977 Surf. Sci. 65, 419.
Cosandey F, Roth L D and Tien J K 1983 Acta Met. 31, 2029.
Harris P J F, Boyes E D and Cairns J A 1983 J. Catal. 82, 127.
Harris P J F 1984 D.Phil. Thesis, University of Oxford.
Harris P J F 1985 Appl. Catal. 16, 439.
McCarroll J J 1975 Surf. Sci. 53, 297.
Somorjai G A 1972 J. Catal. 27, 453.
Wagner R S 1960 Acta Met. 8, 57.
Wynblatt P 1976 Acta Met. 24, 1175.

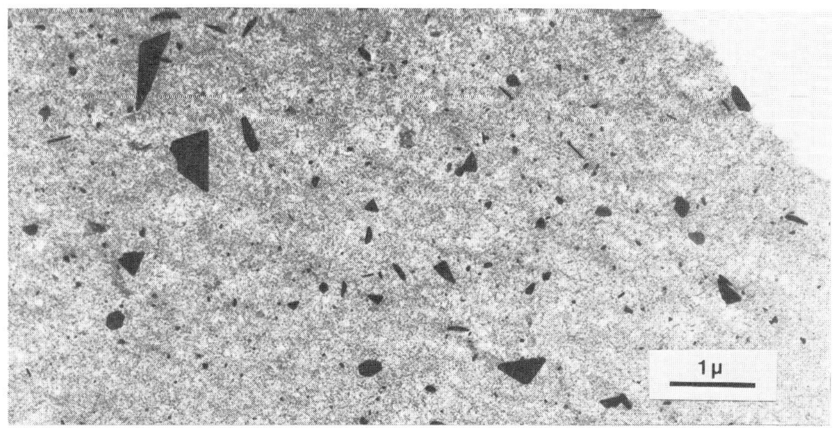


Fig. 1 Pt/alumina heated in air at 800°C for 6 hours.

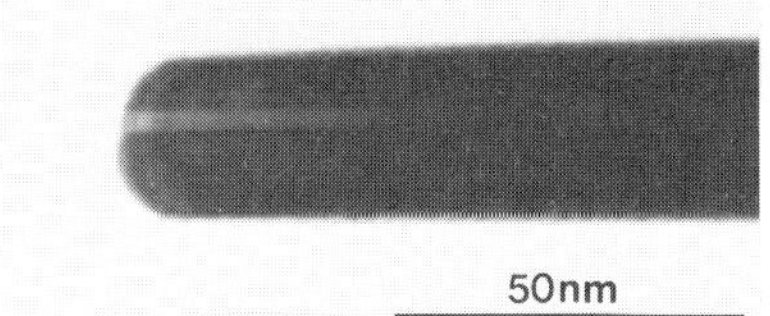


Fig. 2 Triangular plate tilted edge-on showing twin boundaries.

Fig. 3 Tilted triangular plate showing re-entrant feature.

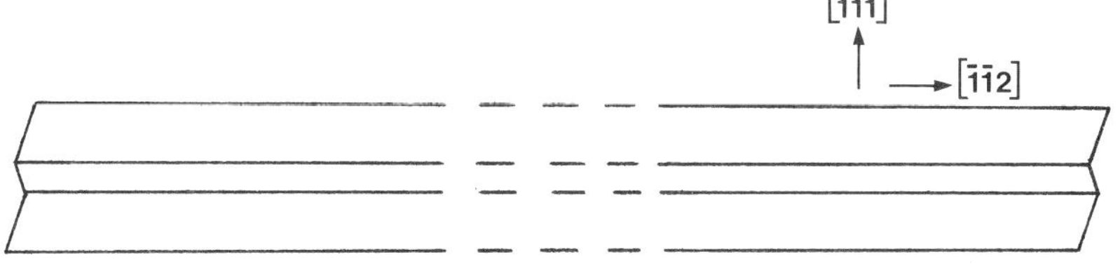


Fig. 4 Cross section of plate-like particle containing two parallel twin boundaries, showing re-entrant edge structure.

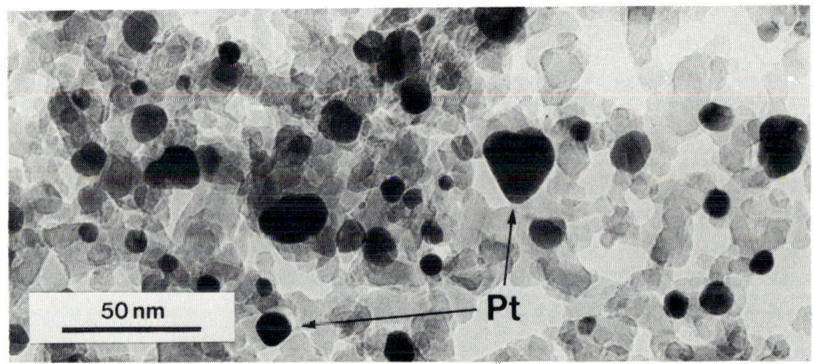


Fig. 5 Pt/alumina heated in air at 700°C for 1 hour.

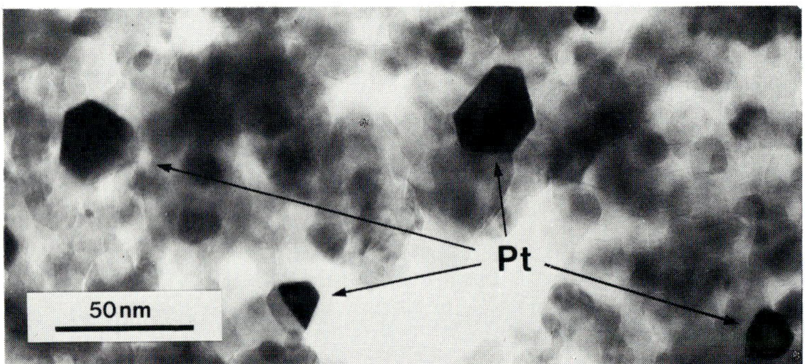


Fig. 6 Pt/alumina heated in air at 700 °C for 1 hour then in vacuum ($10^{-5}$ Torr) at 900°C for 30 minutes.

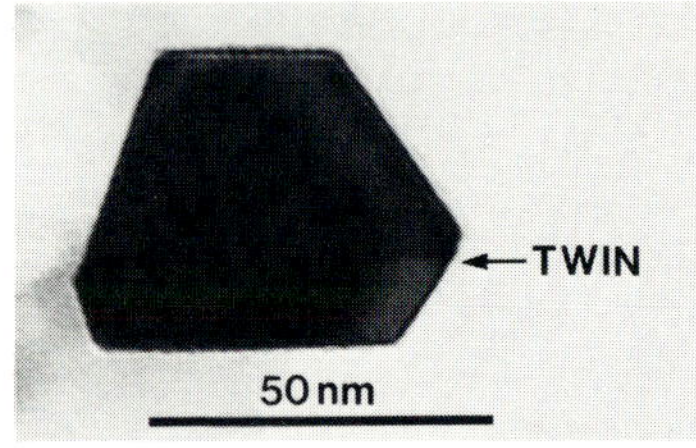


Fig. 7 Strongly faceted particle containing single twin boundary.

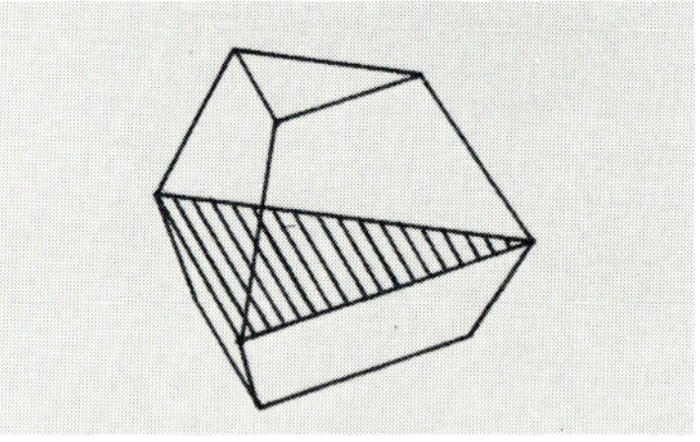

Fig. 8 Approximate structure of particle shown in Fig. 7.

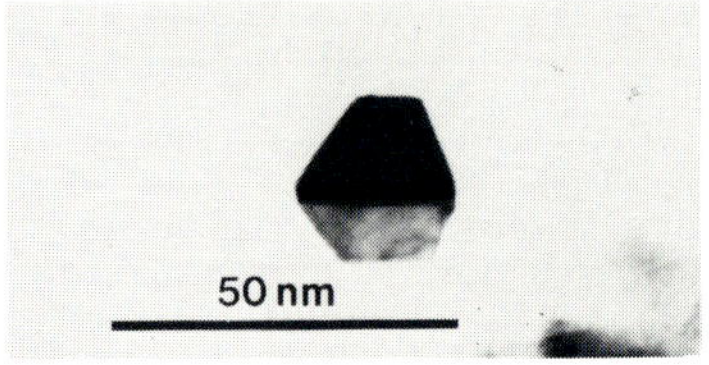


Fig. 9 Faceted single twin.

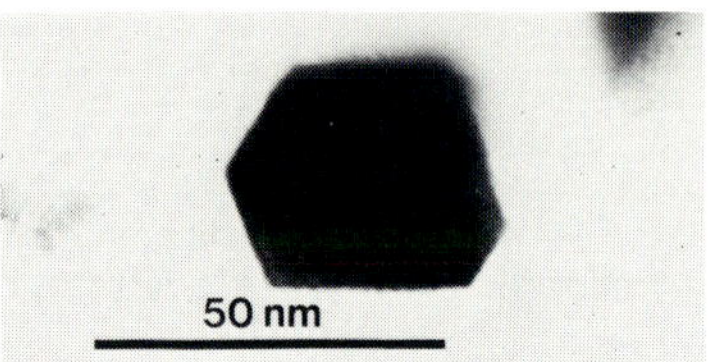


Fig. 10 Faceted single crystal.

# Transmission electron microscopy of metal powder produced by high pressure gas atomisation

R A Ricks

Department of Materials Science and Engineering, University of Surrey, Guildford GU2 5XH.

## 1. Introduction

High pressure inert-gas atomisation, using a twin-fluid atomiser, has been shown recently to be a viable route for the production of rapidly solidified metal and alloy powder (Ricks and Clyne, 1985). Using such equipment, together with atomising gases chosen to maximise heat transfer between the molten droplets and the quenchant gas, very high solidification rates may be expected, comparable to those achieved using "traditional" rapid solidification methods (eg melt spinning) (Clyne et al, 1984).

This paper reports some results concerning the effect of these high solidification rates on the microstructure observed in as-atomised aluminium-alloy powder. It is shown that, for very fine metal droplets (< 1μm) the rate of crystal growth may be so high that complete solute trapping takes place, resulting in a microstructure free from second phases. Alternatively, for larger particles, recalescence effects can alter the degree of segregation in a manner which is qualitatively predictable by a consideration of the heat transfer conditions encountered during atomisation. Several interesting intermediate cases are also reported and discussed.

## 2. Experimental

The aluminium alloy powders used in this investigation were prepared using the high pressure inert-gas atomiser at the University of Surrey (see figure captions for alloy details). Details of the equipment and operating procedure may be found elsewhere (Ricks and Clyne, 1985). For the purposes of this report it is sufficient to state that atomisation was carried out at a gas pressure of 3.5MPa using pure helium as the atomising gas. Analysis of the resultant powder using laser granulometry reveal a typical median diameter of ≃ 15μm with considerable quantities of particles < 5μm. Since the very high cooling rates are associated with these fines, wet sieving techniques using alcohol as a sieving medium were used to obtain sufficient quantities of sub 5 micron powder for analysis by electron microscopy.

Specimens for electron microscopy were prepared by ultramicrotomy of powder embedded in resin. Sections could be cut ≃ 800Å thick which, for the small particles with which this study in concerned, produced large quantities of electron transparent specimen sections. Some problems were encountered with residual stresses contained within the sections and larger particles tended to curl-up, especially when annealed in the electron beam. Despite these problems, ultramicrotomy provides a reliable and quick method for the preparation of TEM specimens of fine powders, with the additional advantage of providing specimens which are not surface-contaminated as is often the case with other specimen preparation techniques.

The thin-foil sections were examined on a Philips EM400T equipped with EDS and EELS for quantitative chemical analysis.

## 3. Heat flow considerations

Heat transfer during atomisation is highly complex but may be simplified by considering a constant velocity difference between the gas and droplet. Using this assumption it has been shown (Clyne et al, 1984b) that a simple analytical expression may be used to calculate the expected value of the interfacial heat transfer coefficient ($h_i$). Values of this parameter for helium are shown in Figure 1 as a function of metal droplet size. The effects of changing this parameter on the expected thermal history of a droplet are illustrated in Figure 2, which shows calculated thermal profiles and the reduction of recalescence effects as $h_i$ increases due to the efficient removal of the evolved latent heat. One of the effects of recalescence will be to reduce the crystal growth velocity as the undercooling is reduced and this will have a profound effect on the microstructure since the effective value of the partition coefficient will not be constant over the entire solidification range. For very high values of $h_i$ (very small droplets) it is expected that growth velocities will be so high that total solute trapping would take place (Clyne et al, 1984a).

## 4. Results

The vast majority of particles investigated above ≃ 2μm in diameter showed the fine cellular microstructure (Figure 3) associated with rapid growth involving solute partitioning. Both EDS and EELS have shown significant solute levels in the intercellular regions as expected and there is some evidence to suggest that enriched solute levels exist in the cells themselves (Ricks et al, 1985). Below this size range most microstructures showed no evidence of segregation and featureless microstructures were observed (Figure 4) as predicted by the simple consideration of heat flow. Two interesting intermediate microstructures were also observed. Figure 5 shows a 1.5μm particle of a Cr and Zr containing particle which had been annealed for a short time in the electron beam. Precipitation of $Al_3Zr$ and $Al_5Cr$ has taken place on one side of the section only. One explanation of this observation is associated with the variation in the value of the partition coefficient during solidification. As the temperature of the interface rises due to recalescence, the level of segregation into the liquid will rise since the interfacial velocity will fall. Thus the last liquid to solidify will be highly solute rich. If at this stage the velocity of the growth front is such that absolute stability is maintained (planar growth front morphology) the last solid to form may aquire this high solute level which would tend to precipitate first when annealed.

Similar observations could also be recorded when the final growth velocity was below that required to maintain absolute stability of the interface. In this case the interface degenerates and produces a cellular microstructure over the last part of the droplet to freeze. An example of this is shown in Figure 6 and the transition to cellular solidification may be clearly seen.

## 5. Disscussion and conclusions

The electron microscopical evidence presented in this paper has been shown to qualitatively support the simple considerations of heat flow pertinent to atomisation as detailed by Clyne at al (1984a,b). Very fine metal droplets (< 1μm) are predicted to have very high interfacial heat transfer coefficients such that recalescence effects are negated and crystal growth is maintained at a very high rate resulting in solute trapping. Predicted heat transfer coefficients associated with larger particles (> 2μm) would not prevent recalescence and the corresponding reduction in interfacial velocity is manifest as an observed degeneration of the growth front to form a fine, cellular solidification structure. Two intermediate cases have been observed where recalescence effects have varied the velocity and morphology of the growth front and hence the value of the partition coefficient. In one case the velocity was still high enough to maintain the criterial for absolute stability of the interface, resulting in an initially featureless microstructure which annealed to give the heterogeneous precipitate dispersion shown in Figure 5. In the second case the interfacial velocity had dropped sufficiently to allow degeneration of the interface resulting in a microstructure partially consisting of a featureless region together with a cellular structure produced towards the end of the droplet solidification.

## Acknowledgements

The author wishes to thank J. Mullervy for careful specimen preparation and acknowledge the helpful assistance of N.J.E. Adkins, T.W. Clyne and G. Von-Bradsky during the course of this investigation.

## References

R A Ricks and T W Clyne (1985) J. Mat. Sci. Letters 4 814

T W Clyne, R A Ricks and P J Goodhew (1984a) Int. J. Rap. Sol. 1 59

T W Clyne, R A Ricks and P J Goodhew (1984b) ibid p85

R A Ricks, P M Budd, P J Goodhew, V L Kohler and T W Clyne, Proc. 3rd Int. Conf. Al-Li Alloys, Oxford (1985) to be published

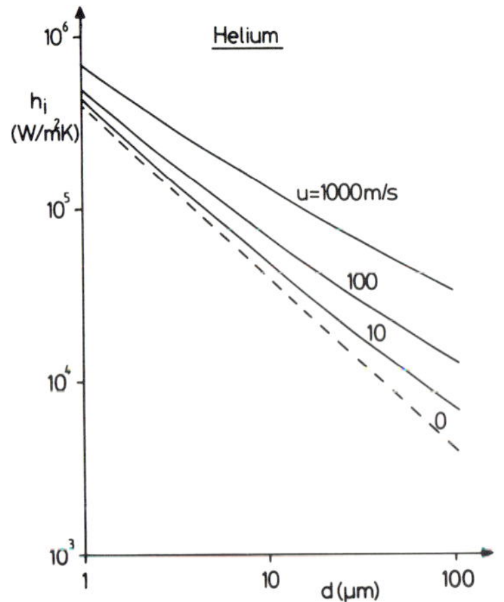


Fig.1 Plot of $h_i$ as a function of particle diameter.

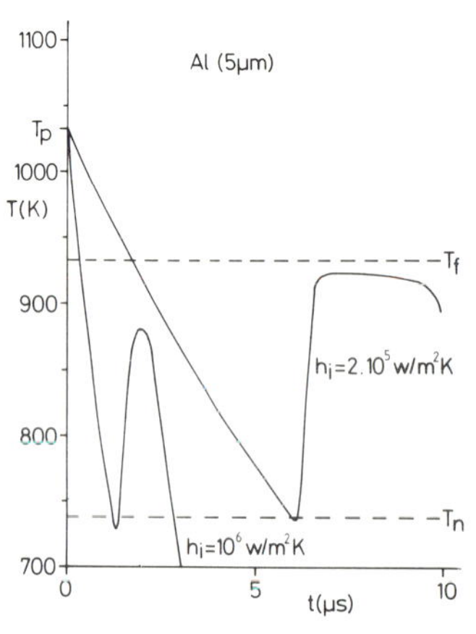


Fig. 2 Thermal histories for 5μm dia. Al droplets for two values of $h_i$.

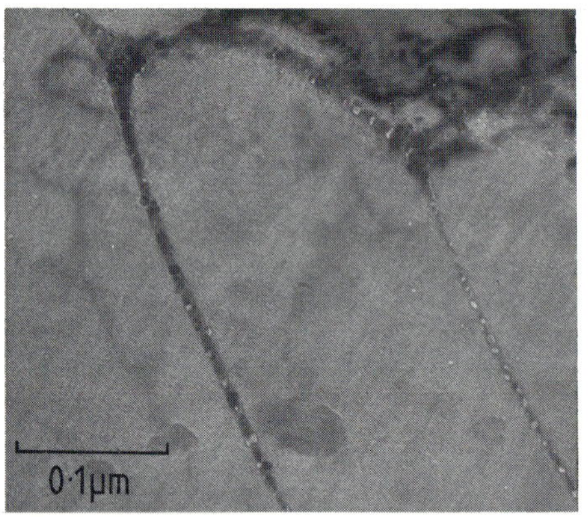


Fig. 3 TEM image showing fine cellular structure in a 4μm dia Al-2.5Li alloy particle.

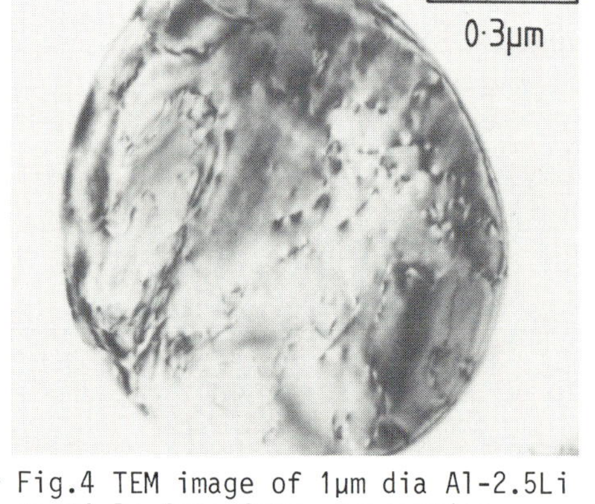


Fig.4 TEM image of 1μm dia Al-2.5Li particle free from segregation.

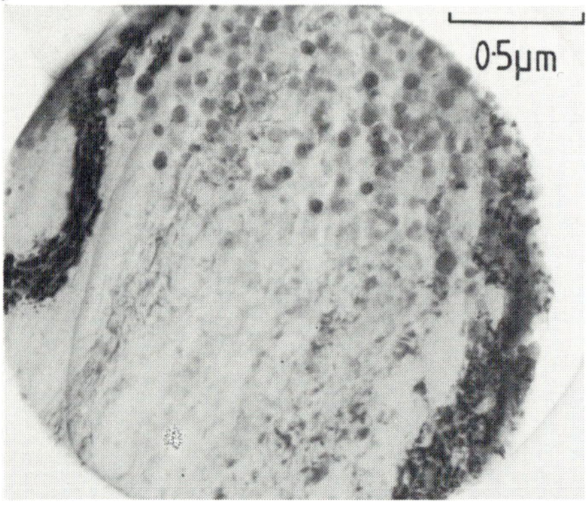

Fig.5 TEM image of 2μm dia Al-Cr-Zr-Mn alloy particle showing heterogeneous precipitation.

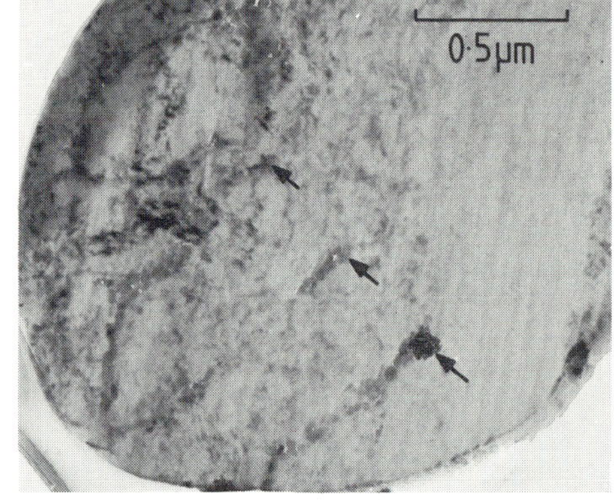


Fig.6 TEM image of 2μm dia. Al-Cr-Zr-Mn alloy showing partially cellular microstructure (arrowed).

*Inst. Phys. Conf. Ser. No 78: Chapter 12*
*Paper presented at EMAG '85, Newcastle upon Tyne, 2–5 September 1985* 

# Electron microscopy of copper bi-metallic systems

B C Smith and P L Gai,
Department of Metallurgy and Science of Materials, University of Oxford, Parks Road, Oxford OX1 3PH.
G Owen and M D Shannon,
ICI Plc, The Heath, Runcorn, Cheshire.

## 1. Introduction

Copper-palladium (Cu-Pd) and copper-ruthenium (Cu-Ru) model bimetallic systems evaporated onto amorphous carbon supports have been studied by a combination of in-situ high voltage electron microscopy (using a gas reaction cell fitted to an AEI-EM7 HVEM), high resolution EM (with a JEOL 200CX HREM) and microanalysis (with a Philips FEG EM400T and JEM 2000FX AEMs). The in-situ studies have provided information on how particles form and sinter, while the microanalysis has shown that there are compositional changes over distances of a few nanometres. Surface structures are revealed by HREM.

## 2. Copper-Palladium

As the thin evaporated film ruptured on heating in vacuo, 10% Hydrogen/He or in 10% CO/He to about 300°C, the Cu-Pd stucture changed from FCC to body centred B2 (CsCl type) structure. At higher temperatures (above 400°C) some coalescence of particles was observed. Fig. 1 shows the general appearance of the samples reduced in (a) 10% $H_2$/He and (b) 10% CO/He. After heating in vacuo the appearance was similar to Fig. 1(a). The faceted rectangular particles in these micrographs are B2 phase with [100] orientation. Those that form in CO have squarer corners and are bounded primarily by (100) faces rather than a mixture of (100), (110) and (111) faces as is the case for those produced in hydrogen. CO adsorbs more strongly on palladium ($-\Delta H$=36 kcal/mole) than on copper ($-\Delta H$=18 kcal/mole). The heat of adsorption for $H_2$ on Pd is similar to that on Cu. In the B2 structure (100) planes contain either all Pd or all Cu atoms, where as the (110) planes contain an equal number of each. Fig. 2 shows a high resolution micrograph of part of a B2 [100] particle. The (100) facets are straight while the (110) facet is irregular and exhibits a number of atomic steps. The crystal structure is essentially defect-free. FCC particles appeared to be less clearly faceted and more defective. Some of them had an hexagonal outline. Because some particles exist with the typical outline of the alternative phase it is believed that phase-transformations occur without altering the external particle shape. From these observations the following phase relationship was deduced: [111]FCC//[110]B2 and [$\bar{2}$11]FCC//[001]B2. No bi-phasic particles were observed and therefore it is assumed that the transformations occur rapidly. Electron diffraction patterns of individual B2 particles exhibited strong diffuse streaks in certain directions (Fig.3), namely <110> streaks in the diffraction pattern (d.p.), in [100] zone, <211> in [110] and <321> in [331] d.p.. The streaking occurs due to soft phonon modes that occur in the B2 lattice (Otsuka et al 1978). In this instance chains of atoms vibrating along the close packed <111> directions give

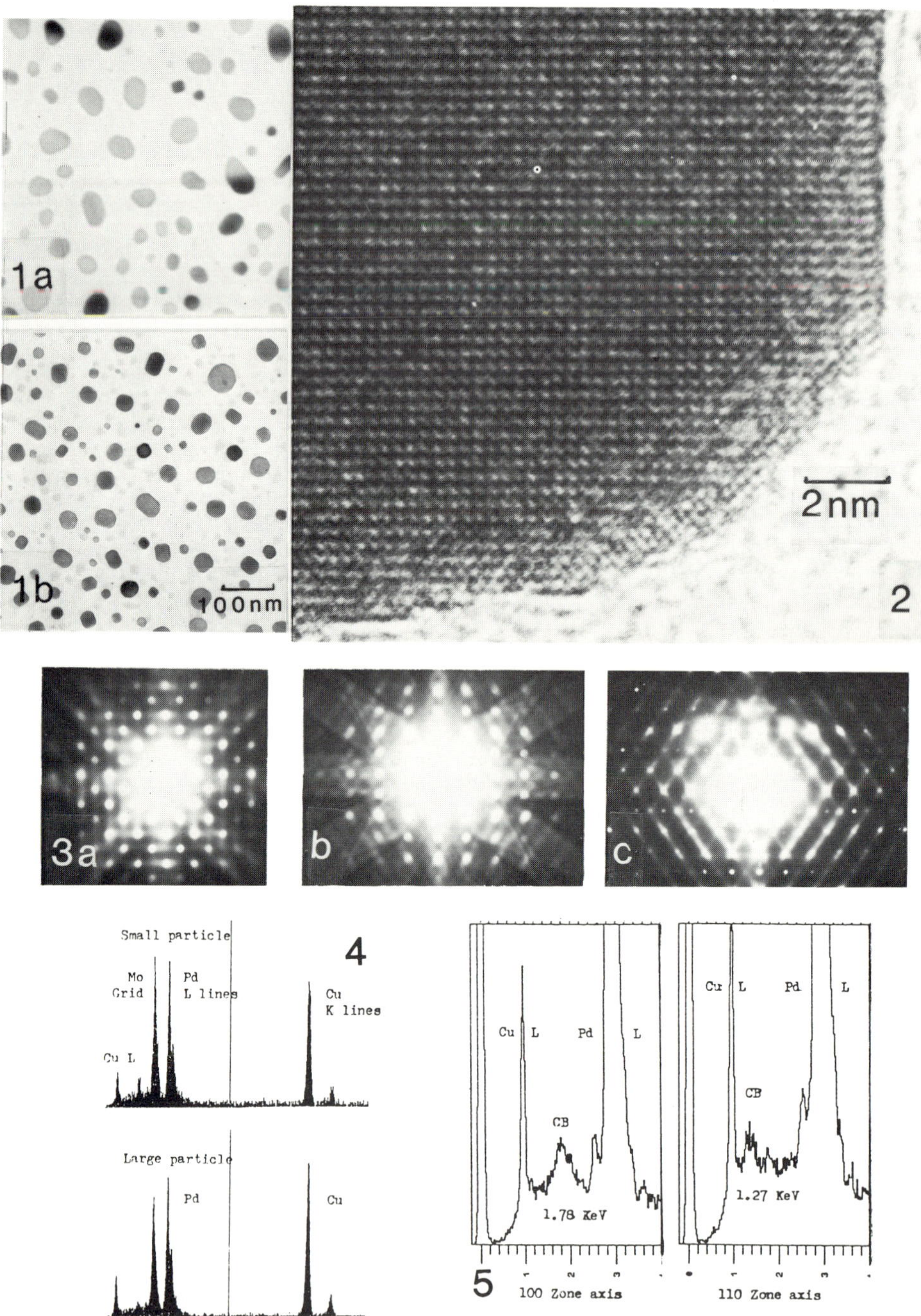


Figs. 1) Cu-Pd/carbon reduced in (a) 10% $H_2$/He (b) 10% CO/He. 2) HREM detail of Cu-Pd B2 [100] particle. 3) Cu-Pd B2 diffraction patterns (a) [100] (b) [110] (c) [331]. 4) Microanalysis of individual Cu-Pd particle. 5) Microanalysis showing coherent bremsstrahlung peaks

rise to {111} rel walls in reciprocal space. The streaks appear when the rel walls intersect the Ewald sphere. Other explanations (e.g. that the streaks are a consequence of arrays of parallel defects) are ruled out by the perfect crystallinity of the lattice as seen in Fig.2.
Microanalysis of the particles (Fig.4) indicated a higher percentage of palladium in the smaller particles (about 50nm) than in the larger ones. Similarly, particle edges were found to be enriched in Pd.
When a particle was tilted exactly on to a low index zone axis peaks due to coherent bremsstrahlung were observed in the microanalysis traces (Fig.5). Upon tilting to a new zone axis the peaks shifted to a different energy. The two spectra shown in Fig. 4 were both obtained from the same particle. For the [100] zone axis the peak is at 1.78 keV but it shifts to 1.27 keV for the [110] zone axis. These values are in reasonable agreement with the predictions made on the basis of the formulae by Spence et al (1983). The actual values depend on the configuration of the microscope and the accelerating voltage used.
In contrast to the above results, heating of the samples in oxygen led to the formation of copper-rich particles , or sheets, surrounded by tiny crystallites of PdO. Electron diffraction patterns indicated that PdO did not have the exact P42/mmc tetragonal structure reported in the literature, since forbidden rings e.g.(111), appeared in the patterns (Fig.6).

3. Copper-Ruthenium

Copper and ruthenium are not known to form a solid solution in the bulk. However, Sinfelt (1977) believes that bimetallic clusters exist where Cu covers the surface of Ru. In our experiments preferential segregation of particular species and evidence of new alloy or bimetallic phases were explored. On heating to about 150° C in vacuo, copper particles were formed. They were identified by electron diffraction and microanalysis. Upon further heating they sintered either by coalescence (high Cu content) or by vapour phase Ostwald ripening (Cu:Ru=1:1). Particles grew along well defined lattice directions. Fig.7 shows a HREM image of Cu particles in Ru-rich specimen. Eventually Ru and an unknown phase (probably with a structure similar to $RuO_2$ were observed in the diffraction patterns. With time the samples oxidised in air and subsequent examination showed the presence of Ru and $Cu_2O$ (Fig.8). Combined STEM and SEM studies (Fig.9) showed that some of the large $Cu_2O$ particles were on the opposite side of the support to the Ru, inducating that Cu had diffused through the amorphous carbon substrate. Similar segregation of Cu and Ru was observed following reduction in 10%$H_2$/He and in 10%CO/He and after oxidation. Fig.10 shows a HREM image of $Ru_2O$ crystallites taken on the Philips EM400T. Work is continuing on identifying copper-rich phases with hitherto unknown metal or oxide structures.
Peaks due to Cu+, CuOH+, Ru+, RuC+, CuRu+ and CuRuC+ were observed in the SIMS (secondary ion mass-spectrometry) spectra of the fresh samples. Copper is at first weakly oxidised, Ru is attached to the carbon support and Cu atoms exist in close proximity to Ru atoms. On heating to 200° C (in-situ), the Ru+, CuRu+ and CuRuC+ peaks become weaker relative to the CU+ peak as Cu diffuses preferentially to the surface while the CuOH+ peak disappears.
These results indicate an interaction of copper with ruthenium in a thin film geometry; forming bimetallic alloy phases which under certain conditions may also contain oxygen.

We thank the SERC for support.

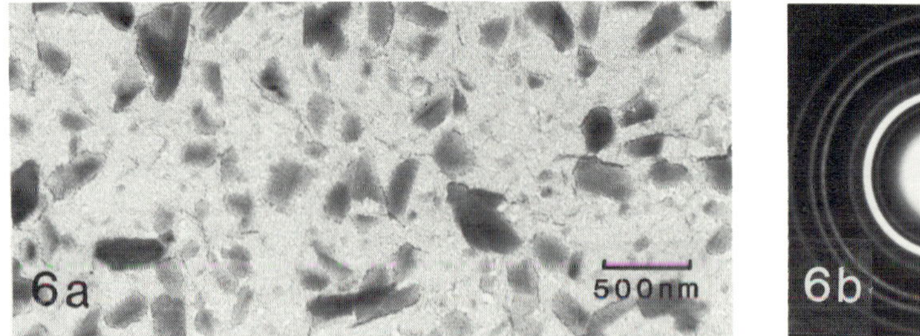


Figs. 6) (a) Cu-Pd in $O_2$ (b) PdO DP showing forbidden (111) ring. 7) HREM Cu particles in Ru-rich specimen. 8) (a) Large $CU_2O$ and small Ru crystallites (b) microanalysis of (a). 9) (a) STEM image (b) SEM image. 10) HREM of $RuO_2$ in Cu-Ru after oxidation.

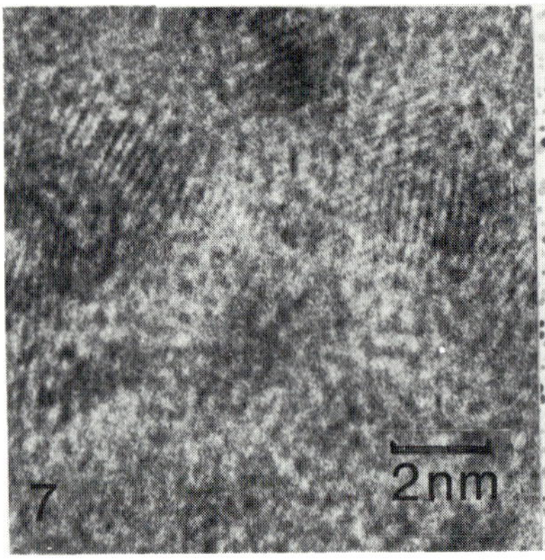

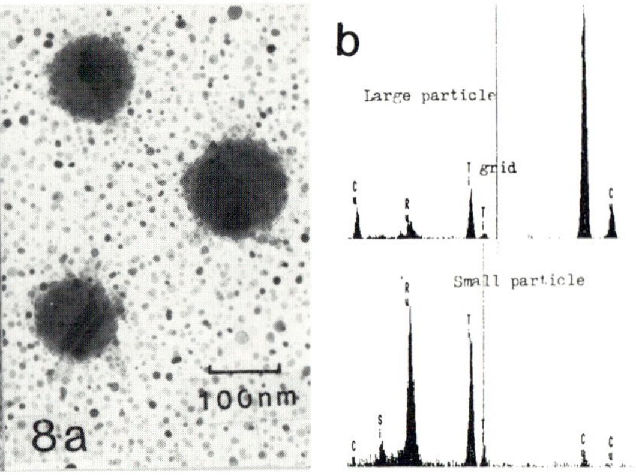

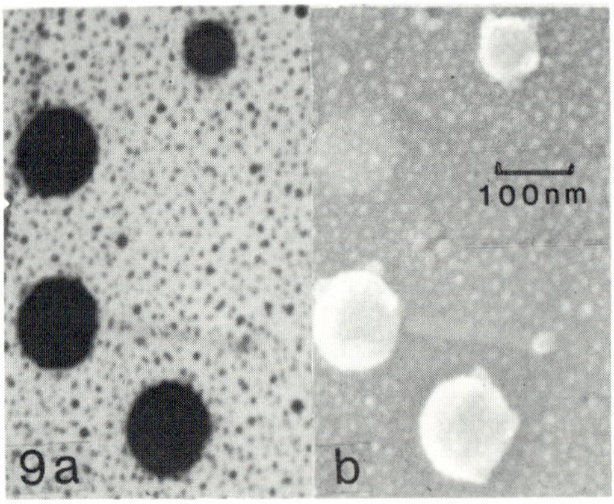

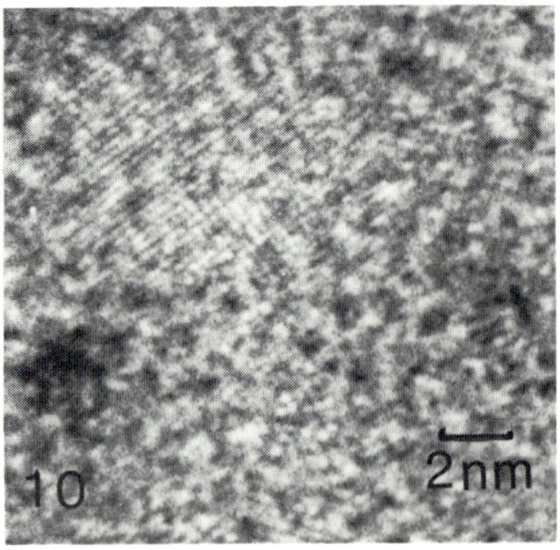

References

Otsuka K, Wayman C M and Kubo H, 1978 Met Trans 9A 1075
Sinfelt J H, 1977 Acc Chem Res 10 15
Spence J C H, Rees G and Yamamoto N, 1983 Phil Mag 48 L39

# Zeolite theta-1—structural studies by TEM

D White, S Ramdas and G R Millward*

BP Research Centre, Chertsey Road, Sunbury-on-Thames, Middlesex TW16 7LN, UK.
* Department of Physical Chemistry, University of Cambridge, Lensfield Road, Cambridge, UK.

## 1. Introduction

Zeolites are crystalline aluminosilicates whose three-dimensional frameworks consist of corner sharing $SiO_4$ and $AlO_4$ tetrahedra. The resultant structures contain intra-crystalline channels and cages which give rise to shape selectivity at the molecular level and provide, inter-alia, high performance catalysts for petroleum or chemical processing. The commercial use of both naturally occurring zeolites and their synthetic counterparts has increased in recent years. This demand for the development of new zeolite catalysts has resulted in the chemists' growing capability to finely tune the synthesis.

In this paper we describe the structural characterisation of a recently discovered, synthetic zeolite, theta-1 (Barri et al. 1984), particularly the use of electron diffraction and High Resolution TEM (HRTEM) to support the full determination of the crystal structure.

## 2. Experimental

### 2.1 Specimen Preparation

A specimen was prepared for characterisation in the SEM by sprinkling some dry zeolite powder on to a droplet of wet carbon cement, followed by conventional gold coating.

For electron diffraction work, a standard, carbon-coated TEM specimen grid was contacted with a sample of the zeolite, leaving many crystals clinging to the carbon film.

Ultrathin specimens for HRTEM were prepared by microtomy techniques. Crystals of theta-1 were first dealuminated as this is known to increase the lifetime under electron irradiation but does not affect the structure (Thomas et al. 1981). Many dealuminated crystals were embedded in resin, to fix the orientation of the rod-like crystals and in particular, to obtain the 'end-on' view. Thin sections were then cut using an ultra-microtome equipped with a diamond knife. Section thickness was estimated from interference colours (Reid, 1975) and those $\leq$500Å thick were picked up on TEM specimen grids. An additional thin carbon film was evaporated onto the specimens to improve conductivity and stability, followed by evacuation for ~1 week to remove excess adsorbed water, prior to examination by HRTEM.

### 2.2 TEM

Electron Diffraction patterns for the major zones were obtained using the Selected Area Electron Diffraction mode of a JEOL 100c TEM, with micro-driven tilt ($\pm$60deg) - rotate goniometer for fine adjustment of single crystal orientation.

Ultrathin sections were examined in a JEOL 2000EX TEM equipped with side entry tilt ($\pm$20deg) - rotate goniometer. Single crystals or small, embedded clusters were tilted into the ⟨001⟩ crystallographic zone, as indicated by the electron diffraction pattern, and focal series images were recorded at X150,000 magnification. Micrographs were recorded on fast film, under minimal exposure conditions to reduce electron irradiation damage to the zeolite structure.

### 2.3 Image Computation

The atomic coordinates from powder X-ray diffraction studies (Highcock et al. 1984) were used as the basis for image calculations. The multiple scattering of electrons in the specimen was simulated using the so-called 'phase grating and multislice' approximation (Cowley, 1975). In addition, the variations in image details as functions of specimen thickness, electron wavelength and lens characteristics were computed for comparison with the observed images.

## 3. Results and Discussion

### 3.1 SEM and Electron Diffraction

SEM photo-micrographs, eg fig. 1, showed the rod-like morphology of the zeolite crystals and the general size of ~1μm long by <0.1μm wide. It was not possible to grow larger crystals, suitable for conventional single crystal X-ray diffraction, but single crystal selected area electron diffraction patterns were obtained. These showed orthorhombic symmetry and revealed unit cell parameters, a=13.84Å, b=17.42Å and in particular, a short c-axis of around 5.0Å which severely restricts structural options in a zeolitic framework. In addition, although systematic absences were consistent with three possible space groups, $Cmc2_1$ was chosen as the most probable from structural considerations. Figure 2 shows, for example, the ⟨010⟩ zone axis pattern, the absence of h0l reflections for l odd indicating a c-glide plane perpendicular to the b-axis. A structural model was then derived that was consistent with these data as well as X-ray powder diffraction, $^{29}$Si solid state NMR and sorption studies (Barri et al. 1984).

### 3.2 HRTEM

Direct confirmation of the proposed framework geometry, projected on to the a-b plane (ie looking down the c-axis - the needle axis) was obtained by HRTEM. Figure 3 shows a cluster made up of around fifteen zeolite crystals, viewed 'end-on'. Some defect structures can be observed which may be twins and/or intergrowths. Figure 4. shows the ultrathin edge of one crystal, together with a simulated image and the schematic proposed structure. Framework oxygen atoms have been omitted for clarity, leaving only the tetrahedral atom sites (T-atoms). The large elliptical 10-membered channels (10-T rings) have diameters of around 6Å and are linked, in the a direction, by 6-T rings and surrounded by edge-sharing 5-T rings.

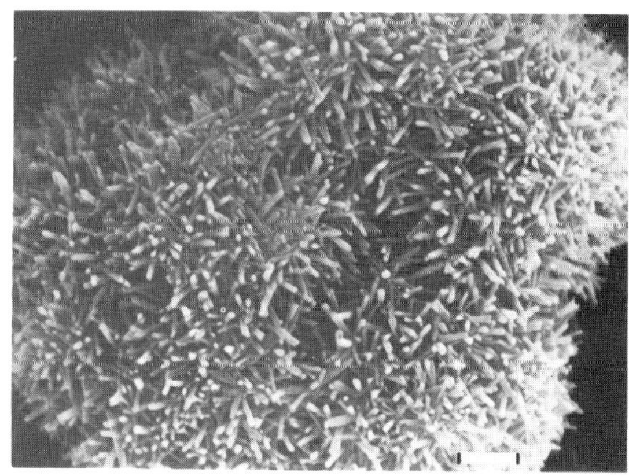

Fig. 1

SEM photomicrograph showing rod-like crystals of theta-1; 1μm scale bar.

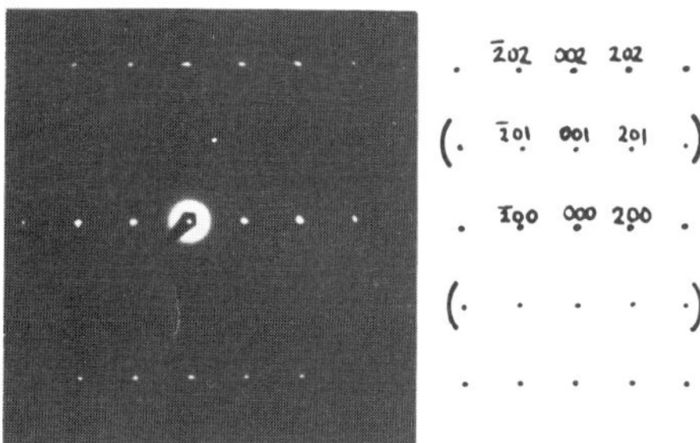


Fig. 2

SAED pattern from single crystal, ⟨010⟩ zone, together with indexing scheme. Bracketted reflections are systematic absences.

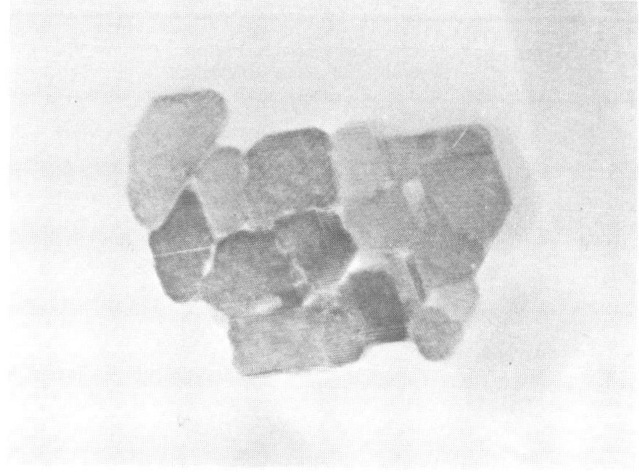

Fig. 3

Section through clustered zeolite crystals in epoxy resin matrix.

400Å

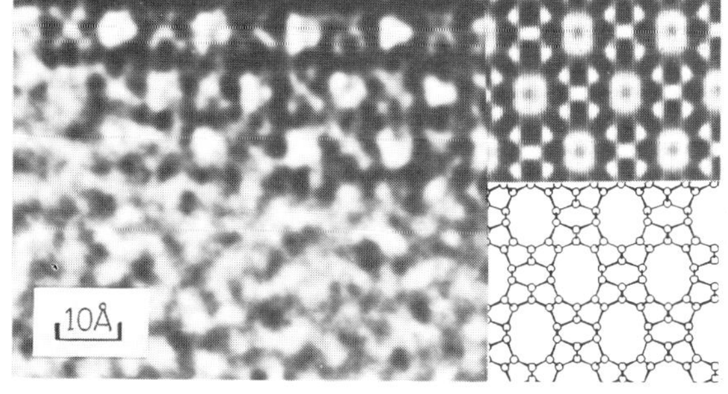


Fig. 4

Ultrathin edge of one crystal with simulated image (upper RHS) and schematic proposed structure (lower RHS).

The full structure-projection is revealed in the photo-image and there is good correspondence between experimental and simulated images. The latter was the 'best match' simulation from comparison of HRTEM-recorded and simulated focal series; the specific simulation used was computed for specimen thickness of 30Å, at 800Å under focus, coefficient of spherical aberration, Cs=2.3mm and electron wavelength $\lambda_{200\ kV}$ = 0.0251Å.

## 4. Conclusions

Theta-1 is the first reported example of a new, topologically distinct, high silica zeolite structure. Electron diffraction and HRTEM observations provided invaluable aids to the determination and confirmation of the crystal structure. The novel, unidimensional 10-T ring channel system should give rise to structure-specific activity with potential catalytic and separation applications. The work illustrates the strengths of electron microscopical characterisation of zeolites, whether for simple size and morphological assessment or to support a full structure determination.

## Acknowledgement

The authors thank British Petroleum plc for permission to publish. We also thank K. Meade for the SEM image and we acknowledge the contributions of our colleagues at Sunbury to aspects of the work outside electron microscopy.

## References

Barri SAI, Smith GW, White D and Young D 1984 Nature 312 533-4
Cowley JM 1975 in Diffraction Physics Pub. North Holland
Highcock RM, Smith GW and Wood D Acta Cryst. in press
Reid N 1975 Practical Methods in Electron Microscopy: Ultra microtomy Ed. AM Glauert Pub. North Holland/American Elsevier p296
Thomas JM, Millward GR, Ramdas S, Bursill LA and Audier M 1981 Faraday Discussion 72 345

# Electron microscopy and analysis of oxynitride ceramics

M H Lewis

Centre for Advanced Materials Technology,
Department of Physics, University of Warwick, U.K.

## 1. Introduction

The new engineering ceramics are normally the products of ultrafine powder-particle sintering with microstructures consisting of a sub-micron phase distribution of high chemical complexity. Hence there is an obvious requirement for application of analytical electron microscopy in defining microstructure and microchemistry with high spatial resolution. This paper will briefly review some of the techniques which are important in analysis of ceramic microstructures with reference to their influence on properties. Examples are derived mainly from a broad programme of research on oxy-nitride ceramics at Warwick over the past decade.

Specific problems in relation to analytical EM of ceramics may be identified :

(i) Preparation of electron-transparent specimens has to be accomplished via ion-beam erosion. With near-theoretically dense ceramics this presents little difficulty; the most successful technique uses a 3 mm dia. metal support ring onto which the ∿100μm thick ceramic slice is glued after diamond lapping. Although there is no visible ion-irradiation damage in oxynitride ceramics there is a surface implantation of an X-ray detectable level of argon ions. The argon concentration is sensitive to crystallography and composition of ceramic phase and is normally greatest within glassy areas. The recent application of 'atom-milling' guns in this laboratory has shown some improvement in specimen thinning rates and susceptibility to implantation but this tentative conclusion has to be supported by controlled experiments.

(ii) Ceramic phases are occasionally susceptible to electron-irradiation

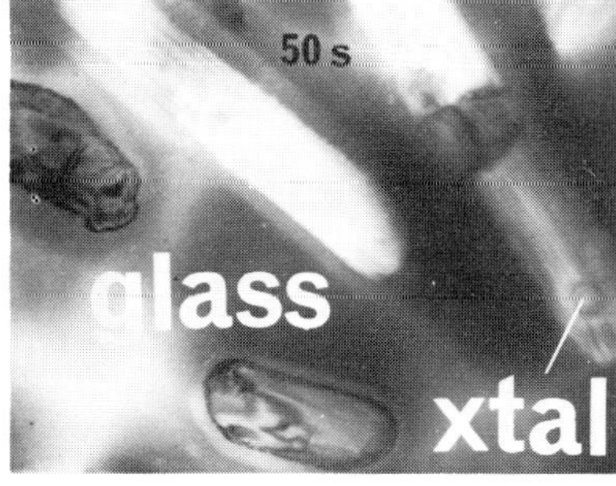


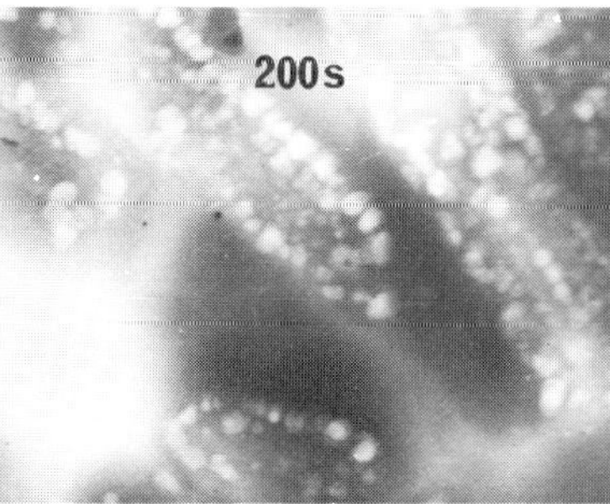


Fig.1 Electron-irradiation damage in a glass-ceramic

damage. Direct displacement damage is not normally a problem for electron energies below 300 keV (there are exceptions with abnormally low displacement thresholds in certain oxides containing non-stoichiometric vacancies - Bell and Lewis 1974). Indirect displacement (radiolysis) damage precludes high-resolution imaging in many silicate phases. Alkali-silicates undergo a sequence of crystalline disordering followed by glass decomposition (Fig.1). Most silicate-like phases within oxynitride ceramics contain di or trivalent metallic ions and are relatively stable but observable damage occurs within a few minutes of imaging. High kV microscopy is an advantage here and there is a marked improvement at 200 keV over conventional 100 keV instruments. The high-covalency nitride or oxynitride phases are resistant to electron-irradiation damage.

(iii) Ceramic phases are largely composed of light elements which have low X-ray fluorescent yields and poor detection sensitivity. This is a special problem in energy-dispersive X-ray microanalysis (EDS) of the elements oxygen and nitrogen in oxynitride ceramics. EDS systems with 'windowless' detectors are in routine use for light element analyses in SEM and at an introductory stage for TEM. The N+O spectra exemplified below illustrates a typical energy-resolution and peak/background which imposes a limit of 5-10 $^{a}/_{o}$ on detection sensitivity, especially for nitrogen. Very thin conducting coatings are necessary on insulating ceramics to minimise low-energy X-ray absorption. Currently available quantitative software is not applicable to the light element range, apart from simple background-stripping procedures, but semiquantitative analysis is possible using standards of appropriate N/O ratio. The use of $LaB_6$ emitters in light-element TEM analysis is an important advantage when high spatial resolution (<20nm) is required in thin sections, to achieve an acceptable spectral accumulation time.

Electron energy-loss spectroscopy (ELS) is an alternative light element analysis technique in TEM. It has a superior spatial resolution, being probe-size limited rather than secondary fluorescence-volume limited as in EDS, but has an inferior detection sensitivity and quantitative capability to windowless EDS using oxynitride standards.

(iv) Comparatively poor back-scattered electron coefficients for light elements reduces the signal/noise limited resolution in SEM images. This is compensated by increased atomic number sensitivity (and hence contrast) in the light element range; an empirical comparison of back-scattered coefficient from various ceramic phases shows a contrast limit equivalent to $\sim$1-2 in atomic no. The use of $LaB_6$ emitters for low kV back-scattered imaging is especially useful for ceramics of mean atomic no. $\lesssim$10.

2. Oxynitride Ceramics

Ceramics which have the greatest potential for high-temperature application have intrinsic properties such as high modulus, hardness, resistance to plastic flow and oxidation, low thermal expansion etc., derived from the high-covalency major phases such as SiC and $Si_3N_4$. Related oxynitride phases in the Si-Al-O-N system have been the basis for development of a range of ceramics, especially in the U.K., known generically as the 'Sialons'. The most prominant member of this family of ceramics is based on the $\beta'$ phase which is a substituted derivative of $\beta Si_3N_4$, of general formula $Si_{3-x}Al_xO_xN_{4-x}$. Alternative major ceramic phases in this system, such as a substituted derivative (O$'$) of $Si_2N_2O$, or one of the polytypoid phases derived from AlN by Si and O substitution have also been studied.

However, they are more difficult to fabricate and have not yet reached the same state of microstructural refinement or property attainment as β´ ceramics.

The limited diffusion rates within high-covalency ceramic phases is detrimental to fabrication via solid-state sintering of component particles. Hence sintering catalysed by a minor liquid phase is the most important fabrication route for the oxynitride ceramics discussed here. Sintering liquids are normally silicates containing variable amounts of dissolved nitrogen. The silicate residue is normally an intergranular glass phase which imposes a limit to high-temperature application at its softening point. The virtue of Si-Al-O-N compositions is not (as commonly believed) the property improvement via solid-solution 'alloying' but rather the flexibility in control of the liquid composition such that sintering is more efficient and the residue may be minimised or controllably crystallised. This microstructural control and property relation is exemplified for β´ ceramics in Section 3.1.

An alternative fabrication process for oxynitride ceramics utilises the 'glass-ceramic' route of homogeneous glass-formation followed by controlled crystallisation. Compositions are limited to glass forming regions of M-Si-Al-O-N systems (where M is a di or trivalent metallic ion) and hence to the intergranular liquid in sintered ceramics. With carefully tailored compositions nearly complete crystallisation of oxide + oxynitride bi-phase ceramics is attainable. The evolution of an oxynitride glass-ceramic microstructure is exemplified in Section 3.2.

## 3. Ceramic Microstructures

### 3.1. Development of high-temperature microstructure in β´ Ceramics

The key to high-temperature ceramic performance under stress, especially in oxidising environment, is the elimination or complete crystallisation of intergranular glass. One approach to the problem is to make initial compositions almost coincident with the β´ solid-solution line and utilise transient sintering liquids derived from additives such as $AlN+SiO_2+MO$. This principle has been used to obtain the hot-pressed monophase β´ microstructures of Fig.2 (Lewis et al 1977). Grain morphology is characteristic of solid-solid interfaces without marked energy anisotropy and typifies the absence of liquid residues. High-resolution lattice imaging may be used in confirmation of interface structure (Fig.2a,b) but suffers from a sampling problem and the geometrical constraint of parallelism between interface and lattice planes in neighbouring crystals. The most important feature in preventing creep-cavitation is the absence of triple-junction residues which have been identified as cavity-nuclei. When of sub-critical size for cavity nucleation, these monophase β´ ceramics exhibit 'superplastic' diffusional creep controlled by grain-boundary diffusion (Karunaratne and Lewis 1980). Diffusion rates vary with grain-boundary segregated impurities which are mainly derived from the sintering liquid and may be detected on intergranular fracture surfaces using Auger electron spectroscopy (Fig.2d).

A different class of β´ ceramic is that fabricated by pressureless-sintering using a comparatively large sintering liquid volume. Low temperature ceramics consist of elongated prismatic β´ crystals within glass matrix of ternary eutectic ($MO-Al_2O_3-SiO_2$) composition. Saturation of the liquid with nitrogen increases the glass softening point but, more important,

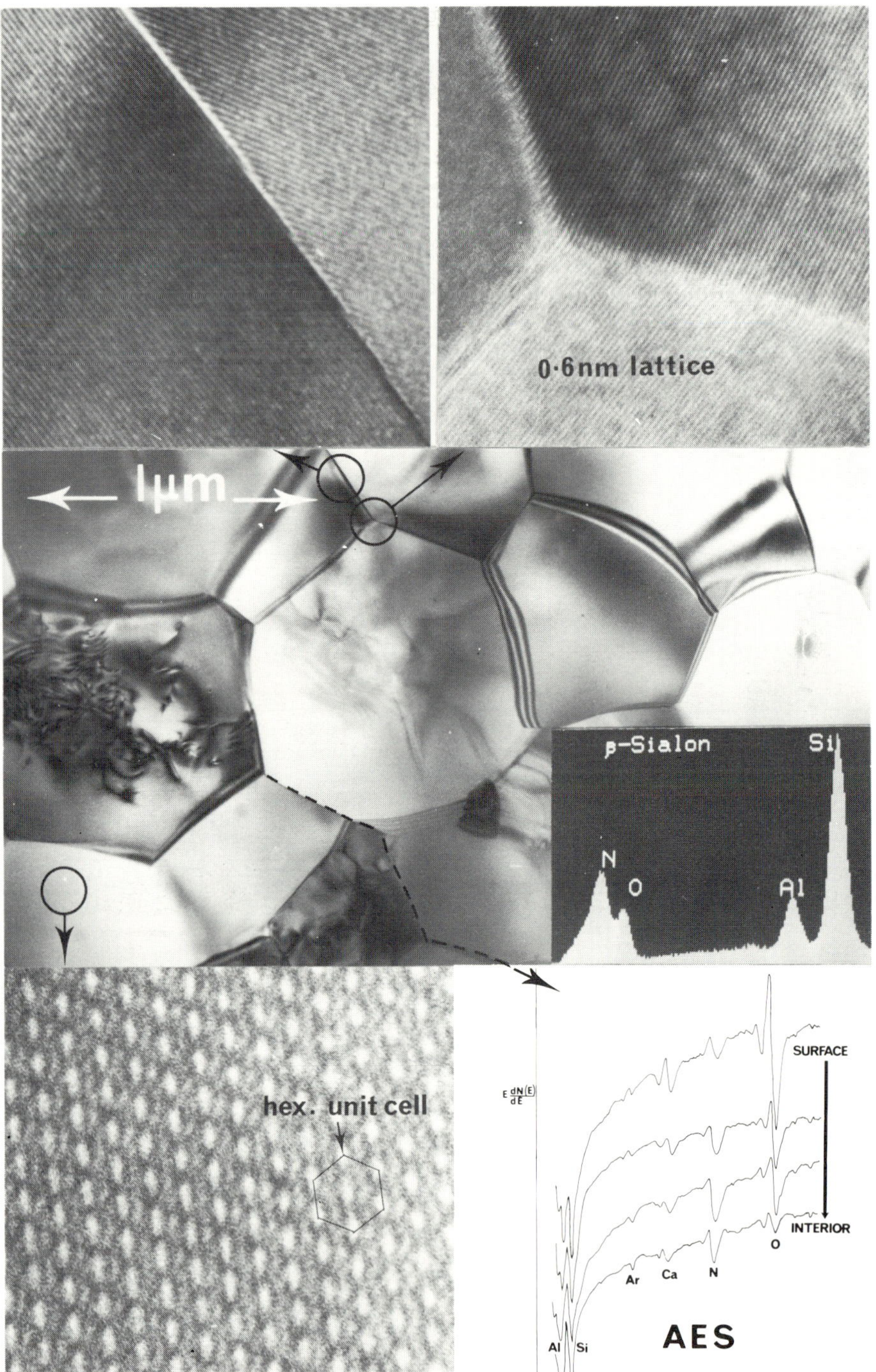


Fig.2 The transmission electron microstructure of a hot-pressed monophase β´ oxynitride ceramic. The lattice-images and Auger electron spectra (AES) from a fracture surface are used in characterising the structural and chemical state of grain-boundaries.

makes it susceptible to crystallisation of non-silicate refractory phases. For example, when MO = $Y_2O_3$ the matrix crystallisation product is $3Y_2O_3.5Al_2O_3$ (YAG) with excess Si,N and some O precipitated as β´. The phase morphology and composition during development of a β´+YAG microstructure are illustrated in Fig.3. Completion of crystallisation of this bi-phase microstructure is essential in attainment of non-cavitating diffusional creep and other high-temperature properties comparable with hot-pressed monophase ceramics (Lewis et al (1980).

3.2. Oxynitride glass ceramics

Synthesis of bulk M-Si-Al-O-N glasses of high nitrogen content may be followed by a crystallising heat-treatment to produce a refractory bi-phase ceramic. For example, a Y-Si-Al-O-N glass may be prepared near to the

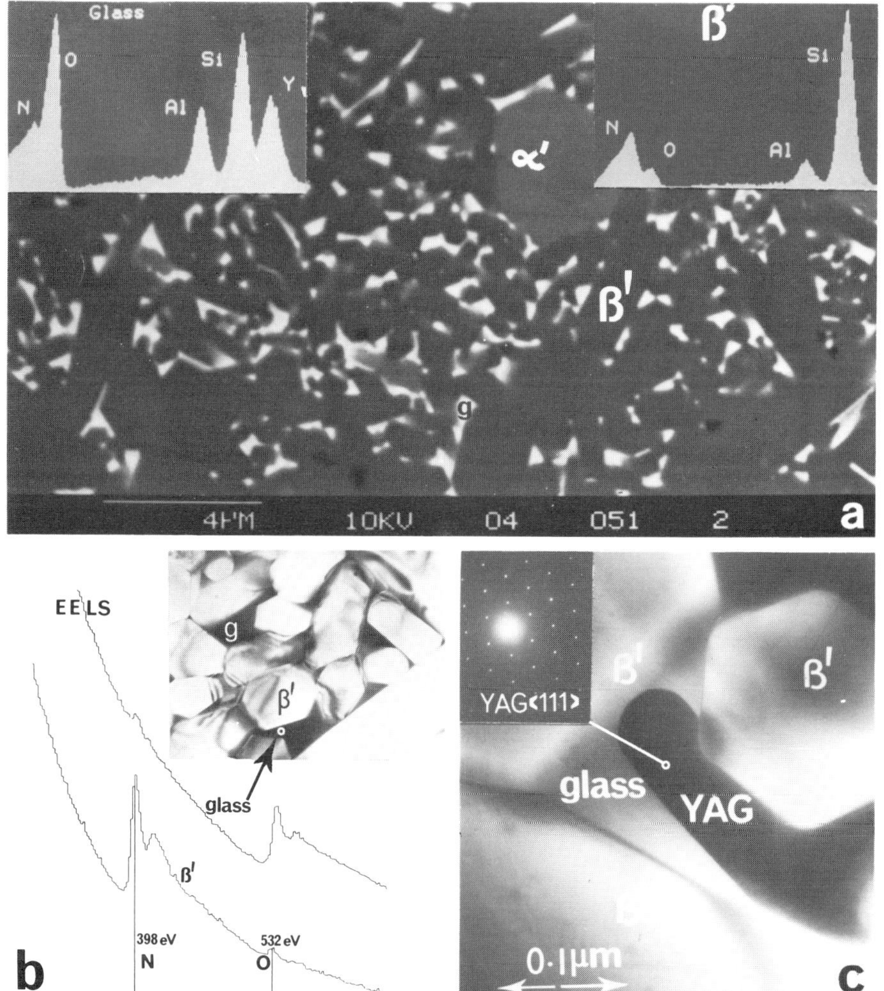


Fig.3 (a) Back-scattered SEM image and windowless EDS from β´ and glass matrix in a pressureless-sintered bi-phase ceramic; (b) Analysis of O/N in glass matrix and β´using ELS;(c)Partial crystallisation of matrix glass as YAG.

Fig.4 Crystallisation of YAG and $Si_2N_2O$ phases in an oxynitride glass-ceramic; (a) back-scattered SEM image, (b) transmission electron image.

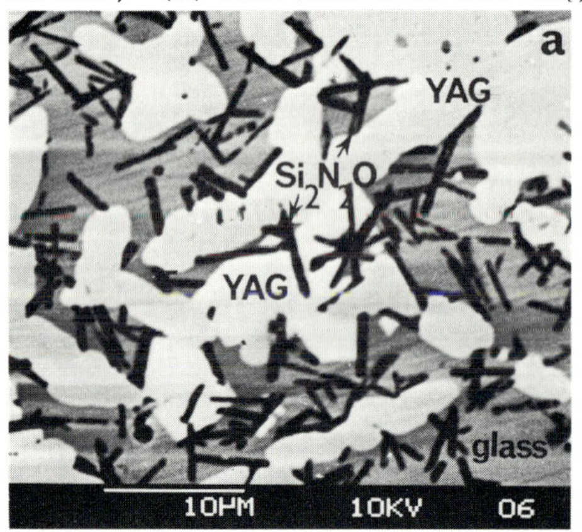


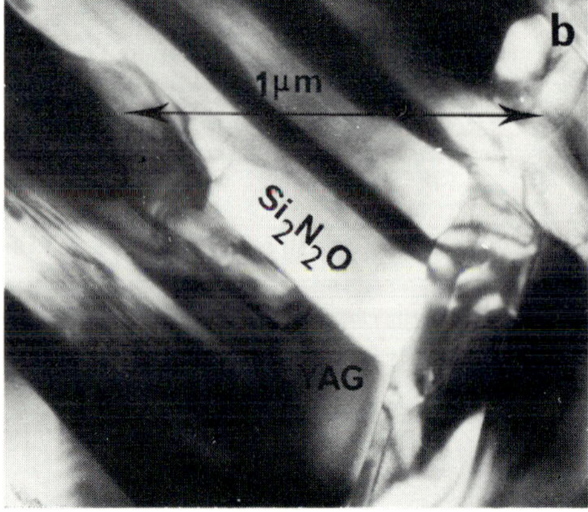


$Si_2N_2O$-YAG compositional tie-line; the parallel development of these phases from the homogeneous glass is illustrated in Fig.4 (Leng-Ward and Lewis 1984). The partitioning of the heavy element (yttrium) between glass and crystallising phases produces good contrast in back-scattered SEM images which may be used in conjunction with windowless EDS for phase analysis. The refinement of microstructure in relation to property development for low and high-temperature application is currently being studied for a range of oxynitride compositions.

## 4. Recent Developments

There are many remaining crystallographic problems involving phases in M-Si-Al-O-N systems in view of the similarity in X-ray scattering factors for the range of light elements. Using the increased electron scattering and dynamical enhancement of weak-reflections it has been possible to provide additional data concerning the ordering of Mg,Si,Al atoms in solid-solutions crystallising from oxynitride glasses (Leng-Ward et al 1984).

Ceramic surface coatings, both on metallic substrates and as diffusion barriers on high-temperature ceramics, such as the oxynitrides, are under development using a variety of deposition techniques, e.g. plasma spraying or CVD. Such films are heterogeneous and frequently contain metastable and non-crystalline domains. It is probable that, in addition to microbeam electron diffraction, direct structure-imaging using HREM will be a powerful technique in coating characterisation. Such techniques have been applied to anodic oxide films in combination with extended electron energy-loss fine structure (EXELFS - Bourdillon et al 1984). This spatially-resolved information is unlikely to provide a complete structure analysis and is most-appropriately used in combination with the 'broad-beam' EXAFS technique and a new technique of 'magic-angle spinning' (MAS) n.m.r.

## References

Bell P S & Lewis M H 1974 Phil.Mag. 29 1175
Bourdillon A, El-Mashri S M & Forty A J 1984 Phil.Mag. 49 341
Karunaratne B S B & Lewis M H 1980 J.Mat.Sci. 15 449
Leng-Ward G, Lewis M H & Wild S 1984 J.Mat.Sci. 19 1726
Leng-Ward G & Lewis M H 1985 Mat.Sci. & Eng. 71 101
Lewis et al 1977 J.Mat.Sci. 12 61
Lewis M H, Bhatti A R, Lumby R J & North B 1980 J.Mat.Sci. 15 103

# A study of the $\alpha' \rightleftharpoons \beta'$ phase transformation in the Ca–SiAlON system

P A Walls

Wolfson Research Group for High-Strength Materials, Department of Metallurgy & Engineering Materials, University of Newcastle upon Tyne

## 1. Introduction

Sialon engineering ceramics are finding ever widening applications from metal cutting tips to engine cylinder liners. The basic sialon materials were developed in the early 1970's. Since then research has been carried out to improve and refine the physical and chemical properties of these materials. For example it has been suggested that the wear resistance of $\beta'$-sialon cutting tips can be improved by incorporation of $\alpha'$-sialon into the microstructure.

## 2. Summary

In the present work the microstructural and compositional nature of such composites is investigated by the use of various electron microscopy techniques on reaction couple samples. Such couples are a useful method of examining reaction mechanisms and to further the understanding of phase behaviour diagrams. This is achieved by studying the compositional variations across the interface using X-ray Digimapping in conjunction with X-ray diffractometry of successive layers. This has revealed the presence of distinct phase regions parallel to the couple interface and S.E. imaging of etched samples and B.E. imaging of polished samples has revealed microstructural details.

## 3. Discussion

Previous work (Hampshire, Park, Thompson and Jack 1978) had shown that the $\alpha' \rightleftharpoons \beta'$ transformations occur when finely ground powders of $\alpha'$ and $Al_2O_3$ reacted to produce $\beta'$, and $\beta'$, AlN, $Ca_3N_2$ mixtures reacted to precipitate $\alpha'$ when fired above 1700°C. However in these specimens the reactions occurred between very small (<10μm) particles and so were difficult to follow using standard TEM techniques. The use of bulk samples of $\alpha'$ and $\beta'$ and the necessary reactants for the transformation in the form of reaction couples, enabled the massive transportation of species to be followed thus providing information on reactions which occur on the micron level.

### 3.1 The $\alpha' \longrightarrow \beta'$ transformation ($\alpha'$-$Al_2O_3$ reaction couples)

Fig. 1 shows the reaction zone of an $\alpha'$-$Al_2O_3$ couple hot pressed above 1700°C, with associated Ca, Si and Al Digimaps, Figs 2, 3 and 4 respectively. On the left of the micrograph a fine grained high calcium content region is observed. This is unreacted $\alpha'$ which contains

≈10c/o of a glassy phase which when liquid acts to densify the α'-sialon.

Reaction Zone

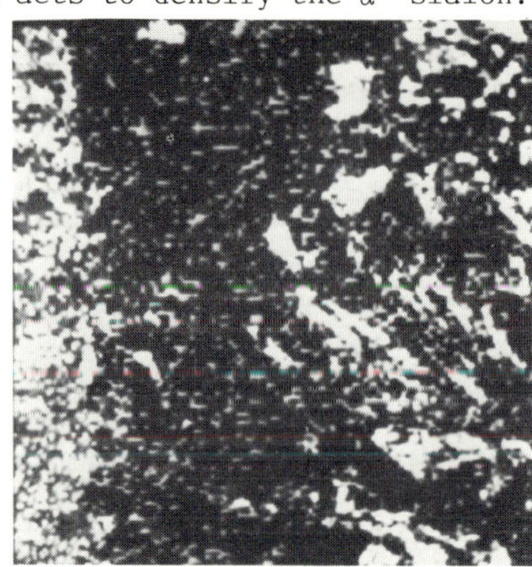

Fig. 2 Ca Digimap

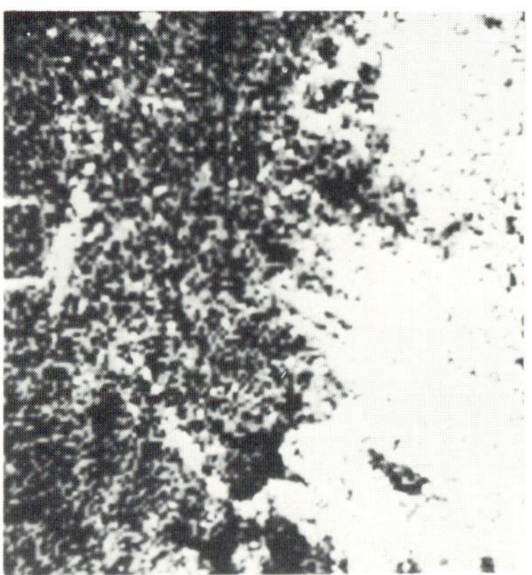

Fig. 3 Si Digimap

Fig. 4 Al Digimap

The central region with a much lower calcium content, slightly lower silicon content and containing approximately the same amount of aluminium as the α' region has not been attacked to such a degree by the HF etchant, indicating a lower glass content material. The region on the right hand side of the micrograph consists of a series of parallel platelet crystallites (seen in section) containing little calcium, a low silicon component but with high aluminium content. These are separated by calcium-rich regions.

X-ray diffractometry across the couple which involved grinding successive layers parallel to the interface showed three distinct regions. These were 15R, β' and α'-sialon. These microstructural and X-ray results agree from a compositional viewpoint in that 15R has the composition $SiAl_4O_2N_4$, β'-sialon cannot accept Ca into its structure and the α' which can accommodate calcium into both its crystal structure and into the intergranular glassy phase.

From these results and with consideration of the sialon behaviour diagram, reaction mechanisms may be suggested.

Initially liquid formation occurs at the interface when the eutectic temperature for the $Al_2O_3$-Ca-sialon glass is reacted. This liquid has an increased Al-O content and so would not be in equilibrium with either the α' or the $Al_2O_3$. A high Al content phase (ie 15R) then precipitates by combination of Al-O and Si-N from the liquid phase leaving the remaining liquid rich in calcium. The 15R observed by X-ray diffraction studies was preferentially orientated such that growth had been predominantly perpendicular to the [001] direction. The stability of the α' is affected by the change in liquid composition causing the former to dissolve increasing the local Si-N concentration. This together with an increased Al content allows precipitation of β'. As more liquid is produced these processes, solution of α' and $Al_2O_3$ and precipitation of 15R and β', proceed as long as Al-O and Si-N transportation is possible through the liquid. This growth mechanism explains why the 15R plates appear to grow away from the Al-O rich region to regions of higher S-N concentration. The β' at the β'- (15R+glass) interface must dissolve and reprecipitate as 15R due to the non-equilibrium compositions of the β' and liquid. Thus expansion of the reaction zone occurs accompanied by shifting of existing phase boundaries.

3.2 β'⟶α' transformation (β'-CaO-AlN reaction couple)

Previous work (Hampshire, Park, Thompson and Jack 1978) has shown the possibility of preparing β' from α', AlN and an appropriate metal nitride eg $Ca_3N_2$ to produce a Ca-α'-sialon. The transformation should be possible by replacing $Ca_3N_2$ by CaO to produce a more oxygen rich α'-sialon.

This transformation has been studied using a reaction couple method, however as three components are involved a "sandwich" method was adopted in which a thin layer of compacted CaO powder is inserted between polished samples of β' and AlN. This, once molten, would act as a carrier for the species involved in the reaction. Due to the large amount of liquid that is produced when firing at 1700°C the couple was pressureless sintered.

Figures 5 and 6 are BSE micrographs indicating the compositional variation across the AlN-CaO and CaO-β' boundaries respectively.

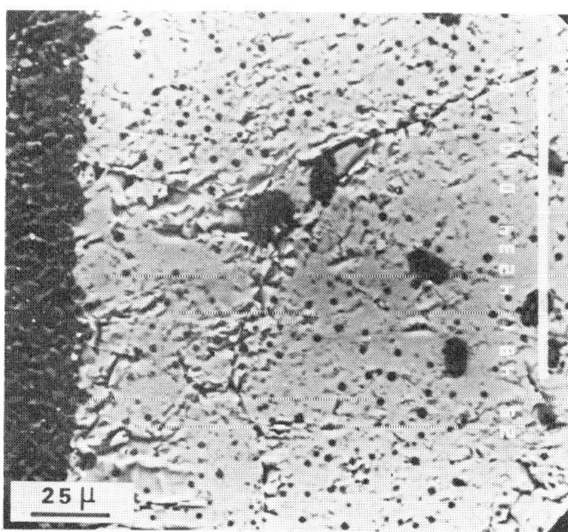


Fig. 5 AlN-CaO Interface

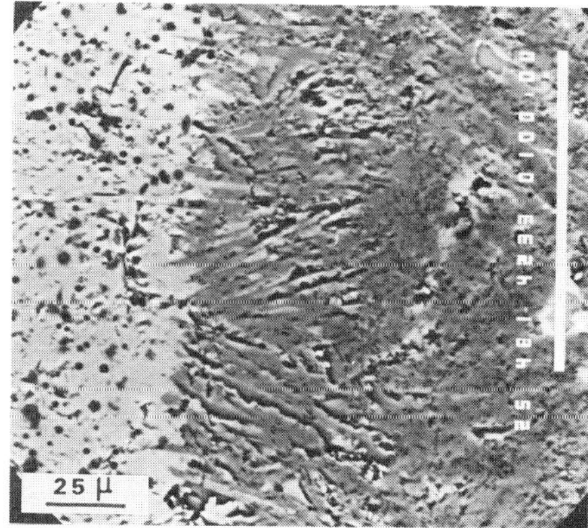


Fig. 6 CaO-β' Interface

Digimapping also showed the presence of the platelet crystallites growing from the β' into a Ca-O rich Si-Al-O-N glass. Particles of AlN (verified by electron microprobe analysis) have become detached from the AlN-CaO interface and are visible as dark regions in the lighter Ca-sialon glass. The size of these AlN particles decreases closer to the β'-CaO interface as they gradually dissolve in the liquid. The platelet crystals observed contain less silicon, more aluminium and have a similar calcium content than the β' which suggests they are α'-sialon. The β' was densified with 10e/o of Ca-sialon liquid and hence shows an even distribution of calcium which is actually distributed at the grain boundaries. In the α' the calcium is contained within the structure.

## 4. Conclusions

The $\alpha' \rightleftharpoons \beta'$ transformations can be followed using bulk specimens in the form of reaction couples. The reaction mechanisms can be deduced from microstructural and compositional information obtained using SEI and BEI with X-ray analysis in the form of Digimaps.

The phase regions observed are consistent with present understanding of the sialon behaviour diagrams and the use of electron microscopy together with X-ray diffraction techniques form a strong tool for the study of sialon reactions.

## 5. References

Hampshire S, Park H K, Thompson D P and Jack K H 1978 Nature 274 pp 880-2

*Inst. Phys. Conf. Ser. No 78: Chapter 13*
*Paper presented at EMAG '85, Newcastle upon Tyne, 2–5 September 1985*


# Interfacial microstructure of SiAlON composites

K M Knowles

Department of Metallurgy and Materials Science,
University of Cambridge, Pembroke Street, Cambridge CB2 3QZ.

## 1. Introduction

During the sintering or hot pressing of sialon ceramics an intergranular glassy film forms which, although beneficial to densification, contributes towards strength degradation at high temperatures through a grain boundary sliding mechanism. The distribution and chemical composition of this grain boundary phase is therefore of both scientific and technological importance. TEM studies of ceramic materials have shown that films as narrow as 1 nm can be detected by techniques such as dark field imaging and lattice fringe imaging (Clarke, 1979), but little attention has been paid to the possibility of obtaining information about the local structure of the amorphous-crystalline boundaries. This is perhaps not surprising given the uncertainties about providing unambiguous structural information about amorphous material even when direct interpretable microscope resolution at the 0.2 nm level is available (Smith et al, 1980). However, recent high resolution electron microscopy of the surfaces of small particles (Marks et al, 1984) and of the silicon-silicon oxide interface (Liliental et al, 1985) is more encouraging, and suggests that images from the amorphous-crystalline boundaries in ceramics can indeed contain interpretable structural information. Here we report initial observations of interfacial microstructures from $\alpha'/\beta'$, $O'/\beta'$ and dual phase Y-Si-O-N ceramic composites and discuss their structural interpretation.

## 2. Experimental details

Ion beam thinned specimens were obtained from samples prepared by hot pressing or reaction sintering of appropriate powders. For example, a mixture of $Si_3N_4$, AlN, $Al_2O_3$ and CaO powders was used to form the $\alpha'/\beta'$ composite. The specimens were carbon coated for initial examination at 100 kV in a JEM 100C electron microscope before transfer to the Cambridge University 600 kV HREM.

## 3. Results and Discussion

The crystal structures of $\alpha'$, $\beta'$ and O' are essentially those of $\alpha$-$Si_3N_4$, $\beta$-$Si_3N_4$ and $Si_2N_2O$, but with substitution of $Si^{4+}$ by $Al^{3+}$ and $N^{3-}$ by $O^{2-}$; $\alpha'$ can also incorporate metal ions such as $Ca^{2+}$ interstitially. In the $\alpha'/\beta'$ and $O'/\beta'$ specimens studied, the larger grains had well developed facets of low index, whereas the smaller grains tended to have a more irregular morphology. For example, Fig. 1 shows an $\alpha'$ grain of about 0.5μm diameter observed down the [0001] zone axis, where the larger facets are $\{10\bar{1}0\}$ and the smaller $\{11\bar{2}0\}$. Grains such as this when viewed near a thin edge clearly offer the best hope of obtaining interpretable images of the transition from the crystalline grain to the amorphous grain boundary phase, since zones are readily obtainable where the interfaces, if atomistically smooth, lie parallel to the electron beam. The situation is then similar to that encountered in the study of native oxide on single crystal silicon (e.g. Bravman et al, 1985; Liliental et al, 1985).

Fig. 2 is an HREM image from the [0001] zone of a β' grain in the O'/β' composite showing two $\{10\bar{1}0\}$ facets. Comparison of this image with the image simulations for β-$Si_3N_4$ by Hiraga et al. (1983) readily shows that in the thinnest regions of the grain the large bright dots correspond to the channels in the structure and the dark hexagons to atom rows. Close inspection of the facet leading away from the thin edge reveals features which imply that it is not atomistically smooth, but instead rough at a level of about 0.3 nm. Indeed, in regions where the grain only one unit cell away is clearly thick, the image at the boundary has an appearance of that from the thinnest regions, which would be consistent with a surface roughening. This image is therefore in principle analysable in the same way as the silicon - silicon oxide interfaces, an example of which is shown for comparison in Fig. 3.

Figs. 4 and 5 show further HREM images, both from α' grains in the [0001] zone. These images are examples of less ideal cases, in that the grains are not faceted on low index planes. The situation is particularly confused in Fig. 4, where the prescence of a one dimensional set of fringes whose spacing correspond to $(10\bar{1}1)$ α' planes masks completely the presence of any intergranular phase. However, in Fig. 5, there is a relatively wide amorphous phase present and as in Fig. 2 the amorphous-crystalline region appears rough at an atomic level. The amorphous phase itself does not exhibit any suggestion of microcrystalline regions, but this is not surprising, given the necessity of having to carbon coat the specimens.

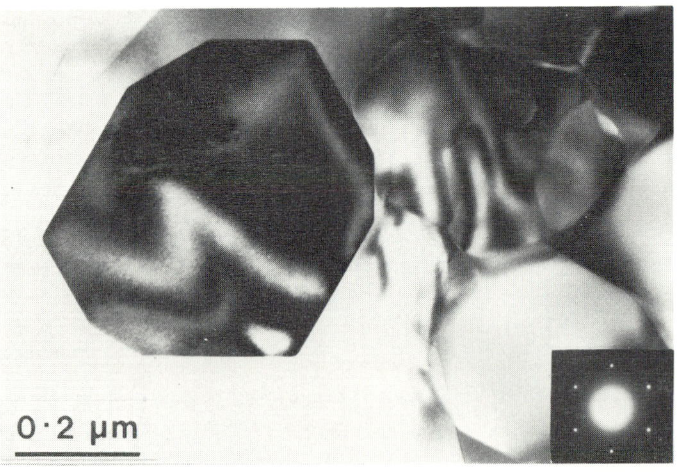


Fig. 1: α' grain in an α'/β' composite showing well defined facets. Part of the [0001] zone diffraction pattern from the grain is shown correctly oriented in the inset.

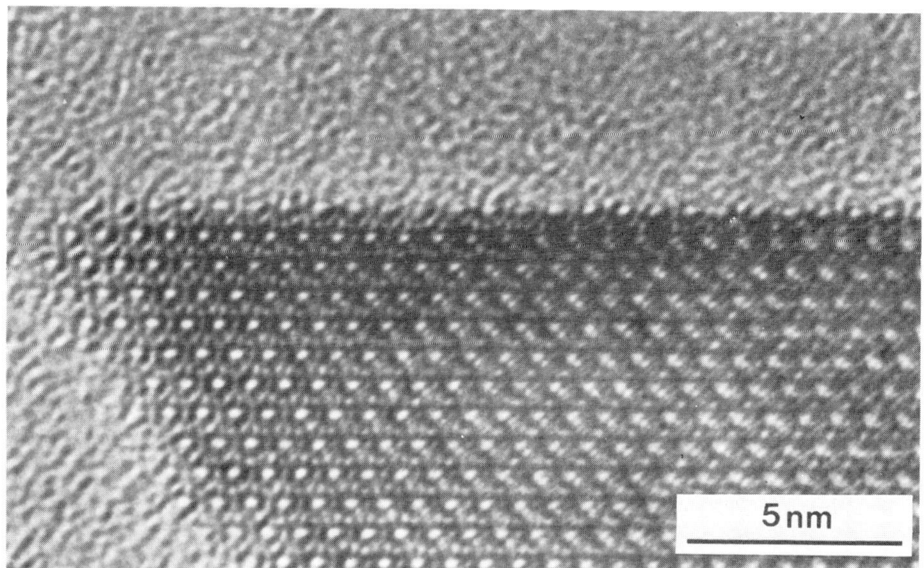


Fig. 2: HREM image of $\{10\bar{1}0\}$ facets in a β' grain. Zone axis [0001]. Foil edge bottom left.

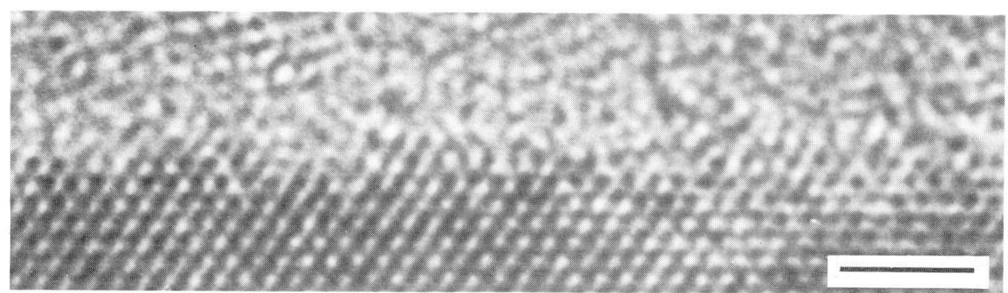

Fig. 3: HREM image of a single crystal silicon - silicon oxide - polycrystalline silicon interface with a ∿1.2 nm oxide layer. The interface is approximately (001) with respect to the silicon. Scale marker 2 nm.

Fig. 4: HREM image of an α' grain down the [0001] zone axis. The $<11\bar{2}0>$ directions are parallel to the edges of the hexagons visible in the grain. Scale marker 5 nm.

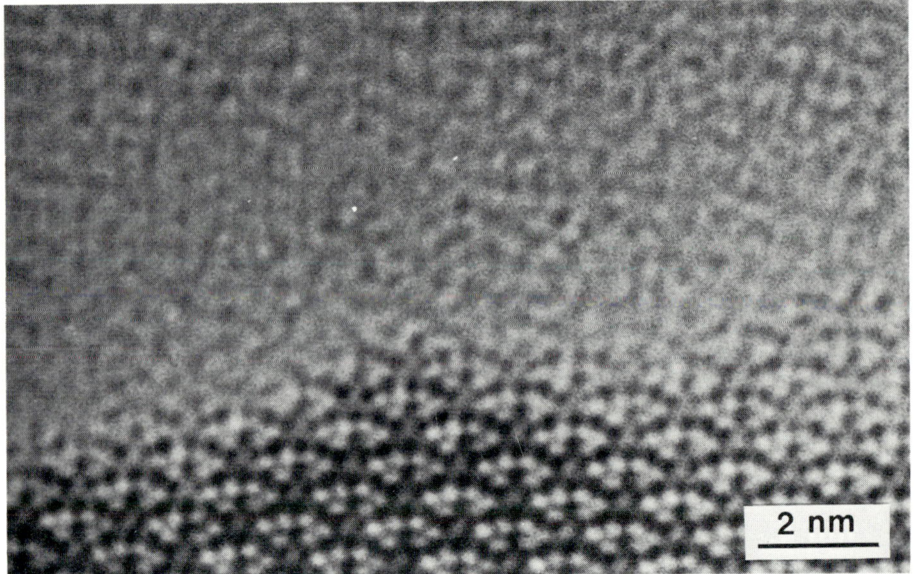


Fig. 5: HREM image of an α' grain down the [0001] zone axis with a grain boundary phase ∿2 nm wide opening up towards a triple point to the right of the photograph. Faint fringes from the adjacent grains are visible.

In contrast to the α'/β' and O'/β' composites, the grains in the dual phase Y-Si-O-N material are less well defined. This composite consists of N-α-wollastonite ($YSiO_2N$), which is known to fault on (001) planes and N-YAM ($Y_4Si_2O_7N_2$). Initial work has shown that the N-YAM twins on (001) and that 'favoured' orientation relationships can occur between grains. This system should therefore be amenable to the study of the wettability of interphase boundaries as a function of orientation relationship.

Acknowledgments

I would like to thank Mr. P. Walls and Dr. W.Y. Sun for their help in specimen preparation. I would also like to thank the SERC, UK for financial support for the Cambridge University HREM and Prof. D. Hull for the provision of laboratory facilities.

References

Bravman J C, Patton G L, Sinclair R and Plummer J D 1985 Mat. Res. Soc. Symp. 37 461

Clarke D 1979 Ultramicroscopy 4 33

Hiraga K, Tsuno K, Shindo D, Hirobayashi M, Hayashi S and Hirai T 1983 Phil. Mag. A 47 483

Liliental Z, Krivanek O L, Goodnick S M and Wilmsen C W 1985 Mat. Res. Soc. Symp. 37 193

Marks L M, Heine V and Smith D J 1984 Phys. Rev. Lett. 52 656

Smith D J, Saxton W O, Cleaver J R A and Catto C J C 1980 J. Microscopy 119 19

*Inst. Phys. Conf. Ser. No 78: Chapter 13*
*Paper presented at EMAG '85, Newcastle upon Tyne, 2–5 September 1985*



# HREM of incoherent $ZrO_2/Al_2O_3$ interfaces

A H Heuer, S P Kraus-Lanteri, P A Labun, and T E Mitchell
Case Western Reserve University
Cleveland, Ohio 44106 USA

## Introduction

Dispersion-toughened ceramics have received much attention in recent years due to their impressive mechanical properties. In particular, $ZrO_2$ additions to $Al_2O_3$ can increase the strength of the material to greater than 1 GPa (1). Typically, small amounts of $ZrO_2$ ($\leq$ 30 vol.%) are dispersed in a fine-grained $Al_2O_3$ matrix; the $ZrO_2$ exists as both intragranular and intergranular, incoherent particles with either monoclinic (m) or tetragonal (t) symmetry. The enhanced strength and toughness arise from the phenomenon known as transformation toughening, involving the martensitic t-$ZrO_2$ to m-$ZrO_2$ transformation. Nucleation of the transformation invariably occurs at the $ZrO_2/Al_2O_3$ interface (2).

High resolution electron microscopy contributes vital information in helping to understand these interfaces. In the past, HREM has been performed on materials of known orientation containing coherent or semi-coherent interfaces. In polycrystalline $ZrO_2$-toughened $Al_2O_3$ (ZTA), the $ZrO_2/Al_2O_3$ interface is not only incoherent but the particle and matrix are of random orientation. In spite of these adverse conditions, HREM is still possible, and yields interesting results.

## Experimental procedure

As-sintered ZTA samples containing 3.8, 10, and 15 vol.% $ZrO_2$ were studied using HREM. Thin foils were prepared by conventional ion-milling techniques. After ion-milling, the thin foils received a short heat treatment (15 min. at 1200°C) to reverse any t $\rightarrow$ m transformation which may have occured during foil preparation. Samples were examined in a JEOL 200CX transmission electron microscope fitted with a top entry $\pm$ 10° tilting stage; the microscope has a $C_s$ of 1.1 mm, and has a point-to-point resolution of 0.236 nm.

## Results and discussion

In order to image the $ZrO_2/Al_2O_3$ interface successfully, each phase must be exactly on zone, while the interface should be parallel to the electron beam. Due to the random distribution of $ZrO_2$ particles, these conditions are difficult to meet; hundreds of particles were investigated, but few were in (or could be tilted to) the proper orientation.

Fig. 1a shows an intragranular, t-$ZrO_2$ particle from a 10 vol.% sample at low magnification. The $Al_2O_3$ matrix is oriented to the $[3\bar{1}\bar{2}1]$ zone, while a single set of {111} planes is visible in the $ZrO_2$, as it is about two degrees off its <110> zone. The {111} $ZrO_2$ planes and $(0\bar{1}11)$ $Al_2O_3$ planes are misoriented by about five degrees; furthermore, the d-spacings of these two sets of planes differ significantly, 0.295 nm for (111) $ZrO_2$ and 0.393 nm for $(0\bar{1}11)$ $Al_2O_3$. In spite of this apparent lattice mismatch, no microcracks or other gross distortions appear at the interface in these HREM images.

Quasi-periodic fringes give rise to some unusual contrast around the particle. At higher magnification (Fig. 1b), the periodicity of the fringes is seen to arise from every fourth $Al_2O_3$ plane (arrowed) stopping short at the interface, as compared to the three preceding planes. As the interface orientation changes, the periodicity changes from every fourth to every third $Al_2O_3$ plane. These arrowed planes are analogous to misfit dislocations commonly observed in semicoherent interfaces. As the orientation changes, so does the apparent configuration of misfit dislocations in the interface structure; they do not appear periodic in Fig. 1c, and are widely spaced in Fig. 1d.

At some orientations, the lattice mismatch is accommodated by a series of ledges one atomic plane in height (Fig. 1e). It is remarkable that features familiar from semicoherent interfaces, i.e. misfit dislocations and ledges, are also a feature of incoherent interfaces in the present system. This must arise in this instance, because the lattice mismatch is in fact small; four $ZrO_2$ (111) planes have a spacing of 1.180 nm, while three $Al_2O_3$ $(0\bar{1}11)$ planes are spaced 1.179 nm.

Near the interface in the $ZrO_2$ particle, some structure images including planes with half of the $(\bar{1}11)$ planar spacing can be discerned. It is believed that the strains arising from the thermal expansion mismatch of the two materials have bent the $ZrO_2$ lattice into a favorable orientation for imaging the (222) planes. This can be seen most clearly in Fig. 1e.

The majority of the intragranular $ZrO_2$ particles in the 10 vol.% material are spherical, but some facetted particles are also present (Fig. 2a). The facetting suggests the existence of a low energy interface between $Al_2O_3$ and $ZrO_2$. The $Al_2O_3$ grain is exactly oriented to the $[10\bar{1}0]$ zone, but the Moiré fringes in the particle unfortunately indicate that the $ZrO_2$ overlaps the $Al_2O_3$. The $ZrO_2$ orientation was determined using optical diffractometry, which showed that (200) planes from t-$ZrO_2$ were parallel to $(1\bar{2}10)$ $Al_2O_3$ planes. Fig. 2b shows that the basal planes of $Al_2O_3$ are also parallel to the facet. Mazerolles, et al. (3) have studied $Al_2O_3$-$ZrO_2(Y_2O_3)$ eutectics and found the following planes of t-$ZrO_2$ and $Al_2O_3$ to be parallel: {001} // {0001}, {100} // $\{2\bar{1}\bar{1}0\}$, {010} // $\{01\bar{1}0\}$; the corresponding directions were also parallel. For the particular facetted particle in Fig. 2b, (200) is parallel to $(1\bar{2}10)$.

Image simulations using O'Keefe's SHRLI (Simulated High Resolution Lattice Imaging) program (4) are currently being performed with various models to obtain a match with experimental images, and thus determine the interface structure. Our strategy is to first create a sharp interface with the ideal $ZrO_2$ and $Al_2O_3$ structures in the observed (random) orientation, and then to relax the atomic positions to best fit the observed images. To date, the perfect lattices of $Al_2O_3$ in the [0001] and $[1\bar{1}01]$ zones have been successfully simulated.

Summary

Incoherent $ZrO_2/Al_2O_3$ interfaces can be successfully imaged with HREM. Lattice mismatch is accommodated by a combination of misfit dislocations and ledges, analogous to those found in semicoherent precipitates. Low energy orientations also exist between $ZrO_2$ and $Al_2O_3$, as attested to by presence of facetted particles.

Acknowledgment

This research was supported by Air Force Grant No. AFSOR-82-0227.

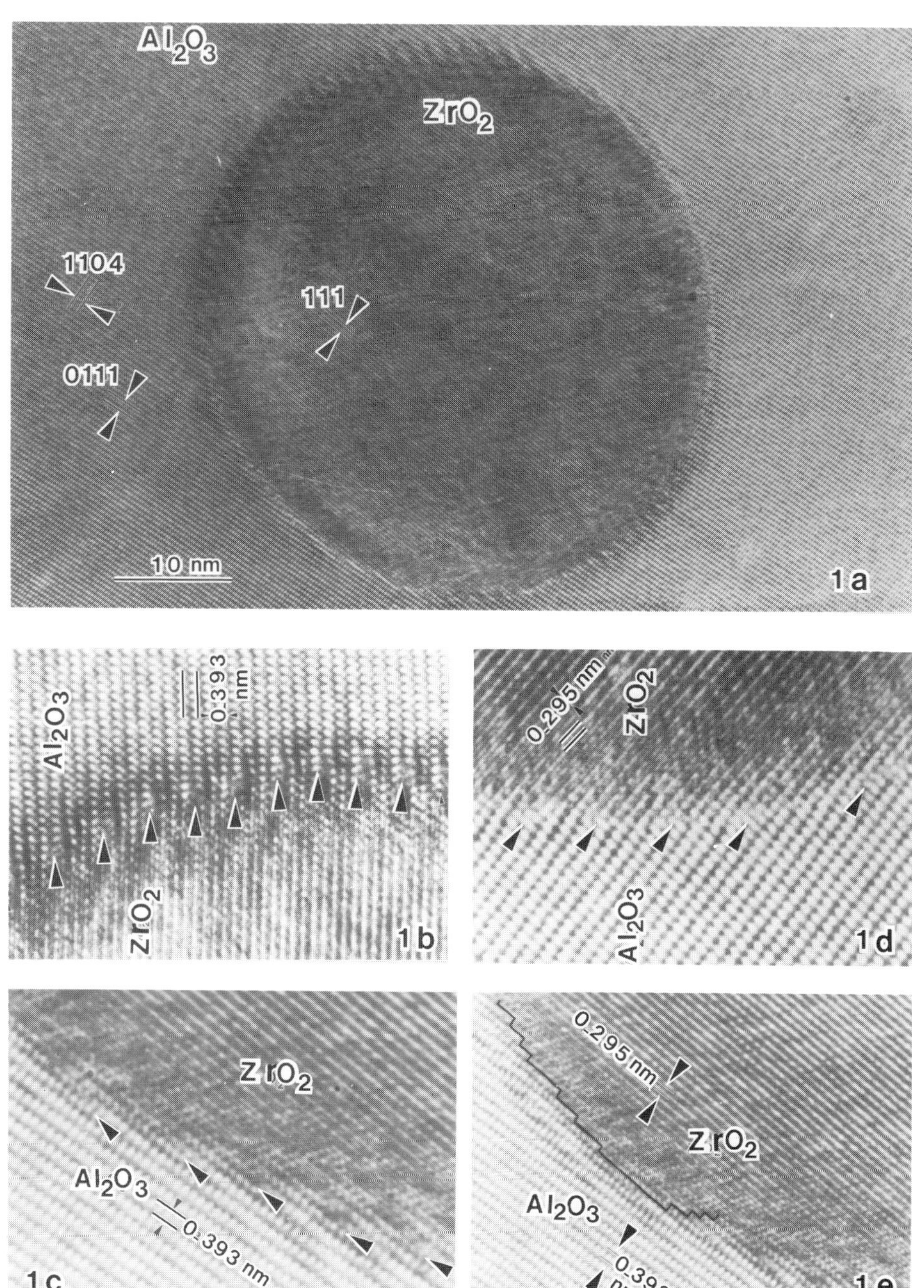


Fig. 1 (a)Intragranular, t-$ZrO_2$ particle with higher magnification images of various areas (b-e). See text for details.

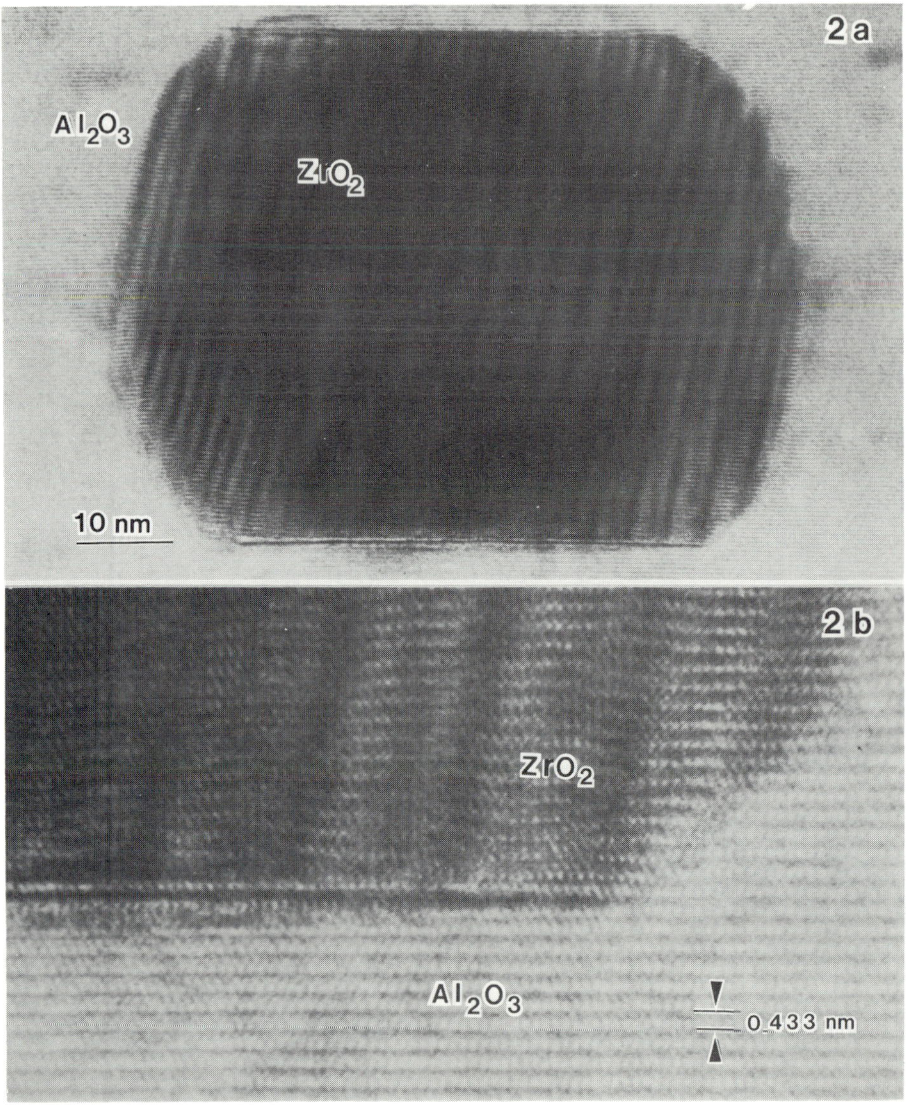

Fig. 2 (a)Intragranular facetted t-$ZrO_2$ particle. (b)Note $Al_2O_3$ basal planes parallel to facet.

References


1. N. Claussen in Advances in Ceramics, Vol. 12, edited by N. Claussen, M. Rühle, and A. H. Heuer, American Ceramic Society (Columbus, Ohio) 1983, p. 325.
2. A. H. Heuer and M. Rühle, in press, Acta Met., 1985.
3. L. Mazerolles, D. Michel, and R. Portier, to be published in J. Am. Ceram. Soc., 1985.
4. M. A. O'Keefe and P. R. Buseck, Trans. Am. Crystallogr. Assoc., 15 24-46 (1979).

# The determination of the mean inner potential of grain boundary films in WC–Co composites by Fresnel techniques

J N Ness, W M Stobbs and T F Page
Department of Metallurgy and Materials Science, Pembroke Street, Cambridge

Thin grain boundary films are found in many ceramic materials and can play an important role in controlling their mechanical and physical properties. In particular, controversy about WC-Co materials relates to whether the WC phase can be contiguous or whether the boundaries always contain a thin wetting film of W-C-Co (Exner and Gurland 1970). The presence of a continuous film would have important consequences for the properties of this material. Observations by Jayaram and Sinclair (1983), using lattice imaging and dark-field techniques (applied to 6-10wt%Co materials) showed that both contiguous carbide grains and thin ~1nm cobalt films were present. These observations contrasted with the STEM analysis of Sharma et al. (1980), on WC-6wt%Co, who found a large increase in cobalt levels at all boundaries, although it was not clear whether this always implied the presence of a second phase at the boundary.

This paper explores the use of defocus or 'Fresnel' imaging techniques applied to boundaries in WC-Co. The technique is not only used in its more normal form for estimating boundary width (e.g. Krivanek et al. 1979; Clarke 1979), but also for estimating the change in mean inner potential across the boundary - and using this to give information on the composition of the boundary. Previous computer simulations (Jepps et al. 1982) have justified the use of the technique for measuring boundary widths. In addition, Rühle and Sass (1984) have used the variation of fringe spacing with defocus to estimate the change in mean inner potential at grain boundary dislocations in NiO. However, they were unable to utilise the intensity information within the fringes, and the purpose of this paper is to extend this technique to include this information.

The change in potential (i.e. the average projected potential of the material, see Goodman and Moodie 1974) at a grain boundary film can be represented by a simple potential well. A variety of boundary profiles were used as bases for the simulation of Fresnel fringe profiles, using a multislice programme (Goodman and Moodie 1974), developed at the University of Melbourne (Bursill and Wood 1978) and modified slightly for this work. No atomic positions were included in the calculations, this being justified by extensive simulations (as will be reported elsewhere), given that the grain boundary widths were not comparable to the lattice spacings and the images were obtained with the grains only weakly diffracting. Various boundary potentials and thicknesses were considered and typical results are shown in figure 1. Fig.1a is a plot of relative central fringe intensity against defocus for a variety of boundary thicknesses and potential changes, for a constant specimen thickness of 25nm.

Fig.1b is a similar plot for the first Fresnel fringe. As can be seen from both figures, the fringe intensities are strongly affected by small changes in potential but are relatively insensitive to similar percentage changes in boundary thickness. The variation in fringe intensity with potential is also sufficiently linear to suggest that measurements of experimental fringe intensities could be reliably used to estimate the potential change at the boundary. Many other simulations were also performed to gauge the effect of other specimen and instrumental parameters, the most important being specimen thickness and beam convergence. However, since both of these are directly measurable, they can be allowed for in the calculations.

The technique requires the boundary to be oriented 'edge-on' to the electron beam, the symmetry of the fringes being used to check this. A through-focal series (usually in steps of 50nm) was taken e.g.fig.2a, using a sufficiently small exposure time (~3s) to minimise the effects of specimen drift whilst still keeping a reasonably small beam convergence. A $LaB_6$ filament was used to provide a fairly coherent and bright illumination source at low beam convergences, for the microscope used (Philips 400ST, operating at 100kV). Beam convergence was measured directly from a carefully exposed diffraction pattern and specimen thickness estimated by a combination of convergent-beam diffraction, weak-beam dark-field imaging and by measuring the distance between contamination spots after tilting through a known angle.

The images were then analysed on a computer-controlled microdensitometer. The image analysis system resident in the computer allowed sections of the boundary images to be averaged, greatly increasing the signal to noise ratio of the resultant profiles. The various intensities and width measurements were then directly taken from the stored profiles. The fringe width measurements could then be used to estimate the boundary thickness, by plotting fringe spacing versus defocus e.g. figure 2b. Errors in this measurement are quite small and have only a minor effect on the subsequent potential estimation, as mentioned above. The estimated boundary thickness was then used as a basis for subsequent simulations, initially in conjunction with an estimate of the boundary potential change and with all the measured specimen and microscope parameters. An iterative procedure enabled the best fit values of potential, beam divergence and specimen thickness to be determined (the last two parameters often needing 'fine-tuning' from the measured values).

Figure 3 shows plots made for the boundary shown in figure 2a in a WC-9wt%Co specimen. Superimposed on the experimental intensities are simulations for potential drops of 4.5 and 3.5eV at the boundary, for a 1.2nm wide film and a measured specimen thickness of 15nm. 3.5eV represents approximately the best fit possible, bearing in mind that there will be inaccuracies in the experimental measurements. Figure 4 shows dark-field images of both the boundary and a large Co-rich area (both imaged with the indicated aperture position). This figure (and related results) suggests that the boundary material is similar to that found in the large areas. In fact, comparable images are produced irrespective of the direction in reciprocal space that is sampled, suggesting that the material may possibly be amorphous (although an amorphous layer on the specimen surface would be expected from the ion-beam thinning process, which is the origin of most of the image intensity in fig.4b). However, a technique such as structure imaging would be necessary to investigate this further.

A drop in potential of 4.5eV would be expected if the boundary contained pure cobalt and the grains were pure WC ($V_{WC}$=33.5eV; $V_{CO}$=29.0eV); whilst the best fit potential of 3.5eV would correspond to a solution of ~20% WC in the cobalt (cobalt is almost insoluble in WC), suggesting that the specimen thickness was marginally underestimated. This figure must thus be taken only as a guide. Also, no account has been taken of inelastic scattering; although this should be low for the small specimen thicknesses examined - even for the heavy elements involved here. However, the results seem to suggest that those boundaries which contain a second phase are very Co-rich, in agreement with previous work. A major advantage of this technique is that, although a relatively large (1.2nm!) boundary was illustrated here, there is no reason why the technique cannot be extended to much thinner boundaries (even monolayers), since image contrast would still be very good and the calculations can be readily modified.

In conclusion, we have illustrated an extension of the Fresnel imaging technique to allow estimation of potential changes at boundaries, and have applied this to a WC-Co ceramic. Results have been presented for a 1.2nm boundary, whose potential drop of 3.5eV (combined with dark-field imaging) indicated that the boundary contained mainly cobalt of possibly amorphous form. The relatively low resolution required makes the technique appealing, and its extension to other ceramics is apparent.

Bursill L A and Wood G J 1978 Phil Mag 38 673
Clarke D R 1979 Ultramicrosopy 4 33
Exner H E and Gurland J 1970 Powder Metall 13 13
Goodman P and Moodie A F 1974 Acta Cryst A34 280
Jayaram V and Sinclair R 1983 J Am Ceram Soc 66 C137
Jepps N W, Page T F and Stobbs W M 1982 Grain Boundaries In Semiconductors (New York: Elsevier) pp45-50
Krivanek O L, Shaw T M and Thomas G 1979 J Appl Phys 50 4223
Rühle M and Sass S L 1984 Phil Mag 49 759
Sharma N K, Ward I D, Fraser H L and Williams W S 1980 J Am Ceram Soc 63 194

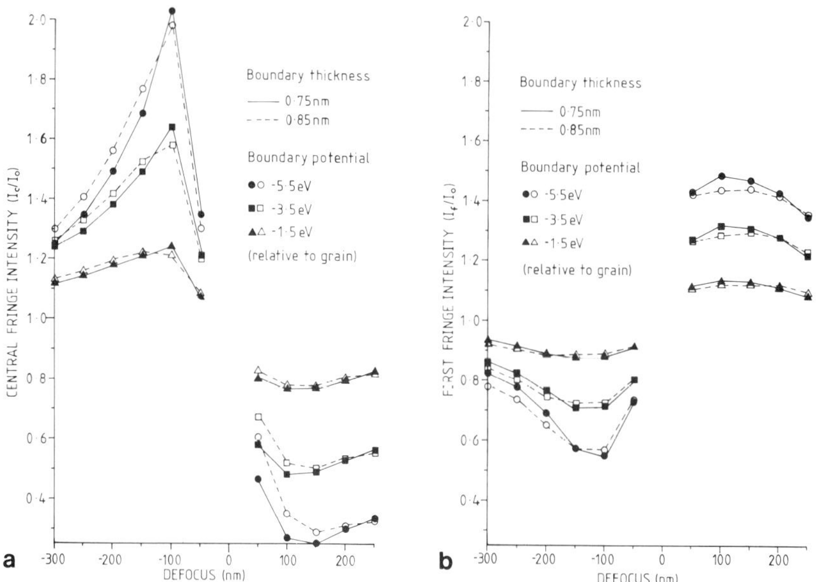


Fig.1. Plots of (a)Central and (b)First Fresnel fringe intensity against defocus for various potential drops and boundary thicknesses.

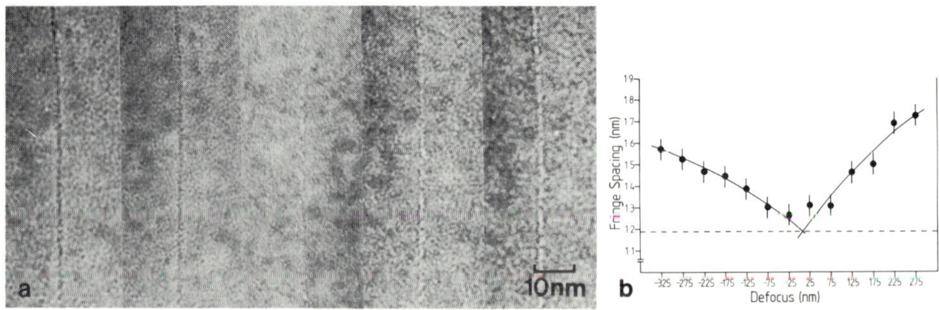


Fig.2. a)Part of a through-focal series for an 'edge-on' boundary in WC-Co.
b)Fresnel fringe spacing versus defocus for the boundary of (a).

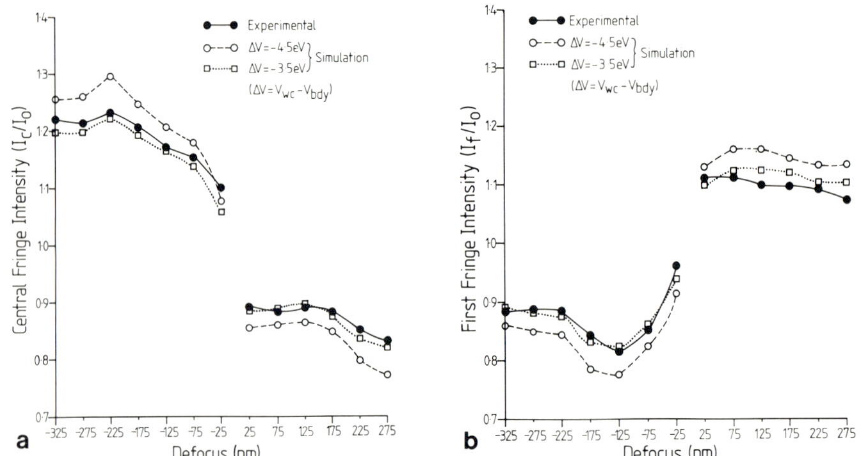


Fig.3. Plots of fringe intensities versus defocus (for the boundary of fig.2) compared to two simulated boundary potentials, and a 1.2nm thickness.

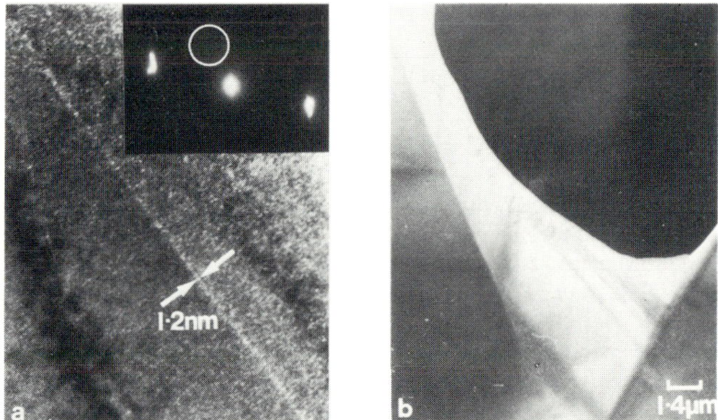


Fig.4. Dark-field images (using the ringed aperture position) of (a)The boundary of fig.2 and (b)A large area of residual cobalt.

*Inst. Phys. Conf. Ser. No 78: Chapter 13*
*Paper presented at EMAG '85, Newcastle upon Tyne, 2–5 September 1985*



# Microscopy and microanalysis of grain boundaries in alumina and zirconia ceramics

E Paul Butler

Alcan International Limited, Southam Road, Banbury, Oxon OX16 7SP.

## 1. Introduction

The properties of dense, sintered technical ceramics are frequently determined by the occurrence and distribution of intergranular phases. These can be either amorphous or crystalline in nature, and arise from the presence of impurities in the original starting powders, from deliberate additions made to assist densification, and from reactions between the ceramic and intergranular components at the sintering temperature. Properties such as fracture strength, creep strength at elevated temperatures, and electrical properties (e.g. ionic conductivity in $ZrO_2$ ceramics) are significantly affected by second phases at boundaries.

Amorphous grain boundary films are commonly found in commercial $Al_2O_3$ and $ZrO_2$ ceramics. This is not unexpected in those aluminas where CaO, MgO, $SiO_2$, etc. are added to coarse-grade powders to act as sintering aids. The additives form a liquid phase at boundaries which provides a high diffusivity path but partial devitrification on cooling to a mixture of amorphous and crystalline phases (Witek & Butler 1984) is also possible. Grain boundary films indicative of liquid phase sintering have been identified even in 99.8% $Al_2O_3$ (Hansen & Phillips 1983). $SiO_2$ is a ubiquitous impurity in $ZrO_2$ powders, and hence low melting point silicate phases readily form during the sintering of $ZrO_2$ ceramics as the $SiO_2$ reacts with the oxide dopants added to stabilise the $ZrO_2$ (MgO, CaO, $Y_2O_3$) together with the other impurities present, such as $Al_2O_3$ picked up from milling. Certain $ZrO_2$ ceramics are precipitation-hardened at 1300-1400°C following sintering at 1750-1800°C, and it is possible for the boundary phase to be still liquid during ageing if it has a composition such that the eutectic temperature is below the ageing temperature. Under these circumstances there can be significant interference with the microstructural evolution close to grain boundaries (Butler & Heuer 1985).

The application of TEM to the characterisation of structure/property relationships in ceramics is not new, but increasing emphasis on optimising structural and electrical performance has necessitated a more scientifically-based understanding of microstructural evolution, especially in grain boundary localities. Microanalytical capabilities are essential to provide the necessary microchemical information from grain boundaries, but as in all multicomponent systems, quantification must be approached with caution: sample preparation, thickness and instrumental parameters may conspire against the simple extraction of meaningful compositional information, and the geometry of grain boundary films or pockets may give rise to incomplete separation of the X-ray signals between the area of interest and the surrounding grains. The ion milling of ceramic samples can result in silicon-rich contaminant layers, and under aggressive milling conditions give additional contamination of the sample with components from gun, support and chamber, together with more pronounced argon ion implantation (Figure 1).

The examples in this contribution have been selected to illustrate a range of grain boundary microstructural features in alumina and zirconia ceramics. All microanalytical data was gathered taking care to minimise specimen contamination from milling and following good experimental practices of quantitative X-ray microanalysis (Williams 1984). $k_{AB}$ factors appropriate to the analyses have been reported separately (Witek & Butler 1984, Butler & Heuer 1985).

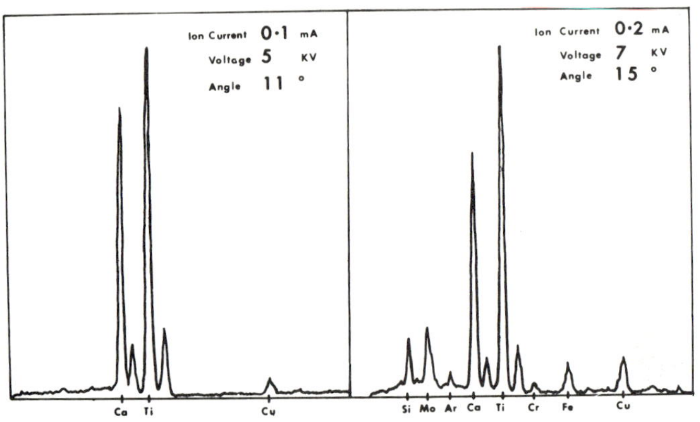


Fig. 1 EDS results for two $CaTiO_3$ foils prepared under standard (left) and aggressive (right) ion milling conditions

## Alumina and Alumina-Zirconia Ceramics

A ceramic material (A1) consisting of Alcoa CT2000 alumina powder with additions (wt %) of 0.36 CaO, 0.97 MgO and 1.67 $SiO_2$ was prepared, cold pressed and conventionally sintered at 1600°C. The influence of $ZrO_2$ additions to this alumina was studied in a second material (A2) containing the same $Al_2O_3$-CaO-MgO-$SiO_2$ mixture to which 15 wt% $ZrO_2$ (doped with 5 wt % $Y_2O_3$) was added in the form of a nitrate sol. A second alumina-zirconia (A3) was prepared from alumina powder with much lower levels of additives/impurities: 0.20 CaO, 0.12 MgO, 0.11 $SiO_2$.

Figure 2(a) shows a typical grain boundary pocket in sample A1, consisting of a dark angular phase A together with a partially devitrified region B containing crystalline needles. Microanalysis of these areas gave the following compositions (mol %):

A: 45 $Al_2O_3$, 0 CaO, 55 MgO, 0 $SiO_2$
B: 20 $Al_2O_3$, 15 CaO, 5 MgO, 60 $SiO_2$

Particle A has a composition close to that of $MgAl_2O_4$ spinel and, according to the $Al_2O_3$-CaO-MgO-$SiO_2$ phase diagram (Levin et al 1964), composition B lies in the anorthite phase field. Anorthite formation is expected as the MgO level is reduced from an $Al_2O_3$-CaO-MgO-$SiO_2$ glass due to local spinel nucleation and growth.

Figure 2(b) is a typical grain boundary pocket in sample A2; the dark rounded regions, e.g. Z, are $ZrO_2$. The composition of glassy area C was as B in Figure 2(a) plus ~ 1 mol % $ZrO_2$ and a similar amount of $Y_2O_3$. The presence of these two constituents dissolved in the grain boundary phase has thus prevented partial devitrification and the formation of anorthite. More detailed work (Witek & Butler 1985) has identified $ZrO_2$ as the important glass former.

The $ZrO_2$ particles in the microstructure of sample A3 were present at both intergranular and intragranular sites (Figure 3a). $ZrO_2$ particles at both types of sites can getter impurities from the ceramic; in the example shown in Figure 3b the particle analysed contains CaO, believed to originate from water washing of the gel precipitate (Butler & Heuer 1982).

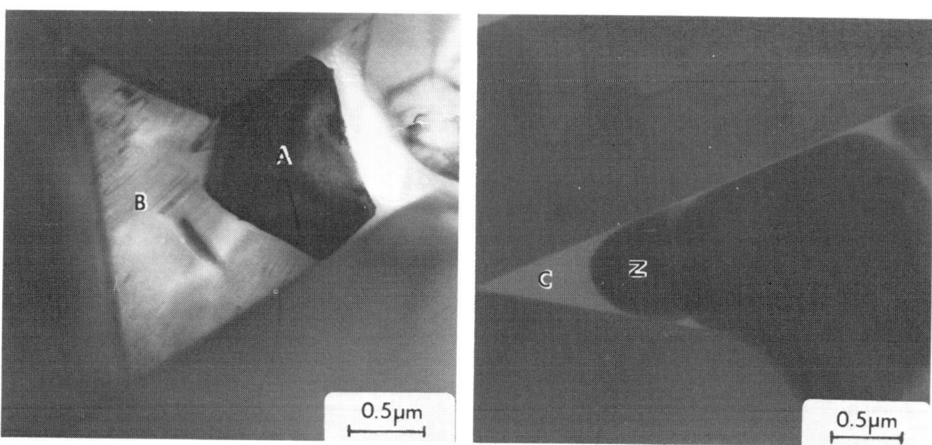


Fig. 2a Crystalline and amorphous grain boundary phases in A1.
2b Amorphous phases plus $ZrO_2$ particles (e.g. Z) in A2.

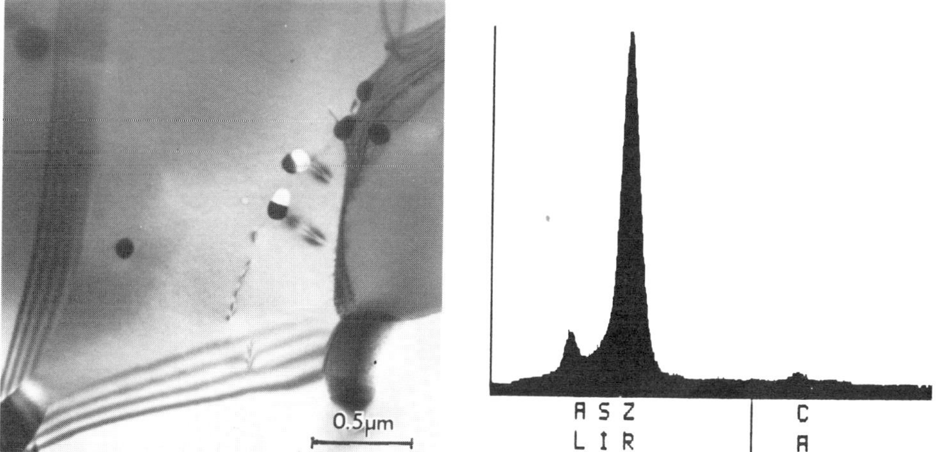


Fig. 3a $ZrO_2$ particles at intergranular and intragranular sites in A3.
3b EDS of intergranular particle showing presence of CaO.

## Zirconia Ceramics

The question of grain boundary phase continuity can be explored using the technique of AC impedance spectroscopy in conjunction with TEM (Butler & Bonanos 1985). Grain boundary films are frequently continuous but not always so; in Figure 4(a) a non-wetting glass is present as discrete 'lenses'. When films are continuous, reactive film migration (RFM) can take place (Butler & Heuer 1985), producing thin regions of cubic symmetry adjacent to boundaries after ageing (Figure 4(b)). In MgO-containing zirconias, forsterite can form at boundaries (Figure 5) which destabilises near grain boundary localities to monoclinic symmetry.

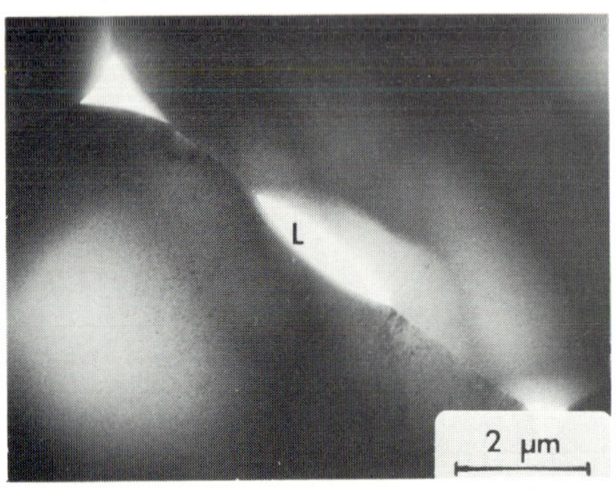


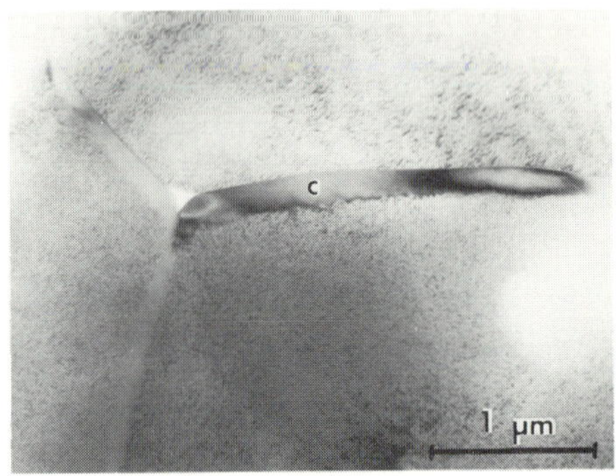


Fig. 4(a) Lens-shaped glassy phase (L) in as-fired Y-PSZ
(b) Cubic zones (c) along boundaries in aged Y-PSZ
Fig. 5 (below) Forsterite particle (F) surrounded by destabilised $ZrO_2$ of m-symmetry in Mg-PSZ:EDS of particle to right

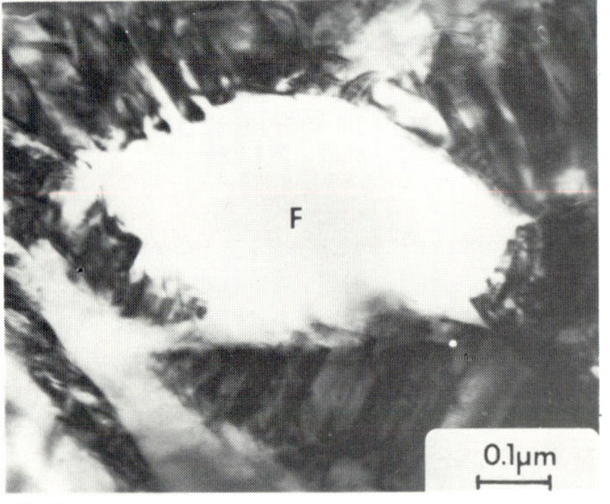


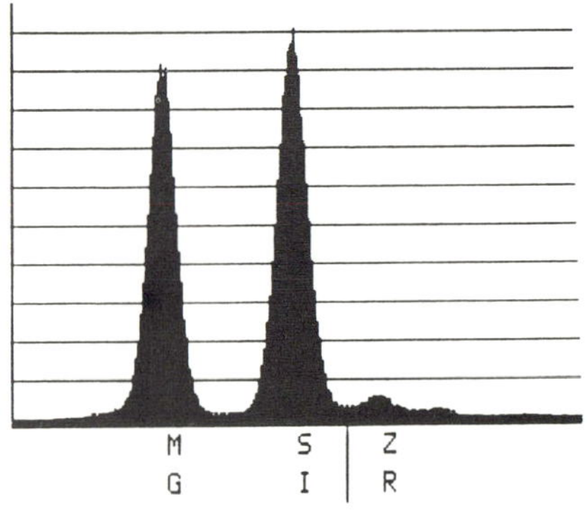


## References

Butler E P and Bonanos N 1985 Mat. Sci. Eng. 71 49
Butler E P and Heuer A H 1982 J. Am. Ceram. Soc. 65 C-206
Butler E P and Heuer A H 1985 J. Am. Ceram. Soc. 68 197
Hansen S C and Phillips D S 1983 Phil. Mag. (A) 47 209
Levin E M, Robbins C R and McMurdie H F 1964 Phase Diagrams for Ceramists ed M K Reser (Columbus, Ohio: American Ceramic Society) Fig 882
Williams D B 1984 Practical Analytical Electron Microscopy in Materials Science (Mahwah, New Jersey: Philips Electron Optics) pp 55-90
Witek S R and Butler E P 1984 Analytical Electron Microscopy - 1984 ed D B Williams and D C Joy (San Francisco: San Francisco Press) pp 318-20
Witek S R and Butler E P 1985 J. Mat. Sci. Lett. (in press)

*Inst. Phys. Conf. Ser. No 78: Chapter 13*
*Paper presented at EMAG '85, Newcastle upon Tyne, 2–5 September 1985* 

# A TEM study of a modified PZT-ferroelectric ceramic

C A Randall*, D J Barber* and R W Whatmore+

*Department of Physics, University of Essex, Colchester CO4 3SQ
+Plessey Co. Ltd., Allen Clark Research Centre, Caswell, Towcester, Northants.

## 1. Introduction:

A modified-PZT hot-pressed ceramic with a composition $Pb(Zr_{.58}Fe_{.20}Nb_{.20}Ti_{.02})_{.995}U_{.005}O_3$, developed by Whatmore et al. (1981) has a Curie temperature of ~ 228°C. Below this temperature it is ferroelectric with the cations displaced to give a dipole polarization in ⟨111⟩ and the cubic (paraelectric) cell is transformed to a rhombohedral cell. This ceramic has found many applications because of the excellent pyroelectric properties which it exhibits.

## 2. Experimental Details

Samples were cut and polished to a thickness approximately 25 μm and then thinned with 5 kV argon ions at an incident angle of 12°. TEM observations were made with a JEOL-200 CX at 120, 160 and 200 kV. Hot and cold stage work was performed at 1 MV with a EM7 microscope.

## 3. Domain Configurations

Periodic wedge domains are essentially twins on {110} with head-to-tail dipole coupling. This type of domain configuration is easily recognised by observing its behaviour using reflections both perpendicular and parallel to the domain wall. Fig. (1) shows dark-field image using a reflection plane parallel to the wall with strong constrast between neighbouring twins, as a consequence of the non-centrosymmetry of the unit cell (Gevers et al[2]). Imaging with reflections perpendicular to the domain wall results in invisibility, the twin displacement vector being parallel to the wall. The fringe contrast at an inclined domain wall shows δ-fringe properties and the nature of the bounding fringes can be used to determine the direction of the dipoles (Gevers et al[3]).

Fine lamellar domains are also frequently observed. These are closely packed and cause diffraction spots in the directions perpendicular to the trace, ⟨110⟩, to be split. The fine lamellar domains are relatively unstable to electron beam heating. Complex configurations of mixed wedge domains and fine lamellar domains are also often found - see fig. (2).

Twinning on {100} is rare compared to {110}. A Landau-type calculation reveals the relative energy of a rhombohedral system twinning on {100} is three times larger than the energy if only {110} walls are allowed.

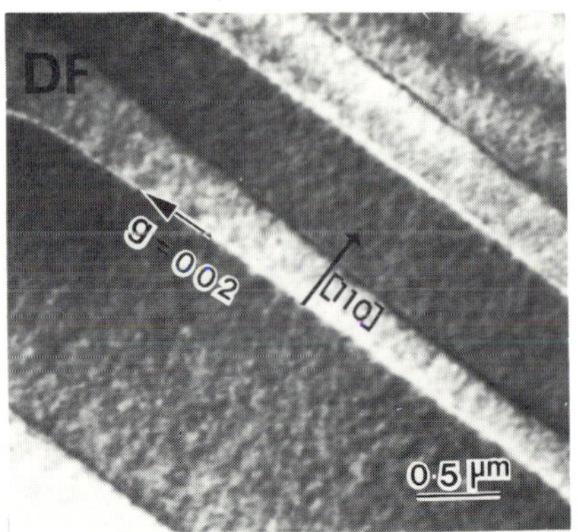


Fig. 1 Dark field image of wedge domains with g = 002 operating.

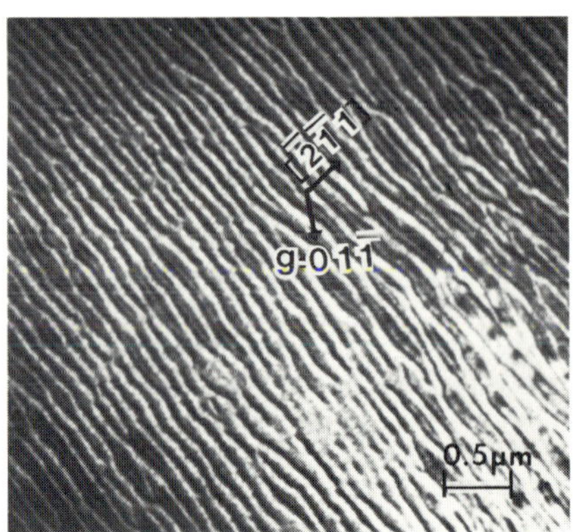


Fig. 2 Fine lamellar domains on ($\bar{1}01$).

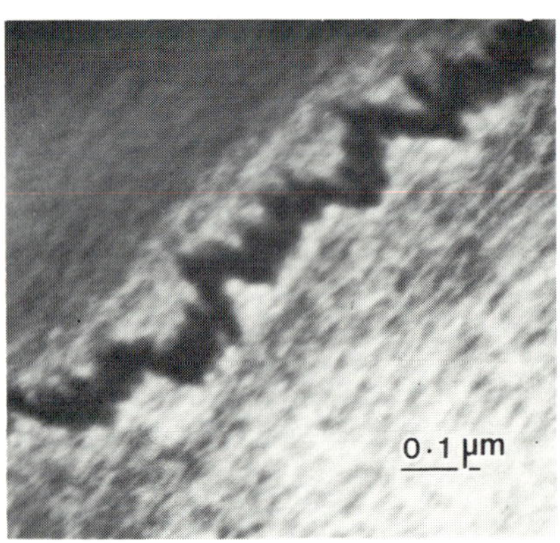


Fig. 3(a) Head-to-head domain.

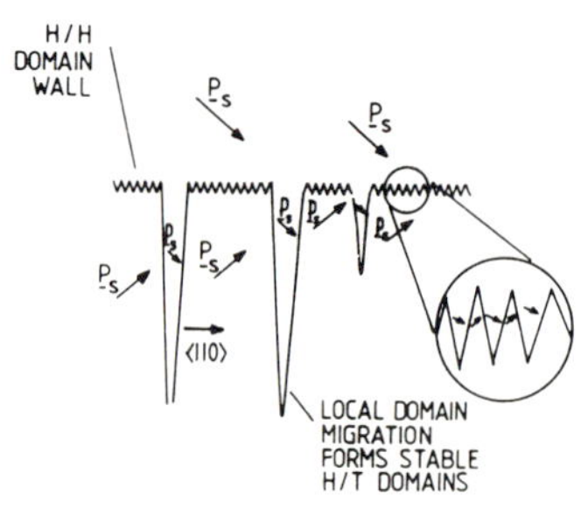


Fig. 3(b) Schematic representation of head-to-head switching to head-to-tail by domain wall migration.

Highly mobile domains with zig-zag boundaries occur and these have head-to-head (or tail-to-tail) dipole coupling. This type of wall has previously been observed in $BaTiO_3$ by S. Yakunin et al.[4]. The zig-zag configuration reduces the the electrostatic self-energy. Elongation of zig-zags by domain wall migration can result in the formation of the wedge-like domains - see fig. 3.

Direct evidence of domain pinning is presented in fig. 4. By using a heating stage it can be demonstrated that the domains use dislocations as nucleation sites. The dislocation that we have shown to belong to the ⟨110⟩ {110} slip system and have predominately screw-character. Dislocation pinning of domain walls act as a source of dielectric loss in ferroelectric ceramics.

## 4. Superlattice Reflections and Diffuse Scattering

(1) F-type reflections e.g. {1/2, 1/2, 1/2}, {3/2,1/2,1/2} etc. (see fig. 5).

(2) Diffuse scattering in $\underline{q}$ = ⟨110⟩ direction with superlattice reflections of type {1/2, 1/2, 0}, {3/2, 1/2, 0}, etc. (see fig. 6).

F-type reflections are seen both above and below the Curie temperature, namely, in the range 0°C - 700°C (above 700°C decomposition takes place). It is thought that at temperatures below the Curie temperature the source of the superlattice spots is two-fold: an octahedral tilt of $a^-a^-a^-$ in the notation of Glazer et al. and also an ordering of the cations, which is only short-range. Above the Curie temperature only the cation ordering remains. With these superlattice spots no contrast effects are seen.

Diffuse scattering in perovskites had been found by Harada et al. in $BaTiO_3$ and in $KNbO_3$ by Comes et al. However, the direction of the scattering here was $\underline{q}$ = ⟨100⟩. In the case of the modified-PZT diffuse scattering is in $\underline{q}$ = ⟨110⟩, with a superlattice spot showing a doubling in this direction. The origin of the diffuse scattering is thought to originate from thermal vibrations (Huller et al.). At this time the origin of the superlattice spots is not known.

Acknowledgements

We express our thanks to Dr. D.R. Tilley for help with Landau calculations and to Dr. S. Bains for X-ray and electrical results.

## References:

Comes R , Lambert M and Guinier A (1968) Stat. Comm. 6 715
Gevers R , Blank H and Amelinckx S (1966) Phys. Stat. Sol. 13 449
Gevers R , Delavignette P , Blank H , Van Landuyt J and Amelinckx S (1964) ibid. 5 595
Glazer A M (1975) Acta Cryst. A31 756
Harada J , Watanabe M , Kodera S and Honjo G (1965) J. Phys. Soc. Japan 20 630
Huller A (1969) Sol. Stat. Comm. 7 589
Whatmore R W and Bell A J (1981) Ferroelectrics 35 155
Yakunin S , Shakmanor V V , Spivak G V and Vasil'eva N V (1972) Sov. Phys. Sol. Stat. 14 No.2

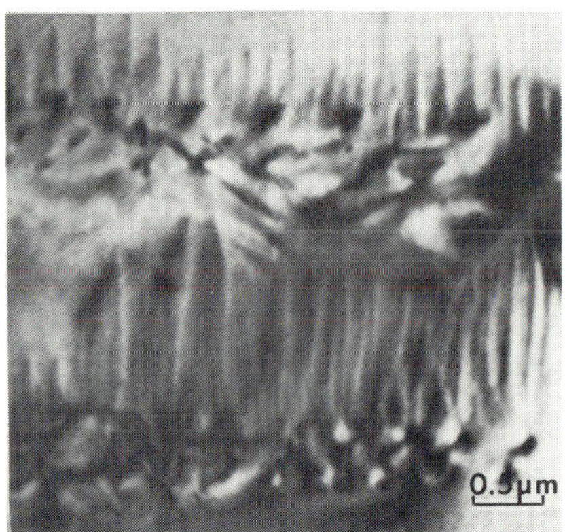


Fig. 4 Slip traces pinning domains in a modified-PZT grain.

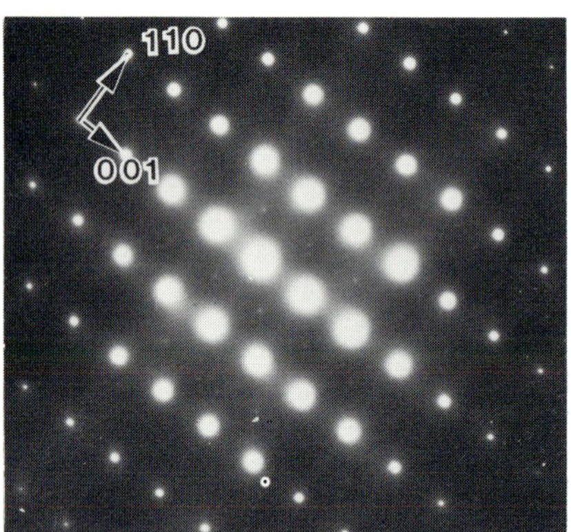


Fig. 5 $\langle 1\bar{1}0\rangle$-Zone - this shows weak F-type superlattice spots.

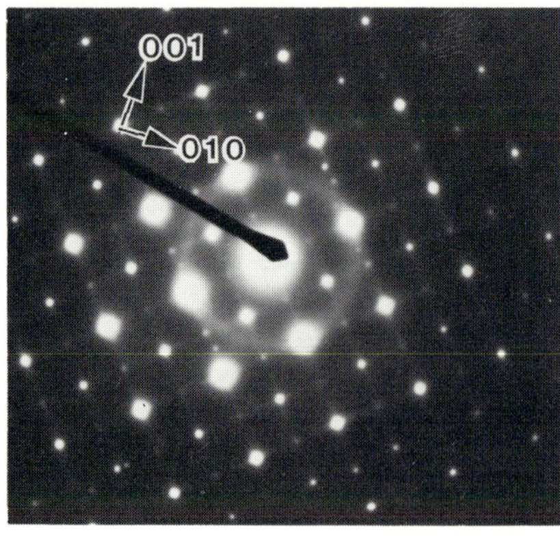


Fig. 6 $\langle 100\rangle$-Zone - weak diffuse scattering in $\underline{q} = \langle 0\bar{1}1\rangle$ and weak superlattice spots.

*Inst. Phys. Conf. Ser. No 78: Chapter 13*
*Paper presented at EMAG '85, Newcastle upon Tyne, 2–5 September 1985*



# A combined RHEED and SEM study of reaction bonded silicon nitride

G J Russell, R Puteh and J S Thorp,
Department of Applied Physics and Electronics,
University of Durham, South Road, Durham, DH1 3LE.

Abstract. Etched surfaces of reaction bonded silicon nitride (RBSN) have been studied by combining the techniques of EDAX and RHEED. Inclusions and impurities were found to be more prevalent in material of low weight gain (∿ 35%). For instance at the surfaces of HF etched samples, particles identified as free Si were observed and were found to be uniformly distributed throughout the material together with small fibrous features which were associated with globules containing iron. Etching in $H_3PO_4$ revealed the additional presence of much larger cylindrical filaments which were rich in calcium and were identified as $CaSi_2$.

## 1. Introduction

Silicon nitride is well established as a leading refractory ceramic partly because of its good high temperature mechanical properties and also because of its potential applications in device passivation and as an electromagnetic window material (Boyer and Moulson 1978, Messier and Wong 1976). However, there is substantial evidence to show that both the mechanical and electrical properties depend on the method of preparation. In an investigation of the mechanisms occurring in the formation of RBSN, Jennings and Richman (1976) conducted a microstructural study of the material and discussed the kinetics of α and β phase formation together with the occurrence of free silicon. Studies in this laboratory have illustrated the importance of the impurity content in relation to both the elastic constants (Thorp and Bushell 1985) and the dielectric behaviour (Thorp et al 1984) and recently Puteh (1984) reported anomalously high values of permittivity and dielectric loss at low weight gains. The present investigation was undertaken in an attempt to identify the origin of these and was directed towards the evaluation of the microstructural properties and impurity contents of both partially and fully nitrided RBSN.

## 2. Experimental

The samples of RBSN supplied by AME Ltd, Gateshead, were examined in their bulk form. Slices with dimensions 8 x 8 x 2 $mm^3$ were cut using a diamond wheel and their large area faces were subsequently mechanically polished with alumina the particle size of which was progressively reduced down to 1 μm. This mechanical polishing served only to provide surfaces which, following subsequent light etching, were sufficiently flat to be suitable for examination by RHEED. The latter technique was carried out using 100 KeV electrons in a JEM 120 TEM. The two different etchants which were employed to remove the polishing damage were concentrated HF(40%) at room temperature and concentrated $H_3PO_4$ at about 160°C for periods of 20 hr and

30 min respectively. The same samples which were studied by RHEED were also examined in the secondary emission mode using a Cambridge S600 SEM in which the elemental composition of individual features was determined by EDAX using a Link Systems 860 type analyser. Due to the highly insulating properties of the fully nitrided material (63% weight gain), it was necessary to sputter coat these samples with Au in order to avoid adverse charging effects in the SEM. In order to substantiate the structural observations of the etched surfaces that were made using RHEED, x-ray diffraction measurements were taken from bulk samples. As the material was comprised of very small grains, it was simply cut into thin rods ∿ 0.5 x 0.5 $mm^2$ in cross section and these were used in the place of powder samples in a Debye-Scherrer camera employing filtered Cu Kα irradiation.

## 3. Results

### 3.1 SEM and EDAX Studies

The most obvious features on surfaces of 35 and 44% weight gain RBSN after etching in HF were the evenly distributed particles which appeared in light contrast in secondary emission images as shown in figure 1a. EDAX studies of these revealed the presence of silicon alone. These features were not so apparent at the surfaces of 63% weight gain samples after the same etching treatment. In addition to these particles observed on HF etched surfaces, fibrous inclusions such as that shown in figure 1b were seen and these were usually associated with globular features located at one of their ends as at A. These globular features were shown by EDAX to contain a significant proportion of iron.

When $H_3PO_4$ at 160°C was used as an alternative etchant, the most important difference in the features at the surface was the appearance of clusters of cylindrical filaments of the order of a few hundred μm in length and about 10 μm in diameter which were particularly evident in low weight gain material. A group of these is shown in figure 2a and another example where two such filaments appear to be "fused" together is provided in figure 2b. These features were quite different from the much smaller ones revealed by HF etching and EDAX spectra taken from the body of these filaments showed that they contained a considerable proportion of Ca while the "weld region" at B contained phosphorous, the source of which was almost certainly the etchant ($H_3PO_4$) used in this case.

(a) Particles rich in Si

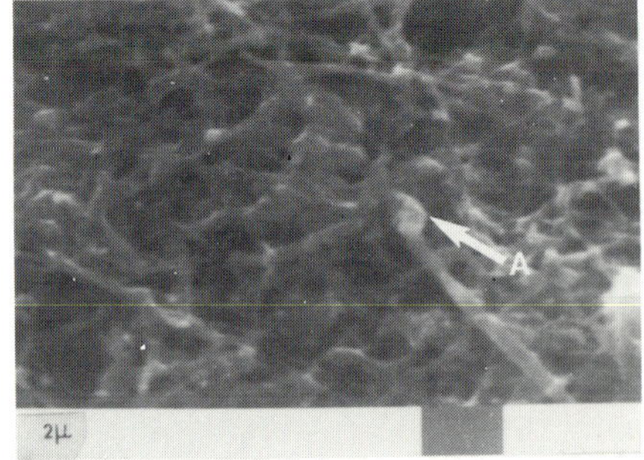


(b) Globule on fibrous feature.

Fig 1. Scanning electron micrographs of HF etched 44% wt. gain RBSN.

In order to ascertain that the impurities detected at the surfaces of etched samples were indeed native to the RBSN, a number of fracture surfaces were also examined in the SEM. While this did not reveal the presence of any fibrous or filament features, EDAX studies of these surfaces did confirm the presence of small quantities of Ca and Fe.

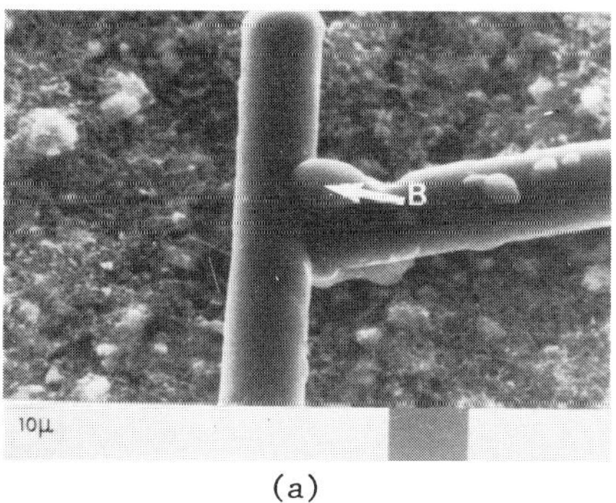


(a)

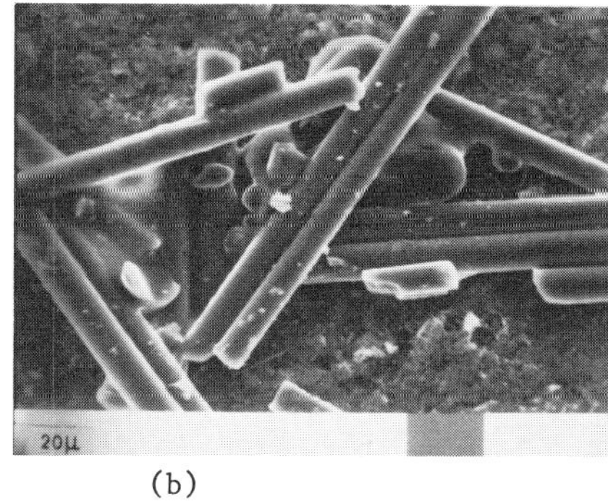


(b)

Fig 2 Scanning electron micrographs of $H_3PO_4$ etched 44% wt. gain RBSN.

3.2 RHEED and x-ray diffraction studies

As stated earlier, all of the samples which were examined in the SEM were also studied by RHEED. The interplanar spacings and estimated relative intensities of diffraction rings from such patterns obtained from 35% weight gain material after etching in HF or $H_3PO_4$ are compared with relevant ASTM index data in table 1. From this it can be seen that for both types of etching, firstly there is good agreement between the measured interplanar spacings and the ASTM index values for $\alpha$-$Si_3N_4$ and secondly there is clear evidence for the existence of additional diffraction rings which may be attributed to silicon. The ASTM data for Ca $Si_2$ is also included in this table though the coincidence of the interplanar spacings of this material with those of $\alpha$ -$Si_3N_4$ and elemental Si, together with the limited accuracy of RHEED relative to x-ray

TABLE 1 Comparison of RHEED data for 35% weight gain RBSN etched in HF and $H_3PO_4$.

| Pattern taken from surface after HF etch | | A. S. T. M. Data | | | | | Pattern taken from surface after $H_3PO_4$ etch | |
|---|---|---|---|---|---|---|---|---|
| | | Si | $\alpha$ -$Si_3N_4$ | | $CaSi_2$ | | | |
| d(A) | rel.int. | d(Å) | d(Å) | $I/I_o$(%) | d(Å) | $I/I_o$(%) | d(Å) | rel.int. |
| | | | 6.69 | 8 | | | 7.03 | W |
| 4.23 | S | | 4.32 | 50 | | | 4.35 | VS |
| | | | 3.88 | 30 | | | 4.00 | S |
| 3.29 | S | | 3.37 | 30 | 3.31 | 12 | 3.30 | VS |
| 3.02 | S | 3.089 | | | 3.08 | 12 | 3.22 | S |
| 2.90 | M | | 2.893 | 85 | | | 2.90 | S |
| | | | 2.599 | 75 | 2.66 | 40) | 2.60 | VS |
| 2.51 | S | | 2.547 | 100 | 2.55 | 28) | | |
| 2.24 | S | | 2.32 | 60 | | | 2.31 | S |
| 2.09 | W | | 2.158 | 30 | 2.13 | 16 | 2.145 | W |
| 1.87 | M | 1.892 | | | 1.92 | 100 | 1.92 | S |
| 1.74 | M | | 1.771 | 25 | | | 1.76 | W |
| | | | | | 1.64 | 16 | 1.63 | W |
| 1.64 | M | 1.613 | 1.596 | 35 | | | 1.61 | W |

TABLE 2 Interplanar spacings (X-ray) for 35% weight gain RBSN compared with ASTM data.

| Expt. d(A) | Results Int. | Si d(Å) | α-$Si_3N_4$ d(Å) | α-$Si_3N_4$ $I/I_o$(%) | β-$Si_3N_4$ d(Å) | β-$Si_3N_4$ $I/I_o$(%) | $CaSi_2$ d(Å) | $CaSi_2$ $I/I_o$(%) | $FeSi_2$ d(Å) | $FeSi_2$ $I/I_o$(%) |
|---|---|---|---|---|---|---|---|---|---|---|
| 4.07 | s | | 4.320 | 50 | | | | | | |
| 3.80 | s | | 3.88 | 30 | | | | | | |
| 3.76 | w | | | | 3.82 | 20 | | | | |
| 3.32 | w | | 3.37 | 30 | | | | | | |
| 3.23 | w | | | | 3.31 | 85 | 3.31 | 12 | | |
| 3.05 | vs | 3.09 | | | | | 3.08 | 12 | | |
| 3.03 | w | | | | | | 3.08 | 12 | | |
| 2.83 | m | | 2.89 | 85 | | | | | | |
| 2.60 | s | | 2.599 | 75 | 2.668 | 100 | 2.66 | 40 | | |
| 2.53 | s | | 2.547 | 100 | | | 2.55 | 28 | | |
| 2.49 | s | | | | 2.492 | 100 | | | | |
| 2.46 | ms | | | | | | | | 2.37 | 60 |
| 2.28 | ms | | 2.32 | 60 | 2.312 | 9 | | | | |
| 2.25 | w | | | | 2.18 | 35 | | | | |
| 2.13 | m | | 2.158 | 30 | | | 2.13 | 16 | | |
| 2.05 | m | | 2.083 | 55 | 1.904 | 5 | | | | |
| 1.89 | vs | | | | 1.892 | 5 | 1.92 | 100 | 1.89 | 30 |
| 1.88 | w | | 1.884 | 8 | | | | | | |
| 1.84 | m | | | | 1.827 | 20 | | | 1.84 | 100 |
| 1.78 | w | | 1.806 | 12 | 1.753 | 70 | | | 1.78 | 15 |

diffraction, do not permit a conclusive association of these additional diffraction rings with Ca $Si_2$ to be made on the basis of RHEED measurements alone. Consequently, the same material was examined by x-ray diffraction in a Debye-Scherrer camera and the measurements taken are recorded in table 2. The better accuracy of the x-ray diffraction measurements enables both the α and β phases of $Si_3N_4$ to be identified. Further, in addition to the lines which can be ascribed to Ca $Si_2$, other reflections indicate the presence of Fe $Si_2$.

## 4. Conclusions

This study has shown that, in addition to voids and free silicon, both iron silicide (Fe $Si_2$) and calcium silicide (Ca $Si_2$) can occur in low weight gain RBSN and exhibit microstructural features similar to those reported by Jennings and Richman (1976) for iron in RBSN. It has recently been shown by Thorp and Rad (1985) that both Fe $Si_2$ and Ca $Si_2$ have significantly higher values of permittivity and dielectric loss than fully nitrided RBSN or free silicon and the anomalous dielectric behaviour of low weight gain silicon nitride is most probably due to the presence of small amounts of these silicide impurities.

## 5. References

Boyer S M and Moulson A J, 1978, J.Mat.Sci, 13, 1637.
Jennings H M and Richman M H, 1976, J.Mat.Sci.11, 2087.
Messier D R and Wong P, 1976, 13th Symposium on Electromagnetic Windows, Georgia Institute of Technology, ed.J.N.Harris, Atlanta, U.S.A.
Puteh R, 1984, Ph.D.Thesis, University of Durham.
Thorp J S and Bushell T G, 1985, J.Mat.Sci 20, 2265.
Thorp J S,Ahmad A B,Kulesza B L J and Bushell T G,1984,J.Mat.Sci,19, 3680.
Thorp J S and Rad N E, 1985, J.Mat.Sci.Lett - accepted for publication.

*Inst. Phys. Conf. Ser. No 78: Chapter 13*
*Paper presented at EMAG '85, Newcastle upon Tyne, 2–5 September 1985* 

# Densification of impure silicon oxynitride by liquid-phase sintering

S A Siddiqi* and A Hendry

Wolfson Laboratory, Department of Metallurgy and Engineering Materials, University of Newcastle upon Tyne, Newcastle upon Tyne, NE1 7RU

* Now at the Centre for Solid State Physics, University of Punjab, Pakistan

## 1. Introduction

A new generation of engineering ceramics based on silicon nitride and known by the generic name 'sialons' has been developed in Newcastle. The applications of sialons range from cutting tools and heat engine components to refractories for metal handling. For each application there is an appropriate sialon phase and a suitable grade of purity which are dictated by the physical and mechanical stresses which the material will experience in service. An important aspect of sialon development therefore has been the search for low-cost manufacturing routes to produce these synthetic ceramics in a usable form. Carbothermal reduction of naturally occurring oxides in nitrogen (Lee and Cutler (1979); Szweda et al (1981)) provides an attractive means of powder preparation of sialons for applications involving low mechanical stresses at operating temperature. The impurities in the silicate used for reduction play an important role in the reactions to produce the sialon phase (Siddiqi and Hendry (1985)) and then provide a potential source of liquid phase for subsequent densification of the powders. It is however these same inherited impurities which degrade the high-temperature strength of the sintered component.

## 2. Results and Discussion

Silicon oxynitride powder was prepared from rice husk ash and carbon containing 2w/o CaO by reacting in nitrogen at 1400°C. The powder used for densification studies therefore contains approximately 2w/o CaO together with a total alkali oxide content ($Na_2O$ and $K_2O$) of 0.05w/o. Hot pressing in a pseudo-isostatic powder bed was used to follow the progress of densification (Hampshire and Jack (1983)) and the pellet shrinkage was measured by the displacement of the press platens monitored by a position transducer. From the observed behaviour it will be demonstrated below that densification proceeds by a solution-reprecipitation mechanism as proposed by Kingery (1959).

Three stages of densification are identified by the Kingery model and are shown schematically in Figure 1 and represented graphically in Figure 2. Initially a liquid forms at densification temperature thus drawing the particles together by capillary pressure and forming a neck (Figure 1)-

Stage 1, particle rearrangement. In the second stage material from the neck dissolves in the liquid, is transported by diffusion away form the neck and is reprecipitated elsewhere in areas of lower stress. The difference in solubility results from differences in stress intensity. The equation governing this important second stage of densification is given in Figure 2. Hence a graph of log ($\Delta V/Vo$) against log t (Figure 2) is linear with a slope of 1/3 or 1/5 for this stage of densification if the liquid sintering model is valid. The third stage of densification (Figure 2) is elimination of closed porosity which is extremely small in extent.

Samples of oxynitride powder were hot-pressed at temperatures in the range 1200 - 1800°C which is consistent with the powder preparation temperature (1400°C) in which liquid-phase transport plays an important role in forming the oxynitride. However, temperatures of 1600°C and above are required before the viscosity of the liquid is sufficiently low to allow complete densification. These experiments indicate that liquid phase sintering occurs.

The results in the second stage of densification give a slope of approximately 1/5 (n = 5) which indicates, after Kingery, that diffusion of material from the inter-particle necks, through the liquid is the rate-controlling step in densification.

Electron microscopy was also used to study the densification mechanism and the development of the sintered microstructure. Figure 3 shows SEM micrographs of successive stages of densification. The initial spherical particles of silicon oxynitride produced by carbothermal reduction of the rice husk ash have begun to form necks on hot-pressing at 1400°C (Figure 3a; c.f. Figure 1) and then elongated angular grains begin to grow with increasing time (Figure 3 b). The morphology of fully dense material prepared by hot pressing at 1600°C is shown in Figure 3 c and the elongated grains are shown by X-ray diffraction and TEM to be silicon oxynitride while the darker regions (heavily etched in Figure 3) are amorphous.

Electron beam microanalysis in TEM, Figure 4, shows that the glassy regions are rich in calcium and potassium inherited from the original raw materials. The precise composition of the glass has not been determined; it is assumed to be a Ca/Si oxynitride glass containing some potassium. It may be concluded that the amorphous phase has solidified from an oxynitride liquid in which, at hot-pressing temperature, all of the impurities in the system are concentrated and which is responsible for densification. Hampshire and Jack (1983) have shown that densification is accompanied by phase transformation in liquid-phase sintering of silicon nitride and sialons with oxide additives but in the present case no phase change occurs. The starting powder is silicon oxynitride and the final dense product is also silicon oxynitride. There has however been a complete change in morphology from spherical particles to elongated grains (Figure 3) due to solution and reprecipitation through the liquid phase and it follows therefore that the driving force for densification must be the reduction in total surface energy of the system by preferred crystal growth.

## 3. References

Hampshire, S. and Jack, K.H., 1983, 'Progress in Nitrogen Ceramics' Ed. F.L. Riley, Noordhof: Leyden, p 225.

Kingery, W.D., 1959, J. Appl. Phys., 30, 301.

Lee, J-G. and Cutler, I.B., 1979, Ceram. Bull., 58, 869.

Siddiqi, S.A. and Hendry, A., 1985, J. Mat. Sci., in the press.

Szweda, A., Hendry, A. and Jack, K.H., 1981, 'Special Ceramics 7', Eds. D. Taylor and P. Papper, Brit. Cer. Soc. Stoke-on-Trent, p 197.

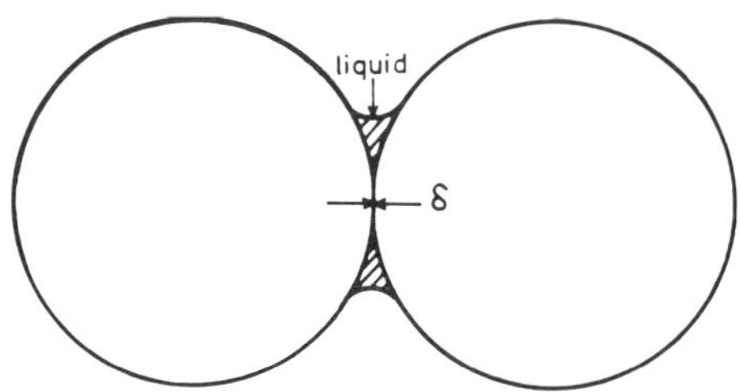


Figure 1. Neck formation in liquid phase sintering.

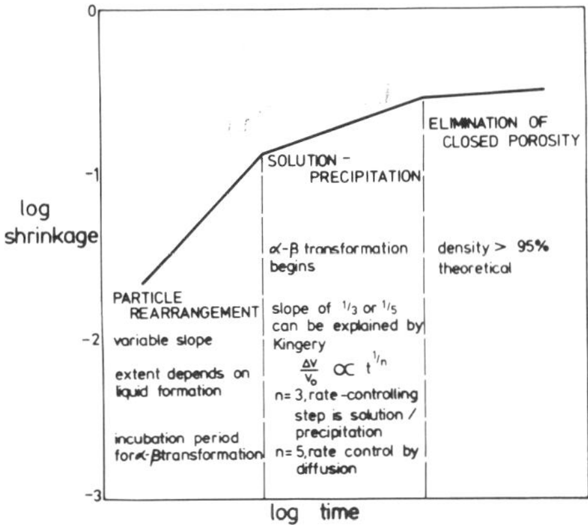


Figure 2. The Kingery model of sintering.

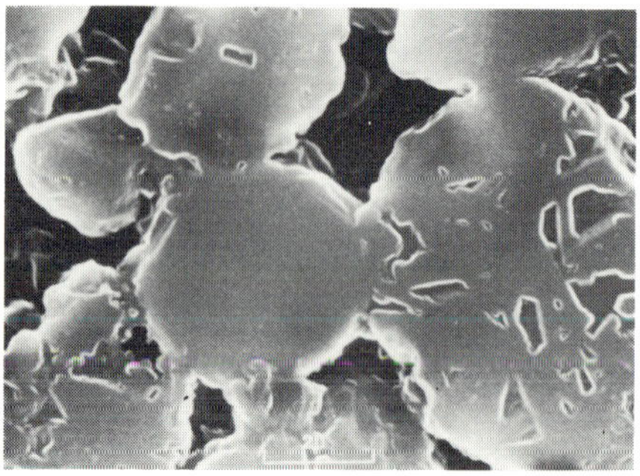

(a)

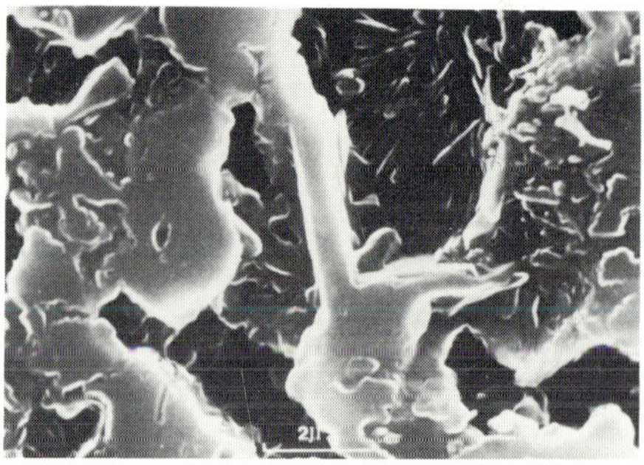

(b)

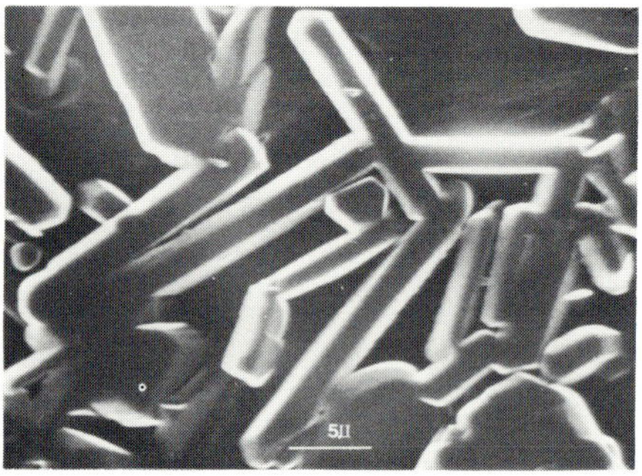

(c)

Figure 3. SEM micrographs of hot-pressed silicon oxynitride, (a) and (b) 1400°C (c) 1600°C.

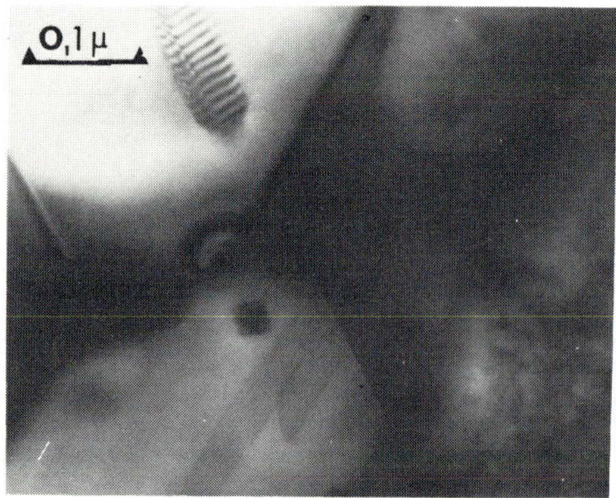


Figure 4: TEM micrograph of hot-pressed silicon oxynitride.

*Inst. Phys. Conf. Ser. No 78: Chapter 13*
*Paper presented at EMAG '85, Newcastle upon Tyne, 2–5 September 1985* 

# Microstructure and composition of plasma sprayed TiC coatings

D Fournier, R G Saint-Jacques, G L'Esperance*, C Brunet** and S Dallaire**

INRS-Energie, Universite du Quebec, Varennes, Quebec, Canada, JOL 2PO
*Ecole Polytechnique, Montreal, Quebec, Canada, H3C 3A7
**IMRI-CNRC, Boucherville, Quebec, Canada, J4B 6Y4

## 1. Introduction

The light element detection capabilities of both EELS and windowless (WL) EDX are known (Zaluzec et al 1984). The main advantage of EELS is the absence of overlap which is observed in WL-EDX when more than one low Z elements are present. We have used EELS to study the carbon depletion of TiC caused by the plasma spraying process. High spatial resolution was needed since the structure of these coatings is not homogeneous.

In plasma spraying of high melting point materials like TiC complete fusion is difficult to obtain. The incomplete fusion followed by a very fast cooling produces an inhomogeneous structure (Safai 1979).

The C/Ti ratio in TiC can vary from 0.48 to 0.97 (Storms 1967). It has already been shown by Auger spectroscopy (Brunet et al 1985) and deduced from X-ray measurement of the lattice parameter (Fournier et al 1985) that during the plasma spraying process the average C content of the material is decreased. However no measurements from individual grains were carried out The object of this paper is to show that EELS is an adequate technique to study the C/Ti ratio of individual grains (even smaller than 0.2 μm).

## 2. Experimental

The powders were projected on polished Al and the coatings were separated by quenching in liquid N. Three mm discs were cut out with an ultrasonic drill. They were subsequently lapped with parallel faces down to 100 μm. Due to their open porosity it was not possible to electropolish them (Wetzig et al 1982). In order to use ion beam milling and not cause relief by a long exposure to the Ar beam, it was necessary to further reduce the thickness at their centre to 5-10 μm. This was done with dimple grinding and polishing on both faces (Gatan apparatus). Microstructure observations were made at 120 keV and 1 MeV.

EELS analysis was carried out on a JEOL 200 FX operated at 200 keV. The convergence and collection angles were respectively 3 and 7 mrad. The plasmons over zero loss intensity ratios ranged between 0.15 and 0.23 (i.e. < 0.3 in order to avoid detrimental effect due to thickness). C/Ti ratios were calcualted from fitting windows of 100 eV under the discontinuities. The background extrapolation beyond the edges was done by hand. The raw edge intensity ratios were converted to elemental ratios using the Sigmak and Sigmal partial cross section models of Egerton (1981).

## 3. Results and discussion

Due to the porous and brittle nature of the TiC coatings, the TEM specimens were delicate to prepare and had no wide transparent regions. Hence HV microscopy was used to look through thick regions and a montage was made to obtain an overall view (Fig. 1). The structure consists of groups of large grains (> 1 μm) surrounded by colonies of smaller grains (< 0.2 μm). The proportions of both sizes of grains are roughly equal. The large grains have the same size as the grains of the original powder (Fournier et al 1985) and hence have not been melted. The small grains have been nucleated from the melted material. Most of them are equiaxed (Fig. 2a). Few regions contain elongated small grains (Fig. 2b). Similar structures are found in splat cooling and in plasma spraying of droplets (Safai 1979). The elongated grains are in contact with unmelted grains and are caused by an unidirectional and faster heat removing in the early stages of solidification. Due to the higher melting point and lower heat conductivity of TiC with respect to metals, no amorphous layer was found.

Microbeam diffraction showed that both sizes of grains are CFC NaCl with a lattice parameter of 4.32 $\pm$ 0.01Å (convergent beam diffraction should give more precise values and indicate any difference). X-R Debye-Scherrer analysis of the original powder yield a value of 4.3270 $\pm$ 0.0005Å. This is the same value as the one of a stoichiometric TiC (Fig. 3). Debye-Scherrer of the coating indicated a somewhat larger value (4.3296Å). According to Storms 1967 (Fig. 3) the TiC having such a lattice parameter can have a C/Ti of 0.90 or 0.76. This last value coincides with Auger spectroscopy (Brunet et al 1985). The carbon depletion observed is caused by oxidation in the case of air spraying and by dissociation in the case of Ar spraying. The C/Ti values indicated above are the average values of C/Ti of both large and small grains.

EELS analysis was performed individually on a large and a small grains (Fig. 4 and 5). The carbon K edge is clearin both cases but it is reduced for the small grain. The qualitative results show that this large grain is nearly stoichiometric (C/Ti ~ 0.92). The maximum ratios usually reported are 0.95 - 0.97 and Auger analysis on the starting powder also gave a value of 0.97 $\pm$ 0.05 (Brunet et al 1985). On the other hand the small grain has a much lower C/Ti value (0.40). This carbon depletion brings this C/Ti down to the bottom limit (0.48) indicated on the phase diagram (Storms 1967).

This preliminary study clearly shows that EELS is an appropriate technique to study the low Z element content in various grains of an inhomogeneous microstructure.

## Acknowledgements

We are grateful to A. Taylor of Argonne for the provision of HV facility. We would like to thank M. Kisker of JEOL-USA for his assistance in EELS. This work was supported by NSERC.

## References

Brunet C, Dallaire S and Saint-Jacques R G 1985 J. Vac. Sci. and Technol. (Nov-Dec) in press

Fournier D, Saint-Jacques R G, Brunet C and Dallaire S 1985 J. Vac. Sci. and Technol. (Nov-Dec) in press
Egerton R F 1981 Proc. EMSA ed G. Bailey (Claitor Publishing) 198
Safai S 1979 Ph.D. thesis State University of New York at Stoney Brook
Storms E K 1967 The refractory carbides (New York; Academic Press) pp 1-9
Wetzig K, Bauer H D, Fisher W and Gille G 1982 Proc. 10th Int. Cong. on Elect. Micr. Hamburg 361
Zaluzec N J, McConville R and Sandborg A 1984 Proc. AEM Workshop Lehigh University (San Francisco Press)

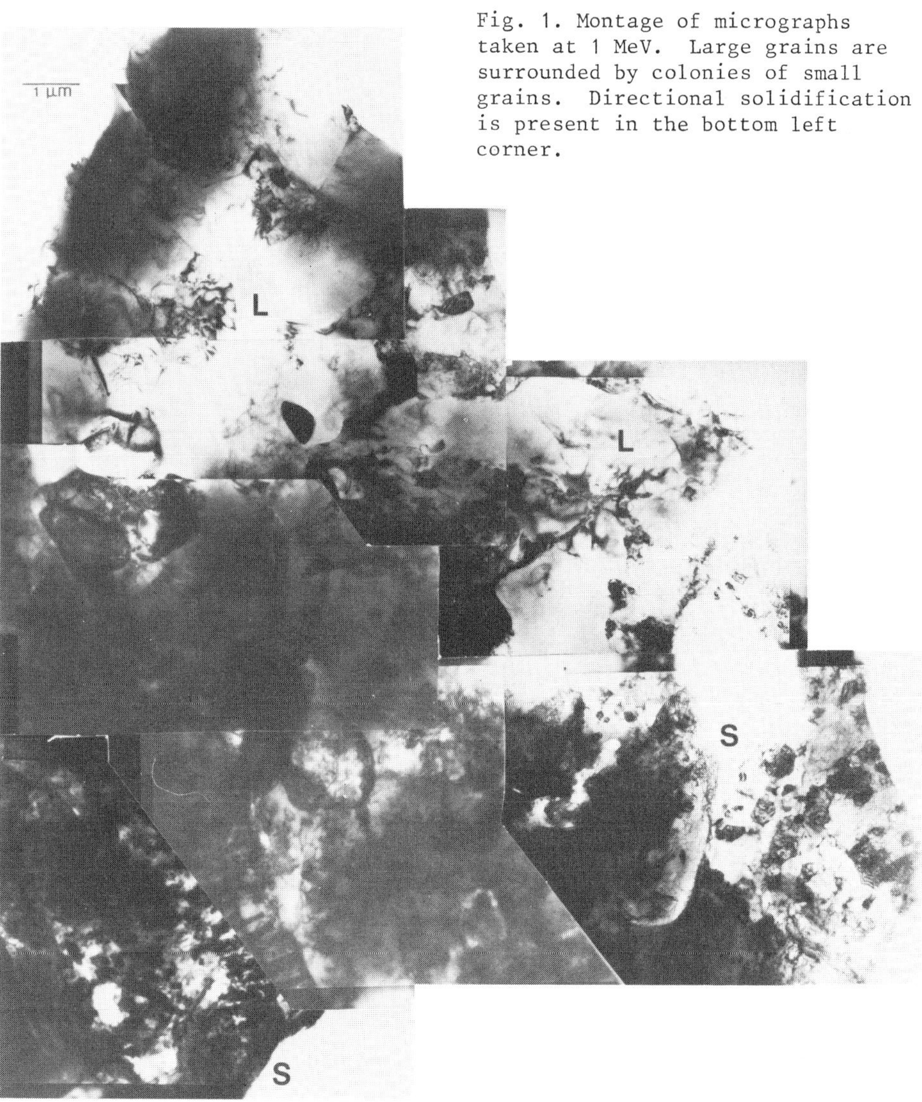


Fig. 1. Montage of micrographs taken at 1 MeV. Large grains are surrounded by colonies of small grains. Directional solidification is present in the bottom left corner.

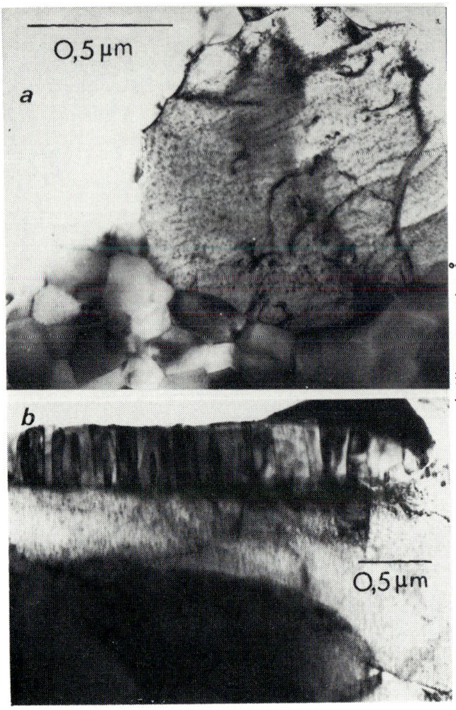

Fig. 2. Micrographs showing
a) equiaxed small grains
b) unidirectionally solidified small grains in contact with large grains.

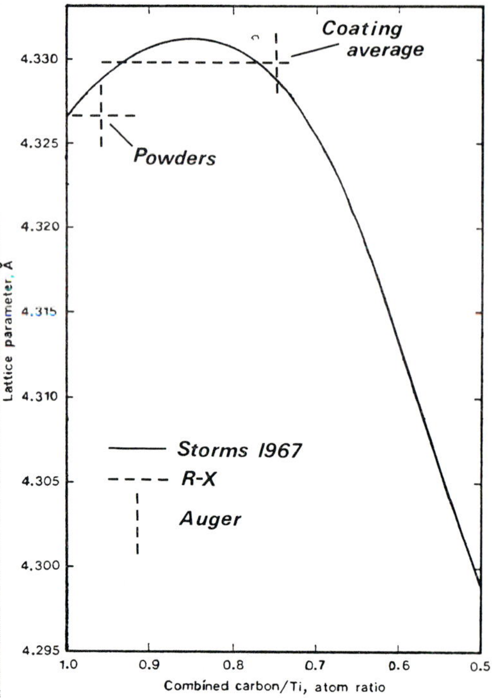

Fig. 3. Variation of the lattice parameter of TiC with C/Ti ratio. Auger and X-R values taken on the original powders and on the coatings are shown.

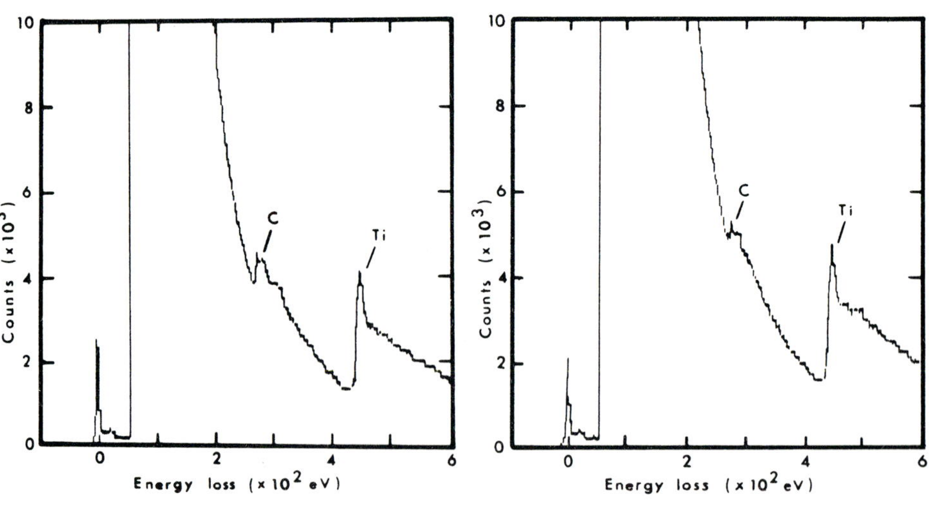


Fig. 4. EELS spectrum of a large grain. The C/Ti ratio is like the one of the starting powder.

Fig. 5. EELS spectrum of a small grain. This material was melted and has lost much carbon.

*Inst. Phys. Conf. Ser. No 78: Chapter 14*
*Paper presented at EMAG '85, Newcastle upon Tyne, 2–5 September 1985* 

# Electron microscopy of carbons—past, present and future

J R Fryer, Chemistry Department, University of Glasgow. Glasgow G12 8QQ Scotland

The use of graphite for refractory crucibles was reported by Agricola(1495-1550) but its identity was not distinguished from molybdenite until the 18th century(Scheele 1742-1786). Davy in 1814 showed that graphite,diamond and amorphous carbon were the same material from their combustion products,however precise identification awaited the invention of X-ray crystallography.In 1922 the hexagonal close packed structure was established by Hull by this technique and also by electron diffraction - Trendelenburg et al (1933) and convergent beam diffraction- Goodman(1976).A rhombohedral modification was reported by Lipson and Stokes(1942)and Friese and Kelly(1963) that was present in variable amounts in graphites.Therefore at this time carbon represented an element which had two important allotropes-diamond and graphite-the major one of which was of variable structure.More stringent requirements were being made of graphite for use in nuclear power and refractories and hence it became the first crystalline material to be investigated by the full range of modern techniques-many of which were in prototype form. Electron microscopy was starting to be applied to material science and image interpretation became a significant problem. It had been predicted that overlapping crystal lattices could produce moiré patterns-Green and Weigle(1948)-and in 1951 Mitsuishi et al reported imaging such fringes in graphite. Dawson and Follett (1959) used moiré fringes to characterise synthetic graphites and crystalline defects whilst Williamson(1960) and Amelinckx and his co-workers(1965) applied diffraction contrast imaging to both basal (i.e. parallel to the layer plane)and non-basal defects.

The reactivity of graphite was of importance to its applications,in particular its oxidation behaviour.A prime innovator in this field was Hennig who developed the technique of etch decoration to reveal expanded monolayer vacancies in graphite single crystals.The graphite crystal was cleaned by heating in vacuum at 900C and then could be etched with a variety of materials,particularly oxygen/chlorine mixtures at 950C,and then decorated with a heavy metal such as gold.The metal particles collected in a circle on the edges of the expanded vacancy as the strong anisotropy of graphite caused oxidation within the layer plane to proceed much faster than perpendicular to the basal plane so the circle of metal represented a hole one atomic layer deep.Hennig's results are reviewed(1966) and by this technique he was able to measure the anisotropy factors as well as detect changes in the graphite crystal caused by the presence of impurity atoms such as boron.This technique for examining the reactivity of graphite was later used by Feates(1968),Montet and Myres(1968) Evans and Thomas(1971) and Yang(1984) who also mentions its application using scanning electron microscopy.

The decoration of un-etched graphite showed the particles aligning at surface steps and clustering at what were thought to be the ends of non-basal dislocations.Baird and Fryer(1974) used different metals and surface treatments and the conclusion was that decoration of graphite is via chemisorbed surface species and not directly on the grahite surface.However,this does not affect the etch decoration results as graphite does not chemisorb oxygen on the perfect basal plane.

Thomas and his co-workers were also examining the reactivity of graphite by optical and electron microscopy of undecorated material to ascertain reaction kinetics(1965) and to reveal non-basal dislocations(1968).They also showed that different metals catalysed the oxidation in different ways according to different anisotropy factors of the graphite to the metals.Another variation was shown by Marsh(1965) who found that atomic oxygen did not show the extent of anisotropy of oxidation as had molecular oxygen.

A difficulty with kinetic studies was that the specimen had to be retrieved from the microscope for subsequent oxidation-although multilayer etch decoration could be used in certain circumstances-Feates(1968).Therefore efforts were made to construct in situ reaction stages in the electron microscope.A simple construction-in a Siemens Elmiskop 1 microscope- involving a gas leak on to the specimen via the objective aperture drive showed the movement of metal catalyst particles on the graphite crystal surface during oxidation reactions-Fig 1-Fryer(1968).A series of non-catalysed reactions were also done to measure reaction rates and activation energies and showed that at low oxygen pressures only the $(10\bar{1}0)$ planes were oxidised and that rate differences between different graphites were a function of different pre-exponential factors and not different activation energies-Fryer(1974).Fig 2 shows graphite being oxidised at 640C with Fig 2a exhibiting typical moire patterns and Fig 2b basal dislocations.Neither feature affecting the oxidation.

A more sophisticated gas reaction stage was made by Feates et al(1970) and cinephotography used to record the images.A large number of catalytic oxidation reactions were examined under carefully controlled conditions of temperature and pressure-Harris et al(1974),Baker(1982).

Recently the industrial interest in graphite oxidation has involved much higher temperatures or more exotic reactants -e.g.plasmas.Most of these involve conditions that would not be easy to simulate within the microscope and it is probable that external reaction chambers would be necessary.Studies external to the microscope have been made showing microconical structures formed by graphite oxidation in a hydrogen plasma by Baird(1984) that are of similar structure to those observed by Jones(1971) in oxygen and nitrogen plasmas.

It is,however,outstanding that virtually no use has been made of the high resolution capability of modern microscopes to examine oxidation reactions in greater detail.

Non-graphitised carbons and graphite formation

The criterion for graphite of an interplanar(0002) spacing of 0.336nm stemmed from X-ray studies,and non-graphitic carbons had larger values with also loss of hkil reflections indicating that the layers were rotated randomly with respect to each other.Such carbons are termed turbostratic and their lack of order rendered microscopy difficult.Dark field examination Hess and Ban(1966) and Rudee(1967) using(0002) reflections provided the first technique for electron microscope characterisation but the real advance was the direct lattice resolution of the(0002)layers-Ban and Hess(1969),Harling and Heckman(1969).The development of the layer planes during graphitisation could now be observed.

Of particular interest was the low temperature formation of graphite on metal surfaces.Hydrocarbons such as methane exposed to a nickel surface at 600C produced flake graphite and also filamentous material-Robertson(1970).Initial characterisation was by dark field microscopy and microprobe analysis but the chemical mechanism did not become clear until lattice imaging showed the relationship between the graphite layers and the catalyst particles-Fig 3-Baird et al(1971,1974)

Controlled atmosphere electron microscopy and conventional microscopy were also used to understand the chemistry of this system-Baker et al(1972),Derbyshire et al (1975).

The development of graphitic structures in both graphitisable and non-graphitisable hydrocarbons and carbon fibres has been studied extensively by lattice imaging.The interpretation was clarified by Johnson and Crawford(1973) and Millward and Thomas (1978).The development of structure in carbon blacks with heat treatment was shown by Marsh et al(1971),in poly--vinylidene chloride by Ban et al(1975) and reviewed-Millward and Jefferson(1978).Decomposition of graphite compounds to show single layers was shown by Iijima(1978) and the general application of lattice imaging to carbons and to mesophase precursors of graphite by Oberlin et al (1975,1980,1983)also the use of the SEM to characterise coke structure has been described by Markovic and Marsh(1983).The(0002)lattice images do not differentiate between graphite and turbostratic material.The$(10\bar{1}0)$(0.21nm) and$(11\bar{2}0)$(0.13nm)have been imaged but there have been no reports of their being used to monitor the formation of the three dimensional graphite structure.

Carbon films are used in microscopy to both support specimens and as phase objects to test optical parameters.Less attention is paid to the structure and composition of carbon films.Their properties have been reviewed-McLintock and Orr(1973) and they concluded that approximately 50% of the evaporated carbon film consists of turbostratic islands of 6-25 Hexagonal rings per layer whilst the remainder is tetrahedral or otherwise distorted carbon.The optical density of carbon films is variable and may differ between areas of a carbon film of the same thickness-Cosslett and Cosslett(1957).In general the island structure within the film is 0.8-1.5nm although larger areas have been reported-Krivanek et al(1978) but the familiar speckle of a carbon film may in part be due to adsorbed gases as nitrogen has been shown to give a similar appearance when adsorbed on a MgO crystal-Moodie and Warble(1974).The surface charge on an evaporated carbon film is negative,however,a charge on an evaporated carbon film is negative,however,a positively charged film can be made by glow discharge in various vapours and the film made hydrophilic-Dubochet et al (1971).Another carbon support is single crystal graphite that can be prepared by cleavage with adhesive tape or by exfoliating the graphite with concentrated sulphuric acid /nitric acid mixture prior to cleaving.The support contrast is a minimum on the bend contours-Heines and Howie(1975).

Heat treating a carbon film in stages to 3000C produced a polycrystalline graphite -Goma and Oberlin(1980) who followed the heat treatment with electron diffraction and lattice imaging.The most dramatic changes occurred above 2000C.

Intercalation compounds

Apart from surface species,graphite and carbons produce a range of compounds with electron donors or acceptors intercalated between the carbon layers.This normally involves expansion of the interlayer distance so that the compound can be detected by electron diffraction but layer distortion also occurs so that diffraction contrast effects are visible even in the decomposed residue compound-Carr(1970).Preparation of the potassium-graphite compound in a microscope hot stage showed the fringe structure around intercalated areas-Saunders(1978).Lattice resolution of the ferric chloride intercalate revealed different packing within the layers-Evans and Thomas(1975) and a graphite iron compound decomposed to give microcrystalline iron throughout the graphite crystal-Saunders and Fryer(1978).Such microdispersed compounds are of catalytic interest and high resolution studies showed the crystalline nature of the iron-Smith et al(1981).

Despite extensive examination of intercalation compounds their general limitation is one low chemical stability.Transition metal and metal oxide intercalates are more stable and the possible microscopy of compounds having the complexity of the metal oxides combined with graphite is stimulating.

Microporosity

Carbons-particularly those of high surface area-are widely used as catalyst supports.In many cases the catalyst acts in conjunction with the carbon so that the microporosity(pores less than 1.5nm diameter) is important to the catalytic process.In a disorded carbon the layer planes can be resolved to show the sizes and shapes of micropores at the edges of the specimen in profile Fig 4-Fryer(1981).The effects of oxidising media can be used in that anisotropic thermal oxidation increases the microporosity and depths of the pores whereas isotropic plasma oxidation has no effect on the microporosity but smooths the surface and reduces mesoporosity(1.5-20nm).

It is anticipated that provided an adsorbed molecule is stable in the microscope it should be possible to resolve its location by high resolution imaging and thus identify active sites on the surface.

Other forms of Carbon

Different forms of carbon have been identified by EELS-Egerton and Whelan(1974) and there has been a thorough high resolution examination of diamond by Hutchison et al(1983).In addition to hexagonal and rhombohedral graphites and cubic diamond there is hexagonal diamond,chaoite-Fryer(1976)-and carbon(IV).These less known forms may be of future importance.

References

Amelinckx S.,Delavignette P.and Heerschap M.1965.in Chemistry and Physics of Carbon 1,(P.L.Walker ed.)Marcel Dekker N.Y.2.
Auguie D.,Oberlin M.,Oberlin A.and Hyvernat P.1980.Carbon 18,337.
Baird T.1984.Proc.8th Eur.Cong.E.M.Budapest.2,1159.
Baird T.and Fryer J.R.1974.Carbon 12,381.
Baird T.,Fryer J.R.and Grant B.1971.Nature 233,329.
Baker R.T.K.1982.J.Catal.78,473.
Baker R.T.K.,Barber M.A.,Harris P.S.,Feates F.S.and Waite R.J. 1972.J.Catal.26,51.
Ban L.L.and Hess W.M.1969.Cities Service Report,October.
Ban L.L.,Crawford D.and Marsh H.1975.Acta Cryst.8,415.
Cosslett A.and Cosslett V.E.1957.Brit.J.Appl.Phys.8,374.
Dawson I.M.and Follett E.A.C.1959.Proc.Roy.Soc.(Lond.)A253,390
Derbyshire F.J.,Presland A.E.B.and Trimm D.L.1972.Carbon 10,113.
Dubochet J.,Ducommun M.,Zollinger M.and Kellenberger E.1971. J.Ultrastruc.Res.35,147.
Egerton R.F.and Whelan M.J.1974.J.Elec.Spectr.3,232.
Evans E.L.and Thomas J.M.1971.3rd Conf.Ind.Carbons and Graphite.London 3.
Evans E.L.and Thomas J.M.1975.J.Solid State Chem.14,99.
Feates F.S.1968 Trans.Farad.Soc.64,3093.
Feates F.S.,Morley H.and Robinson P.S.1970 Proc.7th Int.Cong.E.M. Grenoble 1,295.
Friese E.J.and Kelly A.1963.Phil.Mag.8,1519.
Fryer J.R.1968.Nature 220,1120.
Fryer J.R.1974.Proc.8th Int.Cong.Canberra.1,680.
Fryer J.R.1976.Proc.Carbon Conf.Baden Baden 25.
Green T.A.and Weigle J.1948.Helv.Phys.Acta 21,217.
Goma J.and Oberlin M.1980.Thin Solid Films 65,221.
Goodman P.1976.Acta Cryst.A32,793.
Harling D.F.and Heckman F.A.1969.Mat.Plast.Elast.35,80.
Harris P.S.,Feates F.S.and Reuben B.G.1974.Carbon 12,189.
Hennig G.R.1966.in Chemistry and Physics of Carbon 2,(P.L.Walker Jr.ed.)Marcel Dekker,New York.1.
Hess W.M.and Ban L.L.1966.Proc.6th Int.Cong.E.M.Kyoto.569.
Hines R.L.and Howie A.1975.Phil.Mag.32,257.
Hull A.W.1922.Phys.Rev.20,113.
Hutchison J.L.Barry J.C.and Segall R.L.1983.J.Mat.Sci.18,1421.
Iijima S.1978.Proc.Int.Cong.E.M.Toronto 1,252.
Johnson D.J.and Crawford D.1973.J.Microsc.98,313.
Jones S.S.1971.Carbon 9,259.
Krivanek O.L.,Gaskell P.H.and Howie A.1976.Nature 262,454.
Lipson H.and Stokes A.R.1942.Proc.Roy.Soc.(Lond.)A181,101.
McLintock S.and Orr J.C.1973.in Chemistry and Physics of Carbon 11,(P.L.Walker Jr.and P.A.Thrower eds.)Marcel Dekker New York,243.
Marsh H.,O'Hair T.E.and Reed R.1965.Trans.Farad.Soc.61,285.
Markovic V.and Marsh H.1983.J.Microsc.132,345.
Millward G.R.and Jefferson D.A.1978.in Chemistry and Physics Carbon 14,(P.L.Walker Jr.and P.A.Thrower eds.)Marcel Dekker, New York.
Millward G.R.and Thomas J.M.1979.Carbon 17,1.

Mitsuishi T.,Nagasaki H.and Uyeda R.1951.Proc.Imp.Acad.Japan 27,86.
Montet G.L.and Myres G.E.1968.Carbon 6,627.
Moodie A.F.and Warble C.E.1974.Proc.8th Int.Cong.E.M.Canberra 1,230.
Oberlin A.and Oberlin M.1983.J.Microsc.132,353.
Oberlin A.,Terriere G.and Boulmier J.L.1975.Tanso 80,29.
Robertson S.D.1970.Carbon 8,365.
Rudee M.L.1967.Carbon 5,155.
Saunders K.G.and Fryer J.R.1978.Proc.5th Conf.Ind.Carbons and Graphite.Soc.Chem.Ind.Lond.738.
Saunders K.G.1978.Ph.D Thesis.Glasgow.
Smith D.J.,Fisher R.M.and Freeman L.A.1981.J.Catal.72,51.
Thomas J.M.1965.in Chemistry and Physics of Carbon 1,(P.L.Walker Jr.ed) Marcel Dekker,New York.121.
Trendelenburg F.,Franz E.and Wieland O.1933.Z.Tech.Physik.14, 489.
Williamson G.K.1960.Proc.Roy.Soc.(Lond.)A257,457.
Yang R.T.1984.in Chemistry and Physics of Carbon 19, (P.A.Thrower ed.)Marcel Dekker.New York.163.

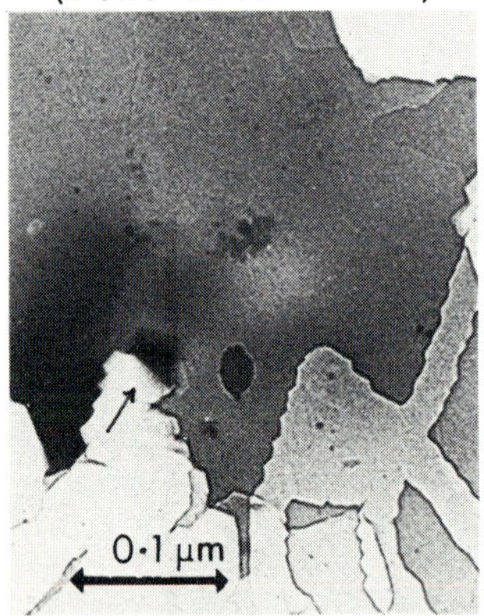


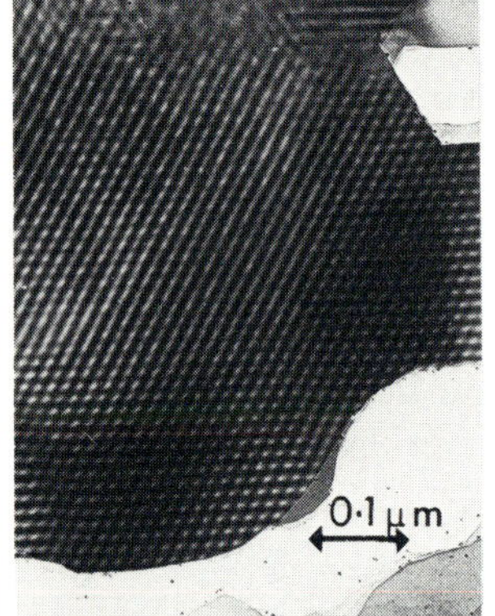


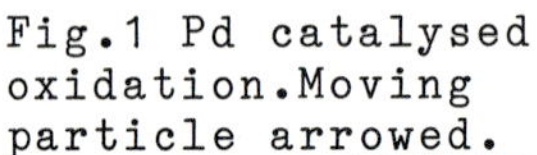

Fig.1 Pd catalysed oxidation.Moving particle arrowed.

Fig.2a Fig.2b
Air oxidation of graphite at 640C in microscope.

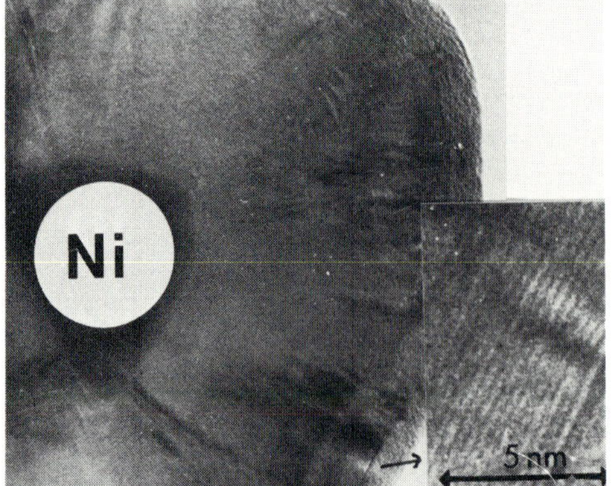


Fig.3 Graphite formed from methane.Lattice shown inset.

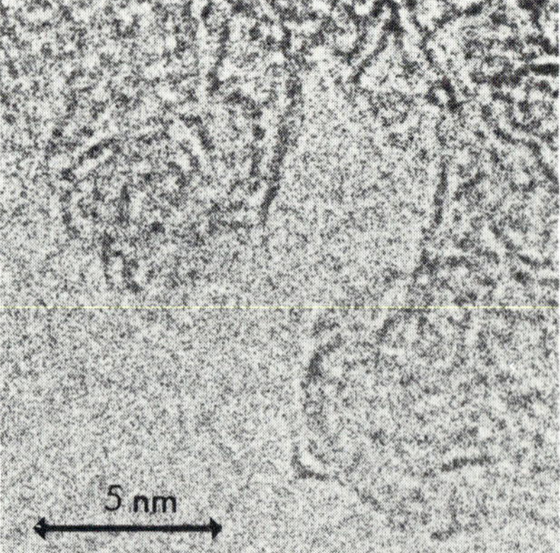


Fig.4 Surface pores of heat treated anthracite carbon.

# Microstructural analysis of metallurgical cokes and intercalated species

M J F Shevlin, J R Fryer, T Baird

Chemistry Dept., University of Glasgow, Glasgow G12 8QQ, Scotland

The presence of recirculating alkali within the blast furnace is believed to be a major factor in the degradation of metallurgical cokes (Hawkins et al, 1974; Abraham and Staffanson 1975; Goleczka et al 1983), leading to a decrease in output and fuel efficienty of the furnace. The presence of alkali gives the possibility of the formation of potassium intercalated species, causing expansion of the coke's semigraphitic layers, resulting in both mechanical and chemical weakening. Penetration of the structure by potassium sets up stresses, caused by differential volume expansion of the carbon lattices which is possibly relieved by crack formation (detrimental to the coke strength), and expansion of the microporous structure makes it more accessible to reactive species at the relatively high temperatures involved.

The microstructure of both blast furnace feed cokes and ex-tuyere cokes have been examined by HREM (figs 1-4). Particular emphasis has been placed on the use of lattice imaging and elemental analysis techniques in investigating the effect of alkali attack on the microstructure, and in assessing the distribution of mineral matter in the coke. The ex-tuyere samples were obtained from the furnace high temperature region in which coke strength is of vital importance and the samples have undergone alkali attack. The feed cokes used were the material as charged to the furnace.

Figures 1 and 2 show examples of two feed cokes of similar rank. Both of these samples exhibit a typical non-graphitised carbon structure, with the lattice spacing of an average of 0.34 nm, of short range order, and very little 0.35 nm (0002) lattice development. Also apparent in both samples are areas of highly ordered crystalline material (arrowed) with lattice spacings 0.32 - 1.8 nm. This material has been shown by EDXRA and elemental mapping techniques to be mineral matter (of an alumino-silicate nature) randomly distributed throughout the sample and inherent in the coke from its formation.

The ex-tuyere samples show two distinct types of microstructure. Figure 3 shows one area of the ex-tuyere coke. The random, "onion" type carbon structure of the feed coke is still apparent although the lattice spacing has been enlarged to around 0.47 nm.

Elemental analysis shows negligible amounts of potassium in the feed cokes, and in contrast very much higher concentrations of potassium in the ex-tuyere samples. The well documented alkali cycle as reported by Lu (1980), together with the observed expansion of the carbon lattices (0.47 nm)

and levels of potassium in the samples are consistent with the formation of an intercalated compound. The (0002) lattice expansion however is not as extensive as would be expected for a species such as $C_8K$, $C_{16}K$ or $C_{24}K$ and is probably a residue compound formed from the intercalated material during the period of removal from the furnace to transfer to the microscope (Carr 1970).

Figure 4 is an example of the other type of microstructure observed throughout the ex-tuyere sample. These areas contain extensively ordered (0002) carbon lattices with spacings of an average of 0.34 nm, no significant expansion of the layer planes is observed within the areas of ordering. The reasons for the occurrence of two microstructural forms are as yet not fully understood but is postulated as being due to one of two processes. Firstly, lattice spacing collapse, of the expanded layer planes, to the original feed coke spacing of 0.34 nm. This phenomenon has been previously reported in graphite (Carr 1970) and is probably due to the highly reactive nature of intercalated species. In this mechanism the coke sample would undergoe potassium attack to form a disordered intercalated species which subsequently forms a residue compound of greater order. The increased order probably being the result of single or repeated intercalation. The second possibility for the occurrence of two microstructural types is the creation of two carbon species during the mesophase in coke production. These two carbon types can then be envisaged as having differing properties with respect to alkali attack and/or heat treatment.

Reproduction of the ex-tuyere microstructure has been successfully carried out in the laboratory. Reaction of feed coke samples with potassium carbonate in an electric furnace has given rise to areas of extensively ordered carbon lattices and substantiated the formation of intercalated species occurring in metallurgical coke.

The separation of the effects of alkali attack and the heat treatment causing ordering of the carbon lattices in the ex-tuyere samples is currently under investigation.

## References

Abraham K P and Staffanson L I 1975 Scand. J. Metallurgy 4 193
Carr K E 1970 Carbon 8 155
Goleczka, Tucker J and Everitt G 1983 ECSC Round Table, Coke, Luxenburg
Hawkins R J, Monte L and Waters J J 1974 Iron and Steelmaking (Quarterly) 3 151
Lu W K 1980 In McMaster Symposium 8

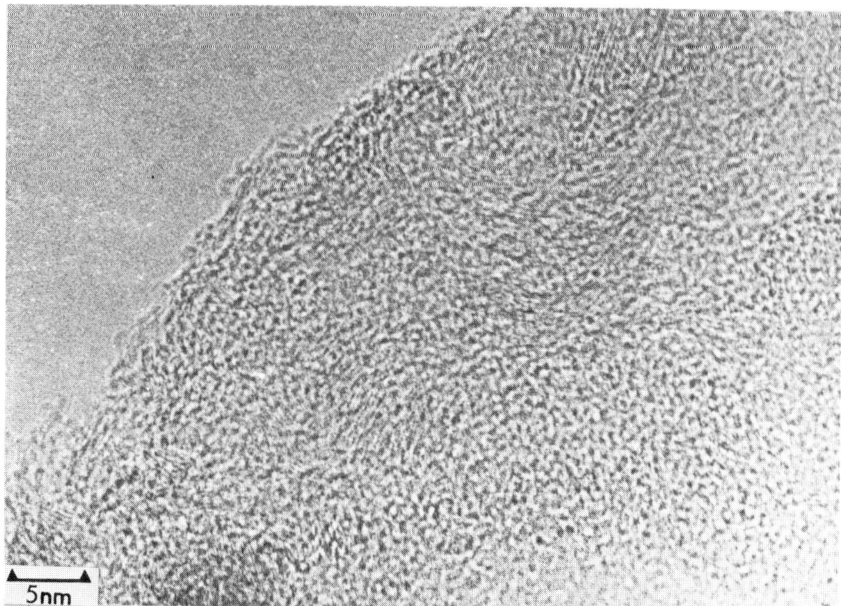


Figure 1.

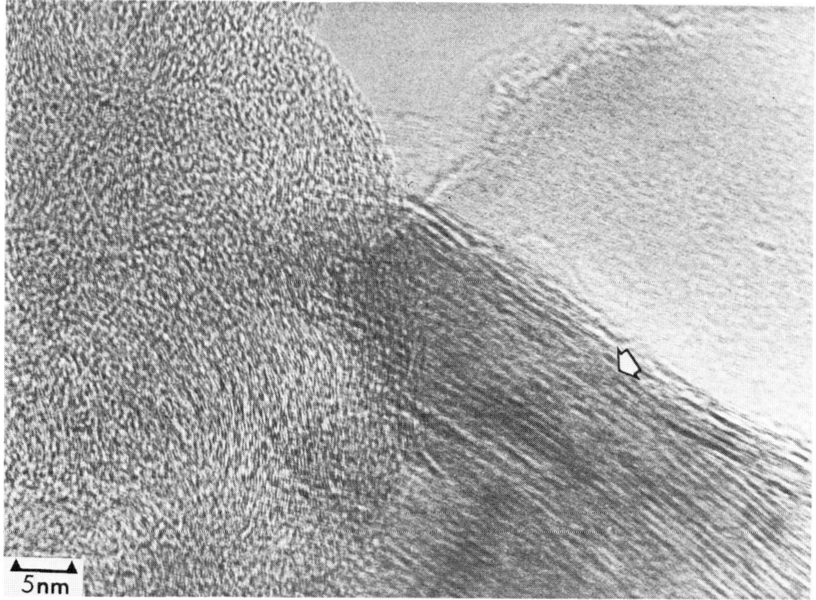


Figure 2.

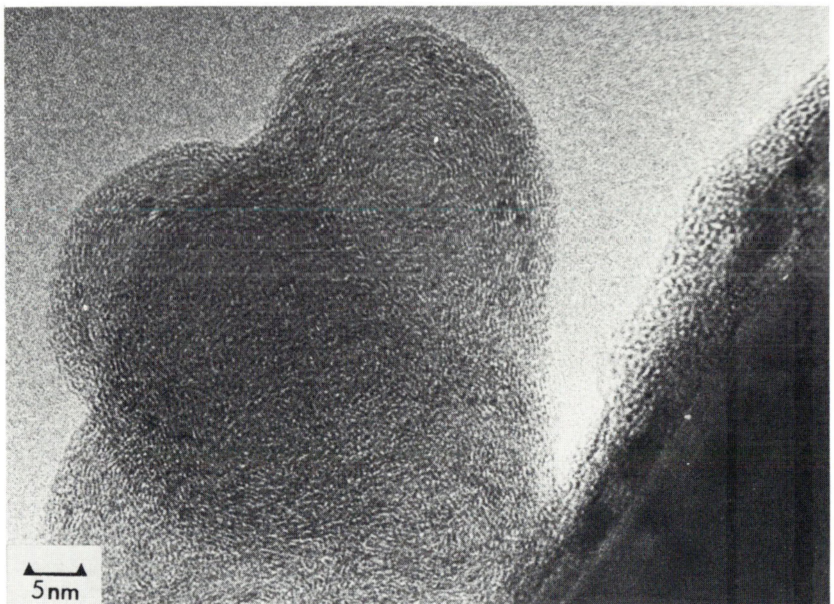


Figure 3.

Figure 4.

Inset showing extreme case of ordering.

*Inst. Phys. Conf. Ser. No 78: Chapter 14*
*Paper presented at EMAG '85, Newcastle upon Tyne, 2–5 September 1985*


# A high resolution examination of graphite–iron chloride intercalates

C P Armstrong, J A Little, W M Stobbs

Department of Metallurgy and Materials Science, Pembroke St, Cambridge

The intercalated graphites have a variety of applications and those based on iron chloride have been considered for use as catalysts in, for example, the Fischer-Tropsch synthesis. Catalytic activity can be increased by changing the intralayer structure of the iron intercalates, for example, by reduction with potassium as in the synthesis of ammonia (Ichikawa et al, 1972). Hence it is of interest to determine the local structure within the iron chloride layers both in the basic intercalate and in related compounds. This is equally relevant given questions raised by Cowley and Ibers' (1956) X-ray structure determination of the graphite- ferric chloride compound. Their results indicated a $30^{o}$ shift of the a-axis of the ferric chloride unit cell with respect to the graphite unit cell and associations between the chlorine and carbon atom positions suggesting distortions of the Cl-Cl distances to values as low as 3.15Å and as high as 4.2Å.

Here we present the preliminary results of an investigation into the use of high resolution electron microscopy for the study of the intralayer structure of the $FeCl_3$ layers in graphite as well as the interlayer stage variability of the compound in relation to mechanisms of intercalation and de-intercalation.

## Experimental

The intercalate was prepared from Madagascan flake graphite and an excess of anhydrous ferric chloride by heating under vacuum at $380^{o}C$ for 24 hours. Specimens for examination in an 'edge-on' orientation in the electron microscope were prepared using a modification of the technique described by Newcomb et al (1985) involving the ion beam thinning of flakes embedded in epoxy resin. The thinned sections were examined in a top entry JEOL 120CX operated at 100kV with an objective lens having a $C_s$ of 0.6mm. In this microscope, under these conditions, transfer is attained down to approximately 2Å for amorphous carbon and the 2.1Å fringes in graphite are readily observed (Donovan et al, 1983). The images were interpreted in relation to simulations of the Cowley and Ibers (1956) structure, and some modifications of it, using computer programs developed by G J Wood and described elsewhere (G J Wood et al, 1984).

## Results

The intercalates produced were, as observed, typically irregularly staged and a micrograph of an intercalated flake obtained at approximately Scherzer defocus is shown in Fig.1. with the supposed positions of the $FeCl_3$ layers and the stage numbers marked. Comparison of a through focal series for the region demonstrated that the apparent first stage structure to the right of the flake is an artefact of Fresnel effects at the edge of the material. Similar conclusions on the staging behaviour for this material have been reached by Thomas et al(1980) and Millward et al (1981) who compared the stability of the ferric chloride intercalate unfavourably with that of a ferrous chloride intercalate for which relatively regular staging was retained. Bifurcations of the staging were also observed (Fig.2.) similar to those described by Thomas et al (1980), as were apparent terminations of graphite layers (A in Fig.3.) presumably associated with de-intercalation and stage rearrangement. Fig.3a. was obtained between Gaussian and Scherzer defocus, Fig.3b. near extended Scherzer defocus. In the former, the $FeCl_3$ layers are light while they are dark in the latter to the left but regain a central bright line in the thicker region to the right at B. Irregularities in the fringe arrangement actually associated with irregularities caused by de-intercalation can, however, be easily misinterpreted, as is emphasized by examination of regions D and E shown in Fig.4a. and Fig.4b. at Scherzer and second broad band defocus.

From these qualitative observations it will be clear that structure instability in the layering makes quantitative assessment of the intralayer structure variations hazardous. A variety of models were, however, simulated for the microscope operating parameters as a function of thickness and defocus. For Fig.5a., the intercalated layer was taken to be iron octahedrally surrounded by chlorine between carbon layers separated by 9.4Å (For simplicity, the carbon and ferric chloride a-axes were considered to be commensurate but rotated $30^{\circ}$ with respect to each other) while for Fig.5b., a mixed double layer was placed in the same gap. For Fig.5c., the 9.4Å gap was left empty. For Fig.5d. and Fig.5e., the gap was given a uniform scattering potential, in the former case, equal to that of Cowley and Ibers' (1956) model (Fig.5a.) and in the latter, equivalent to a large excess of iron. It can be seen that variations in the c direction are virtually indistinguishable for the relatively low thicknesses considered until the scattering potential is raised above that of the graphite (Fig.5e.) when the through focal variation in the fringe spacings changes in the expected manner. The ordered structure, however, shows clear evidence of its nature as viewed perpendicular to the layering and there is some evidence for this. In Fig.6., a variety of spacings are seen (as at regions such as F) along the layers of approximately 2.9Å and 5.1Å (using the average graphite layer spacing of 3.35Å as a standard). If the specimens could be examined with greater confidence at higher thicknesses, the variation in image layer characteristics with defocus could be used to infer more about the structure, for example, whether there are graphite layer strains associated with the chlorine-carbon interactions. While this has been attempted using the techniques described by Wood et al (1984) and Stobbs et al (1984), the results obtained are insufficiently accurate (see Fig.7. and Fig.8.) to allow any modification to the structure described by Cowley and Ibers (1956). It should be noted in particular that although the 2.1Å graphite fringes are readily observed for pristine graphite, the fringes seen in general (crossing the 3.35Å graphite fringes in thinner regions) have a spacing of 3Å (see Fig.3. region C). Since the spacing is associated with 110 reflections for $FeCl_3$, it may be concluded that the intercalated material is covered in de-intercalated $FeCl_3$ and this alone makes structure refinement inadvisable.

One of us (CPA) is grateful to Foseco and the SERC for financial support and we thank Professor D Hull for the provision of laboratory facilities.

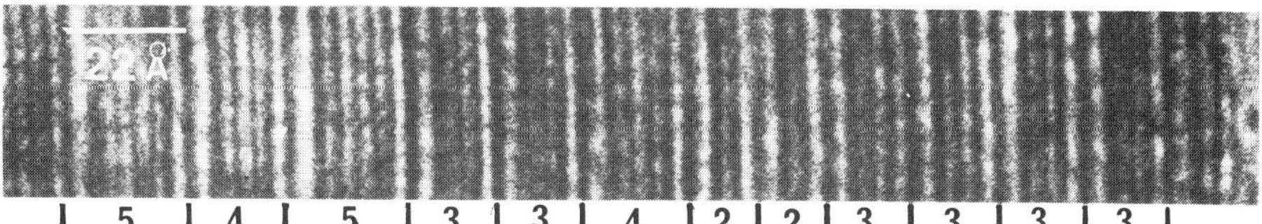


Fig.1 A graphite-ferric chloride flake at near Scherzer defocus.

Fig.2 An example of the bifurcations observed.

Fig.3a and b show apparent terminations of the graphite layers. Fig.3a was obtained between Gaussian and Scherzer defocus, Fig.3b near extended Scherzer.

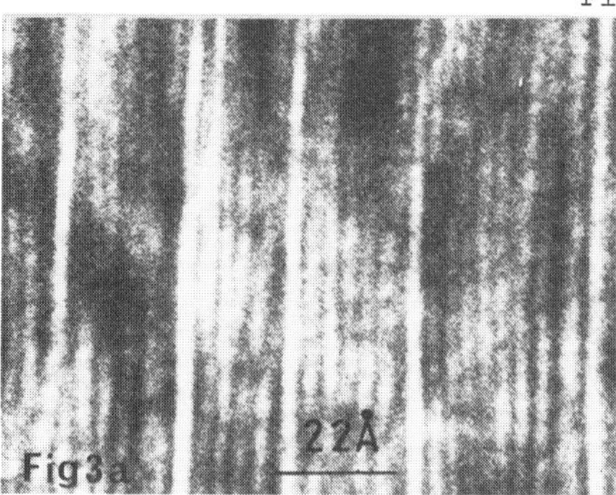


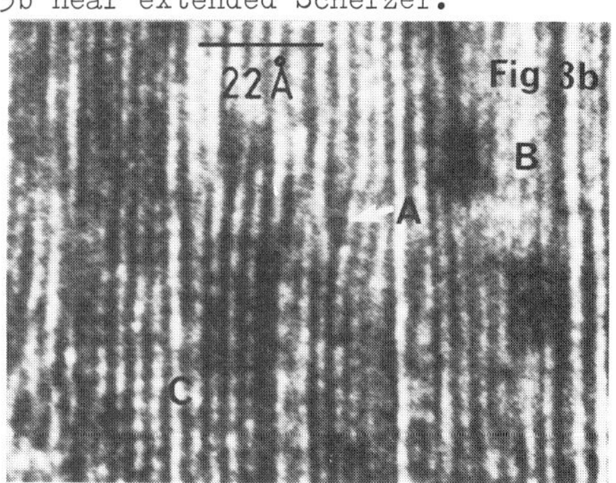


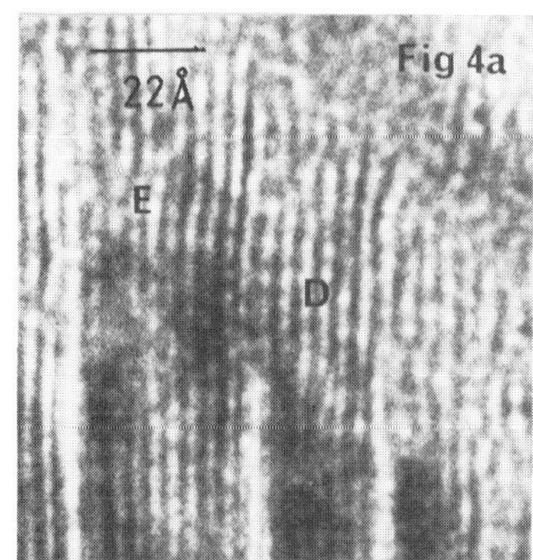


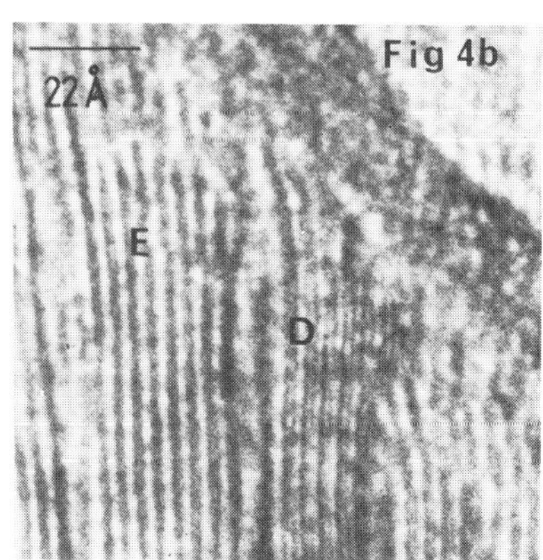


Fig.4a and b show the effects of defocus on the images of irregularities caused by deintercalation. Fig.4a was obtained at Scherzer defocus, Fig.4b at second broad band defocus.

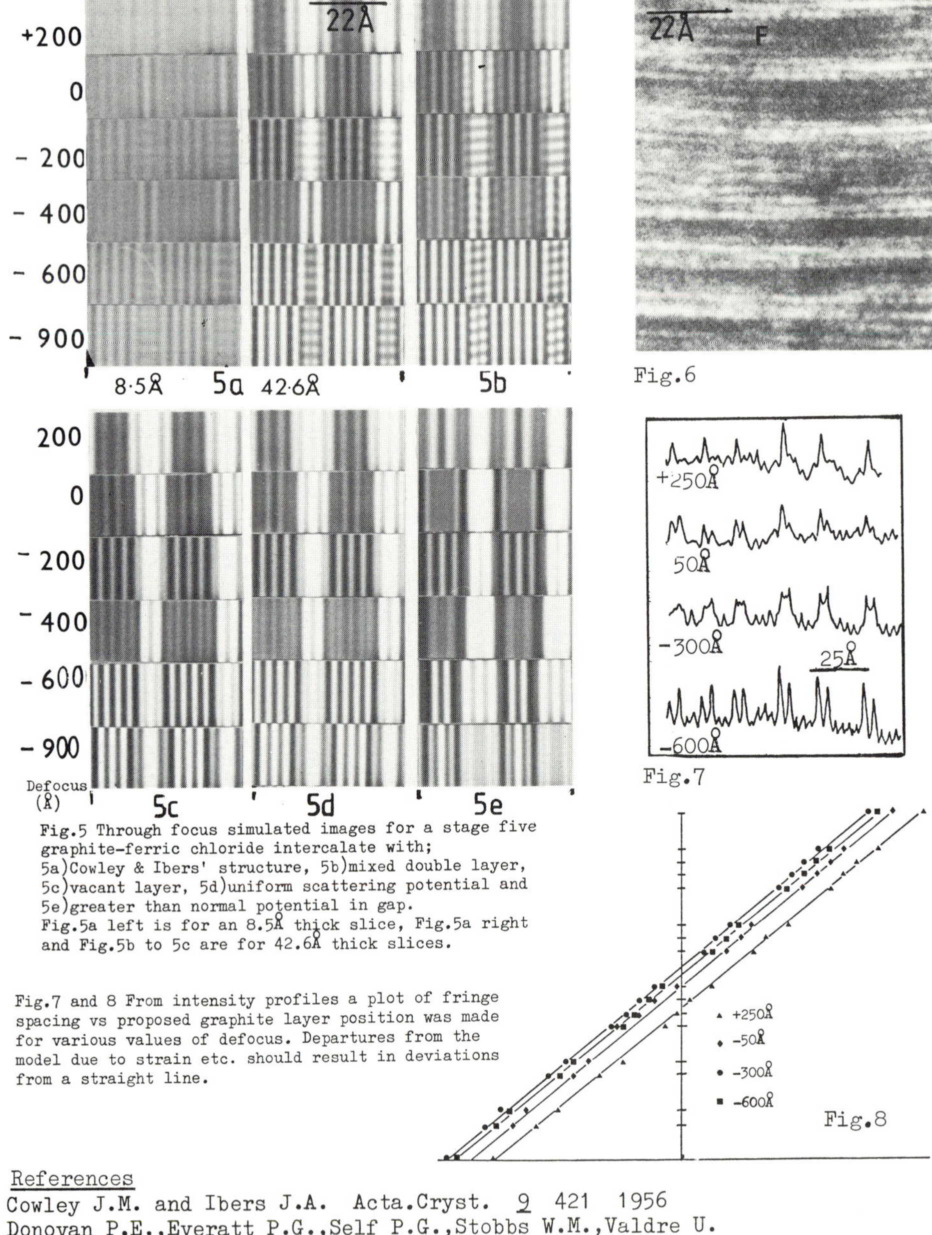


Fig.5 Through focus simulated images for a stage five graphite-ferric chloride intercalate with; 5a)Cowley & Ibers' structure, 5b)mixed double layer, 5c)vacant layer, 5d)uniform scattering potential and 5e)greater than normal potential in gap. Fig.5a left is for an 8.5Å thick slice, Fig.5a right and Fig.5b to 5c are for 42.6Å thick slices.

Fig.7 and 8 From intensity profiles a plot of fringe spacing vs proposed graphite layer position was made for various values of defocus. Departures from the model due to strain etc. should result in deviations from a straight line.

References

Cowley J.M. and Ibers J.A. Acta.Cryst. 9 421 1956

Donovan P.E.,Everatt P.G.,Self P.G.,Stobbs W.M.,Valdre U. J.Phys.E 16 1242 1983

Ichikawa M.,Kondo T.,Kawaso K. J.C.S. Chem.Comm. p176 1972

Millward G.R.,Thomas J.M.,Jones W.,Schlogl R.F. Inst.Phys.Conf.61 (ed. M.J.Goringe) pp.399 1981

Newcomb S.B.,Boothroyd C.B.,Stobbs W.M. J.Microscopy 1985 in press

Stobbs W.M.,Wood G.J.,Smith D.J. Ultramicroscopy 14 145 1984

Thomas J.M.,Millward G.R.,Schlogl R.F.,Boehm H.P. Mat.Res.Bul. 15 671 1980

Wood G.J.,Stobbs W.M.,Smith D.J. Phil.Mag. A50 375 1984

*Inst. Phys. Conf. Ser. No 78: Chapter 14*
*Paper presented at EMAG '85, Newcastle upon Tyne, 2–5 September 1985* 

# A study of the electronic structure of diamond by STEM

J Bruley, L M Brown and S D Berger

Cavendish Laboratory, Madingley Road, Cambridge CB3 0HE

Most studies of the electronic band structure of insulators focus mainly on the optical properties. Electron energy loss spectroscopy (EELS) has been used quite widely to complement optical absorption measurements. Despite the greater sensitivity and higher spectral resolution available optically, EELS shows two main advantages: firstly optical measurements become difficult in the ultraviolet range whereas EELS can provide electronic information up to very high energies. Secondly optical absorption experiments are limited in that only vertical 'direct' electronic transitions are recorded. Very little momentum space information is obtained. Fast electrons however may experience relatively high momentum transfers allowing electronic dispersion to be investigated. As in optical absorption experiments the EELS spectrum may be related to the joint density of states, now linked not only by energy separation but also by momentum separation.

The measurements were made using a scanning transmission electron microscope (STEM) with an incident beam energy of 100 keV. The focused beam was incident along different poles of the crystal and a cone of scattered electrons was collected and analysed with the spectrometer. The spectra show a reproducible anisotropic behaviour in the low loss region, namely losses at 5 - 10 eV, fig. 1. The spectral resolution is limited to 0.5 eV f.w.h.m. by aberrations in the spectrometer. Previous energy-loss studies have been confined to spectra, (e.g. Egerton and Whelan (1974), Festenberg (1969, Whetten (1966)).However, because of its large gap and the intense interest in its optical properties which are controlled by precipitates, diamond is an ideal candidate for STEM, which gives images as well as spectra.

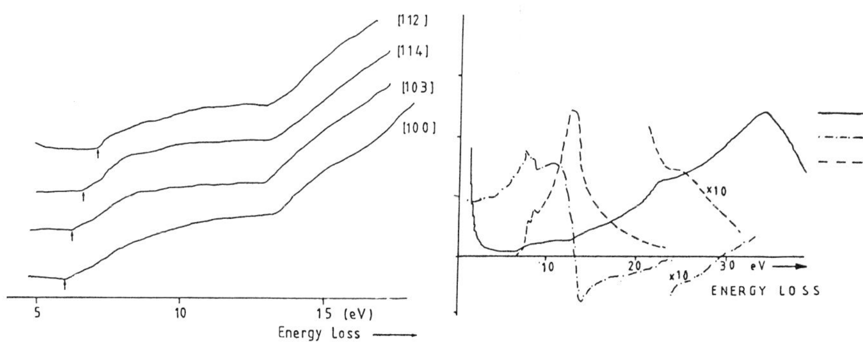


Fig. 1 EELS from various poles of diamond.

Fig. 2 Typical EELS spectrum with permittivity data (Roberts et al).

The standard formula for the loss function (i.e. probability of losing energy $\hbar\omega$ per unit path length) is

$$\left(\frac{1}{\Lambda_e}\right) = \frac{+e^2\ \Delta\omega}{4\pi^3\varepsilon_o\hbar v^2}\int\frac{1}{q^2}\ \mathrm{Im}\left(\frac{-1}{\varepsilon(\omega,q)}\right)\ d^2q \qquad (1)$$

where $v$ is the electron velocity
$q$ is the momentum transfer
$\varepsilon$ is the complex permittivity
$\Lambda_e$ is the mean free path for the loss process

This enables most features of the EELS spectrum of pure diamond to be adequately explained by referring to the optical constants, see fig. 2. The peak at 32 eV corresponds to the bulk plasmon, the hump at 24 eV to the surface plasmon and a possible contribution from an optical transition or from the bulk plasmon of an amorphous carbon layer, and the slope change at 12 eV to a strong interband transition The experimental spectra show very clearly the parabolic shape at the band edge, around 6 eV. The observed anisotropies in the band edge region (fig. 1) can be qualitatively understood with reference to the established band structure calcuations, e.g. Painter (1970).

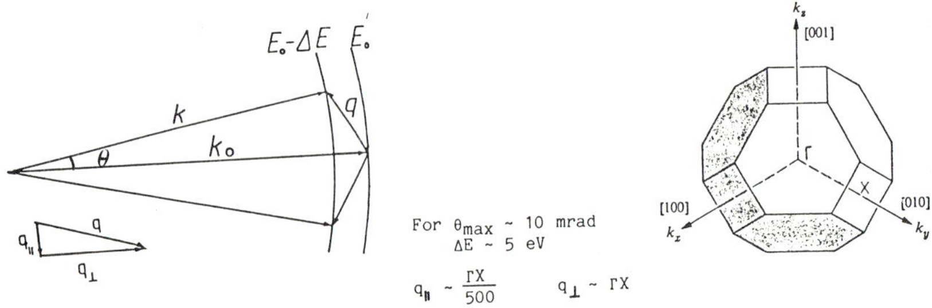


Fig. 3a Allowed cone of momentum transfers, q, for a given energy loss $\Delta E$.

Fig. 3b Brillouin zone of diamond. Conduction band minima lie along [100] directions.

Fig. 3 shows the cone of momentum transfers allowed by the microscope apertures. The momentum changes permissible cover the whole Brillouin zone, but at small energy losses the momentum change parallel to the beam corresponds only to a small fraction of the zone, so that only transitions with momentum changes in a plane perpendicular to the beam can be observed. Of course the transition probability is weighted in the forward direction as can be seen from the $q^{-2}$ term in the integral of equation (1). However, for energy losses just above the (indirect) band gap, only a very restricted range of momentum transfers is possible, namely those connecting extrema in the energy surfaces for the valence and conduction bands. Thus in a first simple analysis of the orientation dependence of the energy gap, the fact that the transition probability is weighted in the forward direction can be neglected. Because in diamond the indirect transitions of minimum energy transfer correspond to momentum transfers in cube planes, the mimimum observed energy gap will be seen in spectra acquired with the cube planes perpendicular to the beam, as is experimentally observed in fig.1.

In thicker specimens there appears a broad hump in the low loss region of the spectrum, fig. 4. This may be interpreted as the loss due to Cherenkov radiation. It has been studied widely and a complete mathematical treatment for the loss function of a relativistic particle passing through a thin dielectric medium has been given by Kröger (1970). A standard calculation (see Schiff, Ch 10) gives the differential mean free path for Cherenkov retardation (i.e. probability of exciting a photon of frequency $\nu$ per unit path length of the fast electron)

$$\frac{1}{\Lambda_{ch}} = \frac{2\pi}{137}\left(1 - \frac{c^2}{v^2\varepsilon(\nu)}\right)\frac{\Delta\nu}{c} \quad \text{if } \frac{c^2}{v^2\varepsilon} < 1 \qquad (2)$$

The result indicates that the mean free path for Cherenkov retardation is of the order of 50 times the mean free path for plasmon excitation, which is consistent with observation in the thick specimens. A plot of $(1 - c^2/\varepsilon(\nu)v^2)$ for diamond shows clearly the broad peak at 10 eV as observed (see fig. 5).

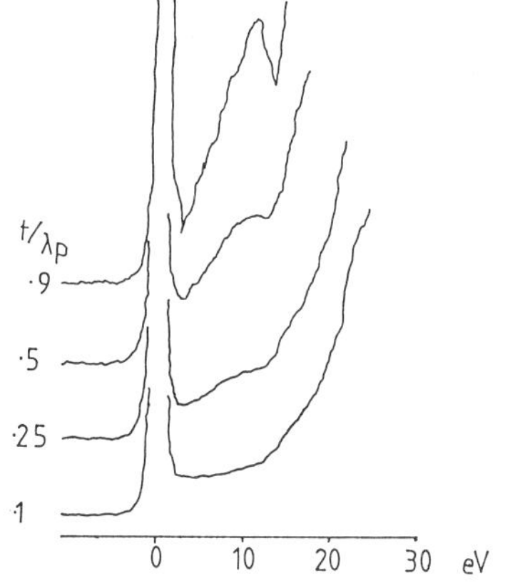


← Fig. 4 Variation of EELS with thickness.

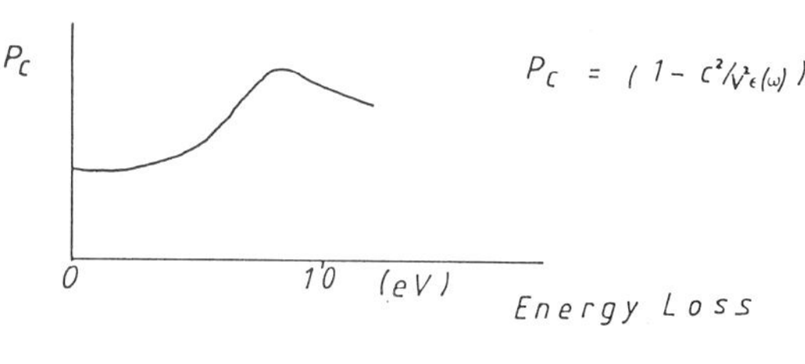


Fig. 5 Probability of Cherenkov radiation excitation.

As the specimen decreases in thickness fig. 4 shows an extinction of the Cherenkov peak. Such an extinction is predicted by Kröger's complicated theoretical formula and may be understood in simple physical terms. As an electron crosses the thin dielectric sample of thickness d it experiences a change in energy. The uncertainty in this energy is related to the film thickness d by the uncertainty principle i.e.

$$\Delta E\ \Delta\tau = h\Delta\nu.\ d/v > h \qquad (3)$$

If this uncertainty in frequency is comparable to the Cherenkov frequency $\nu_o$, no coherent loss process can occur. So we must have $d > v/\nu_o$ for a loss to occur, and this must be compatible with $v\sqrt{\varepsilon}/c > 1$, so $d\nu_o\sqrt{\varepsilon}/c > 1$ is a necessary condition for Cherenkov loss. The critical value of the film thickness is thus the wavelength of Cherenkov radiation in the medium. The critical thickness for the 5 eV photon is ~ 900 Å and so spectra must be collected from specimens thinner than this.

The broad band absorption in the gap region can likewise be understood in terms of the retardation effect brought about as a result of the finite speed of light in a dielectric medium. The effect is called transition radiation and occurs because the electric fields around the electron must change abruptly as it moves across an interface from one medium to the next. The electron momentarily accelerates and radiates, an effect which was first recognized and treated formally by I.M. Frank and V.L. Ginzburg (1946). Using the formulae of Kröger, one finds that the probability of exciting a photon of energy between 4 and 4.5 eV as the electron crosses the interface is approximately 0.02. Again this agrees with the experimental results very well. This radiation loss puts a limit on the detectability of localized states in the gap at about one such state per hundred atoms.

With the advent of the high resolution STEM, it is possible to scan the 1 nm electron probe across the specimen and map the variation in the intensity of absorption within given energy windows selected by the spectrometer below the band gap. Such images show directly the distribution of localized energy states. Fig. 6 demonstrates this clearly. The band gap image is compared with the image formed using the annular dark field detector (ADF). The image shows the difference between insulating clusters D and graphite-like particles, B which appear bright in both images. Also clearly visible are surface states or possibly a thin graphite surface coating. The graphite particles almost certainly result from specimen preparation, which is by burning.

The spatial resolution limit is primarily limited by the impact parameter for such loss processes. For a 5 eV loss this is about 20 nm which roughly agrees with observations.

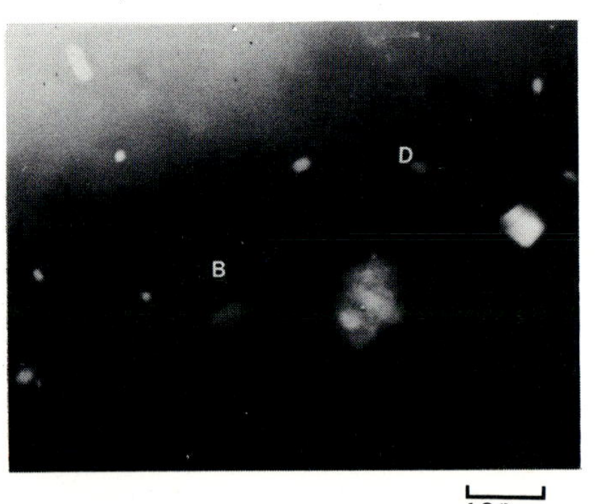


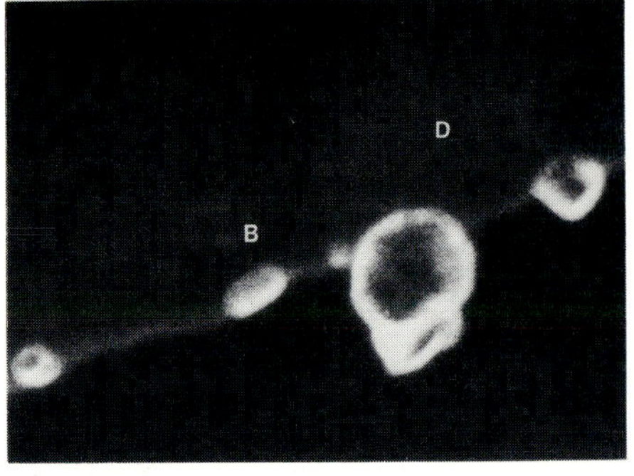


100nm

(a) Annular dark field. (b) Energy filtered, ΔE ~ 4 eV.

Fig. 6 STEM images showing distribution of localised states.

Acknowledgements

We wish to thank Dr. A. Howie FRS for his helpful advice and De Beers for their financial support and supply of samples. Thanks are also due to Professor T. Evans and Dr. M. Seal for their supply of diamond specimens.

References

Egerton, R.F. and Whelan, M.J. Phil. Mag. (1974) 30, 739-749.
Ginzburg, V.L., Frank, I.M. JETP (1946) 16, 1.
Kröger, E. Z. Physik. (1970) 235, 403-421.
Painter, G.S., Ellis, D.E. and Lubinsky, A.R. Phys. Rev. (1971) B4, 3610-3622.
Roberts, A.R. and Walker, W.C. Phys. Rev. (1967) 161, 730-735.
Schiff, L.I. Quantum Mechanics Ch. 10 (1955) (McGraw:Hill).
von Festenberg, C. Z. Physik. (1969) 227, 453-481.
Whetten, N.R. Appl. Phys. Lett. (1966) 8, 135.

*Inst. Phys. Conf. Ser. No 78: Chapter 14*
*Paper presented at EMAG '85, Newcastle upon Tyne, 2–5 September 1985*


# Fractographic studies of carbons

John W Patrick and Alan Walker

Carbon Research Group, Loughborough Consultants Limited
University of Technology, Loughborough, Leics, UK, LE11 3TF

## 1. Introduction

Carbons exhibit brittle behaviour, so that features visible in fracture surfaces are unlikely to be distorted by microplasticity during crack propagation. Fracture surfaces have therefore been examined successfully (Hays et al 1983) using a scanning electron microscope (SEM) to view directly the three-dimensional form of the structural units - building blocks - of which the carbons are composed. Components present in a very wide range of carbons were classified into four broad categories termed flat, lamellar, intermediate and granular according to their appearance in the fracture surface.

Fracture surfaces of carbons also contain information on their mode of fracture. The object of the present paper is to illustrate features observed on fracture surfaces of four bulk carbon products and to show how these contribute to the understanding of the strength of the material.

## 2. Experimental Procedures

The materials studied comprised metallurgical cokes, a briquetted solid fuel, aluminium industry anode carbons and nuclear graphites. Surfaces examined were tensile fracture surfaces produced by diametral compressive loading of 1 cm diameter by 1 cm long specimens using a Tensometer Universal testing machine at a crosshead speed of 0.5 mm/min. Broken specimens were cleaned ultrasonically, mounted on an SEM stub and gold coated before examination in a Cambridge Instruments scanning electron microscope, model 604.

## 3. Results and Discussion

Metallurgical cokes are highly porous carbons, usually with porosity greater than 50%, and with tensile strengths in the range 3 - 6 MN/m$^2$. They differ from the other materials examined in that the solid phase is composed primarily of material which has fused during carbonisation with only 10 - 20% of inert, non-fusing components. Depending on the rank of the coal carbonised, components in all four of the structural unit categories named earlier can be observed in the fracture surfaces. The lamellae in lamellar components are usually aligned circumferentially to the pore surfaces so that crack propagation from pore to pore results in breakage across lamellae (Fig 1a). Granular carbon, however, fails by an intergranular mechanism (Fig 1b). This difference in the mode of fracture implies a difference in the contribution by the components to coke strength and empirical relationships of the form $S = K_0 + \Sigma K_i P_i$ between the coke tensile strength (S) and the proportions (P) of the various structural components present have been obtained. Coke strength has also been related to coke pore structure parameters (Patrick and Stacey 1983) using the Knudsen equation. Future work will involve attempts to combine the two approaches.

The other three materials examined were pitch-bonded carbons, the briquetted solid fuel differing from the anode carbons and the nuclear graphites in that the filler was not precarbonised. Instead, a non-fusing anthracite was used as filler and this was carbonised simultaneously with the binder. Examination of the fracture surfaces of the solid fuel indicated that the material had failed by a mixture of transgranular failure of the filler and by failure at the binder-filler interface. Fractured filler particles had the distinctive rather smooth surface of flat components which bore numerous brittle fracture river patterns (Fig 2a). Such surfaces are characteristic of a carbon formed from a non-fusing precursor. Examples of features formed by transgranular and interfacial failure respectively are shown at A and B in the magnification micrograph in Fig 2b. The proportion of the fracture surface resulting from interfacial failure was assessed by point-counting and a broad relationship between this value and the strength of individual specimens within one batch of briquettes was obtained (Fig 3).

The tensile strengths of the pre-baked anode carbons ranged from 3.9 to 6.9 $MN/m^2$. Thus despite having much lower porosity (20 to 25%) they were little stronger than metallurgical cokes. As with the briquetted solid fuel, fracture of the anode carbons occurred by a mixture of transgranular failure of the lamellar filler particles and failure at the binder-filler interface. The surface features formed are not illustrated but are similar to those shown in Fig 1a and at B in Fig 2b respectively. The interfacial bond was particularly poor at devolatilisation pore surfaces of the petroleum coke particles. For a limited number of experimental carbons from one manufacturer examination of fracture surfaces of specimens of mean strength again showed strength increasing when the proportion of the surface formed by interfacial failure decreased (Fig 4).

The nuclear graphites examined differed from the anode carbons in having significantly higher strength (12 - 36 $MN/m^2$). This was a result of a markedly better quality of the binder-filler bond as evidenced by the predominant transgranular mode of fracture of the filler seen in the fracture surface. Graphite strength has been related to filler particle size and to porosity (Knibbs 1967). Interlamellar fissures in large petroleum coke filler particles provide easy pathways for propagating cracks, especially when the material is stressed perpendicular to the extrusion direction. Higher strength isotropic graphites fail by interlamellar and translamellar fracture of the filler.

References

Hays D, Patrick J W and Walker A 1983 Fuel 62 1079
Knibbs R H 1967 J. Nuclear Materials 24 174
Patrick J W and Stacey A E 1983 ISS Transactions 3 1

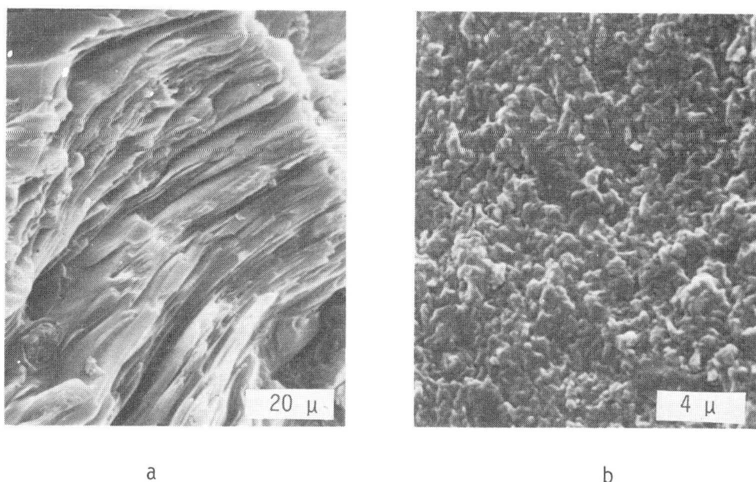


Fig 1. Lamellar (a) and granular (b) components in fracture surfaces of metallurgical coke

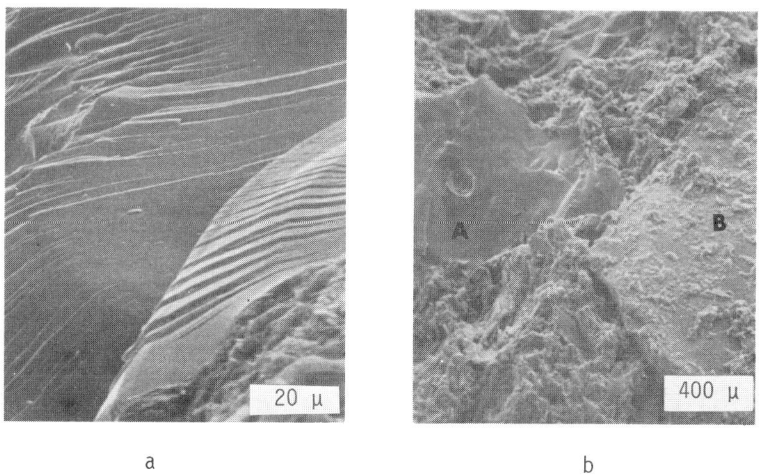


Fig. 2. Fracture surfaces of briquetted fuel showing flat components (a and at A in b) and feature formed by failure at filler-binder interface (B in b)

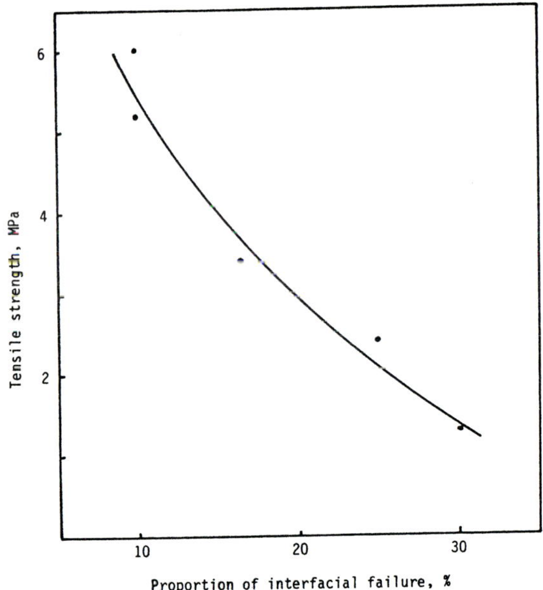


Fig. 3. Influence of failure at binder-filler interface on strength of briquetted fuel

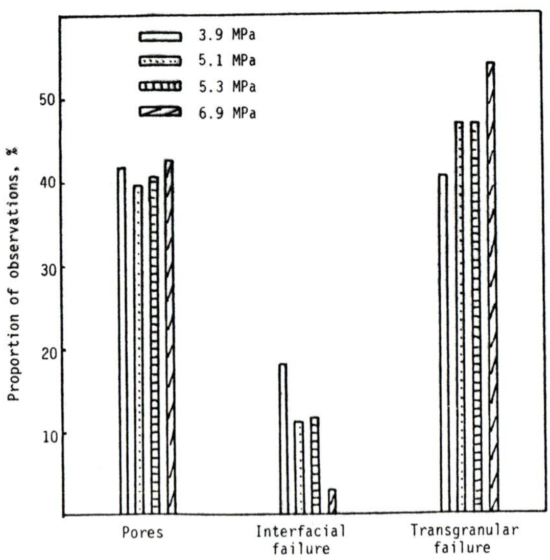


Fig. 4. Variation of strength of anode carbons with extent of interfacial and transgranular failure

# Electron optical studies of shell efflorescence

T Baird[(1)] and N H Tennent[(2)]

(1) Chemistry Department, University of Glasgow, Glasgow G12 8QQ, UK
(2) Glasgow Museums and Art Galleries, Art Gallery and Museum, Kelvingrove, Glasgow G3 8AG, UK

## 1. Introduction

Mollusca collections and other museum specimens are subject to deterioration during prolonged storage through the formation of surface crystalline salts (efflorescence) produced by the action of organic acid vapours present in the storage environment (Fitzhugh and Gettens 1971). In the case of molluscs deterioration occurs via the action on the shell matrix of acetic acid and formic acid vapours emitted by wooden storage cabinets. We have recently identified the efflorescence on numerous types of shells by XRD, IR, TGA and NMR methods (Tennent and Baird 1985); the major products are calcium acetate hemihydrate $Ca(CH_3COO)_2\frac{1}{2}H_2O$, calcium acetate hydrate $Ca(CH_3COO)_2H_2O$ and a novel double salt, calcium acetate-formate hydrate $Ca(CH_3COO)(CHOO)H_2O$. The efflorescence sometimes consisted of a single product or frequently of a mixture of any two or all three of the above in various proportions.

This paper presents the results of TEM and SEM studies on shell deterioration products and the corresponding synthetic reference materials performed with the view to evaluating the usefulness of electron optical methods in this field.

## 2. TEM of Shell Deterioration Products

The samples were prepared for microscopy in the dry state to prevent transformation on recrystallisation of these soluble salts to other products. Possibilities here are changes in the degree of hydration of the salts, e.g. $Ca(CH_3COO)_2\frac{1}{2}H_2O$ to $Ca(CH_3COO)_2H_2O$ and the formation of a pentahydrate from the double salt (Tennent and Baird 1985). Adherence of the products to the carbon film support was poor and gentle grinding was generally necessary to facilitate specimen mounting.

Electron diffraction data obtained from all of the shell products examined were invariably in agreement with the _d_-values for the calcite modification of $CaCO_3$ or for CaO that resulted from further electron beam irradiation. Abnormally low dose viewing occasionally afforded other diffraction data to be recorded but these were usually of little diagnostic significance, generally there being only a few diffraction spots arising from the selected areas. Interestingly it was possible to induce crystallinity by the action of the electron beam on areas that had not previously exhibited evidence of crystallinity, i.e. from areas where rapid initial beam damage had already occurred. When this was observed the resulting diffraction data were again consistent with calcite formation. All the soluble shell products quickly developed a mottled appearance indicative of electron beam degradation (Fig.1) and none of the products possessed morphological characteristics that readily distinguished it from other soluble shell materials.

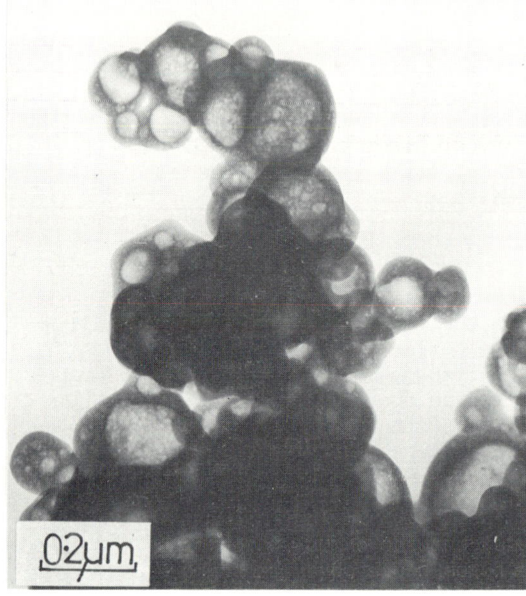


Fig.1. TEM showing beam damage to crystals of $Ca(CH_3COO)(HCOO)H_2O$ soluble shell deterioration product (as identified by XRD, IR and NMR methods).

Insoluble surface material present on some shells, however, exhibited a much more regular morphology than the soluble components and was less susceptible to beam damage (Fig.2). Electron diffraction data here were in accord with the aragonite modification of calcium carbonate. Spicular aragonite crystals grew with their long axis parallel to the [001] direction as was observed in previous work on eggshells (Baird and Solomon 1979). No electron beam induced crystallisation was manifested by the aragonite crystals but CaO could be produced on prolonged beam irradiation as found with the soluble shell products. Confirmation that this material was aragonite was obtained from infrared data. The presence of loose aragonite crystals on some shell surfaces - the shell matrix itself consisting of compact aragonite crystals - appears to be associated with a different mode of deterioration of these shells possibly involving degradation of the organic matrix within the shell.

## 3. TEM of Reference Materials (Calcium Acetate Hemihydrate, Calcium Acetate Hydrate, Calcium Acetate - Formate Hydrate and Calcium Formate)

Dry preparations were used as before. Although calcium formate was never found as a single product on shell surfaces it was included for general comparison with the other carboxylic acid salts.

All the reference materials rapidly developed a mottled appearance similar to that found with the soluble shell deterioration products. Fig.3 taken from calcium acetate hydrate, illustrates the effect produced by the electron beam. Normal viewing conditions gave diffraction data consistent with calcite (and subsequently CaO). Low dose studies afforded higher lattice spacings to be recorded from all the reference samples. Data from calcium formate could generally be equated with the literature d-values whereas the data from the three hydrated salts could not always be related to their known d-spacings, perhaps owing to rapid dehydration in the microscope.

The reference materials, like the shell products, were subject to electron beam induced crystallisation resulting in the formation of calcite. Single crystals of reference materials gave single crystal spot patterns corresponding to calcite in various orientations with respect to the electron beam direction; these induced patterns were observed only after initial loss of crystallinity of the samples had occurred (Fig.4).

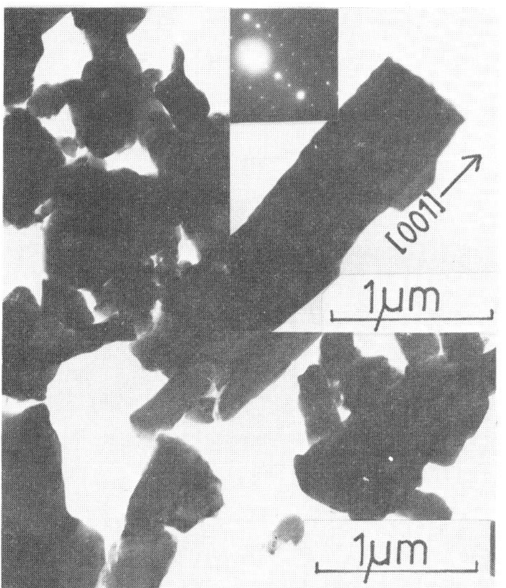


Fig.2. Aragonite crystals removed from a shell surface.

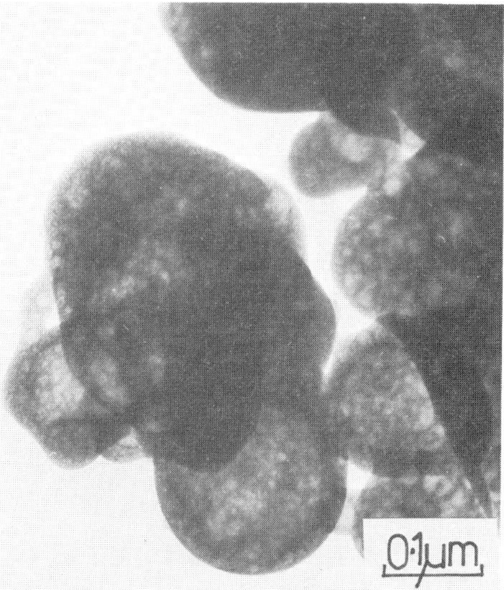


Fig.3. $Ca(CH_3COO)_2H_2O$ reference material.

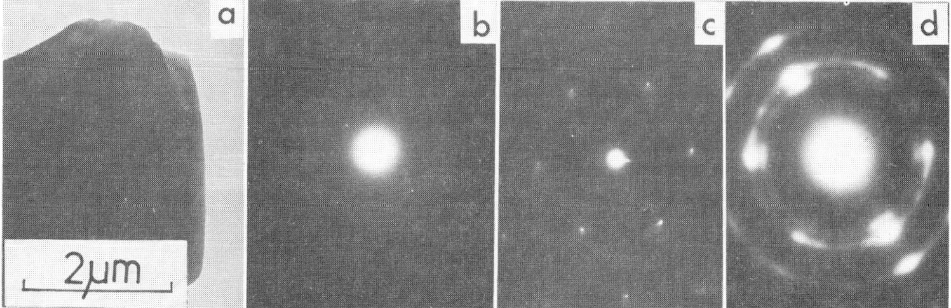


Fig.4. (a) Single crystal of calcium formate. (b) Low dose diffraction pattern from (a). (c) Single crystal diffraction pattern corresponding to $CaCO_3$(calcite) with [001] zone axis obtained from (a) by increasing the electron dose. (d) CaO diffraction pattern produced by severe irradiation of the $CaCO_3$ single crystal.

## 4. SEM of Shell Surfaces Exhibiting Deterioration

SEM was useful for gaining morphological information provided the deterioration was not too severe when shell products masked underlying features. In general efflorescence occurred in patches but there was a marked tendency for crystallisation of the carboxylic acid salts to occur within the grooves associated with shell growth lines (Fig.5). X-ray microanalysis confirmed that calcium was the main elemental constituent of efflorescence.

Shells where loosened aragonite crystals were present revealed different morphological features indicative of the breakdown of the shell matrix rather than the formation of surface salts (Fig.6).

5. Conclusions. TEM when used as the sole technique does not provide an unambiguous identification of shell deterioration products due to the effects of dehydration, beam damage and electron beam induced crystallisation of these materials.

## 6. References

Baird T and Solomon S E 1979 J exp. mar. Biol. Ecol. 36 295-303

Fitzhugh E W and Gettens R J 1971 in Science and Archaeology (ed R Brill) MIT Press Cambridge MA 91-102

Tennent N H and Baird T 1985 Studies in Conservation 30 73-85

Fig.5. SEM illustrating preferential formation of soluble Ca salts within the grooves of shell growth lines.

Fig.6. Fragmented shell surface containing loosened aragonite crystals from the shell matrix.

*Inst. Phys. Conf. Ser. No 78: Chapter 14*
*Paper presented at EMAG '85, Newcastle upon Tyne, 2–5 September 1985*


# Application of TEM and SEM to studies of synthetic copper salts relating to bronze disease

T Baird[(1)], R W Innes[(1)] and N H Tennent[(2)]

(1) Chemistry Department, University of Glasgow, Glasgow G12 8QQ, U.K.
(2) Glasgow Museums and Art Galleries, Art Gallery and Museum, Kelvingrove, Glasgow G3 8AG, U.K.

## 1. Introduction

Copper compounds constitute the main corrosion products found on ancient bronzes and give rise to the surface patinaswhich characterise these artifacts. The most common corrosion products which are found in stable patinas are malachite, $Cu_2(OH)_2CO_3$ and azurite, $Cu_3(OH)_2(CO_3)_2$, whereas the presence of the greenish basic copper chloride polymorphs, atacamite and paratacamite, $CuCl_2 3Cu(OH)_2$, is indicative of active corrosion. Less common, and possibly associated with ancient bronzes found in soils of high alkali carbonate concentrations, is the blue compound, chalconatronite, $Na_2Cu(CO_3)_2 3H_2O$. Other surface products that subsequently arise and of interest in the present studies are cuprite $Cu_2O$ and nantokite, CuCl, the latter being oxidised readily to atacamite (and cuprite).

The feasibility of using TEM to distinguish between the above products was tested in this work and some preliminary studies made of the role of surface features in inducing artificially produced corrosion products on copper surfaces. The corrosion products were made by reacting pretreated copper surfaces with sodium chloride and/or sodium carbonate solutions in various atmospheres (water, air, $CO_2$). Bulk samples of chalconatronite were synthesised by reacting sodium carbonate and copper acetate solutions.

## 2. TEM Studies of Chalconatronite, $Na_2Cu(CO_3)_2 3H_2O$

Synthetic chalconatronite (characterised by IR) was first examined to assess its behaviour in the electron beam. Fig. 1 is typical of the observed morphology. Severe shrinking occurred whilst viewing. Electron diffraction showed the material to be amorphous, this being at variance with corresponding x-ray powder data and indicating irradiation damage. Comparable samples that were decomposed thermally (~ 80°C) outwith the microscope gave morphologically different amorphous products (Fig. 2). A blue compound produced artificially on a copper surface exhibited similar rapid shrinking to the synthetic preparation.

## 3. TEM Studies of Basic Copper Chloride Polymorphs $CuCl_2 3Cu(OH)_2$

Basic copper chlorides produced on copper surfaces were characterised by IR. Unlike chalconatronite rapid decomposition occurred in the beam to give a crystalline material (Fig. 3), identified by electron diffraction as nantokite, CuCl, i.e. a reduction process had occurred. A fragmented thin layer of $Cu_2O$ was usually found associated with the CuCl and this may be the decomposition product of the hydroxide component of basic copper chloride or it may be a $Cu_2O$ surface layer removed from the copper during specimen preparation. Some other diffraction data were observed but remain unassigned.

## 4. SEM Studies of Artificially Produced Corrosion Products

No enhanced corrosion has been observed to date at grain boundaries, surface scratches or on etched copper surfaces as distinct from polished ones. Corrosion products were generally associated with an oxide layer ($Cu_2O$) and initial growth of the products appeared to occur at the interface between the copper and the $Cu_2O$ layer (Fig. 4).

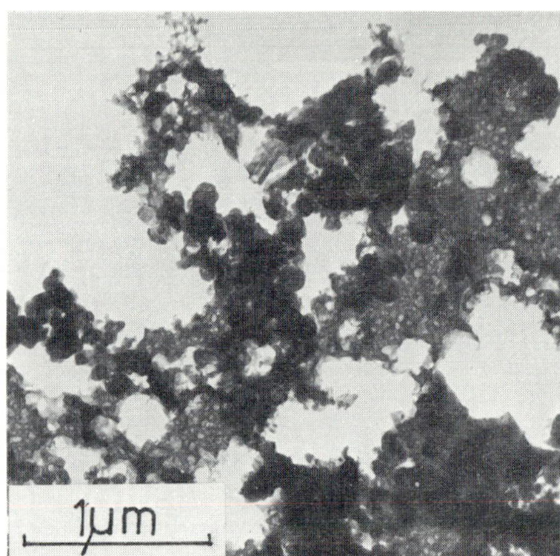


Fig.1 Synthetic chalconatronite morphology

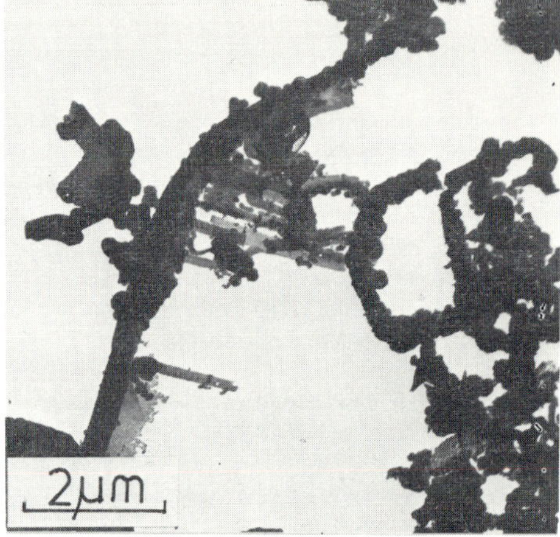


Fig.2 Thermally decomposed chalconatronite

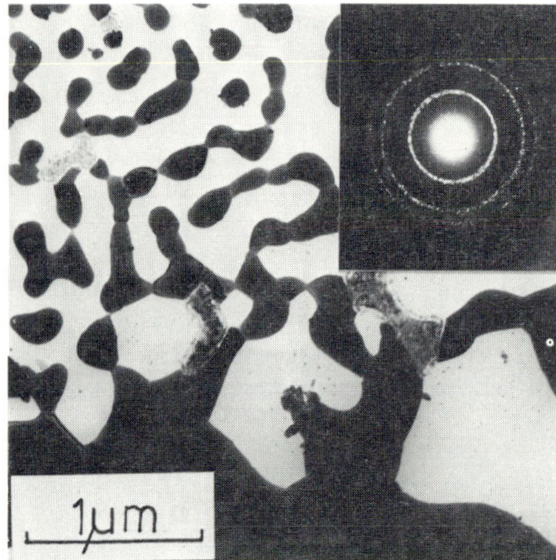


Fig.3 CuCl resulting from beam damage to basic copper chlorides Inset: CuCl diffraction pattern

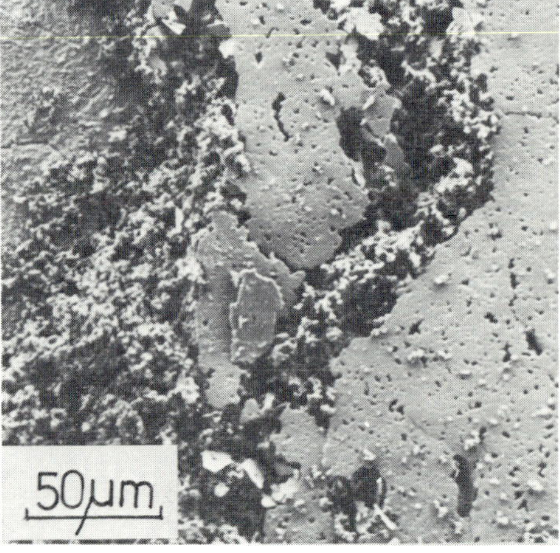


Fig.4 SEM showing the formation of basic copper chlorides at the $Cu/Cu_2O$ interface

*Inst. Phys. Conf. Ser. No 78: Chapter 15*
*Paper presented at EMAG '85, Newcastle upon Tyne, 2–5 September 1985*



# Advantages of a 400 kV high resolution analytical electron microscope and applications

Y BANDO, Y MATSUI, Y KITAMI and Y INOMATA
National Institute for Rescarch in Inorganic Materials,
1-Namiki, Sakura-mura, Niihari-gun, Ibaraki 305, Japan
and
T OIKAWA, S SUZUKI, T HONDA and Y HARADA
JEOL Ltd, 1418 Nakagami, Akishima, Tokyo 196, Japan

## 1. Introduction

The conventional TEM-STEM analytical electron microscope has mainly three disadvantages at 100 and 200 kV. 1). The image resolution in TEM is not high as compared to normal transmission electron microscope. Thus, the compatibility of structure imaging and spectroscopic micro-analysis is not applicable without changing objective polepiece. 2). The peak to the background (P/B) ratio is not high in the energy dispersive x-ray spectroscopy (EDS) and the electron energy loss spectroscopy (EELS). 3). The spatial resolution for microanalysis is not high due to notable electron beam spreading in specimens. In order to improve these observation capabilities, a new analytical electron microscope with a medium voltage of 400 kV (JEM-4000EX) is constructed (Hashimoto et al 1983 and Bando et al 1984a,b and 1985). The present paper describes some of advantages of the 400 kV high-resolution analytical electron microscope in the use of high resolution observation and microanalysis.

## 2. The increase of P/B ratio in EDS and EELS spectra

Fig.1 shows the accelerating voltage dependence of the characteristic x-ray counts, which are normalized at 100 kV. A Si(Li) detector is placed at take-off angle of 65 degrees with solid angle of 0.029 str. The specimens observed are evaporated Al(z=13), Ge(z=32) and Ag(z=47) films with a thickness of about 100 nm. The probe current is kept constantly at about 1 nA at each operating voltage. Counting time is 100 s. The characteristic x-ray counts and the background are integrated over energy window of 400 eV. It is clear that every characteristic x-ray counts decreases very rapidly between 100 and 200 kV, but not so rapidly above 200 kV. This is closely related to the decrease of the ionization cross section with voltages. On the other hand, the x-ray counts of the background also decrease with the increase of the electron energy and it decreases more rapidly than the characteristic x-rays. As a result of

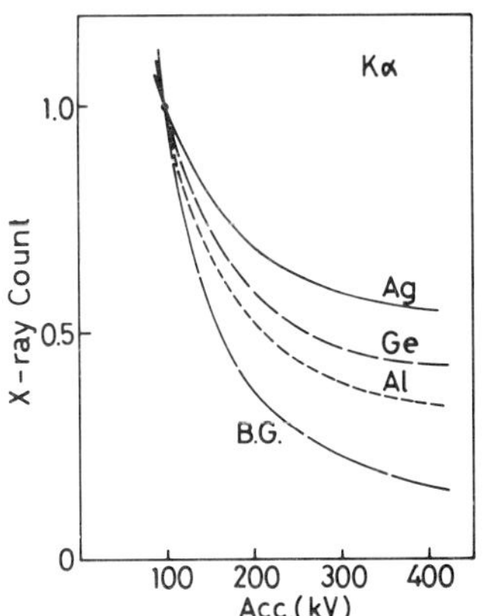


Fig.1 The accelerating voltage dependence of the characteristic x-ray counts and the background with the voltages.

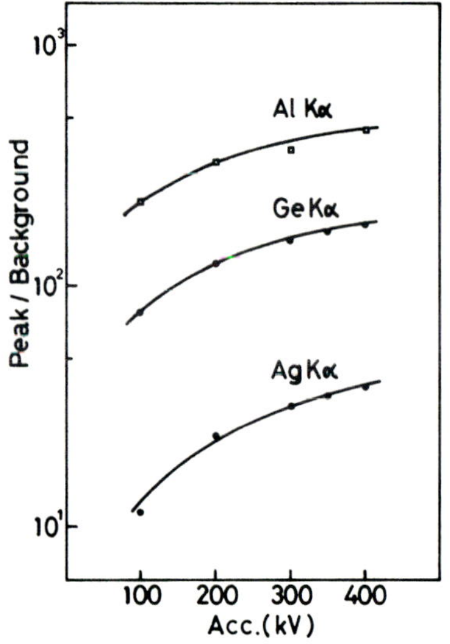


Fig.2 The P/B ratios of the characteritic $Al_{k\alpha}$, $Ge_{k\alpha}$ and $Ag_{k\alpha}$ lines as a function of the accelerating voltages.

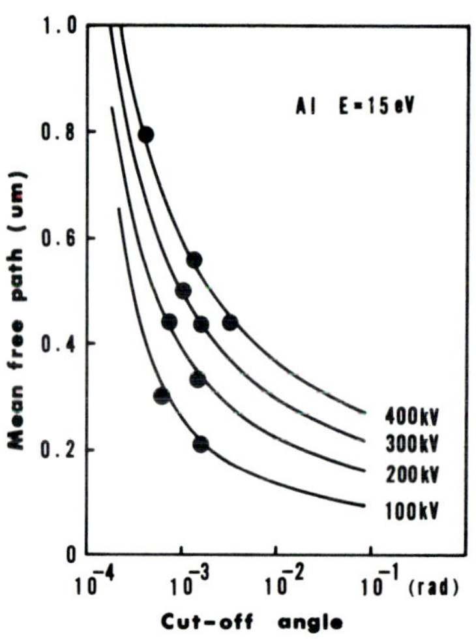


Fig.3 Comparison between measured (dark dots) and calculated (solid lines) mean free path lengths of Al plasmon loss as a function of the cut-off angle and the accelerating voltage.

this, the P/B ratios increase continuously with the increase of the accelerating voltages and the result is shown in Fig.2. The P/B ratios of Al, Ge and Ag at 400 kV are 460, 178 and 38, respectively, which are approximately 1.7, 2.3 and 3.3 times higher than at 100 kV. It is noted that the absolute value of the P/B ratios is larger in the low atomic elements, whereas the dependence of the P/B ratio on the accelerating voltage is smaller in the elements of low atomic numbers (Suzuki et al 1985).

Fig.3 shows a comparison between measured (black dots) and calculated (solid lines) mean free path lengths of Al plasmon loss as a function of the cut-off angle with various accelerating voltages. The measured mean free path lengths increase with the voltages, which are in good agreements with the calculated ones. They are 210, 320, 430 and 560 nm at 100, 200, 300 and 400 kV, respectively , with the cut-off angle of about 1.2 mrad.

Fig.4 shows the accelerating voltage dependence of the P/B ratios of boron K-edge. The specimen is a single crystal of boron with the about 0.5 um thickness and the energy resolution obtained is better than 2 eV. In the boron K-edge spectra, both peaks a and b is due to fine structures corresponding to the density of state of

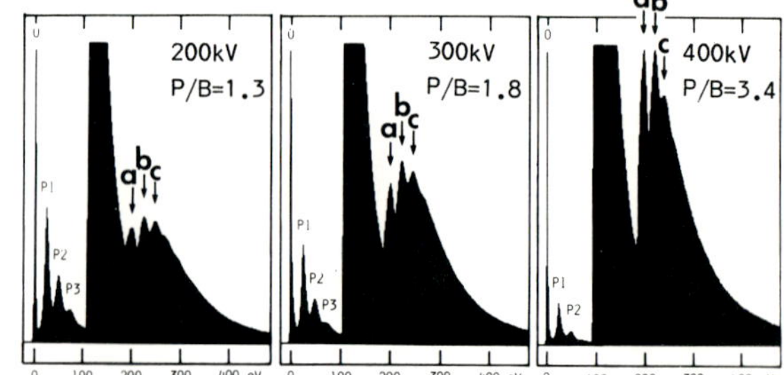


Fig.4 The accelerating voltage dependence of the multiple scattering disturbance on boron K-edge fine structures. The specimen is a single crystal of boron and the energy resolution obtained is better than 2 eV.

the structure, while the peak c is generated by convolution of the multiple scattering. The plasmon loss intensity decreases with the voltages from 200 to 400 kV, and then the P/B ratios increase with the voltage increase. The P/B ratios are 1.3, 1.8 and 3.4 at 200, 300 and 400 kV, respectively. It is also asertained that the peak intensities of a and b increase, while that of c decreases at 400 kV. This is well understood that the background originating from the multiple scattering disturbance decreases with the increase of the voltages (Oikawa et al 1985).

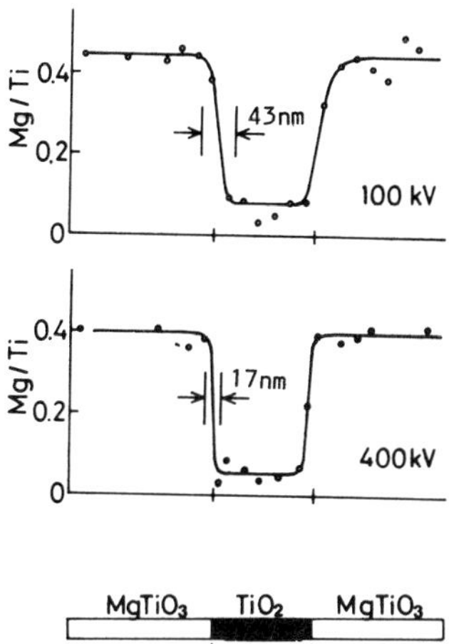


Fig.5 change of x-ray counts of $Mg_{k\alpha}$ to $Ti_{k\alpha}$ across $TiO_2$ precipitates.

## 3. The increase of spatial resolution

Fig.5 shows the result of the spatial resolution test at 100 and 400 kV for the EDS microanalysis. The specimen is $MgTiO_3$ crystal with a needle precipitate of $TiO_2$. The probe size is about 4 nm and the specimen thickness is about 80 nm. The x-ray counts ratio of Mg to Ti changes very sharply at the boundary between $TiO_2$ and $MgTiO_3$. The diffused regions are estimated to be 43 and 17 nm at 100 and 400 kV, respectively, suggesting that the spatial resolution is much improved at 400 kV.

## 4. Compatibility of structure imaging and microanalysis

Fig.6(a) shows a convergent beam electron diffraction pattern of a 12H silicon aluminum oxynitride polytype, and (b) and (c) are corresponding EDS and EELS spectra. The crystal is hexagonal symmetry with lattice parameters a=0.303 and c=3.29 nm. In (a), 6mm smmetry is present in

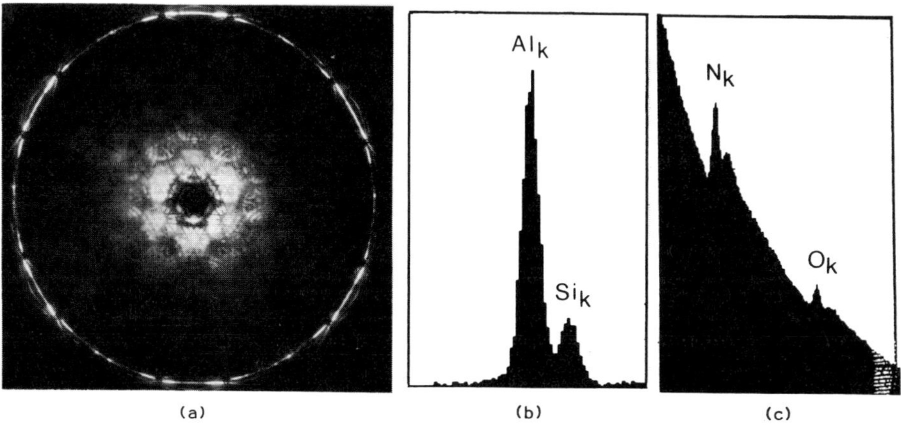


Fig.6 Convergent beam electron diffraction pattern (a), corresponding EDS (b) and EELS (c) spectra of a 12H silicon aluminum oxynitride polytype.

both the bright field disk and the whole pattern. This indicates that the possible space groups are either non-centrosymmetric $P6_3mc$ or centrosymmetric $P6_3/mmc$. From the quantitative analysis of the EDS and the EELS spectra, the chemical composition is determined to be $SiAl_5O_2N_5$. Fig.7(a) shows a structure image of the 12H polytype, obtained subsequently from the same specimen as observed in Fig.6(b) and (c). Each of dark dots and white dots is well resolved with the separation of 0.26 nm normal to the c axis. It should be noted that the structure image reveals symmetries of both the absence of a mirror plane and the presence of a glide plane which are normal and parallel to the c axis, respectively. This means that the space group must be $P6_3mc$. The calculated image is shown in the inset, which agrees well with the observation. The combination of both the structure image and the spectroscopic microanalysis leads to a structure model as shown in Fig.7(b), which is described as a succession of the octahedra (MO), the tetrahedra (MX and $MX_{1.5}$).

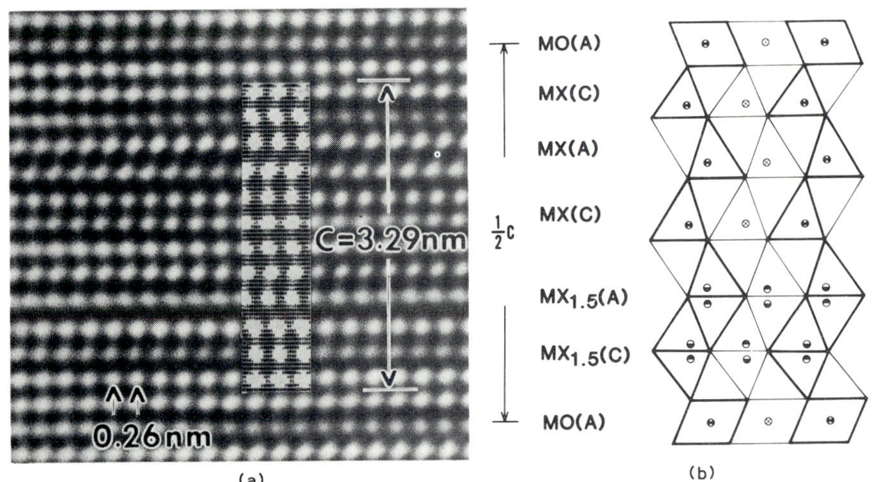


Fig.7 A structure image of the 12H polytype (a) and a proposed crystal structure model (b). The calculated image is inset in (a). The structure consists of a succession of MO octahedra, MX and $MX_{1.5}$ tetrahedra. In $MX_{1.5}$ tetrahedra, half of metal atoms are occupied in order to avoid face sharing between two tetrahedra. A, B and C denotes a stacking sequence.

References

Bando Y, Matsui Y, Kitami, Inomata Y, Ibe K, Honda T and Harada Y 1984a Jpn. J. Appl. Phys. 23 L414

Bando Y, Matsui Y, Kitami Y and Inomata Y 1984b JEOL News, 22 28

Bando Y, Matsui Y, Uemura Y, Oikawa T, Suzuki S, Honda T and Harada Y 1985 Ultramicroscopy in press

Hashimoto H, Endoh H, Honda T, Harada Y, Sakurai S and Etoh T 1983 Proc. 7th Intern. Conf. on High Voltage Electron Microscopy, Berkeley, CA pp81

Oikawa T, Bando Y, Hosoi J, Kokubo Y and Naruse M 1985 Proc. 43rd Annual Meeting of EMSA (Kentucky), pp412

Suzuki S, Bando Y, Kitajima H, Honda T, Harada Y and Kersker M 1985 Proc. 43rd Annual Meeting of EMSA (Kentucky), pp146

# High resolution 400 kV electron microscope

T Honda, K Ibe, S Suzuki, Y Ishida and K Tsuno

JEOL Ltd., 1418 Nakagami, Akishima, Tokyo 196, Japan

Abstract The resolving power of a 400 kV EM was found to be 0.15 nm from an optical diffraction pattern from an amorphous Si image. This value was in good agreement with the one calculated from the information transfer limit of the phase contrast transfer function.

## 1. Introduction

Improvement of the resolving power of an electron microscope is essentially important for modern materials science. And, increase of the accelerating voltage for the electron beam is essential for improving the resolving power because of the beam's short wavelength. In fact it is easy to obtain a 0.2 nm level resolving power in 400 - 1000 kV EMs. But, a resolving power better than that, for instance, 0.15 nm, is very difficult to obtain because of instrumental difficulties. Indeed, few reports have been published on experimental results. This paper discusses experimentally estimated values in comparison with calculated results. Other relevant papers are cited in the references.

## 2. Instrument

The JEM-4000EX, a 400 kV accelerating voltage electron microscope whose objective lens has a 1 mm spherical aberration coefficient, was used for the present study. Several considerations were paid to the fundamental design of this microscope. (1) Minimizing the aberration coefficients of the objective lens by using the finite element method calculation, (2) developing a highly stabilized and bright electron gun, (3) minimizing astigmatism change due to magnification change, (4) improving the stability of the goniometer device for ease of operation and (5) realizing a clean environment around the specimen. As a result, the following were realized: (1) 1 mm $C_s$ and 1.9 mm $C_c$, (2) 2 ppm high voltage stability and $10^7$ A·cm$^{-2}$·str.$^{-1}$ electron beam brightness with a single crystal $LaB_6$ cathode, (3) less than 10 nm astigmatism change over the entire magnification range between 200 kx and 1200 kx, (4) ±15° dual axes specimen tilt angle with a wire type tilt mechanism and (5) $3 \times 10^{-7}$ torr vacuum pressure within the specimen chamber by using an exclusive 140 ℓ·sec$^{-1}$ sputter ion pump.

## 3. Resolution calculation

The resolution was measured by using the optical diffraction (OD) pattern method. To compare an OD pattern with calculated results, the intensity distribution obtained from the phase contrast transfer function (PCTF) was calculated using the following equations:

I: Intensity of OD pattern

$I = C^2$

C: Phase contrast transfer function (PCTF)

$C(k, \alpha, \varepsilon) = B_C \cdot G_\alpha \cdot G_\varepsilon$

$B_C$: Coherent transfer function

$B_C = \cos\left[\frac{\pi}{2} - \frac{\pi}{2\lambda} C_S \lambda^4 k^4 - \pi \Delta f \lambda k^2\right]$

$G_\alpha$: Envelope function for partially coherent illumination

$G_\alpha = \exp[-\pi^2 \alpha^2 (C_S \lambda^2 k^3 - \Delta f k)^2]$

$G_\varepsilon$: Envelope function for chromatic aberration

$G_\varepsilon = \exp[-\frac{1}{2}\pi^2 \lambda^2 (C_C \frac{\Delta V}{V})^2 k^4]$

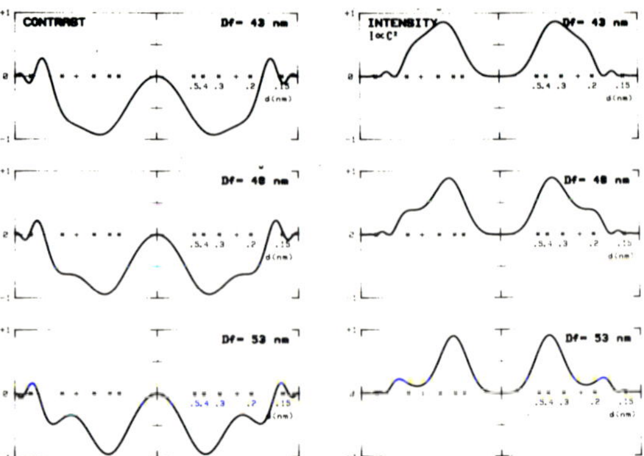

Fig. 1 PCTFs (left) and intensity distributions (right) of OD patterns under defocus conditions around optimum defocus.

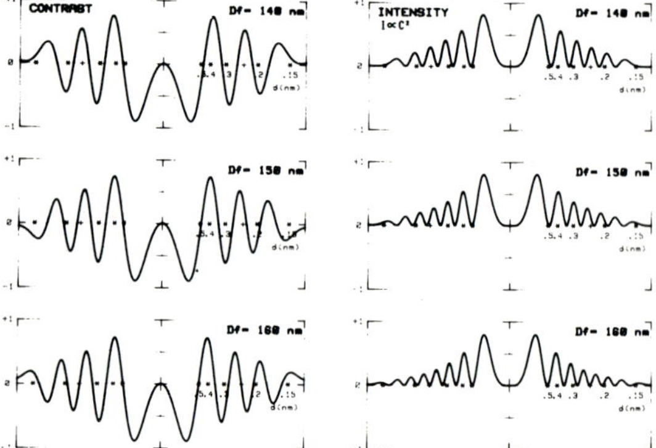

Fig. 2 PCTFs (left) and intensity distributions (right) of OD patterns under defocus conditions around 150 nm defocus.

where $k(=1/d)$ is reciprocal spatial frequency, $\lambda$ wavelength, $\Delta f$ defocus value, $\alpha$ illumination angle, $C_C$ chromatic aberration coefficient, and $\Delta V/V$ stability of accelerating voltage including energy spread of electron beam. The left series of Fig. 1 shows PCTFs under defocus conditions around the optimum defocus (48 nm). For calculation, $\alpha$ of 0.5 mrad and $\Delta V/V$ of 4 ppm (2 ppm acc. volt. stability and 1.5 eV energy spread) were used. The right series of Fig. 1 shows the intensity ($C^2$). The resolving power from the first zero contrast under the optimum defocus condition is in good agreement with a theoretical resolution of 0.168 nm given by $0.65 C_S^{1/4} \lambda^{3/4}$. On the other hand, Fig. 2 shows PCTFs and intensities under defocus conditions about three times the optimum defocus. In this figure, the information transfer limit is estimated to be 0.15 nm ( about 5 % intensity).

4. Specimen

Amorphous Si films of various thickness were made by controlling the sputtering time in the ion beam sputter equipment. On the amorphous Si films, Au particles were evaporated for easy measurement of the OD pattern's camera constant by using their lattice fringes. We used specimen thicknesses between 4 nm and 6 nm, because a specimen more than 10 nm in thickness was too thick for obtaining good image quality, and a specimen of less than 3 nm thickness was susceptible of beam irradiation. Fig. 3 shows an image of a specimen which was used for the present experiment.

5. Results

5-1. Comparison between calculated intensity and OD pattern

Fig. 4 shows an image under the optimum defocus condition. Fig. 5 shows the OD pattern of Fig. 4 together with a calculated intensity distribution curve. The horizontal scale of the curve is calibrated by an Au (111) lattice spacing of 0.235 nm which corresponds to the bright diffraction spots of the OD pattern. The intensity distribution of the OD pattern shows good correspondence to the calculated intensity distribution. The intensity of the first ring which is given by the reversed phase of the PCTF still remains in the OD pattern, but is not clear for detailed discussions.

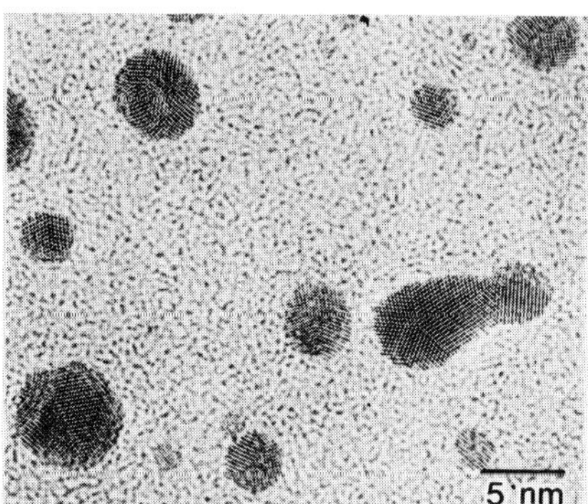

Fig. 3 An amorphous Si film with evaporated Au particles.

5-2. Information transfer limit

Fig. 6 shows an image under 150 nm defocus condition. Fig. 7 shows the OD pattern of Fig. 6, together with a calculated intensity distribution curve. The information transfer limit is estimated to be 0.15 nm from the sixth ring of the OD pattern.

5-3. Structure image of Si(110)

Fig. 8 shows a structure image of Si(110) crystal plane. The calculated atomic arrangement is marked with small circles. The white dots appear to show good correspondence to the atomic arrangement, but the spacing of the dots is slightly wider than 0.136 nm corresponding to (004) reflection.

6. Conclusion

1) A 0.15 nm level point resolution can be ascertained by 5 % intensity of the OD pattern using an amorphous Si film image.
2) The OD pattern of the amorphous Si image under the optimum defocus condition shows a good correspondence with the calculated result of intensity ($I = C^2$). The resolution of the first zero contrast under the optimum defocus was estimated to be 0.168 nm from the OD pattern.
3) The OD pattern of the amorphous Si image under 150 nm defocus condition shows six rings which corresponds with alternated phase of PCTF. The sixth ring of it shows an expected information transfer limit of 0.15 nm.

References

Al-Ali L and Frank J 1980 Optik 56 pp 31 - 40
Boyes E D 1981 E. Microscopy and Analysis 61 pp 27 - 34
Frank J 1973 Optik 38 pp 519 - 536
Hanssen K J and Trepte L 1971 Optik 32 pp 519 - 538
Humphreys C J 1983 Proc. of 7th Int. Conf. HVEM pp 1 - 4
Smith D J, Camps R A, Cosslett V E, Freeman L A and Saxton W O 1983 Ultramicroscopy 9 pp 203 - 214
Watanabe H et al. 1983 Proc. of 7th Int. Conf. HVEM pp 5 - 10
Yagi K, Takayanagi K, Kobayashi K and Nagakura S 1983 Proc. of 7th Int. Conf. HVEM pp 11 - 14

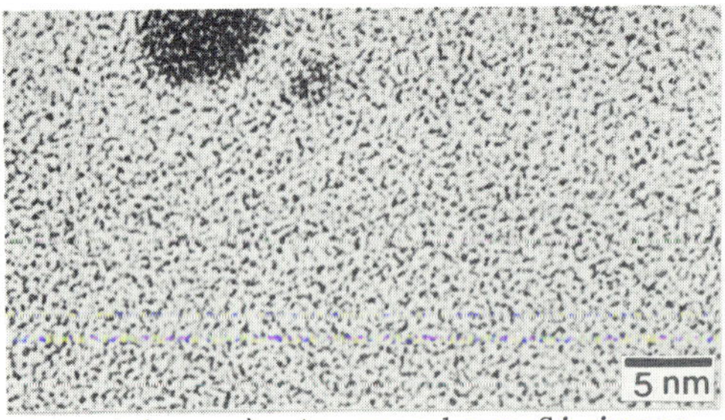

Fig. 4 (upper). An amorphous Si image under optimum defocus condition.

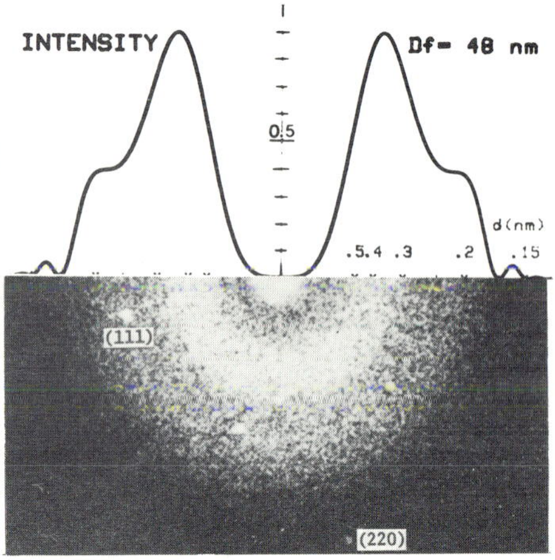

Fig. 5 (right). The OD pattern of Fig. 4 and calculated intensity distribution. Bright spots in the OD pattern correspond to 0.235 nm of Au(111).

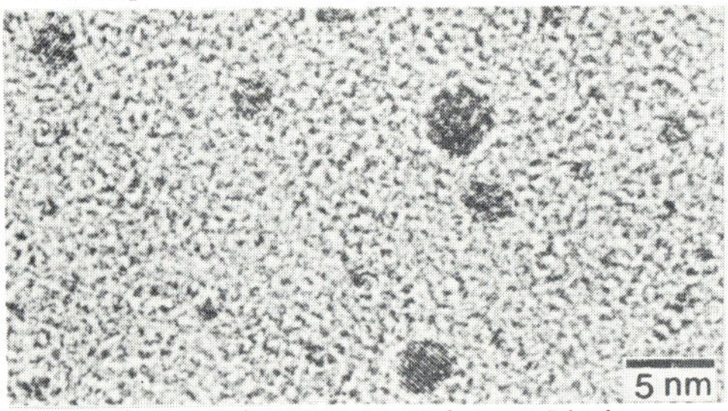

Fig. 6 (upper). An amorphous Si image under 150 nm defocus condition.

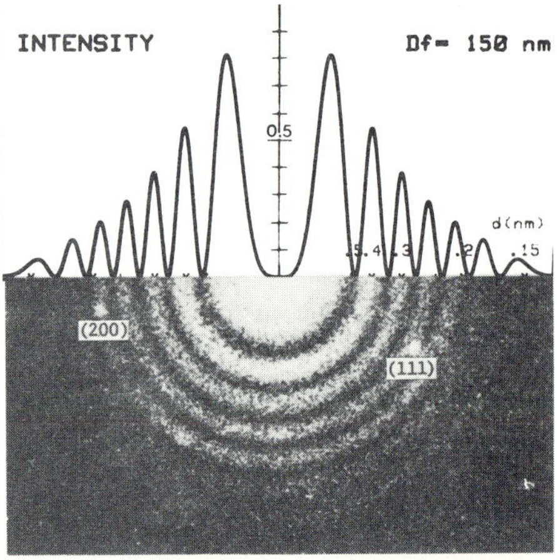

Fig. 7 (right). The OD pattern of Fig. 6 and calculated intensity distribution. Sixth ring of the OD pattern shows good correspondence to calculated intensity of 0.15 nm.

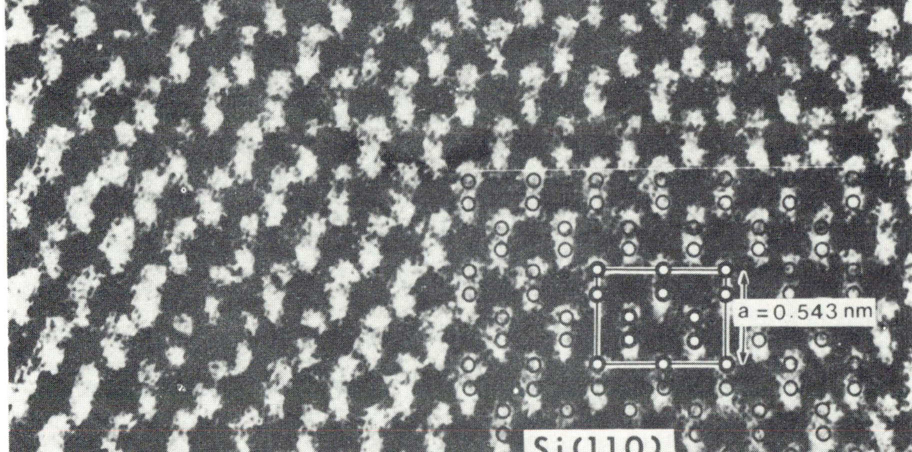


Fig. 8 The structure image of Si(110) plane and calculated atomic arrangement (white circles).

*Inst. Phys. Conf. Ser. No 78: Chapter 15*
*Paper presented at EMAG '85, Newcastle upon Tyne, 2–5 September 1985*



# Medium voltage analytical electron microscopy: Microanalysis versus radiation damage

N J Zaluzec, J F Mansfield, P R Okamoto and N Q Lam
Argonne National Laboratory, MST Division/212, Argonne, IL. 60439, USA

## 1. Introduction

During the last few years the manufacture of analytical microscopes operating at voltages (Vo) as high as 400 kV has progressed steadily. The motivation for this development is enhanced performance in image resolution and microanalysis relative to 100 kV operation. However, in the medium voltage regime the analyst must also be aware of the limitations imposed upon the experiments due to radiation damage effects. These effects can range from sputtering to radiation-induced segregation or electron beam-induced (compositional) mixing, localized to the vicinity of the probe. Since most analytical procedures used today assume that the incident beam does not alter the specimen composition, these deleterious effects can affect the validity of experimental results.

The benefits of higher voltage operation relative to x-ray and electron loss microanalysis have been discussed elsewhere (Zaluzec et al 1983, Zaluzec 1978) and include: increased characteristic signal due to relativisitic effects on the ionization cross-section and gun brightness, increased peak to background ratios due to anisotropic continuum emission, increased spatial resolution and decreased multiple scattering. The disadvantages are twofold. First, the generation of uncollimated radiation (Bentley et al 1979) will be more prolific and hence the precautions and modifications of the basic instrument become more complicated. Second, the effects of radiation damage (displacement versus ionization) become increasingly important as Vo increases. In displacement damage, kinetic energy is directly transfered from the incident electron to atoms within the solid. If the energy transfer is sufficient, then these atoms can be displaced from their lattice sites. At temperatures, where the defects are mobile, segregation to defect sinks can occur and structural or elemental rearrangement may result.

## 2. Results and Discussion

The kinetic energy transfered ($T_T$) to an atom by an electron of kinetic energy $T_E$=eVo, which has scattered through an angle $\phi$ is given by $T_T$ = $[2*T_E*(T_E+2moc^2)*\sin^2(\phi/2)]/[Mc^2]$ where M and mo are the masses of the target atom and electron. Table 1 documents the maximum energy which can be transfered to selected elements at various voltages for $\phi$ = 180°. These values should be compared with the critical displacement energy ($T_d$) required to permanently remove an atom from its site in a lattice and are also given therein. From the preceeding equation one can also calculate the minimum voltage (Vmin) required for $T_T$ to exceed $T_d$; this value serves as a convenient reference point below which one need not consider radiation damage. In addition, the values of $T_d$ change with orientation and structure. Generally, the close-packed directions, <110> in FCC, and <100> in BCC have the lowest $T_d$ values with other directions as much as 2-4 times higher.

The displacement rate of an atom is given by the product of its displacement cross-section ( $\sigma_d$ ) multiplied by the electron current density. Values of $\sigma_d$ (Oen, 1973) are generally in the range of 1 to 40 barns when $T_T > T_d$. The important point here, is that in probe forming systems, although $\sigma_d$ is low, the probe current density can be sufficiently high to yield a significant displacement rate. For example, a current of 1 nA in a 20 nm diameter probe yields a current density of J=318 A/cm$^2$. Using these values, one calculates a displacement rate ($dx/dt = \sigma_d * J$), of 0.002 to 0.08 displacements/atom/sec (dpa/sec). In a $10^3$ sec x-ray analysis, these dpa rates imply that <u>each</u> <u>atom</u> in the probe for which $T_T > T_d$ can be displaced from 2 to 80 times!

The important question, with respect to microanalysis, is how do these values compare with characteristic signal generation? For the case of X-ray analysis, figure 1 compares the calculated number of <u>detected</u> K-shell x-rays/e$^-$/atom from a 8 μm Be window SiLi detector subtending a solid angle ($\Omega$ ) of 0.13 Sr (curve 1), with the maximum number of displacements/e$^-$/atom (curve 2) for Aluminium (fig. 1a) and Nickel (fig. 1b) as a function of voltage. The x-ray parameters (cross-section, etc) used in these calculations are documented elsewhere (Zaluzec, 1984). From this figure one can see that once Vo exceeds Vmin, the displacement rate quickly exceeds the x-ray detection rate. The relatively low x-ray signal is a consequence of the geometrical collection efficiency ( $\varepsilon_X = \Omega/4\pi$ ~1%) of the SiLi detector as well as the x-ray fluorescence yield ($\omega_K$). EELS, on the other hand, does not suffer these effects, having collection efficiencies $\varepsilon_E$ ranging from 30-90% and a effective fluorescence yield of unity. The equivalent plot for EELS would lie above curve #1 by a factor of $\varepsilon_E/(\varepsilon_X * \omega_K)$ ~50X for Al and ~5X for Ni. In alloy systems, subthreshold displacement becomes another factor which must be considered when evaluating the importance of displacement damage. Here, atoms are displaced from their sites at voltages below Vmin. This effect occurs, in alloys, when lower-Z elements, which have undergone electron displacement, transfer their momentum by collisions with higher-Z species and subsequently cause the displacement of that element. This process will basically cause curve #2 to shift horizontally toward lower voltages and has not been included in these calculations since it is composition dependent.

The value for J given above is within the upper limit of some high intensity $LaB_6$ sources, however, for the case of Field Emission systems the current density can reach the range of $10^5$ A/cm$^2$ and the corresponding displacement rates can then reach 50 - 100 dpa/sec! When a material is subjected to such severe conditions, it is highly unlikely that any phases within it remain stable. Observations have ranged from general mass loss experiments (Mochel et al 1983, Thomas 1985) to radiation-induced segregation (RIS) phenomena (Okamoto and Lam, 1985). The mechanism which dominates in a given situation depends on the irradiation conditions. At low defect recombination rates, particuliarly when the probe diameter is less than the specimen thickness, RIS becomes important as the defects have time to migrate away from the irradiation zone. At high displacement rate conditions, beam-induced mixing and/or sputtering will dominate. Unlike sputtering, the implications for AEM in the case of mixing and/or RIS are more subtle and also important for different reasons. First, we have a new high-resolution technique for changing the local composition. Second, we no longer have a nondestructive technique for microanalysis since the composition is changing during the measurement process!

An example of the RIS effect is given in figure 2 which shows a dark-field time sequence depicting the microstructural evolution of an initially two phase Ni-12.7%Al alloy during HVEM irradiation at 700°C at 1 MV. The initially uniform two phase ( $\gamma$ ,$\gamma$ ' $Ni_3Al$ ) material segregates under the action of the incident electron probe forming a central Al-rich precipitate phase ($\gamma$ ') surrounded by a Ni-rich, $\gamma$ ' depleted zone. It has been shown (Okamoto and Lam, 1985) that this effect is due to radial displacement-rate gradients resulting form the Gaussian beam profile. For a focused electron beam, radial variations in the beam intensity (i.e. defect production rate) induce a radial outflux of point defects from the irradiated area. In Ni binary alloys, this defect flux results in a radial influx of oversized solutes like Al into the regions of the irradiated zone where the beam intensity profile (or the defect concentration, $C_d$) is concave downward. The resulting solute concentration in the Ni-Al alloy is proportional to the divergence of the defect gradient $\nabla^2 C_d$; oversize solute enrichment occurs in regions where $\nabla^2 C_d < 0$ and depletion in regions where $\nabla^2 C_d > 0$. The kinetics of this effect is particuliarly strong, increasing as the probe diameter decreases and/or the current density increases, at temperatures where defects are mobile. Similiar results have also been obtained at 300 kV in an Al-1.95%Zn alloy at ~160°C as shown in figure 3. Here a variation in the measured characteristic x-ray intensity ratio of Zn/Al with time indicates a change in composition due to RIS and provides the first experimental evidence that in modern AEM type instruments, appropriate caution must be exercised.

Additional research on these topics is continuing. This work was supported by the U.S. Department of Energy at Argonne National Laboratory.

## 3. References

Bentley J., Zaluzec N., Kenik E., Carpenter R.,SEM/1979/I, pge 581, 1979

Mochel M.E., Humphreys C.J., Eades J.A., Mochel J.M., Petford A.M., App. Phys. Let., Vol 42 (4) pge 392 (1983)

Oen O.S., OakRidge Nat. Lab. Report,#TM-4897, 1973

Okamoto P., Lam N., Proc. Mat. Res. Soc. Sym. Vol. 41 pge 241, 1985

Thomas L.E., Proc. of ASU HREM Symp., (in press Ultramicroscopy) 1985

Zaluzec N.J., Taylor A., Ryan E., Phillipides A., Proc. 7th HVEM meeting, Univ. of Calif.,Berkeley CA. #LBL-16031 CONF-830819 pge 79, 1983

Zaluzec N.J. 9th Int. Cong. on EM, Toronto Canada, pge 548, 1978

Zaluzec N.J. EMSA Bulletin Vol. 14, No. 1,pge 67,Vol. 14 No. 2 pge 61,1984

Zaluzec N.J. Proc. of AEM Workshop July 1984, Lehigh University, Bethelehem Pa, San Francisco Press, pge 279, 1984

Table 1
Maximum Kinetic Energy Transferable to Selected Elements

| Atom | M | 100 kV | 200 kV | 300 kV | 400 kV | Td | Vmin |
|---|---|---|---|---|---|---|---|
| C† | 12.01 | 20.1[eV] | 43.7 | 71.0 | 101.8 | 30 | 145 |
| O | 15.99 | 15.1 | 32.8 | 53.2 | 76.4 | -- | --- |
| Al | 26.98 | 8.9 | 19.5 | 31.6 | 45.3 | 16 | 180 |
| Ti | 47.90 | 5.0 | 11.0 | 17.8 | 25.5 | 15 | 270 |
| Ni | 58.71 | 4.1 | 8.9 | 14.5 | 20.8 | 24 | 450 |
| Mo | 95.94 | 2.5 | 5.5 | 8.9 | 12.7 | 37 | 875 |
| Ag | 107.8 | 2.2 | 4.9 | 7.9 | 11.3 | 28 | 770 |
| Au | 196.9 | 1.2 | 2.7 | 4.3 | 6.2 | 34 | 1300 |

† = Graphite

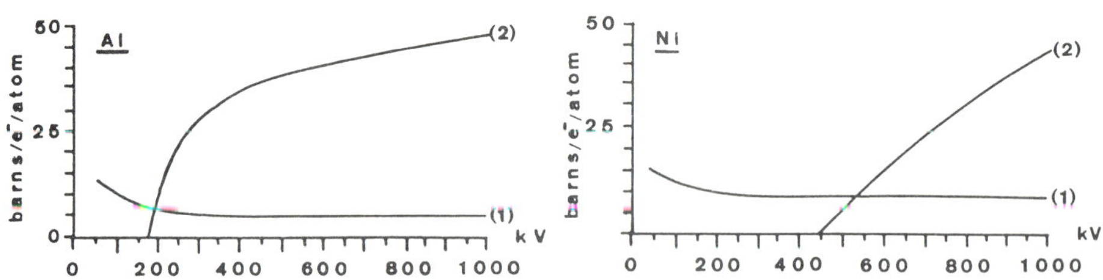


Figure 1. Calculation of the number of detected Kα x-rays/$e^-$/atom (#1) and number of displacements/$e^-$/atom (#2) for Aluminium and Nickel versus Vo.

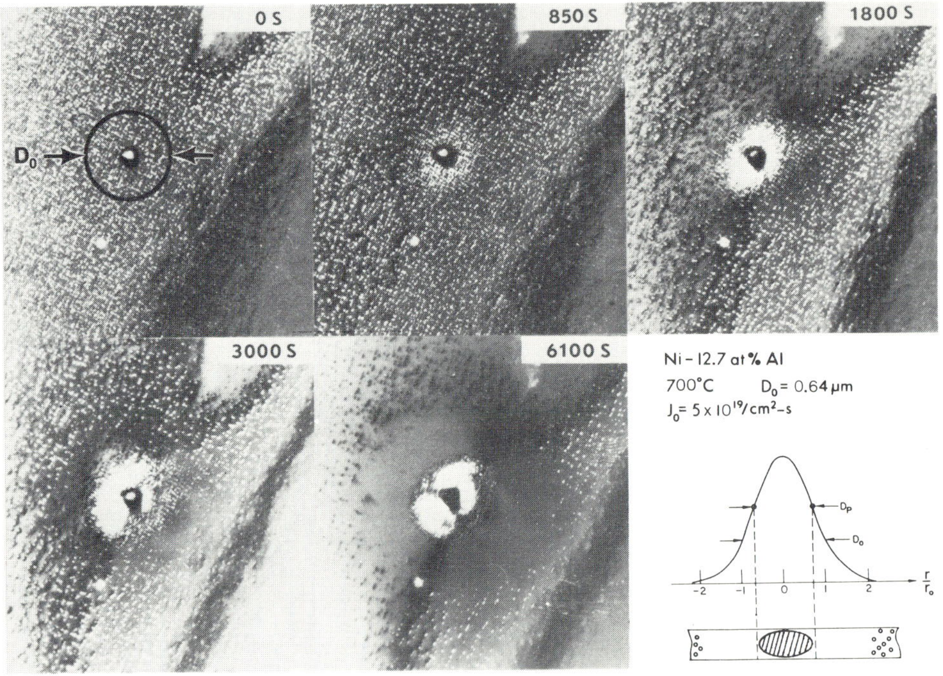


Figure 2.Dark-field (γ'reflection) micrographs showing the development of RIS due to displacement rate gradients in Ni-12.7%Al during 1 MV HVEM irradiation at 700°C. The incident beam diameter is shown by the circle marked Do.

Figure 3. Measured variation of the Zn/Al Kα intensity ratio as a function of time (200 sec intervals) during 300 kV irradiations of Al-1.95%Zn at ~160°C in a medium voltage TEM with a 75 nm diameter probe.

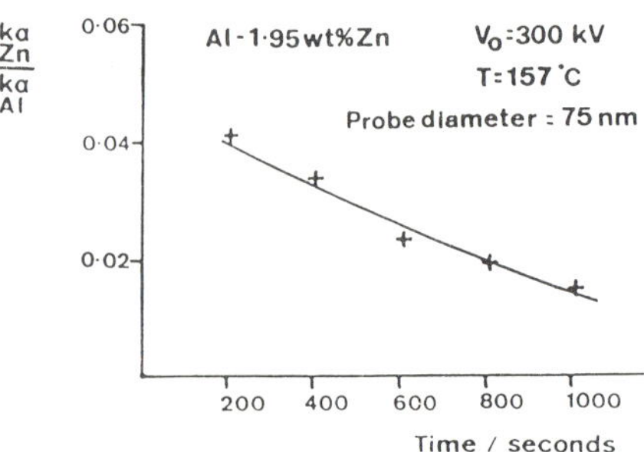

# Author Index

# Subject Index†



† Page numbers refer to the first pages in which the citations occur.